Invasive Birds

Global Trends and Impacts

Invasive Birds

Global Trends and Impacts

Edited by

Colleen T. Downs

Centre for Functional Biodiversity, School of Life Sciences, University of KwaZulu-Natal, Pietermaritzburg, South Africa and Centre for Invasion Biology, University of Stellenbosch, Stellenbosch, South Africa

and

Lorinda A. Hart

Centre for Functional Biodiversity, School of Life Sciences, University of KwaZulu-Natal, Pietermaritzburg, South Africa and Department of Biological Sciences, University of Namibia, Windhoek, Namibia

CABI

CABI is a trading name of CAB International

CABI
Nosworthy Way
Wallingford
Oxfordshire OX10 8DE
UK

Tel: +44 (0)1491 832111
Fax: +44 (0)1491 833508
E-mail: info@cabi.org
Website: www.cabi.org

CABI
745 Atlantic Avenue
8th Floor
Boston, MA 02111
USA

Tel: +1 (617)682-9015
E-mail: cabi-nao@cabi.org

A catalogue record for this book is available from the British Library, London, UK.

Library of Congress Cataloging-in-Publication Data

Names: Downs, Colleen T, editor. | Hart, Lorinda A, editor.
Title: Invasive birds : global trends and impacts / edited by Colleen T. Downs, Lorinda A. Hart.
Description: Wallingford, Oxfordshire ; Boston : CABI, [2020] | Includes bibliographical references and index. | Summary: "Divided into three sections, this book provides detailed information on the 32 global avian invasive species listed by the Invasive Species Specialist Group (ISSG). It also offers an assessment of avian invasions from a biogeographic perspective, and considers the ecological impacts of these invasive species"-- Provided by publisher.
Identifiers: LCCN 2019035836 (print) | LCCN 2019035837 (ebook) | ISBN 9781789242065 (hardback) | ISBN 9781789242072 (ebook) | ISBN 9781789242089 (epub)
Subjects: LCSH: Introduced birds. | Introduced birds--Identification.
Classification: LCC QL677.79.I58 I58 2020 (print) | LCC QL677.79.I58 (ebook) | DDC 598--dc23
LC record available at https://lccn.loc.gov/2019035836
LC ebook record available at https://lccn.loc.gov/2019035837

ISBN-13: 9781789242065 (hardback)
9781789242072 (ePDF)
9781789242089 (ePub)

Commissioning Editor: David Hemming
Editorial Assistant: Lauren Davies
Production Editor: James Bishop

Typeset by SPi, Pondicherry, India
Printed and bound in the UK by Severn, Gloucester

Contents

Contributors

Tim Adriaens, Research Institute for Nature and Forest (INBO), Havenlaan 88 bus 73, B-1000 Brussel, Belgium. Email: tim.adriaens@inbo.be

Craig R. Allen, Nebraska Cooperative Fish and Wildlife Research Unit School of Natural Resources, University of Nebraska-Lincoln, Lincoln, NE 68583, USA. Email: callen3@unl.edu

Michael L. Avery (retired), USDA National Wildlife Research Center, 2906 NW 14th Place, Gainesville, FL 32605, USA. Email: gibs8387@gmail.com

Alexander V. Badyaev, Department of Ecology and Evolutionary Biology, University of Arizona, Tucson, AZ 85721, USA. Email: abadyaev@email.arizona.edu

Fabrice M. Brescia, Institut Agronomique Néo-Calédonien (IAC); Equipe ARBOREAL (AgricultuRE BiOdiversité Et vALorisation), Port-Laguerre, BP73; 98890 Païta, New Caledonia. Email: brescia@iac.nc

Sarah E. Britton, Department of Ecology and Evolutionary Biology, University of Arizona, Tucson, AZ 85721, USA. Email: sbritton@email.arizona.edu

Daniel M. Brooks, Houston Museum of Natural Science, Department of Vertebrate Zoology, 5555 Herman Park Drive, Houston, TX 77019, USA. Email: dbrooks@hmns.org

Jessica L. Burnett, Nebraska Cooperative Fish and Wildlife Research Unit, School of Natural Resources, University of Nebraska-Lincoln, Lincoln, NE 68583, USA. Email: jessicaleighburnett@gmail.com

Matthew Burnett, Centre for Excellence in Invasion Biology, School of Life Sciences, University of KwaZulu-Natal, P/Bag X01, Scottsville, Pietermaritzburg, South Africa, 3209. Email: matthewburnett014@gmail.com

Corey T. Callaghan, School of Biological, Earth and Environmental Sciences; UNSW Sydney, Sydney, New South Wales 2052, Australia. Email: c.callaghan@unsw.edu.au

César Capinha, CIBIO/InBio, Centro de Investigação em Biodiversidade e Recursos Genéticos, Laboratório Associado, Universidade do Porto, Campus Agrário de Vairão, 4485-661 Vairão, Portugal; CIBIO/InBio, Centro de Investigação em Biodiversidade e Recursos Genéticos, Laboratório Associado, Instituto Superior de Agronomia, Universidade de Lisboa, Tapada da Ajuda, 1349-017 Lisbon, Portugal. Email: cesarcapinha@outlook.com

Jen Cele, Centre for Excellence in Invasion Biology, School of Life Sciences, University of KwaZulu-Natal, P/Bag X01, Scottsville, Pietermaritzburg 3209, South Africa. Email: jcele01@gmail.com

Jameson F. Chace, Department of Biology and Biomedical Sciences and Department of Cultural, Environmental and Global Studies, Salve Regina University, 100 Ochre Point Ave, Newport, RI 02840, USA. Email: Jameson.chace@salve.edu

Jocelyn Champagnon, Tour du Valat, Research Institute for the Conservation of Mediterranean Wetlands, Le Sambuc, 13200 Arles, France. Email: champagnon@tourduvalat.org

Alice Cibois, Natural History Museum of Geneva, CP 6434, CH 1211 Geneva 6, Switzerland. Email: alice.cibois@ville-ge.ch

Andrew Close, School of Natural and Environmental Sciences, Newcastle University, Newcastle NE1 7RU, UK. Email: andrew.close@ncl.ac.uk

Alyssa R. Conn, Houston Museum of Natural Science, Department of Vertebrate Zoology, 5555 Herman Park Drive, Houston, TX 77019, USA. Email: alyssa.r.conn@gmail.com

Richard T. Corlett, Xishuangbanna Tropical Botanical Garden, Chinese Academy of Sciences, Menglun, Mengla, Yunnan 666303, China. Email: corlett@xtbg.org.cn.

Adrian J.F.K. Craig, Dept Zoology and Entomology, Life Sciences Building, Barratt Complex, Rhodes University, Grahamstown 6139, South Africa. Email: a.craig@ru.ac.za

Alexander Cruz†, Department of Ecology and Evolutionary Biology, Campus Box 334, University of Colorado, Boulder, CO 80309, USA.

Colleen T. Downs, Centre for Excellence in Invasion Biology, School of Life Sciences, University of KwaZulu-Natal, P/Bag X01, Scottsville, Pietermaritzburg 3209, South Africa. Email: downs@ukzn.ac.za

Amy E. Dunham, Department of BioSciences, Rice University, 6100 Main Street, Houston, TX 77005, USA. Email: aed4@rice.edu

James A. Eaton, Birdtour Asia Ltd, 17 Keats Avenue, Littleover, Derby DE23 4EE, UK. Email: james.birdtourasia@gmail.com

David A. Ehlers Smith, Centre for Excellence in Invasion Biology, School of Life Sciences, University of KwaZulu-Natal, P/Bag X01, Scottsville, Pietermaritzburg 3209, South Africa. Email: ehlerssmithd@ukzn.ac.za

Yvette C. Ehlers Smith, School of Life Sciences, University of KwaZulu-Natal, P/Bag X01, Scottsville, Pietermaritzburg 3209, South Africa. Email: EhlersSmithY@ukzn.ac.za

Mike W. Eichholz, Department of Zoology, Southern Illinois University, Life Science II, Room Number 351, Mail Code 6501, Carbondale, IL 62901, USA. Email: eichholz@siu.edu

Johan Elmberg, Department of Environmental Science and Bioscience, Kristianstad University, Sweden, SE-29188, Kristianstad, Sweden. Email: Johan.Elmberg@hkr.se

Cyril Eraud, Office Français de la Biodiversité Villiers en Bois, France. Email: cyril.eraud@ofb.gouv.fr

Chris J. Feare, WildWings Bird Management, Haslemere, Surrey, UK. Email: feare_wildwings@msn.com

Jason M. Gleditsch, Department of Natural Resources and Environmental Science, University of Illinois at Urbana-Champaign, W-503 Turner, 1102 S Goodwin Ave, Urbana, IL 61801, USA. Email: jason.gleditsch@temple.edu

Guillaume G. Gayet, Unite Mixte de Service Patrimoine Naturel Agence Française pour la Biodiversité, Centre National de la Recherche Scientifique, Muséum national d'Histoire naturelle, 75005 Paris, France. Email: ggayet@mnhn.fr

Matthieu Guillemain, Office Français de la Biodiversité, Unité Avifaune Migratrice, La Tour du Valat, Le Sambuc, 13200 Arles, France. Email: matthieu.guillemain@ofb.gouv.fr

Carlos Gutiérrez-Expósito, Estación Biológica de Doñana (CSIC), Americo Vespucio 26, Isla de la Cartuja, Sevilla 41032, Spain. Email: carlines@ebd.csic.es

Haley E. Hanson, Global and Planetary Health, University of South Florida, 3720 Spectrum Boulevard, Suite 304, Tampa, FL 33612, USA. Email: haleyehanson@gmail.com

Lorinda A. Hart, Centre for Excellence in Invasion Biology, School of Life Sciences, University of KwaZulu-Natal, P/Bag X01, Scottsville, Pietermaritzburg 3209, South Africa; Department of Biological Sciences, University of Namibia, Windhoek 10023, Namibia. Email: lorinda.hart.research@gmail.com

Iain Henderson, National Wildlife Management Centre, Animal and Plant Health Agency, Sand Hutton, York YO41 1LZ, UK. Email: iain.henderson@apha.gsi.gov.uk

Frank Huysentruyt, Research Institute for Nature and Forest (INBO), Havenlaan 88 bus 73, B-1000 Brussel, Belgium. Email: frank.huysentruyt@inbo.be

John G. Innes, Manaaki Whenua Landcare Research, Private Bag 3127, Hamilton 3240, New Zealand. Email: innesj@landcareresearch.ac.nz

Salit Kark, The Biodiversity Research Group, The School of Biological Sciences, Centre for Biodiversity and Conservation Science, The University of Queensland, Brisbane, QLD, 4072 Australia. Email: s.kark@uq.edu.au.

Michael R. Leven, AEC Ltd, 127 Commercial Centre, Palm Springs, Yuen Long, Hong Kong, China. Email: mrl@aechk.hk

Ricardo J. Lopes, CIBIO/InBio, Centro de Investigação em Biodiversidade e Recursos Genéticos, Laboratório Associado, Universidade do Porto, Campus Agrário de Vairão, 4485-661 Vairão, Portugal. Email: riclopes@me.com

Jean-Francois Maillard, Office Français de la Biodiversité, 8 Boulevard Albert Einstein, 44300 Nantes, France. Email: jean-francois.maillard@ofb.gouv.fr

Ndivhuwo Shivambu, Centre for Excellence in Invasion Biology, School of Life Sciences, University of KwaZulu-Natal, P/Bag X01, Scottsville, Pietermaritzburg 3209, South Africa. Email: ndivhuwomaligana@gmail.com

Chloe I. Marcheli, Department of BioSciences, Rice University, 6100 Main St, Houston, TX 77005, USA. Email: cim2@rice.edu

Lynn B. Martin, Global and Planetary Health, University of South Florida, 3720 Spectrum Boulevard, Suite 304, Tampa, FL 33612, USA. Email: lbmartin@usf.edu

Blake A. Mathys, Ohio Dominican University, 1216 Sunbury Road, Columbus, OH 43219, USA. Email: mathysb@ohiodominican.edu

Erik Matthysen, Evolutionary Ecology Group (EVECO), University of Antwerp, Campus Drie Eiken, Universiteitsplein 1, 2610 Wilrijk, Belgium. Email: erik.matthysen@uantwerpen.be

Myriam E. Mermoz, Instituto de Ecología, Genética y Evolución de Buenos Aires (IEGEBA-UBA-CONICET), Facultad de Ciencias Exactas y Naturales, Universidad de Buenos Aires, Piso 4, Pabellón II, Ciudad Universitaria, C1428EGA, Buenos Aires, Argentina. Email: myriammermoz@gmail.com

Jean-Yves Meyer, Délégation à la Recherche, Gouvernement de la Polynésie Française, BP 20981, 98713 Papeete, Tahiti, French Polynesia. Email: jean-yves.meyer@recherche.gov.pf

Dai K.J. Morgan, Applied and Environmental Sciences, NorthTec, 51 Raumanga Valley Road, Private Bag 9019, Whangarei 0110, New Zealand. Email: dmorgan@northtec.ac.nz

Katherine G. Ohman, Houston Museum of Natural Science, Department of Vertebrate Zoology, 5555 Herman Park Drive, Houston, TX 77019, USA. Email: ohmankg@miamioh.edu

Claire Pernollet, Conservatoire d'espaces naturels de Provence-Alpes-Côte d'Azur, 4 avenue Marcel Pagnol, 13100 Aix-en-Provence, France (CEN PACA). Email: clairepernollet@hotmail.com

Murray A. Potter, Wildlife and Ecology Group, School of Agriculture and Environment, Massey University, Palmerston North 4442, New Zealand; Laboratoire d'Excellence Labex-CORAIL, Institut de Recherche pour le Développement (IRD), UMR ENTROPIE (IRD-Université de La Réunion-CNRS), BP A5, Nouméa Cedex 98848, New Caledonia, France. Email: M.Potter@massey.ac.nz

Bill Pranty, Bayonet Point, Florida, USA. Email: billpranty@hotmail.com

Juan Carlos Reboreda, Departamento de Ecología, Genética y Evolución, Facultad de Ciencias Exactas y Naturales, University of Buenos Aires, Pabellón II Ciudad Universitaria, 1428 Buenos Aires, Argentina. Email: reboreda@ege.fcen.uba.ar

Eileen C. Rees, Wildfowl and Wetlands Trust, Slimbridge Wetland Centre, Gloucestershire GL2 7BT, UK. Email: Eileen.Rees@wwt.org.uk

Luís Reino, CIBIO/InBio, Centro de Investigação em Biodiversidade e Recursos Genéticos, Laboratório Associado, Universidade do Porto, Campus Agrário de Vairão, 4485-661 Vairão, Portugal; CIBIO/InBio, Centro de Investigação em Biodiversidade e Recursos Genéticos, Laboratório Associado, Instituto Superior de Agronomia, Universidade de Lisboa, Tapada da Ajuda, 1349-017 Lisbon, Portugal. Email: luisreino@gmail.com

Joana Ribeiro, CIBIO/InBio, Centro de Investigação em Biodiversidade e Recursos Genéticos, Laboratório Associado, Universidade do Porto, Campus Agrário de Vairão, 4485-661 Vairão, Portugal; CIBIO/InBio, Centro de Investigação em Biodiversidade e Recursos Genéticos, Laboratório Associado, Instituto Superior de Agronomia, Universidade de Lisboa, Tapada da Ajuda, 1349-017 Lisbon, Portugal. Email: joanatribeiro@cibio.up.pt; joanateixeiraribeiro@gmail.com

Peter A. Robertson, School of Natural and Environmental Sciences, Newcastle University, Newcastle University, Newcastle NE1 7RU, UK. Email: peter.robertson@ncl.ac.uk

Andrew M. Rogers, The Biodiversity Research Group, The School of Biological Sciences, Centre for Biodiversity and Conservation Science, The University of Queensland, Brisbane, QLD, 4072 Australia. Email: a.munro.rogers@gmail.com

Philip D. Round, Department of Biology, Mahidol University, Bangkok, Thailand. E-Mail: philip.rou@mahidol.ac.th

Joana Santana, CIBIO/InBio, Centro de Investigação em Biodiversidade e Recursos Genéticos, Laboratório Associado, Universidade do Porto, Campus Agrário de Vairão, 4485-661 Vairão, Portugal; CIBIO/InBio, Centro de Investigação em Biodiversidade e Recursos Genéticos, Laboratório Associado, Instituto Superior de Agronomia, Universidade de Lisboa, Tapada da Ajuda, 1349-017 Lisbon, Portugal. Email: joanafsantana@gmail.com

Cavin T. Shivambu, Centre for Excellence in Invasion Biology, School of Life Sciences, University of KwaZulu-Natal, P/Bag X01, Scottsville, Pietermaritzburg 3209, South Africa. Email: shivambucavin@gmail.com

Neftalí Sillero, Centro de Investigação em Ciências Geo-Espaciais (CICGE), Faculdade de Ciências da Universidade do Porto (FCUP), Alameda do Monte da Virgem, 4430-146 Vila Nova de Gaia, Portugal. Email: neftali.sillero@gmail.com

Preshnee Singh, Centre for Excellence in Invasion Biology, School of Life Sciences, University of KwaZulu-Natal, P/Bag X01, Scottsville, Pietermaritzburg 3209, South Africa. Email: preshnee.singh@gmail.com

Pär Söderquist, Department of Environmental Science and Bioscience, Kristianstad University, Faculty of Natural Sciences, SE-29188, Sweden. Email: par.soderquist@kr.se

Karina L. Speziale, Grupo de Investigaciones en Biología de la Conservación, Laboratorio Ecotono, INIBIOMA (CONICET-UNCOMA), Pasaje Gutierrez 1125, Bariloche, Rio Negro 8400, Argentina. Email: karina.speziale@comahue-conicet.gob.ar

Diederik Strubbe, Terrestrial Ecology Unit (TEREC), Ghent University, K.L. Ledeganckstraat 35, 9000 Gent, Belgium. Email: diederik.strubbe@ugent.be

Martin J.P. Sullivan, School of Geography, University of Leeds, Leeds LS2 9JT, UK. Email: M.J.Sullivan@leeds.ac.uk

Jean-Claude Thibault, Muséum National d'Histoire Naturelle, 55 rue Buffon, F-75005 Paris, France. Email: jncldthibault@aol.com

Martin Thibault, Institut Agronomique Néo-Calédonien (IAC), Equipe ARBOREAL (AgricultuRe BiOdiveRsité Et vALorisation) BP 73, 98890 Païta, New Caledonia ; Wildlife and Ecology Group, School of Agriculture and Environment, Massey University, Palmerston North 4442, New Zealand; and Laboratoire d'Excellence Labex-CORAIL, Institut de Recherche pour le Développement (IRD), UMR ENTROPIE (IRD-Université de La Réunion-CNRS), BP A5, Nouméa Cedex 98848, New Caledonia, France. Email: thibault.mn@gmail.com

Berndt J. van Rensburg, School of Biological Sciences, The University of Queensland, Brisbane, QLD 4072, Australia; and Department of Zoology, University of Johannesburg, Auckland Park, Johannesburg 2006, South Africa. Email: b.vanrensburg@uq.edu.au

Eric Vidal, Laboratoire d'Excellence Labex-CORAIL, Institut de Recherche pour le Développement (IRD), UMR ENTROPIE (IRD-Université de La Réunion-CNRS), BP A5, Nouméa Cedex 98848, New Caledonia, France; and Institut Méditerranéen de Biodiversité et d'Ecologie marine et continentale (IMBE), Aix Marseille Université, CNRS, IRD, Avignon Université, Centre IRD Nouméa – BP A5, 98848 Nouméa Cedex, New-Caledonia. Email: eric.vidal@ird.fr

Joseph R. Waas, School of Science, University of Waikato, Private Bag 3105, Hamilton 3240, New Zealand. Email: waasur@waikato.ac.nz

Amy-Leigh Wilson, School of Life Sciences, University of KwaZulu-Natal, P/Bag X01, Scottsville, Pietermaritzburg 3209, South Africa. Email: wilsonamyleigh@gmail.com

Katherine Winston, Houston Museum of Natural Science, 5555 Herman Park Drive, Houston, TX 77019, USA.

Kevin A. Wood, Wildfowl and Wetlands Trust, Slimbridge Wetland Centre, Gloucestershire GL2 7BT, UK. Email: kevin.wood@wwt.org.uk

Pierre Yésou, 3 rue de Châteaulin, 44000 Nantes, France. Email: p.yesou@gmail.com

Ding Li Yong, BirdLife International (Asia), Tanglin International Centre, 354 Tanglin Road, Singapore 247672. Email: dingli.yong@birdlife.org

Jaime E. Zolik, Global and Planetary Health, University of South Florida, 3720 Spectrum Boulevard, Suite 304, Tampa, FL 33612, USA. Email: zolikj@mail.usf.edu

Acknowledgements

Everyone likes birds. What wild creature is more accessible to our eyes and ears, as close to us and everyone in the world, as universal as a bird?
Sir David Attenborough

We could never have imagined that three years ago, when this book about globally invasive bird species was conceptualised, that we would embark on such a journey. The first phase of our expedition saw experts in their field present current species-specific accounts of 'globally' recognised avian invasive species. We were surprised at the limited range of some of these species and the lack of information for a few of them. During the second leg of our voyage each continent and its invasive avifauna were reviewed. The history of a country is often reflected in the invasive species represented today. However, with increased urbanisation and global trade, it will be interesting to see how these assemblages change. How countries and their citizens manage and monitor these species and how their competitive impacts affect local avifauna will become increasingly important and are discussed in the final chapters of this book.

Despite the challenges associated with such an undertaking, we were driven by our goal to bring avian invasive biology back to basics. There are many excellent resources available on the topic of biological invasions, but through our research and personal experiences we began to appreciate the lack of basic knowledge and sometimes misinformation regarding some invasive species - particularly in a global context. It is our hope that this book will form part of the foundation of future research, providing a consolidated review, but also inspiring exploration in areas where research is still needed. We are grateful to our fellow travellers and friends we have made along the way and look forward to embarking on new adventures.

We want to express our most sincere gratitude to the contributors who made this book possible, as well as the team at CABI for their contributions. In particular, we are grateful to James Bishop and Lauren Davis from CABI for their valuable assistance and guidance with the editorial process. A comprehensive peer-review process was undertaken of all contributions submitted for this book, and we are most grateful for the constructive comments that improved the respective original drafts chapters.

We are most grateful to the many authors (more than 80) that have contributed to its success. Each chapter was written with dedication and care. We are also extremely grateful to our families who sacrificed countless hours and provided invaluable support during the compilation of this book. We are appreciative of everyone who contributed images, expertise, and comments. This project was a greater undertaking than we anticipated and was truly a team effort.

Prof Colleen T. Downs and Dr Lorinda A. Hart
Pietermaritzburg, South Africa, 2020

Foreword

Imagine a pond without ducks and swans, a garden without robins and song thrushes, a park without sparrows and pigeons, and perhaps, also, a fridge without eggs. Since antiquity, these descendants of soaring dinosaurs have captured humans' imagination for wilderness and freedom. Simply put, we cannot imagine living in a world without birds. Introductions of alien birds, and other domesticated animals, have accompanied the footprint of early humans. Fowls were in canoes while the Polynesian islanders colonized the vast span of the Pacific. Domesticated in South Asia, the red jungle fowl was introduced via the ancient trade route to Greece around 500 bce and has now reached a global population of nearly 20 billion. Parrots, starlings, mynas and canaries, with their chanting acoustics, have been popular in homes and palaces of Eurasia for centuries. Shakespeare writes in Henry IV, 'I'll have a starling shall be taught to speak.' Subsequently, early European settlers such as Eugene Schieffelin, Cecil John Rhodes and Eastham Guild attempted to 'acclimatise' the colonies using the birds of Shakespeare. John Long (1981) recorded more than 1000 introductions of about 420 bird species mainly during the European Diaspora from the 18th to the 20th century; this has been expanded in the Global Avian Invasion Atlas (Dyer *et al.*, 2017), which documents 971 alien bird species introduced to 230 countries since 6000 bce.

Following Charles Elton's (1958) book, invasion ecology has grown into a mainstream research field on the patterns and processes of human-mediated translocation of alien organisms. As probably the most iconic taxa in conservation, birds provide an ideal natural experiment to test many theories and hypotheses in invasion ecology. Biodiversity conservation requires a solid ground to justify the action of management. Unfortunately, this remains controversial for introduced and invasive birds. This is partly because most exotic birds were introduced deliberately by acclimatization societies or through the pet trade. Such human-mediated introductions add difficulties when we try to identify the environmental determinants and biological traits that are responsible for avian species' invasiveness and the invasibility of recipient ecosystems. Invasion scientists are often divided in their justification of management and control action. For instance, although exotic birds could impose a strong impact on recipient ecosystems, many native birds that are undergoing a rapid range shift from human activities and environmental changes could cause more damage. The impact also often varies from case to case with no general principles. Research priorities should reflect such diversity of opinions and the wind of change.

Biological invasion is a complicated process and can be considered as species breaking geographical, environmental and biotic barriers along the invasion pathway. This naturally classifies species into: (i) those belonging to the regional native species assemblage; (ii) those transported, mainly intentionally, for different purposes; (iii) those having established viable feral populations; and (iv) those starting to spread and expand their range into the recipient areas. Knowledge gained so far has concentrated on the early stages of introduction and naturalization. For instance, the purposes of avian introductions are limited (e.g. food, hunting and ornamental) and primarily deliberate. As such, the taxonomy, localities and time of introductions are selected purposefully, with comparably large and widespread species from only a few families introduced significantly more frequently (e.g. Phasianidae, Passeridae, Psittacidae, Anatidae and Columbidae). For instance, early introductions of game birds of Galliformes in the 1800s were surpassed by the introduction of cage birds in the 1900s (e.g. parrots of Psittaciformes) (Blackburn *et al.*, 2009). The most important predictor of the establishment success is the propagule pressure, specifically the number of individuals released, although its importance often differs among different taxa and cases, signalling the role of other important factors, such as habitat, diet breadth and relative brain size, as well as climate matching. Moving to the last stage of invasion (i.e. spread), we are facing increasing knowledge gaps (Hui and Richardson, 2017), albeit with a few well-observed patterns (Blackburn *et al.*, 2009): (i) a lag phase experienced by most invasive birds before rapid range expansion; (ii) spatial variation in the spreading velocity; (iii) range contraction in those boom-and-bust species; and (iv) the small extent of spread of most invasive birds.

Detailed examinations of current knowledge have revealed a number of research priorities. At the introduction stage, we must distinguish those species that have been selected (or preferred) by humans for translocation from those species that preferred humans. The success of species that are intentionally introduced (i.e. those selected by humans) is probably very much dependent on propagule pressure (also reflecting cultural trends), while the success of species that are introduced unintentionally probably depends on the niche breadth of the species and environmental suitability of the recipient areas (e.g. Indian House Crow, *Corvus splendens*, requires a low propagule size to establish and also benefit from the human-dominated and disturbed landscapes). At the naturalization stage, more emphasis is perhaps needed for those species that failed to establish. For instance, the Yellow Canary

(*Serinus flaviventris*) persisted for more than 100 years in Mauritius only to be wiped out by a single hurricane. The Song Thrush (*Turdus philomelos*) was successfully introduced in 1890 in South Africa but became extinct 45 years later. Showing which traits favour introduction success is not equal to showing that introduced species without these traits will most likely fail. At the spreading stage, we need to identify the determinants of which established species will spread, unveil the mechanisms behind these boom-and-bust species, identify factors determining spread extent, quantify the structure of species distribution, and reveal the mechanisms behind range dynamics.

Key questions on the geographical range dynamics of introduced birds include the following (Hui and Richardson, 2017):

1. *Native and alien concordance.* The range dynamics of the regional avifauna present a complicated picture. Some species show a conserved range, while others are expanding, retracting or shifting their current ranges. In general, the range dynamics of native species reflect the spatial dynamics of their suitable habitat. In contrast, the range dynamics of introduced species depict spread into potentially suitable habitat. For instance, two invasive species in South Africa (Common Starling, *Sturnus vulgaris*, and Common Myna, *Acridotheres tristis*) are expanding their ranges northwards, completely against the flow of most natives (towards the south or west). Two kinds of concordance between native and introduced species can be examined. First, whether expanding invasives share common traits with expanding natives (trait concordance): identifying such traits could be important to understand how species with and without such traits respond differently to the regional environmental changes, which helps conservation managers to pinpoint those species that are more vulnerable to environmental changes. Second, whether introduced and native species expand or shift their ranges into same areas, while withdrawing from other areas (locality concordance): this would highlight areas for conservation management and the environmental factors characterising these areas.
2. *Forms of dispersal.* To produce efficient conservation management, we need to assess the connectivity of spreading populations via dispersal. First, we need to pinpoint the environmental conditions that are responsible for enhanced dispersal capacity in introduced species, and thus their tendency to spread and become invasive. Second, many invasive birds exhibit a fat-tailed dispersal kernel (i.e. the frequency distribution of dispersal distance). When estimating such dispersal kernels, the 'tail' often contains much uncertainty because long-distance dispersal is extremely rare, yet the rate of spread (i.e. the speed of range expansion) is largely determined by these rare events of long-distance dispersal.
3. *Variations in spreading rates.* Although dispersal is an important determinant of the spreading rate, it is not the sole factor. The velocity of advancing range margins also depends on the ability to establish front populations as stepping stones for further spread and traits affecting the rate of producing propagules (e.g. age of maturity, clutch size and breeding success). The spreading rate of introduced birds often has high temporal and spatial variation. A comprehensive understanding of the mechanisms and environmental factors behind this high variation provides an estimate of the rate of response to environmental changes. A main feature of the temporal variation in the spreading rates of introduced birds is the two-phase range expansion, or the existence of a lag phase in some introduced species before the rapid range expansion. Theoretical advances are needed to understand the population structure and dynamics that cause the lag phase. It is also important for conservation management to identity the key environmental factors and the key population structure (e.g. whether the lag is caused by the Allee effect threshold of population density or certain levels of genetic diversity required). By manipulating certain environmental conditions and control strategies to reduce the density under the threshold, we will be able to prevent the rapid expansion of birds.
4. *Human geography*. Our planet has rapidly moving from nature- to human-dominated environments. The role of humans in environmental changes has changed from enduring during the early-mid Holocene (10,000–3000 bp) to an innovative response (from the 17th to the early 20th century) to dominating (especially post-1950). Natural processes have been significantly altered by human activities through, for instance, agriculture, hydrology and urbanization. The way that we alter and interfere with the functioning and processes of natural ecosystems differs dramatically from the self-regulation processes in the pristine natural ecosystem, suggesting a completely different mode of invasibility in human-dominated novel ecosystems. Regional variations in culture, history, legislation and governance will further compartmentalize the response to biological invasions, inevitably affecting how regional ecosystems respond to the new arrivals. Clearly, human geography has become one of the most dominant forces in global change biology and plays a key role in forecasting the future distributions of introduced birds. Many bird species (native and introduced) evolved via allopatric speciation in isolated populations. Little is known about how these species interact in the long term in highly disturbed and connected novel systems, and how these transformed novel ecosystems will change over time. This calls for a systematic assessment of the exposure, sensitivity and adaptive capacity of introduced (and native) birds to human-dominated novel ecosystems.

Plagued by such great knowledge gaps in our understanding of avian invasion and its management, it is such a pleasure to see this massive volume of 45 chapters, edited by Colleen T. Downs and Lorinda A. Hart, with contributions from 78 established ornithologists worldwide, in an attempt to address some of the questions and data gaps in global trends and impacts of alien invasive birds. The book covers the biology, ecology, impact and management of 34 common alien invasive species, with reviews on the history and context of avian introductions and invasions in five major regions (Oceania, Africa, Europe (including the Middle East, Asia and South America), as well as management challenges and the potential of citizen science for monitoring alien birds. The book pitches at the introductory level and is ideal for readers to gain a quick and comprehensive view of the current status of global avian invasions. It has brought the records and research of avian invasion one step ahead of other alien invasive animal taxa. Many chapters contain distribution maps and data tables on the diet and morphology of the species, providing a good reference for the species and its management issues. Each chapter also contains a rich list of references that could help readers dive further into the topic. I hope that readers will use this book as a generic reference on avian invasions and read it with specific questions in mind.

Cang Hui
Centre for Invasion Biology
Stellenbosch University
Stellenbosch 7602
South Africa

References

Blackburn, T.M., Lockwood, J.L. and Cassey, P. (2009) *Avian Invasions: The Ecology and Evolution of Exotic Birds*. Oxford University Press, Oxford.

Dyer, E.E., Redding, D.W. and Blackburn, T.M. (2017) The global avian invasions atlas, a database of alien bird distributions worldwide. *Scientific Data* 4, 170041.

Elton, C.S. (1958) *The Ecology of Invasions by Animals and Plants*. Methuen, London.

Hui, C. and Richardson, D.M. (2017) *Invasion Dynamics*. Oxford University Press, Oxford.

Long, J.L. (1981) *Introduced Birds of the World: The Worldwide History, Distribution and Influence of Birds Introduced to New Environments*. Universe Books, London.

1 Introduction

Lorinda A. Hart[1,2]* and Colleen T. Downs[1]

[1]*Centre for Excellence in Invasion Biology, School of Life Sciences, University of KwaZulu-Natal, Pietermaritzburg 3209, South Africa;* [2]*Department of Biological Sciences, University of Namibia, Windhoek 10023, Namibia*

Citation: Hart, L.A. and Downs, C.T. (2020) Introduction. In: Downs, C.T. and Hart, L.A. (eds) *Invasive Birds: Global Trends and Impacts*. CAB International, Wallingford, UK, pp. 1–8.

Birds evolved from small theropod dinosaurs during the middle to late Jurassic (Padian and Chiappe, 1998) to their current diverse group as we know them today. It is largely accepted that there are around 11,000 described bird species globally (del Hoyo *et al.*, 2014, 2016), although recent molecular and morphological analysis using evolutionary species concepts suggest there could be as many as 18,000 (Barrowclough *et al.*, 2016). The Aves are an extremely diverse group, occupying a range of habitats and dietary niches. Most birds possess the ability of flight, making them highly mobile, with some species undertaking impressive migrations. The Arctic Tern (*Sterna paradisaea*), for example, flies 19,000 km one way between its Arctic breeding and Antarctic overwintering grounds (Åkesson and Hedenström, 2004). The possession of feathers is a unique avian feature that facilitates flight. However, feathers also play a functional role in thermoregulation and behavioural displays (Starr and Taggart, 2004). Birds and their feathers have also captured the attention of mankind for centuries.

Nearly 46% of extant bird species are utilized in some way by humans (Butchart, 2008). These include usage in fashion, weapons (e.g. arrows), stationary, household goods (e.g. down bedding, feather dusters), ornaments, medicine, cultural rituals and fuel, to name a few (Doughty, 1972; Butchart, 2008). Beautifully coloured and melodious birds are particularly popular as pets, which has led to a booming pet trade comprising 37% of extant bird species (Butchart, 2008; Fig. 1.1). While this trade is particularly prolific today due to globalized transport systems (Hulme, 2009), the transport and trade of exotic birds dates back to ancient times. One of the best-documented cases of human-mediated bird introduction in the western hemisphere is that of Emperor Auitzotl who, during 1486–1502, introduced the Great-tailed Grackle (*Quiscalus mexicanus*) to Mexico where it persists today (Haemig, 2012). The Aztecs were fond of keeping exotic birds, from emperors who employed up to 300 workers to look after their vast aviaries and zoos to poorer classes who kept birds as pets (Haemig, 1978). The exotic bird trade is evidenced in many cultural histories for example the Inca, Maya, Lapita and Paquimé, (Haemig, 1978; Hurles *et al.*, 2003; Somerville *et al.*, 2010). The earliest transport of birds was probably linked to their use as a food source (Blackburn *et al.*, 2009). It is estimated that approximately 8000 years ago, the ancestor of the chicken (*Gallus gallus domesticus*), the Red Junglefowl (*G. gallus*), was brought to China and later domesticated (West and Zhou, 1989). Today, the Red Junglefowl is recognized as one of the worst global avian invaders by the Invasive Species Specialist Group (ISSG) from the International Union for Conservation of Nature (IUCN Invasive Species Specialist Group, 2015).

More recently, most bird introductions have coincided with the mass emigration and colonization period of the Europeans from 1850 onwards (Blackburn *et al.*, 2009). Of all known introductions, Hawaii, New Zealand, the USA and Australia account for 40% of these; the last three all being former British colonies (Blackburn and Duncan, 2001). In 1847, The Zoological Society of London steadily introduced species for scientific research, among other reasons (McDowall, 1994). Cecil John Rhodes, a British businessman who became Prime Minister of South Africa in 1890, repeatedly tried to introduce British birds to South Africa (Picker and Griffiths, 2013). While many species failed, one notable success was the Common Starling (*Sturnus vulgaris*) in 1897, when he released a flock of 18 individuals (Hockey *et al.*, 2005; Picker and Griffiths, 2013). The American introduction of these birds was also driven by a 'cultural longing' and it is said that there was a drive to introduce all bird species from Shakespeare's works to the New World (Adeney, 2001; Linz *et al.*, 2007). Today, Common Starlings are considered one of the world's worst invaders and cause significant

*Corresponding author: lorinda.hart.research@gmail.com

Fig. 1.1. Bird market in Afghanistan. (©Photograph: Afghanistan Matters – Flickr: Bird Market, CC BY 2.0, https://commons.wikimedia.org/w/index.php?curid=28546595.)

negative economic and ecological impacts (Linz *et al.*, 2007). Ironically, Common Starlings have also been introduced to serve as biocontrol agents of insect pests and have inadvertently become pests themselves (Feare and Craig, 1998; Yap and Sodhi, 2004; Blackburn *et al.*, 2009). Perhaps linked to the European sense of nostalgia, many of the introduced bird species were favoured for hunting, providing food and a source of entertainment for early colonists (McDowall, 1994; Blackburn *et al.*, 2009). Even scavengers were introduced to facilitate the removal of carcases (Lever, 2005). However, not all introductions have been intentional. For example, the House Crow (*Corvus splendens*) was probably a stowaway on ships travelling to many destinations around the world (Ryall, 2002; Leven and Corlett, 2004; Picker and Griffiths, 2013), and in some cases invasive birds have naturally expanded their ranges into neighbouring regions (Yap and Sodhi, 2004).

In Asia, the religious release of animals in Buddhism or Taoism is believed to generate good karma (Agoramoorthy and Hsu, 2007) and is recognized as one of the important invasion pathways of exotic species for the continent (Liu *et al.*, 2012). Despite up to 90% mortality reported for released birds in some areas (Shiu and Stokes, 2008), it is estimated that 18 invasive bird species have established in Hong Kong since 1860 (Leven and Corlett, 2004). The history of this practice is unclear, but Daoist texts from around the 3rd century (possibly earlier) are credited with some of the first descriptions of animal ritual release (Shiu and Stokes, 2008). It has been suggested that animal release was a cultural practice in China that was incorporated into Chinese Buddhism, as it resonated with Buddhist ideologies (Shiu and Stokes, 2008). Under this pretext, the tradition soon spread to other Asian countries including Japan, Korea, Sri Lanka, Thailand, Cambodia and more recently Taiwan (Shiu and Stokes, 2008). In Taiwan, it is estimated that 93% of the population practice the releasing of animals (Agoramoorthy and Hsu, 2007). Its popularity in Taiwan has spilled over into other religions, including Protestants and Catholics (Severinghaus and Chi, 1999). Approximately US$6 million is spent annually on some 200 million animals ranging from insects to vertebrates (Agoramoorthy and Hsu, 2007). This practice has also reached western countries, with ritual releases reported in Australia and Canada (Shiu and Stokes, 2008). Linked to this practice is the pet trade supplying these animals (Fig. 1.1), some of which are recognized on the ISSG's list of global invaders (Shieh *et al.*, 2006).

Avian invasive success is lower than that of mammals: 64% of established exotic mammals become invasive, while this is only 34% in birds (Jeschke, 2008). Following introduction into a novel habitat, there are several factors and ecological processes that come into play to determine the effective establishment and invasive potential of a species. These can be grouped into location traits (e.g. enemy-release and climate-matching hypotheses, presence of brood parasites), species characteristics (e.g. breeding biology, behavioural flexibility, diet, taxonomy, juvenile development, migratory strategy, body size and genetic variability), and introduction or event factors (e.g. timing of release and introduction effort – the number of individuals released and the number of release events) (Dean, 2000; Sol and Lefebvre, 2000; Kolar and Lodge, 2001; Butler, 2003; Hayes and Barry, 2008; Shwartz *et al.*, 2009). Of course, these are just a few of the known drivers that facilitate invasive success, and by no means act exclusively or even consistently across taxa or locations (Hayes and Barry, 2008).

The ISSG has identified 31 bird species on the Global Invasive Species Database (GISD) that pose a threat to biodiversity (Table 1.1) (IUCN Invasive Species Specialist Group, 2015). Nearly half of these belong to the passerines (order Passeriformes), which is the largest avian clade and is followed by the waterfowl group (order Anseriformes; Table 1.1). A detailed account for each of these species and others is presented in Section 1 (Chapters 2–35, this volume).

Table 1.1. List of invasive birds for which species accounts are presented in Section 1 of this book.

Common name	Species	Order	Family	Impact mechanism[b]	Impact category of concern[c]	Assessment confidence rating
Swamp Harrier	*Circus approximans*	Accipitriformes	Accipitridae	Pred	MO	Low
Mallard Duck	*Anas platyrhynchos*	Anseriformes	Anatidae	Hybr	MR	High
Greylag Goose	*Anser anser*	Anseriformes	Anatidae		DD	
Canada Goose	*Branta canadensis*	Anseriformes	Anatidae	Graz	MO	High
Mute Swan	*Cygnus olor*	Anseriformes	Anatidae	Comp, Graz	MO	High
Ruddy Duck	*Oxyura jamaicensis*	Anseriformes	Anatidae	Hybr	MO	High
Egyptian Goose[a]	*Alopochen aegyptiaca*	Anseriformes	Anatidae	Comp, Chem	MN	Medium
Cattle Egret	*Bubulcus ibis*	Ciconiiformes	Ardeidae	Pred	MO	Medium
Rock Dove	*Columba livia*	Columbiformes	Columbidae	Dis	MO	Low
Eurasian Collared Dove	*Streptopelia decaocto*	Columbiformes	Columbidae	Comp	MN	High
Chukar Partridge	*Alectoris chukar*	Galliformes	Phasianidae	Hybr	MO	High
Red Junglefowl	*Gallus gallus*	Galliformes	Phasianidae	Comp	MR	Low
Green Junglefowl	*Gallus varius*	Galliformes	Phasianidae	Comp	MR	Low
Purple Swamphen	*Porphyrio porphyrio*	Gruiformes	Rallidae		NR	
Grey-headed Swamphen[a]	*Porphyrio poliocephalus*	Gruiformes	Rallidae	Comp, Pred, Graz	MN	High
Jungle Myna	*Acridotheres fuscus*	Passeriformes	Sturnidae		DD	
Common Myna	*Acridotheres tristis*	Passeriformes	Sturnidae	Comp, Pred	MO	High
Common Starling	*Sturnus vulgaris*	Passeriformes	Sturnidae	Comp	MO	High
House Finch	*Carpodacus mexicanus*	Passeriformes	Fringillidae	Comp, Dis	MN	High
House Crow	*Corvus splendens*	Passeriformes	Corvidae	Comp, Pred	MO	High
Common Waxbill	*Estrilda astrild*	Passeriformes	Estrildidae		DD	
Australian Magpie	*Gymnorhina tibicen*	Passeriformes	Cracticidae	Pred	MN	High
Red-billed Leiothrix	*Leiothrix lutea*	Passeriformes	Sylviidae	Comp, Int	MN	High
Brown-headed Cowbird	*Molothrus ater*	Passeriformes	Icteridae		NR	
Shiny Cowbird	*Molothrus bonariensis*	Passeriformes	Icteridae	Para	MO	High
House Sparrow	*Passer domesticus*	Passeriformes	Passeridae	Comp	MO	High
Great Kiskadee	*Pitangus sulphuratus*	Passeriformes	Tyrannidae	Pred	MV	Low
Red-vented Bulbul	*Pycnonotus cafer*	Passeriformes	Pycnonotidae	Comp, Int	MO	High, medium
Red-whiskered Bulbul	*Pycnonotus jocosus*	Passeriformes	Pycnonotidae	Pred	MR	Low
Warbling White-eye	*Zosterops japonicus*	Passeriformes	Zosteropidae	Comp, Dis, Int	MO	High, low, high
Scaly-breasted Munia[a]	*Lonchura punctulata*	Passeriformes	Estrildidae	Comp, Int	MC	Low
Northern Red Bishop[a]	*Euplectes franciscanus*	Passeriformes	Ploceidae		DD	
African Sacred Ibis[a]	*Threskiornis aethiopicus*	Pelecaniformes	Threskiornithidae	Pred	MN	High
Monk Parakeet	*Myiopsitta monachus*	Psittaciformes	Psittacidae	Comp	MN	Medium
Ring-necked Parakeet	*Psittacula krameri*	Psittaciformes	Psittacidae	Comp	MO	High
Great Horned Owl	*Bubo virginianus*	Strigiformes	Strigidae	Pred	MV	Low

[a]Species not identified by ISSG on the Global Invasive Species Database (GISD, 2019). Environmental Impact Classification for Alien Taxa (EICAT) impact categories, mechanisms and confidence of assessments assigned according to Evans *et al.* (2016).
[b]Chem, chemical impact on ecosystem; Comp, competition; Dis, transmission of disease to native species; Graz, grazing/herbivory/browsing; Hybr, hybridization; Int, interaction with other alien species; Para, parasitism; Pred, predation.
[c]DD, data deficient; MC, minimal; MN, minor; MO, moderate; MR, major; MV, massive; NR, species not represented in the analysis of Evans *et al.* (2016).

To better determine avian invasion patterns and processes, a Global Avian Invasions Atlas (GAVIA) was developed by Dyer *et al.* (2017). Their dataset covers 230 countries and administrative areas from 6000 bce to 2014 ad and comprises nearly 28,000 distribution records for 971 introduced bird species (Dyer *et al.*, 2017). Of these, 43% have established a population somewhere (Dyer *et al.*, 2017). The Psittacidae (parrot family; 131 species) and Anatidae (ducks, geese and swans; 92 species) have the greatest numbers of invasive species records (Dyer *et al.*, 2017). House Sparrows (*Passer domesticus*), Common Mynas (*Acridotheres tristis*), Rock Doves (*Columba livia*), Ring-necked Parakeets (*Psittacula krameri*), Common Pheasants (*Phasianus colchicus*), Common Starlings (*Sturnus vulgaris*) and Java Sparrows (*Padda oryzivora*) have the greatest number of records, all exceeding 500 counts each (Dyer *et al.*, 2017). Indeed, most of these species fall within families that have been particularly successful invaders, i.e. the Phasianidae, Passeridae, Psittacidae, Anatidae and Columbidae (Blackburn and Duncan, 2001). Invasive and emerging invasive avian species are discussed by geographical region in Section 2 (Chapters 36–41).

The lack of consistency in invasive species impact assessment protocols allowing for global comparisons prompted the development and formalization of the Environmental Impact Classification for Alien Taxa (EICAT), a standardized protocol that classifies the magnitude of invasive species environmental impacts (Blackburn *et al.*, 2014; Hawkins *et al.*, 2015). EICAT assigns 12 categories to assess the impacts of alien species. These comprise: competition, predation, hybridization, disease transmission, interactions with other alien species, parasitism, biofouling, grazing/herbivory/browsing, poisoning/toxicity, and chemical, physical and structural impacts on ecosystems (Hawkins *et al.*, 2015). Using this technique, a global review of 415 bird species with self-sustaining alien populations showed that most species had a low impact, with only five having a major impact, and a spectrum of influences in between (Evans *et al.*, 2016). Most importantly, these trends were based on only 30% of the species for which there was evidence, with many cases having low confidence scores. This highlights the lack of research in this area and the potential for new trends to emerge as more data become available (Evans *et al.*, 2016).

The Madagascar Turtle Dove (*Nesoenas picturata*), Australian Masked Owl (*Tyto novaehollandiae*), Barn Owl (*Tyto alba*), Great Horned Owl (*Bubo virginianus*) and the Great Kiskadee (*Pitangus sulphuratus*) are the only species listed as having a massive impact (Evans *et al.*, 2016), yet the ISSG only lists the latter two species (Table 1.1). Most ISSG listed species are recognized as having minor to moderate impacts (Table 1.1). This discrepancy in species impact allocation further highlights the need for detailed research, as well as the dissemination of this information among organizations and researchers to make unified and informed decisions. This has implications for management and research priorities on a global scale. Finally, although some species are listed as data deficient, this is not always the case; for example, Jungle Mynas (*Acridotheres fuscus*) are known dispersers of invasive plants (Long, 1981; Aravind *et al.*, 2010; Palita *et al.*, 2011).

The human population continues to expand, and coupled with this, natural habitats are being transformed at unprecedented rates. It is predicted that urban areas will expand by 120 million hectares between 2000 and 2030 (McDonald *et al.*, 2018). Urbanization often leads to homogenization of species, with predominantly more invasive species present in urban areas compared with more natural environments (van Rensburg *et al.*, 2009). Urban areas present an abundance of food and nesting sites, which facilitate avian invasions (Yap and Sodhi, 2004; Clergeau and Vergnes, 2011; Strubbe and Matthysen, 2011). Indeed, urban environments have been linked to the successful establishment and spread of Ring-necked Parakeets (Fig. 1.2), for example, where exotic trees and garden bird-feeding stations are utilized (Clergeau and Vergnes, 2011; Czajka *et al.*, 2011). Invasive birds are generally commensal with humans, who not only provide favourable environments but also directly facilitate their dispersal (Dean, 2000; Tabak *et al.*, 2017).

Invasive species are recognized as one of the leading causes of extinction, particularly of birds (Gurevitch and Padil, 2004; Bird Life International, 2008). Invasive species modify the evolutionary pathway of native species through competitive exclusion (Strubbe and Matthysen, 2009; Hernández-Brito *et al.*, 2014; Grandi *et al.*, 2018), hybridization and introgression (Gaertner *et al.*, 2016), spread of disease (Weber, 1979; Crowl *et al.*, 2008) and predation (Mooney and Cleland, 2001). Generally, invasive species thrive in urban areas where native species tend to be fewer. There is evidence that this is not due to a competitive edge but rather to the ability of exotic species to exploit novel environments and food sources (Sol *et al.*, 2012). Urban birds also have greater problem-solving abilities, which enable them to exploit resources that native species rarely utilize (Sol *et al.*, 2011). However,

Fig. 1.2. A Rose-ringed Parakeet (*Psittacula krameri*) pair feeding on exotic pecan nuts *Carya illinoensis* in a suburban garden in Pretoria, South Africa (©Photograph: H. Jordaan.)

this is not always the case, as evidenced by the sometimes lethal eviction of threatened greater noctule bats (*Nyctalus lasiopterus*) from tree cavities by invasive Ring-necked Parakeets in an urban park in Spain (Hernández-Brito *et al.*, 2018; Sohns, 2018). In Italy, these parakeets also occupy nests favoured by cavity-nesting Common Swifts (*Apus apus*), which then make use of suboptimal nests and suffer greater breeding failure, ultimately reducing their breeding population while the parakeet population increases (Grandi *et al.*, 2018). In Belgium, they also outcompete native Nuthatches (*Sitta europaea*) for nest cavities (Strubbe and Matthysen, 2009). The competitive nature of invasive species can benefit common native species, which take advantage of their aggressive anti-predator behaviour, while rarer species vying for the same resources are more negatively impacted (Hernández-Brito *et al.*, 2014). Green spaces within urban environments are significant refuges for native species and promote biodiversity conservation within an urban landscape (Goddard *et al.*, 2010). While competitive interactions may be reduced in cities (where native species are fewer), these conflicts are probably more significant in green spaces and bordering suburban and natural areas where native wildlife is more prevalent and into which invasive species expand (van Rensburg *et al.*, 2009). Chapter 42 assesses the competition between invasive and native bird species in more detail.

Invasive birds not only have the potential to impact natural ecosystems negatively but are also associated with the spread of disease to humans, damage to property and crops, generating noise and becoming a nuisance (Long, 1981; Kumschick *et al.*, 2011). Pigeons alone are hosts to at least 60 pathogens (Haag-Wackernagel and Moch, 2004). This is probably due to the dense flocks that frequently form within urban areas, resulting in the accumulation of waste and facilitating the transmission of parasites and infections among individuals. In Europe, the impacts of mammals are equally distributed between their effect on economies and the environment; however, birds have twice the impact on the environment than on the economy (Kumschick and Nentwig, 2010). Invasive species in the USA cause losses of US$120 billion each year, with pigeons and starlings contributing US$2.2 billion dollars annually (Pimentel *et al.*, 2005; Pimental, 2007). This excludes losses from an additional 52 harmful exotic species in the USA (Temple, 1992). Nearly half of the country's endangered and threatened species are vulnerable due to pressures from these invasive species (Wilcove *et al.*, 1998). In Great Britain, the economy suffers losses of US$2.24 billion, with nearly US$2 million caused by four goose and swan species and two parakeet species (Williams *et al.*, 2010). Pigeons in the UK are estimated to cause US$2 million in damages (Pimentel *et al.*, 2000, 2001).

In response to the economic and ecological damage attributed to invasive birds, various control efforts have been undertaken. Some of these methods include shooting, nest destruction, limiting resources, using avicides, bioacoustic scaring, sterilization and live trapping (Brook *et al.*, 2003; Yap and Sodhi, 2004; Iriarte, 2005; Feare, 2010). Total eradication of invasive bird populations is probably unrealistic in many cases, owing in part to the difficulty of detecting individuals when population sizes drop and because there is a constant influx from neighbouring areas or new introductions (Brook *et al.*, 2003). Additionally, the control methods used are not always suitable for all locations (Yap and Sodhi, 2004). Invasive species removal can have unexpected knock-on effects, particularly in cases where well-established invaders have replaced a native species' functional role (Zavaleta *et al.*, 2001). However, reducing invasive populations invariably translates into less damage and ecological impact, as well as limiting potential dispersal. Avian control invariably requires a multi-faceted approach in which both the species and the habitat are managed (Yap and Sodhi, 2004). Chapter 43 assesses control methods and successes of controlling invasive birds.

The support for invasive species control varies among taxa, the control methods used, cultures, population demographics, personal experience and the environments people live in (Veitch and Clout, 2001; Bremner and Park, 2007; Coates, 2007). Even the scientific community is divided on the best course of action, with 'denialists' advocating for invasive species (Russell and Blackburn, 2017), while 'eco-xenophobes' and 'preservationists' (Rotherham, 2010; Bhagwat, 2018) are against them. If birds are perceived as agricultural pests or vectors of harmful diseases, for example, they are more likely to gain support for eradication as opposed to beautifully coloured parrots frequenting gardens and perceived as harmless to humans (Veitch and Clout, 2001; Yap and Sodhi, 2004). Today, social and public media provide gateways for groups to sensationalize or vilify invasive species and the groups trying to manage them, often driving public perceptions and support both for and against species removal (Veitch and Clout, 2001). Public education and outreach thus form an integral part of an invasive species control programme and its success (Stafford, 2010). Citizen scientists can also actively contribute to invasive bird research through atlassing (e.g. using the websites eBird, https://ebird.org/home, or BirdLasser, www.birdlasser.com/, accessed 15 October 2019), for example, or by reporting sightings and behaviours. This not only generates interest and support for research but also provides valuable data. The role of citizen scientists in invasive and exotic avifaunal studies is discussed further in Chapter 44.

Globalization continues to intensify and expedite the rate of species invasions due to increased trade, technology and travel (Meyerson and Mooney, 2007). This book aims to synthesize the global knowledge of widespread invasive bird species. Section 1 presents 34 species accounts of globally invasive bird species, predominantly identified by the ISSG GISD but also by others. Section 2 considers globally invasive bird species at a continental scale including the main introduction pathways and methods of control used. Finally, Section 3 presents some aspects of global management and impacts of these species. Regardless of opinion and personal campaigns, it is clear that there is still much to be learnt and that what we have learnt may alter in time as biological systems change. We, too, must respond similarly and be willing to adapt based on the evidence available. These responses must be ethical and respectful, and will no doubt be the source of many future discussions. It is hoped that this book will, at least in part, facilitate these discussions and further future research.

1.1 References

Adeney, J. (2001) European Starling (*Sturnus vulgaris*). Introduced Species Summary Project, Columbia University, New York. Available at: www.columbia.edu/itc/cerc/danoff-burg/invasion_bio/inv_spp_summ/Sturnus_vulgaris.html (accessed 15 October 2018).

Agoramoorthy, G. and Hsu, M.J. (2007) Ritual releasing of wild animals threatens island ecology. *Human Ecology* 35, 251–254.

Åkesson, S. and Hedenström, A. (2004) Migration. In: Sutherland, W.J., Newton, I. and Green, R.E. (eds) *Bird Ecology and Conservation: A Handbook of Techniques*. Oxford University Press, New York, pp. 161–178.

Aravind, N., Rao, D., Ganeshaiah, K., Shaanker, R.U. and Poulsen, J.G. (2010) Impact of the invasive plant, *Lantana camara*, on bird assemblages at Malé Mahadeshwara Reserve Forest, South India. *Tropical Ecology* 51, 325–338.

Barrowclough, G.F., Cracraft, J., Klicka, J. and Zink, R.M. (2016) How many kinds of birds are there and why does it matter? *PLoS One*, 11, e0166307.

Bhagwat, S. (2018) Management of nonnative invasive species in the Anthropocene. In: DellaSala, D.A. and Goldstein M.I. (eds) *Encyclopedia of the Anthropocene*, Vol. 3. Elsevier, Oxford, pp. 409–417.

BirdLife International (2008) Invasive alien species have been implicated in nearly half of recent bird extinctions. In: *BirdLife State of the World's Birds*. Available at: www.birdlife.org/datazone/sowb/casestudy/127 (accessed 15 October 2019).

Blackburn, T.M. and Duncan, R.P. (2001) Establishment patterns of exotic birds are constrained by non-random patterns in introduction. *Journal of Biogeography* 28, 927–939.

Blackburn, T.M., Lockwood, J.L. and Cassey, P. (2009) *Avian Invasions: The Ecology and Evolution of Exotic Birds*. Oxford University Press, Oxford.

Blackburn, T.M., Essl, F., Evans, T., Hulme, P.E., Jeschke, J.M., *et al.* (2014) A unified classification of alien species based on the magnitude of their environmental impacts. *PLoS Biology* 12, e1001850.

Bremner, A. and Park, K. (2007) Public attitudes to the management of invasive non-native species in Scotland. *Biological Conservation* 139, 306–314.

Brook, B.W., Sodhi, N.S., Malcolm, C.K.S. and Lim, H.C. (2003) Abundance and projected control of invasive house crows in Singapore. *Journal of Wildlife Management* 67, 808–817.

Butchart, S.H. (2008) Red List Indices to measure the sustainability of species use and impacts of invasive alien species. *Bird Conservation International* 18, S245–S262.

Butler, C.J. (2003) Population biology of the introduced Rose-ringed Parakeet *Psittacula krameri* in the UK. PhD thesis, University of Oxford, Oxford.

Clergeau, P. and Vergnes, A. (2011) Bird feeders may sustain feral Rose-ringed Parakeets *Psittacula krameri* in temperate Europe. *Wildlife Biology* 17, 248–252.

Coates, P. (2007) *American Perceptions of Immigrant and Invasive Species: Strangers on the Land*. University of California Press, Berkeley, Los Angeles/London.

Crowl, T.A., Crist, T.O., Parmenter, R.R., Belovsky, G. and Lugo, A.E. (2008) The spread of invasive species and infectious disease as drivers of ecosystem change. *Frontiers in Ecology and the Environment* 6, 238–246.

Czajka, C., Braun, M.P. and Wink, M. (2011) Resource use by non-native Ring-necked Parakeets (*Psittacula krameri*) and native starlings (*Sturnus vulgaris*) in central Europe. *Open Ornithology Journal* 4, 17–22.

Dean, W.R.J. (2000) Alien birds in southern Africa: what factors determine success? *South African Journal of Science* 96, 9–14.

del Hoyo, J., Collar, N.J., Christie, D.A., Elliott, A. and Fishpool, L.D.C. (2014) *HBW and BirdLife International Illustrated Checklist of the Birds of the World*, Vol. 1: *Non-passerines*. Lynx Edicions and BirdLife International, Barcelona, Spain/Cambridge.

del Hoyo, J., Collar, N.J., Christie, D.A., Elliott, A., Fishpool, L.D.C., *et al.*. (2016) *HBW and BirdLife International Illustrated Checklist of the Birds of the World*, Vol. 1: *Passerines*. Lynx Edicions and BirdLife International, Barcelona, Spain/Cambridge.

Doughty, R.W. (1972) Concern for fashionable feathers. *Forest History Newsletter* 16, 4–11.

Dyer, E.E., Redding, D.W. and Blackburn, T.M. (2017) The global avian invasions atlas, a database of alien bird distributions worldwide. *Scientific Data* 4, 170041.

Evans, T., Kumschick, S. and Blackburn, T.M. (2016) Application of the Environmental Impact Classification for Alien Taxa (EICAT) to a global assessment of alien bird impacts. *Diversity and Distributions* 22, 919–931.

Feare, C. and Craig, A. (1998) *Starlings and Mynas*. Helm, London.

Feare, C.J. (2010) The use of Starlicide® in preliminary trials to control invasive Common Myna *Acridotheres tristis* populations on St Helena and Ascension Islands, Atlantic Ocean. *Conservation Evidence* 7, 52–61.

Gaertner, M., Larson, B.M., Irlich, U.M., Holmes, P.M., Stafford, L., van Wilgen, B.W. and Richardson, D.M. (2016) Managing invasive species in cities: a framework from Cape Town, South Africa. *Landscape and Urban Planning* 151, 1–9.

Goddard, M.A., Dougill, A.J. and Benton, T.G. (2010) Scaling up from gardens: biodiversity conservation in urban environments. *Trends in Ecology and Evolution* 25, 90–98.

Grandi, G., Menchetti, M. and Mori, E. (2018) Vertical segregation by breeding ring-necked parakeets *Psittacula krameri* in northern Italy. *Urban Ecosystems*, 21, 1011–1017.

Gurevitch, J. and Padil, D.K. (2004) Are invasive species a major cause of extinctions? *Trends in Ecology and Evolution* 19, 470–474.

Haag-Wackernagel, D. and Moch, H. (2004) Health hazards posed by feral pigeons. *Journal of Infection* 48, 307–313.

Haemig, P.D. (1978) Aztec emperor Auitzoti and the Great-tailed Grackle. *Biotropica* 10, 11–17.

Haemig, P.D. (2012) Introduction of the Great-Tailed Grackle (*Quiscalus mexicanus*) by Aztec Emperor Auitzotl: provenance of the historical account. *Auk* 129, 70–75.

Hawkins, C.L., Bacher, S., Essl, F., Hulme, P.E., Jeschke, J.M., *et al.* (2015) Framework and guidelines for implementing the proposed IUCN Environmental Impact Classification for Alien Taxa (EICAT). *Diversity and Distributions* 21, 1360–1363.

Hayes, K.R. and Barry, S.C. (2008) Are there any consistent predictors of invasion success? *Biological Invasions* 10, 483–506.

Hernández-Brito, D., Carrete, M., Popa-Lisseanu, A.G., Ibáñez, C. and Tella, J.L. (2014) Crowding in the city: losing and winning competitors of an invasive bird. *PLoS One* 9, e100593.

Hernández-Brito, D., Carrete, M., Ibáñez, C., Juste, J. and Tella, J.L. (2018) Nest-site competition and killing by invasive parakeets cause the decline of a threatened bat population. *Royal Society Open Science* 5, 172477.

Hockey, P.A.R., Dean, W.R.J. and Ryan, P.G. (2005) *Roberts Birds of Southern Africa*, 7th edn. Trustees of the John Voelker Bird Book Fund, Cape Town.

Hulme, P.E. (2009) Trade, transport and trouble: managing invasive species pathways in an era of globalization. *Journal of Applied Ecology* 46, 10–18.

Hurles, M.E., Matisoo-Smith, E., Gray, R.D. and Penny, D. (2003) Untangling Oceanic settlement: the edge of the knowable. *Trends in Ecology and Evolution* 18, 531–540.

Iriarte, J. (2005) Invasive vertebrate species in Chile and their control and monitoring by governmental agencies. *Revista Chilena de Historia Natural* 78, 143–151.

IUCN Invasive Species Specialist Group (2015) Global Invasive Species Database. Available at: www.iucngisd.org/gisd/search.php (accessed 15 October 2019).

Jeschke, J.M. (2008) Across islands and continents, mammals are more successful invaders than birds. *Diversity and Distributions* 14, 913–916.

Kolar, C.S. and Lodge, D.M. (2001) Progress in invasion biology: predicting invaders. *Trends in Ecology and Evolution* 16, 199–204.

Kumschick, S. and Nentwig, W. (2010) Some alien birds have as severe an impact as the most effectual alien mammals in Europe. *Biological Conservation* 143, 2757–2762.

Kumschick, S., Alba, C., Hufbauer, R.A. and Nentwig, W. (2011) Weak or strong invaders? A comparison of impact between the native and invaded ranges of mammals and birds alien to Europe. *Diversity and Distributions* 17, 663–672.

Leven, M.R. and Corlett, R.T. (2004) Invasive birds in Hong Kong, China. *Ornithological Science* 3, 43–55.

Lever, C. (2005) *Naturalised Birds of the World*. Poyser, London.

Linz, G.M., Homan, H.J., Gaulker, S.M., Penry, L.B. and Bleier, W.J. (2007) European Starlings: a review of an invasive species with far-reaching impacts. *Managing Vertebrate Invasive Species* 24, 378–386.

Liu, X., McGarrity, M.E. and Li, Y. (2012) The influence of traditional Buddhist wildlife release on biological invasions. *Conservation Letters* 5, 107–114.

Long, J.L. (1981) *Introduced Birds of the World: The Worldwide History, Distribution and Influence of Birds Introduced to New Environments*. Universe Books, New York.

McDonald, R.I., Güneralp, B., Huang, C.-W., Seto, K.C. and You, M. (2018) Conservation priorities to protect vertebrate endemics from global urban expansion. *Biological Conservation* 224, 290–299.

McDowall, R.M. (1994) *Gamekeepers for the Nation: The Story of New Zealand's Acclimatisation Societies 1861–1990*. Canterbury University Press, Christchurch, New Zealand.

Meyerson, L.A. and Mooney, H.A. (2007) Invasive alien species in an era of globalization. *Frontiers in Ecology and the Environment* 5, 199–208.

Mooney, H.A. and Cleland, E.E. (2001) The evolutionary impact of invasive species. *Proceedings of the National Academy of Sciences USA* 98, 5446–5451.

Padian, K. and Chiappe, L.M. (1998) The origin and early evolution of birds. *Biological Reviews* 73, 1–42.

Palita, S.K., Ponkshe, A.V. and Dhar, U. (2011) Habitat enrichment and its impact on avian diversity: a study at GBPIHED, Kosi-Katarmal, Uttarakhand, India. *Current Science* 100, 1681–1689.

Picker, M. and Griffiths, C. (2013) *Alien and Invasive Animals: A South African Perspective*. Penguin Random House, Cape Town.

Pimental, D. (2007) Environmental and economic costs of vertebrate species invasions into the United States. In: Witmer, G.W., Pitt, W.C. and Fagerstone, K.A. (eds) *Managing Vertebrate Invasive Species: Proceedings of an International Symposium*. USDA/APHIS Wildlife Services, National Wildlife Research Center, Fort Collins, Colorado, pp. 2–8.

Pimentel, D., Lach, L., Zuniga, R. and Morrison, D. (2000) Environmental and economic costs of nonindigenous species in the United States. *BioScience* 50, 53–65.

Pimentel, D., McNair, S., Janecka, J., Wightman, J., Simmonds, C., *et al.* (2001) Economic and environmental threats of alien plant, animal, and microbe invasions. *Agriculture, Ecosystems and Environment* 84, 1–20.

Pimentel, D., Zuniga, R. and Morrison, D. (2005) Update on the environmental and economic costs associated with alien-invasive species in the United States. *Ecological Economics* 52, 273–288.

Rotherham, I. (2010) Eco-xenophobia – responding to our natural aliens. *Ecos* 31, 2–10.

Russell, J.C. and Blackburn, T.M. (2017) Invasive alien species: denialism, disagreement, definitions, and dialogue. *Trends in Ecology & Evolution* 32, 312–314.

Ryall, C. (2002) Further records of range extension in the House Crow *Corvus splendens*. *Bulletin of the British Ornithologists' Club* 122, 231–240.

Severinghaus, L.L. and Chi, L. (1999) Prayer animal release in Taiwan. *Biological Conservation* 89, 301–304.

Shieh, B.-S., Lin, Y.-H., Lee, T.-W., Chang, C.-C. and Cheng, K.-T. (2006) Pet trade as sources of introduced bird species in Taiwan. *Taiwania* 51, 81–86.

Shiu, H. and Stokes, L. (2008) Buddhist animal release practices: historic, environmental, public health and economic concerns. *Contemporary Buddhism* 9, 181–196.

Shwartz, A., Strubbe, D., Butler, C.J., Matthysen, E. and Kark, S. (2009) The effect of enemy-release and climate conditions on invasive birds: a regional test using the Rose-ringed Parakeet (*Psittacula krameri*) as a case study. *Diversity and Distributions* 15, 310–318.

Sohns, C. (2018) Bloody battles as birds maul European bats. *Nature* 557, 282.

Sol, D. and Lefebvre, L. (2000) Behavioural flexibility predicts invasion success in birds introduced to New Zealand. *Oikos* 90, 599–605.

Sol, D., Griffin, A.S., Bartomeus, I. and Boyce, H. (2011) Exploring or avoiding novel food resources? The novelty conflict in an invasive bird. *PLoS One* 6, e19535.

Sol, D., Bartomeus, I. and Griffin, A.S. (2012) The paradox of invasion in birds: competitive superiority or ecological opportunism? *Oecologia* 169, 553–564.

Somerville, A.D., Nelson, B.A. and Knudson, K.J. (2010) Isotopic investigation of pre-Hispanic macaw breeding in Northwest Mexico. *Journal of Anthropological Archaeology* 29, 125–135.

Stafford, L. (2010) Mallard strategy for South Africa. Available at: https://www.wingshooters.co.za/pdf/NationalMallardStrategy-May10.pdf (accessed 26 March 2020).

Starr, C. and Taggart, R. (2004) *Biology: The Unity and Diversity of Life*. Thomson Brooks/Cole, Pacific Grove, California.

Strubbe, D. and Matthysen, E. (2009) Experimental evidence for nest-site competition between invasive Ring-necked Parakeets (*Psittacula krameri*) and native nuthatches (*Sitta europaea*). *Biological Conservation* 142, 1588–1594.

Strubbe, D. and Matthysen, E. (2011) A radiotelemetry study of habitat use by the exotic Ring-necked Parakeet *Psittacula krameri* in Belgium. *Ibis* 153, 180–184.

Tabak, M.A., Piaggio, A.J., Miller, R.S., Sweitzer, R.A. and Ernest, H.B. (2017) Anthropogenic factors predict movement of an invasive species. *Ecosphere* 8, e01844.

Temple, S.A. (1992) Exotic birds: a growing problem with no easy solution. *Auk* 109, 395–397.

van Rensburg, B.J., Peacock, D.S. and Robertson, M.P. (2009) Biotic homogenization and alien bird species along an urban gradient in South Africa. *Landscape and Urban Planning* 92, 233–241.

Veitch, C.R. and Clout, M.N. (2001) Human dimensions in the management of invasive species in New Zealand. In: McNeely, J.A. (ed.) *The Great Reshuffling: Human Dimensions of Invasive Alien Species.* IUCN, Gland, Switzerland/Cambridge, pp. 63–71.

Weber, W.J. (1979) *Health Hazards from Pigeons, Starlings and English Sparrows: Diseases and Parasites Associated with Pigeons, Starlings, and English Sparrows which Affect Domestic Animals.* Thomson Publications, Fresno, California.

West, B. and Zhou, B.-X. (1989) Did chickens go north? New evidence for domestication. *World's Poultry Science Journal* 45, 205–218.

Wilcove, D.S., Rothstein, D., Bubow, J., Phillips, A. and Losos, E. (1998) Quantifying threats to imperiled species in the United States. *Bioscience* 48, 607–615.

Williams, F., Eschen, R., Harris, A., Djeddour, D., Pratt, C., *et al.* (2010) *The Economic Cost of Invasive Non-native Species on Great Britain.* CABI Project No. VM10066.

Yap, C.A.M. and Sodhi, N.S. (2004) Southeast Asian invasive birds: ecology, impact and management. *Ornithological Science* 3, 57–67.

Zavaleta, E.S., Hobbs, R.J. and Mooney, H.A. (2001) Viewing invasive species removal in a whole-ecosystem context. *Trends in Ecology & Evolution* 16, 454–459.

2 Common Starling (*Sturnus vulgaris* Linnaeus, 1758)

Adrian J.F.K. Craig*

Department of Zoology and Entomology, Rhodes University, Grahamstown, 6139, South Africa

Citation: Craig, A.J.F.K. (2020) Common Starling (*Sturnus vulgaris* Linnaeus, 1758). In: Downs, C.T. and Hart, L.A. (eds) *Invasive Birds: Global Trends and Impacts*. CAB International, Wallingford, UK, pp. 9–24.

2.1 Common Names

Common Starling, European Starling, Eurasian Starling.

2.2 Nomenclature

Two independent molecular phylogenies of the Common Starling (*Sturnus vulgaris* Linnaeus, 1758) come to congruent conclusions, with the genus *Sturnus* restricted to the two species, *S. vulgaris* and *S. unicolor*, which form the sister group to other Asian starlings and mynas (Lovette *et al.*, 2008; Zuccon *et al.*, 2008).

2.3 Distribution

The natural breeding distribution of the Common Starling is from Iceland, the Faroe and Shetland Islands, the Azores and Canary Islands, the British Isles, Scandinavia, France and northern Spain, eastwards through Europe and Russia, to Lake Baikal, northern and eastern Kazakhstan, western Mongolia and western China (Xinjiang). Southern breeding limits are Turkey, Iraq, northern Iran, Turkmenistan, Afghanistan, Pakistan, Kashmir, eastern Uzbekistan and Tajikistan (Fig. 2.1). Migrants in the non-breeding season move south to North Africa, the Middle East, Arabia, Iraq, southern Iran, north-western India and north-eastern China (Craig and Feare, 2009). There is the occasional visitor to coastal China (Hebei), the Korean Peninsula, Japan, Sakhalin and Taiwan (Brazil, 2009).

Long (1981) stated that the Common Starling had been introduced successfully in the USA, Jamaica, South Africa, Australia, New Zealand and the Chatham Islands. It had then colonized the Bahamas, Bermuda, Canada, Alaska, Mexico, Fiji and many small islands off Australia and New Zealand. It was possibly introduced successfully to Vanuatu (New Hebrides) and Tonga but was apparently unsuccessful in Cuba, Venezuela and Ulan-Ude (USSR). Lever (2005) gave its naturalized range as the USA, Canada, Mexico, the West Indies, South Africa, Australia, Lord Howe Island, Norfolk Island, New Zealand and Chatham, Auckland, Campbell and Macquarie Islands, Kermadec Islands, Fiji, Tonga and possibly Vanuatu. He did not discuss failed introductions.

Common Starlings were apparently introduced to Argentina (Peris *et al.*, 2005) and have since been recorded in neighbouring Uruguay (Mazzulla, 2013) and in Brazil (Cavitione e Silva *et al.*, 2017). From South Africa, they have invaded adjoining areas of Namibia (Cunningham, 2016) and Lesotho (Kopij, 2001a).

2.4 Description

Male Common Starlings in fresh plumage have pale buff to whitish tips on all of the body feathers, producing a speckled appearance (Fig. 2.2A). The head is black with purplish-green iridescence, the wings and tail brown with some gloss, and there are narrow buff margins to the feathers. The chin and throat are blackish with some purple gloss, the breast and upper belly dark brown and glossed green, but they have purple gloss on the flanks. The belly and undertail coverts are matt brown with broad whitish margins. The iris is dark brown, the bill blackish and the legs brown. In the breeding season, the pale tips to most feathers have been lost, so that the male bird appears dark and glossy, with a strong purple gloss on the head and throat and a green gloss on the mantle, rump and breast; the elongated throat and upper breast feathers are frequently erected in display. The bill is a striking yellow colour, with the base steel blue, and the legs are pink (Craig and Feare, 2009).

*Corresponding author: a.craig@ru.ac.za

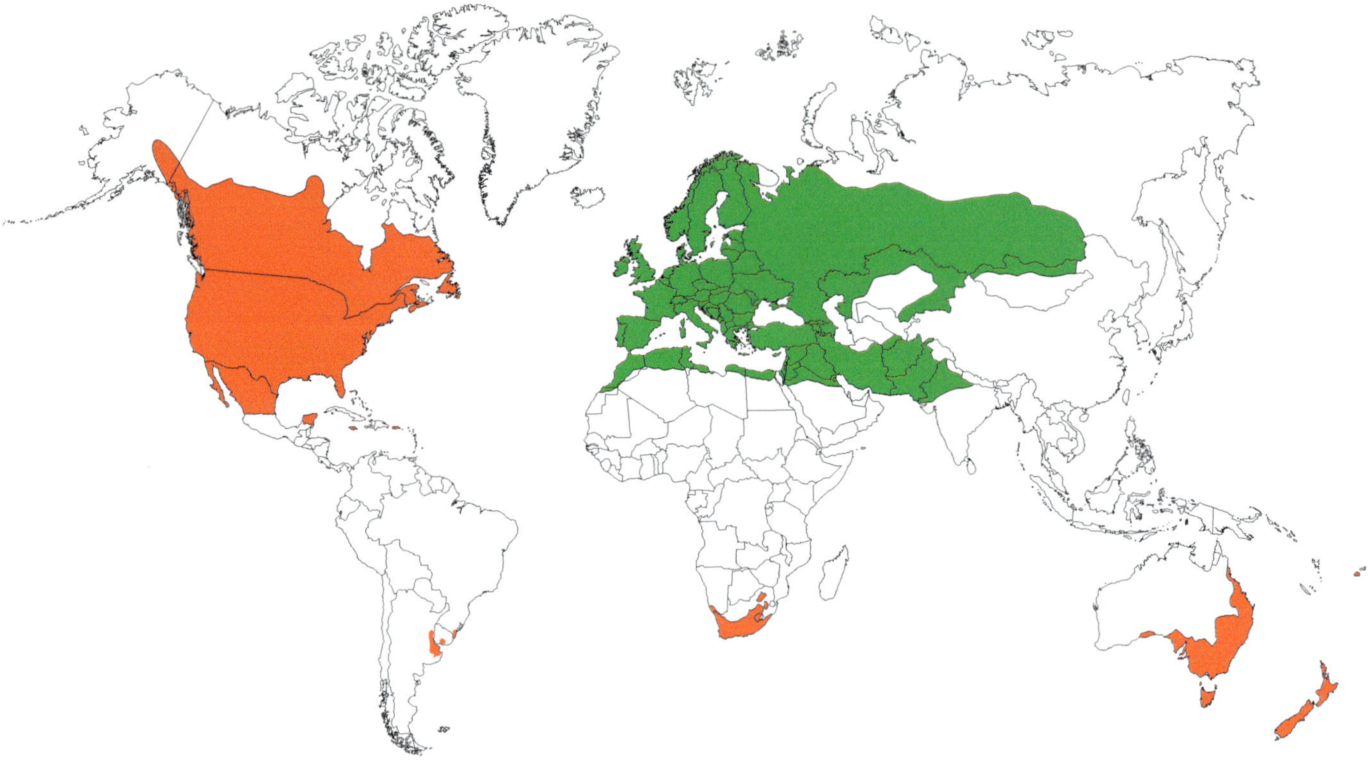

Fig. 2.1. Global distribution of the Common Starling (*Sturnus vulgaris*) showing the natural (green) and invaded (red) ranges.

Fig. 2.2. Adult Common Starlings in fresh non-breeding dress. (A) Male with brown iris. (B) Female with pale outer ring to iris. (©Photographs: Lynette Rudman).

Female Common Starling plumage is similar to that of the male, but the pale tips to the feathers are more persistent and the plumage is generally less glossy (Fig. 2.2B). The iris is dark brown, with a clear pale ring on either the inner or outer margin, while the bill of breeding birds is yellow with a pinkish base. Juvenile birds initially are grey-brown above, with buffy edges to the feathers of the wings and tail, a whitish throat, breast feathers that are whitish at the base with grey-brown tips, and dark shafts and tips to the belly feathers. The bill is dull brownish-black, the iris initially grey, becoming brown, and the legs pinkish-brown. During the moult to adult plumage, a very blotchy appearance is produced, and the head feathers are the last to be replaced (Craig and Feare, 2009).

Both adult and juvenile Common Starlings have a complete moult after the breeding season in both native and introduced populations, lasting for 80–100 days (Kessel, 1957; Cooper and Underhill, 1991; Rothery *et al.*, 2001). There may be an incomplete or interrupted moult in migrant populations or late-breeding birds (Evans, 1986).

In all Common Starling populations, the sexes are most easily distinguished by the difference in colour at the base of the lower mandible during the breeding season (Kessel, 1951; Delvingt, 1961a; Wydoski, 1964; Coleman, 1973), with iris coloration the next most reliable criterion (Smith *et al.*, 2005). The hackles (elongated feathers of the throat) are notably narrow and elongated in male birds but short and stubby in first-year females. However, first-year males and adult females cannot be separated on the appearance of these feathers (Kessel, 1951; Delvingt, 1961a; Coleman, 1973). Whereas the nestling sex ratio in the USA showed a slight female bias, the adult population was 59% male, suggesting higher female mortality (Davis, 1959). In the UK, Bradbury *et al.* (1997) also found a significant female bias in 108 broods containing 350 1-week-old chicks. Coulson (1960) calculated the mean annual mortality of British starlings as 53%, and suggested that differential mortality in the first year (39% for male birds, 70% for females) would explain the imbalance in the adult sex ratio. He noted that there was increased mortality during the breeding season, and ascribed this difference in first-year survival to the higher proportion of first-year females breeding. In the Czech Republic, first-year mortality was estimated at 68%, but by the third year, annual mortality was less than 40%, with some birds expected to survive for 15 years (Beklová, 1972). Longevity records reported were 16 years in the UK, 17 years in the USA and 21 years in Germany (Feare, 1984).

In South Africa, white rectrices were noted in 28 subadult Common Starlings, but only once in an adult (Skead, 2006). It is not clear whether these feathers are replaced by normal-coloured rectrices at the first complete moult or whether these birds suffer higher mortality at an early stage; white rectrices are present in some subadults but extremely rare in adult birds in the native range (C.J. Feare, personal communication). There is a record of three albino nestlings in England, with a normal adult in attendance at the nest (Garner, 1997).

There is some seasonal variation in Common Starling body mass, but male birds are consistently heavier on average than females (Europe: Eble, 1963; South Africa: Cooper and Underhill, 1991; New Zealand: Coleman and Robson, 1975).

Whereas 13 subspecies of Common Starling are recognized over its breeding range, distinguished by plumage coloration and other morphological differences, the introduced populations show only minor changes in size in New Zealand (Ross and Baker, 1982) and in wing morphology in the USA (Bitton and Graham, 2015) and Australia (Cardilini *et al.*, 2016; Phair *et al.*, 2018), probably reflecting random local variations in small, isolated populations. Little genetic variation was found within the populations in New Zealand (Ross, 1983), South Africa (Phair *et al.*, 2018) and the USA (Cabe, 1993), but some genetic differentiation was evident in Australia (Phair *et al.*, 2018).

2.5 Diet

The Common Starling is essentially omnivorous and opportunistic in food selection, feeding on a wide variety of plant and animal material. Vertebrate food items include small reptiles and amphibians and the eggs of other birds, while plant foods include fruit and seeds of a wide range of wild and cultivated plant species, such as yew (*Taxus* spp.), oak (*Quercus* spp.), apple (*Malus* spp.), pear (*Pyrus* spp.), cherry and plum (*Prunus* spp.), rowan (*Sorbus* spp.), elder (*Sambucus* spp.), nightshade (*Solanum* spp.), bryony (*Bryonia* spp.), buckthorn (*Hippophae* spp.), olive (*Olea* spp.), grape (*Vitus* spp.) and grains such as sorghum, wheat, oats, barley, millet and maize; nectar may also be taken from flowers (e.g. *Aloe* and *Erythrina* spp.). Often, the bulk of the food is insects, both adults and larvae, such as craneflies (Tipulidae), butterflies and moths (Lepidoptera), mayflies (Ephemeroptera), dragonflies (Odonata), lacewings (Neuroptera), grasshoppers and crickets (Orthoptera), caddis flies (Trichoptera), flies (Diptera), ants, bees and wasps (Hymenoptera) and beetles (Coleoptera). Other invertebrate food includes small crabs (Decapoda), spiders (Araneae), harvestmen (Opiliones), millipedes (Diplopoda), centipedes (Chilopoda), woodlice (Isopoda), earthworms (Oligochaeta) and snails (Gastropoda). The birds will also scavenge items discarded by humans, and readily take pellets fed to domestic stock and pets (Craig and Feare, 2009).

Dunnet (1955) found that in Scotland most Common Starling food was taken from the upper layers of the soil, with nestlings fed largely (81%) on leatherjackets (larvae of *Tipula* spp.) and earthworms (13%). For resident birds in Halle, Germany, animal food made up 70% of the stomach contents by volume over the year (over 90% in winter), while plant food comprised more than 50% only in June, and was over 20% for July–October. The animal component was mainly insects – Coleoptera (137 beetle species identified), Diptera (25 species), Lepidoptera (20 species, many agricultural pests) and Odonata – and small snails. Plants (27 species) were mostly noted as seeds, but plums, grapes and berries were also identifiable (Eble, 1963). Birds collected from a winter roost in England took mostly vegetable matter during periods of snow (grain of oats and wheat) and household waste. Insect food was mainly larvae of Tipulidae, as well as Coleoptera and Diptera, and also earthworms and snails (Taitt, 1973).

Common Starlings take nectar from indigenous plants in South Africa (e.g. *Aloe* and *Erythrina* spp.; Skead, 1995) and Australia (e.g. *Banksia* spp.; French *et al.*, 2005) and from exotic plants (e.g. *Erythrina* spp. in the USA; Feare, 1993). However, starlings are generally unable to digest sucrose (Martínez del Rio *et al.*, 1988), which presumably restricts their use of nectar to plant groups where other sugars dominate.

Near Algiers, 56% of the diet of Common Starlings by item was insects, particularly harvester ants, as well as many beetles; olives were a minor element (Djennas-Merrar *et al.*, 2016).

In the Western Cape, South Africa, Winterbottom and Liversidge (1954) noted Common Starlings feeding on hairy caterpillars (Lepidoptera), Orthoptera, fruit and arils of the introduced coastal wattle (*Acacia cyclops*). On the Cape Peninsula, birds fed in the intertidal zone both on washed-up kelp and on seaweed attached to rocks, even when the tide was surging there (Skead, 1966). In central South Africa, adult rather than larval Coleoptera, larvae of Lepidoptera, berries, small fruit and seeds were found in the stomachs of Common Starlings (Kopij, 2000).

In North America, when the Common Starling diet was calculated by dry weight of stomach contents, plant material comprised 61%, but numerically animal matter made up 66% of the items, and as it is more digestible, it is likely to be the more important component. The peak of animal food was in June (late summer), with more plant food as large roosts built up in autumn (Fischl and Caccamise, 1987). Dipteran prey in the USA included warble flies (*Hypoderma* spp.), which are responsible for nasal bots in cattle (Bauer, 1978). Dominance by the adults may prevent juvenile birds from foraging in prime areas (Maccarone, 1987a).

In New Zealand, Common Starling nestlings were fed predominantly insects (Coleoptera, Hemiptera, Diptera and Lepidoptera), as well as spiders, isopods, snails and earthworms, with cherries, maize and grass seeds in some gizzards (Moeed, 1975). At Christchurch Airport, where the starlings fed primarily by probing the grass mat, they took earthworms, spiders, and insect larvae (Coleoptera, Lepidoptera and Diptera) (Moeed, 1976). In rural Hawke's Bay, almost 40% of the insects taken by starlings were pest species, with their most common prey being Coleoptera, and at times Diptera, Hemiptera, Orthoptera and Dermaptera; other animal food included lycosid spiders, millipedes, centipedes, earthworms and snails. Plant food included apple, grape, pea, pear, tomato and asparagus (Moeed, 1980).

2.6 Introduction and Invasion Pathways

2.6.1 North America

Gebhardt (1959) stated that there had been unsuccessful attempts to introduce Common Starlings to the USA in the 1870s and earlier. However, it is generally agreed that the first successful introduction was of about 100 birds in Central Park, New York City, in 1890–1891 (Gebhardt, 1954; Cabe, 1993; Linz *et al.*, 2018). They had crossed the Appalachian Mountains by 1921, reached the Mississippi River in 1938 and were in California on the West Coast of the USA in 1942. From there, they moved north to Washington, Oregon State and then British Columbia in Canada in 1947; over a mere 7 years, a winter roost in Vancouver, BC, swelled to 40,000 birds (Myres, 1958). Starlings arrived in Alaska in 1952 (Lever, 2005). Unsuccessful attempts were made to introduce Common Starlings to Canada from 1875, but they invaded Ontario in 1914 and soon spread to adjoining provinces, reaching Alberta in the west by 1934 (Lever, 2005). Common Starlings may have crossed into Mexico from Texas by 1935; by the early 1970s, they were found in Guanajuato, northern Veracruz and Yucatan, and are still expanding southwards (Lever, 2005).

A 1940 Common Starling population estimate for North America was 50 million birds; by 1993, it was considered to exceed 200 million (Cabe, 1993). Whereas in other regions introduced starlings were essentially sedentary, in North America the northern populations retained a migratory habit, with their migration routes determined by local topography. The breeding range was apparently expanded by these migrants, and by the dispersal of first-year and non-breeding second-year birds (Kessel, 1953; Cabe, 1999). This may account in part for their explosive spread in the region (Lever, 2005). There is little genetic difference among North American Common Starling populations (Cabe, 1993), which supports the idea of a small founder population and regular dispersal.

2.6.2 West Indies

Common Starlings were released in Jamaica in 1903–1904 and are now common locally; they were also established on Grand Bahama and the Biminis, and had been recorded on Puerto Rico, Cuba, the Virgin Islands and the Cayman Islands (Lever, 2005). According to BirdLife International (2018), Common Starlings are currently present on all these islands, and on Hispaniola (Haiti and the Dominican Republic), the Turks and Caicos Islands, and Bermuda.

2.6.3 Argentina

The first reports of Common Starlings came from Buenos Aires in 1987, presumed to be either a deliberate release or escapees from an aviary. By 1993, small flocks were noted 200 km away, and in 2001 they were reported 400–500 km from the city. Most nests were in cavities excavated by woodpeckers (Peris *et al.*, 2005). Expansion into the pampas followed conversion of grasslands for agriculture and establishment of trees. The range now covers 67,000 km^2 with an average range expansion of 22 km/year. Urban areas serve as centres, and newly colonized sites are always close to small settled areas with trees for roosting and nesting (Ibañez *et al.*, 2016a, 2017).

2.6.4 Uruguay

Mazzulla (2013) reported that the first sightings of Common Starlings were in Montevideo, with small numbers recorded regularly. They were assumed to be escapees but could have invaded from Argentina.

2.6.5 Brazil

Common Starlings were reported in 2014 from the southernmost state, Rio Grande do Sul, and a flock including juvenile birds was seen in 2017 (Cavitione e Silva *et al.*, 2017). The source of the birds is unknown.

2.6.6 St Helena

Brooke *et al.* (1995) reported that there had been an unsuccessful attempt to introduce the Common Starling to this island in 1852.

2.6.7 South Africa

Although Winterbottom and Liversidge (1954) and Liversidge (1985) accepted the claim by Meinertzhagen (1952) that he had provided Cecil Rhodes with 18 Common Starlings to take to Cape Town in 1899, Brooke *et al.* (1986) noted that this did not match the timing of Rhodes' travels, and 1897 seemed to be the probable date of arrival. There is no evidence that Meinertzhagen was involved, and the number of birds introduced by Rhodes is unknown. After colonizing the Cape Peninsula, the birds spread to the north, reaching the Berg River 140 km from Cape Town by 1928, and were found north of Clanwilliam (250 km from

Cape Town) by 1952. Expanding eastwards, Common Starlings reached the Hottentot Holland Mountains (50 km) by 1910, George in 1948 (500 km) and East London (1050 km) by 1966. The rate of expansion was notably slower than in the USA (Gebhardt, 1954, 1959; Winterbottom and Liversidge, 1954; Liversidge, 1962; Skead, 1995). In the former Transkei region, Common Starlings were found in Kei Mouth in 1971 and in Umtata in 1981 (Quickelberge, 1989). The first sighting in Durban, KwaZulu-Natal Province, was in 1973 (Cyrus and Robson, 1980), but by 1993, there were regular records along the southern coast of this province (Harrison *et al.*, 1997). Occupation of the dry interior was limited, but by 1970, birds were breeding at the mouth of the Orange River on the border with Namibia (Brown, 1985). For the Free State Province, there was a breeding record on the southern border in 1986 (Earlé and Grobler, 1987), but only irregular visitors to Bloemfontein in the centre of the province until 1996, with the first breeding records in the summer of 1997/98 (Kopij, 2001b). In the Western Cape Province, Common Starlings are now present throughout the region, and they were the sixth most frequently recorded species from 1982 to 1986 (Hockey *et al.*, 1989). From recent Southern African Bird Atlas Project data, Ivanova and Symes (2018) noted that the abundance of Common Starlings was consistent with earlier records but that range expansion was still occurring. However, in some cases, new territory was occupied, while the birds failed to persist in some previously occupied areas. Regular occurrence in Gauteng (1400 km north-east of Cape Town) started in the 21st century.

Comparison of British and South African Common Starling populations indicated that both show great spatial and temporal variation, best described by the rule 'stay if conditions are good/ disperse if conditions are bad' (Hui *et al.*, 2012). There appears to be gene flow throughout the South African population, with genetic diversity at the range margins maintained by long-range dispersal events (Berthouly-Salazar *et al.*, 2013).

2.6.8 Lesotho

The first recorded nesting was in Roma in 1991; by 2000, there were at least 20 pairs (Kopij, 2001a).

2.6.9 Namibia

Whereas the first assessment of Common Starling occurrence noted the birds as restricted to Oranjemund, where they had been reported since 1970 (Brown, 1985), recent reports from 70 km south of Grünau, Hohenfels in the east and Lüderitz on the coast suggest a considerable range expansion (Cunningham, 2016).

2.6.10 Australia

Several hundred Common Starlings were released in Victoria, South Australia and New South Wales between 1856 and 1880; there was apparently an unsuccessful introduction to Queensland in the 1860s. Tasmania was colonized by birds from New Zealand, released in Hobart during the late 19th century – various dates have been reported. The birds expanded their range through the coastal areas and adjacent interior, and reached southern Queensland by 1920, while the arid interior represented a barrier, and westward spread along the south coast was slow. The birds are still vagrants in the Northern Territory and over most of Western Australia (Higgins *et al.*, 2006).

2.6.11 New Zealand

Hundreds of birds were released by acclimation societies on both North and South Island between 1862 and 1883; they were reportedly also introduced to Chatham Island during this period. After an initial increase in numbers, by the 1920s they were described as abundant in most parts of the country. Populations reportedly declined in the 1940s when the insecticide dichlorodiphenyltrichloroethane (DDT) was widely used but recovered after it was banned (Higgins *et al.*, 2006). According to Lever (2005), Common Starlings reached Campbell Island by 1907, the Kermadecs by 1910, Macquarie by 1930 and the Antipodes by 1952.

2.6.12 Pacific Islands

By 1951, Common Starlings were well established in Fiji; although it has been suggested that they arrived as immigrants from the Kermadec Islands, it is more probable that they were introduced deliberately, perhaps around 1930 or even earlier (Lever, 2005). In the Tonga group, Rinke (1987) noted that Common Starlings were regular on the main island of Tongatapu, and reported up to 100 birds on 'Eua, where they had first arrived in 1974. It is not clear whether they had been introduced or had invaded from Fiji or the Kermadec Islands. On Vanuatu, birds were reported in the 1950s (Lever, 2005) but are currently considered vagrants there (BirdLife International, 2018).

2.7 Breeding Behaviour

In the northern hemisphere, the Common Starling breeding season is normally March–July with some local and annual variation, while in the southern hemisphere (introduced populations), it is September–December. The birds are often double-brooded, and some males may be polygynous with up to five mates in a single season. In Europe, nests may be clustered in colonies, although many birds also breed singly; in the introduced populations, solitary nesting seems to be the rule. The nest is typically placed in a cavity in a tree, cliff, building or other structure, or in a nest box; occasionally, holes in the ground are used, or nests may even be placed in shrubs or on the ground. Nests are bulky structures of dry grass, conifer needles, twigs, string and other material, while the cup is lined with softer material such as feathers, moss and hair; often, green leaves and flowers may be added by the male. The clutch of four to six plain blue eggs is incubated by both male (25%) and female (75%) by day, but only by the female at night. The incubation period is 11–14 days and the nestling period about 21 days, with the young fed by both parents and for at least 5 days after leaving the nest. Polygynous males usually assist only the first female, so that subsequent mates rear their young unaided. Intraspecific brood parasitism occurs at a low frequency, and parasitic females generally remove a host egg before laying their own (Craig and Feare, 2009).

Polygyny has been found in both the Common Starling native range and in North America. Kessel (1950), in New York, found a male defending three nesting sites, and acquiring three mates during one season. In Belgium, Pinxten *et al.* (1989) observed that about 30% of males in the early phase of the breeding season tried to attract a second female; on average, 20% succeeded. Older males were usually more successful in acquiring more than one female. Komdeur *et al.* (2005) showed that older males were not only more likely to be polygynous but generally had higher breeding success, while older females nested earlier and laid larger clutches with higher hatching success.

Brood parasitism has been recorded both in the Common Starling native range (Yom-Tov *et al.*, 1974) and in North America. In Scotland, it appeared that parasitism was influenced by nest availability and desertion by females at early nests (Evans, 1988). Early clutches were most often parasitized in the USA, and females tended to remove an egg before adding their own to the clutch (Lombardo *et al.*, 1989). Intraspecific brood parasitism might be an option for a female losing a clutch through predation or disturbance (Feare, 1991). The rate of nest parasitism in Belgium varied in different years, averaging 15% of first clutches but only 2% of second clutches; parasites were likely to be females that had lost their own clutch. They usually removed an egg, adding only one to the clutch (Pinxten *et al.*, 1990). Communal breeding, in which two females laid eggs in the same nest (all fathered by the male in attendance) and then all three birds fed the young, has been recorded, but is clearly exceptional in this species (Pinxten *et al.*, 1994). Replacement males during incubation were likely to destroy the eggs, whereas at an earlier stage they would father some of the young and accept the eggs already in the nest (Smith *et al.*, 1996). To date, there are no reports of polygyny or brood parasitism from South Africa (Hockey *et al.*, 2005) or Australasia (Higgins *et al.*, 2006).

2.7.1 Scotland

Over 3 years, Common Starling first broods were more successful than second broods, with the fledging rate from eggs laid being 79%, 85% and 81% compared with 64%, 74% and 78% for second broods. Synchrony of egg-laying was very marked for first clutches, whereas second clutches in the same nest sites were much less synchronized (Dunnet, 1955).

2.7.2 Belgium

Older Common Starling males arrived first, in February, while from April second-year birds arrived; males arriving in May were unlikely to find a mate. In this colony, about half the sexually mature females were second-year birds; females over 3 years old laid eggs a few days earlier (Verheyen, 1969).

2.7.3 Norway

At 69°N, Common Starlings were sedentary and double-brooded, with egg-laying in May and the first brood fledged in late June (Lundberg, 1987).

2.7.4 Sweden

The Common Starling population at 64°N is migratory and single-brooded, although a decline in food availability showed the same timing as for resident Norwegian birds. Eggs are laid in May, and the young fledge in late June (Lundberg, 1987). The timing of breeding is perhaps determined by a circannual rhythm at these latitudes (Lundberg and Eriksson, 1984).

2.7.5 Finland

The Common Starling population is migratory, with birds arriving mid-March to mid-April. Egg-laying then follows from the last week in April to the first week in May. Laying is very closely synchronized for these first clutches. Mean clutch size in different years varied from 4.6 to 5.8 eggs; hatching success was 87% and 2.6–5.2 young fledged from each clutch (mean 3.6). The decrease in population has been ascribed to a reduction in stock and loss of grazing areas (Korpimäki, 1978).

2.7.6 North America

Common Starling nesting may start in late February in the south, and in late March in northern USA and Canada, with egg-laying from March to June. North of 48°N, second broods are not usually attempted. For the first brood, eggs are laid over 3–4 days, with incubation starting once the last egg is laid. The interval between the first and second clutches is usually 40–44 days. Intermediate clutches, between typical first and second broods, may be late-arriving migrants, replacement clutches for early nest losses or first-year females. Breeding success, in terms of eggs laid that produce fledged young, ranged from 57% to 83% for first broods and from 47% to 71% for second broods (Cabe, 1993).

At a Common Starling nest-box colony in Ontario, Canada, there were two distinct breeding periods in April and June (with some ‘intermediate’ clutches). There was a marked synchrony in timing of first clutches, with earlier breeding in a warm spring; 92% of first broods were followed by a second brood, but no intermediate birds were double-brooded. First clutches averaged 5.6 eggs and second clutches 5.0 eggs, with better hatching success in first clutches. Breeding success from eggs to fledglings was 83% for first clutches, 72% for second clutches and 71% for intermediate clutches (Collins and

De Vos, 1966). To the west (Vancouver, British Columbia), eggs were laid in April, and the birds continued to use communal roosts until the last egg was laid. The hatching success was 84% in first clutches and 69% in second clutches. Of the eggs laid, 76% produced fledged young from first broods and 71% from second broods (Johnson and Cowan, 1974).

For birds in Arizona, Common Starling nests were placed in holes in saguaro cacti (*Carnegiea gigantea*) and in the crowns of palm trees. Egg-laying extended from March to May, with the average clutch size being 4.4 eggs, the brood size 3.7 and 2.3 young fledged per brood. First broods were more successful (Royall, 1966). In Ohio, there was a low occupancy rate of nest boxes by Common Starlings, with more boxes occupied by native species, so that the invaders apparently favoured other cavities (Seamans *et al.*, 2015). Kessel (1957) reported that 76% of 1094 eggs produced fledged young in the USA.

2.7.7 South Africa

Common Starling laying dates are September–December (Hockey *et al.*, 2005), and birds sometimes nest in burrows made by African Pied Starlings (*Lamprotornis bicolor*) (Vincent, 1949; Skead, 1995), but on Dassen Island they nested in low shrubs and even on the ground (Cooper and Underhill, 1991). Some birds also nested on an offshore shipwreck (Brooke, 1995). Breeding success on Dassen Island averaged 2.3 nestlings per breeding attempt, with an average clutch size of 3.8 and egg-laying highly synchronized (Cooper and Underhill, 1991). Second broods have not been reported.

2.7.8 Lesotho

Breeding occurred in August–December, with regular second broods. Nest sites were re-used in successive years; 16 nests were in buildings and 12 in holes in exotic trees (Kopij, 2009).

2.7.9 Australia

Egg-laying was from late July to December, with a peak in September–October. The mean clutch size was 4.6 eggs. For clutches with a full history (n = 185 eggs), 58% of eggs hatched and 28% produced fledged young, with a mean of 1.3 young per brood. Many nests were destroyed by people (Higgins *et al.*, 2006). From nest record cards, natural tree hollows were used as often as other sites (Kentish and Peace, 1987).

2.7.10 New Zealand

Common Starling egg-laying is from September to December, with a peak in October and with late broods and second clutches in November. The mean clutch size is 4.5 eggs. In a study on North Island, 52% of eggs laid (n = 17,326) hatched successfully and 33% produced fledglings, with a mean of 1.6 young per brood (Flux and Flux, 1981; Higgins *et al.*, 2006). Artificial nest sites (nest boxes) placed in hedgerows were more attractive to Common Starlings than nest boxes attached to buildings. Boxes with round entrances and those highest above ground were favoured (Coleman, 1974). Later, Moeed and Dawson (1979) found that the breeding success of starlings in nest boxes was influenced by the size of the base (optimal 31 cm^2), the depth of the box and the width of the entrance. When nest sites were in short supply, vicious fights to the death occurred; in one nest-box colony, more than 40 birds were killed in same-sex fights (Flux and Flux, 1992). Bull and Flux (2006) compared breeding at sites in different regions and found that laying of the first clutch was closely synchronized at each site. The mean clutch size was consistently larger at the most southerly locality, and breeding success was also greater; this might be accounted for by a richer food supply. Young fledged as a proportion of eggs laid varied widely, from 19.1% to 60.9 %. There was a variable frequency of second broods.

2.8 Habitat

Common Starlings favour open habitats near short grass areas where most foraging is done (Craig and Feare, 2009). Occupied areas include highly modified suburban and agricultural landscapes, with access to cavities for nesting; these can be in trees or in buildings. Non-breeding birds exploit a wide range of habitats from moorland to saltmarshes and the intertidal zone, stubble fields, orchards, refuse dumps and sewage-treatment works. Roost sites may be in trees, reed beds or on buildings and other structures such as under bridges. It is primarily a bird of lowland areas but occurs at up to 1500 m in Switzerland and 2500 m in the Himalayas (Craig and Feare, 2009). In rural Germany, there was a seasonal shift to meadows and pastures in late summer and then to orchards as fruit ripened, while overwintering birds often roosted in woodlands but fed in close association with people (Eble, 1963). Occurrence of the species in South Africa was best predicted by human population density, irrigated area and heterogeneous vegetation, whereas primary productivity and cultivated area were negatively correlated with Common Starling distribution (Hugo and Van Rensburg, 2009). Although in a French city the birds were evenly distributed throughout different zones, breeding was less successful in the central urban area than in the peri-urban and suburban zones (Mennechez and Clergeau, 2006). Green roofs in cities may offer a new habitat to Common Starlings and other species (Belcher *et al.*, 2018). Starlings are adaptable, and where food and nest sites are available, they have colonized desert environments in the USA and Mexico (Johnston López *et al.*, 2015). Liversidge (1985) noted that, in South Africa, newly arrived Common Starlings were often associated with palm trees.

Range expansion of Common Starlings in Spain from 1970 to 1980 was linked to areas where cereals were cultivated (Motis *et al.*, 1983). Expansion slowed after the Common Starling met the Spotless Starling (*Sturnus unicolor*) (Ferrer *et al.*, 1991). In Argentina, Common Starlings have colonized the pampas as areas have been converted to agriculture, with trees and homesteads established (Zufiaurre *et al.*, 2016).

Supplementary feeding in gardens is seen as favouring exotic species such as Common Starlings in New Zealand (Galbraith *et al.*, 2015, 2017). Generally, human activities induce habitat changes that suit this species better than most native bird species.

2.9 Impacts

Bird strikes – collisions between birds and aircraft – are a concern at many airports. Although starlings are relatively small birds, a flock of Common Starlings was the cause of a major crash in 1960 in Boston, Massachusetts, which resulted in the death of 62 passengers and crew in a Lockheed Electra propeller-driven passenger aircraft (Linz *et al.*, 2018). More recently, in the Netherlands, a C130 Hercules military aircraft struck a flock of starlings on landing and caught fire on the runway, resulting in the death of 34 people. In the most recent review of bird strikes in the USA involving civilian aircraft, DeVault *et al.* (2018) ranked the Common Starling in position 6 with 698 strikes reported for the years 2010–2015. Of these incidents, 28 caused damage costing an estimated US$697,000. Naturally large birds such as the Canada Goose (*Branta canadensis*) present a much more serious problem (more than 100 damaging strikes, costing more than US$10 million). For military aircraft, data from the US Navy and Airforce ranked the Common Starling 32nd overall, with 13% of the impacts causing damage (at the least requiring cleaning of the aircraft engines), but only 3% 'substantial damage', which was defined as repairs costing over US$50,000 (Pfeiffer *et al.*, 2018). Starlings are typically attracted to airfields by the short grass areas where they can forage, and Marateo *et al.* (2015) noted that keeping the grass height above 30 cm effectively discourages this species. Experiments with a 'sonic net' (spatially controlled noise that overlaps with bird communication frequencies) deployed at an airfield in Virginia, resulted in an 80% reduction in the presence of birds at this location (Swaddle *et al.*, 2016). This could be an effective approach to prevent bird strikes, with less danger of habituation. Paz (1987) noted that huge flocks of migrant starlings visiting Israel in winter were a significant hazard to low-flying aircraft.

As potential vectors of diseases that can affect other bird species, domestic animals and humans, Common Starlings could have a significant impact. They have been demonstrated to carry avian malaria (*Plasmodium relictum*) and the causative organism of psittacosis (*Chlamydophila psittaci*) (Dunn and Hatcher, 2015), as well as *Escherichia coli* and *Campylobacter*, *Mycobacterium* and *Salmonella* spp. (Linz *et al.*, 2018). The role of starlings relative to other sources of infection is not clear in most cases, but Carlson *et al.* (2011) showed that reducing starling numbers significantly reduced the prevalence of *Salmonella* in cattle feed and water troughs, and Shwiff *et al.* (2018) found that survey data indicated higher veterinary costs at dairies with flocks of starlings numbering over 1000 birds.

Fruit damaged by Common Starlings in the USA included apples, blueberries, cherries, figs, grapes, peaches and strawberries. Early-ripening cherry cultivars were particularly vulnerable, and starlings were rated as the most significant bird pest by grape producers. However, starlings were not considered serious pests in cereal or oilseed crops. Damage to sweetcorn has been reported but not assessed critically (Linz *et al.*, 2018). In 1993, the estimate of damage caused by Common Starlings in the USA, including agriculture, disease transmission and damage to structures, was US$800 million; a 2005 estimate for agriculture alone was US$800 million (Shwiff *et al.*, 2018).

On dairy farms, Common Starlings feed on pellets provided for the cows (Germany: Höttner *et al.*, 2015; USA: Carlson *et al.*, 2018) and can cause a significant loss of condition, and thus reduced milk production, in the herd (Carlson *et al.*, 2018). In Ohio, a study showed that outdoor feeding of the cows and regular removal and stacking of manure (daily, or after each milking) increased the numbers of starlings in attendance. The distance to the closest active starling roost was also an important factor (Medhanie *et al.*, 2015). Up to 250,000 starlings were recorded at a feedlot for cattle in Kansas, and it was estimated that over 60 days in winter, 1000 starlings could consume 1.5 t of cattle feed (Linz *et al.*, 2018).

Common Starlings were introduced to Australasia in the hope that they would control insect pests and their larvae, and in New Zealand were encouraged by the erection of nest boxes on farms, a practice that continued into the 1980s (Higgins *et al.*, 2006). However, damage to orchards was soon recorded. In a review of bird pests in Australia, Bomford and Sinclair (2002) recorded the impact of Common Starlings as damaging fruit crops (cherries and other stone fruit, olives, grapes and blueberries), spreading invasive weeds, taking food for livestock in feedlots, competing with native birds, transmitting diseases, fouling buildings, causing noise and fouling at roost sites. This species was a major cause of financial loss to cherry and grape growers, and at feedlots, eating and fouling the livestock rations. Starlings can also assist the spread of diseases of poultry, such as avian influenza and Newcastle disease.

In Algeria, in one orchard, 87% of damage to olives was caused by Common Starlings (Berraï *et al.*, 2017). Cherries are particularly vulnerable to damage by starlings. Wagner (1963) noted that, in Switzerland, juvenile birds were responsible for most of the damage to cherry orchards, and this occurred in years when there were many second broods, which in turn was associated with early breeding. There is a recent report of starling damage to blueberries in Argentina (Ibañez *et al.*, 2016b).

Dispersal of invasive plant species by Common Starlings has been highlighted in Australia (e.g. *Olea europaea*; Spenneman and Allen, 2000) and South Africa (e.g. *Acacia cyclops*; Glyphis *et al.*, 1981). Gebhardt (1959) noted that, in New Zealand, starlings fed on introduced bumble bees, which had been imported to pollinate pasture plants such as clover.

As nest sites are a limited resource for birds that cannot excavate their own nest cavities, in all regions there has been a focus on competition between Common Starlings and other hole-nesting species. The endangered Purple Martin (*Progne subis*) in the USA was nesting primarily in nest boxes, generally close to human habitation. Brown (1981) noted that in such situations, competition with starlings was a major factor in nest site loss, and therefore active management of these colonies was necessary. In Canada, a 10-year review of potential competition for nest sites between Common Starlings and the Mountain Bluebird (*Sialia currucoides*) and the Tree Swallow (*Tachycineta bicolor*) concluded that where cavities are abundant, the presence of starlings has little influence on these two native species (Koch *et al.*, 2012). On an island off New Zealand, Lawrence *et al.* (2017) found that Common Starlings regularly usurped nests of the endangered Chatham Island Robin (*Petroica traversi*) in cavities or those placed high up. However, when re-nesting, the robins consistently chose lower nest sites, which were not favoured by starlings. An unusual record from South Africa described a Little Swift (*Apus affinis*) nest usurped by Common Starlings, which bred successfully (Ward, 2006). On Tasmania, the Eastern Rosella (*Platycercus eximius*) and the Green Rosella (*Platycercus calenodicus*) were victims of nest usurpation by Common Starlings, and the decline in these species and other hole-nesting parrots on the island was concomitant with the widespread establishment of the starling (first introduced in 1880); however, other factors such as habitat change were not evaluated (Green, 1983).

Many observers have reported starlings aggressively displacing indigenous woodpecker species and taking over holes excavated by woodpeckers. Ingold (1989) suggested that competition with Common Starlings was important for Red-bellied Woodpeckers (*Melanerpes carolinus*), which lost 52% of their nest cavities, whereas the larger, more aggressive Red-headed Woodpecker (*Melanerpes erythrocephalus*) had only 7% of its nest sites usurped. However, differences in nest-hole characteristics of the two species also suggested some preference by the starlings. After Common Starlings had arrived in Arizona, in 1946, they began nesting in saguaro cacti in holes excavated by woodpeckers. Kerpez and Smith (1990) noted that starlings displaced Gila Woodpeckers (*Melanerpes uropygialis*), which were reduced in numbers, but there was no apparent impact on the larger Northern Flicker (*Colaptes auratus*). Further north in Washington State, Tomasevic and Marzluff (2017) reported that three of 18 Northern Flicker nests were usurped by Common Starlings, and two of 17 Red-breasted Sapsucker (*Sphyrapicus ruber*) nests. However, they also noted that 116 of 120 Common Starling nests were on buildings, whereas of 208 woodpecker nests, only 13 were on man-made structures. For the Red-headed Woodpecker in southern Ontario, Canada, nest survival was clearly lower where there was interference from Common Starlings, and as cavity take-overs were most likely early in the season, the woodpeckers would have little chance of re-nesting (Frei *et al.*, 2015). Nevertheless, when Koenig (2003) reviewed the evidence for Common Starlings reducing the breeding success of 27 native North American species, he concluded that negative effects could be identified in only five species and that declines in their populations could at least in part be attributed to factors other than competition with starlings. Koenig (2003) remarked that these results call into question our ability to predict the effects of exotic species on native species. Considering woodpeckers in both Eurasia and North America, Jackson and Jackson (2016) found that the species whose cavities were most often usurped by Common Starlings were widespread generalists, similar in size to the starlings. Such take-overs occurred most often in human-dominated areas with open habitats where starlings foraged regularly. They argued that habitat loss and fragmentation was the greatest problem for woodpeckers but suggested that Common Starlings might be a significant threat to island endemics. In Argentina, Ifran and Fiorini (2010) speculated that the density of Common Starlings was perhaps too low for serious competition for holes with woodpeckers to be evident. During a study of the Great Spotted Woodpecker (*Dendrocopos major*) in the UK, interference competition by starlings caused nest failures, but as starling populations declined after 1985, woodpecker numbers increased. However, in many areas, Great Spotted Woodpeckers continue to increase in the absence of starlings, so there are likely to be other factors involved (Smith, 2005).

Both Joubert (1945) and van der Merwe (1984) observed Common Starlings displacing Olive Woodpeckers (*Dendropicos griseocephalus*) from nest holes in garden situations in South Africa. However, this is an unusual habitat for the Olive Woodpecker, a forest species that would seldom encounter Common Starlings (Hockey *et al.*, 2005).

South Africa is the only region where introduced Common Starlings are regularly sympatric with indigenous starling species. The Red-winged Starling (*Onychognathus morio*) has adapted to urban life and has been nesting on buildings for more than 100 years (Hockey *et al.*, 2005). At a coastal locality, it was reported that the first nesting of Common Starlings was at a site previously used by Red-winged Starlings (Cooke, 1974), but the latter species typically nests on open ledges rather than in cavities. On the campus of Roma University, Lesotho, the Red-winged Starling appeared to dominate at favoured nesting sites, but direct interactions were not observed, and both species nested annually in separate territories (Kopij, 2009). Sympatry with the Common Myna (*Acridotheres tristis*) is a recent development in South Africa, and the interaction between these two species has not been studied. In New Zealand, Common Mynas regularly displaced Common Starlings from nest sites, and significantly reduced their breeding success in some regions (Higgins *et al.*, 2006).

Foraging Common Starlings in Argentina were most often found in mixed-species flocks, and they were the losers in the few aggressive encounters recorded (Ifran and Fiorini, 2010). Palacio *et al.* (2016) suggested that the greatest ecological overlap with potential competing species in Argentina would be with the Bay-winged Cowbird (*Molothrus badius*), Chalk-browed Mockingbird (*Mimus saturninus*) and House Sparrow (*Passer domesticus*). The Ring-necked Parakeet (*Psittacula krameri*) is a recent invader in many European cities, and this was the most aggressive species at feeders in winter in France. The Common Starling overlapped most with the parakeets in its visits to bird feeders in Paris and was always displaced; it may now also suffer from competition for nest sites with parakeets in urban areas in the Mediterranean region (Le Louarn *et al.*, 2016).

Although it is omnivorous, the Common Starling would not usually be considered a nest predator. However, in the Azores, in a mixed tern colony, starling predation on eggs accounted for 73% and 90% of egg losses over two seasons; for the rare Roseate Tern (*Sterna dougallii*) this was serious, and control measures were proposed (Neves *et al.*, 2006).

2.10 Control

Common Starling control measures in all regions have included frightening devices such as propane exploders and pyrotechnics, and broadcasting distress calls and alarm calls. Generally, such methods produce only short-term effects, and for large areas the cost may outweigh their effectiveness. Chemical repellents have also shown no long-term effects. Exclusion netting is highly effective and may be economically justified for some grape varieties in the USA. Habitat modification, such as thinning trees used for roosting, can also be used to manage starling numbers in critical areas (Linz *et al.*, 2018).

Lethal control of Common Starlings in both the USA and Australia has employed shooting, traps and poison. Starlicide is a toxin that can be used in bait, and while it is also toxic to other bird species, secondary poisoning from carcases is unlikely (Linz *et al.*, 2018). Sodium lauryl sulfate, a surfactant that destroys the insulating properties of feathers, has been used in the USA for spraying at roosts in winter. The results have been variable, and roosts need to be checked in advance for non-target species (Linz *et al.*, 2018).

Exclusion netting for Common Starlings is considered costly, but very effective, for high-value grape and cherry varieties in Australia. However, installing electrified wires is unlikely to reduce damage significantly, and scaring devices are ineffectual. Attempts at population reduction by shooting and poisoning were considered ineffective in terms of damage mitigation but were also likely to affect non-target species (Bomford and Sinclair, 2002). The Western Australian government has run a control programme since 1975 to prevent Common Starlings from crossing the Nullabor Plain and establishing populations in the south-west of the state. Consistent control by live trapping, shooting, netting (and even poison) has kept the numbers low, and a cost–benefit analysis indicated that an annual expenditure of 1.2 million Australian dollars could save an estimated 40 million Australian dollars in agricultural damage alone (Campbell *et al.*, 2016). Where natural cavities are not available, providing wooden nest boxes may concentrate the birds at sites that can be monitored to destroy any breeding attempts (Campbell *et al.*, 2012).

Chemical repellents have been tested for use with Common Starlings in feedlots in North America. Anthranilate is palatable to cattle but aversive to starlings and was effective in both the dimethyl and methyl form in initial trials (Mason *et al.*, 1989, 1991). However, it is not currently recommended as a cost-effective method on cattle farms (Schroeder and Lee, 2015) and trials in Australia suggested that high doses were required (Bomford and Sinclair, 2002). Experiments at tern colonies in the Azores using methiocarb (3,5-dimethyl-4(methylio) phenyl methylcarbamate) found that egg predation by gulls was reduced, but there was little effect on Common Starlings (Neves *et al.*, 2006), so here too chemical repellents proved ineffective. Medhanie *et al.* (2015) suggested that on dairy farms modifying the feeding situation of the cows, and changing manure management practices, were the most effective long-term solutions. Carlson *et al.* (2018) also found that a pellet size greater than 0.95 cm in diameter was acceptable to the cows but reduced Common Starling consumption by 75%.

Helium balloons with eyespots have been reported as effective in deterring Common Starlings from visiting vineyards (Caskin *et al.*, 2018), but it is likely that habituation will occur. However, the use of falconry to protect grapes and strawberries is highly effective; drones may provide a future option (Caskin *et al.*, 2018).

In France, there had been a great increase in the wintering population of Common Starlings, with birds from large roost sites causing significant damage in orchards and to stored forage. An assessment after 10 years of spraying starling roosts with toxic chemicals found that the financial and ecological costs did not justify the relatively local reduction in damage. Alternative proposals included early intervention to prevent the establishment of roosts in critical areas, and a diversity of methods for protecting crops and forage (Clergeau, 1990).

No control measures for Common Starlings have been implemented in South Africa (Hockey *et al.*, 2005), or in New Zealand where this species is viewed in a more positive light than elsewhere in its introduced range (Higgins *et al.*, 2006).

2.11 Roosts

Where Common Starling population densities are high, enormous roosts containing thousands or even millions of birds may be formed in the non-breeding season. Some such roosts are temporary, whereas others may be occupied for extended periods, even throughout the year by small numbers of birds; a few birds even continue to roost at their nest sites throughout the winter (Clergeau, 1981, 1983; Feare, 1996). Foraging groups emerging from communal roosts can be a major source of damage to agricultural products, while the roost sites can cause serious fouling of buildings, streets and vehicles in towns, as well as noise pollution. Common Starlings may roost in association with other species (e.g. Icteridae in North America: Linz *et al.*, 2018; indigenous Sturnidae and Ploceidae in South Africa: Hockey *et al.*, 2005). When roosting starlings in London gathered on the minute hand of Big Ben and stopped the clock in 1947, there were questions in Parliament and the episode was satirized in *The Goon Show*, a radio comedy show (Feare, 1996).

Delvingt (1961b) mapped Common Starling roosts in Belgium and showed that most were within 15 km of pastures, the prime foraging areas. Some had been in constant use over many years (more than 10 years for most, with one known for 100 years), while others were abandoned after a few months, but the assembly points where Common Starlings gathered before flying to a roost were often maintained even when the roosting site changed. He noted many visits by both diurnal and nocturnal bird predators. Changes in roost site were apparently related to thermal benefits; in France, Clergeau and Simonnet (1996) found that urban roost sites in winter were significantly warmer than those in rural areas, while starling roosts in rural Brittany were typically associated with large areas of pasture and arable lands (Clergeau and Fourcy, 2005). Departure from the assembly points to the roost site was dependent on local environmental conditions, especially light levels (Davis and Lussenhop, 1970; Clergeau, 1983).

Observations at British Common Starling roosts suggested that the heavier adult males were at the centre, with lighter birds on the periphery (Summers *et al.*, 1987), and captive studies also showed that dominance determined the position of individual birds (Feare *et al.*, 1995). Clergeau (1991) had suggested that the structure of the roost might depend on particular groupings of birds, and subsequently Hausberger *et al.* (2008) found evidence that the song dialect shared by starlings from the same geographical area led to associations within the roosting site. However, at an urban roost site in Detroit, Michigan, membership changed frequently; this site was used mostly by adult birds in winter, while in summer subadults were the dominant group (Thompson and Coutlee, 1963).

Common Starling flocks approaching a roost site carry out extraordinary aerial manoeuvres, and a mathematical model has been developed to explain how the birds avoid collisions (Ballerini *et al.*, 2008a,b). It is generally thought that these movements are intended to confuse raptors; Gersdorf (1966) described how large flocks repulsed birds of prey, even on occasion forcing them down into water where they drowned.

Tracking studies in North America found that individual Common Starlings were more faithful to their daily activity centres than to roost sites, and joined major roosts on a temporary basis to exploit nearby food sources (Morrison and Caccamise, 1985; Maccarone, 1987b). Birds tracked from a winter roost had a home range of 40 km^2; one bird flew 32 km in 40 min. Most individuals returned to the same general feeding area each day but did not occupy the same spots within the roost (Bray *et al.*, 1975). More recent studies, summarized by Linz *et al.* (2018), found that birds captured at dairies and feedlots showed considerable site fidelity to their feeding areas, and little interchange with sites more than 4 km away. However, this depended on habitat heterogeneity around the roost site (up to 22 km from feeding areas); one bird was located 68 km from its capture site. Studies in three urban areas showed that some birds stayed in town, but others made daily trips of up to 19 km to nearby towns and rural areas.

2.12 Notes

Both male and female Common Starlings sing, and their vocalizations have been the focus of recent research in communication (e.g. Henry *et al.*, 2015), including neurophysiological studies of the brain (e.g. George *et al.*, 2012). Mimicry of other species, and even of the human voice (e.g. West *et al.*, 1983), has been documented from both native and introduced populations (Hindmarsh, 1984; Hausberger *et al.*, 1991). A Common Starling owned by the composer Mozart from 1784 to 1787 may be the source of some pieces such as 'A Musical Joke', K522 (West and King, 1990). Like many other birds, starlings have significant vision in the ultraviolet range, and there is experimental evidence that plumage features visible under UV may influence mate selection by females (Bennett *et al.*, 1997).

The Common Starling is an enthusiastic participant in the remarkable behaviour known as 'anting', in which ants, typically members of the family Formicidae, which release formic acid when disturbed, are picked up by a bird and passed rapidly through the plumage. Sometimes, the ants are eaten afterwards but may be simply discarded. Anting is well known in the species native range (e.g. Poulsen, 1956; Simmons, 1957) and has also been observed in Australia (e.g. Chisholm, 1944), North America (e.g. Potter, 1970) and South Africa (Craig, 1999). A study of captive birds indicated that anting develops at an early age (Querengässer, 1973), but there is no agreement on a possible functional explanation for such behaviour.

The use of green, aromatic herbs in Common Starling nests attracted the interest of several research groups. In the USA, Clark and Mason (1985) had shown that this nest material had potential insecticidal effects and in the laboratory could inhibit feeding or emergence of mites, a regular problem for starling nestlings (Clark and Mason, 1988). However, Fauth *et al.* (1991) argued that this green material played a role in mate selection rather than ectoparasite control. Gwinner (1997), in Germany, stressed the role of such plants in courtship, and later showed that while the aromatic plants reduced bacteria in the nest, they did not affect the numbers or activity of mites. However, in nests with herbs the nestlings appeared to cope better with ectoparasites (Gwinner *et al.*, 2000; Gwinner and Berger, 2005). Wolfs *et al.* (2012) proposed that nestling development is also affected by a balance between ectoparasitic mites and their predators, so that the entire nest fauna needed to be considered.

Feare (1984) described the Common Starling as a problem bird in the UK, but in later years the situation changed dramatically. Today, this is one of the bird species that has shown a consistent decline in numbers over more than 30 years in the UK, marked by a steady decline in juvenile survival matching a reduction in the area under pasture, and increased use of insecticides (Robinson *et al.*, 2006). Over the period 1965–2000 in the UK, there was some reduction in adult survival, but a marked decline in juvenile survival from the 1970s, coinciding with a reduction in dairy cattle (Freeman *et al.*, 2007). The same trend has been evident in most European countries. Whereas Berthold (1968) was seeking to explain the dramatic increase in starling numbers in Germany, which he attributed to the species' ability to take advantage of the milder climate and urbanization through reducing migration and increasing the proportion of first-year birds breeding, within a few decades massive declines were reported from many northern regions. In the Netherlands, over the period 1960–2012, there was an increase in the overall population from 1960 to 1978, a stable population from 1978 to 1990 but then a steady decline. Although there was no consistent trend in reproductive output and adult survival fluctuated irregularly, juvenile survival declined significantly, and this was attributed to changes in agricultural land use (Versluijs *et al.*, 2016). Smith and Bruun (2002) had shown that the breeding success for Swedish starlings was greatest where they had easy access to pastures, which increased nestling survival, whereas other factors such as clutch size and the timing of breeding were not affected. For Finland, starling populations were stable from 1952 to 1970, but a massive decline coincided with a reduction in cattle husbandry and the accompanying loss of pastures (Rintala *et al.*, 2003). Heldbjerg *et al.* (2016) reported that, in Denmark, the population of Common Starlings had declined by 60% between 1976 and 2015. In the 1990s, starling density was correlated with dairy cattle abundance, and the decline from 2001 closely matched the loss of high-intensity grazing, as animal husbandry in Denmark moved indoors. Even in Bulgaria, recent census data have shown declining populations of Common Starlings (Spasov *et al.*, 2017).

Similarly, in North America, the total population of Common Starlings was estimated at 200 million birds in the 1990s, but by 2017 had declined to 140 million (Linz *et al.*, 2018). This was clearly reflected in the annual regional counts in the USA (e.g. Mississippi: Twedt and Pardieck, 2017; see also the raw count data at www.pwrc.usgs.gov/BBS/RawData/) and across Canada between 1967 and 2000 (Downes and Collins, 2003).

2.13 References

Ballerini, M., Cabibbo, N., Candelier, R., Cavagna, A., Cisbani, E., *et al.* (2008a) Empirical investigation of starling flocks: a benchmark study in collective animal behaviour. *Animal Behaviour* 76, 201–215.

Ballerini, M., Cabibbo, N., Candelier, R., Cavagna, A., Cisbani, E., *et al.* (2008b) Interaction ruling animal collective behavior depends on topological rather than metric distance: evidence from a field study. *Proceedings of the National Academy of Sciences USA* 105, 1232–1237.

Bauer, M.S. (1978) Warble-infested cattle attacked by Starlings. *Modern Veterinary Practice* 59, 536–538.

Beklová, M. (1972) Age structure and mortality of the Czechoslovakian populations of *Turdus merula* L. 1758, *Sturnus vulgaris* L. 1758 and *Parus major* L. 1758. *Zoologické Listy* 21, 337–346.

Belcher, R.N., Fornasari, L., Menz, S. and Schroepfer, T. (2018) Birds use of vegetated and non-vegetated high-density buildings – a case study of Milan. *Journal of Urban Ecology* 4, 1–6.

Bennett, A.T.D., Cuthill, I.C., Partridge, J.C. and Lunau, K. (1997) Ultraviolet plumage colors predict mate preferences in starlings. *Proceedings of the National Academy of Sciences USA* 94, 8618–8621.

Berraï, H., Marniche, F., Chaouia, C., Merrar, K. and Doumandji, S. (2017) Estimated damages due to the European Starling *Sturnus vulgaris* in Béjaia's olives groves (North Algeria). *Advances in Environmental Biology* 11, 47–53.

Berthold, P. (1968) Die Massenvermehrung des Stars *Sturnus vulgaris* in fortpflanzungsphysiologischer Sicht. *Journal für Ornithologie* 109, 11–16.

Berthouly-Salazar, C., Hui, C., Blackburn, T.M., Gaboriaud, C., van Rensburg, B.J., *et al.* (2013) Long-distance dispersal maximizes evolutionary potential during rapid geographic range expansion. *Molecular Ecology* 22, 5793–5804.

BirdLife International (2018) Species factsheet: *Sturnus vulgaris*. Available at: http://datazone.birdlife.org/species/factsheet/common-starling-sturnus-vulgaris (accessed 16 October 2019).

Bitton, P.-P. and Graham, B.A. (2015) Change in wing morphology of the European starling during and after colonization of North America. *Journal of Zoology* 295, 254–260.

Bomford, M. and Sinclair, R. (2002) Australian research on bird pests: impact, management and future directions. *Emu* 102, 29–45.

Bradbury, R.B., Cotton, P.A., Wright, J. and Griffiths, R. (1997) Nestling sex ratio in the European Starling *Sturnus vulgaris*. *Journal of Avian Biology* 28, 255–258.

Bray, O.E., Larsen, K.H. and Mott, D.F. (1975) Winter movements and activities of radio-equipped starlings. *Journal of Wildlife Management* 39, 795–801.

Brazil, M. (2009) *Field Guide to the Birds of East Asia*. Christopher Helm, London.

Brooke, R.K. (1995) The wreck of the Otori Maru No. 8 as a bird breeding site. *Promerops* 218, 9.

Brooke, R.K., Lloyd, P.H. and de Villiers, A.L. (1986) Alien and translocated terrestrial vertebrates in South Africa. In: Macdonald, I.A.W., Kruger, F.J. and Ferrar, A.A. (eds) *The Ecology and Management of Biological Invasions in Southern Africa*. Oxford University Press, Cape Town, pp. 63–74.

Brooke, R.K., Lockwood, J.L. and Moulton, M.P. (1995) Patterns of success in passeriform bird introductions on Saint Helena. *Oecologia* 103, 337–342.

Brown, C.J. (1985) Invasive alien birds in South West Africa/Namibia. In: Brown, C.J., Macdonald, I.A.W. and Brown, S.E. (eds) *Invasive Alien Organisms in South West Africa Namibia*. South African National Scientific Programs Report No. 119, pp. 41–43.

Brown, C.R. (1981) The impact of Starlings on Purple Martin populations in unmanaged colonies. *American Birds* 35, 266–268.

Bull, P.C. and Flux, J.E.C. (2006) Breeding dates and productivity of starlings (*Sturnus vulgaris*) in northern, central, and southern New Zealand. *Notornis* 53, 208–214.

Cabe, P.R. (1993) European Starling. In: Poole, A. and Gill, F. (eds) *The Birds of North America*. Academy of Natural Sciences of Philadelphia, Philadelphia. 48, 1–23.

Cabe, P.S. (1999) Dispersal and population structure in the European Starling. *Condor* 101, 451–454.

Campbell S., Roberts, E.J., Craemer, R., Pacioni, C., Rollins, L. and Woolnough, A.P. (2016) Assessing the economic benefits of starling detection and control to Western Australia. *Australasian Journal of Environmental Management* 23, 81–99.

Campbell, S., Powell, C., Parr, R., Rose, K., Martin, G. and Woolnough, A. (2012) Can artificial nest-cavities be used as a management tool to assist the control of Common Starlings (*Sturnus vulgaris*)? *Emu* 112, 255–260.

Cardilini, A.P.A., Buchanan, A.L., Sherman, C.D.H., Cassey, P. and Symonds, M.R.E. (2016) Tests of ecogeographical relationships in a non-native species: what rules avian morphology? *Oecologia* 181, 783–793.

Carlson, J.C., Engeman, R.M. and Hyatt, D.R. (2011) Efficacy of European starling control to reduce *Salmonella enterica* contamination in a concentrated animal feeding operation in the Texas panhandle. *BMC Veterinary Research* 7, 9.

Carlson, J.C., Stahl, R.S., DeLiberto, S.T., Wagner, J.J., Engle, T.E., *et al.* (2018) Nutritional depletion of total mixed rations by European starlings: projected effects on dairy cow performance and potential intervention strategies to mitigate damage. *Journal of Dairy Science* 101, 1777–1784.

Caskin, T., Cirillo, K. and Progar, A.A. (2018) Physical management of pest birds in agricultural settings. Washington State University, Extension FS294E. Available at: https://research.wsulibs.wsu.edu/xmlui/bitstream/handle/2376/12901/FS294E.pdf?sequence=1&isAllowed=y (accessed 26 March 2020).

Cavitione e Silva, F., Pinto, J.M., Mäder, A. and de Souza, V.A.T. (2017) First records of European Starling *Sturnus vulgaris* in Brazil. *Revista Brasileira de Ornitologia* 25, 297–298.

Chisholm, A.H. (1944) The problem of "Anting". *Ibis* 86, 389–405.

Clark, L. and Mason, J.R. (1985) Use of nest material as insecticidal and anti-pathogenic agents by the European Starling. *Oecologia* 67, 169–176.

Clark, L. and Mason, J.R. (1988) Effect of biologically active plants used as nest material and the derived benefit to starling nestlings. *Oecologia* 77, 174–180.

Clergeau, P. (1981) Dynamique des dortoirs d'étourneaux *Sturnus vulgaris* dans le bassin de Rennes. *Alauda* 49, 13–24.

Clergeau, P. (1983) Quelques aspects des comportements hivernaux des étourneaux rejoignant leurs dortoirs nocturnes. *Alauda* 51,124–145.

Clergeau, P. (1990) Réflexions sur le problème "étourneau" et sur le choix des moyens de lutte. *La Défense des Végétaux* 263, 1–7.

Clergeau, P. (1991) Nouvelles hypotheses de structure de dortoirs d'étourneaux *Sturnus vulgaris* et conséquence sur la gestion de l'espèce. *Bulletin de la Société Zoologique de la France* 116, 289–295.

Clergeau, P. and Fourcy, D. (2005) Effects of landscape homogeneity on starling roost distribution. *Agriculture, Ecosystems & Environment* 110, 300–306.

Clergeau, P. and Simonnet, E. (1996) Microclimate in communal roost sites of starlings (*Sturnus vulgaris*). *Journal für Ornithologie* 137, 358–361.

Coleman, J.D. (1973) Determination of the sex and age of starlings in Canterbury, New Zealand. *Notornis* 20, 324–329.

Coleman, J.D. (1974) The use of artificial nest sites erected for starlings in Canterbury, New Zealand. *New Zealand Journal of Zoology* 1, 349–354.

Coleman, J.D. and Robson, A.B. (1975) Variations in body weight, fat-free weights and fat deposition of starlings in New Zealand. *Proceedings of the New Zealand Ecological Society* 22, 7–13.

Collins, V.B. and de Vos, A. (1966) A nesting study of the Starling near Guelph, Ontario. *Auk* 83, 623–636.

Cooke, C.K. (1974) [Letter]. *Bokmakierie* 26, 53.

Cooper, J. and Underhill, L.G. (1991) Breeding, mass and primary moult of European Starlings *Sturnus vulgaris* at Dassen Island, South Africa. *Ostrich* 62, 1–7.

Coulson, J.C. (1960) A study of the mortality of the starling based on ringing recoveries. *Journal of Animal Ecology* 29, 251–271.

Craig, A.J.F.K. (1999) Anting in Afrotropical birds: a review. *Ostrich* 70, 203–207.

Craig, A.J. and Feare, C.J. (2009) Family Sturnidae (starlings). In: Del Hoyo, J., Elliot, A. and Christie, D.A. (eds) *Handbook of the Birds of the World*. Vol. 14. *Bush-shrikes to Old World Sparrows*. Lynx Edicions, Barcelona, Spain, pp. 654–758.

Cunningham, P. (2016) Is Namibia on the brink of being invaded by Common Starlings *Sturnus vulgaris*? *Biodiversity Observations* 7, 1–2.

Cyrus, D. and Robson, N. (1980) *Bird Atlas of Natal*. University of Natal Press, Pietermaritzburg, South Africa.

Davis, D.E. (1959) The sex and age structure of roosting starlings. *Ecology* 40, 136–139.

Davis, G.J. and Lussenhop, J.F. (1970) Roosting of Starlings (*Sturnus vulgaris*): a function of light and time. *Animal Behaviour* 18, 362–365.

Delvingt, W. (1961a) Détermination de l'age et du sexe des étourneaux résidant ou séjournant en Belgique. *Le Gerfaut* 51, 53–63.

Delvingt, W. (1961b) Les dortoirs d'étourneaux, *Sturnus vulgaris* L., de Belgique en 1959–1960. *Le Gerfaut* 51, 121–147.

DeVault, T.L., Blackwell, B.F., Seamans, T.W., Begier, M.J., Kougher, J.D., *et al.* (2018) Estimating interspecific economic risk of bird strikes with aircraft. *Wildlife Society Bulletin* 42, 94–101.

Djennas-Merrar, K., Berraï, H., Marniche, F. and Doumandji, S. (2016) Fall-winter diet of the starling (*Sturnus vulgaris*) between foraging areas and resting areas near Algiers. *Advances in Environmental Biology* 10, 11–18.

Downes, C.M. and Collins, B.T. (2003) The Canadian Breeding Bird Survey, 1967–2000. *Canadian Wildlife Service Progress Notes* 219, 1–39.

Dunn, A.M. and Hatcher, M.J. (2015) Parasites and biological invasions: parallels, interactions and control. *Trends in Parasitology* 31, 189–199.

Dunnet, G.M. (1955) The breeding of the Starling *Sturnus vulgaris* in relation to its food supply. *Ibis* 97, 619–662.

Earlé, R. and Grobler, N. (1987) *First Atlas of Bird Distribution in the Orange Free State*. National Museum, Bloemfontein, South Africa.

Eble, H. (1963) Ernährungsbiologische Untersuchungen an *Turdus philomelos Brehm*, *Turdus merula* L. und *Sturnus vulgaris* L. *Wissenschaftliche Zeitschrift der Martin-Luther-Universität Halle-Wittenburg Mathematik-Naturwissenschaften* 12, 211–234.

Evans, P.G.H. (1986) Ecological aspects of wing moult in the European Starling *Sturnus vulgaris*. *Ibis* 128, 558–561.

Evans, P.G.H. (1988) Intraspecific nest parasitism in the European starling *Sturnus vulgaris*. *Animal Behaviour* 36, 1282–1294.

Fauth, P.T., Krementz, D.G. and Hines, J.E. (1991) Ectoparasitism and the role of green nesting material in the European starling. *Oecologia* 88, 22–29.

Feare, C.J. (1984) *The Starling*. Oxford University Press, Oxford.

Feare, C.J. (1991) Intraspecific nest parasitism in Starlings *Sturnus vulgaris*: effects of disturbance on laying females. *Ibis* 133, 75–79.

Feare, C.J. (1993) Nectar feeding by European starlings. *Wilson Bulletin* 105, 194.

Feare, C.J. (1996) Studies of West Palearctic birds 196. Common Starling *Sturnus vulgaris*. *British Birds* 89, 549–568.

Feare, C.J., Gill, E.L., McKay, H.V. and Bishop, J.D. (1995) Is the distribution of Starlings *Sturnus vulgaris* within roosts determined by competition? *Ibis* 137, 379–382.

Ferrer, X., Motis, A. and Peris, S.J. (1991) Changes in the breeding range of starlings in the Iberian Peninsula during the last 30 years: competition as a limiting factor. *Journal of Biogeography* 18, 631–636.

Fischl, J. and Caccamise, D.F. (1987) Relationships of diet and roosting behavior in the European starling. *American Midland Naturalist* 117, 395–404.

Flux, J.E.C. and Flux, M.M. (1981) Population dynamics and age structure of Starlings (*Sturnus vulgaris*) in New Zealand. *New Zealand Journal of Ecology* 4, 65–72.

Flux, J.E.C. and Flux, M.M. (1992) Nature red in claw: how and why starlings kill each other. *Notornis* 39, 293–300.

Freeman, S.N., Robinson, R.A., Clark, J.A., Griffin, B.M. and Adams, S.Y. (2007) Changing demography and population decline in the Common Starling *Sturnus vulgaris*: a multisite approach to Integrated Population Monitoring. *Ibis* 149, 587–596.

Frei, B., Nocera, J.J. and Fyles, J.W. (2015) Interspecific competition and nest survival of the threatened Red-headed Woodpecker. *Journal of Ornithology* 156, 743–753.

French, K., Major, R. and Hely, K. (2005) Use of native and exotic garden plants by suburban nectarivorous birds. *Biological Conservation* 121, 545–559.

Galbraith, J.A., Beggs, J.R., Jones, D.N. and Stanley, M.C. (2015) Supplementary feeding restructures urban bird communities. *Proceedings of the National Academy of Sciences USA* 112, E2648–E2657.

Galbraith, J.A., Jones, D.N., Beggs, J.R., Parry, K. and Stanley, M.C. (2017) Urban bird feeders dominated by a few species and individuals. *Frontiers in Ecology and Evolution* 5, 81.

Garner, M.S. (1997) Apparent whole brood of Common Starlings exhibiting albinism. *British Birds* 90, 192–193.

Gebhardt, E. (1954) Die gegenwärtige Verbreitung von Haussperling, Star und Buchfink in Südafrika. *Journal für Ornithologie* 95, 58–60.

Gebhardt, E. (1959) Europäische Vögel in überseeischen Ländern. *Bonner Zoologische Beiträge* 10, 310–342.

George, I., Cousillas, H., Richard, J.-P., and Hausberger, M. (2012) Experience with adults shapes multisensory representation of social familiarity in the brain of a songbird. *PLoS One* 7, e38764.

Gersdorf, E. (1966) Beobachtungen über das Verhalten von Vogelschwärmen. *Zeitschrift für Tierpsychologie* 23, 37–43.

Glyphis, J.P., Milton, S.J. and Siegfried, W.R. (1981) Dispersal of *Acacia cyclops* by birds. *Oecologia* 48, 138–141.

Green, R.H. (1983) The decline of the Eastern Rosella and other Psittaciformes in Tasmania concomitant with the establishment of the introduced European Starling. *Records of the Queen Victoria Museum* 82, 1–4.

Gwinner, H. (1997) The function of green plants in nests of European starlings (*Sturnus vulgaris*). *Behaviour* 134, 337–351.

Gwinner, H. and Berger, S. (2005) European starlings: nestling condition, parasites and green nest material during the breeding season. *Journal of Ornithology* 146, 365–371.

Gwinner, H., Oltrogge, M., Trost, L. and Nienaber, U. (2000) Green plants in starling nests: effects on nestlings. *Animal Behaviour* 59, 301–309.

Harrison, J.A., Allan, D.G., Underhill, L.G., Brown, C.J., Tree, A.J., Parker, V. and Herremans, M. (eds) (1997) *The Atlas of Southern African Birds*. Vol. 2. BirdLife South Africa, Johannesburg, South Africa.

Hausberger, M., Bigot, E. and Clergeau, P. (2008) Dialect use in large assemblies: a study in European starling *Sturnus vulgaris* roosts. *Journal of Avian Biology* 39, 672–682.

Hausberger, M., Jenkins, P.F. and Keene, J. (1991) Species-specificity and mimicry in bird song: are they paradoxes? A reevaluation of song mimicry in the European starling. *Behaviour* 117, 53–81.

Heldbjerg, H., Fox, A.D., Levinc, G. and Nyegaard, T. (2016) The decline of the Starling *Sturnus vulgaris* in Denmark is related to changes in grassland extent and intensity of cattle grazing. *Agriculture, Ecosystems and Environment* 230, 24–31.

Henry, L., Craig, A.J.F.K., Lemasson, A. and Hausberger, M. (2015) Social coordination in animal vocal interactions. Is there any evidence of turn taking? The starling as an animal model. *Frontiers in Psychology* 6, 1416.

Higgins, P.J., Peter, J.M. and Cowling, S.J. (eds) (2006) Common Starling. In: *Handbook of Australian, New Zealand and Antarctic birds*, Vol. 7B. *Dunnock to Starlings*. Oxford University Press, Melbourne, Australia, pp. 1906–1934.

Hindmarsh, A.M. (1984) Vocal mimicry in starlings. *Behaviour* 90, 302–324.

Hockey, P.A.R., Dean, W.R.J. and Ryan, P.G. (eds) (2005) *Roberts Birds of Southern Africa*, 7th edn. John Voelcker Bird Book Fund, Cape Town.

Hockey, P.A.R., Underhill, L.G., Neatherway, M. and Ryan, P.G. (1989) Atlas of the birds of the Southwestern Cape. Cape Bird Club, Cape Town.

Höttner, K., Dibbert, R. and Sanftleben, P. (2015) Starenplage in Milchviehbetrieben in Mecklenburg-Vorpommern. *Der Praktische Tierarzt* 96, 1040–1048.

Hugo, S. and van Rensburg, B.J. (2009) Alien and native birds in South Africa: patterns, processes and conservation. *Biological Invasions* 11, 2291–2302.

Hui, C., Roura-Pascual, N., Brotons, L., Robinson, R.A. and Evans, K.L. (2012) Flexible dispersal strategies in native and non-native ranges: environmental quality and the 'good–stay, bad–disperse' rule. *Ecography* 35, 1024–1032.

Ibañez, L.M., Andreucci, F. and Montalti, D. (2016a) Primo registro de daño a cultivo de frutales par el estornino pinto (*Sturnus vulgaris*) (Passeriformes, Sturnidae) en Argentina. *Acta Zoologica Lilloana* 60, 177–180.

Ibañez, L.M., Girini, J.M., Palacio, F.X., Fiorini, V.D. and Montalti, D. (2017) Interacciones entre el estornino pinto (*Sturnus vulgaris*) y aves nativas de Argentina por el uso de cavidades. *Revista Mexicana de Biodiversidad* 88, 477–479.

Ibañez, L.M., Zalba, S., Scorolli, A., Forcelli, D. and Montalti, D. (2016b) El estornino pinto en Argentian: desafíos y prioridades. *Revista del Museo de la Plata*, 1, 159–169.

Ifran, N.R. and Fiorini, V.D. (2010) European Starling (*Sturnus vulgaris*): population density and interactions with native species in Buenos Aires urban parks. *Ornitologia Neotropical* 21, 507–518.

Ingold, D.J. (1989) Nesting phenology and competition for nest sites among Red-headed and Red-bellied Woodpeckers and European Starlings. *Auk* 106, 209–217.

Ivanova, I.M. and Symes, C.T. (2018) Common starling *Sturnus vulgaris* expansion in South Africa. *Biodiversity Observations* 9.9, 1–6.

Jackson, J.A. and Jackson, B.J.S. (2016) Dynamics of starling–woodpecker interactions. *Ornis Hungarica*, 24, 1–41.

Johnson, S.R. and Cowan, I.M. (1974) Thermal adaptation as a factor affecting colonizing success of introduced Sturnidae (Aves) in North America. *Canadian Journal of Zoology* 52, 1559–1576.

Johnston López, K., Duarte, A.M. and Gámez, R.A.C. (2015) Urban birds in the Sonoran Desert: estimating population density from point counts. *Huitzil, Revista Mexicana de Ornitología* 16, 37–47.

Joubert, H.J. (1945) Starlings and others. *Ostrich* 16, 214–216.

Kentish, B. and Peace, H. (1987) Where, when and what are the Starlings doing in Australia? *RAOU Newsletter* 73, 7–9.

Kerpez, T.A. and Smith, N.S. (1990) Competition between European Starlings and native woodpeckers for nest cavities in saguaros. *Auk* 107, 367–375.

Kessel, B. (1950) Observations on the polygamy and territorial behaviour of a male Starling. *Bird-Banding* 21, 112–114.

Kessel, B. (1951) Criteria for sexing and aging European starlings (*Sturnus vulgaris*). *Bird-Banding* 22, 16–23.

Kessel, B. (1953) Distribution and migration of the European Starling in North America. *Condor* 55, 49–67.

Kessel, B. (1957) A study of the breeding biology of the European Starling (*Sturnus vulgaris* L.) in North America. *American Midland Naturalist* 58, 257–331.

Koch, A.J., Martin, K. and Aitken, K.E.H. (2012) The relationship between introduced European Starlings and the reproductive activities of Mountain Bluebirds and Tree Swallows in British Columbia, Canada. *Ibis* 154, 590–600.

Koenig, W.D. (2003) European Starlings and their effect on native cavity-nesting birds. *Conservation Biology* 17, 1134–1140.

Komdeur, J., Oorebeek, M., van Overveld, T. and Cuthill, I.C. (2005) Mutual ornamentation, age, and reproductive performance in the European starling. *Behavioral Ecology* 16, 805–817.

Kopij, G. (2000) Diet of granivorous and frugivorous passerines in South African grassland. *South African Journal of Wildlife Research* 30, 135–138.

Kopij, G. (2001a) *Birds of Roma Valley, Lesotho*. National University of Lesotho, Roma, Lesotho.

Kopij, G. (2001b) *Atlas of Birds of Bloemfontein*. National University of Lesotho, Roma, Lesotho, and Free State Bird Club, Bloemfontein, South Africa.

Kopij, G. (2009) Segregation in sympatrically nesting Red-winged Starling *Onychognathus morio* (L.) and European Starling *Sturnus vulgaris* L. *Polish Journal of Ecology* 57, 589–595.

Korpimäki, E. (1978) Breeding biology of the Starling *Sturnus vulgaris* in western Finland. *Ornis Fennica* 55, 93–104.

Lawrence, C., Paris, D., Briskie, J.V. and Massaro, M. (2017) When the neighbourhood goes bad: can endangered black robins adjust nest site selection in response to the risk of an invasive predator? *Animal Conservation* 20, 321–330.

Le Louarn, M., Couillens, B., Deschamps-Cottin, M. and Clergeau, P. (2016) Interference competition between an invasive parakeet and native bird species at feeding sites. *Journal of Ethology* 34, 291–298.

Lever. C. (2005) *Naturalised Birds of the World*. T. & A.D. Poyser, London.

Linz, G., Johnson, R. and Thiele, J. (2018) European Starlings. In: Pitt, W.C., Beasley, J.C. and Witmer, G.W. (eds) *Ecology and Management of Terrestrial Vertebrate Invasive Species in the United States*. CRC Press, Boca Raton, Florida, pp. 311–332.

Liversidge, R. (1962) The spread of the European Starling in the Eastern Cape. *Ostrich* 33, 13–16.

Liversidge, R. (1985) Alien bird species introduced into South Africa. In: Bunning, L.J. (ed.) *Proceedings of the Birds and Man Symposium*, 10–15 April 1983, Johannesburg, South Africa, pp. 31–44.

Lombardo, M.P., Power, H.W., Stouffer, P.C., Romagnano, L.C. and Hoffenberg, A.S. (1989) Egg removal and intraspecific brood parasitism in the European starling (*Sturnus vulgaris*). *Behavioral Ecology and Sociobiology* 24, 217–223.

Long, J.L. (1981) *Introduced Birds of the World*. David and Charles, London.

Lovette, I.J., McCleery, B.V., Talaba, A.L. and Rubenstein, D.R. (2008) A complete species-level molecular phylogeny for the "Eurasian" starlings (Sturnidae: *Sturnus*, *Acridotheres* and allies): recent diversification in a highly social and dispersive avian group. *Molecular Phylogenetics and Evolution* 47, 251–260.

Lundberg, P. (1987) Breeding seasons of north Scandinavian starlings (*Sturnus vulgaris*): constrained by food or time? *Journal of Animal Ecology* 56, 847–855.

Lundberg, P. and Eriksson, L.-O. (1984) Postjuvenile moult in two northern Scandinavian Starling *Sturnus vulgaris* populations – evidence for difference in the circannual time-program. *Ornis Scandinavica* 15, 105–109.

Maccarone, A.D. (1987a) Age-class differences in the use of food sources by European Starlings. *Wilson Bulletin* 99, 699–704.

Maccarone, A.D. (1987b) Evidence for resource-based communal roosting by European Starlings. *Bird Behaviour* 7, 49–57.

Marateo, G., Grilli, P., Bouzas, N., Ferretti, V., Juárez, M. and Soave, G.E. (2015) Habitat use by birds in airports: a case study and its implications for bird management in South American airports. *Applied Ecology and Environmental Research* 13, 799–808.

Martínez del Rio, C., Stevens, B.R., Daneke, D.E. and Andreadis, P.T. (1988) Physiological correlates of preference and aversion for sugars in three species of birds. *Physiological Zoology* 61, 222–229.

Mason, J.R., Adams, M.A. and Clark, L. (1989) Anthranilate repellency to starlings: chemical correlates and sensory perception. *Journal of Wildlife Management* 53, 55–64.

Mason, J.R., Avery, M.L., Glahn, J.F., Otis, D.L., Matteson, R.E. and Nelms, C.O. (1991) Evaluation of methyl anthranilate and starch-plated dimethyl anthranilate as bird repellent feed additives. *Journal of Wildlife Management* 55, 182–187.

Mazzulla, J. (2013) Primeros registros de estornino pinto *Sturnus vulgaris* (Linnaeus, 1758) (Aves, Passeriformes, Sturnidae) en Uruguay. *Novedades Ornitológicas* 3, 13–17.

Medhanie, G.A, Pearl, D.L., McEwen, S.A., Guerin, M.T., Jardine, C.M. and LeJeune, J.T. (2015) Dairy cattle management factors that influence on-farm density of European starlings in Ohio, 2007–2009. *Preventative Veterinary Medicine* 120, 162–168.

Meinertzhagen, R. (1952) An historical note on the Starling *Sturnus vulgaris* L. at the Cape of Good Hope. *Bulletin of the British Ornithologists' Club* 72, 47.

Mennechez, G. and Clergeau, P. (2006) Effect of urbanisation on habitat generalists: starlings not so flexible? *Acta Oecologica* 30, 182–191.

Moeed, A. (1975) Diets of nestling Starlings and Mynas at Havelock North, Hawke's Bay. *Notornis* 22, 291–294.

Moeed, A. (1976) Birds and their food resources at Christchurch International Airport, New Zealand. *New Zealand Journal of Zoology* 3, 373–390.

Moeed, A. (1980) Diets of adult and nestling starlings (*Sturnus vulgaris*) in Hawke's Bay, New Zealand. *New Zealand Journal of Zoology* 7, 247–256.

Moeed, A. and Dawson, D.G. (1979) Breeding of starlings (*Sturnus vulgaris*) in nest boxes of various types. *New Zealand Journal of Zoology* 6, 613–618.

Morrison, D.W. and Caccamise, D.F. (1985) Ephemeral roosts and stable patches? A radiotelemetry study of communally roosting starlings. *Auk* 102, 793–804.

Motis, A., Mestre, P. and Martínez, A. (1983) La colonizacion y expansion del estornino pinto (*Sturnus vulgaris* L.) y del estornino negro (*Sturnus unicolor* TEMM.) en Cataluña (ne de la Peninsula Iberica). *Miscellenania Zoologica* 7, 131–137.

Myres, M.T. (1958) The European Starling in British Columbia 1947–1957. *British Columbia Provincial Museum Occasional Paper* 11, 1–60.

Neves, V.C., Panagiotakopoulos, S. and Furness, R.W. (2006) A control taste aversion experiment on predators of roseate tern (*Sterna dougallii*) eggs. *European Journal of Wildlife Research* 52, 259–264.

Palacio, F.X., Maragliano, R.E. and Montalti, D. (2016) Functional role of the invasive European Starling, *Sturnus vulgaris*, in Argentina. *Emu* 116, 387–393.

Paz, U. (1987) *The Birds of Israel*. Christopher Helm, London.

Peris, S., Soave, G., Camperi, A., Darrieu, C. and Aramburu, R. (2005) Range expansion of the European Starling *Sturnus vulgaris* in Argentina. *Ardeola* 52, 359–364.

Pfeiffer, M.B., Blackwell, B.F. and DeVault, T.L. (2018) Quantification of avian hazards to military aircraft and implications for wildlife management. *PloS One* 13, e0206599.

Phair, D.J., Le Roux, J.J., Berthouly-Salazar, C.B., Visser, V., Janse van Vuuren, B., *et al.* (2018) Context-dependent spatial sorting of dispersal-related traits in the invasive starlings (*Sturnus vulgaris*) of South Africa and Australia. *bioRxiv* 342451.

Pinxten, R., Eens, M. and Verheyen, R.F. (1989) Polygyny in the European starling. *Behaviour* 111, 234–255.

Pinxten, R., Eens, M. and Verheyen, R.F. (1990) Conspecific nest parasitism in the European Starling. *Ardea* 79, 15–30.

Pinxten, R., Eens, M. and Verheyen, R.F. (1994) Communal breeding in the European Starling: evidence from DNA fingerprinting. *Auk* 111, 482–486.

Potter, E.F. (1970) Anting in wild birds, its frequency and probable purpose. *Auk* 87, 692–713.

Poulsen, H. (1956) A study of anting behaviour in birds. *Dansk Ornithologisk Forenings Tidsskrift* 50, 267–298.

Querengässer, A. (1973) Über das Einemsen von Singvögeln und die Reifung dieses Verhaltens. *Journal für Ornithologie* 114, 96–117.

Quickelberge, C.D. (1989) *Birds of the Transkei*. Durban Natural History Museum, Durban, South Africa.

Rinke, D. (1987) The avifauna of 'Eua and its off-shore islet Kalau, Kingdom of Tonga. *Emu* 87, 26–34.

Rintala, J., Tiainen, J. and Pakkala, T. (2003) Population trends of the Finnish starling *Sturnus vulgaris*, 1952–1998, as inferred from annual ringing totals. *Annales Zoologici Fennici* 40, 365–385.

Robinson, R.A., Siriwardena, G.M. and Crick, H.Q.P. (2006) The population decline of the starling, *Sturnus vulgaris*, in Great Britain: patterns and causes. *Acta Zoologica Sinica* 52 (Suppl.), 550–553.

Ross, H.A. (1983) Genetic differentiation of starling (*Sturnus vulgaris:* Aves) populations in New Zealand and Great Britain. *Journal of Zoology* 201, 351–362.

Ross, H.A. and Baker, A.J. (1982) Variation in the size and shape of introduced starlings, *Sturnus vulgaris* (Aves: Sturnidae), in New Zealand. *Canadian Journal of Zoology* 60, 3316–3325.

Rothery, P., Wyllie, I., Newton, I, Dawson, A. and Osborn, D. (2001) The timing and duration of moult in adult Starlings *Sturnus vulgaris* in east-central England. *Ibis* 143, 435–441.

Royall, W.C. (1966) Breeding of the Starling in central Arizona. *Condor* 68, 196–205.

Schroeder, J.W. and Lee, C.D. (2015) Managing common bird challenges on dairy and livestock operations. *NDSU Extension Service* April 2015, 1–11.

Seamans, T.W., Blackwell, B.F. and Tyson, L.A. (2015) Low occupancy rates of artificial nest cavities by European Starlings. *Ohio Journal of Science* 115, 53–55.

Shwiff, A., Shwiff, S., Holderieath, J., Haden-Chomphosy, W. and Anderson, A. (2018) Economics of invasive species damage and damage management. In: Pitt, W.C., Beasley, J.C. and Witmer, G.W. (eds) *Ecology and Management of Terrestrial Vertebrate Invasive Species in the United States*. CRC Press, Boca Raton, Florida, pp. 35–59.

Simmons, K.E.L. (1957) A review of the anting-behaviour of passerine birds. *British Birds* 50, 401–424.

Skead, C.J. (1995) *Life-history Notes on East Cape Bird Species 1940–1990*. Algoa Regional Services Council, Port Elizabeth, South Africa.

Skead, C.J. (2006) The Common (European) Starling *Sturnus vulgaris* at Summerstrand, Port Elizabeth, 1991–2004. VI. The occurrence of white tail feathers (rectrices) *Diaz Diary* 34, 16–17.

Skead, D.M. (1966) Birds frequenting the intertidal zone of the Cape Peninsula. *Ostrich* 37, 10–16.

Smith, E.L., Cuthill, I.C., Griffiths, R., Greenwood, V.J., Goldsmith, A.R. and Evans, J.E. (2005) Sexing Starlings *Sturnus vulgaris* using iris colour. *Ringing & Migration* 22, 193–197.

Smith, H.G. and Bruun, M. (2002) The effect of pasture on starling (*Sturnus vulgaris*) breeding success and population density in a heterogeneous agricultural landscape in southern Sweden. *Agriculture, Ecosystems & Environment* 92, 107–114.

Smith, H.G., Wennerberg, L. and von Schantz, T. (1996) Adoption or infanticide: options of replacement males in the European starling. *Behavioural Ecology and Sociobiology*, 38, 191–197.

Smith, K.W. (2005) Has the reduction in nest-site competition from Starlings *Sturnus vulgaris* been a factor in the recent increase of Great Spotted Woodpecker *Dendrocopos major* numbers in Britain? *Bird Study* 52, 307–313.

Spasov, S., Hristov, I., Eaton, M. and Nikolov, S.C. (2017) Population trends of common birds in Bulgaria: is their status improving after the EU accession? *Acta Zoologica Bulgarica*, 69, 95–104.

Spenneman, D.H.R. and Allen, L.R. (2000) The avian dispersal of olives *Olea europaea*: implications for Australia. *Emu* 100, 264–273.

Summers, R.W., Westlake, G.E. and Feare, C.J. (1987) Differences in the physical condition of starlings at the centre and periphery of roosts. *Ibis* 129, 96–102.

Swaddle, J.P., Moseley, D.L., Hinders, M.K. and Smith, E.P. (2016) A sonic net excludes birds from an airfield: implications for reducing bird strike and crop losses. *Ecological Applications*, 26, 339–345.

Taitt, M.J. (1973) Winter food and feeding requirements of the Starling. *Bird Study* 20, 226–236.

Thompson, W.L. and Coutlee, E.L. (1963) The biology and population structure of Starlings at an urban roost. *Wilson Bulletin* 75, 358–372.

Tomasevic, J.A. and Marzluff J.M. (2017) Cavity nesting birds along an urban-wildland gradient: is human facilitation structuring the bird community? *Urban Ecosystems*, 20, 435–448.

Twedt, D.J. and Pardieck, K.L. (2017) North American Breeding Bird Survey in Mississippi. *The Mississippi Kite* 47, 70–83.

van der Merwe, F. (1984) 'n Nalatenskap van Rhodes: Europese spreeus verdring gryskopspegte. *African Wildlife* 38, 152–157.

Verheyen, R.F. (1969) Arrival at the breeding area, maturity for breeding and onset of laying according to age in the starling population (*Sturnus v. vulgaris* L.) of lower and middle Belgium. *Gerfaut* 59, 378–384.

Versluijs, M., van Turnhout, C.A.M., Kleijn, D. and van der Jeugd, H.P. (2016) Demographic changes underpinning the population decline of Starlings *Sturnus vulgaris* in The Netherlands. *Ardea* 104, 153–165.

Vincent, A.W. (1949) On the breeding habits of some African birds. *Ibis* 91, 483–507.

Wagner, G. (1963) Zum Brutverlauf beim Star, *Sturnus vulgaris*. *Ornithologischer Beobachter* 60, 205–213.

Ward, D. (2006) Common Starling usurps Little Swift nest. *Bee-eater* 57, 83.

West, M. J. and King, A.P. (1990) Mozart's starling. *American Scientist* 78, 106–114.

West, M.J., Stroud, A.N. and King, A.P. (1983) Mimicry of the human voice by European Starlings: the role of social interaction. *Wilson Bulletin* 95, 635–640.

Winterbottom, J.M. and Liversidge, R. (1954) The European Starling in the South West Cape. *Ostrich* 25, 89–96.

Wolfs, P.H.J., Lesna, I.K., Sabelis, M.W. and Komdeur, J. (2012) Trophic structure of arthropods in Starling nests matter to blood parasites and thereby to nestling development. *Journal of Ornithology* 153, 913–919.

Wydoski, R.S. (1964) Seasonal changes in the color of Starling bills. *Auk* 81, 542–550.

Yom-Tov, Y., Dunnet, G.M. and Anderson, A. (1974) Intraspecific nest parasitism in the Starling *Sturnus vulgaris*. *Ibis* 116, 87–90.

Zuccon, D., Pasquet, E. and Ericson, P.G.P. (2008) Phylogenetic relationships among Palearctic-Oriental starlings and mynas (genera *Sturnus* and *Acridotheres*: Sturnidae). *Zoologica Scripta* 37, 469–481.

Zufiaurre, E., Abba, A., Bilenca, D. and Codesido, M. (2016) Role of landscape elements on recent distributional expansion of European Starlings (*Sturnus vulgaris*) in agroecosystems of the pampas, Argentina. *Wilson Journal of Ornithology* 128, 306–313.

3 Common Myna (*Acridotheres tristis* Linnaeus, 1766)

Lorinda A. Hart[1,2]*, Andrew M. Rogers[3] and Berndt J. van Rensburg[3,4]

[1]*School of Life Sciences, University of KwaZulu-Natal, Pietermaritzburg 3209, South Africa;* [2]*Department of Biological Sciences, University of Namibia, Windhoek 10023, Namibia;* [3]*School of Biological Sciences, The University of Queensland, Brisbane, QLD 4072, Australia;* [4]*Department of Zoology, University of Johannesburg, Auckland Park, Johannesburg 2006, South Africa*

Citation: Hart, L.A., Rogers, A. and van Rensburg, B.J. (2020) Common Myna (*Acridotheres tristis* Linnaeus, 1766). In: Downs, C.T. and Hart, L.A. (eds) *Invasive Birds: Global Trends and Impacts*. CAB International, Wallingford, UK, pp. 25–32.

3.1 Common Names

Given the vast distribution of the Common Myna, there are many common names recognized for this species. These include the English names Common Myna, Calcutta Myna, House Myna, Indian Myna and Talking Myna. More colloquially, in New Zealand and Australia, they are referred to as Tasmanian Starlings, Tickbirds, Chocolate Birds, Chocolate Dollar-birds, White Wings, Flying Cane-toads, Rat-with-wings or Thynne Birds (presumably after Minister Thynne who is believed to have introduced the birds) (Higgins *et al*., 2006, Gray and Fraser, 2013). Other names include but are not limited to: Hirtenmaina (German), Manu Kavamani (Māori, Cook Islands), Martin Triste (French), Miná Común (Spanish), Maina (Danish), Pihamaina (Finnish), Brun Majna (Swedish), Maina Comune (Italian), Majna Brunatna (Polish), Indiese Spreeu (Afrikaans) and Treurmaina (Dutch).

3.2 Nomenclature

Classification of the Common Myna (*Acridotheres tristis* Linnaeus, 1766) has been revised and debated on several occasions. Traditional classifications were based on distributional and morphological differences (Sibley and Monroe, 1990; Feare and Craig, 1998). More recently molecular studies focusing on the phylogenetic relationships among Palearctic-Oriental Sturnidae, notably the genera *Sturnus* (starlings) and *Acridotheres* (mynas), have been conducted. Christidis and Boles (2008) acknowledged that there were various ways in which to represent the lineage based on data available at the time. Mynas could either be retained as a genus or, as they proposed, form part of the genus *Sturnus*, thus becoming *Sturnus tristis* (Christidis and Boles, 2008). However, more recent and extensive genetic analysis supports the retention of the genus *Acridotheres* (Lovette *et al*., 2008; Zuccon *et al*., 2008). Based on this and the continued use by leading organizations such as BirdLife International and the International Union for Conservation of Nature (IUCN), we have chosen to retain the classification of *Acridotheres tristis* in this chapter.

Two subspecies of Common Myna are recognized, namely: *A. t. tristis* Linnaeus, 1766, found throughout the continental range of this species, and *A. t. melanosternus* Legge, 1879, endemic to the island country of Sri Lanka (CAB International, 2018; Craig *et al*., 2019). The subspecies *A. t. melanostemus* has darker plumage and is smaller than *A. t. tristis* (Higgins *et al*., 2006). A third subspecies, *A. t. tristoides* Hodgson, 1836, has been proposed and is represented in Assam (India), Nepal and Myanmar (Craig, 2005). However, some classify this subspecies as a member of the polytypic *A. t. tristis* (Feare and Craig, 1998; Craig *et al*., 2019). Subspecies *A. t. tristis* is paler and duller than birds identified as *A. t. tristoides* on the under parts and less red on the back, but there is no apparent size difference (Craig, 2005).

3.3 Distribution

The Common Myna occurs on all continents except South America and Antarctica (Fig. 3.1) (Peacock *et al*., 2007). Their native range spans India and south and central Asia. Countries include Afghanistan, Bangladesh, Bhutan, Cambodia, China, India, Iran, Kazakhstan, Kyrgyzstan, Laos, Malaysia, Myanmar, Nepal, Pakistan, Singapore, Sri Lanka, Tajikistan, Thailand, Turkmenistan, Uzbekistan and Vietnam.

*Corresponding author: lorinda.hart.research@gmail.com

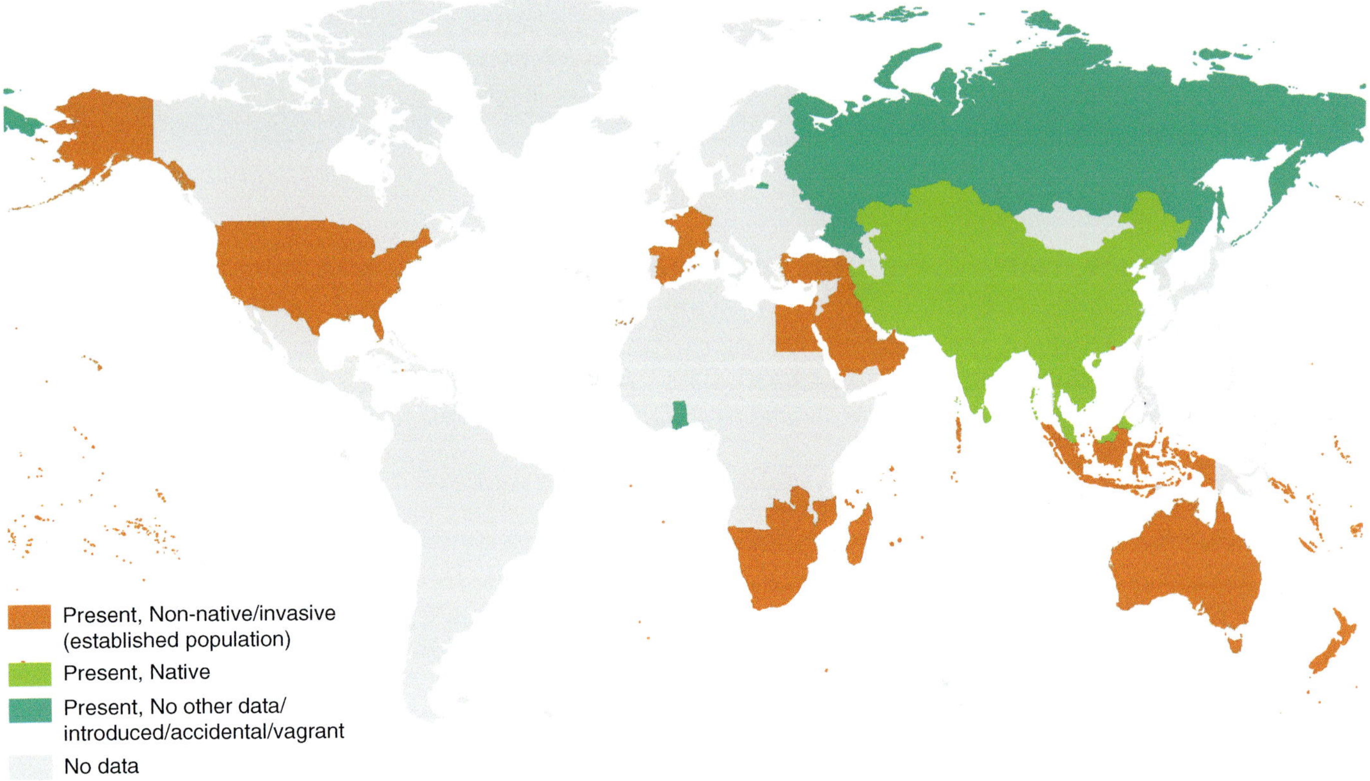

Fig. 3.1. Global distribution of the Common Myna (*Acridotheres tristis*) showing the natural (green) and invaded (red) ranges.

Common Mynas have been introduced in southern Africa, New Zealand, Australia, the USA, the Middle East and on numerous islands around the world. Countries and islands include: American Samoa, Australia, Brunei, Bahrain, British Indian Ocean Territory, Botswana, Comoros, Cook Islands, Egypt, Eswatini, Fiji, France, French Polynesia, Hong Kong, Indonesia, Iraq, Israel, Kiribati, Kuwait, Lesotho, Madagascar, Maldives, Mauritius, Mayotte, Mozambique, Namibia, New Caledonia, New Zealand, Oman, Qatar, Reunion, Saint Helena, Ascension and Tristan da Cunha, Samoa, Saudi Arabia, Seychelles, Singapore, Solomon Islands, South Africa, Spain, Tonga, Turkey, United Arab Emirates, the USA, US Minor Outlying Islands (Vanuatu, Wallis and Futuna), Zambia and Zimbabwe.

Common mynas are listed as both indigenous and invasive in Singapore, as they most likely arrived by natural range expansion due to the deforestation of Peninsular Malaysia (Gibson-Hill, 1950, Craig *et al.*, 2019, Chong *et al.* 2012). Common Mynas are reported as vagrants in the Russian Federation and Ghana.

Table 3.1. Morphological traits (mean ± SD) and ratios of Common Mynas in South Africa. (From Berthouly-Salazar *et al.*, 2012.)

Traits	Male (n = 217)	Female (n =172)
Bill depth (mm)	8.2 ± 0.4	7.9 ± 0.4
Bill length (mm)	19.2 ± 0.9	18.7 ± 1.0
Bill width (mm)[a]	7.8 ± 0.5	7.5 ± 0.5
Head length (mm)	35.2 ± 0.9	34.4 ± 1.0
Tail length (mm)[a]	92.1 ± 5.5	86.9 ± 5.6
Tarsus length (mm)	38.1 ± 1.5	36.9 ± 1.3
Wing length (mm)	144.5 ± 4.0	139.1 ± 3.7
Weight (g)	124.7 ± 11.0	113.4 ± 8.9
Wing:tail ratio	0.64 ± 0.04	0.62 ± 0.04
Bill length:width ratio	2.50 ± 0.19	2.52 ± 0.20
Head:body length ratio	3.54 ± 0.29	3.30 ± 0.25
Tarsus:body length ratio	3.27 ± 0.27	3.08 ± 0.21
Wing loading	0.01 ± 0.02	0.01 ± 0.02

[a]Traits displaying non-significant dimorphism between the sexes.

3.4 Description

3.4.1 Morphology

The Common Myna is 25 ± 2 cm (mean ± SD) in length and weighs 113 ± 31 g, with a wingspan of 12–14 cm. A detailed account of the morphological traits is presented in Table 3.1. There is little difference between the sexes, although males are slightly larger. Birds generally stand on the ground with an upright posture, although they run lower to the ground. In adults, the head and collar are glossy black/purple, and the body and back are predominantly brown (Fig. 3.2A, B). The wings are dark brownish black with white primary coverts and primary feather bases (Craig, 2005). These wing

Fig. 3.2. (A, B) Adult Common Mynas with varying yellow eye patches. (C) A juvenile Common Myna. (D) A nest site in a roof; note the blue plastic in the nesting material. (E) Common Myna in flight displaying white wing patches, coverts and vent. (F) Common Mynas resting on a horse's back. (©Photo credits: H. Jordaan (A–D); J. Hart (E); N. Puntis (F). All images taken in South Africa.)

patches are visible in flight (Fig. 3.2E). The undertail coverts and tip of the tail are also white (Feare and Craig, 1998). The bill, legs, feet and bare patches of skin around the eyes are yellow (Fig. 3.2A, B) (Feare and Craig, 1998). Eyes are mottled white, reddish brown or brown (Craig, 2005). Variation in iris colour and the number of white spots in eyes has been associated with ageing and seasons (Feare *et al.*, 2015a). Juveniles have brown heads and are paler than adults (Fig. 3.2C) (Craig, 2005). Baldness in Common Mynas is reported wherever they are found, and in extreme cases the head and neck lack feathers completely, exposing yellow skin (Kasambe *et al.*, 2010). These are colloquially termed 'King Mynas' and, although there are several proposed theories, the exact cause of this condition is unknown.

In some of their non-native ranges, Common Myna populations show slight differences in morphometric parameters (Berthouly-Salazar *et al.*, 2012). In South Africa, birds on the front edge of the invading population showed morphological variation in traits facilitating dispersal; for example, wing length was greater in females (the dispersing sex) when compared with female birds from the core range (Berthouly-Salazar *et al.*, 2012). In New Zealand, males and females showed varying responses to environmental parameters; males were larger in warmer locations than in cold locations, while only females varied along an altitudinal gradient (Baker and Moeed, 1979). Both sexes showed morphological differences in response to precipitation (Baker and Moeed, 1979). In South Africa, environmental factors have also been shown to influence morphological traits (Berthouly-Salazar *et al.*, 2012). Common Mynas also display morphological variation between introduced populations, which is probably due to bottlenecks in small founding populations and the short time that has elapsed since introductions occurred (Gibson *et al.*, 1984).

3.4.2 Physiology

Understanding the physiological responses of an introduced species to the environment can provide more accurate parameters for modelling predictive dispersal rates and managing the species. A study in South Africa showed that Common

Myna body temperature is affected by ambient temperature, as body temperatures were lowered during the night, presumably to conserve energy (Senoge, 2015). Body temperatures were also significantly lower in June (winter) than in September (spring), with resting metabolic rates being significantly higher in winter (Senoge, 2015). This indicates that in colder conditions the birds have higher energetic demands (Senoge, 2015) and supports studies suggesting that mynas avoid cold areas and use urban areas to find shelter from the elements (Pell and Tidemann, 1997).

3.5 Diet

Common Mynas have an omnivorous diet, eating fruits, grains, invertebrates, crustaceans, food scraps, nectar, small vertebrates, eggs and chicks of other bird species (Feare and Craig 1998; Craig, 2005; Markula *et al.*, 2009), Green Turtle (*Chelonia mydas*) eggs and carrion (Saavedra, 2009). Common Mynas generally forage on the ground but will forage opportunistically on flowering and fruiting trees. Within their native range, their diet is predominantly insectivorous (Ali *et al.*, 2010); however, they feed more on nectar in the spring, presumably to increase their energy intake before breeding (Sengupta, 1974, 1976; Ahmad and Sahi, 2012). They also scavenge around dumps, farms and city streets. Diets are habitat dependent, with more insects consumed by birds foraging in agricultural fields than in non-agricultural locations (Ali *et al.*, 2010). Nestlings are exclusively fed insects for the first 10 days (Markula *et al.*, 2009), with a more varied diet introduced later during chick development, which includes eggshells (Moeed, 1975). A recent laboratory study showed that Common Mynas prefer a lipid-rich diet followed by a protein-rich and lastly carbohydrate-based diet (Gumede and Downs, 2019). However, within an urban environment, protein-based foods are preferred and can result in intraspecific aggression, suggesting that these environments may be protein deficient (Machovsky-Capuska *et al.*, 2016). Additionally, Common Mynas prefer glucose sugar over sucrose and fructose (Gumede and Downs, 2018). Diet preferences of Common Mynas have implications for the agricultural industry, as farmers growing lipid-rich crops are more likely to suffer crop damage and losses to this species, for example (Gumede and Downs, 2019).

3.6 Introduction and Invasion Pathways

The Common Myna has been introduced or moved both intentionally and unintentionally starting in the mid-18th century (Lever, 1987). Introduced populations have an initial lag period, after which their rate of spread increases (Peacock *et al.*, 2007). Common Mynas are typically sedentary, but can travel longer distances, with records of 7–40 km (Peneaux and Griffin, 2015). A ring recovery of an adult myna showed that it had travelled 381 km over 13 months (Oschadleus, 2001). The success of Common Mynas as an introduced species is in part attributable to their ability to exploit resources available in human-modified environments, and to compete with native species. Their innovative problem-solving abilities and use of novel resources no doubt further facilitate their establishment in new environments, particularly in urban areas (Sol *et al.*, 2011, 2012b). In its invasive range, the Common Myna reaches its highest densities in urban and suburban environments (van Rensburg *et al.*, 2009). The two most common introduction pathways for this species include intentional release as a biocontrol agent and use in the pet trade.

3.6.1 Biocontrol

The literal translation of *Acridotheres* is 'grasshopper hunter' (Clinning, 1989) and in some regions Common Mynas were introduced as biocontrol agents, particularly of locusts (Peacock *et al.*, 2007). In Australia and New Zealand, acclimatization societies initially facilitated this process (Bull *et al.*, 1985). Countries where Common Mynas were introduced to control insects include: Ascension Island (Hughes, 2013), Mauritius and Reunion (Cheke and Hume, 2008), Seychelles, Madagascar and the Comoros (Benson, 1960), the Andaman Islands and the Nicobars (Ali and Ripley, 1972), Fiji (Long, 1981; Lever, 1987), the Hawaiian Islands, Australia and New Zealand (Hone, 1978), St Helena (McCulloch, 2004), and Tahiti and Raratonga (Lever, 1987). On Ascension Island, they were introduced to control ticks on cattle (Saavedra, 2009).

3.6.2 Pet trade

The keeping of Common Mynas as a caged bird has been the source of most recent introductions and continues to facilitate the range expansion of the species, both in areas adjacent to its historical native range and in its invasive range (Watling, 2001; Nagle, 2006). The Common Myna is kept as a caged bird in part due to its ability to mimic human speech. The release or escape of caged birds has been attributed to the introductions in the Chagos Group (Bourne, 1971), South Africa (Peacock *et al.*, 2007), Vanuatu (Medway and Marshall, 1975), Hong Kong and Taiwan (Gilbert *et al.*, 2012), and Israel (Holzapfel *et al.*, 2006).

3.7 Breeding Behaviour

Common Mynas generally breed in the spring and summer months. Pairs line a nesting cavity with material that includes grass, feathers and, in some cases, plastic (Fig. 3.2D), and lay eggs in a cup created in the nesting material. Common Mynas have a strong homing ability, with pairs forming life-long bonds and often reusing nest sites (Peneaux and Griffin, 2015). One nest that was 4–5 years old contained beetle wings, cigarette butts and an estimated 720 pieces of a variety of anthropogenic rubbish in addition to natural nesting materials (Peacock, 2016). The nesting cavity can be in almost any structure that provides some cover including, but not limited to, under eaves, in roofs, in nest boxes, in sufficiently dense trees, in tree cavities, in palm tree crowns and in cliff clefts. The clutch size ranges from three to six pale greenish-blue eggs measuring

2.9 × 2.2 cm (Craig, 2005). Both sexes brood and care for the young (Massam, 2001). Incubation ranges from 13–15 days, and the average number of fledglings is 2 ± 1.5 (n = 39) (CAB International, 2018). Fledging occurs on average after 25 days, but can range from 20 to 32 days, and one pair may lay two or three clutches in a season. Sexual maturity occurs at 1 year, and their lifespan is on average 4 years for a wild bird (Markula *et al.*, 2009). Nests are parasitized by Great Spotted Cuckoos (*Clamator glandarius*) in South Africa (Craig, 2005).

3.8 Habitat

The Common Myna's native range falls predominantly within the tropics and subtropics, and although they can tolerate a wide variety of climates, they prefer warm conditions (CAB International, 2018). Where cold conditions are present, they find refuge in and near buildings (Pell and Tidemann, 1997; CAB International, 2018). They occur in a diverse range of habitats including flood plains, grasslands, cultivated areas, plantations and desert oases (Feare and Craig, 1998). They evolved in open woodland (Sengupta, 1968), but today in both its native and invasive range, the Common Myna is considered a human commensal species. In its non-native range, it is largely adapted to modified habitats including open woodland, agricultural landscapes, and suburban and urban environments. In Australia and South Africa, they do not appear to invade very far into intact forests with high tree density, although they will make use of edge habitats of forests or urban reserves, particularly for breeding (Brooke *et al.*, 1986; Pell and Tidemann, 1997; Craig, 2005). With this species being competitively dominant in man-made habitats, land transformation coupled with a growing human population is likely to increase the available habitat and foraging areas for such commensal alien birds (Richardson *et al.*, 2000). It is expected that with such landscape-level changes, together with known spatial congruence between areas of high human activity and high indigenous species richness at macro-ecological scales (see e.g. Chown *et al.*, 2003, and van Rensburg *et al.*, 2004, for studies on birds), in future, alien bird species are expected to constitute a further economic, agricultural and environmental burden.

3.9 Impacts

3.9.1 Agricultural pests

The wide dietary tolerance of the Common Myna makes it a potential pest of fruit crops and cereals of economic importance (Feare and Craig, 1998). This is particularly true when insect prey is scarce (Martin, 1996). They also damage subsistence farms and gardens (Parkes, 2006; Saavedra, 2009), which has implications particularly for impoverished communities.

3.9.2 Seed dispersal of invasive and exotic plants

Common Mynas have the potential to assist in the spread of undesirable invasive and exotic plants (Feare and Craig, 1998). For example, in Hawaii and on Ascension Island, they disperse common lantana (*Lantana camara*) (Pimentel *et al.*, 2000; Saavedra, 2009). On Mangaia Island, Common Mynas feed on red passionfruit (*Passiflora rubra*), chillies (*Capsicum frutescens*), and pawpaws (*Carica papaya*) and probably facilitate their dispersal (Parkes, 2006).

3.9.3 Aggression, competition and predation

One of the Sanskrit names for the Common Myna is 'Kalahapriya', which translates to 'quarrelsome' or 'fond of contention' (Dave, 2005), as mynas are very vocal and often aggressive. The presence of Common Mynas has mixed impacts, depending on their location. In New Zealand, native species abundance increased following the control of Common Mynas (Tindall *et al.*, 2007). The continued spread of Common Mynas in southern Africa has led to public concern, particularly regarding interspecific competition (which has been observed with a range of indigenous species) and predation (Peacock *et al.*, 2007). The recent arrival of Common Mynas in many important conservation areas is worrying and warrants further monitoring (Peacock *et al.*, 2007).

In Australia, the evidence of significant impacts of the Common Myna are mixed. Pell and Tidemann (1997) found that Common Mynas used tree hollows for nesting that would have otherwise been used by native species. Indeed, they successfully and aggressively competed for nesting sites with two native parrots. In a separate study, long-term abundance of three cavity-nesting and eight small bird species was negatively affected by the establishment of Common Mynas (Grarock *et al.*, 2012). While Grarock *et al.* (2013) found a negative relationship in the spatial distribution of Common Myna nests and Crimson Rosella (*Platycercus elegans*), they concluded that this pattern may also be because of habitat change. In general, it appears that aggression around food resources by Common Mynas is no more than that displayed by native species, and these interactions are mostly restricted to urban areas (Crisp and Lill, 2006; Lowe *et al.*, 2011; Haythorpe *et al.*, 2012, 2014; Sol *et al.*, 2012a).

Their impacts on islands are likely to be greater, as many of these locations have a high degree of endemism, are historically predator free and are important breeding areas for many species. Their detrimental effects on many endemic Seychelles species are well documented. They compete with the endangered Seychelles Magpie Robin (*Copsychus sechellarum*) by occupying nest boxes, disrupting incubation, and predating eggs and chicks (Komdeur, 1996; Canning, 2011). Similar competition occurs with the critically endangered Seychelles Flycatcher (*Terpsiphone corvina*) (Feare, 2010) and vulnerable Seychelles White-eye (*Zosterops modestus*) (Payet, 2007). Attacks and injury to Seychelles Fodies (*Foudia sechellarum*) have also been observed (Feare, 2010).

Common Mynas have also been observed predating seabird eggs in the Seychelles (Feare *et al.*, 2015b), Cook Islands (Parkes, 2006) and Ascension Island (Dhami and Nagle, 2009), to name a few. On Ascension Island, the removal of feral cats saw a 25% increase in the predation of Sooty Tern (*Onychoprion fuscatus*) eggs by Common Mynas (Saavedra, 2009), which was more damage than that caused by rats (Dhami and Nagle, 2009).

Common Mynas are also believed to have a detrimental effect on native insect species, given that these make up the bulk of their diet (Feare and Craig, 1998), and potentially, to a lesser degree, any vertebrate prey.

3.9.4 Noise, nuisance and property damage

Common Mynas roost communally and can be a nuisance as they are very noisy, particularly at dawn and dusk. Their roosts also result in large faecal deposits, foul odours and tree limb damage (Yap *et al.*, 2002). They steal food from plates at restaurants (CAB International, 2018), feed on offerings and sacrifices (Dave, 2005), and consume food put out for domestic stock (Parkes, 2006). Common Mynas mob people and animals, particularly near nest sites. They often nest on building structures, in gutters and on the roofs of houses (Saavedra, 2009), which results in the contamination of water collected from these sources (Parkes, 2006). They also nest on telecommunication structures (Parkes, 2006), and in a more unique scenario, Common Mynas affected the electrical system at a USA base, which not only posed a security risk but was also a great financial expense to repair (Saavedra, 2009).

3.9.5 Disease transmission

Common Mynas carry avian diseases such as avian malaria (Clark *et al.*, 2015). They can spread mites such as *Ornithonyssus bursa* (causing dermatitis in humans) and other diseases including psittacosis, salmonellosis and arbovirusal infections (CAB International, 2018). Disease spread has been linked to their droppings and nest sites, as well as their scavenging habits, particularly at landfills and on carrion (Saavedra, 2009; CAB International, 2018). Common Mynas have potentially transferred parasite lineages from their native regions to areas where they have been introduced and are carriers of novel parasites in their invasive ranges (Ishtiaq *et al.*, 2005).

3.9.6 Positive impacts

As Common Mynas feed on fruits and nectar, they are potential pollinators and dispersers of indigenous seeds. Additionally, they remove carrion from the environment (Saavedra, 2009). They predate on invasive bird species eggs and chicks (Saavedra, 2009), but the effect and degree of this is not known. In their native range, they are regarded as important biocontrol agents of agricultural pests (Ahmad and Sahi, 2012), with up to 92% of their diet consisting of insects (Ali *et al.*, 2010). Additionally, they have a commensal relationship with Fork-tailed Drongos (*Dicrurus adsimilis*) in India, where drongos had greater food intake when associated with myna flocks than birds that foraged in isolation (Veena and Lokesha, 1993). Nest boxes previously occupied by Common Mynas have been used by Spotted Owlets (*Athene brama*) in their native range, where these birds make use of their nesting material (Kler, 2004; Kler and Kumar, 2012).

3.10 Control

Control in the form of shooting has been used in South Africa, New Zealand, the Cook Islands and Tokelau (Nagle, 2006; Tindall *et al.*, 2007). Wide-scale eradication on continental countries is probably not possible, but problem birds and groups are controlled as needed (Picker and Griffiths, 2013). Control in Australia has mostly been using live traps, but trapping did not remove individuals faster than the replacement rate in Canberra (Grarock *et al.*, 2014). On Atiu Island, a bounty of NZ$2 per bird and live trapping further assisted the eradication of nearly 13,000 birds (Schwartz, 2014). In total, more than 26,000 birds have been culled, ridding the Cook Islands of this species using poisons, shooting and live trapping (see Chapter 43, this volume). Common Mynas were eradicated from Kiribati by shooting, as numbers were low and warranted a targeted approach (Butler, 2015). The Seychelles has also successfully eradicated Common Mynas from some islands using various trapping techniques, nest disturbance and shooting (Canning, 2011). On Fregate Island, 745 birds and 42 eggs were removed to achieve total eradication (Canning, 2011). The use of Starlicide poisoned baits on St Helena and Ascension Islands had mixed results, but ultimately reduced Common Myna numbers (Feare, 2010). Live trapping on Ascension Island also reduced the population by 56% (Saavedra, 2009).

Common Mynas are preyed on opportunistically by birds of prey. Birds and their eggs may also be eaten by snakes (Ali and Ripley, 1972; CAB International, 2018). Other predators known to raid nests or feed on adults include: House Crows (*Corvus splendens*), domestic and feral cats (*Felis catus*) and Javan Mongooses (*Herpestes javanicus*) (Lin, 2007). Common Mynas are eaten by people on some of the Pacific Islands (Lin, 2007).

3.11 Uses

The species is used as a pet or caged bird, probably due to its ability to mimic human speech and its diverse vocalizations. It has been used as a biocontrol for insects, although its effectiveness is unclear.

3.12 References

Ahmad, F. and Sahi, D.N. (2012) Food and feeding habits of the Common Myna, *Acridotheres tristis* (Family: Sturnidae) at Jammu, Jammu and Kashmir. *Ecoscan* 6, 69–72.

Ali, A.M.S., Asokan, S., Manikannan, R. and Radhakrishnan, P. (2010) Evaluation on the role of Common Myna (*Acridotheres tristis*) in agricultural crop protection in Cauvery Delta region. *Indian Journal of Environmental Education* 10, 37–44.

Ali, S. and Ripley, S.D. (1972) *Handbook of the Birds of India and Pakistan*, Vol. 5: *Larks to Grey Hypocolius*. Oxford University Press, Delhi, India.

Baker, A.J. and Moeed, A. (1979) Evolution in the introduced New Zealand populations of the Common Myna, *Acridotheres tristis* (Aves: Sturnidae). *Canadian Journal of Zoology* 57, 570–584.

Benson, C.W. (1960) The birds of the Comoro Islands: results of the British Ornithologists' Union Centenary Expedition, 1958. *Ibis* 103, 5–106.

Berthouly-Salazar, C., van Rensburg, B.J., Le Roux, J.J., van Vuuren, B.J. and Hui, C. (2012) Spatial sorting drives morphological variation in the invasive bird, *Acridotheris tristis*. *PLoS One* 7, e38145.

Bourne, W.R.P. (1971) The birds of the Chagos Group, Indian Ocean. *Atoll Research Bulletin* 149, 175–207.

Brooke, R.K., Lloyd, P.H. and de Villiers, A.L. (1986) Alien and translocated terrestrial vertebrates in South Africa. In: Macdonald, I.A.W., Kruger, F.J. and Ferrar, A.A. (eds) *The Ecology and Management of Biological Invasions in Southern Africa*. Oxford University Press, Cape Town, pp. 63–74.

Bull, P.C., Gaze, P.D. and Robertson, C.J.R. (1985) *The Atlas of Bird Distribution in New Zealand*. Ornithological Society of New Zealand, Wellington.

Butler, D. (2015) Final Report – Eradication of Myna from Kiribati. Ministry of Environment, Lands and Agricultural Development, Government of Kiribati, Kiribati.

CAB International (2018) *Acridotheres tristis* (common myna). In: *Invasive Species Compendium*. CAB International, Wallingford, UK. Available at: www.cabi.org/isc/datasheet/2994 (accessed 22 October 2019).

Canning, G.J.P. (2011) Eradication of the invasive Common Myna, *Acridotheres tristis*, from Fregate Island, Seychelles. *Phelsuma*, 19, 43–53.

Cheke, A. and Hume, J. (2008) *Lost Land of the Dodo*. T. & A.D. Poyser, London.

Chong, K.Y., Teo S., Kurukulasuriya, B., Chung, Y.F. Rajathurai, S., Lim, H.C. and Tan, H.T. (2012) Decadal changes in urban bird abundance in Singapore. *Raffles Bulletin of Zoology* 25, pp. 189–196.

Chown, S.L., van Rensburg, B.J., Gaston, K.J., Rodrigues, A.S.L. and van Jaarsveld, A.S. (2003) Energy, species richness, and human population size: conservation implications at a national scale. *Ecological Applications* 13, 1233–1241.

Christidis, L. and Boles, W. (2008) *Systematics and Taxonomy of Australian Birds*. CSIRO Publishing, Melbourne, Australia.

Clark, N.J., Olsson-Pons, S., Ishtiaq, F. and Clegg, S.M. (2015) Specialist enemies, generalist weapons and the potential spread of exotic pathogens: malaria parasites in a highly invasive bird. *International Journal for Parasitology* 45, 891–899.

Clinning, C. (1989) *Southern African Bird Names Explained*. The Southern African Ornithological Society, Johannesburg, South Africa.

Craig, A., Feare, C. and Garcia, E.F.J. (2019) Common Myna (*Acridotheres tristis*). In: del Hoyo, J., Elliott, A., Sargatal, J., Christie, D.A. and de Juana, E. (eds) *Handbook of the Birds of the World Alive*. Lynx Edicions, Barcelona, Spain. Available at: www.hbw.com/species (accessed 22 October 2019).

Craig, A.J.F.K. (2005) Common Myna. In: Hockey, P.A.R., Dean, W.R.J. and Ryan, P.G. (eds) *Roberts Birds of Southern Africa*, 7th edn. Trustees of the John Voelker Bird Book Fund, Cape Town, p. 972.

Crisp, H. and Lill, A. (2006) City slickers: habitat use and foraging in urban Common Mynas *Acridotheres tristis*. *Corella* 30, 9–15.

Dave, K.N. (2005) *Birds in Sanskrit Literature*. Motilal Banarsidass Private Ltd., Delhi, India.

Dhami, M. and Nagle, B. (2009) Review of the biology and ecology of the Common Myna (Acridotheres tristis) and some implications for management of this invasive species. Pacific Invasives Initiative, The University of Auckland, Auckland, New Zealand.

Feare, C.J. (2010) The use of Starlicide in preliminary trials to control invasive Common Myna *Acridotheres tristis* populations on St Helena and Ascension islands, Atlantic Ocean. *Conservation Evidence* 7, 52–61.

Feare, C. and Craig, A. (1998) *Starlings and Mynas*. Helm, London.

Feare, C.J., Edwards, H., Taylor, J.A., Greenwell, P., Larose, C.S., Mokhoko, E. and Dine, M. (2015a) Stars in their eyes: iris colour and pattern in Common Mynas *Acridotheres tristis* on Denis and North Islands, Seychelles. *Bulletin of the British Ornithologists' Club* 135, 61–68.

Feare, C.J., Lebarbenchon, C., Dietrich, M. and Larose, C.S. (2015b). Predation of seabird eggs by Common Mynas *Acridotheres tristis* on Bird Island, Seychelles, and its broader implications. *Bulletin of the African Bird Club* 22, 162–170.

Gibson, A.R., Baker, A.J. and Moeed, A. (1984) Morphometric variation in introduced populations of the Common Myna (*Acridotheres tristis*): an application of the jackknife to principal component analysis. *Systematic Biology* 33, 408–421.

Gibson-Hill, C.A. (1950) Myna matters. *Malayan Nature Journal* 5, 58–75.

Gilbert, M., Sokha, C., Joyner, P.H., Thomson, R.L. and Poole, C. (2012) Characterizing the trade of wild birds for merit release in Phnom Penh, Cambodia and associated risks to health and ecology. *Biological Conservation* 153, 10–16.

Grarock, K., Tidemann, C.R., Wood, J. and Lindenmayer, D.B. (2012) Is it benign or is it a pariah? Empirical evidence for the impact of the Common Myna (*Acridotheres tristis*) on Australian birds. *PLoS One* 7, e40622.

Grarock, K., Lindenmayer, D.B., Wood, J.T. and Tidemann, C.R. (2013) Does human-induced habitat modification influence the impact of introduced species? A case study on cavity-nesting by the introduced Common Myna (*Acridotheres tristis*) and two Australian native parrots. *Environmental Management* 52, 958–970.

Grarock, K., Tidemann, C.R., Wood, J.T. and Lindenmayer, D.B. (2014) Understanding basic species population dynamics for effective control: a case study on community-led culling of the Common Myna (*Acridotheres tristis*). *Biological Invasions* 16, 1427–1440.

Gray, J. and Fraser, I. (2013) *Australian Bird Names: A Complete Guide*. CSIRO Publishing, Clayton, Australia.

Gumede, S.T. and Downs, C.T. (2018) Sugar preference of invasive Common Mynas (*Sturnus tristis*). *Journal of Ornithology* 160, 71–78.

Gumede, S.T. and Downs, C.T. (2019) Preliminary observations suggest Common Myna prefer lipid over protein and carbohydrate diets in a food choice experiment. *Ostrich* (in press).

Haythorpe, K.M., Burke, D. and Sulikowski, D. (2014) The native versus alien dichotomy: relative impact of native Noisy Miners and introduced Common Mynas. *Biological Invasions* 16, 1659–1674.

Haythorpe, K.M., Sulikowski, D. and Burke, D. (2012) Relative levels of food aggression displayed by Common Mynas when foraging with other bird species in suburbia. *Emu* 112, 129–136.

Higgins, P.J., Peter, J.M. and Cowling, S.J. (2006) *Handbook of Australian, New Zealand & Antarctic Birds*. Oxford University Press, Melbourne, Australia.

Holzapfel, C.L., Levin, N.O., Hatzofe, O.H. and Kark, S.A. (2006) Colonisation of the Middle East by the invasive Common Myna *Acridotheres tristis* L., with special reference to Israel. *Sandgrouse* 28, 44–51.

Hone, J. (1978) Introduction and spread of the Common Myna in New South Wales. *Emu* 78, 227–230.

Hughes, B.J. (2013) Breeding and population ecology of Sooty Terns on Ascension Island. PhD thesis, University of Birmingham, Birmingham, UK.

Ishtiaq, F., Beadell, J.S., Baker, A.J., Rahmani, A.R., Jhala, Y.V. and Fleischer, R.C. (2005) Prevalence and evolutionary relationships of haematozoan parasites in native versus introduced populations of Common Myna *Acridotheres tristis*. *Proceedings of the Royal Society B: Biological Sciences* 273, 587–594.

Kasambe, R., Wadatkar, J., Joshi, A., Wagh, G. and Balpande, T. (2010) Baldness in Common Mynas *Acridotheres tristis* and Rose-ringed Parakeets *Psittacula krameri* in India. *Wetlands* 14, 323–334.

Kler, T. (2004) Successful breeding of Common Myna in wooden nest boxes and nest box acceptance and occupation by Spotted Owlet. *Pestology* 27, 37–39.

Kler, T.K. and Kumar, M. (2012) Breeding of Spotted Owlet *Athene brama* in nest boxes and conservation aspects. *Journal of the Bombay Natural History Society* 109, 135–139.

Komdeur, J. (1996) Breeding of the Seychelles Magpie Robin *Copsychus sechellarum* and implications for its conservation. *Ibis* 138, 485–498.

Lever, C. (1987) *Naturalised Birds of the World.* Longman Scientific and Technical, London.

Lin, T. (2007) *Acridotheres tristis* Common Myna. Animal Diversity Web. Available at: https://animaldiversity.org/ (accessed 20 October 2019).

Long, J.L. (1981) *Introduced Birds of the World: The Worldwide History, Distribution and Influence of Birds Introduced to New Environments.* Universe Books, New York.

Lovette, I.J., McCleery, B.V., Talaba, A.L. and Rubenstein, D.R. (2008) A complete species-level molecular phylogeny for the "Eurasian" starlings (Sturnidae: Sturnus, Acridotheres, and allies): recent diversification in a highly social and dispersive avian group. *Molecular Phylogenetics and Evolution* 47, 251–260.

Lowe, K.A., Taylor, C.E. and Major, R.E. (2011) Do Common Mynas significantly compete with native birds in urban environments? *Journal of Ornithology* 152, 909–921.

Machovsky-Capuska, G.E., Senior, A.M., Zantis, S.P., Barna, K., Cowieson, A.J., *et al.* (2016) Dietary protein selection in a free-ranging urban population of Common Myna birds. *Behavioral Ecology* 27, 219–227.

Markula, A., Hannan-Jones, M. and Csurhes, S. (2009) *Pest Animal Risk Assessment: Indian Myna* Acridotheres tristis. Department of Employment, Economic Development and Innovation, Brisbane, Australia.

Martin, W.K. (1996) The current and potential distribution of the Common Myna *Acridotheres tristis* in Australia. *Emu* 96, 166–173.

Massam, M. (2001) Common Myna (Farmnote No. 61/2001). Department of Agriculture, Kensington, Australia.

McCulloch, N. (2004) *A Guide to the Birds of St Helena and Ascension Islands.* Royal Society for the Protection of Birds, Sandy, UK.

Medway, L. and Marshall, A.G. (1975) Terrestrial vertebrates of the New Hebrides: origin and distribution. *Philosophical Transactions of the Royal Society of London B: Biological Sciences* 272, 423–365.

Moeed, A. (1975) Diet of nestling starlings and mynas at Havelock North, Hawke's Bay. *Notornis* 22, 291–294.

Nagle, B. (2006) Protection of Tokelau Fakaofo from Myna Bird (*Acridotheres* spp.) invasion. Report of a feasibility study, 12–16 May 2006. Pacific Invasives Initiative, Auckland, New Zealand.

Oschadleus, H.D. (2001) Report on recoveries received at SAFRING: July 2000–June 2001. *Afring News* 30, 74–81.

Parkes, J. (2006) Feasibility plan to eradicate Common Mynas (*Acridotheres tristis*) from Mangaia Island, Cook Islands. Landcare Research Contract Report: LC0506/184, Lincoln, New Zealand. Available at: http://www.issg.org/cii/Electronic%20references/pii/project_docs/cooks_myna/mangaia_myna_feasibility_plan_2006.pdf (accessed 26 March 2020).

Payet, H.E. (2007) The introduced Seychelles White-eye (*Zosterops modestus*) population of Fregate Island, Seychelles: territories and aspects of population dynamics, biology and ecology. Masters thesis, University of Réunion, Saint-Denis, France

Peacock, D.S., van Rensburg, B.J. and Robertson, M.P. (2007) The distribution and spread of the invasive alien Common Myna, *Acridotheres tristis* L. (Aves: Sturnidae), in southern Africa. *South African Journal of Science* 103, 465–474.

Peacock, F. (2016) Recycling mynas. Available at: https://faansiepeacock.com/recycling-mynas (accessed 20 October 2019).

Pell, A.S. and Tidemann, C.R. (1997) The ecology of the Common Myna in urban nature reserves in the Australian Capital Territory. *Emu* 97, 141–149.

Peneaux, C. and Griffin, A. (2015) Opportunistic observations of travel distances in Common Mynas (*Acridotheres tristis*). *Canberra Bird Notes* 40, 228–234.

Picker, M. and Griffiths, C. (2013) *Alien and Invasive Animals: A South African Perspective.* Penguin Random House, Cape Town.

Pimentel, D., Lach, L., Zuniga, R. and Morrison, D. (2000) Environmental and economic costs of nonindigenous species in the United States. *BioScience* 50, 53–65.

Richardson, D.M., Bond, W.J., Dean, W.R.J., Higgins, S.I., Midgley, G.F., *et al.* (2000) Invasive alien species and global change: a South African perspective. In: Mooney, H.A. and Hobbs, R.J. (eds) *Invasive Species in a Changing World.* Island Press, Washington, pp. 303–349.

Saavedra, S. (2009) First control campaign for Common Myna (*Acridotheres tristis*) in Ascension Island 2009. Live Arico Invasive Species Department, Canary Islands, Spain.

Schwartz, D. (2014) Hunting down the Common Myna on the island of Atiu in the Cook Islands. Available at: https://www.abc.net.au/news/2014-05-25/hunting-down-the-common-myna-in-the-cook-islands/5475868 (accessed 26 March 2020).

Sengupta, S. (1968) Studies of the Common Myna *Acridotheres tristis tristis* (Linnaeus). *Proceedings of the Zoological Society* 21, 1–27.

Sengupta, S. (1974) Anticipation of weather change by the Common Myna. *Science Reporter* 11, 37–38.

Sengupta, S. (1976) Food and feeding ecology of the Common Myna, *Acridotheres tristis* (Linn.). *Proceedings of the National Academy of Sciences of India* 42, 338–345.

Senoge, N.D. (2015) Eco-physiological aspects of invasive Common Myna. MSc thesis, University of KwaZulu-Natal, Pietermaritzburg, South Africa.

Sibley, C.G. and Monroe, B.L. (1990) *Distribution and Taxonomy of Birds of the World.* Yale University Press, New Haven/London.

Sol, D., Bartomeus, I. and Griffin, A.S. (2012a) The paradox of invasion in birds: competitive superiority or ecological opportunism? *Oecologia* 169, 553–564.

Sol, D., Griffin, A.S. and Bartomeus, I. (2012b) Consumer and motor innovation in the Common Myna: the role of motivation and emotional responses. *Animal Behaviour* 83, 179–188.

Sol, D., Griffin, A.S., Bartomeus, I. and Boyce, H. (2011) Exploring or avoiding novel food resources? The novelty conflict in an invasive bird. *PLoS One* 6, e19535.

Tindall, S.D., Ralph, C.J. and Clout, M.N. (2007) Changes in bird abundance following Common Myna control on a New Zealand island. *Pacific Conservation Biology* 13, 202–212.

van Rensburg, B.J., Erasmus, B.F., van Jaarsveld, A.S., Gaston, K.J. and Chown, S.L. (2004) Conservation during times of change: interactions between birds, climate, and people in South Africa. *South African Journal of Science* 100, 266–272.

van Rensburg, B.J., Peacock, D.S. and Robertson, M.P. (2009) Biotic homogenization and alien bird species along an urban gradient in South Africa. *Landscape and Urban Planning* 92, 233–241.

Veena, T. and Lokesha, R. (1993) Association of drongos with myna flocks: are drongos benefitted? *Journal of Biosciences* 18, 111–119.

Watling, D. (2001) Common Mynah – *Acridotheres tristis*. In: *A Guide to the Birds of Fiji and Western Polynesia: Including American Samoa, Niue, Samoa, Tokelau, Tonga, Tuvalu and Wallis and Futuna.* Environmental Consultants, Suva, Fiji, pp. 143–144.

Yap, C.A.M., Sodhi, N.S. and Brook, B.W. (2002) Roost characteristics of invasive mynas in Singapore. *Journal of Wildlife Management* 66, 1118–1127.

Zuccon, D., Pasquet, E. and Ericson, P.G. (2008) Phylogenetic relationships among Palearctic-Oriental starlings and mynas (genera *Sturnus* and *Acridotheres*: Sturnidae). *Zoologica Scripta* 37, 469–481.

4 Jungle Myna (*Acridotheres fuscus* Wagler, 1827)

Lorinda A. Hart*

School of Life Sciences, University of KwaZulu-Natal, Pietermaritzburg 3209, South Africa; Department of Biological Sciences, University of Namibia, Windhoek 10023, Namibia

Citation: Hart, L.A. (2020) Jungle Myna (*Acridotheres fuscus* Wagler, 1827). In: Downs, C.T. and Hart, L.A. (eds) *Invasive Birds: Global Trends and Impacts*. CAB International, Wallingford, UK, pp. 33–39.

4.1 Common Names

There are many common names listed in a variety of languages for this species (Lepage, 2004). Some examples include Jungle Myna, Buffalo Myna, Indian Jungle Myna, Jungle Mynah (English); Mainate de jungle, Martin brun, Martin forestier (French); Morihakka (Japanese); Burung Tiong Hutan, Burung Tiong Sawah (Malay); nók îaŋ khwaay (Thai); and Kattu Naganavaai (Tamil).

4.2 Nomenclature

Jungle Mynas (*Acridotheres fuscus* Wagler, 1827) are members of the family Sturnidae. Four races of Jungle Myna are recognized, with the nominate population:

- *A. f. fuscus* Wagler, 1827, present in the Himalayas (Pakistan to Assam, Rajasthan and Orissa).
- *A. f. mahrattensis* Sykes, 1832, found in the western peninsula of India.
- *A. f. fumidus* Ripley, 1950, in north-eastern Assam, and
- *A. f. torquatus* Davison, 1892, found from Myanmar to the Thai-Malay peninsula, excluding the southern tip (Clements *et al.*, 2018).

4.3 Distribution

Jungle Mynas are native to India and parts of south-east Asia (Fig. 4.1). Countries include: Bangladesh, Bhutan, India, Laos (probably vagrant or misidentified), Malaysia, Myanmar, Nepal, Pakistan, Thailand and Singapore (Lepage, 2004; CABI, 2018).

Jungle Mynas have been introduced in American Samoa, Cook Islands (rare), Fiji, Indonesia, Japan, Samoa, Taiwan, Tokelau (New Zealand), Tonga, and Wallis and Futuna Islands (France) (Fig. 4.1) (Eguchi and Amano, 2004a; Lepage, 2004; McAllan and Hobcroft, 2005; Theuerkauf *et al.*, 2010; CABI, 2018). Two observations (probable escaped birds) have been reported in Miami-Dade County in Florida (EDDMapS, 2019).

Common Mynas are sometimes confused with Jungle Mynas; however, their females are 33 g and males 49 g larger than Jungle Mynas (Ali and Ripley, 1972). Jungle Mynas are generally a sleeker bird, lack the yellow eye ring present in Common Mynas, and have a small crest at the base of their beaks. Such misidentification could explain the two sightings of Jungle Mynas in Laos, where Common Mynas are present (see eBird species report, Sullivan *et al.*, 2009). Additionally, the invasion history for some countries is uncertain due to species confusion (Trail, 1994).

Continental Jungle Mynas, Javan Mynas (*Acridotheres javanicus*) and Pale-bellied Mynas (*A. cinereus*) are genetically very similar, suggesting a recent divergence (Lovette *et al.*, 2008). It is possible that Jungle Mynas and Javan Mynas interbreed in areas of Malaysia, Thailand, Myanmar and Singapore where the latter has been introduced (Kang *et al.*, 1994). In 2003, in areas of Malaysia where Javan Mynas have been introduced, Jungle Mynas have since disappeared, with possible hybrids observed (Wells, 2010). This supports the theory that Javan Mynas have genetically swamped Jungle Mynas in this area, although further research is needed for confirmation (Wells, 2010). As such, the Jungle Myna faces extinction in the Malaysian Peninsula and is considered locally endangered (Wells, 2010).

4.4 Description

Jungle Mynas typically forage in pairs or small flocks of 10–30 birds (Long, 1981). However, larger flocks of 50–100 can congregate at feeding sites, such as landfills (Trail, 1994). They nest alone or communally, and non-breeding birds often roost

*Corresponding author: lorinda.hart.research@gmail.com

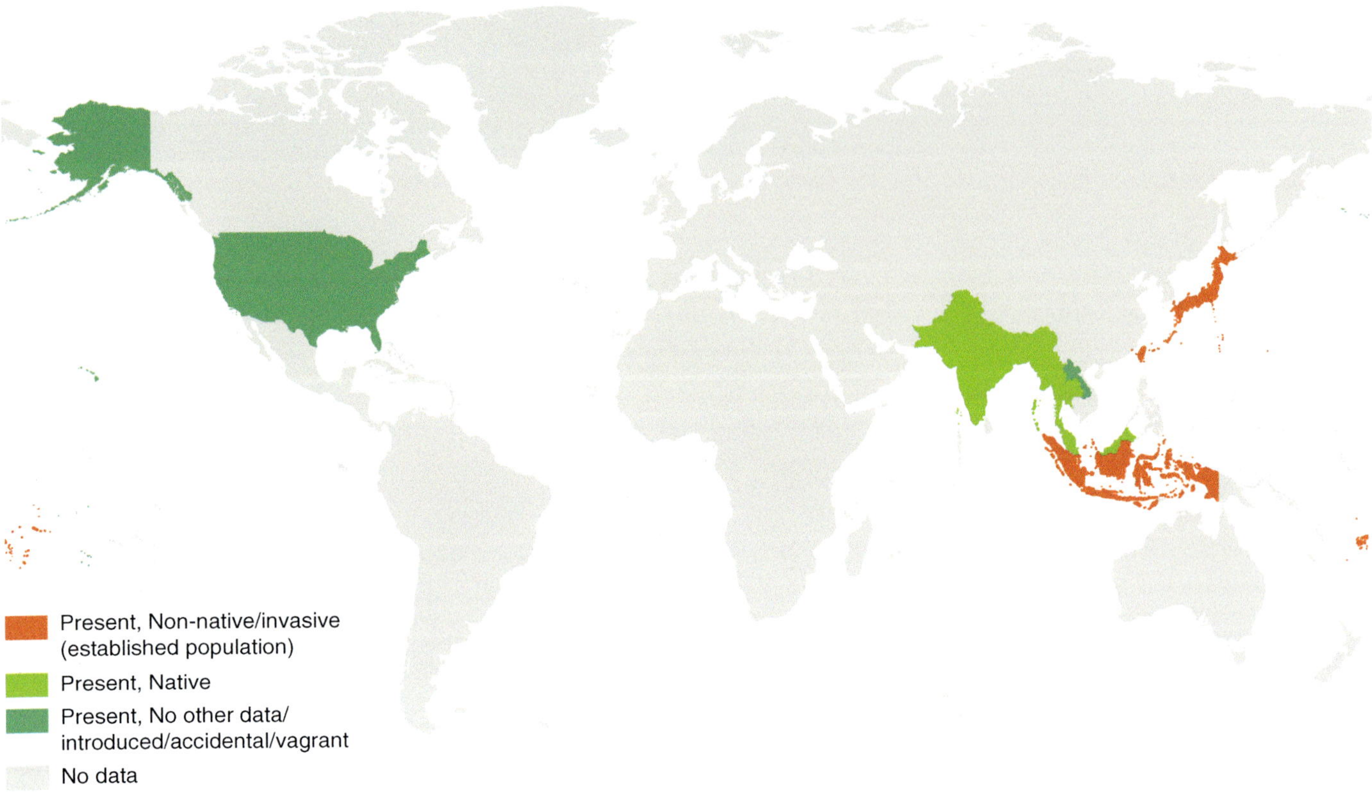

Fig. 4.1. Global distribution of the Jungle Myna (*Acridotheres fuscus*) showing the natural (green) and invaded (red) ranges.

in mixed-species communal roosts (Wells, 2010). They are approximately 22–24 cm tall (Feare and Craig, 1998). The sexes are alike in plumage, but females are smaller and typically weigh 78 g, while males weigh 94 g (Ali and Ripley, 1972). They have grey/brown bodies, with darker brown/black heads, outer wings and tails (Fig. 4.2B, D). The bases of wing primaries are white, with patches conspicuous when the wings are spread, and tail feathers are tipped white (Feare and Craig, 1998). The beaks are predominantly orange-yellow, with the base black and a crest of feathers is present where the forehead and beak meet (Fig. 4.2B, C) (Feare and Craig, 1998). Birds are generally paler on the underparts with a white vent (Fig. 4.2A, F) (Feare and Craig, 1998). Juvenile beaks are solid yellow and the overall plumage is browner (Fig. 4.2A) (Wells, 2010). The legs and eyes are yellow, although there is a race in southern India that presents with a blue-white iris (Fig. 4.2C) (Feare and Craig, 1998).

The four races of Jungle Myna vary in colour as follows: *A. f. fuscus* has a pale creamy vent and yellow iris, *A. f. fumidus* has a greyish vent, *A. f. mahrattensis* has a blue iris and *A. f. torquatus* has a salmon-buff throat and half collar around the neck (Davison, 1892; Marien, 1950; Amadon *et al.*, 1962).

4.5 Diet

Jungle Mynas are omnivorous, typically feeding on nectar (Fig 4.2F) (Raju *et al.*, 2005), fruits, seeds (including cereal grains, groundnuts (*Arachis hypogaea*), coriander (*Coriandrum* spp.) and cloves (*Syzygium* spp.)), and a variety of insects and their larvae (Long, 1981; Narang and Lamba, 1984; Feare and Craig, 1998). They also consume plant matter, centipedes and spiders (Narang and Lamba, 1984). Some interesting records include feeding on maggots at a carcass (Samson and Ramakrishnan, 2017) and the predation of a field mouse (Johnsingh, 1979). Jungle Mynas are also cleaner birds, feeding on ticks and flies while perched on a host (Fig. 4.2E) (Sazima, 2011). They also take advantage of moving animals by feeding on flushed insects (Wells, 2010). Some fruit species consumed by Jungle Mynas include: mulberries (*Morus alba*), a variety of figs (*Ficus* spp.), Asian barberry (*Berberis asiatica*), Himalayan firethorn (*Pyracantha crenulata*) and chillies (*Capsicum* spp.) (Feare and Craig, 1998; Palita *et al.*, 2011). *Erythrina* spp. are popular nectar trees, but other species include *Bombax*, *Butea* and *Careya* spp. (Feare and Craig, 1998; Wells, 2010). They also feed on *Grevillea* spp. flowers and damage groundnut plumules (Watling, 1975; Feare and Craig, 1998). Jungle Mynas are also scavengers, particularly in urban and transformed landscapes.

4.6 Introduction and Invasion Pathways

Jungle Mynas arrived by airplane on Tutuila Island, American Samoa, with four birds noted at the airport in 1986 (Trail, 1994; Freifeld, 1999). They were reportedly purposefully introduced

Fig. 4.2. Jungle Myna images. (**A**) An immature bird, in Kolkata, India. (©Photograph: J.M. Garg, CC BY-SA 3.0, https://creativecommons.org/licenses/by-sa/3.0). (**B**) Adult with blue iris in outskirts of Bengaluru, India. (©Photograph: Vimal Rajyaguru, CC BY-SA 4.0, https://creativecommons.org/licenses/by-sa/4.0). (**C**) An adult exiting a pipe nest in Kolkata, India. (©Photograph: J.M. Garg, CC BY-SA 3.0, https://creativecommons.org/licenses/by-sa/3.0; image cropped). (**D**) Adults fighting in Nepal, with the white wing patches visible (©Photograph: Dick Daniels, http://carolinabirds.org/, CC BY-SA 3.0 https://creativecommons.org/licenses/by-sa/3.0). (**E**) Foraging alongside a Black Drongo (*Dicrurus macrocercus*) on a buffalo in Kolkata, India. (©Photograph: J.M. Garg, CC BY-SA 3.0, https://creativecommons.org/licenses/by-sa/3.0). (**F**) Probing a Kapok (*Ceiba pentandra*) flower in Kolkata, India. (©Photograph: J.M. Garg, CC BY-SA 3.0, https://commons.wikimedia.org/w/index.php?curid=3002564.).

to Fiji *c.*1900 to control agricultural pests such as armyworm caterpillars (Watling, 1982; Clunie, 1999) and grasshoppers in sugarcane fields (Lever, 1987). Their initial introduction to Western Samoa is unclear due to probable incorrect identification of species (Trail, 1994). The first records were possibly noted around 1965, with a confirmed, established population present by 1978 (Watling, 1978; Gill *et al.*, 1993).

It has been postulated that the arrival of Jungle Mynas to Samoa and Tonga was unassisted, possibly storm driven (Lever, 1987). However, Gill *et al.* (1993) stated that they almost certainly would have been assisted in some way, as their establishment occurred in Apia, an unlikely starting point for natural dispersal to Samoa. The first records for Tonga date back to 1980–1982 (Rinke, 1991). An estimated population of 100 Jungle

Mynas was reported only on Futuna Island during 2010 surveys; however, it is not known how they got there (Theuerkauf *et al.*, 2010). More recently, the population is estimated to be approximately 1000 individuals (Thibault *et al.*, 2015).

In Japan, Jungle Mynas are kept for aesthetic reasons, such as pets (Eguchi and Amano, 2004b), and were probably introduced by accidental escape (Eguchi and Amano, 2004a) or deliberate release, possibly for religious reasons (Shiu and Stokes, 2008). Similarly, in Taiwan, it is estimated that 93% of the population practice the releasing of animals predominantly as Buddhists or Taoists generating good karma (Agoramoorthy and Hsu, 2007), but other religions, including Protestants and Catholics, also practice this ritual (Severinghaus and Chi, 1999). Whether or not this is the original introduction pathway in these countries is unclear, but their persistence and spread has no doubt been fuelled by the continued trade and release of this species.

4.7 Breeding Behaviour

Jungle Mynas are secondary cavity nesters. Nest building, incubation and chick rearing is carried out by both parents (Feare and Craig, 1998). They lay three to six slightly glossy blue-green eggs, 26.0–32.8 × 19.0–23.0 mm ($n = 124$) (Schönwetter, 1983; Wells, 2010). They nest inside terraced wall holes (Palita *et al.*, 2011), tree cavities, at the base of palm fronds (Freifeld, 1999), in holes in cliff faces, in concrete embankment seepage outlets, in house eaves, in roof thatch and under roof tiles, although roof nesting is less common in this species (Watling, 1982; Feare and Craig, 1998; Wells, 2010). At bridge and wall drainage holes, large breeding colonies may form and return annually (Feare and Craig, 1998). In one case, communal nesting was observed in the stone wall of a railway tunnel, with nests just under 1 m apart (Mukherjee, 1970). Communal nesting has also been reported in a baobab tree (*Adansonia digitata*), with nests ranging from 5 to 10 m high (Unnithan and Unnithan, 1998). Nest heights are variable, with old woodpecker holes generally ranging from 2 to 6 m (Feare and Craig, 1998), while nest heights range from 6 to 20 m in man-made structures (Wells, 2010). Nests in cavities form neat cups, while more exposed nests, such as those in palm fronds, are a messier collection of nesting material, with a neatly lined cup at the centre (Wells, 2010). Plant matter (leaves, pine needles, twigs, roots and soft wood fibres), feathers and small pieces of rubbish are all used as nesting material (Mukherjee, 1970; Wells, 2010). More unusual nesting materials recorded include cassette tape ribbon (Unnithan and Unnithan, 1998), snake skin, tissue paper and onion peel (Ali and Abdulali, 1941). Nests are reused by myna pairs, who typically raise two broods per season (Wells, 2010). Ring-necked Parakeets (*Psittacula krameri*) have been observed to use vacated Jungle Myna tree nest holes (Unnithan and Unnithan, 1998).

4.8 Habitat

Typically, Jungle Mynas are found in lowlands and foothills, but they are present at altitudes of up to 2000 m in the Himalayas (Feare and Craig, 1998). Within some of its native Indian range, Jungle Mynas are more common in human habitats than in natural habitats, probably because of food and nest site abundance (Palita *et al.*, 2011). In Malaysia, they are absent from forests and urban centres (Wells, 2010). In their invasive range, Jungle Mynas have been observed in secondary forests, but not in primary or well-established forests (Trail, 1994). Due to their ground-foraging habits, they prefer open or short vegetation areas, including but not limited to: open-cast mining land, road verges, forest edges, paddy fields, gardens, agricultural fields, coastal plains, grazing fields and deciduous woods with open areas (Feare and Craig, 1998; Wells, 2010). Jungle Mynas are generally commensal with humans, and in their invasive range on the Pacific Islands, they are reported in villages, urbanized areas and agricultural landscapes (Freifeld, 1999).

Where Jungle Mynas are established in American Samoa, they appear to inhibit the spread of Common Mynas (Trail, 1994). Conversely, in the capital of Samoa, Apia, Common Mynas have replaced Jungle Mynas as the more common species (Gill, 1999). However, as Common Mynas are considered more commensal with humans than Jungle Mynas (Ali and Ripley, 1972; Watling, 1975), Jungle Mynas could remain more dominant in open natural and agricultural landscapes (Gill, 1999). Transect surveys on Upolu Island (where Apia is based) yielded 15,755–30,979 Common Mynas as opposed to 90,390–119,530 Jungle Mynas (Nagle and Associates, 2015). Similarly, in Fiji, Common Mynas dominate city centres while Jungle Mynas dominate periurban areas (Watling, 1975). In agricultural landscapes where most overlap occurs between these two species, Common Mynas forage near the farm buildings while Jungle Mynas typically feed in the fields (Watling, 1975).

4.9 Impacts

4.9.1 Negative impacts

Jungle Mynas are reported to cause damage to fruit orchards and subsistence gardens in both their native and introduced ranges (Long, 1981; Nagle, 2006). Long (1981) reported 30–40% losses of groundnuts. They also damage groundnut plumules (Watling, 1975) and clove plants (Narang and Lamba, 1984). Jungle Mynas feed on lantana (*Lantana camara*), Himalayan firethorn, chillies and mulberries (to name a few) and are thus potential dispersers of invasive plants (Long, 1981; Aravind *et al.*, 2010; Palita *et al.*, 2011). Jungle Mynas are often observed near pig stalls on the Pacific Islands, where they raid livestock food (Nagle, 2006; Thibault *et al.*, 2015).

One of the prominent concerns is that Jungle Mynas will outcompete native species (Lever, 1987), but formal research is lacking and observations remain scant on the direct effect of these birds on native avifauna. In native India, following an aggressive attack, they successfully outcompeted Lesser Golden-backed Woodpeckers (*Dinopium benghalense*) for a nesting hole (Jha, 2001). They have also been observed in two aggressive interactions with Samoan Starlings (*Aplonis atrifusca*), with the starlings winning most likely due to their larger size (Watling, 1978). However, the general absence of starlings where mynas are present suggests that the sheer numbers of the latter species could be enough to deter Samoan Starlings, which are not a flocking species (Freifeld, 1999).

Jungle Mynas roost and nest communally. Although not formally listed as one of the major impacts of this species, problems associated with this behaviour could be similar to that experienced with Common Mynas. These include noise, nuisance, property damage and spread of disease (Yap *et al.*, 2002; Parkes, 2006).

4.9.2 Positive impacts

The forehead feather crest of Jungle Mynas acts as an efficient brush for cross-pollination when feeding on nectar (Ali and Ripley, 1972; Raju *et al.*, 2005) and they serve as seed dispersers (Long, 1981). Jungle Mynas have a commensal relationship with Drongos (*Dicrurus adsimils*) in native India, whereby drongo foraging success increases when feeding alongside myna flocks in fallow lands (Fig. 4.2E) (Veena and Lokesha, 1993). Jungle Mynas are also considered important biocontrol agents in tea plantations in India (Sinu, 2011). Indeed, Narang and Lamba (1984) advocated their protection by law in India, stating that they did more good than harm. Furthermore, they remove ectoparasites and other insect pests from a variety of domestic and wild animals (Sazima, 2011). Their scavenging habits can also be considered a valuable ecosystem service.

4.10 Control

The control of Jungle Mynas requires prevention, control, eradication and continued monitoring (Nagle and Associates, 2015). Unfortunately, many Pacific Islands lack systematic baggage checks and security measures such as dogs and scans (Theuerkauf *et al.*, 2010). Restrictions on the importation of this species also remain limited in some areas, and therefore inadequate biosecurity measures are in place for the prevention of new introductions (Theuerkauf *et al.*, 2010). The offering of a bounty of NZ$200 per bird in Tokelau and 20c in Samoa was unsuccessful in reducing the population of mynas and thus other strategies were employed (Nagle, 2006; Nagle and Associates, 2015). On Tokelau Fakaofo Atoll, cage and nest box traps, as well as bread baited with the narcotic alphachloralose are used to control mynas (Nagle, 2006). In Samoa, the Ministry for Natural Resources and Environment initially used traps in 2004 (Doherty, 2006), but in 2008 poison trials were approved and implemented (Esera, 2012). Starlicide DRC-1339 baited bread has been used, but with little success (Nagle and Associates, 2015). A conservative estimate of 2489–3627 mynas (Jungle and Common) removed weekly is needed in Samoa for one year to reduce populations (Nagle and Associates, 2015). Thus, a multi-faceted strategy including baiting, reduction of suitable habitat and food sources, nest and roost site destruction, trapping and shooting was proposed (Nagle and Associates, 2015).

Natural events may also decrease Jungle Myna populations on islands, as was the case in Western Samoa following a cyclone in 1991 (Gill *et al.*, 1993). However, birds remained prevalent in urban areas, probably shielded from strong winds by buildings (Gill *et al.*, 1993). Ultimately, the key to successfully eradicating a species is early detection. Jungle Mynas were reported in Kiribati, and as numbers were still low, targeted control could be implemented to remove the birds before they became established (Butler, 2015). These birds probably arrived by boat from nearby Pacific Islands where established populations occur (Butler, 2015).

4.11 Uses

Historically imported for pest control (Watling, 1982; Lever, 1987; Clunie, 1999), today Jungle Mynas are predominantly used in the pet trade or for religious ceremonies (Eguchi and Amano, 2004b; Agoramoorthy and Hsu, 2007). They are also eaten by people (BirdLife International, 2019).

4.12 Notes

Jungle Mynas have been observed to use millipedes during anting, most likely as a form of pest control, as millipede secretions deter many arthropods (Clunie, 1976; Wee, 2008).

4.13 References

Agoramoorthy, G. and Hsu, M.J. (2007) Ritual releasing of wild animals threatens island ecology. *Human Ecology* 35, 251–254.

Ali, S. and Abdulali, H. (1941) *The Birds of Bombay and Salsette*. Prince of Wales Museum (Natural History Section), Mumbai, India.

Ali, S. and Ripley, S.D. (1972) *Handbook of the Birds of India & Pakistan*, Vol. 5. Oxford University Press, Mumbai, India.

Amadon, D., Blake, E.R., Greenway, J.C. Jr, Mayr, E., Moreau, R.E. and Vaurie, C. (eds) (1962) *Check-list of Birds of the World*. Museum of Comparative Zoology, Cambridge, Massachusetts.

Aravind, N., Rao, D., Ganeshaiah, K., Shaanker, R.U. and Poulsen, J.G. (2010) Impact of the invasive plant, *Lantana camara*, on bird assemblages at Malé Mahadeshwara Reserve Forest, South India. *Tropical Ecology* 51, 325–338.

BirdLife International (2019) Species factsheet: *Acridotheres fuscus*. Available at: http://datazone.birdlife.org/species/factsheet/jungle-myna-acridotheres-fuscus (accessed 23 October 2019).

Butler, D. (2015) Final Report – Eradication of Myna from Kiribati. Ministry of Environment, Lands and Agricultural Development, Government of Kiribati, Kiribati, pp. 1–25.

CABI (2018) *Acridotheres fuscus* (jungle myna). In: *Invasive Species Compendium*. CAB International, Wallingford, UK. Available at: www.cabi.org/isc/datasheet/95671 (accessed 23 October 2019).

Clements, J.F., Schulenberg, T.S., Iliff, M.J., Roberson, D., Fredericks, T.A., *et al.* (2018) The eBird/Clements checklist of birds of the world: v2018. Available at: www.birds.cornell.edu/clementschecklist/download/ (accessed 23 October 2019).

Clunie, F. (1976) Jungle Mynah "anting" with a millipede. *Notornis* 23, 77.

Clunie, F. (1999) *Birds of the Fiji Bush*. Fiji Museum, Suva.

Davison, W. (1892) Descriptions of some new species of birds from the eastern coast of the Malayan Peninsula. *Ibis* 34, 99–103.

Doherty, N. (2006) Information booklet for the "National Control of the Myna" (*Acridotheres* spp.). Ministry of Natural Resources and Environment, Samoa.

EDDMapS (2019) Early Detection and Distribution Mapping System. Center for Invasive Species and Ecosystem Health, University of Georgia, Athens, Georgia. Available at: www.eddmaps.org/ (accessed 23 October 2019).

Eguchi, K. and Amano, H.E. (2004a) Invasive birds in Japan. *Global Environmental Research* 8, 29–39.

Eguchi, K. and Amano, H.E. (2004b) Spread of exotic birds in Japan. *Ornithological Science* 3, 3–11.

Esera, U. (2012) Pacific islands discuss ways to eradicate myna birds. Talamua Media and Publications. Available at: www.talamua.com/pacific-islands-discuss-ways-to-eradicate-myna-birds/ (accessed 23 October 2019).

Feare, C. and Craig, A. (1998) *Starlings and Mynas*. Helm, London.

Freifeld, H. (1999) Habitat relationships of forest birds on Tutuila Island, American Samoa. *Journal of Biogeography* 26, 1191–1213.

Gill, B., Lovegrove, T. and Hay, J. (1993) More myna matters – notes on introduced passerines in Western Samoa. *Notornis* 40, 300–302.

Gill, B.J. (1999) A myna increase – notes on introduced mynas (*Acridotheres*) and bulbuls (*Pycnonotus*) in Western Samoa. *Notornis* 46, 268–269.

Jha, A. (2001) Competition between Jungle Myna *Acridotheres fuscus* and Lesser Goldenbacked Woodpecker *Dinopium benghalense* for a nest hole. *Journal of the Bombay Natural History Society* 98, 115.

Johnsingh, A.J.T. (1979) A note on the predation of Jungle Myna *Acridotheres fuscus* Wagler on field mouse. *Journal of the Bombay Natural History Society* 761, 159.

Kang, C., Chan, W., Feare, C., Wells, D. and Kang, N. (1994) The relationship between mynas: An application of DNA sequencing. *Journal of Ornithology* 135, 33.

Lepage, D. (2004) Avibase: The World Bird Database, Bird Studies Canada. Available at: https://avibase.bsc-eoc.org (accessed 23 October 2019).

Lever, C. (1987) *Naturalised Birds of the World*. Longman Scientific and Technical, London.

Long, J.L. (1981) *Introduced Birds of the World: The Worldwide History, Distribution and Influence of Birds Introduced to New Environments*. Universe Books, New York.

Lovette, I.J., McCleery, B.V., Talaba, A.L. and Rubenstein, D.R. (2008) A complete species-level molecular phylogeny for the "Eurasian" starlings (Sturnidae: Sturnus, Acridotheres, and allies): recent diversification in a highly social and dispersive avian group. *Molecular Phylogenetics and Evolution* 47, 251–260.

Marien, D. (1950) Notes on some Asiatic Sturnidae (birds). *Journal of the Bombay Natural History Society* 49, 471–487.

McAllan, I.A.W. and Hobcroft, D. (2005) The further spread of introduced birds in Samoa. *Notornis* 52, 16–20.

Mukherjee, R. (1970) Jungle Mynas and their nests. *Newsletter for Birdwatchers* 10, 11.

Nagle, B. (2006) Protection of Tokelau Fakaofo from Myna Bird (*Acridotheres* spp.) invasion. Report of a feasibility study, 12–16 May 2006. Pacific Invasives Initiative, Auckland, New Zealand.

Nagle, W.G. and Associates (2015) Managing Myna Birds (*Acridotheres fuscus* and *A. tristis*) in the Independent State of Samoa. Division of Environment and Conservation, Ministry of Natural Resources and Environment, Clyde, New Zealand.

Narang, M. and Lamba, B.S. (1984) Contribution to the food habits of some Indian Mynas (Aves). *Records of Zoological Survey of India, Calcutta, Occasional Paper* 44, 1–76.

Palita, S.K., Ponkshe, A.V. and Dhar, U. (2011) Habitat enrichment and its impact on avian diversity: a study at GBPIHED, Kosi-Katarmal, Uttarakhand, India. *Current Science* 100, 1681–1689.

Parkes, J. (2006) Feasibility plan to eradicate Common Mynas (*Acridotheres tristis*) from Mangaia Island, Cook Islands. Landcare Research Contract Report: LC0506/184, Lincoln, New Zealand.

Raju, A.J., Rao, S.P. and Rangaiah, K. (2005) Pollination by bats and birds in the obligate outcrosser *Bombax ceiba* L. (Bombacaceae), a tropical dry season flowering tree species in the Eastern Ghats forests of India. *Ornithological Science* 4, 81–87.

Rinke, D. (1991) Birds of 'Ata and Late, and additional notes on the avifauna of Niuafo'ou, Kingdom of Tonga. *Notornis* 38, 131–151.

Samson, A. and Ramakrishnan, B. (2017) Scavenging mode of vertebrate scavengers on domestic buffalos *Bubalus bubalis* (Linnaeus, 1785) killed by tiger *Panthera tigris* and natural deaths in southern India. *Podoces* 12, 22–26.

Sazima, I. (2011) Cleaner birds: a worldwide overview. *Revista Brasileira de Ornitologia* 19, 32–47.

Schönwetter, M. (1983) *Handbuch der Oologie*. Academie-Verlag, Berlin.

Severinghaus, L.L. and Chi, L. (1999) Prayer animal release in Taiwan. *Biological Conservation* 89, 301–304.

Shiu, H. and Stokes, L. (2008) Buddhist animal release practices: historic, environmental, public health and economic concerns. *Contemporary Buddhism* 9, 181–196.

Sinu, P.A. (2011) Avian pest control in tea plantations of sub-Himalayan plains of northeast India: mixed-species foraging flock matters. *Biological Control* 58, 362–366.

Sullivan, B.L., Wood, C.L., Iliff, M.J., Bonney, R.E., Fink, D. and Kelling, S. (2009) eBird: a citizen-based bird observation network in the biological sciences. *Biological Conservation* 142, 2282–2292.

Theuerkauf, J., Jourdan, H., Rouys, S., Gula, R., Gajewska, M., *et al.* (2010) Inventory of alien birds and mammals in the Wallis and Futuna Archipelago. *Biological Invasions* 12, 2975–2978.

Thibault, J.-C., Cibois, A. and Meyer, J.-Y. (2015) Birds of Uvea (Wallis), Futuna and Alofi Islands (South-West Pacific): an update. *Notornis* 62, 30–37.

Trail, P.W. (1994) Distribution and status of mynas in American Samoa. *Elepaio* 54, 19–21.

Unnithan, S. and Unnithan, G.V.K. (1998) Breeding of southern Jungle Myna *Acridotheres fuscus mahrattensis* (Sykes) at Karanja. *Journal of the Bombay Natural History Society* 95, 346.

Veena, T. and Lokesha, R. (1993) Association of drongos with myna flocks: are drongos benefitted? *Journal of Biosciences* 18, 111–119.

Watling, D. (1975) Observations on the ecological separation of two introduced congeneric mynahs (*Acridotheres*) in Fiji. *Notornis* 22, 37–53.

Watling, D. (1978) A myna matter. *Notornis* 25, 148–166.
Watling, D. (1982) *Birds of Fiji, Tonga and Samoa*. Millwood Press, Wellington, New Zealand.
Wee, Y. (2008) Anting in Singapore birds. *Nature in Singapore* 1, 23–25.
Wells, D.R. (2010) *The Birds of the Thai-Malay Peninsula*, Vol. 2: *Passerines*. Christopher Helm, London.
Yap, C.A.M., Sodhi, N.S. and Brook, B.W. (2002) Roost characteristics of invasive mynas in Singapore. *Journal of Wildlife Management* 66, 1118–1127.

5 Red-vented Bulbul (*Pycnonotus cafer* Linnaeus, 1766)

Martin Thibault[1,2,3]*, Murray A. Potter[2], Eric Vidal[3,4], Fabrice M. Brescia[1] and Daniel M. Brooks[5]

[1]*Institut Agronomique Néo-Calédonien (IAC), Equipe ARBOREAL (AgricultuRe BiOdiveRsité Et vALorisation) BP 73, 98890 Païta, New Caledonia, France;* [2]*Wildlife and Ecology Group, School of Agriculture and Environment, Massey University, Palmerston North 4442, New Zealand;* [3]*Laboratoire d'Excellence Labex-CORAIL, Institut de Recherche pour le Développement (IRD), UMR ENTROPIE (IRD-Université de La Réunion-CNRS), BP A5, Nouméa Cedex 98848, New Caledonia, France;* [4]*Institut Méditerranéen de Biodiversité et d'Ecologie Marine et Continentale (IMBE), Aix Marseille Université, CNRS, IRD, Avignon Université Centre IRD Nouméa – BP A5 98848 Nouméa Cedex, New Caledonia, France;* [5]*Houston Museum of Natural Science, Department of Vertebrate Zoology, Houston, TX 77019, USA*

Citation: Thibault, M., Potter, M.A., Vidal, E., Brescia, F.M. and Brooks, D.M. (2020) Red-vented Bulbul (*Pycnonotus cafer* Linnaeus, 1766). In: Downs, C.T. and Hart, L.A. (eds) *Invasive Birds: Global Trends and Impacts*. CAB International, Wallingford, UK, pp. 40–52.

5.1 Common Name

Red-vented Bulbul.

5.2 Nomenclature

The Red-vented Bulbul (*Pycnonotus cafer* Linnaeus, 1766) has eight subspecies (Table 5.1). These relate mostly to different geographical ranges, but some have morphologic differences (Dickinson *et al*., 2002).

5.3 Description

The Red-vented Bulbul measures approximately 21 cm in length (Berger, 1972) and its weight can vary between 26 and 45 g (Long, 1981). The morphometric parameters for this species are presented in Table 5.2. It may live for up to 11 years (Walker, 2008). Its feathers are predominantly dark brown (Fig. 5.1). Those on the back and breast have white tips. It has a distinctive erectile black crest on the top of the head, deep crimson subcaudal feathers, a white rump that is highly visible in flight, and a tail that is long and blackish with a prominent white tip (Berger, 1981; Pratt *et al*., 1987). The males tend to be slightly larger than the females but otherwise there is no sexual dimorphism (Stuart and Stuart, 1999). The immature bird looks like the adult but with paler feathering with some brownish edging, and an orange or pinkish, rather than red, vent. The morphological differences of the eight subspecies are presented in Table 5.1.

5.4 Distribution

The Red-vented Bulbul is native to the Indian subcontinent, South-east Asia and Malay Peninsula (Fig. 5.2) (Long, 1981). Its native range extends from eastern Pakistan to eastern China and Vietnam, and from northern India to Sri Lanka. Historically, the species was also present in Bangladesh, Bhutan, Myanmar and Nepal.

The Red-vented Bulbul was introduced into 19 countries and became established in 17 of them. This alien range comprises two main geographical areas: islands of the Pacific Ocean and littoral countries of the Middle East. There are also established populations in North America (e.g. Houston, Texas), in

*Corresponding author: thibault.mn@gmail.com

Table 5.1. The eight subspecies of *Pycnonotus cafer* and their native distributions and morphological characteristics.

Subspecies	Morphological characteristic	Distribution	Descriptor
P. c. humayuni	Pale brown	South-eastern Pakistan (Sind), north-western and north-central India	Deignan, 1951
P. c. intermedius	Black hood extending to mid-breast	Western Himalayas from Pakistan, Jammu and Kashmir east at least to Nepal	Blyth, 1846
P. c. bengalensis	Dark hood with brown auriculars	Central and eastern Himalayas and Gangetic Plain, and Bangladesh	Blyth, 1845
P. c. stanfordi	Black hood extending to breast	South Assam	Deignan, 1949
P. c. melanchimus	Black crown	South Myanmar	Deignan, 1949
P. c. wetmorei	Pale brown with dark hood	North-eastern Peninsular India	Deignan, 1960
P. c. cafer	Dark brown with scaly pattern, dark head	Southern India	Linnaeus, 1766
P. c. haemorrhousus	Dark mantle with narrow, pale edges	Sri Lanka	J.F. Gmelin, 1789

Table 5.2. Morphometric parameters of male and female *Pycnonotus cafer bengalensis*, measured on adults from an introduced population in New Caledonia. (Thibault *et al.*, unpublished data.)

	Females			Males		
Character	Mean	SD	*n*	Mean	SD	*n*
Body mass (g)	38.05	4.54	27	43.69	5.01	31
Total length (cm)	19.75	0.97	13	19.98	0.87	18
Tail length (cm)	8.74	0.35	27	9.04	0.59	31
Head length (mm)	15.83	1.08	12	16.36	1.41	31
Crest length (mm)	18.01	0.87	12	18.02	2.59	18
Mandible length (mm)	12.15	1.20	27	12.50	0.87	31
Mandible width (mm)	5.65	0.27	13	5.70	0.31	18
Mandible height (mm)	5.61	0.24	13	5.98	0.31	18
Right wing length (cm)	9.31	0.25	27	9.63	0.57	30
Right tarsus length (mm)	24.66	2.31	27	24.92	2.11	30
Third digit length (mm)	16.71	1.66	13	17.87	1.52	17
Crimson patch length (cm)	4.00	0.40	13	3.91	0.47	17

Fig. 5.1. An adult Red-vented Bulbul. (©Photograph: Coralie Thouzeau-Fonseca, IAC.)

the Canary Islands (e.g. Fuerteventura) and in southern Spain (e.g. Malaga). It is now present on at least 37 islands and in seven continental locations, and is anticipated to continue its range expansion on several archipelagos (Thibault *et al.*, 2018a).

5.5 Introduction and Invasion Pathways

5.5.1 Success of introductions

The Red-vented Bulbul is currently considered to be established in 38 of the 46 locations where it has been recorded historically (Table 5.3). Up-to-date information is lacking for islets of the Ailinglaplap archipelago (Marshall Islands) ('Eua, Savai'i and Ailinglaplap). Red-vented Bulbuls were reported in Melbourne in 1918 and 1942 (Lendon, 1952; Watling, 1978), but the species has not been reported there since and is considered 'eradicated' in Australia in the global invasive species database (www.issg.org/database). It was observed on five islands in the Hawaii Archipelago between 1982 and 1989, but it seems that it failed to establish beyond Oahu (Walker, 2008). It was eradicated from Auckland, New Zealand, in 1955 (Watling, 1978), 3 years after the first observation in 1952 (Turbott, 1956).

5.5.2 Introduction history

The oldest documented observation of *P. cafer* outside its native range is from Fiji, *c.*1903 (Parham, 1955; Watling, 1978). It is very likely that the dispersal of the species started with Indian workers who travelled from India to Fiji during this period (Ali and Ripley, 1996; Watling, 1978). A study conducted on Viti Levu between 1970 and 1973 revealed that the Red-vented Bulbul was already present in all major habitat types of the island, from sea level to 1320 m above sea level (Gorman, 1975). In 1978, it was also present at low densities on Ovalau, Wakaya and Beqa Islands, and on the western coast of Taveuni island (Watling, 1978). Today, the Red-vented Bulbul has spread to Vanua Levu, Gau, Kadavu, Nairi island, Waya islet, and recently to Fulaga islet.

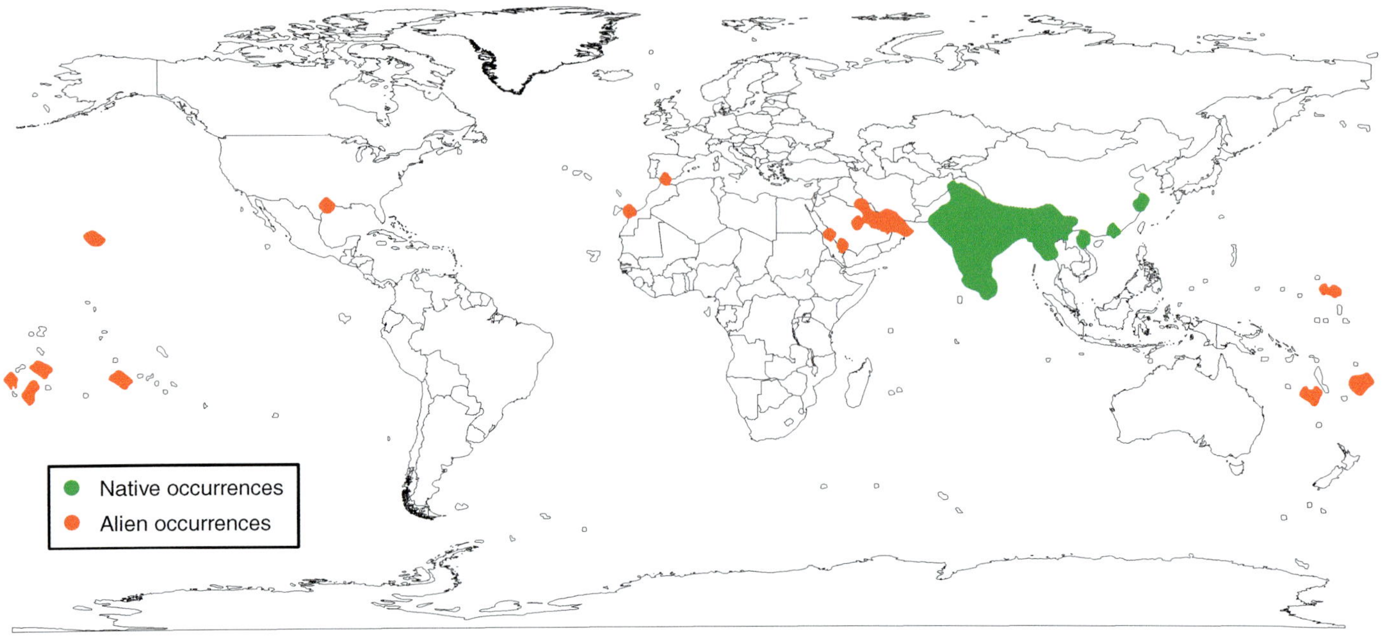

Fig. 5.2. Global distribution of the Red-vented Bulbul showing natural (green) and alien (red) distributions. (Adapted from Thibault *et al.*, 2018a.)

Table 5.3. Year of first observation of non-native populations of the Red-vented Bulbul.

Country	First observation	Reference(s)
Fiji	1903	Parham (1955); Watling (1978)
Tonga	1928	Watling (1978)
Western Samoa	1943	Watling (1978); Dhondt (1976a); DuPont (1972)
American Samoa	After 1957	Clapp and Sibley (1966)
USA (Hawaii)	1966	Berger (1975)
Qatar	1971	Nation *et al.* (1997); Warr (1986)
United Arab Emirates	1974	Pedersen and Aspinall (2015)
French Polynesia	1979	Meyer (1996)
Kuwait	1981	Gregory (2005)
New Caledonia	1983	Gill *et al.* (1995)
Bahrain	1986	Khamis (2010)
Oman	1987	J. Eriksen (personal communication)
Saudi Arabia	1980s	J. Babbington (personal communication)
USA (Texas)	1996	Brooks, 2013)
Spain	1998	Ministerio de Agricultura, Alimentacion y Medio Ambiente (2013)
Marshall Islands	2000	Vander Velde (2002)
Iran	2007	Azin *et al.* (2008)

According to Watling (1978), a pair of Red-vented Bulbuls was recorded on Niuafo'ou Island (Kingdom of Tonga) in 1928/1929. The species was then introduced on Tongatapu in the 1940s (Watling, 1978). It is assumed to have promptly spread from Tongatapu to 'Eua. It is now also present on Vava'u and Foa islands, where it was not recorded by the aforementioned authors. It was then recorded in Western Samoa in 1943 (Dhondt, 1976a), probably brought from Fiji (Watling, 1978). According to DuPont (1972), the first record of the species in Savai'i, the second main island of Western Samoa, was in 1970. In 1974, the species was widely distributed in Upolu and was observed in small numbers in two areas on Savai'i (Dhondt, 1976b). In 2005, McAllan and Hobcroft considered that the species had spread to all the inhabited areas of Upolu, Samoa, regardless of elevation or rainfall. Clapp and Sibley (1966) suspected that the Red-vented Bulbul was introduced on to Tutuila (American Samoa) around 1957, although the species was first officially recorded in Pago Pago in 1963. It appears to have arrived by aircraft and become established on the Tafuna Plain, a cultivated area next to the international airport (Freifeld, 1999). Soon after, in 1966, the Red-vented Bulbul was observed in Hawaii (Berger, 1975). It was first recorded in October 1966 near Waipahu, on O'ahu (Walker, 2008). Until 1982, the species was restricted to the south-east of the Island (Williams and Giddings, 1984). It arrived in French Polynesia in 1979, on Tahiti (Meyer, 1996). It quickly became abundant in coastal and lowland areas in the mainland and on the peninsula, even in areas up to 1500 m above sea level (Monnet *et al.*, 1993). It is highly suspected that its introduction started with escaped or released caged birds. The species is now present in Bora-Bora, Moorea, Raiatea, Tahaa, Huahineiti and Huahine Nui, Tetiaroa, and is currently spreading to other islands and islets in the archipelago (T. Ghestemme, Société d'Ornithologie de Polynésie

(MANU), personal communication). In New Caledonia, intentional release of caged birds is strongly suspected to have occurred in 1983 in Nouméa to avoid prosecution for possession (Gill *et al.*, 1995). Nowadays, the species is still restricted to the mainland's southern province but continues to increase its range. Finally, in terms of Pacific Island occupation, the species was first observed in the Marshall Islands in 2000 by Vander Velde (2002). The species was initially restricted to urbanized areas in Majuro and then spread to agricultural areas of Laura (Vander Velde, 2002). It remains unknown exactly how the species got to Majuro, but it has been suggested that a pair built a nest in a container on a ship from Oahu (Vander Velde, 2002). The Red-vented Bulbul has since spread to Ailinglaplap Island (ebird, 2019: https://ebird.org, accessed 18 November 2019).

The species has also spread to the Arabian Peninsula, especially around the Persian Gulf. The Red-vented Bulbul was recorded in Qatar in 1971 (Nation *et al.*, 1997). It is suspected that there was more than one introduction event, possibly from birds escaping from cages. Two further inland occurrences were recorded – one in May 1992 and the other in March 1995 (Nation *et al.*, 1997). The species was then observed in Abu Dhabi in the United Arab Emirates in 1974 (Pedersen and Aspinall, 2015). Once again, its introduction was probably due to escaped or released caged birds (Khan, 1993). Local populations bred, and by 1996, a self-sustaining population of between 1000 and 5000 pairs had established (Jennings, 2010). The species is now present in many inland and coastal cities in the north of the country. During the spring of 1981, Red-vented Bulbuls were observed in the grounds of a hospital in the south of Kuwait City (Gregory, 2005). In 2003, the resident population was reported to comprise 100 breeding pairs (Jennings, 2010). This population appears to have decreased since then, and currently has a very restricted distribution (M. Pope, personal communication). The Red-vented Bulbul was first recorded in Bahrain in 1986 (Khamis, 2010), having probably escaped from cages. By 2003, the population comprised 250 pairs, was on the increase and was considered established. It was first observed in Saudi Arabia in the 1980s around the Riyadh Airbase (Jem Babbington, personal communication). Again, escaped caged birds were the likely source of this introduction. Additional introductions probably occurred because the bird is still sold in local markets. Anecdotal reports suggest that it was locally abundant near the airbase in the early 1990s, but it has since declined and today is considered to be a scarce breeding resident. The first record in Oman was an inland observation of a single bird on 2 January 1987 (J. Eriksen, personal communication). The species became increasingly common and dispersed across the capital, and some individuals were observed concurrently in northern Oman (J. Eriksen, personal communication). Ten to 12 Red-vented Bulbuls were observed on Kish Island, Iran, Persian Gulf, in March 2007 (Azin *et al.*, 2008). A survey was conducted immediately, revealing the presence of the species in two locations on the island, situated 2 km apart. As this is the only record on Iranian Islands and coastal areas, Azin *et al.* (2008) suggested that the introduction was human-induced.

In addition to the alien populations in the Pacific Islands and the Middle East, the Red-vented Bulbul has established in Houston, Texas. They were first observed in the Greenspoint area in the mid-1990s, having arrived as caged birds (Brooks, 2013). Two populations are recorded in Houston and have been monitored continuously since 2008.

Further afield, *P. cafer* has been present in Torremolinos, Malaga, Spain, since 1998 (Ministerio de Agricultura, Alimentacion y Medio Ambiente, 2013) and at Corralejo on the island of Fuerteventura, Canary Islands, where several ornithologists have recorded its presence since 2014 (ebird, 2019: https://ebird.org).

5.5.3 Suspected expansion along urban corridors

Monitoring of the expansion of the introduced population of Red-vented Bulbuls in New Caledonia over the past 10 years has provided reliable information on the dispersal strategy of this species on an oceanic island (Thibault *et al.*, 2019). First, the population established and started expanding into inhabited areas, apparently using urban corridors as major dispersal pathways. Three complementary research strategies were used to monitor their spread. The first involved constant monitoring at the edge of the distribution, data collection from community-participation point-count monitoring of terrestrial songbirds, and a survey conducted at borders between occupied areas and natural forests. Second, an important density gradient was described from the city centre (historical introduction) to rural habitats at the edge of the current distribution. Third, the distribution monitoring suggested that the expansion of the Red-vented Bulbul's range accelerated when the species started to disperse beyond city boundaries. It took 25 years for the Red-vented Bulbul to establish in Nouméa, the capital of New Caledonia, and to move 40 km north. In comparison, its range has expanded an additional 35 km in just 4 years (Thibault *et al.*, 2019). It is very likely that the species dispersal in Houston, Texas, is following a similar pattern (D. Brooks, personal communication).

5.6 Diet

5.6.1 Foraging behaviour

Foraging was the most frequent behavioural activity (33%) in an alien population of Red-vented Bulbuls, with perching/resting (22%) and calling (14%) accounting for an additional one-third of all activities (Brooks, 2013). The Red-vented Bulbul is one of the few animal species that cannot synthesize ascorbic acid and is thus susceptible to scurvy (Roy and Guha, 1958). This explains why it feeds mainly on fruit and berries (Islam and Williams, 2000; Brooks, 2013). Consumption of flowers, buds, insects and small reptiles is also reported (Vander Velde, 2002; Brooks, 2013). According to Bates' and co-workers (2014) study, the preferred foraging substrate for Red-vented Bulbuls is possibly related to its short slender legs that force it to hop from place to place. It is used to foraging in the presence of some conspecifics but with only a few heterospecifics around. Competitive interactions with heterospecific species was reported and could potentially pose a threat to some native frugivorous bird species (see section 5.10).

5.6.2 Colour preference

In a study in French Polynesia, the authors concluded that fruit selection in some species of birds, including Red-vented Bulbuls, might be driven more by colour preference than by fruit abundance (Spotswood *et al.*, 2013). A food colour-preference experiment showed that they significantly preferred red fruits over yellow, green and a control colour (Thouzeau-Fonseca, 2013; Thibault *et al.*, 2019).

5.6.3 Diet composition

A quantitative assessment of the diet of Red-vented Bulbuls was conducted recently following the classical method of gut content analysis (Thibault *et al.*, 2019). As suspected according to the literature, the diet mostly comprises plant items, which were found in 92% of the cadavers analysed (n = 134). Animal remains were found in 50% of the digestive tracts of Red-vented Bulbuls examined. This study contributed to the list of both plant and animal species known to be consumed by the species (Thibault *et al.*, 2018a). Seeds and entire fruits accounted for more than 70% of plant remains found, and most could be identified to species level. The 134 individuals analysed had consumed more than 16 plant species belonging to 14 families. Among these, only two were native and one endemic; the rest (80%) were introduced species, and five of these were considered invasive. Moreover, 31% of the species of fruit consumed by the Red-vented Bulbuls that were found around the city of Nouméa were cultivated plants. The maximum length of the largest seed swallowed by bulbuls was 14 mm.

All identified animals in the digestive tracts of Red-vented Bulbuls in this study were arthropods (Thibault *et al.*, 2018a). A total of 22 families belonging to ten orders were identified from chitin remains. Functionally, 88% of the insects consumed were primary consumers. Seasonal variation in the annual diet of the Red-vented Bulbul is highly suspected but has not been assessed specifically. The consumption of small skinks (e.g. *Emoia cyanura* and *Caledoniscincus* spp.) and geckos (e.g. *Hemidactylus frenatus*) has been reported in the literature and observed in the field, but no corresponding remains were found in this study.

Similarly, in the Houston population of Red-vented Bulbuls, 45% of the 20 species of identified plants consumed were exotic species found within the native range of the bulbul, whereas six (30%) were exotic species found outside the native range and five (25%) were native Texas plants (Brooks, 2013).

5.7 Breeding Behaviour

The breeding characteristics of the Red-vented Bulbul are summarized in Table 5.4. The observed number of clutches per year is generally one, although a second may be possible if the first clutch fails early in the season (Watling, 1983). Both sexes take part in nest construction (Prajapati *et al.*, 2011). The nest is cup-shaped (Vander Velde, 2002) and made of plant matter (Brooks, 2013) and spider webs. Sometimes, they will nest in plastic dishes. Nests are constructed about 3 m above ground

Table 5.4. Descriptive and breeding traits of Red-vented Bulbuls.

Characteristic	Value	Reference
Adult height (cm)	21	Berger (1972)
Adult weight (g)	26–45	Long (1981)
Lifespan (years)	≤11	Walker (2008)
Clutches per year	1	Watling (1983)
Clutch size (mean ± SD)	3 ± 1	Zia *et al.* (2014)
Incubation length (days; mean ± SD)	13 ± 1	Watling (1983)
Incubation success (%)	28	Watling (1983)
Days as nestling	12	Watling (1983)
Nestling survival rate (%)	50	Watling (1983)
Fledging (days)	14	Watling (1983)
No. fledged per pair	0.33	Watling (1983)
Flocks of young	Yes	Watling (1983)
Sexual dimorphism	No	Watling (1983)
Parental investment	Both	Watling (1983)

Fig. 5.3. A nest and eggs of Red-vented Bulbuls. (©Photograph: Fabrice Brescia, IAC.)

(Vijayan, 1980; Watling, 1983; Brooks, 2013), often in forks of branches of shrubs (Vijayan, 1980; Zia *et al.*, 2014). The nest is about 10 cm wide, 6 cm height and 5 cm deep (Vijayan, 1980; Watling, 1983; Prajapati *et al.*, 2011). The eggs of the Red-vented Bulbul are oval shaped and pinkish-white with small red-brown dots, and the mean clutch size is three (Fig. 5.3) (Zia *et al.*, 2014). The incubation period is around 12–14 days (Berger, 1981; Watling, 1983; Duncan *et al.*, 2001). Chicks stay in the nest for approximately 12 days and remain around the nest for about 2 weeks after fledging. The mean number of fledglings produced per nesting attempt was estimated to be 0.33 in Fiji, with an incubation success rate of 28% and a nestling survival rate of about 50% (Watling, 1983).

5.8 Habitat

The Red-vented Bulbul can live in diverse habitat types. It is found in open areas, dry scrub, forest (natural or plantation), plains and cropland, as well as urban areas (Vander Velde, 2002). It is preferentially present in anthropogenic environments (urban areas, gardens, parks, farms), savannah areas, shrub vegetation, and more rarely on the edge of forest (Vander Velde, 2002). Virtually all (96%) of the alien bulbuls in Houston were found in residential suburbs, often with well-planted gardens (Brooks, 2013). In this study, bulbuls perched in 36 different species of plants: 44% were native to Texas, 42% were exotic plants found within the native range of bulbuls and 14% were exotic plants found outside their native range (Brooks, 2013). Tree species that alien bulbuls most frequently used for perching were bamboo (*Bambusa* spp.) and crepe myrtle (*Lagerstroemia indica*), along with edible fig (*Ficus carica*) and tallow (*Sapium sebiferum*) (Brooks, 2013). Bulbuls also perched on abiotic anthropogenic structures, especially telephone and utility wires, along with bird baths and water fountains (Brooks, 2013).

Bird densities across habitats

In New Caledonia, Red-vented Bulbuls are extremely abundant in Nouméa (over 200 birds/km^2) and present in much lower densities (30 birds/km^2) 50 km away. Together with these estimates of bird densities at different distances from Nouméa, a detection probability curve is presented for the species in inhabited contexts (Fig. 5.4). This curve shows that the probability of detecting a Red-vented Bulbul falls to 50% when the individual is 25 m from the observer.

5.9 Impacts

The Red-vented Bulbul is commonly blamed for three categories of negative impacts (described below), mostly related to its diverse diet comprising fruits and berries (Islam and Williams, 2000; Brooks, 2013), and flowers, buds, insects and small reptiles (Vander Velde, 2002; Brooks, 2013). Of the 165 reports of plants eaten by the Red-vented Bulbul in the literature, 50% concern the degradation of cultivated plants and 35% relate to seed dispersal (Thibault *et al.*, 2018a). The remaining 17% (26 species from 17 families) were reports of consumption without consideration of the impacts.

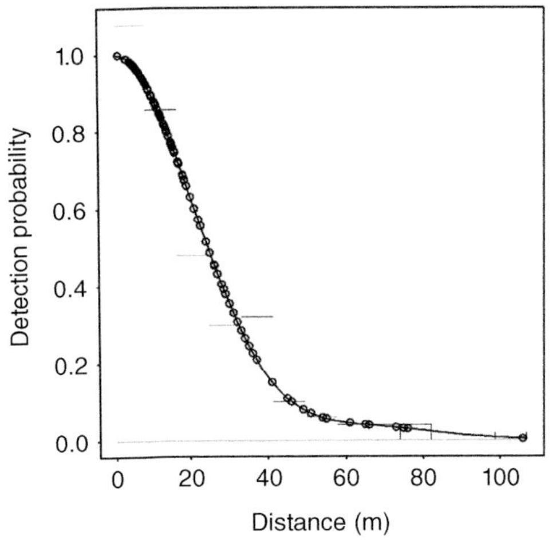

Fig. 5.4. Probability of detecting a Red-vented Bulbul individual as a function of distance (m) from the transect in inhabited areas. (From Thibault *et al.*, 2019.)

Damage to cultivated plants is the most frequently reported impact of the Red-vented Bulbul, within both its native and alien ranges (Fig. 5.5). However, information from its alien range comes from just four locations. In contrast, the references reporting the Red-vented Bulbul as a problematic seed disperser come from eight locations in six countries, and negative interactions with local fauna are reported for 17 species from 11 locations.

In 2018, a review of the literature on alien Red-vented Bulbul populations, their impacts, population trends and management found no quantitative support for considering this species in the '100 world's worst invasive species list' (Thibault *et al.*, 2018a). Indeed, applying the available quantitative information from the literature to the Environmental Impact Classification of Alien Taxa would have led to classifying the species as 'data deficient', 'minimal concern' or 'minor concern', depending on the alien location considered. This is due to the lack of quantitative evaluations of impacts caused by Red-vented Bulbuls in their alien range.

A risk assessment was conducted in 2019 regarding the rencent entry of the Red-vented Bulbul within the European territory. Following a framework proposed by the European Commission's section dedicated to biological invasions, the assessment concluded in a "moderate overall risk with medium confidence". Once again, the medium conficence level attributed to this conclusion was attributed to a lack of quantitative evaluations of both the impacts of and habitat suitability for the species.

5.9.1 Damage to cultivated plants

When reviewing the literature on consumption of cultivated plant species by the Red-vented Bulbul, 52 plant species belonging to 25 families were identified (Thibault *et al.*, 2018a). Among these, 67% (35 species) were cultivated for food production and 33% (17 species) for ornamental purposes.

The location where alien Red-vented Bulbuls have been blamed for the most significant damage to cultivated plants is Oahu, Hawaii, where Walker (2008) reported them consuming several species of fruits, vegetables and flowers, leading to considerable economic losses. The estimated value of the damage to Oahu's orchid industry in one year (1989) was US$300,000 (Fox, 2011) when the Red-vented Bulbul, together with the Warbling or Japanese White-eye (*Zosterops japonicus*), reportedly destroyed up to 75% of Hawaiian orchid and *Anthurium* plantations (Cummings *et al.*, 1994). In New Caledonia, significant impacts have been recorded for some crops and plant nurseries (Metzdorf and Brescia, 2008) with up to 35% losses (Caplong and Barjon, 2010). An open-field experiment suggested that the Red-vented Bulbul was responsible for nearly 18% loss of tomato crops (Thibault *et al.*, 2019). Pecked fruits were generally medium-sized (50–70 mm width), orange to red in colour (Fig. 5.6) and had relatively high sugar concentrations (4–6°Bx). The Red-vented Bulbul is not considered an agricultural pest in Fiji (Watling, 1979) or in Houston, Texas, where it was found to consume mainly introduced tropical plant species (Brooks, 2013).

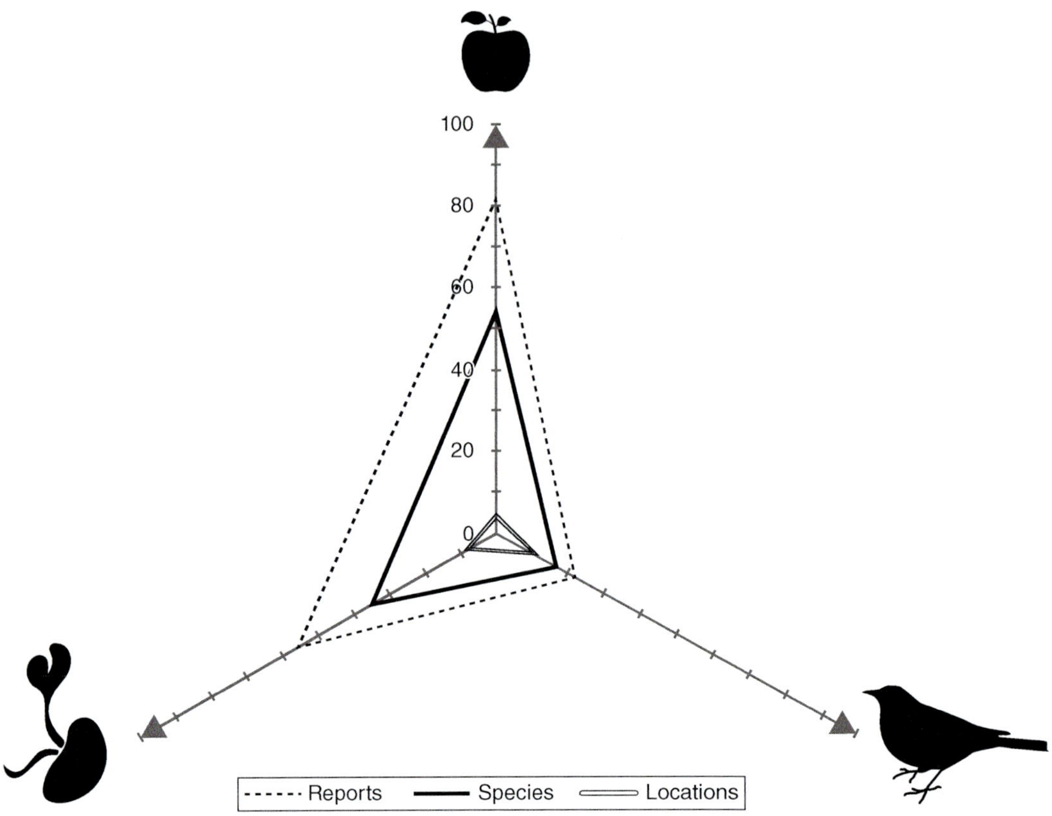

Fig. 5.5. Numbers of reports in the literature, species and location associated with three categories of impact from introduced Red-vented Bulbuls: damage to fruit production, disturbance of native fauna and dispersal of noxious plant seeds. (Adapted from Thibault *et al.*, 2018a.)

Fig. 5.6. A) Cultivated tomato pecked by Red-vented Bulbuls. B) An individual feeding on a cultivated papaya. (©Photographs: (A) Martin Thibault, IAC and (B) Coralie Thouzeau-Fonseca, IAC.)

5.9.2 Dispersal of noxious plant seeds

Thibault *et al.* (2018a) found 56 reports of problematic seed dispersal by the Red-vented Bulbul from six countries inside its alien range. The Red-vented Bulbul can spread the seeds of at least 33 plant species from 25 families. Among these species, 30% were considered non-native (ten species) and 42% invasive (14 species) in the places where the corresponding studies were conducted. They found records of only one endemic (*Coprosma taitensis*, Tahiti) and eight native species that were spread by bulbuls (Spotswood *et al.*, 2012).

The Red-vented Bulbul is considered a major vector of the invasive tree *Miconia calvescens* in Tahiti (Meyer, 1996) and can potentially disperse seven other alien plant species in French Polynesia, including *Lantana camara* (Spotswood *et al.*, 2012, 2013). Its ability to disperse *M. calvescens* and *L. camara* is not unique to the Red-vented Bulbul, as many other avian species, both alien and native, also disperse seeds of these plants. The propensity of the Red-vented Bulbul to disperse seeds of these plants varies from island to island and generally depends on the community of introduced plants and bird species present. For example, the introduced Silvereye (*Zosterops lateralis*) also disperses these seeds in Tahiti, but in Moorea, the endemic Fruit Dove (*Ptilinopus purpuralis*) predominantly disperses seeds of these alien plants. In Fiji, the Red-vented Bulbul contributes to the spread of primary colonist weeds (Watling, 1979). In New Caledonia, the Red-vented Bulbul spreads seeds of another invasive species, *Schinus terebinthifolius* and is often observed feeding on its fruits (Fig. 5.7).

A recent study in New Caledonia was dedicated to describing the seed disperser effectiveness of the Red-vented Bulbul (Thibault *et al.*, 2018c). Their mean gut passage time of a fruit consumed was between 20 and 30 min. When coupled with bird movement data, this corresponded to a median dispersal distance of approximately 100 m around a tree. Germination experiments conducted on both the endemic *Myrtastrum rufopunctata* and the invasive *S. terebinthifolius* showed contradicting trends. Consumption by the Red-vented Bulbul accelerated the germination speed and enhanced the germination rate of *S. terebinthifolius* but lowered the germination capacity of the endemic *M. rufopunctatum*. Even if this observation could be partly explained by differences in fruit characteristics, such effects of favouring an invasive species could drive an 'invasional meltdown' and should be described in depth (Thibault *et al.*, 2018c).

5.9.3 Interactions with native fauna

According to Thibault *et al.* (2018a), 15 species of bird, one reptile and one insect suffer negative interactions with alien Red-vented Bulbuls. Only one study has addressed the issue of how the aggressive behaviour of the Red-vented Bulbul affected other bird species (Pernetta and Watling, 1978). In Fiji, several authors have reported Red-vented Bulbuls displaying aggressive behaviour and competition for food resources towards other passerine species (Clunie, 1976; Pernetta and Watling, 1978; Williams, 2011). However, Watling (1979) suspected that the observed association of native bird species with forest was mainly due to habitat loss rather than the aggressive behaviour of the Red-vented Bulbul in Fiji. Similarly, alien bulbuls in Houston, Texas, showed limited agonistic behaviour towards smaller more passive species when foraging, whereas larger species too close to nesting bulbuls stimulated mobbing behaviour; overall bulbuls showed little aggression towards native birds (Brooks, 2013).

Fig. 5.7. A Red-vented Bulbul feeding on the red berries of the invasive *Schinus terebinthifolius* in New Caledonia. (©Photograph: Coralie Thouzeau-Fonseca, IAC.)

In New Caledonia, the presence and abundance of Red-vented Bulbuls in inhabited areas was correlated with a lower abundance of nine native songbird species but had no effect on the three introduced species that were tested (Fig. 5.8) (Thibault *et al.*, 2018b). In this study, the temporal trend in the abundance of native songbirds suggested that the Red-vented Bulbul contributed to a niche contraction rather than a direct depletion of these species. On Tutuila, American Samoa, Sherman and Fall (2010) observed that bulbuls competed for access to food resources with two passerine species.

Insect and skink predation by Red-vented Bulbuls is mentioned in several studies (Vander Velde, 2002; Walker, 2008; Brooks, 2013). On Oahu, Hawaii, direct predation of the monarch butterfly (*Danaus plexippus*) by the Red-vented Bulbul led to an induced colour selection against the orange morph in the monarch (Stimson and Berman, 1990). After 10 years, the same authors reported a predation transfer to the larvae, leading to an overall decline in abundance of the butterfly (Stimson and Kasuya, 2000). In Tahiti, Red-vented Bulbuls are considered a threat to the Tahiti Monarch (*Pomarea nigra*), an endemic and critically endangered passerine, through competition for nest sites and territory (Blanvillain *et al.*, 2003).

In the Middle East, cross-breeding between the exotic Red-vented Bulbul and the three closely related native species (White-cheeked Bulbul (*Pycnonotus leucogenys*), White-eared Bulbul (*P. leucotis*) and Yellow-vented Bulbul (*P. xanthopygos*)) is often reported as a potential threat for native bulbuls (Khan, 1993; Nation *et al.*, 1997; Gregory, 2005; Azin *et al.*, 2008; Khamis, 2010).

5.9.4 Positive effects

Red-vented Bulbuls feed on a variety of native plant species (Trail, 1994; Sherman and Fall, 2010), and dispersal of native seeds is the only service that has been explored in the Red-vented Bulbul's alien range (Spotswood *et al.*, 2012). As an example, the endemic *M. rufopunctatum* was found to be among the most frequently consumed species in New Caledonia during the study of Thibault *et al.* (2018c). Interestingly, in a village-scale survey led by Daigneault and Brown (2013) in Viti Levu, Fiji, 47% of the respondents reported that the Red-vented Bulbul was good for their community for three main

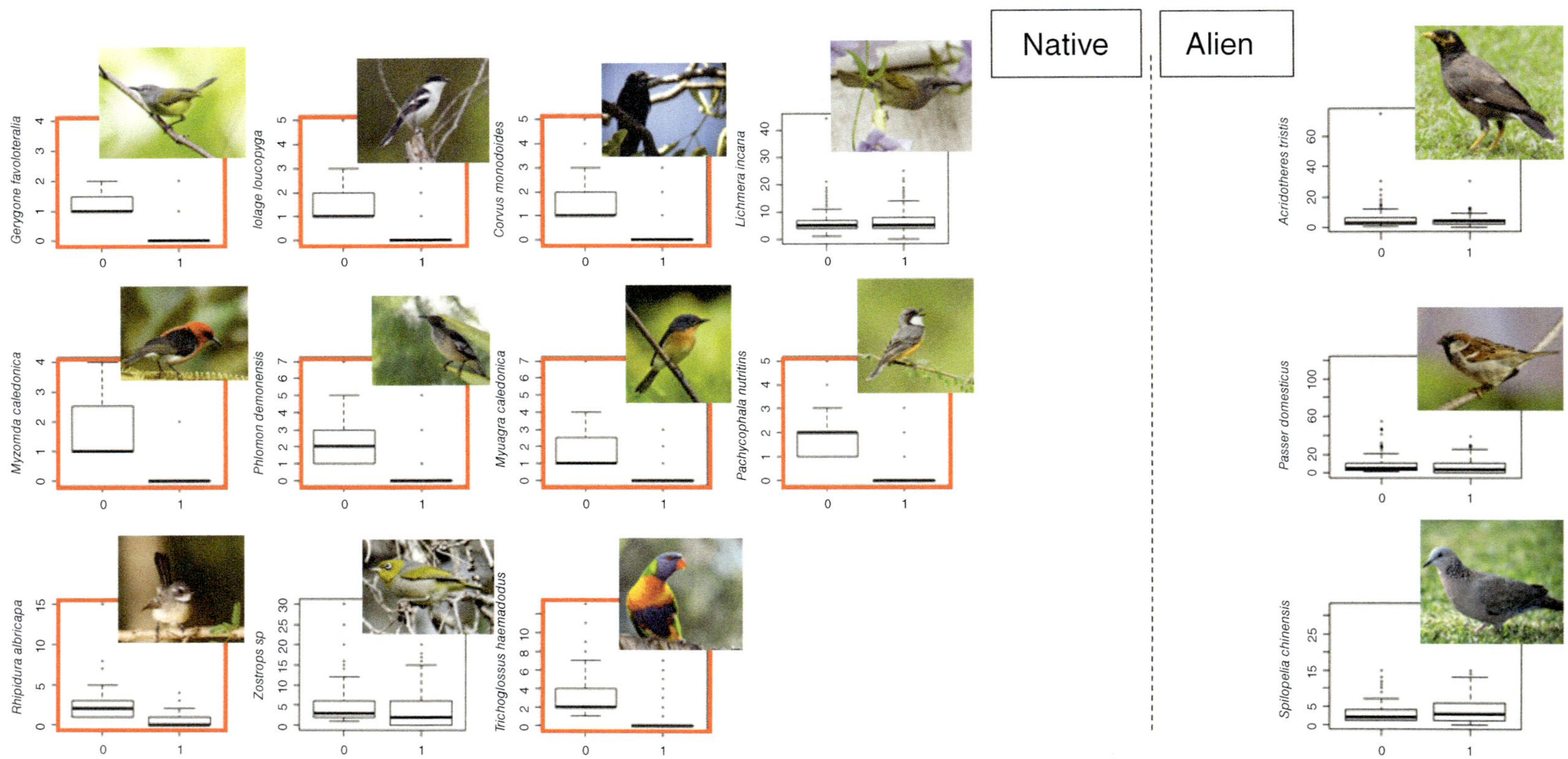

Fig. 5.8. Mean abundance of 14 songbirds in inhabited areas of New Caledonia, depending on whether the Red-vented Bulbul was present or not. Species highlighted in red were less abundant when the Red-vented Bulbul was present. (Adapted from Thibault *et al.*, 2018b.)

reasons: (i) it was effective at insect control; (ii) it reduced mongoose attacks on chickens as a result of the chickens eavesdropping on bulbul alarm calls; and (iii) village focus groups responded that Red-vented Bulbuls were occasionally eaten by villagers.

5.10 Control

To the best of our knowledge, predation of Red-vented Bulbuls has not been studied specifically. Predation by domestic and feral cats (*Felis catus*) is very likely, considering both the low height at which the Red-vented Bulbul nest is built (3 m on average) and their preferred habitat. Raptors are also suspected to prey on Red-vented Bulbuls. During a trapping experiment in New Caledonia, a Swamp Harrier (*Circus approximans*) was recorded by a camera trap while attempting to catch Red-vented Bulbuls (M. Thibault, personal communication).

The Red-vented Bulbul is considered an invasive species and environmental pest under the law in Australia (Tasmanian Government, 2010), Fiji (Minister of Primary Industries, 1985), French Polynesia (Direction de l'Environnement de la Polynésie Française, 2017), Hawaii (Division of Forestry and Wildlife, 2014), New Caledonia (Direction du Développement Economique et de l'Environnement de la Province Nord, 2008; Direction de l'Environnement de la Province Sud, 2016), New Zealand (Ministry of Primary Industries, 2017), South Africa (Department of Environmental Affairs, 2016) and Spain (Ministerio de Agricultura, Alimentacion y Medio Ambiente, 2013). In these countries, transportation, trade and possession of this species are forbidden, and hunting is authorized. No mention of this species as a pest or an invasive species is available from other countries.

Thibault *et al.* (2018a) listed only three examples of management action taken against the Red-vented Bulbul in its alien range documented in the published scientific literature. The first was the successful eradication programme implemented in New Zealand between 1952 and 1955 (Turbott, 1956). This programme allowed the early detection and shooting of bulbuls due to a reward associated with a call for information, and led to an announcement of eradication in 1955 (Watling, 1978). This management strategy remains in place in New Zealand and helped prevent establishment after three more recent introduction events (September 2006, February 2013 and May 2016).

The second location where management actions have been implemented against the Red-vented Bulbul is the island of Tahiti, French Polynesia. In Tahiti, an experimental management programme that was not focused on Red-vented Bulbul management specifically but rather on Tahiti Monarch conservation aimed to control alien birds. Control campaigns were implemented twice, in 2012 and 2013 (Saavedra, 2012, 2013), against the Red-vented Bulbul and the Common Myna (*Acridotheres tristis*). For these, poisoning, shooting and trapping were used simultaneously. These actions resulted in the extermination of 1035 Red-vented Bulbuls in 2012 and 849 in 2013, and led to an increase in the breeding success of the Tahiti Monarch (Saavedra, 2013). Elsewhere in the French Polynesian Archipelago, Red-vented Bulbul removal programmes are in progress in Bora-Bora, Makatea and Nuku Hiva, three islands where the species is still rare but that are located near uninvaded parts of the archipelago.

Third, a test conducted in Hawaii on bird repellent showed that ziram (zinc dimethyldithiocarbamate), methiocarb and methyl anthranilate reduced the consumption of treated papaya mash by Red-vented Bulbuls (Cummings *et al.*, 1994). In an open-field test, the same authors showed that methiocarb significantly reduced damage on orchids.

In contrast, a cost–benefit analysis conducted in Fiji on controlling the Red-vented Bulbul recommended 'taking no action against the bulbul until such time as other benefits and/or means of control have been field tested' (Daigneault and Brown, 2013). This response could be considered a fourth management option.

Nevertheless, locally, unpublished actions exist. In New Caledonia, where expansion of the bulbul population is ongoing, birds are shot, where possible, when they are detected in new areas. In 2016, a Red-vented Bulbul was located at 'La Foa, Pocquereux' by the Institut Agronomique néo-Calédonien (IAC), in the current northern edge of their distribution, resulting in the very first official control action organized in New Caledonia. Following this, a collaboration was established between several institutions to implement a 'quick reaction' protocol that aims to locate and shoot Red-vented Bulbuls that are observed near or beyond the edge of the current distribution range.

In order to reduce the increasing levels of crop damage, and to meet a social demand reported by New Caledonian farmers, locals and agriculture authorities, a management method inspired by the example of Red-whiskered Bulbul (*Pycnonotus jocosus*) control in La Réunion was investigated by the IAC (Thouzeau-Fonseca, 2013). The method consisted of using magpie-type traps and decoys to capture bulbuls. The results show that reddish fruit baits attract Red-vented Bulbuls efficiently. The use of a decoy bird is essential to enhance trapping success, but frequent accidental releases by inadequately trained users should be avoided when trapping outside the actual distributional range to prevent any release in the field.

5.11 Uses

In Assam in north-eastern India, Red-vented Bulbul fights were part of a traditional and religious annual celebration (Fig. 5.9) until this was banned in January 2016. Wild bulbuls were trapped, kept in cages and prepared for the fights, and were finally released if they won (Shalet, 2016).

Red-vented Bulbuls are in high demand for the Asian songbird trade, both in Asia and in many places where people from this region have emigrated. When the captive birds held in other countries escape, this is probably the primary source of introduced birds. This is the case in several countries of the Arabian Peninsula, where bulbuls are sold as cage birds in local markets (J. Baddington, personal communication) and on the internet, as well as in Houston, Texas (Brooks, 2013), where they are also believed to have arrived on large cargo barges travelling from Asia.

Fig. 5.9. A Red-vented Bulbul fight in Assam, India. (©Photograph: Manash Pratim Gogoi.)

5.12 Notes

Dispersal of the Red-vented Bulbul is strongly linked to human activities. The long and close relationship with humans certainly fostered bulbul range expansion at both global and local scales. It led to the transportation of caged individuals across the Pacific Ocean by Indian migrants from the early 20th century first, then by boat and airplane during the 1950s. Within invaded territories such as the New Caledonia mainland, it is assumed that human transportation of caged birds accelerates the dispersal rate of the species.

Part of the information presented here comes from a PhD thesis published recently (Thibault, 2018). In this thesis, an attempt was made to model the potential global distribution of the Red-vented Bulbul according to available distribution and climate data (Thibault *et al.*, unpublished data). The preliminary results of this study suggest that many tropical and subtropical islands (e.g. Mayotte, Madagascar, Indian Ocean Islands) appear to be climatically suitable for the species. When applying Intergovernmental Panel on Climate Change (IPCC) climate change scenarios for 2050 to this potential global distribution, two results are causes for alarm. First, two locations could become climatically suitable for Red-vented Bulbuls in the future (Guinea Gulf and South America). Second, the climate requirements of the Red-vented Bulbul overlap those of other invasive bird species native to the same area (e.g. Common Myna, Red-whiskered Bulbul). Therefore, expected climate change may increase the overlap of invasive bird species that tolerate specific climate parameters. This could favour the establishment and multi-invasion of several territories.

5.13 References

Ali, S. and Ripley, S.D. (1996) *Handbook of the Birds of India and Pakistan: Together with those of Bangladesh, Nepal, Bhutan and Sri Lanka*. Oxford University Press, Delhi, India.

Azin, F., Nosrati, S.M. and Amini, H. (2008) Occurrence of Red-vented Bulbul *Pycnonotus cafer* on Kish Island, Northeastern Persian Gulf, Iran. *Podoces* 3, 105–107.

Bates, J.H., Spotswood, E.N. and Russel, J.C. (2014) Foraging behaviour and habitat partitioning in sympatric invasive birds in French Polynesia. *Notornis* 61, 35–42.

Berger, A.J. (1972) *Hawaiian Birdlife*, 1st edn. University Press of Hawaii, Honolulu, Hawaii.

Berger, A.J. (1975) Red-whiskered and Red-vented bulbuls on Oahu. *Elepaio* 36, 16–19.

Berger, A.J. (1981) *Hawaiian Birdlife*, 2nd edn. University Press of Hawaii, Honolulu, Hawaii.

Blanvillain, C., Salducci, J.M., Tutururai, G. and Maeura, M. (2003) Impact of introduced birds on the recovery of the Tahiti Flycatcher (*Pomarea nigra*), a critically endangered forest bird of Tahiti. *Biological Conservation* 109, 197–205.

Brooks, D.M. (2013) Ecology, behavior, and reproduction of an introduced population of Red-vented Bulbuls (*Pycnonotus cafer*) in Houston, Texas. *Wilson Journal of Ornithology* 125, 800–808.

Caplong, P. and Barjon, F. (2010) Le Bulbul à ventre rouge, une star méconnue en Nouvelle Calédonie. *La Calédonie Agricole* 123, 22–25.

Clapp, R.B. and Sibley, F.C. (1966) Notes on the birds of Tutuila, American Samoa. *Notornis* 13, 157–164.

Clunie, F. (1976) Red-vented Bulbul uses Vanikoro Broadbill nest. *Notornis* 23, 263–263.

Cummings, J.L., Mason, J.R., Otis, D.L., Davis, J.E. and Ohashi, T.J. (1994) Evaluation of methiocarb, ziram, and methyl anthranilate as bird repellents applied to dendrobium orchids. *Wildlife Society Bulletin* 22, 633–638.

Daigneault, A. and Brown, P. (2013) Invasive species management in the Pacific using survey data and benefit-cost analysis. In: *57th Australian Agricultural and Resource Economic Conference*, 5–8 February, Sydney, NSW, Australia. Australian Agricultural and Resource Economics Society. Available at: https://pdfs.semanticscholar.org/74ff/2c5bd929bdae567470430c185a357dcbb07b.pdf

Department of Environmental Affairs (2016) Biodiversity (NEMBA) Act 10 of 2004, List 4: National List of Invasive Bird Species. Department of Environmental Affairs, Pretoria, South Africa. Available at: http://invasives.org.za/legislation/what-does-the-law-say (accessed 10 April 2017).

Dhondt, A. (1976a) Bird Notes From The Kingdom of Tonga. *Notornis* 23, 4–7.

Dhondt, A. (1976b) Bird Observations in Western Samoa. *Notornis* 23, 29–43.

Dickinson, E.C., Dekker, R.W.R.J., Eck, S. and Somadikarta, S. (2002) Systematic notes on Asian birds. 26. Types of the Pycnonotidae. *Zoologische Verhandelingen* 340, 115–160.

Direction de l'Environnement de la Polynésie Française (2017) Section 2: Désignation des espèces menaçant la biodiversité. In: *Code de l'Environnement de la Polynésie Française*. Papeete, Tahiti, French Polynesia, pp. 23–27. Available at: www.2dattitude.org/ressources/k2d/pdf/1/1D/1D05/1D05-01/1D05-01-01.pdf (accessed 15 April 2017).

Direction de l'Environnement de la Province Sud (2016) Code de l'environnement de la Province Sud, Livre II, Titre V, Article 250–2, IV: Liste des espèces animales exotiques envahissantes. Nouméa, New Caledonia. https://eprovince-sud.nc/sites/default/files/2016-04-Code_environnement_province_Sud_Avril_2016.pdf (accessed 15 April 2017).

Direction du Développement Économique et de L'environnement de la Province Nord (2008) Code de l'environnement de la Province Nord, Livre II, Titre VI, Chapter 1, Annexe á l'article 261-1: Liste des espèces envahissantes en province Nord, Koné, New Caledonia. Available at: https://www.province-nord.nc/sites/default/files/parutions/code_environnement_pn_chg_10-06-2016.pdf (accessed 18 November 2019).

Division of Forestry and Wildlife (2014) Chapter 124. Exhibit 4: Introduced wild birds other than game birds which have become established in the wild. Division of Forestry and Wildlife, Honolulu, Hawaii. Available at: http://dlnr.hawaii.gov/dofaw/files/2013/09/Chap124a-Ex.pdf (accessed 15 April 2017).

Duncan, R.P., Bomford, M., Forsyth, D.M. and Conibear, L. (2001) High predictability in introduction outcomes and the geographical range size of introduced Australian birds: a role for climate. *Journal of Animal Ecology* 70, 621–632.

DuPont, J.E. (1972) Notes from Western Samoa, including the description of a new parrot-finch (*Erythrura*). *Wilson Bulletin*, 84, 375–376.

Fox, L. (2011) Red vented bulbul can alter ecosystems. *The Maui News* 1/9/2011, C6.

Freifeld, H.B. (1999) Habitat relationships of forest birds on Tutuila Island, American Samoa. *Journal of Biogeography* 26, 1191–1213.

Gill, B.J., Hunt, G.R. and Sirgouant, S. (1995) Red-vented Bulbuls (*Pycnonotus cafer*) in New Caledonia. *Notornis* 42, 214–215.

Gorman, M.L. (1975) Habitats of The land-birds of Viti Levu, Fiji Islands. *Ibis* 117, 152–161.

Gregory, G. (2005) *The Birds of the State of Kuwait*. Cadeby Books, Grimsby, UK.

Islam, K. and Williams, R.N. (2000) Red-vented Bulbul (*Pycnonotus cafer*) and Red-whiskered Bulbul (*Pycnonotus jocosus*). In: Poole, A. and Gill, F. (eds) *Birds of North America: Life Histories for the 21st Century*. Cornell Lab of Ornithology, Ithaca, New York, and the Academy of Natural Sciences, Philadelphia, Pennsylvania. Available at: http://bna.birds.cornell.edu/bna/species/520bdoi:10.2173/bna.520 (accessed 5 February 2020).

Jennings, M.C. (2010) Atlas of the breeding birds of Arabia. *Fauna of Arabia* 25, 1–751.

Khamis, A. (2010) Alien species in Bahrain: the overseen threats. *BNHS Newsletter* 1, 20–26.

Khan, M.A.R. (1993) The introduced but naturalized avifauna of the United Arab Emirates. *Journal of the Bombay Natural History Society* 90, 437–445.

Lendon, A. (1952) Bulbuls in Melbourne. *Emu* 52, 67–68.

Long, J.L. (1981) *Introduced Birds of the World: The Worldwide History, Distribution and Influence of Birds Introduced to New Environments*. Universe Books, New York.

McAllan, I.A.W. and Hobcroft, D. (2005) The further spread of introduced birds in Samoa. *Notornis* 52, 16–20.

Metzdorf, N. and Brescia, F. (2008) Impact de l'avifaune et des roussettes sur les productions fruitières en Nouvelle-Calédonie. Etude préliminaire – état des lieux et recommandations. IAC/Axe2, Paita, Peru.

Meyer, J.-Y. (1996) Status of *Miconia calvescens* (Melastomataceae), a dominant invasive tree in the Society Islands (French Polynesia). *Pacific Science* 50, 66–67.

Minister of Primary Industries (1985) Laws of Fiji, Chapter 170: Birds and game protection. Minister of Primary Industries, Viti-Levu, Fiji. Available at: www.paclii.org/fj/legis/consol_act_OK/bagpa231/ (accessed 15 April 2017).

Ministerio de Agricultura, Alimentacion y Medio Ambiente (2013) *Pycnonotus cafer*. Catálogo español de especies exóticas invasoras. Ministerio de Agricultura, Alimentacion y Medio Ambiante, Madrid.

Ministry of Primary Industries (2017) Biosecurity Act 1993. Wellington, New Zealand. Available at: www.legislation.govt.nz/act/public/1993/0095/latest/whole.html#DLM315361 (accessed 15 April 2017).

Monnet, C., Thibault, J.C. and Varney, A. (1993) Stability and changes during the twentieth century in the breeding landbirds of Tahiti (Polynesia). *Bird Conservation International* 3, 261–280.

Nation, B., Nation, H. and Hooper, A. (1997) Birds new to Qatar. *Sandgrouse* 19, 56–62.

Parham, B.E.V. (1955) Birds as pests in Fiji. *Journal of Fijian Agriculture* 25, 9–14.

Pedersen, T. and Aspinall, S.J. (compilers) (2015) EBRC annotated checklist of the birds of the United Arab Emirates. Ornithological Society of the Middle East, the Caucasus and Central Asia, Sandy, UK.

Pernetta, J.C. and Watling, D. (1978) The introduced and native terrestrial vertebrates of Fiji. *Pacific Science* 32, 223–244.

Prajapati, S.H., Patel, C.D., Parmar, R.V. and Patel, M.I. (2011) Breeding performance of Red-vented Bulbul (*Pycnonotus cafer*). *Life Sciences Leaflets* 11, 298–304.

Pratt, H.D., Bruner, P.L. and Berrett, D.G. (1987) *A Field Guide to the Birds of Hawaii and the Tropical Pacific*. Princeton University Press, Princeton, New Jersey.

Roy, R.N. and Guha, B.C. (1958) Production of experimental scurvy in a bird species. *Nature* 182, 1689–1690.

Saavedra, S. (2012) First control campaign for common myna and red vented bulbul on Tahiti. Internal Report, SOP Manu, Tahiti.

Saavedra, S. (2013) Second control campaign for common myna and red vented bulbul on Tahiti. Internal Report, SOP Manu, Tahiti.

Shalet, J. (2016) India: Guwahati High Court stays bulbul fight, protests erupt in Assam. *International Business Times* 14 January 2016. Available at: www.ibtimes.co.uk/india-guwahati-high-court-stays-bulbul-fight-protests-errupt-assam-1537937 (accessed 25 January 2016).

Sherman, J. and Fall, P. (2010) Observations on feeding frequencies among native and exotic birds and fruit bats at *Erythrina variegata* and *Dysoxylum* trees on American Samoa. Australian National University Press, Canberra.

Spotswood, E.N., Meyer, J.-Y. and Bartolome, J.W. (2012) An invasive tree alters the structure of seed dispersal networks between birds and plants in French Polynesia. *Journal of Biogeography* 39, 2007–2020.

Spotswood, E.N., Meyer, J.-Y. and Bartolome, J.W. (2013) Preference for an invasive fruit trumps fruit abundance in selection by an introduced bird in the Society Islands, French Polynesia. *Biological Invasions* 15, 2147–2156.

Stimson, J. and Berman, M. (1990) Predator induced colour polymorphism in *Danaus plexippus* L. (Lepidoptera: Nymphalidae) in Hawaii. *Heredity* 65, 401–406.

Stimson, J. and Kasuya, M. (2000) Decline in the frequency of the white morph of the monarch butterfly (*Danaus plexippus plexippus* L., Nymphalidae) on Oahu, Hawaii. *Journal of the Lepidopterists Society* 54, 29–32.

Stuart, C. and Stuart, T. (1999) *Birds of Africa: from Seabirds to Seed-eaters*. MIT Press, Cambridge, Massachusetts.

Tasmanian Government (2010) Wildlife (General) Regulations, S.R. 2010, NO. 113 – Schedule 5: Restricted animals. Tasmanian Government, Hobart, Tasmania. Available at: www.austlii.edu.au/au/legis/tas/num_reg/wr20102010n113319/sch5.html (accessed 15 April 2017).

Thibault, M. (2018) The red-vented bulbul (*Pycnonotus cafer*): invasion dynamics and ecological impacts of an introduced pest bird in New Caledonia and implications for management. PhD thesis, Massey University, Palmerston North, New Zealand.

Thibault, M., Vidal, E., Potter, M.A., Dyer, E. and Brescia, F. (2018a) The red-vented bulbul (*Pycnonotus cafer*): serious pest or understudied invader? *Biological Invasions* 20, 121–136.

Thibault, M., Vidal, E., Potter, M.A., Sanchez, T. and Brescia, F. (2018b) The invasive Red-vented bulbul (*Pycnonotus cafer*) outcompetes native birds in a tropical biodiversity hotspot. *PLoS One* 13, e0192249.

Thibault, M., Masse, F., Pujapujane, A., Lannuzel, G., Bordez, L., *et al.* (2018c) "Liaisons dangereuses": the invasive red-vented bulbul (*Pycnonotus cafer*), a disperser of exotic plant species in New Caledonia. *Ecology and Evolution* 8, 9259–9269.

Thibault, M., Vidal, E., Potter, M.A., Masse, F., Pujapujane, A., *et al.* (2019) Invasion by the red-vented bulbul: an overview of recent studies in New Caledonia. In: Veitch, C.R., Clout, M.N., Martin, A., Russell, J. and West, C. (eds) *Island Invasives: Scaling up to Meet the Challenge*. International Union for Conservation of Nature, Gland, Switzerland, pp. 309–316.

Thouzeau-Fonseca, C. (2013) Contribution à la gestion d'un oiseau introduit envahissant en Nouvelle-Calédonie, le bulbul à ventre rouge (*Pycnonotus cafer* L.): étude des modalités de piégeage et du comportement alimentaire en milieu naturel. Master II, Institut National d'Horticulture et du Paysage, AgroParisTech/IAC, Païta, New Caledonia.

Trail, P.W. (1994) *The Phenology of Rainforest plants in Tutuila, American Samoa*. Department of Marine and Wildlife Resources Biological Report Series, Pago Pago, American Samoa.

Turbott, E.G. (1956) Bulbuls in Auckland. *Notornis* 6, 185–192.

Vander Velde, N. (2002) The Red-vented bulbul has come to Micronesia. *Aliens* 16, 13–14.

Vijayan, V.S. (1980) Breeding biology of bulbuls, *Pycnonotus cafer* and *Pycnonotus luteolus* (Class: Aves, Family: Pycnonotidae) with special reference to their ecological isolation. *Journal of the Bombay Natural History Society* 75, 1090–1117.

Walker, R. (2008) The Red-vented bulbul, Superbird? *Elepaio* 68, 71–78.

Warr, F.E. (1986) The birds of Qatar. *Journal of Qatar Natural History Society* 4, 38–51.

Watling, D. (1978) Observations on the naturalised distribution of the red-vented bulbul in the pacific, with special reference to the Fiji Islands. *Notornis* 25, 109–117.

Watling, D. (1979) Bulbul gets a clean bill. *New Scientist* 81, 963–965.

Watling, D. (1983) The breeding biology of the Red-vented Bulbul *Pycnonotus cafer* in Fiji. *Emu* 83, 173–180.

Williams, G. (2011) *100 Alien Invaders*. Bradt Travel Guides, Chesham, UK.

Williams, R.N. and Giddings, L.V. (1984) Differential range expansion and population growth of bulbuls in Hawaii. *Wilson Bulletin* 96, 647–655.

Zia, U., Ansari, M.S., Akhter, S. and Rakha, B.A. (2014) Breeding biology of Red Vented Bulbul (*Pycnonotus cafer*) in the area of Rawalpindi/Islamabad. *Journal of Animal and Plant Sciences* 24, 656–659.

6 Red-whiskered Bulbul (*Pycnonotus jocosus* Linnaeus, 1758)

Lorinda A. Hart*

School of Life Sciences, University of KwaZulu-Natal, Pietermaritzburg 3209, South Africa; Department of Biological Sciences, University of Namibia, Windhoek 10023, Namibia

Citation: Hart, L.A. (2020). Red-whiskered Bulbul (*Pycnonotus jocosus* Linnaeus, 1758). In: Downs, C.T. and Hart, L.A. (eds) *Invasive Birds: Global Trends and Impacts*. CAB International, Wallingford, UK, pp. 53–59.

6.1 Common Names

The Red-whiskered Bulbul's name has been translated into many different languages (Lepage, 2004). Some examples include: Red-whiskered Bulbul, Crested Bulbul, Red-eared Bulbul, Persian Nightingale (English); Bulbul de Bigoti Vermell, Bulbul Orfeu (Catalan); Hóng-ěr Bēi (Pinyin, Chinese); Bulbul de Bigotes Rojos, Bulbul Orfeo (Spanish); Bulbul Orphée (French); Sivappu Meesai Chinnaan (Tamil); Nók Kroŋ, Nók Kroŋ Hŭa-cùk, Nók Prítlĕew (Thai); and Merl Konde (Creole).

6.2 Nomenclature

Nine subspecies/races of Red-whiskered Bulbul (*Pycnonotus jocosus* Linnaeus, 1758) are recognized:

- *P. j. abuensis* Whistler, 1931, present in north-western India, excluding arid areas.
- *P. j. fuscicaudatus* Gould, 1866, found in western and central India.
- *P. j. pyrrhotis* Bonaparte, 1850, which occupies northern India and Nepal.
- *P. j. emeria* Linnaeus, 1758, located in eastern India, south-western Thailand, Bangladesh and northern, western and southern Myanmar.
- *P. j. whistleri* Deignan, 1948, which occupies the Andaman Islands.
- *P. j. monticola* Horsfield, 1840, present in the eastern Himalayas, north-eastern India, southern Tibet, northern Myanmar and southern China.
- *P. j. pattani* Deignan 1948, found in the far south of Myanmar, Thailand, northern peninsular Malaysia, Laos and southern Indochina.
- *P. j. hainanensis* Hachisuka, 1939, found in northern Vietnam and southern China.
- *P. j. jocosus* Linnaeus, 1758, located in south-eastern China (Deignan, 1948; Fishpool and Tobias, 2017).

6.3 Distribution

The Red-whiskered Bulbul's native range spans tropical Asia from Pakistan throughout India, and southwest to China. Islands and countries include: Andaman Islands, Bangladesh, Bhutan, Cambodia, China, India, Laos, Myanmar, northern Malaysia, Nepal, Thailand and Vietnam (Fig. 6.1).

The Red-whiskered Bulbul has established invasive populations in Australia, Borneo, the Canary Islands, Hawaii, Hong Kong (China), Indonesia, Japan, Juan de Nova Island, southern Malaysia, Mauritius, Nicobar Islands, Oman, Réunion, Saudi Arabia, Singapore, Spain, Taiwan, the United Arab Emirates and the USA (Fig. 6.1) (Leven and Corlett, 2004; Russell, 2008; Walther, 2011; Clements *et al.*, 2018).

Red-whiskered Bulbuls have been reported in South Africa and Zimbabwe but are not currently considered an established invader of continental Africa (Lepage, 2004; van Wilgen and Wilson, 2018). They have also been observed in Madagascar (Bertrand, 2000) and reported as non-breeding in Bahrain (Jennings, 2010).

6.4 Description

Red-whiskered Bulbuls are typically seen in pairs, with flocks observed at fruiting and roost trees (Ali, 1943; Carleton and

*Corresponding author: lorinda.hart.research@gmail.com

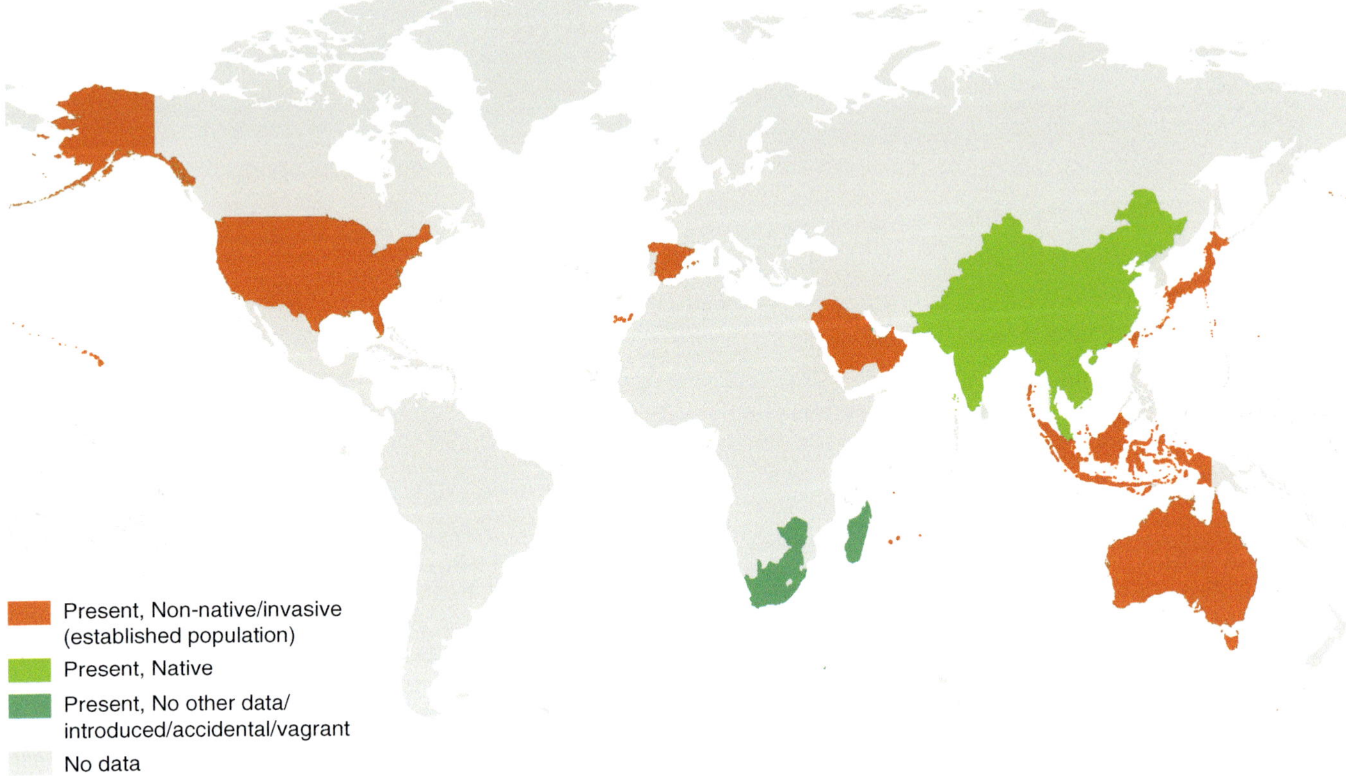

Fig. 6.1. Global distribution of the Red-whiskered Bulbul (*Pycnonotus jocosus*) showing the natural (green) and invaded (red) ranges.

Owre, 1975; Wood, 1999). They are 18–20.5 cm long and weigh 25–31 g (Fishpool and Tobias, 2017). The head is dark brown-black, the belly and breast are off-white, the flanks are tan and the back and tail are brown, with white tips on the tail (Fig. 6.2B) (Fishpool and Tobias, 2017; CABI, 2018). The chin and throat are white and separated from the duller breast by a brown-black collar (Fig. 6.2C) (Fishpool and Tobias, 2017). They have a dark crest (approximately 2 cm), which is often upright (except during flight) and a striking white cheek patch (Fig. 6.2C) (Sinclair, 2013; Fishpool and Tobias, 2017). They have black legs, eyes and beaks, with bristles at the base (Fishpool and Tobias, 2017; CABI, 2018). The sexes are alike in appearance, with males larger than females (Amiot *et al.*, 2007). Adult birds have a red vent and a small red 'whisker' patch below and towards the back of the eye, which is lacking in juveniles (Fig. 6.2A) (Ali, 1943; Sinclair, 2013). Juveniles are also browner, have smaller crests, and their vents appear pink-orange (Fig. 6.2A) (Fishpool and Tobias, 2017; CABI, 2018).

On Réunion Island, Red-whiskered Bulbuls have undergone rapid morphological divergence in fewer than ten generations (Amiot *et al.*, 2007). Nine characteristics differed between windward and leeward populations, particularly bill morphology, which suggests that diet differences between these locations could be a driver (Amiot *et al.*, 2007).

The nine Red-whiskered Bulbul races can differ in the hue of their upper plumage, the presence of white tail tips, crest length, breast band size or bill shape (Fig. 6.2B–D) (Deignan, 1948).

6.5 Diet

As with most bulbuls, Red-whiskered Bulbuls feed primarily on fruit but will also feed on grains, insects and their larvae, arachnids, flower buds, nectar, leaves and kitchen waste (Ali and Ripley, 1972; Corlett, 1998; Higgins *et al.*, 2006; Walther, 2011). This species reportedly also predates young birds and eggs (Clergeau and Mandon-Dalger, 2001; Lever, 2005), although this is uncommon and remains to be quantified (Thibault *et al.*, 2002). Their proportionately large gape allows them to feed on a wide range of fruits and their diet flexibility facilitates success in seasonally fruiting locations (Corlett, 1998). It has been suggested that increased diet protein content reduces incubation time and egg size in Red-whiskered Bulbuls (Mazumdar and Kumar, 2007b) and thus protein-rich food may be preferred during breeding. A variety of invertebrate prey and their larvae make up the bulk of a chick's diet, which decreases with nestling age as plant foods increase (Li *et al.*, 2015).

6.6 Introduction and Invasion Pathways

The rate of establishment and spread of Red-whiskered Bulbuls is slower on continents than on islands: 3 km/yr versus 5–30 km/yr, respectively (Clergeau and Mandon-Dalger, 2001). This is probably because of reduced competition on islands, as invasion rates were faster on islands where bird diversity was lower

Fig. 6.2. Red-whiskered Bulbul.
(A) An immature bird in India, lacking the red 'whisker' patch and with a pale vent (©Photograph: J.M. Garg, CC BY-SA 3.0, https://creativecommons.org/licenses/by-sa/3.0).
(B) Adult birds in Vietnam, showing a reduced collar (©Photograph: Sam Thuong).
(C) An adult bird in India, with a prominent collar (©Photograph: Adarshajoisa, CC BY 3.0, https://creativecommons.org/licenses/by/3.0).
(D) An adult in India, collecting nesting material (©Photograph: Shiva Shankar, CC BY-SA 2.0, https://creativecommons.org/licenses/by-sa/2.0).
(E) Chicks and egg in a nest in Nīoi (*Eugenia koolauensis*) in Hawaii (©Photograph: David Eickhoff, CC BY 2.0, https://creativecommons.org/licenses/by/2.0).
(F) An urban nest site in India (©Photograph: Reji Jacob, CC BY-SA 3.0, https://creativecommons.org/licenses/by-sa/3.0).
(A, B, E and F cropped from original images.)

(Clergeau and Mandon-Dalger, 2001). Habitat suitability and slope also influence the rate and direction of spread (Mandon-Dalger *et al.*, 1999; Clergeau and Mandon-Dalger, 2001).

Red-whiskered Bulbuls are tentatively recognized as established invaders in Hong Kong (Leven and Corlett, 2004). They arrived during the colonial period and their absence in forests in this and neighbouring areas suggests that they are not native to the area, but their introduction pathway remains uncertain (Kwok and Corlett, 2000; Leven and Corlett, 2004). In Taiwan, Red-whiskered Bulbuls are commonly sold as they are popular cage birds and, as with many Asian countries, they are also used during religious releases (Severinghaus and Chi, 1999; Agoramoorthy and Hsu, 2007; Shiu and Stokes, 2008; Walther, 2011). The first wild Taiwanese individual was noted in 1985, with breeding being inferred from young birds observed from 1999 to 2004 (Shieh *et al.*, 2006). However, breeding was only confirmed in 2010 (Walther, 2011). Singapore populations are attributed to cage escapees (Yap and Sodhi, 2004).

The Red-whiskered Bulbul population in Florida, USA arose from a small group of birds (five to ten pairs) that escaped in 1960 (Fisk, 1966; Carleton and Owre, 1975). The first successful breeding was recorded in 1961 (Owre, 1973, cited in Mo, 2015). These birds were popular and were protected by residents in Florida (Carleton and Owre, 1975). No roosts in Florida have been reported for Red-whiskered Bulbuls since the study by Carleton and Owre (1975), with birds maintaining a small population within a relatively small range of 41.7 km^2 (Pranty, 2010). In California, USA birds were detected in 1968 and probably also arose from aviary escapees (Clark, 1976; Lever, 2005).

Red-whiskered Bulbuls were introduced to Mauritius in 1892 and from there were accidentally introduced to Réunion in 1972 (Carié, 1910). Six birds, also from Mauritius, were introduced to Assumption Island in the Seychelles in 1977 and within 9 years grew to an estimated 200 pairs spread throughout the island (Roberts, 1988). The population reached several thousand individuals prior to eradication in 2014 (Uranie, 2015). Of concern was that these individuals would fly to Aldabra Atoll, a UNESCO World Heritage Site, 27 km away (Roberts, 1988). Indeed, in 2012, one bird was reported in Aldabra (Uranie, 2015). The British introduced these birds to the Nicobar Islands in the 1800s (Sankaran, 1998) and they were accidentally introduced to Hawaii in 1965 (Stone, 1985).

Mo (2015) provides a detailed history and current distribution of this species in Australia. The first importation of Red-whiskered Bulbuls into the country occurred in 1865 for an aviary in the Botanical Gardens (Higgins *et al.*, 2006). The Zoological and Acclimatisation Society tried to establish this species through deliberate releases, which were successful during the 20th century (Lever, 2005; Mo, 2015). These populations have probably also been supplemented by cage escapees (Gregory-Smith, 1983). Today, four extant populations are recognized in Australia, occurring predominantly in the south-east (Mo, 2015).

6.7 Breeding Behaviour

Red-whiskered Bulbuls breed primarily from February to August in their native range (Ali, 1943; Li *et al.*, 2015). Both sexes partake in nest building (Fig. 6.2D), incubation and rearing young (Ali, 1943). Nest sites are reused annually (Mazumdar and Kumar, 2007a). However, one study reports that, in southern China, nest building and incubation are carried out by the female only (Liu, 1992). Incubation is almost continuous throughout the night, with roughly half of daylight hours spent incubating (Mazumdar and Kumar, 2007a; Li *et al.*, 2015). Nests are cup-shaped and consist of a variety of natural and man-made materials, including: fine twigs and roots,

Casuarina needles, spider webs, metal wire, bark, paper, leaves, thread and grass (Fig. 6.2F) (Ali, 1943; Mazumdar and Kumar, 2007a; Mo, 2015). Nesting material differs among locations, with birds opportunistically using the materials available (Carleton and Owre, 1975). Nests are made in low trees, shrubs, creepers, garden hedges, thatch roofs, lamp shades, and woven palm leaf walls and roofs of huts (Ali, 1943; Mazumdar and Kumar, 2007a). In south-west China, nests were located in cultivated landscapes and were absent from tropical rain forests, with up to 50 different plant species used (Li *et al.*, 2015). In Florida, birds nest exclusively in the suburbs in a variety of hedges, shrubs and trees (Carleton and Owre, 1975).

When comparing Red-whiskered Bulbul nest sites in the city with agricultural and natural sites in India, it was found that nest locations were more clumped in the city and that there were 50% fewer (Mazumdar and Kumar, 2007a). City nests were around 1.5 m higher (4 m versus 2.5 m) and nest cups were approximately 1 cm shallower than nests in non-urban environments (Mazumdar and Kumar, 2007a). Nests were also more exposed in cities, took 1–2 days longer to construct and consisted of proportionately more anthropogenic material (Mazumdar and Kumar, 2007a). Nests in cultivated landscapes in China are similar to those observed in the same habitat type in India (Li *et al.*, 2015). However, there are differences between the Indian and Chinese populations' breeding biology (Li *et al.*, 2015). In India, they lay two to four eggs, which are incubated for 15–16 days (Ali, 1943) and chicks fledge after 12–13 days (Mazumdar and Kumar, 2007a). In south-western China, both incubation and nestling period are approximately 11 days (Li *et al.*, 2015).

Red-whiskered Bulbuls can have two to three broods per season (Ali, 1943). Eggs have an off-white base colour with pinkish-brown spots that are denser at the obtuse end of the egg (Fig. 6.2E). They are 22.0 ± 2.24 mm long (mean ± SD) by 15.5 ± 1.46 mm wide and weigh 2.81 ± 0.25 g (Mazumdar and Kumar, 2007a; Walther, 2011; Li *et al.*, 2015). Ali (1943) described their egg and chick mortality as 'amazingly heavy'. Indeed, in China, overall nesting success was only 34.2% (Li *et al.*, 2015). However, in India, overall nesting success was 72.2% in the city and 80.5% in more natural habitats, with chick mortality of 17% and 10.4%, respectively (Mazumdar and Kumar, 2007a). Nest failure could be due to infertility, storms, parent desertion, parasites, disease or predation by a variety of birds, lizards, snakes and mammals (Carleton and Owre, 1975; Walther, 2011; Li *et al.*, 2015).

6.8 Habitat

Red-whiskered Bulbuls occupy a range of habitats typically below 2000 m elevation, including scrublands, wooded habitats, forest and mangrove edges, reed beds, parks, secondary forests, gardens, agricultural landscapes, and urban and suburban areas (Clergeau and Mandon-Dalger, 2001; Walther, 2011; Fishpool and Tobias, 2017). Densely vegetated areas, such as continuous forest, appear to be avoided, with their preferred habitat consisting of well-watered, open wooded areas (Robson, 2000; Clergeau and Mandon-Dalger, 2001; Fishpool and Tobias, 2017). In India, they prefer humid habitats and avoid the dry areas to the north-west of the country (Ali, 1943). In southern China, their abundance is greatest in sparse bush and valley farmland (Liu, 1992). On Réunion Island, they are more abundant in areas where invasive plants supply year-round fruit than in seasonally fruiting stands (Mandon-Dalger *et al.*, 2004). In Australia, they are present in both suburban locations and surrounding natural vegetation (Mo, 2015). On the Mascarene Islands, they are more abundant in areas of human habitation, where native avifauna are scarcer (Clergeau and Mandon-Dalger, 2001). Similarly, in Hong Kong and Japan, Red-whiskered Bulbuls are more abundant in suburban areas (Chan, 2004; Eguchi and Amano, 2004a). In Florida, they are almost exclusively associated with the suburbs (Carleton and Owre, 1975).

Red-whiskered Bulbuls roost communally, especially during the non-breeding season (Carleton and Owre, 1975; Wood, 1999). Roost trees shift, seemingly in response to the level of protection they offer during different seasons; e.g. pine trees (*Pinus* spp.) are used during warmer months while more densely foliaged figs (*Ficus* spp.) provide roosts during colder months (Carleton and Owre, 1975). They also hawk insects near to and from roost trees prior to nightfall, and therefore roost tree species could also be selected based on their associated insect assemblages (Carleton and Owre, 1975).

6.9 Impacts

6.9.1 Negative impacts

One of the main concerns of invasive Red-whiskered Bulbuls is their role as dispersers of invasive alien plants, which has been recorded throughout much of their range (Mo, 2015). In so doing, they have altered plant communities (Simberloff and Von Holle, 1999), particularly as they move between degraded and intact landscapes (Linnebjerg *et al.*, 2009). On Réunion Island, invasive fruits make up 80% of their diet (Mandon-Dalger *et al.*, 2004), with some of these species benefitting from more rapid and increased germination following gut passage (Linnebjerg *et al.*, 2009). Examples of invasive plants that benefit from Red-whiskered Bulbuls include (among others) Brazilian pepper trees (*Schinus terebinthifolius*), camphor trees (*Cinnamomum camphora*), Sri Lankan privet (*Ligustrum robustum*), lantana (*Lantana camara*), blackberry bramble (*Rubus fruticosus*), many fig species, and Koster's curse (*Clidemia hirta*) (Carleton and Owre, 1975; Panetta and McKee, 1997; Mandon-Dalger *et al.*, 2004; Tassin *et al.*, 2007; Linnebjerg *et al.*, 2009, 2010; Walther, 2011; Mo, 2015). They have also been observed to roost in lantana thickets (Wood, 1999).

Red-whiskered Bulbuls also cause damage to fruit orchards and vegetable gardens, feed on and impact indigenous invertebrate populations, predate young birds, damage eggs and compete with indigenous avifauna for resources (Clergeau and Mandon-Dalger, 2001; Yap and Sodhi, 2004; Lever, 2005). Additionally, Red-whiskered Bulbuls feed on seedlings and growing shoots, causing damage to young and growing plants (Carleton and Owre, 1975). Laboratory trials and field observations found that smaller drupes or larger soft fruits were easily eaten by Red-whiskered Bulbuls (Carleton and Owre, 1975). While citrus (*Citrus* spp.) and mangoes (*Mangifera indica*) are readily consumed by the birds, they relied on other species to break the skin of the fruit open (Carleton and Owre, 1975). Whole fruits were left alone by birds in cages with no other food available for 24 h, until the skins were pierced (Carleton and Owre, 1975).

Red-whiskered Bulbul intra- and interspecies territorial aggression increases during their nesting period, although territorial behaviour is observed year-round (Sotthibandhu, 2003; Mazumdar and Kumar, 2007a). In Hong Kong, they have displaced the indigenous Chinese Bulbul (*Pycnonotus sinensis*) from suburban habitats to more natural habitats (Chan, 2004). Due to their aggressive and competitive nature, they have had a negative impact on many endemic and endangered species, particularly on islands (Eguchi and Amano, 2004b; Lever, 2005; Fishpool and Tobias, 2017). For example, they have displaced the Nicobar Bulbul (*Hypsipetes nicobariensis*) on the Nicobar Islands (Sankaran, 1998).

The effect of Red-whiskered Bulbuls on indigenous invertebrate populations is raised as a potential concern, but this remains to be quantified. However, in Mauritius, the loss of large, native golden orb weaving spiders (*Nephila* spp.) has been attributed to bulbul predation (Islam and Williams, 2000; Linnebjerg *et al.*, 2010).

Red-whiskered Bulbuls are very popular as pet birds and have been hybridized with several other bulbul species to enhance favourable qualities (Techachoochert and Round, 2013). This poses a genetic risk to indigenous Red-whiskered Bulbul populations should these captive birds escape and breed with wild populations. It also highlights a potential risk of invasive Red-whiskered Bulbuls or their hybrids interbreeding with indigenous bulbul species.

Red-whiskered Bulbuls are possible reservoirs of avian malaria in Hawaii (Lever, 2005) and a new avian malaria parasite, *Plasmodium* (*Novyella*) *jiangi*, was isolated from Red-whiskered Bulbuls in south-east China (He and Huang, 1993). Additionally, one individual tested positive for subtype H9 of the avian influenza virus in Vietnam (Thinh *et al.*, 2012). External parasites such as lice have been recorded on Red-whiskered Bulbuls in India, including amblyceran lice *Myrsidea pycnonoti*, *M. eurysternus* and *Menacanthus eurysternus* and ischnoceran lice *Brueelia guldum* (Arya *et al.*, 2010; Saxena *et al.*, 2012). It is likely that other parasites and diseases are hosted by this species and, given their interspecies roosting behaviour and widespread abundance, they could be important vectors in the transmission, prevalence and spread of avian pathogens.

6.9.2 Positive impacts

Red-whiskered Bulbuls are one of the most important seed dispersers in their native and invasive range for a wide variety of fleshy-fruited species (Corlett, 1998; Clergeau and Mandon-Dalger, 2001). In Hong Kong shrublands, they consume fruits from at least 61 taxa (Corlett, 1998). Cheng (1963, cited by Yap and Sodhi, 2004) reported that Red-whiskered Bulbuls benefit agriculture, due to pest control or pollination (Gregory-Smith, 1983). As they also feed on nectar and forage for insects occurring on some flowers (Carleton and Owre, 1975), they are probably pollinators of many plant species as well (Raju *et al.*, 2005).

In Australia, the indigenous Pallid Cuckoo (*Cuculus pallidus*) uses Red-whiskered Bulbuls as hosts (Gregory-Smith, 1983).

6.10 Control

While snakes, cats, rats and birds may predate on Red-whiskered Bulbuls (Carleton and Owre, 1975), formal control measures would need to be implemented to have any meaningful impact on invasive populations. To prevent these bulbuls from establishing in California, a decoy stuffed bird mounted on a pole surrounded with speakers playing the Red-whiskered Bulbul call was used to lure birds within shooting or detection range (Clark, 1976). At the time 75 birds were shot, but vagrant sightings continued (Clark, 1976). This species is now prohibited in California (Clark, 1976).

No formal control programme is in place for Red-whiskered Bulbuls in Australia, as their impact on Australian flora and fauna has yet to be quantified (Mo, 2015). However, localized culling is reported in response to agricultural crop damage (Higgins *et al.*, 2006). While control of invasive species is generally more difficult on continents, their isolated and contained distributions in Australia should make control possible (Mo, 2015).

Eradication has been achieved on some islands. In Mayotte, Red-whiskered Bulbuls were eliminated in the 1990s (Sinclair, 2013). In the Seychelles, a vagrant individual was removed from Aldabra in 2013, and a total of 5279 birds were successfully eradicated from Assumption Island in 2014, using targeted mist-netting and shooting over a 3-year period (Bunbury *et al.*, 2013; Uranie, 2015). On the Mascarene Islands, farmers use pesticides and bird lime to control Red-whiskered Bulbuls, but unfortunately these untargeted approaches also have a detrimental effect on indigenous birdlife (Clergeau and Mandon-Dalger, 2001). Réunion Island has established a working group to manage Red-whiskered Bulbuls by means of assessing risks and using management practices to reduce conflict, instead of eliminating this species (Clergeau *et al.*, 2006). However, culling by means of trapping, using decoy birds and fruit bait, has been implemented to reduce numbers as these birds damage agricultural crops on the island (Amiot *et al.*, 2007).

Based on invasion rates of this species, it has been recommended that effective control is implemented within the first 5 years of introduction (3 years on tropical islands), and thus early detection and implementation of control programmes is key (Clergeau and Mandon-Dalger, 2001).

6.11 Uses

Red-whiskered Bulbuls are commonly bought and released during religious ceremonies and kept as pets (Severinghaus and Chi, 1999; Agoramoorthy and Hsu, 2007; Shiu and Stokes, 2008; Walther, 2011). Their popularity as pets not only stems from their appealing song and appearance, but Ali

Fig. 6.3. Poaching using a caller bird in a cage (©Photograph: Lip Kee Yap, CC BY-SA 2.0, https://creativecommons.org/licenses/by-sa/2.0.)

(1943) reports that this species is easily tamed and will fly great distances to follow their owner.

6.12 Notes

Although not globally threatened, the Red-whiskered Bulbul faces local extinctions within its native range (Fig. 6.3) (Techachoochert and Round, 2013; Fishpool and Tobias, 2017). In Thailand, this species is perhaps the most popular indigenous caged bird and, as such, illegal trapping has resulted in this species becoming locally threatened and rare (Techachoochert and Round, 2013). Estimates of captive birds range in the millions, based on reports of over 100 Red-whiskered Bulbul clubs in Thailand; one of which has around 50,000 members owning an average of five to ten birds each (Techachoochert and Round, 2013). The Thai government has inadvertently further promoted their popularity by hosting bulbul singing competitions (Techachoochert and Round, 2013). Prize-winning birds on average fetch 200,000 Baht (approximately US$6300), with a maximum of 1.6 million Baht (US$50,500) reported (Techachoochert and Round, 2013).

6.13 References

Agoramoorthy, G. and Hsu, M.J. (2007) Ritual releasing of wild animals threatens island ecology. *Human Ecology* 35, 251–254.

Ali, S. (1943) *The Book of Indian Birds*. The Bombay Natural History Society, Mumbai, India.

Ali, S. and Ripley, S.D. (1972) *Handbook of the Birds of India & Pakistan*, Vol. 5. Oxford University Press, Delhi, India.

Amiot, C., Lorvelec, O., Mandon-Dalger, I., Sardella, A., Lequilliec, P. and Clergeau, P. (2007) Rapid morphological divergence of introduced Red-whiskered Bulbuls *Pycnonotus jocosus* in contrasting environments. *Ibis* 149, 482–489.

Arya, G., Bansal, N., Khan, V., Ahmad, A. and Saxena, A. (2010) Population characteristics of *Phthiraptera* occurring on Red Whiskered Bulbul (*Pycnonotus jocosus*). *Journal of Applied and Natural Science* 2, 263–265.

Bertrand, G. (2000) Compte-rendu ornithologique des espèces observées à Juan de Nova (Canal du Mozambique). *Bulletin Phaethon* 12, 57–61.

Bunbury, N., Mahoune, T., Raguain, H., Richards, H. and Fleischer-Dogley, F. (2013) Red-whiskered Bulbul eradicated from Aldabra. *Aliens: The Invasive Species Bulletin* 33, 7–8.

CABI (2018) *Pycnonotus jocosus (Red-whiskered Bulbul)*. In: *Invasive Species Compendium*. CAB International, Wallingford, UK. Available at: www.cabi.org/isc/datasheet/67471 (accessed 23 October 2019).

Carié, P. (1910) Note sur l'acclimatation du Bulbul (*Octocompsa jocosa* L.) à l'île Maurice. *Bulletin de la Société Nationale d''cclimatation* 57, 462–464.

Carleton, A.R. and Owre, O.T. (1975) The Red-whiskered Bulbul in Florida: 1960–71. *Auk* 92, 40–57.

Chan, E. (2004) Space partitioning by two common bulbuls in Hong Kong. *Porcupine* 31, 8–9.

Cheng, T. (1963) *China's Economic Fauna: Birds*. Science Publication, Beijing.

Clark, D.O. (1976) An overview of depredating bird damage control in California. In: *Proceedings of the Seventh Bird Control Seminar*, Bowling Green State University, Ohio, 9–11 November. Internet Center for Wildlife Damage Management, University of Nebraska, Lincoln, Nebraska, pp. 21–27.

Clements, J.F., Schulenberg, T.S., Iliff, M.J., Roberson, D., Fredericks, T.A., *et al.* (2018) The eBird/Clements checklist of birds of the world: v2018. Available at: www.birds.cornell.edu/clementschecklist/download/ (accessed 23 October 2019).

Clergeau, P. and Mandon-Dalger, I. (2001) Fast colonization of an introduced bird: the case of *Pycnonotus jocosus* on the Mascarene Islands. *Biotropica* 33, 542–546.

Clergeau, P., Lay, G.L. and Mandon-Dalger, I. (2006) Intégrer les analyses géographiques, écologiques et sociales pour gérer la faune sauvage [Integrating geographic, ecological and social analyses in managing wildlife]. In: Legay, J.M. (ed.) *L'Interdisciplinarité Dans les Sciences de la Vie*. Édition Cemagref, Antony Cedex, France, pp. 103–113.

Corlett, R.T. (1998) Frugivory and seed dispersal by birds in Hong Kong shrubland. *Forktail* 13, 23–27.

Deignan, H.G. (1948) The races of the Red-whiskered Bulbul, *Pycnonotus jocosus* (Linnaeus). *Journal of the Washington Academy of Sciences* 38, 279–281.

Eguchi, K. and Amano, H.E. (2004a) Invasive birds in Japan. *Global Environmental Research* 8, 29–39.

Eguchi, K. and Amano, H.E. (2004b) Spread of exotic birds in Japan. *Ornithological Science* 3, 3–11.

Fishpool, L. and Tobias, J. (2017) Red-whiskered Bulbul (*Pycnonotus jocosus*). Lynx Edicions, Barcelona, Spain. Available at: www.hbw.com/species (accessed 23 October 2019).

Fisk, E.J. (1966) A happy newcomer in a fruitful land. *Florida Naturalist* 39, 10–11.

Gregory-Smith, R. (1983) Introduced species. *Canberra Bird Notes* 8, 70–89.

He, J. and Huang, J. (1993) A new species of avian malaria parasite from *Pycnonotus jocosus* (Sporozoa: Plasmodiidae). *Acta Zootaxonomica Sinica* 18, 129–133.

Higgins, P.J., Peter, J.M. and Cowling, S.J. (2006) *Handbook of Australian, New Zealand and Antarctic Birds*. Oxford University Press, Melbourne, Australia.

Islam, K. and Williams, R.N. (2000) Red-whiskered Bulbul (*Pycnonotus jocosus*). In: Poole, A.F. (ed.) *The Birds of North America*. Cornell Lab of Ornithology, Ithaca, New York. Available at: https://birdsna.org/Species-Account (accessed 23 October 2019).

Jennings, M.C. (2010) *Fauna of Arabia*, Vol. 25: *Atlas of the Breeding Birds of Arabia*. King Abdulaziz City for Science and Technology, Riyadh.

Kwok, H. and Corlett, R.T. (2000) The bird communities of a natural secondary forest and a *Lophostemon confertus* plantation in Hong Kong, South China. *Forest Ecology and Management* 130, 227–234.

Lepage, D. (2004) Avibase: The World Bird Database, Bird Studies Canada. Available at: https://avibase.bsc-eoc.org (accessed 23 October 2019).

Leven, M.R. and Corlett, R.T. (2004) Invasive birds in Hong Kong, China. *Ornithological Science* 3, 43–55.

Lever, C. (2005) *Naturalised Birds of the World*. T. & A.D. Poyser, London.

Li, H., Zhang, M.-X., Yang, X.-J., Cui, L.-W. and Quan, R.-C. (2015) The breeding biology of Red-Whiskered Bulbul (*Pycnonotus jocosus*) in Xishuangbanna, southwest China. *Zoological Research* 36, 233–240.

Linnebjerg, J.F., Hansen, D.M. and Olesen, J.M. (2009) Gut passage effect of the introduced Red-whiskered Bulbul (*Pycnonotus jocosus*) on germination of invasive plant species in Mauritius. *Austral Ecology* 34, 272–277.

Linnebjerg, J.F., Hansen, D.M., Bunbury, N. and Olesen, J.M. (2010) Diet composition of the invasive Red-whiskered Bulbul *Pycnonotus jocosus* in Mauritius. *Journal of Tropical Ecology* 26, 347–350.

Liu, X. (1992) A preliminary study on the ecology of Red-whiskered Bulbul. *Journal of Guangxi Academy of Sciences* 1, 54–62.

Mandon-Dalger, I., Le Corre, M., Clergeau, P., Probst, J.M. and Besnard, N. (1999) Modalités de la colonisation de l'Île de la Réunion par le Bulbul Orphée (*Pycnonotus jocosus*). *Revue d'Ecologie: La Terre et La Vie* 54, 283–295.

Mandon-Dalger, I., Clergeau, P., Tassin, J., Riviere, J.-N. and Gatti, S. (2004) Relationships between alien plants and an alien bird species on Réunion Island. *Journal of Tropical Ecology* 20, 635–642.

Mazumdar, A. and Kumar, P. (2007a) Nesting ecology of the Red-whiskered Bulbul at city center and periphery in Lucknow, Northern India. *Berkut* 16, 98–102.

Mazumdar, A. and Kumar, P. (2007b) Role of protein diet in the regulation of clutch size, incubation period, and egg size of the Red-whiskered Bulbul. *Berkut* 16, 164–165.

Mo, M. (2015) The Red-whiskered Bulbul *Pycnonotus jocosus* in Australia – a global perspective, history of introduction, current status and potential impacts. *Australian Zoologist* 37, 461–471.

Owre, O.T. (1973) A consideration of the exotic Avifauna of southeastern Florida. *Wilson Bulletin* 85, 491–500.

Panetta, F. and McKee, J. (1997) Recruitment of the invasive ornamental, *Schinus terebinthifolius*, is dependent upon frugivores. *Australian Journal of Ecology* 22, 432–438.

Pranty, B. (2010) Status and current range of Red-whiskered Bulbuls (*Pycnonotus jocosus*) in Florida. *Florida Field Naturalist* 38, 146–149.

Raju, A.J., Rao, S.P. and Rangaiah, K. (2005) Pollination by bats and birds in the obligate outcrosser *Bombax ceiba* L. (Bombacaceae), a tropical dry season flowering tree species in the Eastern Ghats forests of India. *Ornithological Science* 4, 81–87.

Roberts, P. (1988) Introduced birds on Assumption Island – a threat to Aldabra. *Oryx* 22, 15–17.

Robson, C. (2000) *A Field Guide to the Birds of South-East Asia*. New Holland, London.

Russell, J. (2008) Juan de Nova Island Scientific Mission 7th July–28th August 2008. Université de la Réunion, Reunion.

Sankaran, R. (1998) An annotated list of the endemic avifauna of the Nicobar Islands. *Forktail* 13, 17–22.

Saxena, A.K., Arya, G. and Bansal, N. (2012) Egg laying site and oviposition pattern of two phthirapteran species parasitizing Red Whiskered Bulbul (*Pycnonotus jocosus*). *Türkiye Parazitoloji Dergisi* 36, 166–168.

Severinghaus, L.L. and Chi, L. (1999) Prayer animal release in Taiwan. *Biological Conservation* 89, 301–304.

Shieh, B.-S., Lin, Y.-H., Lee, T.-W., Chang, C.-C. and Cheng, K.-T. (2006) Pet trade as sources of introduced bird species in Taiwan. *Taiwania* 51, 81–86.

Shiu, H. and Stokes, L. (2008) Buddhist animal release practices: historic, environmental, public health and economic concerns. *Contemporary Buddhism* 9, 181–196.

Simberloff, D. and Von Holle, B. (1999) Positive interactions of nonindigenous species: invasional meltdown? *Biological Invasions* 1, 21–32.

Sinclair, I. (2013) *Birds of the Indian Ocean Islands*. Penguin Random House, South Africa.

Sotthibandhu, S. (2003) Territorial defense of the Red-whiskered Bulbul, *Pycnonotus jocosus* (Pycnonotidae), in a semi-wild habitat of the bird farm. *Songklanakarin Journal of Science and Technology* 25, 553–563.

Stone, C.P. (1985) Alien animals in Hawaii's native ecosystems: toward controlling the adverse effects of introduced vertebrates. In: Stone, C.P. and Scott, J.M. (eds) *Hawaii's Terrestrial Ecosystems: Preservation and Management*. Cooperative National Park Resources Studies Unit, University of Hawaii, Honolulu, Hawaii, pp. 251–297.

Tassin, J., Rivière, J.N. and Clergeau, P. (2007) Reproductive versus vegetative recruitment of the invasive tree *Schinus terebenthifolius*: implications for restoration on Réunion Island. *Restoration Ecology* 15, 412–419.

Techachoochert, S. and Round, P.D. (2013) Red-whiskered Bulbul: are trapping and unregulated avicultural practices pushing this species towards extinction in Thailand? *BirdingASIA* 20, 49–52.

Thibault, J.-C., Martin, J.-L., Penloup, A. and Meyer, J.-Y. (2002) Understanding the decline and extinction of monarchs (Aves) in Polynesian Islands. *Biological Conservation* 108, 161–174.

Thinh, T.V., Gilbert, M., Bunpapong, N., Amonsin, A., Nguyen, D.T., *et al.* (2012) Avian influenza viruses in wild land birds in northern Vietnam. *Journal of Wildlife Diseases* 48, 195–200.

Uranie, S. (2015) Eradication success – Seychelles wins war against invasive Red-whiskered Bulbul. *Seychelles News Agency*. Available at: www.seychellesnewsagency.com/articles/2332/Eradication+success++Seychelles+wins+war+against+invasive+red-whiskered+bulbul (accessed 23 October 2019).

van Wilgen, B.W. and Wilson, J.R. (eds) (2018) *The Status of Biological Invasions and Their Management in South Africa in 2017*. South African National Biodiversity Institute, Kirstenbosch and DST-NRF Centre of Excellence for Invasion Biology, Stellenbosch, South Africa.

Walther, B.A. (2011) First documented nesting of the Red-whiskered Bulbul *Pycnonotus jocosus* in Taiwan. *Taiwan Journal of Biodiversity* 13, 121–133.

Wood, K. (1999) Roosting habits of the Red-whiskered Bulbul at Wollongong, New South Wales. *Australian Field Ornithology* 18, 75–80.

Yap, C.A. and Sodhi, N.S. (2004) Southeast Asian invasive birds: ecology, impact and management. *Ornithological Science* 3, 57–67.

7 Great Kiskadee (*Pitangus sulphuratus* Linnaeus, 1766)

Blake A. Mathys*
Ohio Dominican University, Columbus, OH 43219, USA

Citation: Mathys, B. (2020) Great Kiskadee (*Pitangus sulphuratus* Linnaeus, 1766). In Downs, C.T. and Hart, L.A. (eds) *Invasive Birds: Global Trends and Impacts*. CAB International, Wallingford, UK, pp. 60–62.

7.1 Common Names

Great Kiskadee, Derby Flycatcher.

7.2 Distribution

The Great Kiskadee (*Pitangus sulphuratus* Linnaeus, 1766) occurs naturally in the New World tropics and subtropics, from southern Texas through Central America and south to Argentina. It was introduced to Bermuda, Tobago (ffrench, 1991) and Rio Negro in central Argentina (Sibley and Monroe, 1990) from where it may have expanded further south into Chubut due to tree planting in formerly open habitats (Leveque, 1979). The Bermuda population is the most well-established and by far the largest and best known of the non-native populations.

7.3 Description

The Great Kiskadee is boldly marked, with a black-and-white striped head; it has a bright yellow crown patch that is usually concealed, but can be raised when the bird is agitated, distressed or displaying (Fig. 7.2). It has a bright yellow body below, with brownish back and rufous patches in the wings. Mean body mass ranges from 55 to 74 g in the native range (Brush and Fitzpatrick, 2002), with Bermuda individuals having mean mass of 62.7 g based on 84 adult individuals captured over 4 years (Mathys and Lockwood, 2009). Sexes are similar in appearance and have a broad overlap in body dimensions, with males tending to be slightly larger (Brush and Fitzpatrick, 2002).

7.4 Diet

The Great Kiskadee is a generalist omnivore; it consumes a wide range of animal prey (including many arthropods and vertebrates, such as fish and lizards), fruits and human-derived food (e.g. dog food, bird seed) whenever available (Gorena, 1997; Brush and Fitzpatrick, 2002).

7.5 Introduction and Invasion Pathways

The Great Kiskadee was introduced to Bermuda to control invasive anole (*Anolis* spp.) lizards. It was hoped that a reduction in anole populations would increase the population of their prey, ladybird beetles. More beetles would then be available to consume scale insects (order Hemiptera), which were negatively impacting the endemic Bermuda cedar (*Juniperus bermudiana*). To this end, a total of 200 Great Kiskadees were captured near Port of Spain, Trinidad, and released on Bermuda in 1957. The population quickly increased (Crowell, 1962) and was estimated at 60,000 in 1976 (Crowell and Crowell, 1976), and the kiskadee is now at least the third most common full-time avian resident of Bermuda (Long, 1981). Bermuda contains a relatively small land area, only 53.2 km^2, and the kiskadee spread to every part of the islands soon after introduction and continues to be found in all Bermuda habitats. Due to Bermuda's isolation (approximately 1000 km from any other land) and the kiskadee's non-migratory habits, it is unlikely that any individuals will ever naturally disperse from the Bermuda population. However, ship-assisted movements, especially to mainland North America, could be possible, as this species regularly spends time near water and on boats, including building nests on boats (B. Mathys, personal observation).

*Corresponding author: mathysb@ohiodominican.edu

Fig. 7.1. Global distribution of the Great Kiskadee (*Pitangus sulphuratus*) showing the natural (green) and invaded (red) ranges.

Fig. 7.2. An adult Great Kiskadee. (©Photograph: Blake Mathys.)

7.6 Breeding Behaviour

Nesting of the Great Kiskadee has been reported in nearly every month in Trinidad (ffrench, 1991) but is usually concentrated in the early to middle parts of the year (Hilty, 2003; Stiles and Skutch, 1989). No significant studies have been carried out concerning breeding in its introduced range, although it is probably concentrated in the northern hemisphere spring and early summer (April–July), similar to what is seen for kiskadees in Texas (Brush and Fitzpatrick, 2002). Nests are made primarily of grass and are large, domed structures with an entrance on one side. No extensive breeding studies have been carried out for any kiskadee population; it is known that parents are socially monogamous, clutch size ranges from two to five eggs, and both parents provision the nestlings and fledglings (Brush and Fitzpatrick, 2002). It is not known whether any differences exist between native and non-native populations in any aspect of breeding behaviour.

7.7 Habitat

In its native range, the Great Kiskadee is found in a variety of habitats, although it prefers more open areas and is at relatively low densities or absent in thick forests. It occupies all available habitats on Bermuda. It is regularly found near human habitations, including throughout cities, in both the native and introduced ranges (Brush and Fitzpatrick, 2002).

7.8 Impacts

The desired control of *Anolis* spp. lizards in Bermuda was not realized. While Great Kiskadees do consume some lizards, their generalism prevents them from having much impact on the lizard populations. *Anolis* lizards make up less than 10% of the items consumed (B. Mathys, unpublished data; Long, 1981), with Samuel (1975) finding no lizard remains in the stomach contents of 82 individuals collected during the summers of 1973 and 1974. Great Kiskadees have been implicated in negative effects on native and long-established Bermudian organisms, such as the Eastern Bluebird (*Sialia sialis*), Grey Catbird (*Dumetella carolinensis*), Northern Cardinal (*Cardinalis cardinalis*), White-eyed Vireo (*Vireo griseus*) and the endemic Bermuda Skink (*Plestiodon longirostris*) (Long, 1981; Davenport *et al.*, 2001), all through direct predation or harassment. The Great Kiskadee has also been suggested as one of the causes of the putative extinction of the Bermuda cicada (*Tibicen bermudiana*), which already had a reduced population due to the decline in the Bermuda cedar (Gehrman, 2012). Due to the Great Kiskadee's willingness to consume a diversity of food items, it is likely that it competes with native and non-native passerine birds for food and may also compete for nesting sites with the non-native Common Starling (*Sturnus vulgaris*) and perhaps other species. The Great Kiskadee has also been suggested as one cause of the 'seed rain' that has distributed seeds of invasive plants throughout Bermuda (Wolsak *et al.*, 2018).

7.9 Control

No dedicated control attempts have been implemented, although the idea was mentioned by Samuel (1975). David Wingate did attempt to control Great Kiskadee populations (through shooting) on Nonsuch Island in hopes of saving the Bermuda cicada, but his efforts were unsuccessful (Gehrman, 2012). The species' high density and presence in all habitats in Bermuda make extirpation unlikely. Bermuda's depauperate fauna contain few potential predators, although domestic cats (*Felis catus*) (Brush and Fitzpatrick, 2002) and introduced rats (*Rattus rattus* and *Rattus norvegicus*) may kill some nestlings and adults.

7.10 Uses

Aside from giving Bermuda a more 'tropical' feel, the Great Kiskadee has no practical uses.

7.11 Notes

The success of the Great Kiskadee on Bermuda can largely be traced to this species' opportunistic omnivory and Bermuda's coverage with ideal kiskadee habitat. Additionally, there are no resident confamilials on Bermuda (although some do migrate through the islands) and few other bird species generally, meaning that the Great Kiskadee probably experiences limited competition for food and other resources (Crowell, 1962). Mathys and Lockwood (2009) found that Great Kiskadees have evolved in the time since they were introduced from Trinidad, increasing in body size. The Great Kiskadee is one of several tropical tyrannid flycatchers that have very similar coloration, which has been suggested to be the result of a mimicry complex (Prum and Samuelson, 2016). None of the other similar species has been successfully introduced outside of their native range.

7.12 References

Brush, T. and Fitzpatrick, J. W. (2002) Great Kiskadee (*Pitangus sulphuratus*). In: Poole, A.F. and Gill, F.B. (eds) *Birds of North America*. Cornell Lab of Ornithology, Ithaca, New York. Available at: https://doi.org/10.2173/bna.622 (accessed 28 October 2019).

Crowell, K. (1962) Reduced interspecific competition among the birds of Bermuda. *Ecology* 43, 75–88.

Crowell, K., and Crowell, M. (1976) Bermuda's abundant, beleaguered birds. *Natural History* 85, 48–56.

Davenport, J., Hills, J., Glasspool, A. and Ward, J. (2001) Threats to the critically endangered endemic Bermudian skin *Eumeces longirostris*. *Oryx* 35, 332–339.

ffrench, R. (1991) *A Guide to the Birds of Trinidad and Tobago*, 2nd edn. Cornell University Press, New York.

Gehrman, E. (2012) *Rare Birds: The Extraordinary Tale of the Bermuda Petrel and the Man who Brought it Back from Extinction*. Beacon Press, Boston, Massachusetts.

Gorena, R.L. (1997) Notes on the feeding habits and prey of adult great kiskadees. *Bulletin of the Texas Ornithological Society* 30, 18–19.

Hilty, S. (2003) *Birds of Venezuela*, 2nd edn. Princeton University Press, Princeton, New Jersey.

Leveque, R. (1979) Nidification du Tyran bentevi *Pitangus sulphuratus* dans le Chubut (Argentine). *Alauda* 47, 116.

Long, J. (1981) *Introduced Birds of the World*. Universe Books, New York.

Mathys, B.A. and Lockwood, J.L. (2009) Rapid evolution of great kiskadees on Bermuda: an assessment of the ability of the island rule to predict the direction of contemporary evolution in exotic vertebrates. *Journal of Biogeography* 36, 2204–2211.

Prum, R.O. and Samuelson, L. (2016) Mimicry cycles, traps, and chains: the coevolution of toucan and kiskadee mimicry. *American Naturalist* 187, 753–764.

Samuel, D.E. (1975) The kiskadee in Bermuda. *Newsletter of the Bermuda Biological Station* 4, 2.

Sibley, C.G. and Monroe, B.L. (1990) *Distribution and Taxonomy of Birds of the World*. Yale University Press, New Haven, Connecticut.

Stiles, F.G. and Skutch, A.F. (1989) *A Guide to the Birds of Costa Rica*. Comstock Publishing Associates, Cornell, New York.

Wolsak, S., Wingate, D. and Cronk, Q. (2018) Environmental change in the terrestrial vegetation of Bermuda: revisiting Harshberger (1905). *Brittonia* 70, 257–275.

8 Red-billed Leiothrix (*Leiothrix lutea* Scopoli, 1786)

Lorinda A. Hart*
School of Life Sciences, University of KwaZulu-Natal, Pietermaritzburg 3209, South Africa; Department of Biological Sciences, University of Namibia, Windhoek 10023, Namibia

Citation: Hart, L.A. (2020) Red-billed Leiothrix (*Leiothrix lutea* Scopoli, 1786). In: Downs, C.T. and Hart, L.A. (eds) *Invasive Birds: Global Trends and Impacts*. CAB International, Wallingford, UK, pp. 63–68.

8.1 Common Names

Of the many common names for this species (Lepage, 2004), a few examples include: Japanese Hill Robin, Japanese Nightingale, Chinese Nightingale, Red-billed Mesia, Doubtful Leiothrix, Japanese Honeycreeper Hill-robin, Pekin Nightingale, Pekin Robin, Pekin Robin, Pekin Nightingale, Rimba Paruh Merah (Malay), Léiothrix Jaune (French), Leiótrix Piquirrojo (Spanish), Rotschnabel-Sonnenvogel (German), Usignolo del Giappone (Italian), Soushichou (Japanese), Japanse Nachtegaal (Dutch).

8.2 Nomenclature

The Red-billed Leiothrix (*Leiothrix lutea* Scopoli, 1786) is a member of the family Leiotrichidae, which includes laughing thrushes and their allies (Collar *et al.*, 2019). There are five subspecies of Red-billed Leiothrix:

- *L. l. kumaiensis* (Whistler, 1943) occurring in the north-western Himalayas from north-eastern Pakistan to northern India;
- *L. l. calipyga* (Hodgson, 1837) ranging from Nepal to Bhutan, occurring in north-eastern India, southern China and extreme north-western Myanmar;
- *L. l. yunnanensis* (Rothschild, 1921) found in north-eastern Myanmar and southern China;
- *L. l. lutea* (Scopoli, 1786) occurring in south-central and eastern China; and
- *L. l. kwangtungensis* (Stresemann, 1923) found in south and south-eastern China and northern Vietnam (Collar *et al.*, 2019).

There is only one other species in this genus, the Silver-eared Mesia (*Leiothrix argentauris*), which co-occurs within parts of their native range (Ali and Ripley, 1972; Male *et al.*, 1998).

8.3 Distribution

Red-billed Leiothrix are widespread throughout China, with populations spanning Myanmar to Pakistan largely restricted to the Himalayas. Their native range includes Bangladesh, Bhutan, China, India, Myanmar, Nepal, Pakistan (rare or vagrant) and Vietnam (Fig. 8.1).

The introduced range of the Red-billed Leiothrix includes France, Hawaii, Italy, Germany, Japan, Réunion, the USA, Portugal and Spain (Fig. 8.1). A free-living adult male was trapped in 2013 in Cuba and marked the first record for this species in the West Indies (Castaneda *et al.*, 2017). A sighting of one individual among a flock of Silver-eared Mesia was reported in Malaysia in 1988 (Sullivan *et al.*, 2009).

8.4 Description

The Red-billed Leiothrix is a small (14–15 cm) olive-grey babbler with a distinctly forked tail and weighs 18–28 g (Collar *et al.*, 2019). They generally form small flocks, especially during the non-breeding months, with up to 100 individuals present (Fisher and Baldwin, 1947; Male *et al.*, 1998). The crown is more olive, in contrast to an overall greyer body (Fig. 8.2). A broad, buff-yellow eye ring extends to a bright red bill (Male *et al.*, 1998). They have a yellow throat, darkening to an orange breast, with a dull yellow to grey belly and yellow vent (Male *et al.*, 1998; Collar *et al.*, 2019). Primary and secondary wing feathers are black with dark orange to yellow edges, and a red square is formed near the shoulder when the wing is closed (Male *et al.*, 1998; Collar *et al.*, 2019). Tails are predominantly grey-black with olive-grey uppertail coverts, sometimes bordered by a thin white line (Male *et al.*, 1998). This white line is less pronounced or absent in females (Karsten, 2002). Legs are horn-brown to pinkish in colour. Pagani-Núñez *et al.* (2013) found no biometric differences between the sexes. However, differences in the hue, chroma and brightness of the throat, crown and breast were more

*Corresponding author: lorinda.hart.research@gmail.com

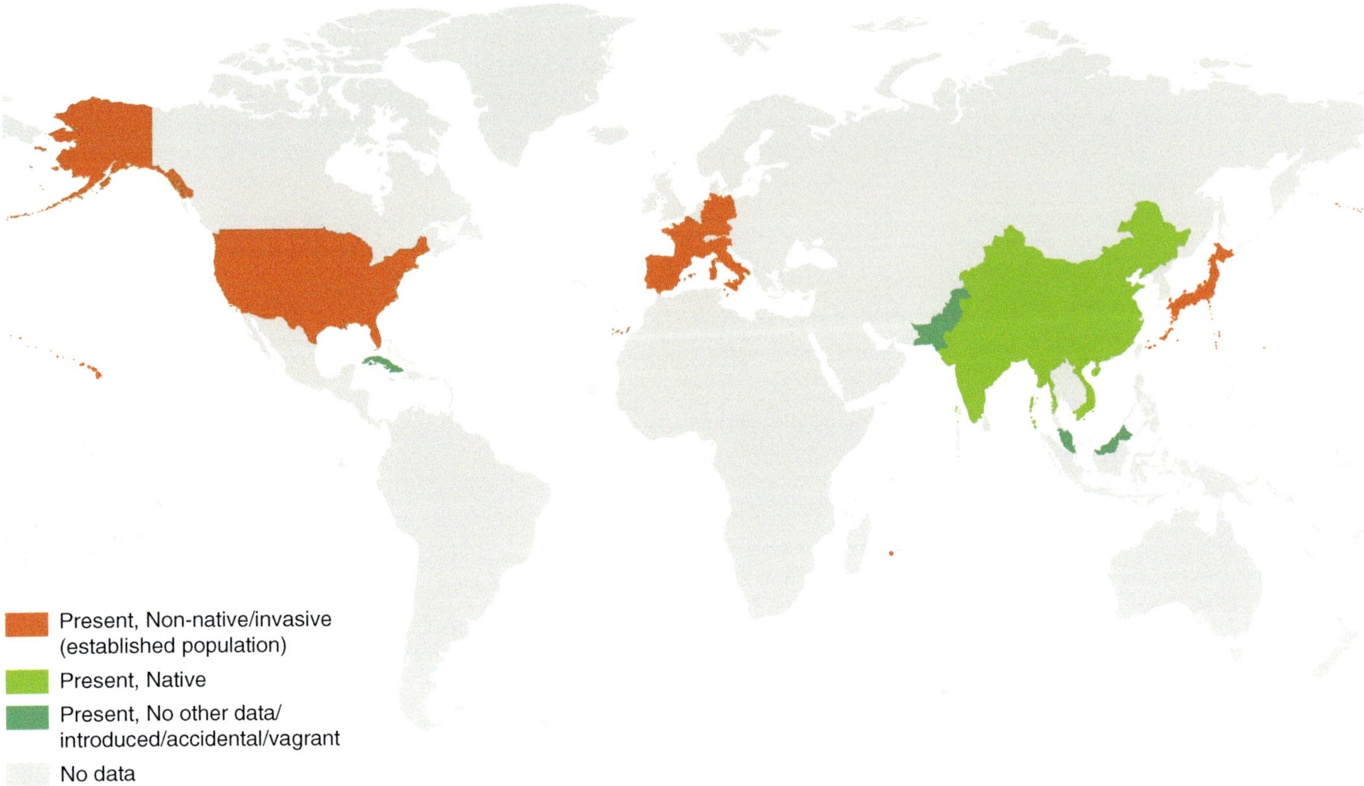

Fig. 8.1. Global distribution of the Red-billed Leiothrix (*Leiothrix lutea*) showing the natural (green) and invaded (red) ranges.

pronounced in males. Male chin to belly areas were more orange than in females (Pagani-Núñez *et al.*, 2013). Additionally, males have a more distinctive black stripe from the bill bordering the yellow throat (Sinclair, 2013). A trained ear can also sex them by their calls (Gibson, 1978). Juveniles are more dull and greyer in appearance, lacking bright chests, heads and bills compared with adults (Sinclair, 2013). Male *et al.* (1998) report that juveniles have black bills.

A fair amount of research has been conducted on the visual capabilities of Red-billed Leiothrix (e.g. Maier and Bowmaker, 1993; Maier, 1992, 1994). These birds can see in the ultraviolet range, and thus, although seemingly dull and cryptic in the undergrowth to the human eye, an entirely different image may be perceived by this species.

8.5 Diet

Red-billed Leiothrix feed on insects and their larvae, spiders, snails, fruits and seeds (Fisher and Baldwin, 1947; Gibson, 1978; Collar *et al.*, 2019). They also probe flowers, although this is rare (1%; Ralph and Noon, 1986). Although Red-billed Leiothrix feed on seeds, they are generally passed undigested (Gibson, 1978; Karsten, 2002). Birds ingest grit to facilitate seed digestion, as they lack crops and muscular stomachs (Gibson, 1978; Karsten, 2002). Red-billed Leiothrix use a 'jumping' foraging technique to feed on aerial insects just above bush canopies (Amano and Eguchi, 2002a), and glean insects and their larvae from branches and foliage (Eguchi and Masuda, 1994; Ralph *et al.*, 1998). Estimates of the percentage of fruit consumed by this species varies from 38% (Ralph and Noon, 1986) to 40–60% (Fisher and Baldwin, 1947) and probably reflects fruit availability and diet preferences at different spatial and temporal scales.

In captivity, they feed on a wide variety of foods including dry dog food pellets, scrambled egg, bread, a variety of fruits and insects, cooked chicken, plant tips and shoots, soft plants, cottage cheese and commercially produced bird feeds (Gibson, 1978; Karsten, 2002). They do not eat red meat, hard beetles or 'legless' organisms (e.g. maggots), and ignore very small insects such as aphids (Gibson, 1978). A great quantity and diversity of insects must be offered to bring birds into breeding condition (Karsten, 2002). In captivity, birds from a few weeks old were recorded to consume dry bird droppings, despite being offered a varied and rich diet (Gibson, 1978). Adults have also been observed dipping food in artificial nectar solutions or water on hot days prior to feeding chicks, presumably to prevent dehydration (Karsten, 2002). In captivity, adults and their three chicks can consume up to 300 insects daily, with parent birds generally consuming less non-insect foods during the breeding season (Karsten, 2002).

8.6 Introduction and Invasion Pathways

Red-billed Leiothrix populations in Japan (cage escapees from the early 1980s) and Spain (first recorded in 1992) continue to expand (Amano and Eguchi, 2002a; Eguchi and Amano, 2004;

Fig. 8.2. Adult Red-billed Leiothrix in Vietnam. (A) Front view. (B) Back view. (©Photographs: Sam Thuong.)

Herrando *et al.*, 2010). Some Spanish populations may have arisen from the southward migration of invasive populations in France (Sanz-azkue *et al.*, 2014). French and Italian populations were probably also escapees or deliberate releases that occurred in the 1990s (Kumschick and Nentwig, 2010). These birds were also historically released in Australia, Tahiti, Colombia, Taiwan and England, but no populations established there (Long, 1981; Dyer et al., 2017; Brazil, 2009). Red-billed Leiothrix arrived in Hawaii in 1911 (Fisher and Baldwin, 1947), with deliberate introductions started in c. 1928 and in subsequent years on various islands (Caum, 1933). Populations have fluctuated significantly on the different Hawaiian Islands, but reasons for these trends remain unclear (Male *et al.*, 1998). In Portugal, they were rarely seen exotics in 1997 (Costa *et al.*, 1997). Later, in 2007, they were first noted in central Portugal in an area rarely visited by birdwatchers, where they have since established and increased in numbers (Matias, 2010).

8.7 Breeding Behaviour

The Red-billed Leiothrix builds an open-cup nest in dense foliage and lays three to four eggs (Amano and Eguchi, 2002b; Eguchi and Amano, 2004). Both parents build the nest and incubate and care for the young (Gibson, 1978). Eggs hatch after 11–12 days, with chicks fed insects exclusively for one week; the chicks leave the nest after approximately 10–12 days (Fisher and Baldwin, 1947; Gibson, 1978). It is parasitized by the Common Cuckoo (*Cuculus canorus*) in its native Indian range (Yang *et al.*, 2014). Consequently, Red-billed Leiothrix lay eggs that range from white to blue with a variety of deep red spot patterns as an anti-parasitic measure (Yang *et al.*, 2014). Eggs are approximately 2.06 × 1.59 cm and weigh around 2.6 g (Fisher and Baldwin, 1947). In Hawaii, where populations have been free of brood parasites for over a century, egg rejection behaviour is still equally strong, but they display lower interclutch variation and intraclutch consistency in egg colour (Yang *et al.*, 2014). This suggests that egg colour is under natural selection (Yang *et al.*, 2014). One captive pair produced seven consecutive clutches with a total of 26 eggs, but breeding success was only 20–25% (Gibson, 1978). In the wild, second clutches are observed, but it is uncertain if these are replacement clutches or if the species is a multi-brooder (Male *et al.*, 1998).

Nests are 9.7 ± 1.0 cm (mean ± SD) wide and 9.3 ± 1.2 cm deep, with an inner diameter and depth of roughly 5 × 5 cm and consist of bamboo leaves, plant roots, moss, grass, occasional animal hair, sheep wool and plastic chords (Fisher and Baldwin, 1947; van Riper, 1977; Male *et al.*, 1998; Amano and Eguchi, 2002b). Feathers are not reported as nesting material, despite being available (Gibson, 1978). The deep cups may provide concealment, as only the tail and head protrude during incubation (Gibson, 1978).

In native China, they breed from April to October (Etchecopar and Hue, 1983, cited by Male *et al.*, 1998). In Japan, breeding occurs from April to September (Amano and Eguchi, 2002a), in central Portugal from April to August (Pereira *et al.*, 2017) and in Hawaii from May to August (Ralph *et al.*, 1998), but early nests are recorded during March (Fisher and Baldwin, 1947). In captivity, birds form strong pair bonds year-round and display intraspecific aggression towards other pairs (although this could dissipate in larger enclosures) (Gibson, 1978). In the wild, Red-billed Leiothrix home ranges overlap significantly (Male *et al.*, 1998).

8.8 Habitat

Exotic birds typically prefer or flourish in disturbed, degraded, human-altered landscapes (Case, 1996). However, this is not the case for the invasive Red-billed Leiothrix. These birds occupy habitats with dense understoreys and sufficient water, including various indigenous and exotic forests, open woodlands and shrublands (Fisher and Baldwin, 1947; Scott *et al.*, 1986; Herrando *et al.*, 2010). Water is important not only for drinking but also for frequent bathing in this species (Fisher and Baldwin, 1947; Gibson, 1978). They also occur in secondary forests near human settlements and, in their native range, they typically occur between 900 and 3000 m above sea level (a.s.l.) (de Schauensee, 1984, cited by Amano and Eguchi, 2002a; Long, 1987). In Japan, they occur in deciduous broad-leaved forests with dense undergrowth above 1000 m a.s.l. (Eguchi and Masuda, 1994).

Although this species is a forest generalist, it is also observed in tea plantations (Chettri *et al.*, 2018) and has been observed in Hawaiian suburbs where they feed on figs (*Ficus* spp.) (Male *et al.*, 1998). In Hawaii, Red-billed Leiothrix were 92% more abundant in discontinuous forest that had dense understoreys with abundantly fruiting bushes as opposed to continuous forest with bare understoreys (Ralph *et al.*, 1998). Red-billed Leiothrix avoid open pastures, degraded forests and windswept, exposed ranges (Fisher and Baldwin, 1947; Scott *et al.*, 1986). These understoreys are also utilized for nesting, with upper tree canopies barely used by this species (Male *et al.*, 1998). In captivity, dense foliage and water must be available to encourage breeding (Gibson, 1978). This propensity for dense foliage makes these birds difficult to see, despite their distinctive appearance and appearing in small flocks (Gibson, 1978; Sinclair, 2013). They rarely fly more than 15 m, unless they are in open habitats (Fisher and Baldwin, 1947). Gibson (1978) noted that when birds flew across open spaces, their flight was weak, but the agility and speed with which they navigated dense foliage was impressive.

Originally from subtropical climates, they now also occur in areas with Mediterranean climates (Herrando *et al.*, 2010). They generally exhibit some altitudinal migration, moving to lower altitudes during colder months or during breeding season (Fisher and Baldwin, 1947; Male *et al.*, 1998; Amano and Eguchi, 2002a). Their altitudinal preference varies on the Hawaiian Islands, again probably due to the vegetation density and fruit availability associated with these (Fisher and Baldwin, 1947). Although they occur from sea level, abundance is greater at 1000–1200 m a.s.l., with their upper limit approximately 2400–2700 m a.s.l. (Fisher and Baldwin, 1947; Scott *et al.*, 1986). Mortalities increase from around 2700 m a.s.l.; however, birds are still observed at 4200 m a.s.l., despite significant die-offs at this altitude (Fisher and Baldwin, 1947; Montgomery and Howarth, 1980). Altitudinal preference could be linked to thermoregulation, as it has been suggested that Red-billed Leiothrix avoid extremely hot and cold climates, although a link has also been found with water and food availability (Fisher and Baldwin, 1947; Scott *et al.*, 1986). Huddling has been observed in cold climates in captivity, with any accommodating species (Gibson, 1978). Tolerance of below-freezing temperatures is also reported in captive birds, provided they have wind protection and access to unfrozen food and water, without which death occurs within 24 h (Karsten, 2002).

8.9 Impacts

A global analysis identified Red-billed Leiothrix as one of the species with the most detrimental influence on local communities, with disease, competition and interactions with other non-native species identified as negative impacts (Martin-Albarracin *et al.*, 2015).

8.9.1 Diseases and parasites

Red-billed Leiothrix have been infected with the highly pathogenic avian influenza H5N1 subtype in association with captivity (Rappole and Hubálek, 2006) and have been implicated in the spread of disease to humans and/or animals where they have been introduced (Chan, 2006). The intestinal protozoan *Coccidia* sp. has also been observed in captive birds, although these infections are harmless to healthy, well-kept individuals (Gibson, 1978). The intestinal bacteria *Yersinia enterocolitica* has also been detected in one captive bird (Hacking and Sileo, 1974). Red-billed Leiothrix carry avian malaria and avian pox in Hawaii (Fisher and Baldwin, 1947; van Riper and van Riper, 1985; Male *et al.*, 1998). They seem unaffected by diseases that have had devastating effects on indigenous island species (Ralph *et al.*, 1998). Indeed, Gibson (1978) commented that Red-billed Leiothrix were among 'the healthiest and disease-free birds' in his collection.

Red-billed Leiothrix also carry both external and internal parasites such as *Isospora* spp. (McQuistion and Buice, 1996), the nematode *Procyrnea mansion* (Ehrsam *et al.*, 1985), spinning mites *Neocheyletiella media* (Kniest and Hoffman, 1983) and a variety of other mites including *Cytodites* spp., *Ornithocheyletia* spp. and *Ornithocheyla* spp. among others (Goff, 1980, 1983, 1987).

8.9.2 Competition

In Japan, the invasion of Red-billed Leiothrix does not appear to have affected indigenous bird communities (Eguchi and Masuda, 1994). While the Japanese Bush Warbler (*Cettia diphone*) and various tits (*Parus* spp.) overlap in feeding range with the Red-billed Leiothrix, there is no evidence of competition, as the latter utilizes a different feeding microhabitat by foraging on lower canopy aerial insects (Amano and Eguchi, 2002a). Additionally, Japanese Bush Warblers and Red-billed Leiothrix also differ in nest microhabitat, despite both species nesting exclusively in dwarf bamboo (Sasamorpha borealis) thickets (Amano and Eguchi, 2002b). Red-billed Leiothrix nest in the upper leaves at approximately 2 m, while warblers nest lower down (Amano and Eguchi, 2002b). In one study, there were two to three times more Red-billed Leiothrix nests than warbler nests, and the possibility of high nest densities increasing predation risk (i.e. indirect impacts) on cohabiting species should not be discounted (Amano and Eguchi, 2002b).

In Spain, there also appears to be no negative impact or alteration in native bird assemblages following leiothrix invasion (Vall-llosera *et al.*, 2016). However, Red-billed Leiothrix do have a high isotopic niche overlap with the European Robin (*Erithacus rubecula*) and to a lesser degree with Blackcaps (*Sylvia atricapilla*) in Spain, suggesting that there is potential for resource competition (Pagani-Núñez *et al.*, 2018). In an experimental study, Red-billed Leiothrix were initiators of aggressive interactions, and while apparently more dominant through victorious outcomes, they were not more aggressive than European native species (Pereira *et al.*, 2018). In Portugal, their competitive superiority is attributed to their more efficient foraging morphology when compared with indigenous European Robins and Blackcaps, rather than occupying a vacant niche (Pereira *et al.*, 2017). In Italy, they are a vocally dominant species, calling loudly year-round, and it is predicted that these alterations in the soundscape will result in changes of both invasive and indigenous species' calling repertoires (Farina *et al.*, 2013).

In Hawaii, the invasive White-rumped Shama (*Copsychus malabaricus*) has seemingly had a negative effect on Red-billed Leiothrix populations, as the latter has declined significantly on islands where the former has been introduced (Ralph, 1991). In captivity, birds display high levels of intraspecific aggression, particularly when breeding, but tolerate other species at feeding bowls (Gibson, 1978).

8.9.3 Seed Dispersal

Red-billed Leiothrix are known to disperse a wide variety of plants (Starr *et al.*, 2003; Male *et al.*, 1998). However, their small gape limits the plant species on which they can feed (Pejchar, 2015). On Hawaii, one study found that while they dispersed both invasive and native plant seeds, they predominantly fed on two indigenous small-seeded species, *Rubus hawaiensis* and *Vaccinium calycinum*, which constituted 92% of their total dispersed seeds (Pejchar, 2015). Thus, their role as dispersers may be limited (Pejchar, 2015).

8.10 Control

No formal control measures are reported for this species, but predation on birds and eggs in their invasive range is reported by rats (*Rattus* spp.), feral cats (*Felis catus*) and birds of prey (Male *et al.*, 1998).

8.11 Uses

Red-billed Leiothrix have been popular cage birds for at least the last 100 years, due to their varied call repertoire, hardy and active nature, attractive appearance and intelligence (Male *et al.*, 1998; Karsten, 2002; Collar *et al.*, 2019). They can learn to incorporate new notes into their calls from other birds or even electronic recordings, and live up to 20 years in captivity (Karsten, 2002). Due to habitat destruction and its popularity in the cage-bird trade, the population is believed to be in decline within its natural range (Collar *et al.*, 2019). In 1997, the Red-billed Leiothrix was listed on CITES (Convention for International Trade in Endangered Species) Appendix II to restrict its trade (Butchart *et al.*, 2019). However, it remains a popularly traded species, especially in China (Li and Jiang, 2014). This could be because wild birds have more diverse and varied calls (Karsten, 2002), which are sought after in bird-calling competitions (Jepson, 2008).

This species is also released during religious ceremonies (Chan, 2006). The continued release and trade of this species may lead to new invasions in the future.

8.12 Notes

Anting has been observed in this species in captivity, whereby birds will rub ants vigorously over their feathers before consuming them (Gibson, 1978).

8.13 References

Ali, S. and Ripley, S.D. (1972) *Handbook of the Birds of India and Pakistan*, Vol. 5. Oxford University Press, Mumbai, India.

Amano, H.E. and Eguchi, K. (2002a) Foraging niches of introduced Red-billed Leiothrix and native species in Japan. *Ornithological Science* 1, 123–131.

Amano, H.E. and Eguchi, K. (2002b) Nest-site selection of the Red-billed Leiothrix and Japanese Bush Warbler in Japan. *Ornithological Science* 1, 101–110.

Brazil, M. (2009) *Birds of East Asia: China, Taiwan, Korea, Japan and Russia*. Christopher Helm, London.

Butchart, S., Stattersfield, A. and Ekstrom, J. (2019) BirdLife International Species Factsheet: *Leiothrix lutea*. Available at: http://datazone.birdlife.org/species/factsheet/22716522 (accessed 23 October 2019).

Case, T.J. (1996) Global patterns in the establishment and distribution of exotic birds. *Biological Conservation* 78, 69–96.

Castaneda, Y.R., Wiley, J.W. and Garrido, O.H. (2017) Additional records of Lazuli Bunting (*Passerina amoena*) and first records of several wild-caught exotic birds for Cuba. *Journal of Caribbean Ornithology* 30, 134–142.

Caum, E.L. (1933) The exotic birds of Hawaii. *Bishop Museum Occasional Papers* 10, 1–55.

Chan, S.W. (2006) Religious release of birds in Hong Kong. MPhil, University of Hong Kong, Pokfulam, Hong Kong.

Chettri, A., Sharma, K., Dewan, S. and Acharya, B.K. (2018) Bird diversity of tea plantations in Darjeeling Hills, Eastern Himalaya, India. *Biodiversitas* 19, 1066–1073.

Collar, N., Robson, C. and de Juana, E. (2019) Red-billed Leiothrix (*Leiothrix lutea*). In: del Hoyo, J., Elliott, A., Sargatal, J., Christie, D.A. and de Juana, E. (eds) *Handbook of the Birds of the World Alive*. Lynx Edicions, Barcelona, Spain. Available at: www.hbw.com/node/59657 (accessed 23 October 2019).

Costa, H., Elias, G.L. and Farinha, J.C. (1997) Exotic birds in Portugal. *British Birds* 90, 562–567.

de Schauensee, R.M. (1984) *The Birds of China*. Oxford University Press, Oxford.

Dyer, E.E., Redding, D.W. and Blackburn, T.M. (2017) The global avian invasions atlas, a database of alien bird distributions worldwide. *Scientific Data* 4, 170041.

Eguchi, K. and Amano, H.E. (2004) Invasive birds in Japan. *Global Environmental Research* 8, 29–39.

Eguchi, K. and Masuda, T. (1994) A report on the habitats of Peking Robin *Leiothrix lutea* in Kyushu. *Japanese Journal of Ornithology* 43, 91–100.

Ehrsam, H., Spillmann, S. and Wolff, K. (1985) Nematodes (*Procyrnea mansioni*, Spirurida) as the cause of changes in the stomach of the Chinese Nightingale (*Leiothrix lutea*) and Rothschild's Mynah bird (*Leucopsar rothschildi*). *Schweizer Archiv für Tierheilkunde* 127, 665–670.

Etchecopar, R.D. and Hue, F. (1983) *Les Oiseaux de Chine de Mongolie et de Core*. Societe Novelle des Editions Boubee, Paris.

Farina, A., Pieretti, N. and Morganti, N. (2013) Acoustic patterns of an invasive species: the Red-billed Leiothrix (*Leiothrix lutea* Scopoli 1786) in a Mediterranean shrubland. *Bioacoustics* 22, 175–194.

Fisher, H.I. and Baldwin, P.H. (1947) Notes on the Red-billed Leiothrix in Hawaii. *Pacific Science* 1, 45–51.

Gibson, L. (1978) The Red-billed Leiothrix: a four year study. *Aviculture Magazine* 84, 4–17.

Goff, M.L. (1980) Mites (Chelicerata: Acari) parasitic on birds in Hawaii Volcanoes National Park. Technical Report, No. 29. Cooperative National Park Resources Studies Unit and University of Hawaii, Hawaii.

Goff, M.L. (1983) Notes and exhibitions. *Proceedings of the Hawaiian Entomological Society* 24, 158.

Goff, M.L. (1987) A catalog of Acari of the Hawaiian Islands. Research Extension Series 075. Hawaii Institute of Tropical Agriculture and Human Resources, Honolulu, Hawaii.

Hacking, M. and Sileo, L. (1974) *Yersinia enterocolitica* and *Yersinia pseudotuberculosis* from wildlife in Ontario. *Journal of Wildlife Diseases* 10, 452–457.

Herrando, S., Llimona, F., Brotons, L. and Quesada, J. (2010) A new exotic bird in Europe: recent spread and potential range of Red-billed Leiothrix *Leiothrix lutea* in Catalonia (northeast Iberian Peninsula). *Bird Study* 57, 226–235.

Jepson, P. (2008) Orange-headed Thrush *Zoothera citrina* and the avian X-factor. *Birding Asia* 9, 58–61.

Karsten, P. (2002) Pekin Robins. Information on their care and breeding. *AFA Watchbird* 29, 56–68.

Kniest, F.M. and Hoffman, J.R. (1983) Spinning mites (*Neocheyletiella media* Fain, 1972) (Acari: Cheyletiellidae) on feathers of the domestic Japanese Nightingale (*Leiothrix lutea* (L.)). *International Journal of Acarology* 9, 63–65.

Kumschick, S. and Nentwig, W. (2010) Some alien birds have as severe an impact as the most effectual alien mammals in Europe. *Biological Conservation* 143, 2757–2762.

Lepage, D. (2004) *Avibase: The World Bird Database, Bird Studies Canada*. Available at: https://avibase.bsc-eoc.org (accessed 23 October 2019).

Li, L. and Jiang, Z. (2014) International trade of CITES listed bird species in China. *PLoS One*, 9, e85012.

Long, J.L. (1981) *Introduced Birds of the World: The Worldwide History, Distribution and Influence of Birds Introduced to New Environments*. Universe Books, New York.

Long, Z. (1987) Leiothrix. In: Chen, T., Long, Z. and Zheng, B. (eds) *Fauna Sinica*. Science Press, Beijing, pp. 154–162.

Maier, E. (1992) Spectral sensitivities including the ultraviolet of the passeriform bird *Leiothrix lutea*. *Journal of Comparative Physiology A* 170, 709–714.

Maier, E. (1994) Ultraviolet vision in a passeriform bird: from receptor spectral sensitivity to overall spectral sensitivity in *Leiothrix lutea*. *Vision Research* 34, 1415–1418.

Maier, E. and Bowmaker, J. (1993) Colour vision in the passeriform bird, *Leiothrix lutea*: correlation of visual pigment absorbance and oil droplet transmission with spectral sensitivity. *Journal of Comparative Physiology A* 172, 295–301.

Male, T.D., Fancy, S.G. and Ralph, C.J. (1998) Red-billed Leiothrix. In: Poole, A.F. and Gill, F.B. (eds) *The Birds of North America*, No. 359. The Birds of North America, Philadelphia, Pennsylvania, pp. 1–12.

Martin-Albarracin, V.L., Amico, G.C., Simberloff, D. and Nuñez, M.A. (2015) Impact of non-native birds on native ecosystems: a global analysis. *PLoS One* 10, e0143070.

Matias, R. (2010) Aves exóticas em Portugal: Anos de 2005–2008. *Anuário Ornitológico* 7, 95–108.

McQuistion, T. and Buice, R. (1996) A new species of *Isospora* (Apicomplexa) from captive Pekin Robins, *Leiothrix lutea* (Passeriformes: Sylviidae), from the Dallas Zoo. *Acta Protozoologica* 35, 73–75.

Montgomery, S.L. and Howarth, F.G. (1980) Records of mummified Leiothrix from the summits of Mauna Loa and Mauna Kea. *Elepaio* 41, 30–31.

Pagani-Núñez, E., Senar, J.C. and Tella, J.L. (2013) Red-billed Leiothrix sexing is a matter of colour. *Bird Study* 60, 285–288.

Pagani-Núñez, E., Renom, M., Furquet, C., Rodríguez, J., Llimona, F. and Senar, J.C. (2018) Isotopic niche overlap between the invasive leiothrix and potential native competitors. *Animal Biodiversity and Conservation* 41, 427–434.

Pejchar, L. (2015) Introduced birds incompletely replace seed dispersal by a native frugivore. *AoB Plants* 7, plv072.

Pereira, P.F., Godinho, C., Vila-Viçosa, M.J., Mota, P.G. and Lourenço, R. (2017) Competitive advantages of the Red-billed Leiothrix (*Leiothrix lutea*) invading a passerine community in Europe. *Biological Invasions* 19, 1421–1430.

Pereira, P.F., Lourenço, R. and Mota, P.G. (2018) Behavioural dominance of the invasive Red-billed Leiothrix (*Leiothrix lutea*) over European native passerine-birds in a feeding context. *Behaviour* 155, 55–67.

Ralph, C.J. (1991) Population dynamics of land bird populations on Oahu, Hawaii: fifty years of introductions and competition. *Acta XX Congressus Internationalis Ornithologici* 3, 1444–1457.

Ralph, C.J. and Noon, B.R. (1986) Foraging interactions of small Hawaiian forest birds. *Acta XIX Congressus Internationalis Ornithologici* 2, 1992–2006.

Ralph, C.J., Fancy, S.G. and Male, T.D. (1998) Demography of an introduced Red-billed Leiothrix population in Hawaii. *Condor* 100, 468–473.

Rappole, J.H. and Hubálek, Z. (2006) Birds and influenza H5N1 virus movement to and within North America. *Emerging Infectious Diseases* 12, 1486–1492.

Sanz-azkue, I., Michelena, P.L., Esparcia, J.F. and Belzunce, J.A. (2014) Observación de Leiotrix Piquirrojo *Leiothrix lutea* (Scopoli, 1786) en Irún. Nueva especie de ave exótica en Gipuzkoa [Red-billed Leoithrix *Leiothrix lutea* (Scopoli, 1786) at Irún. A new exotic species in Gipuzkoa]. *Munibe (Ciencias Naturales-Natur Zientziak)* 62, 145–151.

Scott, J.M., Mountainspring, S., Ramsey, F.L. and Kepler, C.B. (1986) Introduced species accounts. In: Raitt, R.J. and Thompson, J.P. (eds) *Forest Bird Communities of the Hawaiian Islands: Their Dynamics, Ecology, and Conservation*. Studies in Avian Biology, No. 9. Cooper Ornithological Society, Los Angeles, California, pp. 183–330.

Sinclair, I. (2013) *Birds of the Indian Ocean Islands*. Penguin Random House, South Africa.

Starr, F., Starr, K. and Loope, L. (2003) *Morella faya*. United States Geological Survey – Biological Resources Division, Haleakala Field Station, Maui, Hawaii.

Sullivan, B.L., Wood, C.L., Iliff, M.J., Bonney, R.E., Fink, D. and Kelling, S. (2009) eBird: a citizen-based bird observation network in the biological sciences. *Biological Conservation* 142, 2282–2292.

Vall-llosera, M., Llimona, F., de Cáceres, M., Sales, S. and Sol, D. (2016) Competition, niche opportunities and the successful invasion of natural habitats. *Biological Invasions* 18, 3535–3546.

van Riper, C. (1977) The use of sheep wool in nest construction by Hawaiian birds. *Auk* 94, 646–651.

van Riper, S. and van Riper C. III, (1985) A summary of known parasites and diseases recorded from the avifauna of the Hawaiian Islands. In: Stone, C.P. and Scott, J.M. (eds) *Hawaii's Terrestrial Ecosystems: Preservation and Management*. University of Hawaii Press, Honolulu, Hawaii, pp. 298–371.

Yang, C., Liu, Y., Zeng, L. and Liang, W. (2014) Egg color variation, but not egg rejection behavior, changes in a cuckoo host breeding in the absence of brood parasitism. *Ecology and Evolution* 4, 2239–2246.

9 Ring-necked Parakeet (*Psittacula krameri* Scopoli, 1769)

Diederik Strubbe[1*] and Erik Matthysen[2]
[1]Terrestrial Ecology Unit (TEREC), Ghent University, K.L. Ledeganckstraat 35, 9000 Gent, Belgium; [2]Evolutionary Ecology Group (EVECO), University of Antwerp, Campus Drie Eiken, Universiteitsplein 1, Wilrijk 2610, Belgium

Citation: Strubbe, D. and Matthysen, E. (2020) Ring-necked Parakeet (*Psittacula krameri* Scopoli, 1769). In: Downs, C.T. and Hart, L.A. (eds) *Invasive Birds: Global Trends and Impacts*. CAB International, Wallingford, UK, pp. 69–75.

9.1 Common Names

Ring-necked Parakeet, Rose-ringed Parakeet.

9.2 Distribution

Ring-necked Parakeets (*Psittacula krameri* Scopoli, 1769) have an extensive natural distribution, spanning two disjunct ranges across parts of Africa and Asia. In Africa, their distribution largely overlaps with the Sahel zone, stretching from Senegal to parts of Ethiopia. Across Asia, the species occurs from western Pakistan, across of all India, Bangladesh and Sri Lanka, up to parts of Nepal and central Myanmar.

The species has established breeding populations in numerous countries outside of its natural range. In Europe, Ring-necked Parakeets have successfully invaded parts of Belgium, the Netherlands, France, Germany, Greece, Italy, Portugal, Spain, Turkey and the UK. Breeding populations have been present in Austria and Switzerland as well, but these have become extinct (Pârâu *et al.*, 2016). Based on demographic data spanning about 90 different populations, Pârâu *et al.* (2016) estimated that at least 85,000 Ring-necked Parakeets are present in Europe, although this likely is an underestimate. Across the Mediterranean, Ring-necked Parakeets are present in Lebanon, Jordan, Israel, the Palestinian territories, Egypt, Algeria and Tunisia (Butler, 2003). In the Middle East, the species can be found in Saudi Arabia, Oman, United Arab Emirates, Dubai, Qatar, Bahrein and Iran. Further east, Ring-necked Parakeets have successfully invaded parts of Japan, and can be found in Macau, Hong Kong and Singapore (Lever, 2005). Recently, emerging populations have also been reported from Thailand and Indonesia (Iqbal, 2017). In the USA, growing populations have been reported mainly from Florida, California and Hawaii (Butler, 2005). Breeding birds have also been observed in Caracas, Venezuela, but the status of this population is currently unclear (Nebot, 1999). Finally, south of its native African distribution range, Ring-necked Parakeets are rapidly expanding in South Africa, parts of Namibia and the island of Mauritius (Jones, 1980; Hart and Downs, 2014; Symes, 2014).

9.3 Description

Ring-necked Parakeets are slim, grass-green parakeets characterized by a long, pointed tail, measuring about 40 cm in total length (including tail). Males, in their third year, acquire a rose-pink and black neck collar, suffused variably with blue nape feathers (Fig. 9.2). Females lack the collar but sport an indistinct emerald-green ring around the neck. The central tail feathers are tipped bluish, with green-yellowish lateral tail feathers. Currently, there are four recognized subspecies, two in Africa and two in Asia. The Asian subspecies are slightly larger than the African Ring-necked Parakeets. They can best be separated from each other based on beak and neck collar characteristics. The subspecies *P. k. borealis* (found in northern parts of the Asiatic range) is characterized by brightly red upper mandibles and partly red lower mandibles, whereas *P. k. manillensis* (found in southern parts of Asiatic range) has largely black lower mandibles. Both Asian subspecies have rather broad, rose-pink hindneck collars. African parakeets have smaller, and sometimes nearly invisible, neck collars but sport a broader black stripe to their chins. The subspecies *P. k. krameri*, which can be

*Corresponding author: diederik.strubbe@ugent.be

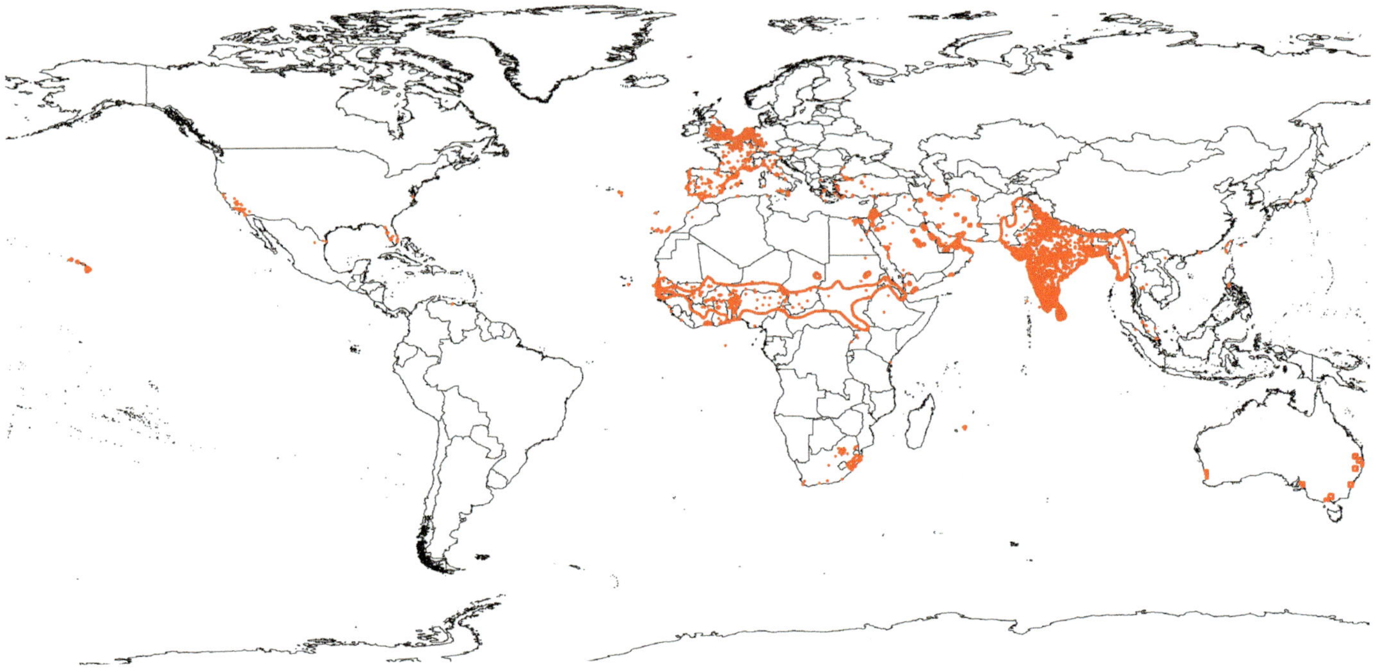

Fig. 9.1. Ring-necked Parakeet occurrence data (red) obtained from the Global Biodiversity Information Facility (www.GBIF.org). The native range outlines are taken from BirdLife International (www.birdlife.org). Selected invasive range occurrences probably reflect self-sustaining populations.

Fig. 9.2. Male Ring-necked Parakeet in Brussels. (©Photograph: Frank Adriaensen.)

found from Senegal and Guinea up to southern Sudan, has a very dark red to black upper, and a black lower mandible. The subspecies *P. k. parvirostris*, which ranges from parts of Sudan east through northern Ethiopia and north-west Somalia, is similar but has a smaller beak and a somewhat greener, less yellowish head and cheeks (Ali and Ripley, 1969; Forshaw, 1978, 2010; Fry and Keith, 1988). Butler and Gosler (2004) found that for Ring-necked Parakeets in the UK, female birds were characterized by a higher number of yellow underwing coverts, while juveniles could reliably be identified on the basis of their more pointed primary feathers.

9.4 Diet

Ring-necked Parakeets have a broad and flexible diet, consisting mainly of fruits, cereal, grain and seeds of all kinds, wild as well as cultivated. In both their native and invasive ranges, they have been observed feeding on a very large number of plants – almost any fruit or seed may be consumed including unripe fruits. Studies examining gut contents found that these parakeets indeed forage exclusively on vegetable matter (Shiels *et al.*, 2018). For example, in India, Ring-necked Parakeet diets included mustard, wheat, maize, rice and wild fruits, while in Africa, a comparable mix of fruit and seed was found, including wild figs and millet but also flowers (Ali and Ripley, 1969; Soni, 1991; Pithon, 1996). Across Europe, Ring-necked Parakeets readily visit backyard bird feeders, and it has been suggested that anthropogenic food provisioning is among the factors contributing to their invasion success (Clergeau and Vergnes, 2011).

9.5 Introduction and Invasion Pathways

Ring-necked Parakeets were probably first brought to Europe by the army of Alexander the Great (356–323 BC) from the Punjab region. By Roman times, parakeets had become well known in Europe, and pet birds were mainly kept as a status symbol. In medieval times, parakeets were attributed religious powers (Verdi, 2007). For example, the painting 'Virgin and Child with Canon Joris van der Paele' by the eminent Flemish painter Jan Van Eyck (*c.*1390–1441) pictures the Virgin and

Child holding a Ring-necked Parakeet (Fig. 9.3). This is probably symbolic for the dogma of the virgin birth of Jesus, as parakeets were believed to naturally greet people with the word 'Ave', and the Dominican Franciscus de Retza, in 1425, famously wondered, 'If a parrot has the power from nature to say Ave, why might not a pure Virgin conceive through the word Ave?' (Hanley, 2007; Verdi, 2007).

The first records of feral Ring-necked Parakeet populations in Europe, however, stem from a much later date. In 1855, breeding pairs were present in Northrepps (UK) and in the 1930s, flocks of Ring-necked Parakeets were observed in Epping Forest and Lilford, UK, but these did not persist (Lever, 2005). For most of the period 1930–1966, importing birds such as Ring-necked Parakeets was forbidden in the UK, but soon after this ban was lifted, the number of feral Ring-necked Parakeets records began to increase, and in 1969, parakeets established the first UK breeding colony in Kent (Butler, 2003; Lever, 2005). From the 1970s onwards, large numbers of Ring-necked Parakeets were caught across their native range and transported for the pet industry, mainly to Europe. Between 1984 (the earliest trade record available from the CITES trade database) and 2005 (before the European Union ban on the trade of birds was implemented), about 110,000 Ring-necked Parakeets were legally imported into Europe from the Asian native range, and about 37,000 from the African native range. Imports from Africa into Europe were mostly received from Senegal, whereas imports from Asia were received from a wider geographic source including India, Pakistan, Sri Lanka and Bangladesh. The source composition of imported Ring-necked Parakeets also varied considerably among countries; for example, all imports into France were from Senegal while all imports to Spain and Italy were from Pakistan. Greater London, Germany and France received larger numbers of Ring-necked Parakeets from Africa than from Asia (Strubbe *et al.*, 2015; Cardador *et al.*, 2016; Reino *et al.*, 2017). Interestingly, a study based on mitochondrial DNA collected from parakeet feathers found that European Ring-necked Parakeet populations comprised mostly individuals of Asian descent, suggesting a higher invasion success for these subspecies compared with the African counterparts (Jackson *et al.*, 2015).

Fig. 9.3. Detail of 'Virgin and Child with Canon Joris van der Paele' by Jan Van Eyck (*c.*1390–1441, Groeningemuseum, Bruges, Belgium). The Virgin and the Child are holding a Ring-necked Parakeet. In medieval times, parakeets were believed to hold religious powers.

Ring-necked Parakeet populations thus stem entirely from pet birds that escaped or were released from captivity. While such novel introductions continue to give rise to new populations, range expansion from already established populations has recently become more important. While Butler (2003) estimated a relatively slow rate of expansion of about 2.3 km per year from the London stronghold, Ring-necked Parakeets have now expanded into most of England, reaching cities such as Manchester and Liverpool (Balmer *et al.*, 2013). Similarly, independently established populations in the Dutch cities of Rotterdam, The Hague and Amsterdam have now largely merged, exchanging individuals between roost sites (Klaassen and van Kleunen, 2012). The same is true for Belgium, where the Brussels population has spread towards other cities and has merged with the Ring-necked Parakeet population present in the city of Lille, northern France. Recent data also indicate ongoing range expansion across the Mediterranean, especially in Spain and Italy where Ring-necked Parakeets increasingly occupy areas along the seaboard as well as inland (Pârâu *et al.*, 2016).

Available data from outside Europe indicate similar patterns; whereas populations were initially founded via escaped or released pet birds, established populations start spreading into suitable habitats surrounding their site of release. Although natal dispersal data are largely lacking, it is clear that Ring-necked Parakeets are strong fliers, covering distances of up to 20 km during their daily commutes between breeding, foraging and roosting sites (Keijl, 2001; Kahl-Dunkel and Werner, 2002). Ringing capture–recapture data obtained from the EURING database show a maximum distance of about 27 km, although median distances are only about 6 km.

9.6 Breeding Behaviour

Ring-necked Parakeets are cavity-nesters and usually breed in tree holes, either naturally formed or excavated by primary cavity-nesters such as woodpeckers. They also easily accept crevices in rocks or buildings, however, when tree cavities are in short supply. Pairs often nest together in close vicinity, forming a loose breeding colony. Across Asia, Ring-necked Parakeets mainly breed from January to April, although breeding may go on until July, with some latitudinal variation in timing of breeding across the subcontinent (Ali and Ripley, 1969). In Africa, the breeding season ranges from August to November (Fry and Keith, 1988). Ring-necked Parakeets normally only breed once a year, although replacement clutches are possible after (early) nest failure. Pairs start forming at the end of the winter, involving extensive courtship preening, whereby the male stands as far back from the female as possible, just making contact with the beak. Preening leads to bouts of courtship feeding, whereby the male offers food while 'saluting' by frequently raising one foot as high as possible, ultimately resulting in copulation. Females occupy nesting cavities long before the first egg is laid and defend their cavities against competitors, especially other female Ring-necked Parakeets. The clutch size ranges from two to six eggs, but mostly three or four are laid with 1–2 days between eggs. Only the female incubates and incubation lasts 22–24 days. Chicks hatch asynchronously, as the female starts incubation as soon as the first egg has been laid (Ali and Ripley, 1969; Forshaw, 1978). Chicks fledge at an age of 6–7 weeks, and after leaving the nest, the chicks are still being fed by their parents, for at least 2 weeks (Braun and Wink, 2013).

Data on Ring-necked Parakeet breeding success, covering 19 different localities across the native Asian range and two invasive populations (the UK and Israel), show that they typically lay about four eggs (Shwartz *et al.*, 2009). Clutch sizes do not differ between native and invasive areas, but in the UK, about half of the eggs fails to hatch. UK Ring-necked Parakeets thus only have about half the number of hatchlings in the nest (about two) compared with the four observed across Asia and in Israel. This lower breeding success can be attributed to climate mismatch, whereby cold European temperatures in late winter to early spring lead to egg infertility. Hatchling survival in the UK was high as almost all hatchlings managed to fledge their nest. Across Asia, in contrast, nest predation by crows (*Corvus* spp.), monitor lizards and snakes took a heavy toll on nestlings, so that an average nest fledged only about 2.5 young. In Israel, a similarly high nestling loss was observed, although the underlying cause was not immediately clear (Shwartz *et al.*, 2009). A separate study in Heidelberg, Germany, also found an average clutch size of four eggs, with 61% of eggs hatching and 79% of all chicks surviving to fledging (Braun and Wink, 2013).

Ring-necked Parakeets in Europe start breeding early, and the first eggs are laid at the end of February (Butler, 2003; Strubbe and Matthysen, 2009). Luna *et al.* (2017) tested whether breeding phenology could affect parakeet invasion success in areas colder than their native range, and found that, compared with breeding dates from the native Asian range, European Ring-necked Parakeets breed earlier than expected. In addition, the discrepancy between expected and actual breeding phenology is greater in northern Europe. In northern European populations, population growth tends to slow down in years that are colder than average. Such a phenological mismatch may also explain why Ring-necked Parakeets from African populations (which are more likely to breed in autumn) have been poor invaders compared with parakeets from Asia. Timing of reproductive phenology can thus be a limiting factor for establishment and range expansion of Ring-necked Parakeets across colder climates.

9.7 Habitat

In their native range, Ring-necked Parakeets are found in a variety of forested habitats from light secondary forest, riparian woodland, mangroves, savannah grasslands and open farmlands with scattered trees to parks and gardens in urban areas. They avoid deserts, treeless wetlands and dense forest, preferring lightly wooded habitats, gardens, orchards and cultivated areas in the vicinity of human habitation (Ali and Ripley, 1969; Forshaw, 1978). Asiatic subspecies are recognized as serious agricultural pests, and Khan *et al.* (2004) mentioned that, in their native Asian range, Ring-necked Parakeets have benefited from agricultural development, as much of the Central Punjab region was originally covered by tropical thorn forest, but the introduction of irrigation and subsequent intensification of agriculture, accompanied by the introduction of larger trees not native to the area, has provided the parakeets with food as well as nesting opportunities.

In Europe, Ring-necked Parakeets are most abundant in areas close to the urban centres where they were released. Radio-tracking (Clergeau and Vergnes, 2011; Strubbe and Matthysen, 2011) and habitat selection studies (Strubbe and Matthysen, 2007; Newson *et al.*, 2011) indicate that Ring-necked Parakeets prefer to forage in city parks and gardens, rather than inside forest fragments or on agricultural lands. In urban areas, bird feeders and (often non-native) ornamental vegetation present Ring-necked Parakeets with abundant food, as well as offering an abundance of suitable nesting sites, because large, old trees are often retained for their aesthetic value. In Germany, Ring-necked Parakeets increasingly breed in holes and crevices within the thermal insulation layers of buildings, where a more favourable microclimate enables them to achieve a higher breeding success compared with natural cavities (Braun, 2004). Strubbe *et al.* (2015) argued that association with human-modified habitats in the native range may enable Ring-necked Parakeets to exploit equivalent human-modified landscapes in Europe, allowing them to colonize areas far colder than their native range. While the bulk of the European Ring-necked Parakeet populations are still centred in urban areas, smaller breeding populations are also present in rural areas (Butler, 2005). This suggests that Ring-necked Parakeets are not necessarily restricted to urban and suburban areas as long as adequate food is available and temperatures do not drop too low (Butler, 2003, 2005).

9.8 Impacts

Invasive Ring-necked Parakeets have been suggested to pose a threat to native species through competition for nesting cavities and for food resources. The most extensive research on competition for cavities with birds was conducted in Brussels, Belgium, including observational, empirical and experimental studies. Observations and experiments led to the conclusion that, at least in urban forested parks, Ring-necked Parakeets compete with native Nuthatches (*Sitta europaea*; Strubbe and Matthysen, 2007, 2009). However, long-term monitoring data in Brussels suggest that competition does not translate into population-level impacts on any native cavity-nester, including the Nuthatch (Weiserbs *et al.*, 2002). Somewhat similarly, a study in the UK showed that several native cavity-nesters shift their nesting cavity preference and timing of breeding in response to competition with introduced Ring-necked Parakeets (British Trust for Ornithology, 2010). However, monitoring programmes in areas where Ring-necked Parakeets have increased did not find any evidence for parakeet impacts (Newson *et al.*, 2011). Multiple studies from the Netherlands, Germany and France have also reported aggressive interactions with native cavity-nesters including bats but failed to find any evidence of population-level impacts (Clergeau and Vergnes, 2009; Czajka *et al.*, 2011; Tamis, 2015). In southern Europe, there is some evidence that Ring-necked Parakeets compete locally with native cavity-nesters such as Common Starlings (*Sturnus vulgaris*; Dodaro and Battisti, 2014) and Lesser Kestrels (*Falco naumanni*; Hernández-Brito *et al.*, 2014a), rats (*Rattus rattus*; Hernández-Brito *et al.*, 2014b) and squirrels (*Sciurus vulgaris*; Mori *et al.*, 2013). Probably the best documented case of impact on a native cavity-nester is from Seville, Spain, where Hernández-Brito *et al.* (2018) reported that Ring-necked Parakeets compete with a threatened bat species, the greater noctule (*Nyctalus lasiopterus*). Both species reportedly share preferences for the same tree cavities. Coinciding with a strong increase in Ring-necked Parakeet numbers, the number of trees occupied by noctules declined by 81%. Parakeets occupied most cavities used previously by noctules, and remaining noctules tried to avoid cavities close to parakeets. Parakeets were highly aggressive towards noctules, trying to occupy their cavities, sometimes resulting in noctule death. This has led to a dramatic population decline but also an unusual aggregation of the occupied trees, probably disrupting the complex social behaviour of this bat species.

There is some evidence for food competition by Ring-necked Parakeets with native species, but the impact on native populations is unknown. An experimental study found that Ring-necked Parakeet presence significantly reduced feeding rates and increased vigilance among native birds (Peck *et al.*, 2014). Similarly, a study conducted in Paris, France, using bird feeders and cameras in several private gardens suggested that Ring-necked Parakeets may be a superior competitor for anthropogenic food resources by showing aggressive behaviour and by hindering food access to the native bird species present at the feeders (Le Louarn *et al.*, 2016).

Despite the Ring-necked Parakeets' reputation as an agricultural pest in its native range, extensive damage in Europe has yet to be reported, and most claims about damage to crops and ornamental trees are anecdotal. For example, in Surrey, UK, parakeets reportedly decimated the grape crop in a vineyard, causing a reduction in wine production from 3000 to 500 bottles of British rosé wine (Butler, 2003). Recently, Mentil *et al.* (2018) conducted one of the first systematic assessments of Ring-necked Parakeet damage to a suburban almond orchard inside the metropolitan area of Rome. They found that of all almonds recorded, about 32% showed signs of damage due to the feeding action of Ring-necked Parakeets. Damage was highest at the edges of the plantation.

9.9 Control

There are a few case studies of successful removal of introduced Ring-necked Parakeets, mainly from islands. Ring-necked Parakeets have been removed from Mahé, Seychelles, after a campaign of several years that relied on citizen scientists to spot parakeets. The birds were then targeted for shooting by a team of professional hunters, who killed 548 birds from 2013 to 2017. Similarly, on the Canary Island of La Palma, an ongoing campaign is eradicating Ring-necked Parakeets from the island, mainly using live traps, which must be checked daily. This method has also been used successfully in the city of Ghent, Belgium, where an emerging population was largely removed between 2015 and 2018 (41 birds from an estimated total of 55). Shooting is the main control method in Kauai, Hawaii. Crop damage appears less severe when shooters patrol cornfields, although the parakeet population does not seem to have noticeably declined so far (Gaudioso *et al.*, 2012). In the UK, the fertility control agent diazacon has been tested and proven effective against captive Ring-necked Parakeets (Lambert *et al.*, 2010). However, a suitable formulation and delivery system is needed before diazacon can be used effectively for parakeet population control. In the species' native range, researchers have tested different techniques to reduce agricultural damage. In Pakistan, this includes multi-mirror reflectors, reflecting ribbons, distress sound players, gas explosions and wind-powered chemical repellents. Of these, reflecting ribbons, wind-powered repellents and distress sounds provided at least some relief against bird damage (H. A. Khan, personal communication).

9.10 Uses

Ring-necked Parakeets are a popular aviary bird. Multiple colour mutations are bred in captivity, which can occasionally be seen in the wild as well, and appear to breed successfully in at least some locations (Grandi *et al.*, 2018).

9.11 Notes

The number of individuals introduced has a strong positive effect on invasion success (Blackburn *et al.*, 2015), and the high number of Ring-necked Parakeets that has been traded worldwide (Cardador *et al.*, 2016), combined with captive breeding in non-native areas, is an important driver of the widespread invasion success of Ring-necked Parakeets. Parakeets in general have larger relative brain sizes (i.e. relative to body mass), helping them to respond to novel conditions and be more successful at establishing themselves in novel environments (Sol *et al.*, 2005). Their flexible diet and habitat generalism probably also contribute to their near-global invasion success (Sol *et al.*, 2012). In urban areas in particular, Ring-necked Parakeets are often well liked by many people, necessitating early, inclusive public engagement and open, responsive communication strategies to minimize social conflicts when management questions arise (Crowley *et al.*, 2019).

9.12 References

Ali, S. and Ripley, S. (1969) *Handbook of the birds of India and Pakistan*. Vol. 3: *Stone Curlews to Owls*. Oxford University Press, Mumbai.

Balmer, D.E., Gillings, S., Caffrey, B., Swann, B., Downie, I. and Fuller R. (2013) *British Trust for Ornithology Bird Atlas 2007–11*: the breeding and wintering birds of Britain and Ireland. British Trust for Ornithology, Thetford, UK.

Blackburn, T.M., Lockwood, J.L. and Cassey, P. (2015) The influence of numbers on invasion success. *Molecular Ecology* 24, 1942–1953.

Braun, M. and Wink, M. (2013) Nestling development of ring-necked parakeets (*Psittacula krameri*) in a nest-box population. *Open Ornithology Journal* 6, 9–24.

Braun, M.P. (2004) Neozoen in urbanen Habitaten: Okologie und Nischenexpansion des Halsbandsittichs (*Psittacula krameri* Scopoli, 1769) in Heidelberg. PhD thesis, Philipps-Universitat Marburg, Marburg, Germany.

British Trust for Ornithology (2010) Impact of ring-necked parakeets on native birds – DEFRA Project WC0732. British Trust for Ornithology, Thetford, UK.

Butler, C. (2003) Population biology of the introduced ring-necked parakeet *Psittacula krameri* in the UK. PhD, thesis, University of Oxford, Oxford.

Butler, C. (2005) Feral parrots in the continental United States and United Kingdom: past, present and future. *Journal of Avian Medicine and Surgery* 19, 142–150.

Butler, C.J. and Gosler, A. (2004) Sexing and ageing Rose-ringed Parakeets *Psittacula krameri* in Britain. *Ringing and Migration* 22, 7–12.

Cardador, L., Carrete, M., Gallardo, B. and Tella, J.L. (2016) Combining trade data and niche modelling improves predictions of the origin and distribution of non-native European populations of a globally invasive species. *Journal of Biogeography* 43, 967–978.

Clergeau, P. and Vergnes, A. (2009) La Perruche à collier *Psittacula krameri* en Ile-de-France. Dispersion et approche des impacts. Report of Muséum National d'Histoire Naturelle, Paris.

Clergeau, P. and Vergnes, A. (2011) Bird feeders may sustain feral Rose-ringed parakeets *Psittacula krameri* in temperate Europe. *Wildlife Biology* 17, 248–252.

Crowley, S.L., Hinchliffe, S. and McDonald, R.A. (2019) The parakeet protectors: understanding opposition to introduced species management. *Journal of Environmental Management* 229, 120–132.

Czajka, C., Braun, M.P. and Wink, M. (2011) Resource use by non-native ring-necked parakeets (*Psittacula krameri*) and native starlings (*Sturnus vulgaris*) in Central Europe. *Open Ornithology Journal* 4, 17–22.

Dodaro, G. and Battisti, C. (2014) Rose-ringed parakeet (*Psittacula krameri*) and starling (*Sturnus vulgaris*) syntopics in a Mediterranean urban park: evidence for competition in nest-site selection? *Belgian Journal of Zoology* 144, 5–14.

Forshaw, J. (1978) *Parrots of the World*. T.F.H. Publications, Neptune, New Jersey.

Forshaw, J. (2010) *Parrots of the World*. CSIRO Publishing, Sydney, Australia.

Fry, C. and Keith, S. (1988) *The Birds of Africa*. Vol. III: *Parrots to Woodpeckers*. Princeton University Press, Princeton, New Jersey.

Gaudioso, J.M., Shiels, A.B., Pitt, W.C. and Bukoski, W.P. (2012) Rose-ringed parakeet impacts on Hawaii's seed crops on the island of Kauai: population estimate and monitoring of movements using radio telemetry. Unpublished Report QA1874, USDA, National Wildlife Research Center, Hilo, Hawaii.

Grandi, G., Menchetti, M., and Mori, E. (2018) Vertical segregation by breeding ring-necked parakeets *Psittacula krameri* in northern Italy. *Urban Ecosystems*, 21, 1011–1017.

Hanley, S. (2007) The optical concerns of Jan van Eyck's painting practise. PhD thesis, University of York, York, UK.

Hart, L. and Downs, C.T. (2014) Public surveys, of rose-ringed parakeets *Psittacula krameri*, in the Durban metropolitan area, South Africa. *African Zoology* 49, 283–289.

Hernández-Brito, D., Carrete, M., Popa-Lisseanu, A.G., Ibáñez, C. and Tella, J.L. (2014a) Crowding in the city: losing and winning competitors of an invasive bird. *PLoS One*, 9, e100593.

Hernández-Brito, D., Luna, Á., Carrete, M. and Tella, J.L. (2014b) Alien rose-ringed parakeets (*Psittacula krameri*) attack black rats (*Rattus rattus*) sometimes resulting in death. *Hystrix* 25, 121–123.

Hernández-Brito, D., Carrete, M., Ibáñez, C., Juste, J. and Tella, J.L. (2018) Nest-site competition and killing by invasive parakeets cause the decline of a threatened bat population. *Royal Society Open Science* 5, 172477.

Iqbal, M. (2017) First record of Rose-ringed Parakeet *Psittacula krameri* for Indonesia. *BirdingAsia* 25, 73–74.

Jackson, H., Strubbe, D., Tollington, S., Prys-Jones, R., Matthysen, E. and Groombridge, J.J. (2015) Ancestral origins and invasion pathways in a globally invasive bird correlate with climate and influences from bird trade. *Molecular Ecology* 24, 4269–4285.

Jones, C.G. (1980) Parrot on the way to extinction. *Oryx* 15, 350–354.

Kahl-Dunkel, A. and Werner, R. (2002) Winterverbreitung des Halsbandsittichs *Psittacula krameri* in Koln. *Die Vogelwelt* 123, 17–20.

Keijl, G. (2001) Halsbandparkieten *Psittacula krameri* in Amsterdam, 1976–2000. *Limosa* 74, 29–32.

Khan, H., Beg, M. and Khab, A. (2004) Breeding habits of the rose-ringed parakeet (*Psittacula krameri*) in the cultivations of Central Punjab. *Pakistan Journal of Zoology* 36, 133–138.

Klaassen, O. and van Kleunen, A. (2012) Halsbandparkieten in Nederland in de winter van 2011/2012. Verslag van de slaapplaatstellingen en adviezen voor toekomstige monitoring. SOVON, Nijmengen, The Netherlands. Available at: https://www.sovon.nl/sites/default/files/doc/Rap_2012-26_Halsbandparkieten_2011-12.pdf (accessed 25 October 2019).

Lambert, M., Massel, G., Yoder, C. and Cowan, D. (2010) An evaluation of diazacon as a potential contraceptive in non-native Rose-Ringed Parakeets. USDA National Wildlife Research Center – Staff Publications, no. 934. Available at: https://digitalcommons.unl.edu/icwdm_usdanwrc/934 (accessed 25 October 2019).

Le Louarn, M., Couillens, B., Deschamps-Cottin, M. and Clergeau, P. (2016) Interference competition between an invasive parakeet and native bird species at feeding sites. *Journal of Ethology* 34, 291–298.

Lever, C. (2005) *Naturalised Birds of the World*. T. and A.D. Poyser, Chicago, Illinois.

Luna, A., Franz, D., Strubbe, D., Shwartz, A., Braun, M.P., *et al.* (2017) Reproductive timing as a constraint on invasion success in the Ring-necked parakeet (*Psittacula krameri*). *Biological Invasions* 19, 2247–2259.

Mentil, L., Battisti, C. and Carpaneto, G.M. (2018) The impact of *Psittacula krameri* (Scopoli, 1769) on orchards: first quantitative evidence for Southern Europe. *Belgian Journal of Zoology* 148, 129–134.

Mori, E., Ancillotto, L., Menchetti, M., Romeo, C. and Ferrari, N. (2013) Italian red squirrels and introduced parakeets: victims or perpetrators? *Hystrix* 24, 195–196.

Nebot, J. (1999) First report on the rose-ringed parakeet (*Psittacula krameri*) in Venezuela and preliminary observations on its behaviour. *Ornithologia Neotropical* 10, 115–117.

Newson, S.E., Johnston, A., Parrott, D. and Leech, D.I. (2011) Evaluating the population-level impact of an invasive species, Ring-necked Parakeet *Psittacula krameri*, on native avifauna. *Ibis* 153, 509–516.

Pârâu, L.G., Strubbe, D., Mori, E., Menchetti, M., Ancillotto, L. *et al.* (2016) Rose-ringed Parakeet populations and numbers in Europe: a complete overview. *Open Ornithology Journal* 9, 1–13.

Peck, H.L., Pringle, H.E., Marshall, H.H., Owens, I.P.F., and Lord, A.M. (2014) Experimental evidence of impacts of an invasive parakeet on foraging behavior of native birds. *Behavioral Ecology* 25, 582–590.

Pithon, J. (1996) The status and ecology of the Ring-necked Parakeet *Psittacula krameri* in Great-Britain. PhD thesis, University of York, York, UK.

Reino, L., Figueira, R., Beja, P., Araújo, M.B., Capinha, C. and Strubbe, D. (2017) Networks of global bird invasion altered by regional trade ban. *Science Advances* 3, e1700783.

Shiels, A.B., Bukoski, W.P. and Siers, S.R. (2018) Diets of Kauai's invasive rose-ringed parakeet (*Psittacula krameri*): evidence of seed predation and dispersal in a human-altered landscape. *Biological Invasions* 20, 1449–1457.

Shwartz, A., Strubbe, D., Butler, C.J., Matthysen, E. and Kark, S. (2009) The effect of enemy-release and climate conditions on invasive birds: a regional test using the rose-ringed parakeet (*Psittacula krameri*) as a case study. *Diversity and Distributions* 15, 310–318.

Sol, D., Duncan, R.P., Blackburn, T.M., Cassey, P. and Lefebvre, L. (2005) Big brains, enhanced cognition, and response of birds to novel environments. *Proceedings of the National Academy of Sciences USA* 102, 5460–5465.

Sol, D., Maspons, J., Vall-Llosera, M., Bartomeus, I., García-Peña, G.E., *et al.* (2012) Unraveling the life history of successful invaders. *Science* 337, 580–583.

Soni, A.C. (1991) Food habit and morphometrics of the rose-ring parakeet *Psittacula krameri* Scopoli. MSc thesis, Anand Agricultural University, Anand, Gujarat, India.

Strubbe, D. and Matthysen, E. (2007) Invasive ring-necked parakeets *Psittacula krameri* in Belgium: habitat selection and impact on native birds. *Ecography* 30, 578–588.

Strubbe, D. and Matthysen, E. (2009) Experimental evidence for nest-site competition between invasive ring-necked parakeets (*Psittacula krameri*) and native nuthatches (*Sitta europaea*). *Biological Conservation* 142, 1588–1594.

Strubbe, D. and Matthysen, E. (2011) A radiotelemetry study of habitat use by the exotic Ring-necked Parakeet *Psittacula krameri* in Belgium. *Ibis* 153, 180–184.

Strubbe, D., Jackson, H., Groombridge, J., and Matthysen, E. (2015) Invasion success of a global avian invader is explained by within-taxon niche structure and association with humans in the native range. *Diversity and Distributions* 21, 675–685.

Symes, C.T. (2014) Founder populations and the current status of exotic parrots in South Africa. *Ostrich* 85, 235–244.

Tamis, W. (2015) Review van de wetenschappelijke literatuur over de effecten van de uitheemse soorten Halsbandparkiet, Grote Alexanderparkiet en Monniksparkiet. Een update vanaf 2010. Centrum voor Milieuwetenschappen CML, Universiteit Leiden, Leiden, The Netherlands.

Verdi, P. (2007) *The Parrot in Art: from Durer to Elisabeth Butterworth*. Scala Publishers, London.

Weiserbs, A., Jacob, J. and Rotsaert, G. (2002) Evaluation de l'incidence de développement des populations de perruches sur les habitats et les espèces indigènes en Région bruxelloise. Final report. Convention entre l'IBGE et AVES, Brussels, Belgium.

10 Monk Parakeet (*Myiopsitta monachus* Boddaert, 1783)

Michael L. Avery*

USDA APHIS National Wildlife Research Center, Gainesville, FL 32641, USA

Citation: Avery, M.L. (2020) Monk Parakeet (*Myiopsitta monachus* Boddaert, 1783). In: Downs, C.T. and Hart, L.A. (eds) *Invasive Birds: Global Trends and Impacts*. CAB International, Wallingford, UK, pp. 76–84.

10.1 Common Names

Monk Parakeet, Quaker Parakeet, Quaker Parrot, Quaker Conure, Grey-headed Parakeet.

10.2 Distribution

The natural distribution of the Monk Parakeet (*Myiopsitta monachus* Boddaert, 1783) extends from southern Bolivia, through Paraguay, southern Brazil and Uruguay to southern Argentina (Fig. 10.1). It is documented as invasive in a range of countries, and of these, there are at least 17 countries where they are breeding in the wild, as detailed below.

- *Belgium:* Monk Parakeets were imported in limited numbers in the 1980s. Currently, they are nesting at several locations in Brussels (Nixon, 2018).
- *Brazil:* Monk Parakeets are native to southern Brazil, but populations are now established in other parts of the country, such as Rio de Janeiro, through releases and escapes of pet birds (Amorim and Piacentini, 2006; Viana *et al.*, 2016).
- *Cayman Islands:* Monk Parakeets were introduced to Grand Cayman Island in 1987, and they adapted readily to the new location. The growing Monk Parakeet population caused concern over impacts to agriculture and the electric utility service, which prompted initiation of a control programme. Control efforts were successful but could not be sustained to eradication, and the population rebounded (Godbeer, 2014).
- *Chile:* Monk Parakeets were released in 1972 by residents of Santiago (Iriarte *et al.*, 2005). Since then, the species has spread to many other parts of the country, aided by intentional releases and escapes (Briceño *et al.*, 2017).
- *Denmark:* Small numbers of Monk Parakeets have bred at one location (Køge Bugt) 'for many years' (Fox *et al.*, 2015).
- *France:* Populations of Monk Parakeets in southern France (the cities of Toulon and Marseilles) continue to maintain themselves (Dubois and Cugnasse, 2015; GT IBMA, 2018).
- *Greece*: A Monk Parakeet nesting colony of at least 21 birds was documented in 2010 in Athens (Kalodimos, 2013).
- *Israel:* Monk Parakeets were first detected near Tel Aviv in 1995. The population has grown exponentially, and these parakeets now occupy urban and agricultural areas (Postigo *et al.*, 2017).
- *Italy:* Italy is second only to Spain in Europe for CITES (Convention on International Trade in Endangered Species; https://trade.cites.org/, accessed 30 October 2019) imports of Monk Parakeets. The species has been in Italy for decades and now breeds throughout the country (Mori *et al.*, 2013).
- *Mexico:* Monk Parakeets were first reported in the late 1990s (MacGregor-Fors *et al.*, 2011; Hobson *et al.*, 2017). In the past decade, Monk Parakeets have been seen with increasing frequency in many parts of the country coincident with large numbers of imports (Hobson *et al.*, 2017).
- *Morocco:* In recent years, nesting of Monk Parakeets has been documented in several cities including Casablanca, Melilla, Tangier and Marrakech (MaghrebOrnitho, 2018).
- *Netherlands:* Recent surveys (2011–2013) documented Monk Parakeet breeding in eastern and southwest parts of the country (van Kleunen *et al.*, 2014).
- *Portugal:* Observations in the cities of Lisbon and Porto 'indicate the successful reproduction' by Monk Parakeets (Matias, 2012). The Monk Parakeet population in Porto is well established and spreading (da Silva Carneiro, 2017).
- *Puerto Rico:* Monk Parakeets are widespread and are increasing throughout the island (Falcón and Tremblay, 2018), and the species is regularly recorded on annual Audubon Christmas Bird Counts (http://netapp.audubon.org/CBCObservation/, accessed 30 October 2019).

*Corresponding author (retired): gibs8387@gmail.com

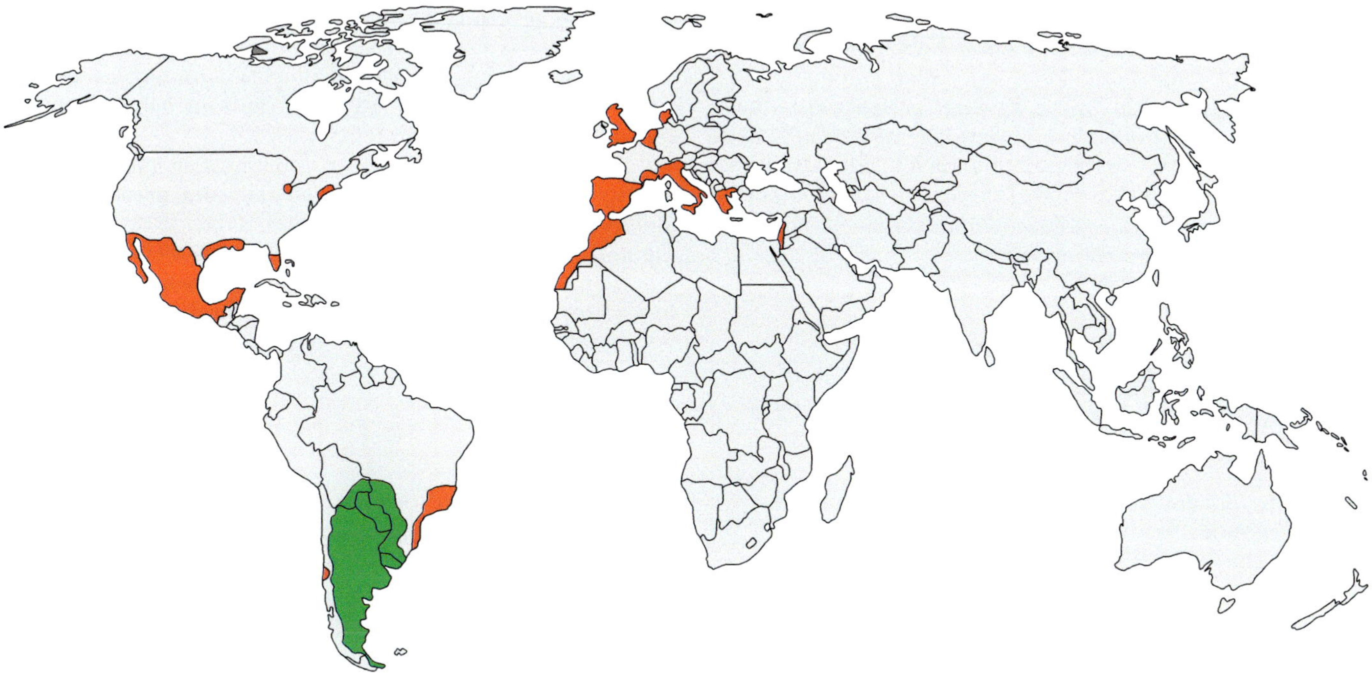

Fig. 10.1. Global distribution of the Monk Parakeet showing native (green) and invasive (red) breeding ranges, based on recent records. (World outline map by www.freeworldmaps.net.)

- *Spain:* Thousands of Monk Parakeets were imported in the 1980s and 1990s. Currently, the species is common in many locations, including the Canary Islands (Sol *et al.*, 1997; Rodríguez-Pastor *et al.*, 2012; Souviron-Priego *et al.*, 2018).
- *UK:* Small numbers of free-living Monk Parakeets have been reported since 1987 (Tayleur, 2010). Serious efforts began in 2011 to eradicate the species in the UK (Carrington, 2014).
- *USA:* Monk Parakeets were first sighted in 1967 in New York and in 1969 in Florida (Neidermyer and Hickey, 1977). Through the pet trade, Monk Parakeets quickly occupied other parts of the country, and the US population expanded exponentially through 2003 (Avery and Shiels, 2018). Breeding populations currently occur in seven states: Florida, Louisiana, Texas, Connecticut, New York, New Jersey and Illinois. Reports from other states are common.

There are more than 13 countries where Monk Parakeet breeding in the wild is not verified. These include the following:

- *Australia:* 'To date, *M. monachus* has not naturalised in Queensland. However, it is kept in considerable numbers as a pet and escape/release is inevitable. ...it seems reasonable to predict that *M. monachus* will eventually naturalise in Queensland' (Csurhes, 2016). The Monk Parakeet is among the non-native caged-bird species most frequently reported missing in Australia (Vall-llosera and Cassey, 2017).
- *Bahamas:* Monk Parakeets are regularly recorded on Audubon Christmas Bird Counts, and are frequently reported on blogs by tourists (e.g. www.smartertravel.com/bahamas-birdwatching-trip-report/, accessed 30 October 2019).
- *Canada:* There have been infrequent sightings of Monk Parakeets in southern Canada of presumably escaped pet birds but no established population (Christie, 1992; Crins, 2004).
- *Czech Republic:* There have been occasional sightings of Monk Parakeets, but no verified instance of nesting (Hudec, 2015).
- *Germany:* According to Bauer and Woog (2008), there are no longer any nesting colonies of Monk Parakeets in Germany.
- *Guadeloupe:* Monk Parakeets are considered to be 'rare' by Raffaele and Wiley (2014), but show up on eBird sites (e.g. two birds on 28 November 2018; https://ebird.org/newzealand/region/caribbean, accessed 30 October 2019).
- *Japan:* There are sporadic sightings of escaped pet birds, plus two instances of nesting but no accompanying documentation or explanation (Eguchi and Amano, 2004).
- *Singapore:* Occasional observations of free-flying birds include nest-building activity (Lim, 2009; Kwong, 2013).
- *South Africa:* There have been just two records (in 1980 and 2012) of free-flying Monk Parakeets (Symes, 2014).
- *Switzerland:* Isolated breeding attempts have been reported but with no indication of success (Wittenberg, 2005).
- *Thailand:* Monk Parakeets are among the many bird species found in Bangkok bird markets (Chng and Eaton 2016).
- *United Arab Emirates:* Escaped pet Monk Parakeets have been sighted in Dubai (Aspinall and Porter, 2011).
- *Venezuela:* Feral Monk Parakeets have been reported previously (Nebot, 1999), but the current status of the species in unclear.

Other countries listed as invaded areas are Austria, Slovakia, Virgin Islands and Gibraltar (BirdLife International, 2019).

No records of wild Monk Parakeet presence have been found for China, Finland, Hong Kong, Malaysia, Malta, Norway, Panama and Sweden, despite each of these countries being listed by the CITES Trade Database as having imported the species (Table 10.1), although minimally in the cases of Finland ($n = 5$) and Norway ($n = 4$).

10.3 Description

The Monk Parakeet is a medium-sized parrot (110–130 g body mass, approximately 28 cm total length; Fig. 10.2). Avery *et al.* (2012) examined 845 parakeets collected in south Florida to document body size, moult and reproductive biology within the invasive population. Adult males on average were 1.5–3.5% larger than adult females, except during the breeding season (March–May) when the body mass of females increased slightly due to egg development. Body measurements placed the Florida birds in the subspecies *M. m. monachus*, the largest of the four Monk Parakeet subspecies (Spreyer and Bucher, 1998).

Adult males and females are identical in plumage. The plumage is green on the back and tail, and greyish on the underside. The wings are mostly dull green with blue outer flight feathers (Spreyer and Bucher, 1998). Females initiate moult of wing feathers sooner than males (Avery *et al.*, 2012). Replacement of primary feathers starts in April and extends into October among Florida Monk Parakeets.

10.4 Diet

Monk Parakeets eat a wide variety of fruit, seeds, buds and flowers. In its South American native range, the species is regarded as a major pest to crops such as sorghum, sunflower and

Table 10.1. Monk Parakeet importation based on importer reports from the CITES Trade Database (1981–2014), and current status of the species in each country.

	No. imported					
Importing country	Total	1981–1993	1994–2005	2006–2014	Current status[a]	Reference
Belgium	380	380	0	0	B	Nixon (2018)
Canada	10	0	10	0	P	Christie (1992); Crins (2004)
Chile	5880	0	5880	0	B	Iriarte *et al.* (2005)
China	100	0	0	100	U	
Czech Republic	410	0	410	0	P	Hudec (2015)
Denmark	291	291	0	0	B	Fox *et al.* (2015)
Finland	5	5	0	0	U	
France	4270	3000	1270	0	B	Dubois and Cugnasse (2015)
Germany	5,038	4,761	277	0	P	Bauer and Woog (2008)
Greece	630	0	630	0	B	Kalodimos (2013)
Hong Kong	2020	40	0	1980	U	
Israel	100	0	100	0	B	Postigo *et al.* (2017)
Italy	29,187	17,919	11,268	0	B	Mori *et al.* (2013)
Japan	1,166	960	204	2	P	Eguchi and Amano (2004)
Malaysia	55	0	35	20	U	
Malta	26	0	26	0	U	
Mexico	576,818	0	3,052	573,766	B	MacGregor-Fors *et al.* (2011)
Netherlands	250	250	0	0	B	van Kleunen *et al.* (2014)
Norway	4	4	0	0	U	
Panama	75	0	75	0	U	
Portugal	14,167	900	13,267	0	B	Matias (2012)
Singapore	8,510	60	2,250	6,200	P	Kwong (2013)
South Africa	1,360	820	540	0	P	Symes (2014)
Spain	161,899	80,400	81,499	0	B	Souviron-Priego *et al.* (2018)
Sweden	240	240	0	0	U	
Thailand	140	0	140	0	P	Chng and Eaton (2016)
United Arab Emirates	480	0	240	240	P	Aspinall and Porter (2011)
UK	2,448	1927	521	0	B	Tayleur (2010)
USA	161,510	161,510	0	0	B	Neidermyer and Hickey (1977)
Total no. birds	**977,469**	**273,467**	**121,694**	**582,308**		
No. importers	**29**	**17**	**20**	**7**		

[a]B, breeding; P, present; U, unknown.

rice. They sometimes feed on tropical fruit crops in the USA (Tillman *et al.*, 2001). The species is flexible and adaptable in its diet. For example, in Chicago, Monk Parakeets usually feed on plant buds, weeds, and fruits and berries of ornamental shrubs and trees (South and Pruett-Jones, 2000), but in winter (December–February), they feed extensively on bird seed at backyard feeders (Hyman and Pruett-Jones, 1995). In Spain, Monk Parakeets feed on a variety of crops (Senar *et al.*, 2016), and in the Cayman Islands, invasive Monk Parakeets damage mango crops (Godbeer, 2014). Monk Parakeets feeding on wheat have been reported in Israel (Postigo *et al.*, 2017). In urban areas of their invasive range, they often feed on anthropogenic food waste (Fig. 10.3; L. Hart, personal observation).

Fig. 10.2. An adult Monk Parakeet in Mato Grosso, Brazil. (©Photograph: Bernard DuPont, https://creativecommons.org/licenses/by-sa/2.0/deed.en.)

10.5 Introduction and Invasion Pathways

The Monk Parakeet has been very popular in the caged-bird trade since the 1960s. Its current status as an invasive species is solely due to its availability and popularity as a pet. Hundreds of thousands of Monk Parakeets were sent around the world from Uruguay, Paraguay and Argentina as part of the commercial pet trade. Genetic evidence positively links invasive populations of Monk Parakeets in Spain and the USA to the international caged-bird trade (Russello *et al.*, 2008; Edelaar *et al.*, 2015). The CITES Trade Database is a unique resource for examining the volume and geographical scope of commercial trade in dozens of species (e.g. Cardador *et al.* 2017; Hobson *et al.* 2017). Analyses of Monk Parakeet import records for 1981–2014 were conducted using only the data ascribed to 'importer reports' because the export and import reports are often dissimilar in a given year for the same species (Hobson *et al.*, 2017). Data were restricted to wild-caught birds (source code 'W') exported for commercial use (purpose code 'T'). It was found that trade trends have been strongly influenced by national controls in key import markets. In 1992, the USA passed the Wild Bird Conservation Act, which sharply reduced the number of parrots and other wild birds imported to the USA. In 2005, the European Union banned the import of wild birds due to concerns about transmission of highly pathogenic avian influenza virus (European Commission, 2005). Thereafter, imports shifted to Mexico and Asian countries (Cardador *et al.*, 2017). Hobson *et al.* (2017) provided a detailed accounting of the Monk Parakeet in Mexico and analysed the implications of Mexico's ascendance as world import leader of the species in the wake of the European import restrictions.

Overall, from 1981 to 2014, 29 countries reported importing Monk Parakeets (Table 10.1). During 1981–1993, the USA was leading importer (161,510 birds) of Monk Parakeets, accounting for 59% of the world total. Other major importing countries included Spain, Italy, Germany and France. With the passage of the 1992 Wild Bird Conservation Act, the USA imported no Monk Parakeets after 1993. Spain (81,499 birds)

Fig. 10.3. Adult Monk Parakeets feeding on bread left on a street in Barcelona, Spain. (©Photographs: Lorinda Hart.)

accounted for 57% of the total world imports during 1994–2005, followed by Portugal, Italy and Chile. After the European ban on wild bird trade, CITES lists just seven importing countries, with Mexico (573,766 birds) responsible for 98.4% of the total during 2006–2014. In fact, Mexico alone has accounted for 59% of the total Monk Parakeet imports recorded by CITES.

10.6 Breeding Behaviour

Monk Parakeets are monogamous. One clutch of four to eight eggs is produced annually during the well-defined spring breeding season. The female incubates the eggs and broods the nestlings, while the male contributes nest materials and brings food to the female. Monk Parakeets are unique among psittacines as they use sticks and twigs to construct bulky nests, which house from one to many nesting chambers (Figs 10.4 and 10.5) (Spreyer and Bucher, 1998). The nest structure is the focus of the parakeets' social system; the birds occupy their nests year-round.

Breeding adults and non-breeding subadults defend and maintain their nests throughout the year, and instances of non-breeding parakeets helping to feed nestlings or recent fledglings have been documented (Bucher *et al.*, 1991; Eberhard, 1998). Nesting season and the moult cycle of the species in North America are shifted 6 months with respect to the native range (Avery *et al.*, 2012). In Spain, Monk Parakeets generally nest in palm trees, especially *Phoenix* spp. (Sol *et al.*, 1997; Rodríguez-Pastor *et al.*, 2012; L. Hart, personal observation).

10.7 Habitat

Monk Parakeets in their native range typically inhabit open woodlands, savannahs and agricultural landscapes. *Eucalyptus* trees are a favourite nesting substrate.

Fig. 10.4. Nesting Monk Parakeets in the USA. (©Photographer: E.A. Tillman, courtesy of USDA/APHIS.)

In their introduced range, Monk Parakeets are primarily found in urban/suburban areas, but expansion into agricultural landscapes is occurring in some areas (Postigo *et al.*, 2017). It is a common visitor at bird feeders and exploits ornamental plantings for food and nest sites. Preferred nest substrates also include anthropogenic structures such as electric utility structures, light poles and cell towers (Fig. 10.4 and 10.6).

10.8 Impacts

There are no positive impacts of Monk Parakeets as an invasive species and no known negative impacts on native species in its introduced range.

Monk Parakeets damage agriculture crops in the USA (Tillman *et al.*, 2001), Cayman Islands (Godbeer, 2014), Spain (Senar *et al.*, 2016) and Israel (Postigo *et al.*, 2017). Nest construction on electric utility facilities causes power outages and maintenance problems (Avery *et al.*, 2006; Godbeer, 2014; Reed *et al.*, 2014).

There is some evidence that Monk Parakeets facilitate the dispersal of the bacterium *Cryptosporidium* spp., which can cause illness in humans (Briceño *et al.*, 2017), but no direct link has yet been reported between parakeets and illness related to *Cryptosporidium* spp.

10.9 Control

10.9.1 Control methods

The life of the Monk Parakeet is centred on its nest structure where breeding takes place and where it roosts at night. Consequently, management actions to control Monk Parakeet populations usually target the nest structure. Formerly, in the native range, managers applied paste containing toxicants such as carbofuran to nest openings so that parakeets entering the

Fig. 10.5. Nests of Monk Parakeets in a palm tree in Barcelona, Spain. (©Photograph: Lorinda Hart.)

nest would die from ingesting the toxicant as they preened the paste from their feathers (Linz *et al.*, 2015). Such lethal measures have not been employed in the non-native range.

Nest destruction is frequently used by utility companies to reduce the risk of power outages. Monk Parakeets quickly reoccupy the site, however, resulting in short-term relief only (Avery and Lindsay, 2016). Trapping birds at their nest has proven effective as a management tool. Approaching the nest after sunset reduces the likelihood that the birds will bolt. To reach the nests, a long-handled net is useful, as is a truck with an articulating arm to raise the trapper safely to the proper height. This approach enabled authorities in the Cayman Islands to reduce the Monk Parakeet population by 86% (Godbeer, 2014). There is little doubt that Monk Parakeet populations can be extirpated with a persistent, integrated management effort (trapping, shooting and toxic baiting). The necessary methods exist, but public opinion could make any such management programme difficult, if not impossible, to implement.

Installation of alternative nest platforms adjacent to distribution poles to encourage parakeets to switch nesting sites has been applied with limited success to address persistent, isolated problems at specific locations (Menzer, 2006). This approach is probably not cost-effective or practical on a large scale.

Extensive aviary and field trials have demonstrated that reproductive inhibition using diazacon as an oral contraceptive can be a safe, effective tool for reducing the growth of Monk Parakeet populations (Fig. 10.7) (Avery *et al.*, 2008).

Fig. 10.6. Nesting Monk Parakeets on electrical infrastructures. (©Photographer: E.A. Tillman, courtesy of USDA/APHIS.)

Fig. 10.7. Monk Parakeets eating seed treated with an oral contraceptive at bait trays in the USA. (©Photographer: E.A. Tillman, courtesy of USDA/APHIS.)

10.9.2 Natural predators

Nest predation is a substantial source of mortality for Monk Parakeets in their native range (Navarro *et al.*, 1992). Fish Crows (*Corvus ossifragus*) occasionally attack Monk Parakeet nests in Florida (Avery and Shiels, 2018), and Monk Parakeets have been preyed on by Great Horned Owls (*Bubo virginianus*) in Connecticut and by Peregrine Falcons (*Falco peregrinus*) in New York (Moscatello, 2003, cited in Burgio *et al.*, 2016).

10.9.3 Conclusions

The Monk Parakeet is among the most successful invasive bird species in the world. Unlike many other invasive birds, such as the Ring-necked Parakeet (*Psittacula krameri*) and the Common Myna (*Acridotheres tristis*), there is no evidence that invasive Monk Parakeets compete with or otherwise negatively affect native species. Their economic impact on agricultural crops has been limited to date, but reports from Spain (Senar *et al.*, 2016) and Israel (Postigo *et al.*, 2017) suggest that serious problems might be developing in some countries. The unique nest-building behaviour of Monk Parakeets does create serious problems for electric utility companies (Avery *et al.*, 2006) and in at least one instance prompted an eradication effort (Godbeer, 2014). Several factors contribute to the success of the Monk Parakeet as an invasive species:

1. *There is no need for cavities.* The construction of large nest structures with sticks and branches distinguishes the Monk Parakeet from all other psittaciform species, and this behaviour is a principal reason for their success as an invader. Monk Parakeets are not dependent on availability of natural cavities for nesting and do not have to compete for such a limited resource. Because of their unique, flexible behaviour, they are not constrained and can build nests on a variety of man-made and natural substrates.
2. *Dietary flexibility.* The Monk Parakeet's flexible behaviour extends to their diet. They readily adapt to local conditions in subtropical and temperate environments where they feed on seeds, flowers and fruits from the variety of locally available native and exotic plants. Monk Parakeets exploit backyard bird feeders, which become particularly important food sources during winter in places like New York and Chicago. The Monk Parakeet diet extends to fruit, vegetable and grain crops. Ultimately, the Monk Parakeet is flexible, adaptable and opportunistic, which are highly advantageous traits for an invading species.
3. *Socio-ecology.* The Monk Parakeet is also a highly social animal, with the nest structure at the centre of their social activity. A single nest structure can include numerous individual nest chambers, each of which is occupied by a breeding pair and offspring. There is evidence that non-breeding parakeets assist the breeding pair in nest maintenance and predator detection (Bucher *et al.*, 1991; Eberhard, 1998). A study of captive birds in a large flight pen revealed that the Monk Parakeet social structure is built on pair bonds (not always male–female) and linear dominance hierarchies (Hobson *et al.*, 2014). The Monk Parakeet's social system, grounded in resilient pair-wise relationships, could facilitate a population's recovery from major disruption or large-scale natural disaster (e.g. Sevenair, 2012), which in turn would contribute to the invasion success of the species.

10.10 Uses

The Monk Parakeet is a very popular caged bird. When they are released or escape, Monk Parakeets in urban and suburban locales can provide enjoyment as interesting, charismatic visitors to backyard bird feeders. For researchers, the Monk Parakeet is an excellent study species for investigating animal social behaviour and information networks (e.g. Hobson *et al.*, 2014).

10.11 Acknowledgements

Thanks to Eric Tillman for review comments and for providing photographs. Michael Moulton kindly facilitated access to reference materials.

10.12 References

Amorim, J.F. and de Queiroz Piacentini, V. (2006) Novos registros de aves raras em Santa Catarina, Sul do Brasil, incluindo os primeiros registros documentados de algumas espécies para o Estado. *Revista Brasileira de Ornitologia* 14, 145–149.

Aspinall, S. and Porter, R.F. (2011) *Birds of the United Arab Emirates*. Christopher Helm, London.

Avery, M.L. and Lindsay, J.R. (2016) Monk Parakeets. Wildlife Damage Management Technical Series. USDA, APHIS, WS National Wildlife Research Center, Fort Collins, Colorado. https://www.aphis.usda.gov/wildlife_damage/reports/Wildlife%20Damage%20Management%20Technical%20Series/Monk-Parakeet.pdf

Avery, M.L. and Shiels, A.B. (2018) Monk and rose-ringed parakeets. In: Pitt, W.C., Beasley, J. and Witmer, G.W. (eds) *Ecology and Management of Terrestrial Vertebrate Invasive Species in the United States*. CRC Press, Boca Raton, Florida, pp. 333–357.

Avery, M.L., Lindsay, J.R., Newman, J.R., Pruett-Jones, S. and Tillman, E.A. (2006) Reducing monk parakeet impacts to electric utility facilities in south Florida. In: Feare, C.J. and Cowan, D.P. (eds) *Advances in Vertebrate Pest Management*, Vol. IV. Filander Verlag, Fürth, Germany, pp. 125–136.

Avery, M.L., Tillman, E.A., Keacher, K.L., Arnett, J.E. and Lundy, K.J. (2012) Biology of invasive monk parakeets in south Florida. *Wilson Journal of Ornithology* 124, 581–588.

Avery, M.L., Yoder, C.A. and Tillman, E.A. (2008) Diazacon inhibits reproduction in invasive monk parakeet populations. *Journal of Wildlife Management* 72, 1449–1452.

Bauer, H.-G. and Woog, F. (2008) Non-native and naturalized bird species (neozoa) in Germany, part I: occurrence, population size and status. *Vogelwarte* 46, 157–194.

BirdLife International (2019) Species factsheet: *Myiopsitta monachus*. Available at: http://datazone.birdlife.org/species/factsheet/monk-parakeet-myiopsitta-monachus (accessed 25 October 2019).

Briceño, C., Surot, D., González-Acuña, D., Martínez, F.M. and Fredes, F. (2017) Parasitic survey on introduced monk parakeets (*Myiopsitta monachus*) in Santiago, Chile. *Revista Brasileira de Parasitologia Veterinária* 26, 129–135.

Bucher, E.H., Martín, L.F., Martella, M.B. and Navarro, J.L. (1991) Social behavior and population dynamics of the monk parakeet. *Proceedings of the International Ornithological Congress* 20, 681–689.

Burgio, K.R., van Rees, C.B., Block, K.E., Pyle, P., Patten, M.A., *et al.* (2016) Monk Parakeet (*Myiopsitta monachus*). In: Rodewald, P.G. (ed.) *The Birds of North America*. Cornell Lab of Ornithology, Ithaca, New York. Available at: https://birdsna.org/Species-Account/bna/species/monpar/ (accessed 25 October 2019).

Cardador, L., Lattuada, M., Strubbe, D., Tella, J.L., Reino, L., *et al.* (2017) Regional bans on wild-bird trade modify invasion risks at a global scale. *Conservation Letters* 10, 717–725.

Carrington, D. (2014) Eradication efforts bring UK's monk parakeet numbers down to last 50. *The Guardian*, 24 September. Available at: https://www.theguardian.com/environment/2014/sep/24/monk-parakeets-parrots-uk (accessed 25 October 2019).

Chng, S.C.L. and Eaton, J.A. (2016) Snapshot of an on-going trade: an inventory of birds for sale in Chatuchak weekend market, Bangkok, Thailand. *BirdingAsia* 25, 24–29.

Christie, D. (1992) Nature news summer 1992. *N. B. Naturalist* 19, 41–43.

Crins, W.J. (2004) Ontario bird records committee report for 2003. *Ontario Birds* 22, 54–74.

Csurhes, S. (2016) Invasive animal risk assessment: Monk/Quaker Parakeet *Myiopsitta monachus*. Department of Agriculture and Fisheries, Queensland Government, Queensland, Australia. Available at: www.daf.qld.gov.au/__data/assets/pdf_file/0005/65597/IPA-Monk-Parakeet-Risk-Assessment.pdf (accessed 25 October 2019).

da Silva Carneiro, I.F. (2017) New neighbours to gossip about: people's attitudes towards alien parakeets in Porto. MSc thesis, Universidade do Porto, Porto, Portugal.

Dubois, P.J. and Cugnasse, J.-M. (2015) Les populations d'oiseaux allochtones en France en 2014 (3e enquête nationale) [Third invasive bird species Survey in France in 2014]. *Ornithos* 22, 72–91.

Eberhard, J.R. (1998) Breeding biology of the monk parakeet. *Wilson Bulletin* 110, 463–473.

Edelaar, P., Roques, S., Hobson, E.A., Gonçalves da Silva, A., Avery, M.L., *et al.* (2015) Shared genetic diversity across the global invasive range of the monk parakeet suggests a common restricted geographic origin and the possibility of convergent selection. *Molecular Ecology* 24, 2164–2176.

Eguchi, K. and Amano, H.E. (2004) Invasive birds in Japan. *Global Environmental Research* 8, 29–39.

European Commission (2005) Avian influenza: EU bans imports of captive live birds from third countries. European Commission press release, 25 October 2005. Available at: https://europa.eu/rapid/press-release_IP-05-1351_en.htm (accessed 25 October 2019).

Falcón, W. and Tremblay, R.L. (2018) From the cage to the wild: introductions of psittaciformes to Puerto Rico. *PeerJ* 6, e5669.

Fox, A.D., Heldbjerg, H. and Nyegaard, T. (2015) Invasive alien birds in Denmark. *Dansk Ornitologisk Forening Tidsskrift* 109, 193–205.

Godbeer, K.D. (2014) Cayman's invasive monk parakeet. *PsittaScene* 26.2, 12–15.

GT IBMA (2018) *Myiopsitta monachus*. Information base on biological invasions in aquatic environments. National Working Group Biological Invasions in Aquatic Environments. IUCN France and French Agency for Biodiversity. Available at: www.gt-ibma.eu/espece/myiopsitta-monachus/ (accessed 20 November 2018).

Hobson, E.A., Avery, M.L. and Wright, T.F. (2014) The socioecology of monk parakeets: insights into parrot social complexity. *Auk* 131, 756–775.

Hobson, E.A., Smith-Vidaurre, G. and Salinas-Melgoza, A. (2017) History of nonnative monk parakeets in Mexico. *PLoS One* 12, e0184771.

Hudec, K. (2015) Non-native species of birds in the Czech Republic. Special supplement. Zooreport. Available at: www.zoobrno.cz/img/UK%20profi%20cerven%202015%20K03.pdf (accessed 25 October 2019).

Hyman, J. and Pruett-Jones, S. (1995) Natural history of the monk parakeet in Hyde Park, Chicago. *Wilson Bulletin* 107, 510–517.

Iriarte, J.A., Lobos, G.A. and Jaksic, F.M. (2005) Invasive vertebrate species in Chile and their control and monitoring by governmental agencies. *Revista Chilena de Historia Natural* 78, 143–154.

Kalodimos, N.P. (2013) First account of a nesting population of monk parakeets, *Myiopsitta monachus*, with nodule-shaped bill lesions in Katehaki, Athens, Greece. *Bird Populations* 12, 1–6.

Kwong, W.C. (2013) Monk parakeets in Singapore. Bird Ecology Study Group site, 13 February 2013. Available at: www.besgroup.org/2013/02/22/monk-parakeets-in-singapore/ (accessed 25 October 2019).

Lim, K.S. (2009) Bird report: September 2009. *Singapore Avifauna* 23, 1–13. Available at: www.nss.org.sg/wildbirdsingapore/SINAV_Vol_23_No_09_Sep_09Edit25022010.pdf (accessed 7 February 2020).

Linz, G.M., Bucher, E.H., Canavelli, S.B., Rodriguez, E. and Avery, M.L. (2015) Limitations of population suppression for protecting crops from bird depredation: a review. *Crop Protection* 76, 46–52.

MacGregor-Fors, I., Calderón-Parra, R., Meléndez-Herrada, A., López-López, S. and Schondube, J.E. (2011) Pretty, but dangerous! Records of non-native monk parakeets (*Myiopsitta monachus*) in Mexico. *Revista Mexicana de Biodiversidad* 82, 1053–1056.

MaghrebOrnitho (2018) Breeding of monk parakeet at Casablanca. Moroccan Birds blog (moroccanbirds.blogspot.com). Available at: www.magornitho.org/2018/03/monk-parakeet-breeding-casablanca/ (accessed 22October 2018).

Matias, R. (2012) Aves exóticas em Portugal: ano de 2011. *Anuário Ornitológico* 9, 57–65.

Menzer, K. (2006) With bird tower, TXU's winging it. Coastal Bend Companion Bird Club & Rescue Mission, September 2006, 10. Available at: http://cbcbirdclub.org/newsletters/2006/2006-09.pdf (accessed 30 October 2019).

Mori, E., Di Febbraro, M., Foresta, M., Melis, P., Romanazzi, E., *et al.* (2013) Assessment of the current distribution of free-living parrots and parakeets (Aves: Psittaciformes) in Italy: a synthesis of published data and new records. *Italian Journal of Zoology* 80, 158–167.

Moscatello, B. (2003) Preliminary review of the status of monk parakeets in New Jersey. *New Jersey Birds* 29, 3–5.

Navarro, J.L., Martella, M.B. and Bucher, E.H. (1992) Breeding season and productivity of monk parakeets in Cordoba, Argentina. *Wilson Bulletin* 104, 413–424.

Nebot, J.C. (1999) First report on the rose-ringed parakeet (*Psittacula krameri*) in Venezuela and preliminary observations on its behaviour. *Ornitologia Neotropical* 10, 115–117.

Neidermyer, W.J. and Hickey, J.J. (1977) The monk parakeet in the United States, 1970–75. *American Birds* 31, 273–278.

Nixon, J. (2018) Brussels parakeets make the world a better place – or don't. TheSupercargo blog, 18 July 2018. Available at: www.thesupercargo.com/brussels-parakeets/ (accessed 8 November 2018).

Postigo, J.L., Shwartz, A., Strubbe, D. and Muñoz, A.R. (2017) Unrelenting spread of the alien monk parakeet *Myiopsitta monachus* in Israel. Is it time to sound the alarm? *Pest Management Science* 73, 349–353.

Raffaele, H.A. and Wiley, J.W. (2014) *Wildlife of the Caribbean*. Princeton University Press, Princeton, New Jersey.

Reed, J.E., McCleery, R.A., Silvy, N.J., Smeins, F.E. and Brightsmith, D.J. (2014) Monk parakeet nest-site selection of electric utility structures in Texas. *Landscape and Urban Planning* 129, 65–72.

Rodríguez-Pastor, R., Senar, J.C., Ortega, A., Faus, J., Uribe, F. and Montalvo, T. (2012) Distribution patterns of invasive Monk parakeets (*Myiopsitta monachus*) in an urban habitat. *Animal Biodiversity and Conservation* 35.1, 107–117.

Russello, M.A., Avery, M.L. and Wright, T.F. (2008) Genetic evidence links invasive monk parakeet populations in the United States to the international pet trade. *BMC Evolutionary Biology* 8, 217.

Senar, J.C., Domènech, J., Arroyo, L., Torre, I. and Gordo, O. (2016) An evaluation of monk parakeet damage to crops in the metropolitan area of Barcelona. *Animal Biodiversity and Conservation* 39, 141–145.

Sevenair, J.P. (2012) Monk parakeet (*Myiopsitta monachus*) nests in metropolitan New Orleans before and after Hurricane Katrina. *Journal of Louisiana Ornithology* 9, 8–15.

Sol, D., Santos, D.M., Feria, E. and Clavell, J. (1997) Habitat selection by the monk parakeet during colonization of a new area in Spain. *Condor* 99, 39–46.

South, J.M. and Pruett-Jones, S. (2000) Patterns of flock size, diet, and vigilance of naturalized monk parakeets in Hyde Park, Chicago. *Condor* 102, 848–854.

Souviron-Priego, L., Román Muñoz, A., Olivero, J., Vargas, M. and Fa, J.E. (2018) The legal international wildlife trade favours invasive species establishment: the monk and ring-necked parakeets in Spain. *Ardeola* 65, 233–246.

Spreyer, M., and Bucher, E. (1998) Monk parakeet (*Myiopsitta monachus*). In: Poole, A. and Gill, F. (eds) *The Birds of North America*, No. 322. The Birds of North America, Inc., Philadelphia, Pennsylvania.

Symes, C.T. (2014) Founder populations and the current status of exotic parrots in South Africa. *Ostrich* 85, 235–244.

Tayleur, J. (2010) A comparison of the establishment, expansion and potential impacts of two introduced parakeets in the United Kingdom. In: *British Ornithologists' Union Proceedings 2008: The Impacts of Non-native Species*, pp. 1–12. Available at: https://www.bou.org.uk/bouprocnet/impacts-of-non-native-species/ (accessed 25 October 2019).

Tillman, E.A., van Doorn, A. and Avery, M.L. (2001) Bird damage to tropical fruit in south Florida. *Wildlife Damage Management Conference* 9, 47–59.

Vall-llosera, M. and Cassey, P. (2017) Leaky doors: private captivity as a prominent source of bird introductions in Australia. *PLoS One* 12, e0172851.

van Kleunen, A., Kampichler, C. and Sierdsema, H. (2014) De verspreiding van Halsbandparkiet en andere in het wild voorkomende papegaaiachtigen (Psittaciformes) in Nederland [Distribution of free-living *Psittaciformes* in The Netherlands]. Sovon-rapport 2014/31. Sovon Dutch Centre for Field Ornithology, The Netherlands (in Dutch with English summary). Available at: www.sovon.nl/sites/default/files/doc/Rap_2014-31_Verspreiding_papegaaiachtigen.pdf (accessed 7 February 2020).

Viana, I.R., Strubbe, D. and Zocche, J.J. (2016) Monk parakeet invasion success: a role for nest thermoregulation and bactericidal potential of plant nest material? *Biological Invasions* 18, 1305–1315.

Wittenberg, R. (ed.) (2005) An inventory of alien species and their threat to biodiversity and economy in Switzerland. CABI Bioscience Switzerland Centre report to the Swiss Agency for Environment, Forests and Landscape. The Environment in Practice, No. 0629. Federal Office for the Environment, Bern, Switzerland.

11 House Sparrow (*Passer domesticus* Linnaeus, 1758)

Haley E. Hanson*, Jaime E. Zolik and Lynn B. Martin*
Global and Planetary Health, University of South Florida, Tampa, FL 33612, USA

Citation: Hanson, H.E., Zolik, J.E. and Martin, L.B. (2020) House Sparrow (*Passer domesticus* Linnaeus, 1758). In: Downs, C.T. and Hart, L.A. (eds) *Invasive Birds: Global Trends and Impacts*. CAB International, Wallingford, UK, pp. 85–96.

11.1 Common Names

House Sparrow, English Sparrow.

11.2 Nomenclature

A range of House Sparrow (*Passer domesticus* Linnaeus, 1758) subspecies are described: *P. d. bactrianus*, *P. d. biblicus*, *P. d. domesticus*, *P. d. hufufae*, *P. d. hycranus*, *P. d. indicus*, *P. d. niloticus*, *P. d. parkini*, *P. d. persicus*, *P. d. rufidorsalis* and *P. d. tingitarus*.

11.3 Distribution

The House Sparrow is one of the most widely dispersed species globally. The native distribution of the House Sparrow includes: Afghanistan, Albania, Algeria, Armenia, Austria, Azerbaijan, Bahrain, Bangladesh, Belarus, Belgium, Bhutan, Bosnia and Herzegovina, Bulgaria, China, Croatia, Cyprus, Czech Republic, Denmark, Egypt, Estonia, Finland, France, Georgia, Germany, Greece, Hungary, India, Iran, Iraq, Ireland, Israel, Jordan, Kazakhstan, Kosovo, Kyrgyzstan, Latvia, Lebanon, Libya, Lithuania, Luxembourg, Macedonia, Moldova, Mongolia, Montenegro, Morocco, Myanmar, Nepal, Netherlands, Norway, Oman, Poland, Portugal, Qatar, Romania, Russia, Saudi Arabia, Serbia, Slovakia, Slovenia, South Sudan, Spain, Sudan, Sweden, Switzerland, Syria, Tajikistan, Tunisia, Turkey, Turkmenistan, Ukraine, United Arab Emirates, the UK, Uzbekistan, and Yemen (Fig. 11.1).

The House Sparrow introduced distribution includes: Angola, Anguilla, Argentina, Aruba, Australia, Barbados, Belize, Botswana, Brazil, British Virgin Islands, Burundi, Cameroon, Canada, Cape Verde, Cayman Islands, Chad, Chile, Columbia, Comoro Islands, Costa Rica, Côte d'Ivoire, Cuba, Curaçao, Democratic Republic of the Congo, Dominican Republic, Ecuador, El Salvador, French Guiana, The Gambia, Ghana, Guadeloupe, Guatemala, Guinea Bissau, Guyana, Haiti, Honduras, Jamaica, Kenya, Lesotho, Liberia, Madagascar, Malawi, Mauritania, Mauritius, Mayotte, Mexico, Mozambique, Namibia, New Caledonia, New Zealand, Nicaragua, Niger, Nigeria, Panama, Papua New Guinea, Paraguay, Peru, Puerto Rico, Republic of the Congo, Rwanda, Réunion, Senegal, Singapore, South Africa, St Kitts and Nevis, St Vincent and the Grenadines, Suriname, Swaziland, Tanzania, Tasmania, The Bahamas, Trinidad and Tobago, Turks and Caicos Islands, Uganda, USA, Uruguay, US Virgin Islands, Vanuatu, Venezuela, Zambia, and Zimbabwe (Fig. 11.1).

It is unknown whether the House Sparrow's distribution was part of a natural expansion or from introduction event in the following: Cambodia, China (parts), Djibouti, Eritrea, Ethiopia, Laos, Thailand, Vietnam and Western Sahara.

11.4 Description

House Sparrows range in size from 140 to 180 mm in length and weigh between 21 and 39.5 g (Summers-Smith, 1988). The species is sexually dimorphic. Male plumage differs from females on the head, throat, breast and wing (Fig. 11.2). Male House Sparrows have grey crests, black lores and black post-ocular stripes with noticeable white post-ocular spots (Fig. 11.2). The black post-ocular stripes fade into a dark brown colour that extends down the neck and around a grey cheek. The abdomen is grey with a noticeable black badge underneath the black bill on the throat that can vary in size (Sánchez-Tójar *et al.*, 2018). The upper back, tail feathers and wings are brown and black. Each wing has one small white wing bar. Female House Sparrows

*Corresponding authors: haleyehanson@gmail.com or lbmartin@usf.edu

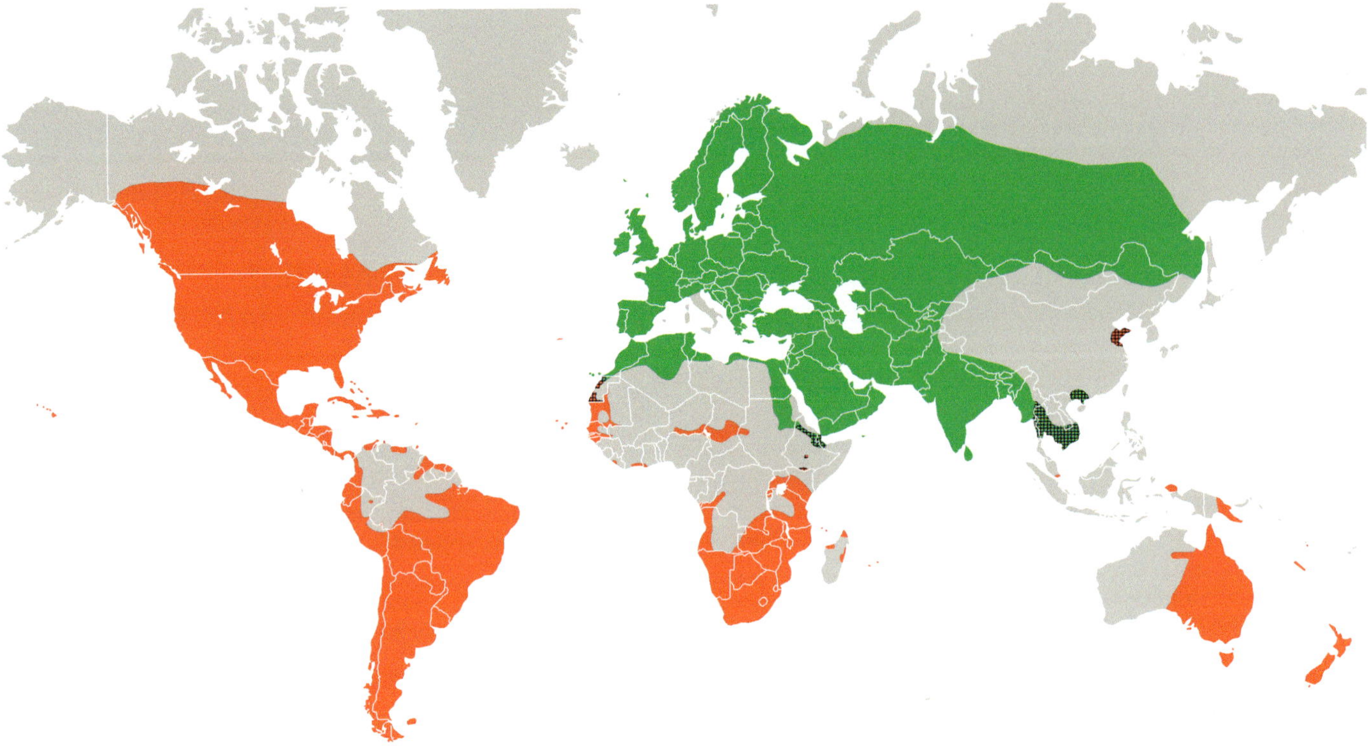

Fig. 11.1. Global distribution of the House Sparrows showing native (green) and introduced (red) distributions. Crosshatched areas represent House Sparrows from unknown origins, with the colour representing presumed status as part of a natural expansion or from an introduction. The introduced range was constructed largely through eBird observations based on three criteria: (i) multiple sightings (more than 20 birds seen in the same area over multiple years); (ii) pictures that could be positively identified; and/or (iii) sightings from sources we could verify had experience with bird identification (i.e. ornithologists, ornithological societies, nature guides, naturalists) (eBird: www.ebird.org (accessed 26 October 2019)). Primary literature was used when available (Azpiroz *et al.*, 2006; Ingels *et al.*, 2007; Waltert *et al.*, 2007; Amonsin *et al.*, 2008; Barlow and Dodman, 2011; Gilbert *et al.*, 2012; Tang *et al.*, 2013; Gedeon *et al.*, 2015; Andrew and Griffith, 2016; Goodman *et al.*, 2017) and several other databases and sources were referenced (Barlow and Wacher, 2006; Browne, 2010; Isenmann *et al.*, 2010; African Bird Club, 2018: www.africanbirdclub.org (accessed 15 October 2018); Global Biodiversity Information Facility, 2018: www.gbif.org/ (accessed 28 September 2018); Liedén and Brouwer, 2018). The map was adapted from Summers-Smith (1988) with supporting information from Anderson (2006). Note that some regions had relatively fewer observations, leading to conservative distribution estimates, particularly in Africa and South America).

are muted in colour overall (Fig. 11.3). They have brown crests and lighter brown post-ocular stripes. Females lack the distinctive black head markings of males. Female cheeks, bills and abdomens are grey-brown, and their upper back, tail and wing feathers are streaked with brown and dark brown. Wing bars are muted compared with males. Colours in both sexes become more pronounced during the breeding season (Summers-Smith, 1963). Juvenile House Sparrows are similar to females and thus difficult to distinguish based on plumage (Anderson, 2006). Subspecies differ slightly in size and plumage (reviewed by Summers-Smith 1988).

11.5 Diet

In more rural habitats, House Sparrows feed predominantly on grains, weed and grass seeds; their diet is up to 97% granivorous in the non-breeding season (Kalmbach, 1940; Hammer, 1948; Simenonov, 1964; Beri *et al.*, 1972; Macmillan, 1981; Anderson, 2006). During the breeding season, there is a marked increase in the consumption of animal matter. Females, in particular, consume more insects during the breeding season than at other times in the year to keep up with resource demands of egg-laying (Pinowska, 1975; Anderson, 2006). In urban environments, diets can differ dramatically from rural populations; fewer cereal grains are available and more human refuse is consumed (Summers-Smith, 1988). Overall, House Sparrows are opportunistic foragers, particularly of human resources.

After hatching, hatchling House Sparrows mostly subsist on insects for the first 3 days; thereafter, the percentage of animal material in their diets decreases (review by Anderson 2006). However, this balance varies depending on food availability. In an urban German population, nestlings were fed a high proportion of food obtained from humans, including dog

Fig. 11.2. A male House Sparrow. ('*Passer domesticus male*' by Lip Kee is licensed under CC BY-SA 2.0, https://creativecommons.org/licenses/by-sa/2.0/.)

Fig. 11.3. A female House Sparrow. ('House Sparrow ('*Passer domesticus*') in Kolkata, West Bengal, India' by J.M. Garg is licensed under CC BY-SA 3.0, https://creativecommons.org/licenses/by-sa/3.0/deed.en.)

food, bird food and bread (Bower, 1999). Interestingly, nestlings are found to have a high degree of plasticity in the activity of their intestinal digestive enzymes when fed different diets during ontogeny (Brzek *et al.*, 2009, 2011). Changes in enzyme production can occur within just 24 h of an experimental diet change, which may be a mechanism that allows individuals to exploit different diets as resource availability fluctuates (Rott *et al.*, 2017).

Introduced populations probably encounter unfamiliar resources more frequently, or at least they probably did at the time of initial arrival to new areas. House Sparrows have different proclivities to eat novel foods, depending on population age. For example, introduced House Sparrows in Panama consumed novel foods quicker than their conspecifics from a longer-established population in New Jersey, USA; feeding rate did not differ between the populations (Martin and Fitzgerald, 2005). Similarly, younger populations in Kenya consumed novel foods more quickly than other introduced populations established for longer (Liebl and Martin, 2014). A proclivity to eat novel foods may allow a wider variety of resources to be exploited, but it also has risks (Martin and Fitzgerald, 2005). One way to minimize risk is through observation of the feeding choices of conspecifics.

Rarely do House Sparrows feed alone. They are most often seen in flocks ranging in size from a few to several thousand birds (Summers-Smith, 1963; Popp, 1988). Flocking, and in particular communal feeding, increases foraging efficiency by decreasing the amount of time spent on vigilance, although this effect is density dependent (Anderson, 2006). Foraging in flocks may have other advantages, including social learning. During feeding, House Sparrows orient themselves to keep other foragers in view, thus stimulating others to feed at the sight of conspecifics foraging (Caraco and Bayham, 1982). Juveniles are highly stimulated to forage, and will eat novel foods upon the sight of adults doing so (Turner, 1964). Adult House Sparrows are also more likely to consume novel foods after observing conspecifics eating those foods (Fryday and Greig-Smith, 1994). This sensitivity to and the propensity to learn from the behaviour of others may enable the consumption of novel foods to spread quickly through introduced populations in new areas.

11.6 Introduction and Invasion Pathways

House Sparrows have been introduced all over the world, both intentionally and unintentionally. In North America, eight pairs were brought from England and intentionally released in Brooklyn, New York, in 1851 in an effort to control cankerworms (Barrows, 1889; Long, 1981). This introduction was followed by the release of approximately 100 birds into Brooklyn in 1852 and 1853, brought via the *Europa* steamship from Liverpool (Barrows, 1889). However, the initial release in 1851 is thought to be the first successful introduction into North America, highlighting the ability of House Sparrows to survive with a small founding population (Barrows, 1889; Moulton *et al.*, 2010). Since this period, there have been many more releases across approximately 100 cities across the USA and Canada, most from European stock or previously established populations across the USA (Long, 1981). These introductions were made either to control insects or owing to the nostalgia of some Europeans for familiar bird species (i.e. naturalization societies) (Long, 1981; Anderson, 2006). From 1853 to 1910, House Sparrows expanded their geographic range in the US at approximately 72 km/year (see Anderson, 2006, for discussion), leading to a generally ubiquitous presence across the country by 1910 (Robbins, 1973). From 1910, they began their spread into Mexico and eventually Central America, where they reached Panama around the 1980s (Robbins, 1973; Anderson, 2006).

Their next introduction took place in Melbourne, Australia in 1862 where, birds shipped from India were released (Andrew and Griffith, 2016). This was followed by a shipment of birds from Great Britain in 1863 and from Germany shortly after (Andrew and Griffith, 2016). House Sparrows were released in Sydney in 1865 and in Adelaide between 1863 and 1865 where they successfully established (Andrew and Griffith, 2016). Additionally, they were introduced to New Zealand between 1859 and 1871 (Long, 1981; Anderson, 2006). These sparrows were intentionally released into both Australia and New Zealand for very similar reasons as their release into North America (Long, 1981).

In South America, House Sparrows were brought to Buenos Aires, Argentina, from Europe in 1872 or 1873 to control psychid moth populations (Long, 1981). From this initial population, House Sparrows were brought into Uraguay, Chile and Peru (Anderson, 2006). House Sparrows from Europe were also introduced into Santiago, Chile, in 1904, and Rio de Janeiro, Brazil, in 1905, brought specifically to Brazil to help keep mosquito populations down (Long, 1981). Their range expansion continues in Brazil and is facilitated by roads, including the Brasilia-Belém and Transamazon highways (Summers-Smith, 1988; Anderson, 2006). They were also introduced into Venezuela, and were reported breeding for the first time in 2006 (Azpiroz *et al.*, 2006). House Sparrows are now found in every country in South America (Fig. 11.1).

In Africa, House Sparrows were intentionally released into Durban, South Africa, between 1890 and 1900, and into East London, South Africa, from Great Britain between 1907 and 1930 (Summers-Smith, 1963; Lever, 1987). After 1948, their rate of expansion increased from 7 to 30–101 km/year, allowing them to establish throughout most of southern Africa to southern Tanzania by the 1980s (Vierke, 1970; Summers-Smith, 1988). They were accidentally introduced to Mombasa, Kenya, most likely as stowaway birds on a cargo ship from South Africa in the 1950s (Anderson, 2006). They have since spread across the country along the Mombasa highway and into Uganda, possibly facilitated by trucks or trains (Schrey *et al.*, 2014). Similarly, House Sparrows have been introduced into Senegal from South Africa in the 1970s, most likely by ship (Lever, 1987; Summers-Smith, 1988, 1992; Anderson, 2006).

In addition to introduction events, House Sparrows have also expanded from their native range into new territory. In Russia, they expanded their range in the 19th century at a rate of between 23.0 and 27.9 km/year as agriculture grew and as the trans-Siberian railway was completed (Summers-Smith, 1963; van den Bosch *et al.*, 1992; Anderson, 2006). Their expansion into China is less well known. They are also expanding in south-eastern Asia into Thailand, Singapore, Vietnam, Cambodia and Laos; however, it is unknown whether this represents an introduction or if it is a naturally occurring range expansion (Fig. 11.1).

11.7 Breeding Behaviour

House Sparrows are quite flexible in nest-site choice. They are most commonly found nesting in anthropogenic structures such as roof cavities, eaves, walls and nest boxes (Anderson, 2006). They also utilize natural cavities or those made by other species, such as holes in trees and cliffs, and have been found excavating their own nesting sites (Philipson, 1938; Pitman, 1961). Once thought to be limited exclusively to cavities, they are now known to nest frequently in vegetation (Sheldon and Griffith, 2017). Once they have selected a site, they exhibit nest-site fidelity during the breeding season, as well as in successive years (Summers-Smith, 1963). During the breeding season, female House Sparrows in England were found at the same nest during successive broods 72% of the time and males 76% of the time (Dawson 1972). Across their range, males have higher nest-site fidelity than females, generally returning to the same site every year, regardless of whether or not they have the same mate (Sappington, 1977; Anderson, 2006; Morrison *et al.*, 2008).

Nest composition appears opportunistic, mostly made of vegetable material such as straw, but also containing clay, sand, cloth and even dung (Heij, 1986). Several studies have found nests containing antimalarial and anti-arthropod parasite secondary compounds including the leaves of margosa or neem (*Azadirachta indica*) and the Krishnachura tree (*Caesalpinia pulcherrima*) (Sengupta and Shrilata, 1997; Anderson, 2006). Both male and females participate in nest building, beginning before egg-laying (Anderson, 2006). Nests in trees are generally spherical, whereas those under cover or inside buildings tend to be a flattened oval without a roof (Heij, 1986). Both types are used throughout the year for roosting (Heij, 1986).

Male House Sparrows select nesting sites and subsequently advertise for mates by vocal and visual displays (Summers-Smith, 1963). Females select males based on the display and by the location of nest sites (Anderson, 2006). Once bonded, the pair remains together for the entire attempt, but may stay together for the entire season or for multiple years (Summers-Smith, 1963; Anderson, 2006). Pairs build nests, defend nests against intruders, feed young and remove faecal sacs (Anderson, 2006). Upon laying eggs, females develop vascularized brood patches, absent of feathers, and incubate the eggs and brood young (Anderson, 2006).

Although socially monogamous, polygyny leading to extra-pair offspring is common in House Sparrows. The rate of extra-pair paternity spans is 1–26% (Stewart *et al.*, 2015). Polygynous males tend to have higher reproductive success, whereas females have lower reproductive success (Veiga, 1990). Females defend their monogamous status by showing aggression towards other females near their nests (Veiga, 1992). The incidence of extra-pair offspring is related to food availability and environmental conditions; however, food supplementation does not increase the number of extra-pair young (Westneat *et al.*, 2015).

House Sparrows often begin breeding during their first year of life. Both first-year males and females have lower reproductive success than older conspecifics, which is attributed to their later onset of breeding than older House Sparrows (Hatch and Westneat, 2007). Timing of breeding varies geographically, with temperate populations initiating a short breeding season lasting from spring to summer, and tropical and subtropical species having longer breeding seasons, not uncommonly lasting throughout the year (Naik and Mistry, 1980; Anderson, 2006). Photoperiod, temperature and rainfall all contribute to the onset of breeding, and House Sparrows may adjust the timing of breeding to match local conditions (Seel, 1968; Krogstad *et al.*, 1996; Anderson, 2006).

House Sparrows exhibit extreme variation in clutch size. Clutch size varies with subspecies, latitude and time of laying

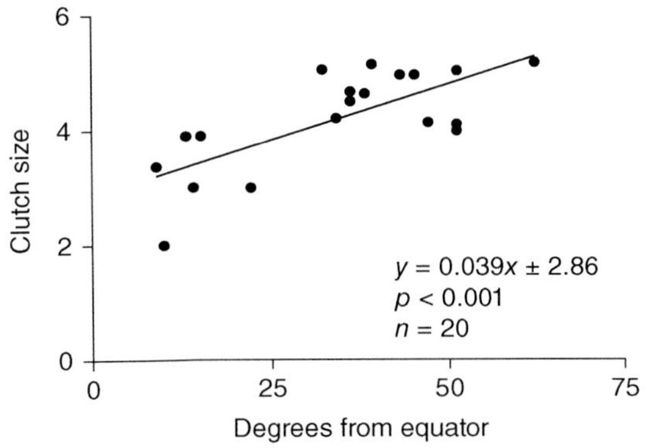

Fig. 11.4. Clutch size decreases towards the equator in House Sparrows; equation and associated statistics are from linear regression of data in Table 11.1.

within season (review by Anderson, 1994, 2006). Clutch size ranges from one to nine eggs, with clutch size increasing with distance from the equator (Fig. 11.4) (Naik and Mistry, 1972; Gil-Delgado *et al.*, 1979; Anderson, 2006). In a comparison of 20 populations (both native and introduced), populations in tropical areas close to the equator were found to have smaller clutches than non-tropical populations, but tropical populations generally had more offspring per year (Table 11.1). Additionally, there is variation in hatching, fledging and nest success among populations (Summers-Smith, 1963; Seel, 1968; North, 1968, 1973; Anderson, 1973, 1994; Mitchell *et al.*, 1973; Sappington, 1977; Murphy, 1978; Pitts, 1979; Fleischer, 1982; Anon., 1986; Thurber, 1986; Singer and Yom-Tov, 1988; Krogstad *et al.*, 1996; Nhlane, 2000). Interestingly, parents from these populations spend different amounts of time in the nest, during both incubation and nestling periods, which could be due to differences in ambient conditions.

11.8 Habitat

Although they are generally associated with human habitation and agriculture, House Sparrows reach peak abundance in urban areas and have a relatively low abundance in undeveloped and natural sites (Summers-Smith, 1988; Magudu and Downs, 2015). However, House Sparrows are sometimes able to sustain populations in remote sites away from human development. For example, they have been found breeding miles away from human settlements in Australia and New Zealand, and maintained a population on a volcanic island following an eruption that destroyed the only human settlement (Wodzicki, 1956; Hobbs, 1958). In the Middle East and Central Asia (Iran and Kazakhstan), *P. d. bactrianus* are not associated with human activity and are found in semi-natural grasslands (Sætre *et al.*, 2012).

House Sparrows persist in a range of ambient temperatures and habitat types in both their native and introduced ranges. They occupy latitudes of 71.1°N (about 5° above the Arctic Circle) and 54.9°S in sub-Antarctic habitat (Anderson, 2006). They thrive in equatorial Ecuador and Kenya and are expanding their range in Amazonia. They are found in Death Valley, California, where the mean maximum daily temperature is above 43°C in the summer (Grinnell, 1919). They even endure conditions in Churchill, Manitoba, where the mean minimum temperature is –16°C (Kendeigh, 1976). They breed in alpine regions ranging up to 4500 m in elevation and also have been seen breeding 640 m below ground in a coal mine (Summers-Smith, 1988; Singh *et al.*, 2013).

11.9 Impacts

Upon introduction to a new area, House Sparrows interact and may thus compete with other species. A well-known example is with Eastern Bluebirds (*Sialia sialis*) over nesting sites in North America. House Sparrows have aggressive interactions with adult and nestling bluebirds, occasionally leading to mortality (Gowaty, 1984). In one study, House Sparrows caused 29.5% of nest failures over 27 years in Green Bay, Wisconsin (Radunzel *et al.*, 1997). Similar aggressive interactions were seen in Brazil with native Monk Parakeets (*Myiopsitta monachus*) (Wagner, 2012); in this instance, multiple House Sparrows were aggressive together to usurp Monk Parakeet nests. House Sparrows can also negatively impact avian community structure and diversity. In Western Mexico, introduced House Sparrow presence was correlated with a loss of native species in eight different habitat types compared with non-invaded areas (MacGregor-Fors *et al.*, 2009). It was suggested that this may be due to House Sparrow aggression towards native species during feeding (MacGregor-Fors *et al.*, 2009).

In addition to competition, House Sparrows have been implicated in the transmission of many different pathogens due to their abundance and close association with humans and livestock. They are thought to play an important role in the transmission of several arthropod-borne viruses including eastern equine encephalitis virus, St Louis encephalitis virus, western equine encephalitis virus, West Nile virus and Buggy Creek virus (Anderson, 2006). In the case of West Nile virus, House Sparrows were implicated as one of the main agents of dispersal of this introduced pathogen across North America (Rappole and Hubálek, 2003). Furthermore, House Sparrows in Nebraska carrying Buggy Creek virus invaded native swallow nesting sites. Compared with Cliff Swallows (*Petrochelidon pyrrhonota*), House Sparrows had eight times the infection prevalence, which may in turn lead to increased exposure to Cliff Swallows (O'Brien *et al.*, 2011). House Sparrow nestlings in particular were highly infectious (O'Brien and Brown, 2012). House Sparrows are also susceptible to some strains of avian influenza virus, and may be involved in its transmission (Boon *et al.*, 2007). In the case of the H9N2 avian influenza strain, House Sparrows successfully spread infections to poultry (Iqbal *et al.*, 2013). Other highly virulent isolates have been obtained from House Sparrows including H5N1 (Perkins and Swayne, 2003; Boon *et al.*, 2007). Upon infection, sparrows shed virus for several days before the onset of symptoms, which suggests that they may play a role in transmission of these isolates (Brown *et al.*, 2009). House Sparrows can become infected with a wide range of bacterial pathogens (reviewed by Anderson 2006). They carry *Salmonella* spp., although their

Table 11.1. Life history characteristics of native and introduced populations of House Sparrows. Data was collected from published sources, except for Panama, Hungary and Brazil, which came from personal communications with individuals who have worked extensively on the species (L.B. Martin, unpublished data).

		Clutch size			Time in nest			Nest success				
Location	Latitude (°N)	Mean	SD	No./year	Eggs/female	Incubation	Nestling	Hatching	Fledging	Nest	Offspring/ pair/year	Reference
Panama	9	3.4	0.6	3.2	10.8	10.5	16.2	0.84	0.86	0.72	7.79	Martin (unpublished data)
Costa Rica	10	2.0	0.6									Fleischer (1982)
Zambia	13	3.9				11.5	17.0					Anon. (1986)
El Salvador	14	3.0										Thurber (1986)
Malawi	15	3.9	1.0			11.6	14.9	0.64	0.73	0.46		Nhlane (2000)
Brazil	22	3.0		2.0	6.0	13.0	13.0					Mallet-Rodrigues (unpublished data)
Israel	32	5.1	0.9	2.3	11.6	15.0	16.0	0.70	0.75	0.52	6.07	Singer and Yom-Tov (1988)
Mississippi	34	4.2	0.9	1.8	7.7	12.2	17.1	0.83	0.77	0.64	4.95	Sappington (1977)
Texas	34	4.2		3.45	14.5	11.3		0.61	0.67	0.41	5.93	Mitchell *et al.* (1973)
Oklahoma	36	4.5		1.7	7.6		13.9	0.50	0.65	0.33	2.46	North (1968)
Tennessee	36	4.7		2.5	11.7	11.8	14.6	0.52	0.65	0.34	3.93	Pitts (1979)
Missouri	38	4.6	0.9	2.6	11.8	14.1	14.8	0.65	0.63	0.41	4.80	Anderson (1973)
Kansas	39	5.1	0.3			10.7		0.53	0.68	0.36		Murphy (1978)
Wisconsin	43	5.0		1.5	7.4	11.7	15.4	0.51	0.61	0.31	2.31	North (1973)
Michigan	45	5.0	1.1			11.6	14.3	0.72	0.78	0.56		Anderson (1994)
Hungary	47	4.1	1.0									Lendvai and Liker (unpublished data)
Britain	51	4.0	0.8	2.1	8.4	10.8	15.0	0.88				Seel (1968)
Britain	51	4.1		2.1	8.6		18.0	0.71	0.50	0.36	3.06	Summers-Smith (1963)
Alberta	51	5.0	0.2			10.7		0.70	0.67	0.46		Murphy (1978)
Norway	62	5.2	0.9	2.5	12.9							Krogstad *et al.* (1996)

role in transmission to humans is debated (Lawson *et al.*, 2014; Rouffaer *et al.*, 2016). They have also been found to carry antibiotic-resistant strains of *Escherichia coli*, and may spread these strains to other taxa (Dolejská *et al.*, 2008; Sacristán *et al.*, 2014). House Sparrows are also able to transmit *Mycoplasma gallisepticum* to House Finches (*Haemorhous mexicanus*; Dhondt *et al.*, 2008).

House Sparrows are also considered an agricultural pest due to their propensity to consume substantial commercially produced grains. In the Czech Republic, House Sparrows caused 2–4% grain loss in a 1040 ha area of wheat and barley farms, with small farms losing up to 12.5% of their yield (Havlín, 1974). In India, they caused a loss of approximately 22.3% in millet fields and 18.6% in sorghum fields (Dhindsa *et al.*, 1984). In New Zealand, they were responsible for a loss of 1–25% in individual wheat fields and 6–19% in barley fields (Dawson, 1970). In addition to grains, House Sparrows can cause substantial losses of sunflowers (up to 51.3% in a field in Karaj, Iran; Khaleghizadeh, 2011) and fruits (Dawson and Bull, 1970; Manzoor *et al.*, 2013). In addition, House Sparrows exploit opportunities to feed on harvested grain, such as livestock feed. This feed has been found to be contaminated with House Sparrow excrement, which may lower its quality, as well as representing a mechanism by which House Sparrows might expose livestock to pathogens, as mentioned above (Daniels *et al.*, 2003).

Other negative impacts by House Sparrows include damage to buildings and property. In Australia, House Sparrows are listed as a species of serious problem for damaging buildings, citing that their nests built in rain gutters can lead to water damage (Bomford and Sinclair, 2002). Other concerns included damage to home gardens (Bomford and Sinclair, 2002).

11.10 Control

Dawson (1903) wrote, 'without question the most deplorable event in the history of American ornithology was the introduction of the English Sparrow.' As such, many manual and chemical methods have been used to control House Sparrow populations, both in their introduced and native range. Poisons, such as strychnine, have been widely used to treat food sources and were effective in killing potentially hundreds of thousands of House Sparrows in Germany in the 1950s (Anderson, 2006). While effective in the short term, the populations rebounded quickly due to an influx of dispersing birds from surrounding areas (Anderson, 2006). Non-lethal chemical methods have also been widely used. For example, quinine sulfate, tannic acid and methiocarb have been found to be aversive in taste to House Sparrows (Greig-Smith, 1987; Moran, 2001; Clapperton *et al.*, 2012; Day *et al.*, 2012). These treatments may become more effective in combination with visual cues such as colouring seeds with colours that are aversive to House Sparrows (Greig-Smith, 1987; Clapperton *et al.*, 2012). Other methods may decrease ability or preference for House Sparrow feeding including the selective breeding of plant characteristics to reduce feeding (Tipton *et al.*, 1970; Khaleghizadeh, 2011). For example in sunflowers, the height of the plant and angle of the head and stem may impact the rate of sparrow damage, and selecting for shorter plants with heads and stems angled more towards the ground may mitigate potential damage (Seiler and Rogers, 1987; Khaleghizadeh, 2011).

Manual methods to control House Sparrows on a small scale include trapping. Automatic nest box traps have been used with success during the breeding season. In one case, 14 House Sparrows were trapped at the same nest box (Stewart, 1973). In a study of effectiveness of trap types, funnel traps and elevator traps performed similarly (Fitzwater, 1981). Trapping is relatively labour intensive and ineffective at managing large numbers of House Sparrows, as they are prone to bycatch and birds become wary of them over time (Dearborn, 1912; Fitzwater, 1981).

Predators also affect House Sparrow populations. Domestic cats (*Felis catus*) predate House Sparrows frequently, in some cases accounting for 50% of House Sparrow mortality in one village (Binder and Scharfman, 2004; Anderson, 2006). Interestingly, cats are found to predate House Sparrows in relatively poor condition compared with those who are killed by other means (Baker *et al.*, 2008). Other predators have been attributed to the decline of native House Sparrows, namely the Eurasian Sparrowhawk (*Accipiter nisus*), as their populations in Europe have increased dramatically, particularly in urban areas (Anderson, 2006; Bell *et al.*, 2010). In urban areas, House Sparrows have been found to be the most common prey item of sparrowhawks (Frimer, 1989; Anderson, 2006). Other avian predators have been increasing in cities and have been found consuming House Sparrows, including Northern Goshawks (*Accipiter gentilis*), Common Kestrels (*Falco tinnunculus*) and Merlins (*Falco columbarius*) (Bosakowski and Smith, 1992; Sodhi and Oliphant, 1993; Kübler *et al.*, 2005; Rutz, 2006; Dulisz *et al.*, 2016).

House Sparrows are declining in their native range, which is probably due to a number of reasons. For example, they avoid nesting in newer buildings built after 1985 (Shaw *et al.*, 2008). New building design and construction methods may preclude sparrows from nesting by including the implementation of different building materials such as hard plastics, instead of wood, which do not break down as readily to form cavities that can be exploited (Fitzwater, 1981; Shaw *et al.*, 2008). Other factors, such as a heavier reliance on pesticides, more industrialized agriculture practices and better grain storage, may play a role in the declining numbers of House Sparrows in their native range (Summers-Smith, 2003).

In North America, there has been a recent push for citizen House Sparrow management, especially from Eastern Bluebird conservation groups (Larson *et al.*, 2016). Members of citizen science nest-monitoring programmes reported that they commonly removed House Sparrow nests being built, destroyed their eggs once laid, and changed the location or type of nest box to preclude them from nesting or disturbing other species (Larson *et al.*, 2016). About half who managed House Sparrows said they used lethal measures including trapping or shooting (Larson *et al.*, 2016). While it is unclear what effects these methods are having on populations, it is an interesting method to consider reducing House Sparrow numbers without prohibitive cost or governmental intervention.

11.11 Uses

Joseph Grinnell documented the arrival of House Sparrows to Death Valley, California, in 1919 and aptly titled his paper 'An experiment in nature' (Grinnell, 1919). Many scientists continue to use them as a sort of natural experiment to study various processes. For example, approximately 10,000 years ago, as humans began domesticating animals and relying more heavily on agriculture, they created a novel niche that allowed House Sparrows to begin their commensal relationship (Sætre *et al.*, 2012; Riyahi *et al.*, 2013). This commensalism has a single origin, most likely beginning in the Middle East, and allowed House Sparrows to greatly increase their range (Sætre *et al.*, 2012). It also presents an opportunity to investigate genes that underwent selection during their transition to commensalism. For example, *COL11A*, a gene that is involved in the regulation of facial development, and *AMY2A*, a gene in the amylase gene family involved in the digestion of starch, were found also to be important in the range expansion and adaptation to novel diets in humans (Ravinet *et al.*, 2018).

In the face of climate change, House Sparrows may be useful in identifying the mechanisms in which they, and potentially other species, alter their behaviour and physiology in response to ambient temperature. For example, smaller body size is correlated with high ambient temperatures in the breeding season rather than winter ambient temperatures in populations in Australia and New Zealand (Andrew *et al.*, 2018). In a study of North American House Sparrows from 33 different locations, House Sparrows differed in all 16 skeletal characteristics studied, and these morphological characteristics were correlated with temperature (Johnston and Selander, 1971). This differentiation occurred in relatively few generations after their introduction, lending support to their ability to adapt to local climatic conditions (Johnston and Selander, 1971). In addition to morphology, higher spring ambient temperatures have been correlated with the dispersal rate of House Sparrows in lower-quality habitats and with the earlier onset of lay date (Pärn *et al.*, 2011). As warming continues, they may continue to expand their range into new areas and alter their reproductive physiology in order to maximize the number of clutches they have each year.

As House Sparrows are generally abundant in urban areas, they have also been used as bioindicators for environmental contaminants, such as air pollution and heavy metals, including lead (Swaileh and Sansur, 2006; Herrera-Dueñas *et al.*, 2014; Cid *et al.*, 2018). Furthermore, as they tend to live near human habitation and are competent hosts of many arboviruses, they have been proposed as sentinel organisms for the detection of West Nile virus (although the efficacy is disputed), St Louis encephalitis virus, eastern equine encephalitis virus and western equine encephalitis virus (Williams *et al.*, 1971; McLean *et al.*, 1983; Darbro and Harrington, 2006; Millins *et al.*, 2011).

11.12 Notes

Introduced House Sparrows exhibit considerable phenotypic variation among populations. Behaviourally, House Sparrows from recently introduced populations are less neophobic, more exploratory and more apt to eat novel foods than their conspecifics from longer-established populations (Martin and Fitzgerald, 2005; Liebl and Martin, 2012, 2014). Compared with native populations, they also disperse more often and over longer distances (Skjelseth *et al.*, 2007). Physiologically, they differ in the regulation of glucocorticoids and immune genes (Liebl and Martin, 2012; Martin *et al.*, 2014, 2015, 2017). Introduced House Sparrows also exhibit phenotypic plasticity in traits such as vigilance during feeding, clutch and egg size, and nestling regulation of digestive enzymes (Brzek *et al.*, 2011; Wetzel *et al.*, 2012; Westneat *et al.*, 2015; Boujja-Miljour *et al.*, 2018). Strong selection for plasticity could facilitate House Sparrow range expansions by allowing them to adapt rapidly to environmental conditions and overcome challenges.

It has also been argued that epigenetic mechanisms may be one way for phenotypic variation and plasticity to be maintained, even in the face of genetic founder effects and bottlenecks that may occur upon introduction. In Kenya, where much phenotypic variation is seen, House Sparrows had lower genetic diversity than conspecifics from other introduced and native areas (Schrey *et al.*, 2011). Interestingly, low genetic diversity was correlated with high epigenetic diversity in Kenya, suggesting that there may be a compensatory relationship (Liebl *et al.*, 2013). Recently, it was proposed that individuals may differ in epigenetic potential, or their capacity for epigenetic mechanisms to work within the genome and give rise to phenotypic plasticity (Kilvitis *et al.*, 2017). Upon testing, introduced populations of House Sparrows had increased epigenetic potential in two of the three Toll-like receptor genes investigated (Hanson *et al.*, unpublished data). Whereas additional studies will need to access the functional relevance, these results together suggest that mechanisms underlying phenotypic plasticity and variation may facilitate House Sparrow range expansions and may be a reason why they are able to successfully establish upon introduction.

11.13 References

Amonsin, A., Choatrakol, C., Lapkuntod, J., Tantilertcharoen, R., Thanawongnuwech, R., *et al.* (2008) Influenza virus (H5N1) in live bird markets and food markets, Thailand. *Emerging Infectious Diseases* 14, 1739–1742.

Anderson, T.R. (1973) A comparative ecological study of the House Sparrow and the Tree Sparrow near Portage des Sioux, Missouri. PhD thesis, St Louis University, St Louis, Missouri.

Anderson, T.R. (1994) Breeding biology of house sparrows in northern lower Michigan. *Wilson Bulletin* 106, 537–548.

Anderson, T.R (2006) *Biology of the Ubiquitous House Sparrow: From Genes to Populations*. Oxford University Press, Oxford.

Andrew, S.C. and Griffith, S.C. (2016) Inaccuracies in the history of a well-known introduction: a case study of the Australian House Sparrow (*Passer domesticus*). *Avian Research* 7, 1–15.

Andrew, S.C., Awasthy, M., Griffith, A.D., Nakagawa, S., Simon, C., *et al.* (2018) Clinal variation in avian body size is better explained by summer maximum temperatures during development than by cold winter temperatures. *Auk* 135, 206–217.

Anon. (1986) Breeding data: House Sparrow *Passer domesticus*. *Zambian Ornithological Society Newsletter* 16, 66–67.

Azpiroz, A., Ascanio, D., Restall, R., Soto, A., Bosque, C. and Rodríguez-Ferraro, A. (2006) Status and distribution of the house Sparrow (*Passer domesticus*) in Venezuela. *Ornithologia Neotropical* 14, 457–460.

Baker, P.J., Molony, S.E., Stone, E., Cuthill, I.C. and Harris, S. (2008) Cats about town: is predation by free-ranging pet cats *Felis catus* likely to affect urban bird populations? *Ibis* 150, 86–99.

Barlow, C. and Dodman, T. (2011) Un oiseau nouveau au Burundi: le Moineau domestique *Passer domesticus*. *Malimbus* 33, 57–58.

Barlow, C. and Wacher, T. (2006) *A Field Guide to Birds of The Gambia and Senegal*. Yale University Press, New Haven, Connecticut.

Barrows, W.B. (1889) The English sparrow (*Passer domesticus*) in North America, especially in its relation to agriculture. *US Department of Agriculture, Division of Economic Ornithology and Mammalogy* 1, 1–442.

Bell, C.P., Baker, S.W., Parkes, N.G., Brooke, M.D.L. and Chamberlain, D.E. (2010) The role of the Eurasian sparrowhawk (*Accipiter nisus*) in the decline of the house sparrow (*Passer domesticus*) in Britain. *Auk* 127, 411–420.

Beri, Y., Raizada, P.M. and Bhatnagar, R.K. (1972) Food in adults of common sparrow *Passer domesticus* Linnaeus. *Entomology Newsletter* 2, 65–66.

Binder, D.K. and Scharfman, H.E. (2004) Brain-derived neurotrophic factor. *Growth Factors* 22, 123–31.

Bomford, M. and Sinclair, R. (2002) Australian research on bird pests: impact, management and future directions. *Emu* 102, 29–45.

Boon, A.C.M., Sandbulte, M.R., Seiler, P., Webby, R.J., Songserm, T., *et al.*(2007) Role of terrestrial wild birds in ecology of influenza A virus (H5N1). *Emerging Infectious Diseases* 13, 1720–1724.

Bosakowski, T. and Smith, D.G. (1992) Comparative diets of sympatric nesting raptors in the eastern deciduous forest biome. *Canadian Journal of Zoology* 70, 984–992.

Boujja-Miljour, H., Leighton, P.A. and Beauchamp, G. (2018) Individual vigilance profiles in flocks of house sparrows (*Passer domesticus*). *Canadian Journal of Zoology* 96, 1016–1023.

Bower, S. (1999) Fortpflanzungsaktivitat, Habitatnutzung und populationsstruktur eines Schwarms von Haussperlingen (*Passer d. domesticus*) im Hamburger Stadtgebiet–Hamburger avifaun. *Hamburger Avifaun* 30, 91–128.

Brown, J.D., Stallknecht, D.E., Berghaus, R.D. and Swayne, D.E. (2009) Infectious and lethal doses of H5N1 highly pathogenic avian influenza virus for house sparrows (*Passer domesticus*) and rock pigeons (*Columbia livia*). *Journal of Veterinary Diagnostic Investigation* 21, 437–445.

Browne, P. (2010) *Atlas of the Birds of Mauritania*. Available at: http://atlasornmau.org (accessed 26 October 2019).

Brzek, P., Kohl, K., Caviedes-Vidal, E. and Karasov, W.H. (2009) Developmental adjustments of house sparrow (*Passer domesticus*) nestlings to diet composition. *Journal of Experimental Biology*, 212, 1284–1293.

Brzek, P., Kohl, K.D., Caviedes-Vidal, E. and Karasov, W.H. (2011) Fully reversible phenotypic plasticity of digestive physiology in young house sparrows: lack of long-term effect of early diet composition. *Journal of Experimental Biology* 214, 2755–2760.

Caraco, T. and Bayham, M.C. (1982) Some geometric aspects of house sparrow flocks. *Animal Behaviour* 30, 990–996.

Cid, F.D., Fernández, N.C., Pérez-Chaca, M.V., Pardo, R., Caviedes-Vidal, E. and Chediack, J.G. (2018) House sparrow biomarkers as lead pollution bioindicators. Evaluation of dose and exposition length on hematological and oxidative stress parameters. *Ecotoxicology and Environmental Safety* 154, 154–161.

Clapperton, B.K., Porter, R.E.R., Day, T.D., Waas, J.R. and Matthews, L.R. (2012) Designer repellents: combining olfactory, visual or taste cues with a secondary repellent to deter free-ranging house sparrows from feeding. *Pest Management Science* 68, 870–877.

Daniels, M.J., Hutchings, M.R. and Greig, A. (2003) The risk of disease transmission to livestock posed by contamination of farm stored feed by wildlife excreta. *Epidemiology and Infection* 130, 561–568.

Darbro, J.M. and Harrington, L.C. (2006) Bird-baited traps for surveillance of West Nile mosquito vectors: effect of bird species, trap height, and mosquito escape rates. *Journal of Medical Entomology* 43, 83–92.

Dawson, D.G. (1970) Estimation of grain loss due to sparrows (*Passer domesticus*) in New Zealand. *New Zealand Journal of Agricultural Research* 13, 681–688.

Dawson, D.G. (1972) The breeding biology of the House Sparrow. PhD thesis, University of Oxford, Oxford, UK.

Dawson, D.G. and Bull, P.C. (1970) A questionnaire survey of bird damage to fruit. *New Zealand Journal of Agricultural Research* 13, 362–371.

Dawson, W.L. (1903) *The Birds of Ohio*. Wheaton Publishing Company, Columbus, Ohio.

Day, T.D., Clapperton, B.K., Porter, R.E.R., Waas, J.R. and Matthews, L.R. (2012) Responses of free-ranging house sparrows to feed containing primary and secondary repellents. *New Zealand Journal of Crop and Horticultural Science* 40, 127–138.

Dearborn, N. (1912) *The English Sparrow as a Pest*. US Department of Agriculture, Washington, DC.

Dhindsa, M.S., Toor, H.S. and Sandhu, P.S. (1984) Community structure of birds damaging pearl millet and sorghum and estimation of grain loss. *Indian Journal of Ecology* 11, 154–159.

Dhondt, A.A., Dhondt, K.V. and McCleery, B.V. (2008) Comparative infectiousness of three passerine bird species after experimental inoculation with *Mycoplasma gallisepticum*. *Avian Pathology* 37, 635–640.

Dolejská, M., Šenk, D., Čížek, A., Rybaříková, J., Sychra, O. and Literák, I. (2008) Antimicrobial resistant *Escherichia coli* isolates in cattle and house sparrows on two Czech dairy farms. *Research in Veterinary Science* 85, 491–494.

Dulisz, B., Nowakowski, J.J. and Górnik, J. (2016) Differences in biometry and body condition of the House Sparrow (*Passer domesticus*) in urban and rural population during breeding season. *Urban Ecosystems* 19, 1307–1324.

Fitzwater, W.D. (1981) Outwitting the House Sparrow [*Passer domesticus* (Linnaeus)]. In: *Proceedings of the Great Plains Wildlife Damage Control Workshop*, Vol. 5. University of Nebraska, Lincoln, Nebraska, pp. 244–251.

Fleischer, R. (1982) Clutch size in Costa Rican House Sparrows. *Journal of Field Ornithology* 53, 280–281.

Frimer, O. (1989) Food and predation in suburban sparrowhawks *Accipiter nisus* during the breeding season. *Dansk Ornitologisk Forenings Tidsskrift* 83, 35–44.

Fryday, S.L. and Greig-Smith, P.W. (1994) The effects of social learning on the food choice of the house sparrow (*Passer domesticus*). *Behaviour* 128, 281–300.

Gedeon, K., Cauldwell, A., Management, E.R. and Regassa, F. (2015) House Sparrow *Passer domesticus* and hybrids with Somali Sparrow *P. castanopterus* in Ethiopia. *Bulletin of the African Bird Club* 22, 70–73.

Gilbert, M., Sokha, C., Joyner, P.H., Thomson, R.L. and Poole, C. (2012) Characterizing the trade of wild birds for merit release in Phnom Penh, Cambodia and associated risks to health and ecology. *Biological Conservation* 153, 10–16.

Gil-Delgado, A., Antonio, J., Pardo, R., Bellot Abad, J.F. and Lucas, I. (1979) Avifauna del naranjal valenciano. II: el Gorrión Común (*Passer domesticus* L.). *Mediterránea. Serie de Estudios sobre Biología Terrestre Mediterránea* 3, 69–99.

Goodman, S.M., Achille, P., Andriniaina, H.A. and Nahina, E. (2017) The distribution and ecology of invasive alien vertebrate species in the greater Toamasina region, central eastern Madagascar. *Malagasy Nature* 12, 95–109.

Gowaty, P.A. (1984) House Sparrows kill Eastern Bluebirds. *Journal of Field Ornithology* 55, 378–380.

Greig-Smith, P.W. (1987) Aversions of starlings and sparrows to unfamiliar, unexpected or unusual flavours and colours in food. *Ethology* 74, 155–163.

Grinnell, J. (1919) The English sparrow has arrived in Death Valley: an experiment in nature. *American Naturalist* 53, 468–472.

Hammer, M. (1948) Investigations on the feeding-habits of the house-sparrow (*Passer domesticus*) and the tree-sparrow (*Passer montanus*). *Danish Review of Game Biology* 1, 1–59.

Hatch, M.I. and Westneat, D.F. (2007) Age-related patterns of reproductive success in house sparrows *Passer domesticus*. *Journal of Avian Biology* 38, 603–611.

Havlín, J. (1974) Vom Haussperling (*Passer domesticus*) und Feldsperling (*Passer montanus*) an reifenden Getreidepflanzen verursachte Schaden. *Zoologické Listy* 23, 241–259.

Heij, C.J. (1986) Nests of house sparrows *Passer domesticus*; composition and occupants. *International Study on Sparrows* 13, 28–34.

Herrera-Dueñas, A., Pineda, J., Antonio, M.T. and Aguirre, J.I. (2014) Oxidative stress of house sparrow as bioindicator of urban pollution. *Ecological Indicators* 42, 6–9.

Hobbs, J.N. (1958) House Sparrow breeding away from man. *Emu* 58, 422–423.

Ingels, J., Studer, P. and Tostain, O. (2007) Distribution and status of the house Sparrow *Passer domesticus* in French Guiana. *Alauda* 75, 405–412.

Iqbal, M., Yaqub, T., Mukhtar, N., Shabbir, M.Z. and McCauley, J.W. (2013) Infectivity and transmissibility of H9N2 avian influenza virus in chickens and wild terrestrial birds. *Veterinary Research* 44, 100.

Isenmann, P., Benmergui, M., Browne, P., Diam Ba, A., Hamallah Diagana, C., *et al.* (2010) *Oiseaux de Mauritanie [Birds of Mauritania]*. Société d'Études Ornithologiques de France, Paris.

Johnston, R.F. and Selander, R.K. (1971) Evolution in the House Sparrow. II. Adaptive differentiation in North American populations. *Evolution* 25, 1-28.

Kalmbach, E.R. (1940) *Economic Status of the English Sparrow in the United States*. US Department of Agriculture, Washington, DC.

Kendeigh, S.C. (1976) Latitudinal trends in the metabolic adjustments of the house sparrow. *Ecology* 57, 509–519.

Khaleghizadeh, A. (2011) Effect of morphological traits of plant, head and seed of sunflower hybrids on house sparrow damage rate. *Crop Protection* 30, 360–367.

Kilvitis, H.J., Martin, L.B., Hanson, H.E. and Schrey, A.W. (2017) Epigenetic potential as a mechanism of phenotypic plasticity in vertebrate range expansions. *Integrative and Comparative Biology* 57, 385–395.

Krogstad, S., Sæther, B.E. and Solberg, E.J. (1996) Environmental and genetic determinants of reproduction in the house sparrow: a transplant experiment. *Journal of Evolutionary Biology* 9, 979–991.

Kübler, S., Kupko, S. and Zeller, U. (2005) The kestrel (*Falco tinnunculus* L.) in Berlin: investigation of breeding biology and feeding ecology. *Journal of Ornithology* 146, 271–278.

Larson, L.R., Cooper, C.B. and Hauber, M.E. (2016) Emotions as drivers of wildlife stewardship behavior: examining citizen science nest monitors' responses to invasive house sparrows. *Human Dimensions of Wildlife* 21, 18–33.

Lawson, B., de Pinna, E., Horton, R.A., Macgregor, S.K., John, S.K., *et al.* (2014) Epidemiological evidence that garden birds are a source of human salmonellosis in England and Wales. *PLoS One* 9, e88968.

Lever, C. (1987) *Naturalized Birds of the World*. Longman Scientific and Technical, New York.

Liebl, A.L. and Martin, L.B. (2012) Exploratory behaviour and stressor hyper-responsiveness facilitate range expansion of an introduced songbird. *Proceedings of the Royal Society B: Biological Science* 279, 4375–4381.

Liebl, A.L. and Martin, L.B. (2014) Living on the edge: range edge birds consume novel foods sooner than established ones. *Behavioral Ecology* 25, 1089–1096.

Liebl, A.L., Schrey, A.W., Richards, C.L. and Martin, L.B. (2013) Patterns of DNA methylation throughout a range expansion of an introduced songbird. *Integrative and Comparative Biology* 53, 351–358.

Liedén, U. and Brouwer, J. (2018) The West African Bird Database (WABDaB). Available at: www.wabdab.org/ (accessed 15 October 2018.)

Long, J.L. (1981) *Introduced Birds of the World*. David & Charles Publishers, London.

MacGregor-Fors, I., Morales-Pérez, L., Quesada, J. and Schondube, J.E. (2009) Relationship between the presence of house sparrows (*Passer domesticus*) and Neotropical bird community structure and diversity. *Biological Invasions* 12, 87–96.

Macmillan, B.W.H. (1981) Food of house sparrows and greenfinches in a mixed farming district, Hawke's Bay, New Zealand. *New Zealand Journal of Zoology* 8, 93–104.

Magudu, K. and Downs, C.T. (2015) The relative abundance of invasive house sparrows (*Passer domesticus*) in an urban environment in South Africa is determined by land use. *African Journal of Wildlife Research* 45, 354–359.

Manzoor, S., Khan, H.A. and Javed, M. (2013) Inhibiting damage of watermelon (*Citrulus lanatus*) against some bird pests in an orchard of Faisalabad, Pakistan. *Journal of Animal and Plant Sciences* 23, 464–468.

Martin, L.B. and Fitzgerald, L. (2005) A taste for novelty in invading house sparrows, *Passer domesticus*. *Behavioral Ecology* 16, 702–707.

Martin, L.B., Coon, C.A.C., Liebl, A.L. and Schrey, A.W. (2014) Surveillance for microbes and range expansion in house sparrows. *Proceedings of the Royal Society B: Biological Sciences*, 281, 20132690.

Martin, L.B., Liebl, A.L. and Kilvitis, H.J. (2015) Covariation in stress and immune gene expression in a range expanding bird. *General and Comparative Endocrinology* 211, 14–19.

Martin, L.B., Kilvitis, H.J., Thiam, M. and Ardia, D.R. (2017) Corticosterone regulation in house sparrows invading Senegal. *General and Comparative Endocrinology* 250, 15–20

McLean, R.G., Mullenix, J., Kerschner, J. and Hamm, J. (1983) The house sparrow (*Passer domesticus*) as a sentinel for St. Louis encephalitis virus. *American Journal of Tropical Medicine and Hygiene* 32, 1120–1129.

Millins, C., Reid, A., Curry, P., Drebot, M.A., Andonova, M., Buck, P. and Leighton, F.A. (2011) Evaluating the use of house sparrow nestlings as sentinels for West Nile Virus in Saskatchewan. *Vector-Borne and Zoonotic Diseases* 11, 53–58.

Mitchell, C.J., Hayes, R.O., Holden, P. and Hughes, T.B. Jr (1973) Nesting activity of the House Sparrow in Hale County, Texas, during 1968. *Ornithological Monographs* 14, 49–59.

Moran, S. (2001) Aversion of the feral pigeon and the house sparrow to pellets and sprouts treated with commercial formulations of methyl anthranilate. *Pest Management Science* 57, 248–252.

Morrison, E.B., Kinnard, T.B., Stewart, I.R.K., Poston, J.P., Hatch, M.I. and Westneat, D.F. (2008) The links between plumage variation and nest site occupancy in male house sparrows. *Condor* 110, 345–353.

Moulton, M.P., Cropper, W.P., Avery, M.L. and Moulton, L.E. (2010) The earliest House Sparrow introductions to North America. *Biological Invasions* 12, 2955–2958.

Murphy, E. (1978) Breeding ecology of House Sparrows: spatial variation. *Condor* 80, 180–193.

Naik, R.M. and Mistry, L. (1972) Breeding season and reproductive rate of *Passer domesticus* (L.) in Baroda, India. In: S. Kendeigh, S.C. and Pinowski, J. (eds) *Productivity, Population Dynamics and Systematics of Granivorous Birds*. Polish Scientific Publishers PWN, Warsaw, pp. 133–140.

Naik, R.M. and Mistry, L. (1980) Breeding season in a tropical population of the House Sparrow. *Journal of Bombay Natural History Society* 75, 1118–1142.

Nhlane, M. (2000) The breeding biology of the House Sparrow *Passer domesticus* at Balantyre, Malawi. *Ostrich* 71, 80–82.

North, C.A. (1968) A study of house sparrow populations and their movements in the vicinity of Stillwater, Oklahoma. PhD thesis, Oklahoma State University, Stillwater, Oklahoma.

North, C.A. (1973) Population dynamics of the House Sparrow *Passer domesticus* (L.) in Wisconsin, USA. In: Kendeigh, S.C. and Pinowska, J. (eds) *Productivity, Population Dynamics, and Systematics of Granivorous Birds*. Polish Scientific Publishers PWN, Warsaw, pp. 195–210.

O'Brien, V.A. and Brown, C.R. (2012) Seasonal variation and age-related correlates of Buggy Creek Virus (Togaviridae) infection in nestling house sparrows. *Journal of Wildlife Diseases* 48, 138–147.

O'Brien, V.A., Moore, A.T., Young, G.R., Komar, N., Reisen, W.K. and Brown, C.R. (2011) An enzootic vector-borne virus is amplified at epizootic levels by an invasive avian host. *Proceedings of the Royal Society B: Biological Sciences* 278, 239–246.

Pärn, H., Ringsby, T.H., Jensen, H. and Sæther, B.E. (2011) Spatial heterogeneity in the effects of climate and density-dependence on dispersal in a house sparrow metapopulation. *Proceedings of the Royal Society B: Biological Sciences* 279, 144–152.

Perkins, L.E.L. and Swayne, D.E. (2003) Varied pathogenicity of a Hong Kong-origin H5N1 avian influenza virus in four passerine species and budgerigars. *Veterinary Pathology* 40, 14–24.

Philipson, W.R. (1938) House-sparrows excavating nest-hole. *British Birds* 32, 17.

Pinowska, B. (1975) Food of female house sparrows (*Passer domesticus* L.) in relation to stages of the nesting cycle. *Polish Ecological Studies* 1, 211–225.

Pitman, C.R.S. (1961) Unusual nesting behaviour of the House Sparrow, *Passer domesticus* (L.). *Bulletin of the British Ornithologists' Club* 81, 148–149.

Pitts, T. (1979) Nesting habits of rural and suburban House Sparrows in northwest Tennessee. *Journal of the Tennessee Academy of Science* 54, 145–148.

Popp, J.W. (1988) Selection of horse dung pats by foraging house sparrows. *Journal of Field Ornithology* 59, 385–388.

Radunzel, L.A., Muschitz, D.M., Bauldry, V.M. and Arcese, P. (1997) A long-term study of the breeding success of eastern bluebirds by year and cavity type. *Journal of Field Ornithology* 68, 7–18.

Rappole, J.H. and Hubálek, Z. (2003) Migratory birds and West Nile virus. *Journal of Applied Microbiology* 94, 47S–58S.

Ravinet, M., Elgvin, T.O., Trier, C., Aliabadian, M., Gavrilov, A. and Sætre, G.-P. (2018) Signatures of human-commensalism in the house sparrow genome. *Proceedings of the Royal Society B: Biological Sciences* 285, 20181246.

Riyahi, S., Hammer, Ø., Arbabi, T., Sánchez, A., Roselaar, C.S., *et al.* (2013) Beak and skull shapes of human commensal and non-commensal house sparrows *Passer domesticus*. *BMC Evolutionary Biology* 13, 200.

Robbins, C.S. (1973) Introduction, spread, and present abundance of the house sparrow in North America. *Ornithological Monographs* 14, 3–9.

Rott, K.H., Caviedes-Vidal, E. and Karasov, W.H. (2017) Intestinal digestive enzyme modulation in house sparrow nestlings occurs within 24 hours of a change in diet composition. *Journal of Experimental Biology* 220, 2733–2742.

Rouffaer, L.O., Lens, L., Haesendonck, R., Teyssier, A., Hudin, N.S., *et al.* (2016) House sparrows do not constitute a significant *Salmonella typhimurium* reservoir across urban gradients in Flanders, Belgium. *PLoS One*, 11, e0155366.

Rutz, C. (2006) Home range size, habitat use, activity patterns and hunting behaviour of urban-breeding Northern Goshawks *Accipiter gentilis*. *Ardea* 94, 185–202.

Sacristán, C., Esperón, F., Herrera-León, S., Iglesias, I., Neves, E., *et al.* (2014) Virulence genes, antibiotic resistance and integrons in *Escherichia coli* strains isolated from synanthropic birds from Spain. *Avian Pathology* 43, 172–175.

Sætre, G., Riyahi, S., Aliabadian, M., Hermansen, J.S., Hogner, S., *et al.* (2012) Single origin of human commensalism in the house sparrow. *Journal of Evolutionary Biology* 25, 788–796.

Sánchez-Tójar, A., Nakagawa, S., Sánchez-Fortún, M., Martin, D.A., Ramani, S., *et al.* (2018) Meta-analysis challenges a textbook example of status signalling and demonstrates publication bias. *eLife Sciences* 7, e37385.

Sappington, J.N. (1977) Breeding biology of house sparrows in north Mississippi. *Wilson Bulletin* 89, 300–309.

Schrey, A.W., Grispo, M., Awad, M., Cook, M.B., McCoy, E.D., *et al.* (2011) Broad-scale latitudinal patterns of genetic diversity among native European and introduced house sparrow (*Passer domesticus*) populations. *Molecular Ecology* 20, 1133–1143.

Schrey, A.W., Liebl, A.L., Richards, C.L. and Martin, L.B. (2014) Range expansion of house sparrows (*Passer domesticus*) in Kenya: evidence of genetic admixture and human-mediated dispersal. *Journal of Heredity* 105, 60–69.

Seel, D.C. (1968) Breeding seasons of the house sparrow and tree sparrow *Passer* spp. at Oxford. *Ibis* 110, 129–144.

Seiler, G.J. and Rogers, C.E. (1987) Influence of sunflower morphological characteristics on achene depredation by birds. *Agriculture, Ecosystems and Environment* 20, 59–70.

Sengupta, S. and Shrilata (1997) House sparrow *Passer domesticus* yses Krishnachura leaves as an antidote to malarial fever. *Emu* 97, 248–249.

Shaw, L.M., Chamberlain, D. and Evans, M. (2008) The house sparrow *Passer domesticus* in urban areas: reviewing a possible link between post-decline distribution and human socioeconomic status. *Journal of Ornithology* 149, 293–299.

Sheldon, E.L. and Griffith, S.C. (2017) A high incidence of non-cavity nesting in an introduced population of house sparrows suggests that the species should not be constrained by cavity-nest site availability. *Avian Research* 8, 29.

Simenonov, S.D. (1964) Über die Nahrung des Haussperlings in der Umgebung von Sofia. *Annuaire de l'Universite de Sofia*, 61, 239–275.

Singer, R. and Yom-Tov, Y. (1988) The breeding biology of the house sparrow *Passer domesticus* in Israel. *Ornis Scandinavica* 19, 139–144.

Singh, R., Kour, D.N., Ahmad, F. and Sahi, D.N. (2013) The causes of decline of house sparrow (*Passer domesticus*, Linnaeus 1758) in urban and suburban areas of Jammu region, J and K. *Munis Entomology and Zoology* 8, 803–811.

Skjelseth, S., Ringsby, T.H., Tufto, J., Jensen, H. and Sæther, B.E. (2007) Dispersal of introduced house sparrows *Passer domesticus*: an experiment. *Proceedings of the Royal Society B: Biological Sciences* 274, 1763–1771.

Sodhi, N.S. and Oliphant, L.W. (1993) Prey selection by urban-breeding Merlins. *Auk* 110, 727–735.

Stewart, I.R.K., Westneat, D.F. and Barrick, R.L. (2015) Food supplementation and extrapair paternity in house sparrows. *American Midland Naturalist* 174, 278–289.

Stewart, P.A. (1973) Replacement of cavity-hunting starlings and house sparrows after removal. *Wilson Bulletin* 85, 291–294.

Summers-Smith, J.D. (1963) *The House Sparrow*, 1st edn. Collins, London.

Summers-Smith, J.D. (1988) *The Sparrows: A Study of the Genus* Passer. A. & C. Black Publishers, London.

Summers-Smith, J.D. (1992) *In Search of Sparrows*. Academic Press. London.

Summers-Smith, J.D. (2003) The decline of the House Sparrow: a review. *British Birds* 96, 439–446.

Swaileh, K.M. and Sansur, R. (2006) Monitoring urban heavy metal pollution using the house sparrow (*Passer domesticus*). *Journal of Environmental Monitoring* 8, 209–213.

Tang, Y., Diao, Y., Yu, C., Gao, X., Ju, X., *et al.* (2013) Characterization of a Tembusu virus isolated from naturally infected house sparrows (*Passer domesticus*) in Northern China. *Transboundary and Emerging Diseases* 60, 152–158.

Thurber, J. (1986) Range expansion of the house sparrow through Guatemala and El Salvador. *American Birds* 40, 341–349.

Tipton, K.W., Floyd, E.H., Marshall, J.G. and McDevitt, J.B. (1970) Resistance of certain grain sorghum hybrids to bird damage in Louisiana. *Agronomy Journal* 62, 211–213.

Turner, E.R.A. (1964) Social feeding in birds. *Behaviour* 24, 1–45.

van den Bosch, F., Hengeveld, R. and Metz, J.A.J. (1992) Analysing the velocity of animal range expansion. *Journal of Biogeography* 19, 135–150.

Veiga, J.P. (1990) Sexual conflict in the house sparrow: interference between polygynously mated females versus asymmetric male investment. *Behavioral Ecology and Sociobiology* 27, 345–350.

Veiga, J.P. (1992) Why are house sparrows predominantly monogamous? A test of hypotheses. *Animal Behaviour* 43, 361–370.

Vierke, J. (1970) Die Besiedlung Sudafrikas durch den Haussperling (*Passer domesticus*). *Journal of Ornithology* 111, 94–103.

Wagner, N. (2012) Occupation of Monk Parakeet (*Myiopsitta monachus*) nest cavities by house sparrows (*Passer domesticus*) in Rio Grande do Sul, Brazil. *Boletín SAO* 20, 72–78.

Waltert, M., Yaokokore-Beibro, K.H., Mühlenberg, M. and Waitkuwait, W.E. (2007) House sparrow *Passer domesticus* colonises Guinea-Bissau. *West African Ornithological Society* 17, 58.

Westneat, D.F., Wright, J. and Dingemanse, N.J. (2015) The biology hidden inside residual within-individual phenotypic variation. *Biological Reviews* 90, 729–743.

Wetzel, D.P., Stewart, I.R.K. and Westneat, D.F. (2012) Heterozygosity predicts clutch and egg size but not plasticity in a house sparrow population with no evidence of inbreeding. *Molecular Ecology* 21, 406–420.

Williams, J.E., Young, O.P., Watts, D.M. and Reed, T.J. (1971) Wild Birds as Eastern (EEE) and Western (WEE) Equine Encephalitis Sentinels. *Journal of Wildlife Diseases* 7, 188–194.

Wodzicki, K. (1956) Breeding of the House Sparrow away from Man in New Zealand. *Emu* 56, 146–147.

12 Shiny Cowbird (*Molothrus bonariensis* Gmelin, 1788)

Myriam E. Mermoz[1]*, Alexander Cruz[2†], Jameson F. Chace[3] and Juan C. Reboreda[1,4]

[1]*Instituto de Ecología, Genética y Evolución de Buenos Aires (IEGEBA-UBA-CONICET), Facultad de Ciencias Exactas y Naturales, Universidad de Buenos Aires, Buenos Aires, Argentina;* [2]*Department of Ecology and Evolutionary Biology, University of Colorado, Boulder CO, USA;* [3]*Department of Biology and Biomedical Sciences and Department of Cultural, Environmental and Global Studies, Salve Regina University, Newport, Rhode Island, USA;* [4]*Departamento de Ecología, Genética y Evolución, Facultad de Ciencias Exactas y Naturales, Universidad de Buenos Aires, Buenos Aires, Argentina*

Citation: Mermoz, M.E., Cruz, A., Chace, J.F. and Reboreda, J.C. (2020) Shiny Cowbird (*Molothrus bonariensis* Gmelin, 1788). In: Downs, C.T. and Hart, L.A. (eds) *Invasive Birds: Global Trends and Impacts*. CAB International, Wallingford, UK, pp. 97–104.

12.1 Common Names

Tordo Renegrido (Argentina), Tordo (Uruguay), Maria-preta, Chopim and Vira-bosta, Gaudério (Brazil), Chamón (Colombia and Costa Rica), Mirlo and Tordo Argentino (Chile), Tordo Lustroso (Puerto Rico), Vacher luisant (Haiti), Pájaro Vaquero (Cuba), and Shiny or Glossy Cowbird (USA).

12.2 Distribution

The native distribution of the Shiny Cowbird (*Molothrus bonariensis* Gmelin, 1788) is from southern Argentina (Chubut Province) to eastern Panama, except most areas above 2000 m above sea level (a.s.l.) and widely forested regions such as the Amazon rainforest (Fig. 12.1). During the early 1900s, they started expanding and invaded the West Indies (Cruz *et al.*, 1985). Shiny Cowbirds were first recorded on the Florida mainland in 1987 and have been considered permanent residents there since 1991 (Post, 1992; Cruz *et al.*, 2000). They also invaded Chile, from Aysén to Atacama (Reed, 1913; Marín, 2000). In 2004, they reached Costa Rica from Panama (Carman and Martínez, 2004) (Fig. 12.1).

12.3 Description

Shiny Cowbirds are sexually dimorphic in plumage coloration and body size. Males are entirely glossy black with blue and purplish iridescence (Fig. 12.2A) (Friedmann, 1929; Ortega, 1998). Females are dull brown in most of the species distribution, but in some areas, they can be dark brown or black (Fig. 12.2B) (Friedmann, 1929; Jaramillo and Burke, 1999; Mahler *et al.*, 2010). The bill is medium length and conical, blackish grey. Males are 15–20% larger than females, and body size varies geographically. Hatchlings have sparse tufts of pale grey down on a yellowish skin, and their oral flanges range from white to yellow (Fraga, 1978).

There are seven subspecies that differ mostly in size (based on Lowther, 2011, unless otherwise indicated):

- *M. b. bonariensis* Gmelin, 1789. This subspecies is present in eastern and southern Brazil, north to Mato Grosso, Maranhão, Piauí and Ceará; eastern Bolivia, Paraguay, Uruguay and Argentina south to Chubut. It was introduced in Chile where it has established from Coquimbo Province

*Corresponding author: myriammermoz@gmail.com
† Deceased 15 May 2018.

Fig. 12.1. Distribution of Shiny Cowbird (*Molothrus bonariensis*) showing the native (green) and invaded (red) ranges.

Fig. 12.2. Photographs of the male (A) and female (B) of the subspecies *Molothrus bonariensis bonariensis*. (©Photographs: Alec Earnshaw.)

south to Valdivia in Los Lagos Province and to Antofagasta Province in the north. The male has a silky violet-black head, back, breast and upper abdomen, with the rest of the body blue-black and the wings a glossy greenish black. The female has a greyish olive-brown head and back, a greyer rump and tail coverts, dusky brown wings and tail, and hair-brown underparts. In north-eastern Argentina and south-eastern Brazil, females of the dark morphs '*melanogyna*' and in Mato Grosso, Brazil '*milleri*' have been observed; these were formerly considered subspecies *M. b. melanogyna* and *M. b. milleri*, respectively. These females are almost as dark as males but lack the strong glossy plumage (Jaramillo and Burke, 1999). Its mean body mass is 46.8 g for females and 51.6 g for males (Reboreda *et al.*, 1996).

- *M. b. cabanisii* Cassin, 1866. This is the largest subspecies. Its distribution includes eastern Panama, and tropical and lower subtropical zones of Colombia west of the eastern Andes and eastern slope of the eastern Andes. It intergrades with *M. b. bonariensis* in south-eastern Colombia and it reached Costa Rica in 2004 (Carman and Martínez, 2004). The male's plumage is similar to that of *M. b. bonariensis*, while females are slightly paler. Its mean body mass 55.6 g for females and 66.7 g for males (Kattan, 1993).
- *M. b. venezuelensis* Stone, 1891. The distribution of this subspecies includes the tropical zone of eastern Colombia, from Zulia Valley south to the eastern Llanos; northern Venezuela south to Llanos to the Apure and Orinoco rivers and south of Orinoco in north-western Amazonas and northern Bolívar. The male is more richly glossed with purple than *M. b. bonariensis*, similar to *M. b. occidentalis*. The female is darker than *M. b. bonariensis*, like *M. b. minimus*.
- *M. b. occidentalis* Berlepsch & Stolzmann, 1892. The distribution of this subspecies includes south-western Ecuador and western Peru, east to the province of Jaén in Cajamarca, and south to Lima and Ica. The males are more bluish than *M. b. bonariensis* and are similar to *M. b. venezuelensis*. The females differ from all other subspecies by pale upperparts, very pale and streaked underparts, and a conspicuous buffy post-ocular stripe.
- *M. b. minimus* Dalmas, 1900. This is the smallest subspecies. It was originally confined to northern Brazil in the region of upper Rio Branco, the Guianas, eastern Venezuela, and Trinidad and Tobago (Cruz *et al.*, 1995). About 1900, it started to expand its range through West Indies, reaching Florida Keys, USA, in 1985 and Florida mainland in 1987. The male is similar to *M. b. bonariensis*. The female has a forehead and crown darker than *M. b. bonariensis* and distinct streaks on the scapulars and interscapulars. The body mass is 34.1–36.4 g for females and 37.9–40.0 g for males, with body lengths of 17.0–17.6 cm for females and 17.1–18.9 cm for males (Lowther, 2011).

- *M. b. aequatorialis* Chapman, 1915. The distribution of this subspecies includes the tropical zone of south-western Colombia, south of Río Patía, and western Ecuador south to Guayaquil. It is larger than *M. b. bonariensis*. The males are more violet and have less blue iridescence than *M. b. bonariensis*, while the females are darker and without the buffy whitish post-ocular streak.
- *M. b. riparius* Griscom and Greenway, 1937. The distribution of this subspecies includes the lower Amazon Valley (to Obidos in northern Brazil on the north bank of the Amazon) west to the Río Ucayali, eastern Peru. The male is similar to *M. b. bonariensis*. The female is slightly blacker on the upperparts and paler on the underparts than *M. b. bonariensis*.

12.4 Diet

Shiny Cowbirds are omnivorous, and their diet includes a variety of arthropods and seeds. The adults feed on the ground, usually in association with cattle and other large grazing mammals. They show seasonal variation in diet (i.e., mainly arthropods during the breeding season, and grains and feed mixtures in feedlots during the winter) (Post *et al.*, 1990; Lowther, 2011). Nestlings need a protein-rich diet for normal development and cannot be reared successfully by hosts that feed them mostly with seeds (Salvador and Salvador, 1986).

12.5 Introduction and Invasion Pathways

Since *c.*1900, the subspecies *M. b. minimus* has expanded its range from Venezuela and Trinidad and Tobago Islands north through the Lesser and then Greater Antilles, and most recently to the USA, Canada and Mexico (Cruz *et al.*, 1995). The expansion sequence was: 1899: Grenadines (Bond, 1956); 1901: Grenada (Bond, 1956); 1916: Barbados (Bond, 1956); 1924: St Vincent (Bond, 1956); 1931: St Lucia (Danforth, 1935); 1934: St Croix (Bond, 1956); 1948: Martinique (Pinchon, 1963); 1955: St John (Robertson, 1962); 1955: Puerto Rico (Grayce, 1957); 1959: Antiqua (Pinchon, 1963); 1959: Marie-Galante (Pinchon, 1963); 1961: Yabucoa, Puerto Rico (Biaggi, 1963); 1965: Guajataca Cliffs, Puerto Rico (Buckley and Buckley, 1970); 1969: Guánica, Puerto Rico (Kepler and Kepler, 1970); 1971: Mona Island (Bond, 1973); 1972: Santo Domingo, Dominican Republic (Bond: 1973); 1978: Neiba, Dominican Republic (Arendt and Vargas Mora, 1984); 1980: Port-au-Prince, Haiti (Cruz *et al.*, 1985); 1981: Monte Cristi, Dominican Republic (Arendt and Vargas Mora, 1984); 1982: Cardenas, Cuba (Cruz *et al.*, 1985); 1985: Lower Matecumbe Key, Florida (Cruz *et al.*, 2000); 1987: Homstead, Florida (Cruz *et al.*, 2000); 1988: Jacksonville, Florida (Cruz *et al.*, 2000); 1989: Cameron, Louisiana (Cruz *et al.*, 2000); 1989: Aurora, North Carolina (Cruz *et al.*, 2000); 1990: Fort Hood, Texas (Cruz *et al.*, 2000); 1993: Goliad, Texas (Cruz *et al.*, 2000); 1993: Jamaica (Cruz *et al.*, 2000); 1993: New Brunswick, Canada (Benoit, 1995); and 1996: Yucatán, Mexico (Kluza, 1998).

During the late 1800s and early 1900s, a similar expansion occurred from Argentina to Chile involving the subspecies *M. b. bonariensis*. The Shiny Cowbird was a popular cage bird in the early 1900s, and large-scale importations took place from Argentina to markets in the central regions of Chile. This species is not distributed in areas higher than 2500 m a.s.l. Therefore, the high Andes Mountains that separate Argentina from Chile constitute a geographical barrier that would have made the natural dispersion of Shiny Cowbirds relatively difficult (i.e. the lower altitudes are around 4000 m a.s.l.). In addition, the rapid expansion of this species in Chile supports the hypothesis that the population originated from caged birds that escaped from captivity (Reed, 1913; Marín, 2000). The Chilean population expanded from central Chile and probably was the source of the apparently isolated population of northern Chile to south-west Peru. Individuals that originated from this population could have reached this area during the humid years of El Niño, when there is an increase in coastal vegetation in central to northern Chile (Marín, 2000; Cossíos, 2010).

Similarly, as a result of anthropogenic habitat transformation, the subspecies *M. b. cabanisii* expanded throughout Panama, reaching Costa Rica in 2004 (Carman and Martínez, 2004).

Information on Shiny Cowbird movements on a large scale is scarce. In the Dominican Republic, 14 out of 196 banded individuals were recovered at a mean distance of 25.8 km and a maximum distance of 130 km (Arendt and Vargas Mora, 1984). In Florida, Shiny Cowbirds are more common in April–June, indicating that part of the population may migrate from the Greater Antilles during the breeding season (Post *et al.*, 1993).

12.6 Breeding Behaviour

The Shiny Cowbird is an extreme generalist brood parasite. In contrast to its North American counterpart, the Brown-headed Cowbird (*Molothrus ater*), it does not avoid parasitizing hosts with domed or close nests. The number of total hosts that have been recorded is 270, and 97 of these reared the parasite young successfully (Lowther, 2018). Shiny Cowbird breeding season varies across its distribution, coinciding with the one of its main hosts. In Argentina and Chile it extends from September to January (Mason, 1986a; Marín, 2000), in Colombia, from June to late December (Kattan, 1993), in Venezuela from June to October (Cruz and Andrews, 1997), in Trinidad from April to August (Cruz *et al.*, 1995), in Puerto Rico from March to August on the coast (Wiley, 1988) and from March to October in the interior (Pérez-Rivera, 1986), and in the Dominican Republic from April to July (Arendt and Vargas Mora, 1984).

During the breeding season, female Shiny Cowbirds use relatively constant areas of approximately 50 ha where they search for and parasitize host nests (Scardamaglia and Reboreda, 2014). Females spend the morning hours watching host behaviour and use the host's nest-building activity as a cue for finding nests (Wiley, 1988). They may also locate host nests by using conspicuously noisy searching (i.e., flushing the host from its nest; Wiley, 1988) or by searching systematically through the habitat (Kattan, 1997; Fiorini and Reboreda, 2006;

Svagelj *et al.*, 2009). Females do not have exclusive territories for parasitism, as several females may overlap in their morning ranges (Scardamaglia and Reboreda, 2014), and multi-parasitism in a single day is common in nests of several hosts (Mermoz and Reboreda, 1999; Gloag *et al.*, 2014). Males have a home range much larger than that of females (i.e., over 150 ha; Scardamaglia and Reboreda, 2014), probably because during the breeding season they simultaneously follow several females, which is consistent with a socially polygynous or promiscuous mating system (Mason, 1987). Females, but not males, are faithful to one communal roost located near their diurnal home range (Scardamaglia *et al.*, 2018).

Most parasitism (over 90%) occurs in a short time window 30 min before sunrise, with the remaining parasitic events occurring up to 30 min after sunrise (Gloag *et al.*, 2013). Females prospect host nests and make at least one nest visit before parasitism (range one to eight visits; Scardamaglia *et al.*, 2017). On the laying day, they fly directly from the roost to the target nest (Scardamaglia and Reboreda, 2014; Scardamaglia *et al.*, 2017, 2018). Video records of parasitic events have shown that they last less than 30 s (Gloag *et al.*, 2013). During this time, the female first tries to peck and puncture host eggs and then lays her egg. In many cases, the female lays her egg at the same time she is being mobbed by the host (Gloag *et al.*, 2013; Fiorini *et al.*, 2014). Host mobbing does not reduce the probability of parasitism but decreases the probability of punctures in host eggs (Gloag *et al.*, 2013). Egg punctures reduce the clutch of the host and therefore the number of host chicks that hatch, thus reducing the competition for food and increasing the survival of the parasite nestlings. This is particularly advantageous when hosts are larger than the parasite (Mermoz and Reboreda, 2003; Astié and Reboreda, 2009; Fiorini *et al.*, 2009; Gloag *et al.*, 2012). Once a female has parasitized a nest, she does not visit it again (Gloag *et al.*, 2014; Scardamaglia *et al.*, 2017), thus avoiding the costs of competition between her own chicks and damaging her own previously laid eggs (Gloag *et al.*, 2014).

Shiny Cowbirds have a very high reproductive potential. The breeding season of *M. b. bonariensis* in the province of Buenos Aires, Argentina, lasted approximately 3 months, from early October to mid-January, and the laying period of individual females in that area is approximately 65 days, with a maximum laying rate of 0.5 eggs/day (de la Colina *et al.*, 2016). This laying rate is similar to that reported for the subspecies *M. b. cabanisii* in Central Colombia (0.66 eggs/day; Kattan, 1993).

Females synchronize parasitism with host laying, but the degree of synchronization varies among hosts. In open nesters it is relatively high (70–80%; Massoni and Reboreda, 1998; Mermoz and Reboreda, 1999; Fiorini and Reboreda, 2006; Astié and Reboreda, 2009), while in cavity nesters it is lower (47–57%; Kattan, 1998; Tuero *et al.*, 2007). Shiny Cowbird eggs have an eggshell that is thicker than expected by egg volume (Spaw and Rohwer, 1987; Mermoz and Ornelas, 2004). This increase in eggshell thickness provides higher resistance against punctures by other parasitic females (Hudson, 1874; Spaw and Rohwer, 1987; Mermoz and Ornelas, 2004; Fiorini *et al.*, 2014) or against hitting by subsequently laid cowbird eggs in multiply parasitized nests (López *et al.*, 2018). It also decreases the probability of damage when eggs are laid from an elevated position (López *et al.*, 2018). Among females, there is a high egg polymorphism, with eggs varying in background colour and spotting pattern (Fig. 12.3) (Friedmann, 1929; Ortega, 1998, Mahler *et al.*, 2008), but individual females lay eggs of consistent appearance (Gloag *et al.*, 2014). The incubation period is 11–13 days and does not differ from that expected by allometric regressions (Mermoz and Ornelas, 2004).

Shiny Cowbird nestlings have long and tremulous begging calls, which are louder than the host's begging calls (Gochfeld, 1978; Lichtenstein, 1997) and highly efficient in stimulating parental feeding, even in non-host species (Gloag and Kacelnik, 2013). In addition, Shiny Cowbird nestlings may adjust their begging intensity depending on intrabrood competition

Fig. 12.3. The variation of eggshell spotting patterns in Shiny Cowbirds. All eggs were collected in Buenos Aires Province, Argentina, and belong to the subspecies *Molothrus bonariensis bonariensis*.

(Tuero *et al.*, 2016). A disadvantage of the louder begging calls is that they may make host nests more detectable to potential predators (Massoni and Reboreda, 1998; Vincenty *et al.*, 2009). Nestlings fledge when they are 10–14 days of age (Fraga, 1978; Mermoz and Reboreda, 1999) and reach independence from their foster parents after 30–35days after fledging (I. Crudele, Buenos Aires, 2017, personal communication).

12.7 Habitat

Shiny Cowbirds occur in open or semi-open habitats and are attracted to deforested areas such as fields and pastures with scattered patches of trees and shrubs, or tree groves near farms, gardens and plantations (Cruz *et al.*, 1985; Jaramillo and Burke, 1999). They also are found in undisturbed habitats such as riparian thickets, savannahs, semi-arid scrub, marshes, open flood plains including oxbows, and mangroves (Post and Wiley, 1977a; Mason, 1985; Cavalcanti and Pimentel, 1988; Jaramillo and Burke, 1999).

12.8 Impacts

Brood parasitism by Shiny Cowbirds reduces the breeding success of their hosts in several ways. First, parasite females peck and puncture host eggs during parasitic and non-parasitic visits, reducing clutch size and increasing the probability of nest abandonment (Post and Wiley, 1977b; Cruz and Andrews, 1997; Kattan, 1998; Massoni and Reboreda, 1998, 2002; Mermoz and Reboreda, 1999; Sackmann and Reboreda, 2003; Cruz *et al.*, 2005; Astié and Reboreda, 2009; Dominguez *et al.*, 2015). Second, parasite eggs may interfere with host incubation lowering the hatching success of the remaining host eggs (Wiley, 1985; Woodworth, 1997; Astié and Reboreda, 2006, Tuero *et al.*, 2007). Third, parasite nestlings may outcompete host nestlings for food, reducing their survival (Wiley, 1985; Woodworth, 1997; Tuero *et al.*, 2007; Duré-Ruiz *et al.*, 2008). In addition, they might lower the post-fledging survival of host young and the future reproductive success of host parents. In general, when hosts are similar or larger in body mass than the parasite, the impact of parasitism is restricted mainly to the puncture of host eggs, but when hosts are smaller than the parasite, the hatchability of host eggs and the survival of host chicks are also reduced (Reboreda *et al.*, 2003).

In areas of long-term sympatry with Shiny Cowbirds, some hosts have developed antiparasitic defences such as recognition and aggression against adults and rejection of parasite eggs, which reduce the costs of brood parasitism (Mason, 1986b; Cruz *et al.*, 1990; Sackmann and Reboreda, 2003; Astié and Reboreda, 2005; Segura and Reboreda, 2012; Gloag *et al.*, 2013; Mermoz *et al.*, 2013; Reboreda *et al.*, 2013). However, in the Antilles, where Shiny Cowbirds have arrived recently, potential new hosts have evolved without contact with the parasite and therefore have no antiparasitic defences. Therefore, the costs of parasitism on host breeding success are particularly high in recently invaded areas. For example, in the Puerto Rican Vireo (*Vireo latimeri*), parasitism reduces the number of hosts fledged per successful nest by 82% (Woodworth, 1997), while in the Yellow-shouldered Blackbird (*Agelaius xanthomus*) the reduction was estimated at 39% (Post and Wiley, 1977b).

Human activities associated with the intensification of agriculture imply the loss of natural habitats and the transformation of habitats more favourable for Shiny Cowbirds because of the anthropogenic sources of food available for them. This has resulted in an increase in Shiny Cowbird densities and in an increase in the frequency of parasitism, which is particularly serious for endangered species, where even moderate rates of parasitism imply an important conservation risk.

In the Antilles, three endemic species, the Bahamas Oriole (*Icterus northropi*; Price *et al.*, 2011), the Yellow-shouldered Blackbird (Cruz *et al.*, 2005) and the Puerto Rican Vireo (Woodworth, 1997) suffer the joint pressure of anthropogenic habitat loss and brood parasitism by the Shiny Cowbird. Two of these species are threatened globally: the Bahamas Oriole (critically endangered) and Yellow-shouldered Blackbird (endangered) (IUCN, 2018). In addition, according to the IUCN (2018), other globally threatened species within the native distribution of Shiny Cowbirds are also negatively affected, such as the endangered Pale-headed Brush-finch (*Atlapetes pallidiceps*; Krabbe *et al.*, 2011), Forbes's Blackbird (*Anumara* (= *Curaeus*) *forbesi*; Studer and Vielliard, 1988), Apolinar's Wren (*Cistothorus apollinari*; Villaneda-Rey and Rosselli, 2011) and the Yellow Cardinal (*Gubernatrix cristata*; Dominguez *et al.*, 2015), and the vulnerable Saffron-cowled Blackbird (*Xanthopsar flavus*; Fraga *et al.*, 1998), the Pampas Meadowlark (*Leistes* (= *Sturnella*) *defilippi*; Azpiroz, 2015) and the Black-and-white Monjita (*Xolmis dominicanus*; Azpiroz, 2015).

With regard to positive impacts, in Chile, Shiny Cowbirds are the principal pollinator of native *Embothrium coccineum* trees that remain isolated due to deforestation (Smith-Ramírez and Armesto, 2003), and in Venezuela's Savannah, they are an important seed disperser of native *Palicourea rigida* trees (Wütherich *et al.*, 2001).

12.9 Control

Shiny Cowbird trapping was established from 1985 to 2003 in south-western Puerto Rico, near the breeding areas of the globally endangered Yellow-shouldered Blackbird, which in 1982 had a population of approximately 300 individuals (Wiley *et al.*, 1991). During that period, 29,981 Shiny Cowbirds were captured, and parasitism on Yellow-shouldered Blackbird nests dropped from 95% (pre-trapping period, 1973–1983) to less than 10% (trapping period, 1985–2003) (Cruz *et al.*, 2005). Since 1985, control of Shiny Cowbirds through adult trapping and egg and chick removal in nests built in artificial structures has been implemented without interruption. However, during the years 2000–2012, the population of Yellow-shouldered Blackbird did not increase and fluctuated between 500 and 1000 individuals (Medina-Miranda *et al.*, 2013), indicating that factors other than Shiny Cowbird parasitism are affecting the recovery of this species.

Another example of Shiny Cowbird control was implemented in the Yunguilla Reserve, Ecuador, to protect the endangered Pale-headed Brush-finch. In association with the shooting of adults over a 7-year period, parasitism rates decreased from 60% to almost zero, and territory occupancy of Pale-headed Brush-finches increased from less than 40 to more than 100 territories (Krabbe *et al.*, 2011). Due to local topography, Shiny Cowbirds entered this reserve every morning from the same direction, and therefore one skilled shooter was very effective in controlling the cowbirds. Eventually, the population of Pale-headed Brush-finches stabilized, and vegetation management was found to increase their reproductive success, although cowbird shooting was maintained (Hartmann *et al.*, 2015).

12.10 Uses

During early 1900, Shiny Cowbirds were widely used as caged birds (Reed, 1913). Currently, they are still used as domestic pets in some rural areas of Latin America (Bezerra *et al.*, 2011).

12.11 Notes

Shiny Cowbirds move widely during the breeding and non-breeding seasons (Friedmann, 1929; Post *et al.*, 1993; Jaramillo and Burke, 1999). They rapidly use human disturbed landscapes, readily invading deforested areas converted to agriculture (Arendt and Vargas Mora, 1984; Cavalcanti and Pimentel, 1988; Marín, 2000).

12.12 References

Arendt, W.J. and Vargas Mora, T.A. (1984) Range expansion of the Shiny Cowbird in the Dominican Republic. *Journal of Field Ornithology* 55, 104–107.

Astié, A.A. and Reboreda, J.C. (2005) Creamy-bellied Thrush defenses against Shiny Cowbird brood parasitism. *Condor* 107, 788–796.

Astié, A.A. and Reboreda, J.C. (2006) Costs of egg punctures and parasitism by shiny cowbirds (*Molothrus bonariensis*) at Creamy-bellied Thrush (*Turdus amaurochalinus*) nests. *Auk* 123, 23–32.

Astié, A.A. and Reboreda, J.C. (2009) Function of egg punctures by Shiny Cowbird in parasitized and unparasitized Creamy-bellied Thrush nests. *Journal of Field Ornithology* 80, 336–343.

Azpiroz, A.B. (2015) Shiny Cowbird (*Molothrus bonariensis*) parasitism records for three globally threatened species from the South American Pampas. *Wilson Journal of Ornithology* 127, 746–752.

Benoit, G. (1995) A day to remember: first Shiny Cowbird in Canada. *Birders Journal* 4, 89–90.

Bezerra, D.M.M., Araujo, H.F.P. and Alves, R.R.D.N. (2011) The use of wild birds by rural communities in the semi-arid region of Rio Grande do Norte State, Brazil. *Bioremediation, Biodiversity and Bioavailability* 5, 117–120.

Biaggi, V. Jr. (1963) Record of the White Pelican and additional information on the Glossy Cowbird from Puerto Rico. *Auk* 80, 198.

Bond, J. (1956) *Check-list of Birds of the West Indies*, 4th ed. Academy of Natural Sciences of Philadelphia, Philadelphia, Pennsylvania.

Bond, J. (1973) *Eighteenth supplement to the check-list of birds of the West Indies (1956)*. Academy of Natural Sciences of Philadelphia, Philadelphia, Pennsylvania.

Buckley, P.A. and Buckley, F.G. (1970) Notes on the distribution of some Puerto Rican birds and the courtship behaviour of White-tailed Tropicbirds. *Condor* 72, 483–486.

Carman, E. and Martínez, D. (2004) Primer registro de *Molothrus bonariensis*, Icteridae (Chamón Parásito) en Costa Rica. *Zeledonia* 8, 36–37.

Cavalcanti, R.B. and Pimentel, T.M. (1988) Shiny Cowbird parasitism in central Brazil. *Condor* 90, 40–43.

Cossíos, E.D. (2010) Vertebrados naturalizados en el Perú: historia y estado del conocimiento. *Revista Peruana de Biología* 17, 179–189.

Cruz A., Manolis, T.D. and Andrews, R.W. (1990a) Reproductive interactions of the Shiny Cowbird *Molothrus bonariensis* and the Yellow-hooded Blackbird *Agelaius icterocephalus* in Trinidad. *Ibis* 132, 436–444.

Cruz, A. and Andrews, R.W. (1997) The breeding biology of the Pied Water-Tyrant and its interactions with the Shiny Cowbird in Venezuela. *Journal of Field Ornithology* 68, 91–97.

Cruz, A., Lopez-Ortiz, R., Ventosa-Febles, E.A., Wiley, J.W., Nakamura, T.K., *et al.* (2005) Ecology and management of Shiny Cowbirds (*Molothrus bonariensis*) and endangered Yellow-shouldered Blackbirds (*Agelaius xanthomus*) in Puerto Rico. *Ornithological Monographs* 57, 38–44.

Cruz, A., Manolis, T.D. and Andrews, R.W. (1990b) Reproductive interactions of the Shiny Cowbird *Molothrus bonariensis* and the Yellow-hooded Blackbird *Agelaius icterocephalus* in Trinidad. *Ibis* 132, 436–444.

Cruz, A., Manolis, T.D. and Andrews, R.W. (1995) History of Shiny Cowbird *Molothrus bonariensis* brood parasitism in Trinidad and Tobago. *Ibis* 137, 317–321.

Cruz, A., Manolis, T.D. and Wiley, J.W. (1985) The Shiny Cowbird: a brood parasite expanding its range in the Caribbean region. *Ornithological Monographs* 36, 607–620.

Cruz, A., Prather, J.W., Post, W. and Wiley, J.W. (2000) The spread of the Shiny and Brown-headed Cowbird into the Florida Region. In: Smith, J.N.M., Cook, T.L., Rothstein, S.I., Robinson S.K. and Sealy, S.G. (eds) *Ecology and Management of Cowbirds and Their Hosts: Studies in the Conservation of North American Birds*. University of Texas, Austin, Texas, pp. 47–57.

Danforth, S.T. (1935) *The Birds of Saint Lucia*. Monographs of the University of Puerto Rico, Series B, Physical and Biological Sciences. University of Puerto Rico, San Juan, Puerto Rico.

de la Colina M. A., Hauber, M.E., Strausberger, B.M., Reboreda, J.C. and Mahler, B. (2016) Molecular tracking of individual host use in the Shiny Cowbird, a generalist brood parasite. *Ecology and Evolution* 6, 4684–4696.

Dominguez, M., Reboreda, J.C. and Mahler, B. (2015) Impact of Shiny Cowbird and botfly parasitism on the reproductive success of the globally endangered Yellow Cardinal *Gubernatrix cristata*. *Bird Conservation International* 25, 294–305.

Duré-Ruiz, N.M., Fernández G.J. and Mermoz, M.E. (2008) Effects of Cowbird parasitism in the brood reduction of the Brown-and-yellow Marshbird *Pseudoleistes virescens*. *Condor* 110, 507–513.

Fiorini, V.D. and Reboreda, J.C. (2006) Cues used by Shiny Cowbirds (*Molothrus bonariensis*) to locate and parasitise Chalk-browed Mockingbird (*Mimus saturninus*) nests. *Behavioral Ecology and Sociobiology* 60, 379–385.

Fiorini, V.D, Tuero, D.T. and Reboreda J.C. (2009) Shiny Cowbirds synchronize parasitism with host laying and puncture host eggs according to host characteristics. *Animal Behaviour* 77, 561–568.

Fiorini, V.D., Gloag, R., Reboreda, J.C. and Kacelnik, A. (2014) Strategic egg destruction by brood parasitic cowbirds? *Animal Behaviour* 93, 229–235

Fraga, R.M. (1978) The Rufous-collared Sparrow as a host of the Shiny Cowbird. *Wilson Bulletin* 90, 271–284.

Fraga, R.M., Casañas, H. and Pugnali, G. (1998) Natural history and conservation of the endangered Saffron-cowled Blackbird *Xanthopsar flavus* in Argentina. *Bird Conservation International* 8, 255–267.

Friedmann, H. (1929) *The Cowbirds: a Study in the Biology of Social Parasitism*. Charles C. Thomas Publisher, Springfield, Illinois.

Gloag, R. and Kacelnik, A. (2013) Host manipulation via begging call structure in the brood-parasitic Shiny Cowbird. *Animal Behaviour* 86, 101–109.

Gloag, R., Tuero D.T., Fiorini, V.D., Reboreda, J.C. and Kacelnik, A. (2012) The economics of nestmate-killing in avian brood parasites: a provisions trade-off. *Behavioral Ecology* 23, 132–140.

Gloag, R., Fiorini V.D., Reboreda, J.C. and Kacelnik, A. (2013) The wages of violence: mobbing by mockingbirds as a frontline defence against brood parasitic cowbirds. *Animal Behaviour* 86, 1023–1029.

Gloag, R., Fiorini V.D., Reboreda, J.C. and Kacelnik, A. (2014) Shiny Cowbirds share foster mothers but not true mothers in multiply parasitized mockingbird nests. *Behavioral Ecology and Sociobiology* 68, 681–689.

Gochfeld, M. (1978) Begging by nestling shiny cowbirds: adaptive or maladaptive. *Living Bird* 17, 41–50.

Grayce, R.L. (1957) Range extensions in Puerto Rico. *Auk* 74, 106.

Hartmann, S.A., Segelbacher, G., Juiña, M.E. and Schaefer, H.M. (2015) Effects of habitat management can vary over time during the recovery of an endangered bird species. *Biological Conservation* 192, 154–160.

Hudson, W.H. (1874) Notes on the procreant instincts of the three species of *Molothrus* found in Buenos Ayres. *Proceeding of the Zoological Society of London* 11, 153–174.

IUCN (2018) The IUCN Red List of Threatened Species, Version 2018–2. Available at: www.iucnredlist.org (accessed 30 October 2018).

Jaramillo, A. and Burke, P. (1999) *New World blackbirds: The Icterids*. Princeton University Press, Princeton, New Jersey.

Kattan, G.H. (1993) Reproductive strategy of a generalist brood parasite, the Shiny Cowbird, in the Cauca Valley, Colombia. PhD thesis, University of Florida, Gainesville, Florida.

Kattan, G.H. (1997) Shiny Cowbirds follow the 'shotgun' strategy of brood parasitism. *Animal Behaviour* 53, 647–654.

Kattan, G.H. (1998) Impact of brood parasitism. Why do House Wrens accept Shiny Cowbird eggs? In: Rothstein, S.I. and Robinson, S.K. (eds) *Parasitic Birds and Their Hosts: Studies in Coevolution*. Oxford University Press, New York, pp. 212–220.

Kepler, C.B. and Kepler, A.K. (1970) Preliminary comparison of bird species diversity and density in Luquillo and Guinica Forests. In: Odum, H. (ed.) A *Tropical Rain Forest*. US Atomic Energy Commission, Oak Ridge, Tennessee, pp. 183–186.

Kluza, D.A. (1998) First record of Shiny Cowbird (*Molothrus bonariensis*) in Yucatán, Mexico. *Wilson Bulletin* 110, 429–430.

Krabbe, N., Juina, M. and Sornoza, A.F. (2011) Marked population increase in Pale-headed Brush-finch *Atlapetes pallidiceps* in response to cowbird control. *Journal of Ornithology* 152, 219–222.

Lichtenstein, G. (1997) Begging behaviour and host exploitation in three species of parasitic cowbirds. PhD thesis, University of Cambridge, Cambridge, UK.

López, A.V., Fiorini, V.D., Ellison, K. and Peer, B.D. (2018) Thick eggshells of brood parasitic cowbirds protect their eggs and damage host eggs during laying. *Behavioral Ecology* 29, 965–973.

Lowther, P.E. (2011) Shiny Cowbird (*Molothrus bonariensis*), version 1.0. In: Schulenberg, T.S. (ed.) *Neotropical Birds*. Cornell Lab of Ornithology, Ithaca, New York, USA. Available at: https://doi.org/10.2173/nb.shicow.01 (accessed 25 October 2018).

Lowther, P.E. (2018) Lists of victims and hosts of the parasitic cowbirds (*Molothrus*). Available at: www.fieldmuseum.org/sites/default/files/cowbird-hosts-05sep2018.pdf (accessed 21 September 2018).

Mahler, B., Lovette, I., Confalonieri, V. and Reboreda, J.C. (2008) Eggshell spotting in brood parasitic Shiny Cowbirds (*Molothrus bonariensis*) is not linked to the female sex chromosome. *Behavioral Ecology and Sociobiology* 62, 1193–1199.

Mahler, B., Lopez, N.I., Di Giacomo, A.G. and Reboreda, J.C. (2010) Increased plumage darkness of female Shiny Cowbirds *Molothrus bonariensis* in the subtropics: an adaptation to bacterial degradation? *Ibis* 152, 775–781.

Marín, M. (2000) The Shiny Cowbird (*Molothrus bonariensis*) in Chile: introduction or dispersion? Its hosts and parasitic trends. *Ornitología Neotropical* 11, 285–296.

Mason, P. (1985) The nesting biology of some passerines of Buenos Aires, Argentina. *Ornithological Monographs* 36, 954–972.

Mason, P. (1986a) Brood parasitism in a host generalist, the Shiny Cowbird: II. Host selection. *Auk* 103, 61–69.

Mason, P. (1986b) Brood parasitism in a host generalist, the Shiny Cowbird: I. The quality of different species as hosts. *Auk* 103, 52–60.

Mason, P. (1987) Pair formation in cowbirds: evidence found for Screaming but not Shiny Cowbirds. *Condor* 89, 349–356.

Massoni, V. and Reboreda, J.C. (1998) Cost of brood parasitism and lack of defenses on the Yellow-winged Blackbird-Shiny Cowbird system. *Behavioral Ecology and Sociobiology* 42, 273–280.

Massoni, V. and Reboreda, J.C. (2002) A neglected cost of brood parasitism: egg punctures by Shiny Cowbirds during inspection of potential host nests. *Condor* 104, 407–412.

Medina-Miranda, R., López-Ortiz, R., Ramos-Álavarez, K., Cruz-Burgos, J., McKenzie, P., *et al.* (eds) (2013) Yellow-shouldered Blackbird/Shiny Cowbird population and habitat viability assessment workshop final report. IUCN SSC Conservation Breeding Specialist Group, Apple Valley, Minnesota.

Mermoz M.E. and Ornelas, J.F. (2004) Phylogenetic analysis of life-history adaptations in parasitic cowbirds. *Behavioral Ecology* 15, 109–119.

Mermoz, M.E and Reboreda, J.C. (1999) Egg-laying behaviour by Shiny Cowbirds parasitizing Brown-and-yellow Marshbirds. *Animal Behaviour* 58, 873–882.

Mermoz, M.E and Reboreda, J.C. (2003) Reproductive success of Shiny Cowbird (*Molothrus bonariensis*) parasitizing the larger Brown-and-yellow Marshbird (*Pseudoleistes virescens*) in Argentina. *Auk* 120, 1128–1139.

Mermoz, M.E., Reboreda, J.C. and Fernández, G.J. (2013) High rates of Shiny Cowbird parasitism on the Brown-and-yellow Marshbird select for complementary host defenses. *Condor* 115, 910–920.

Ortega, C.P. (1998) *Cowbirds and Other Brood Parasites*. University of Arizona Press, Tucson, Arizona.

Pérez-Rivera, R.A. (1986) Parasitism by the Shiny Cowbird in the interior parts of Puerto Rico. *Journal of Field Ornithology* 57, 99–104.

Pinchon, P.R. (1963) *Faune des Antilles Francaise: Les Oiseaux*. Fort-de-France, Martinique.

Post, W. (1992) First Florida specimens of the Shiny Cowbird. *Florida Field Naturalist* 20, 17–18.

Post, W. and Wiley, J.W. (1977a) The Shiny Cowbird in the West Indies. *Condor* 79, 119–121.

Post, W. and Wiley, J.W. (1977b) Reproductive interactions of the Shiny Cowbird and the Yellow-shouldered Blackbird. *Condor* 79, 176–184.

Post, W., Nakamura, T.K. and Cruz, A. (1990) Patterns of Shiny Cowbird parasitism in St. Lucia and southwestern Puerto Rico. *Condor* 92, 461–469.

Post, W., Cruz, A. and McNair, D.B. (1993) The North American invasion pattern of the Shiny Cowbird. *Journal of Field Ornithology* 64, 32–41.

Price, M.R., Lee, V.A. and Hayes, W.K. (2011) Population status, habitat dependence, and reproductive ecology of Bahama Orioles: a critically endangered synanthropic species. *Journal of Field Ornithology* 82, 366–378.

Reboreda, J.C., Clayton, N.S. and Kacelnik, A. (1996) Species and sex differences in hippocampus size in parasitic and non-parasitic cowbirds. *NeuroReport* 7, 505–508.

Reboreda, J.C., Mermoz, M.E., Massoni, V., Astié, A.A. and Rabuffetti, F.L. (2003) Impacto del parasitismo de cría del Tordo Renegrido (*Molothrus bonariensis*) sobre el éxito reproductivo de sus hospedadores. *Hornero* 18, 77–88.

Reboreda, J.C., Fiorini V.D. and de Mársico M.C. (2013) Antiparasitic defences in hosts of South American cowbirds. *Chinese Birds* 4, 57–70.

Reed, C.S. (1913) Datos para la biología del *Molothrus bonariensis*. *Revista Chilena de Historia Natural* 13, 172–179.

Robertson, W.B. Jr (1962) Observation on the birds of St. John, Virgin Islands. *Auk* 79, 44–76.

Sackmann, P. and Reboreda, J.C. (2003) A comparative study of Shiny Cowbird parasitism in two large hosts: Chalk-browed Mockingbird and Rufous-bellied Thrush. *Condor* 105, 728–736.

Salvador, S.A. and Salvador, L.A. (1986) Nota sobre la reproducción del Misto (*Sicalis luteola*) en Córdoba, Argentina. *Hornero* 12, 274–280.

Scardamaglia, R.C. and Reboreda, J.C. (2014) Ranging behavior of female and male Shiny Cowbirds and Screaming Cowbirds while searching for host nests. *Auk* 131, 610–618.

Scardamaglia, R.C., Fiorini, V.D., Kacelnik, A. and Reboreda, J.C. (2017) Planning host exploitation through prospecting visits by parasitic cowbirds. *Behavioral Ecology and Sociobiology* 71, 23.

Scardamaglia, R.C., Kacelnik, A. and Reboreda, J.C. (2018) Temporal and spatial patterns of roosting behaviour by brood parasitic cowbirds. *Ibis* 160, 779–789.

Segura, L.N. and Reboreda, J.C. (2012) Red-crested Cardinal defences against Shiny Cowbird parasitism. *Behaviour* 149, 325–343.

Smith-Ramírez, C. and Armesto, J.J. (2003) Foraging behaviour of bird pollinators on *Embothrium coccineum* (Proteaceae) trees in forest fragments and pastures in southern Chile. *Austral Ecology* 28, 53–60.

Spaw, C.D. and Rohwer, S. (1987) A comparative study of eggshell thickness in cowbirds and other passerines. *Condor* 89, 307–318.

Studer, A. and Vielliard, J. (1988) Premières données étho-écologiques sur l'Ictéridé brésilien *Curaeus forbesi* (Sclater, 1886) (Aves, Passeriformes). *Revue suisse de Zoologie* 95, 1063–1077.

Svagelj, W.S., Fernández, G.J. and Mermoz, M.E. (2009) Effects of nest-site characteristics and parental activity on cowbird parasitism and nest predation in Brown-and-yellow Marshbirds. *Journal of Field Ornithology* 80, 9–18.

Tuero, D.T, Gloag, R. and Reboreda, J.C. (2016) Nest environment modulates begging behavior of a generalist brood parasite. *Behavioral Ecolology* 27, 204–210.

Tuero, D.T., Fiorini, V.D. and Reboreda, J.C. (2007) Effects of Shiny Cowbird parasitism on different components of House Wren reproductive success. *Ibis* 149, 521–527.

Villaneda-Rey, M. and Rosselli, L. (2011) Abundancia del Chamón Parásito (*Molothrus bonariensis*, Icteridae) en 19 humedales de la Sabana de Bogotá, Colombia. *Ornitología Colombiana* 11, 37–48.

Vincenty, M., Tossas, A.G., Bird-Pico, F.J., Lopez-Ortiz, R. and Ventosa-Febles, E.A (2009) Yellow warbler (*Dendroica petechia*) breeding success in relation to Shiny Cowbird (*Molothrus bonariensis*) brood parasitism in Boqueron, Puerto Rico. *Ornitología Neotropical* 20, 523–533.

Wiley, J.W. (1985) Shiny Cowbird parasitism in two avian communities in Puerto Rico. *Condor* 87, 165–176.

Wiley, J.W. (1988) Host selection by the Shiny Cowbird. *Condor* 90, 289–303.

Wiley, J.W., Post, W. and Cruz, A. (1991) Conservation of the Yellow-shouldered Blackbird *Agelaius xanthomus*, an endangered West Indian species. *Biological Conservation* 55, 119–138.

Woodworth, B.L. (1997) Brood parasitism, nest predation, and season-long reproductive success of a tropical island endemic. *Condor* 99, 605–621.

Wüthérich, D., Azócar, A., García-Nuñez, C. and Silva, J.F. (2001) Seed dispersal in *Palicourea rigida*, a common treelet species from neotropical savannas. *Journal of Tropical Ecology* 17, 449–458.

13 Brown-headed Cowbird (*Molothrus ater* Boddaert, 1783)

Amy-Leigh Wilson*
School of Life Sciences, University of KwaZulu-Natal, Pietermaritzburg 3209, South Africa

Citation: Wilson, A.-L. (2020). Brown-headed Cowbird (*Molothrus ater* Boddaert, 1783). In: Downs, C.T. and Hart, L.A. (eds) *Invasive Birds: Global Trends and Impacts*. CAB International, Wallingford, UK, pp. 105–109.

13.1 Common Names

Brown-headed Cowbird, Common Cowbird, Buffalo Bird, Cowbird, Tordo cabeza café (Spanish), Vacher à Tête Brune (French).

13.2 Nomenclature

The Brown-headed Cowbird (*Molothrus ater* Boddaert, 1783) belongs to the family Icteridae. Three subspecies of Brown-headed Cowbirds are recognized: *M. a. obscurus* J.F. Gmelin, 1789; *M. a. artemisiae* Grinnell, 1909 and *M. a. ater* Boddaert, 1783. These subspecies vary in their ranges: *M. a. obscurus* is found in the extreme south-western parts of the USA, *M. a. artemisiae* in the north-western parts of the USA and *M. a. ater* in the eastern parts of the USA (Ortega, 1998; Peterjohn *et al.*, 2000).

13.3 Distribution

Brown-headed Cowbirds are not globally threatened and are common to locally abundant throughout most of their range (Fraga and Garcia, 2019). Brown-headed Cowbirds were originally limited to North America's mid-continental prairies (Fig. 13.1) (Powell, 2017; CABI, 2018). However, during the 19th and 20th centuries, the species greatly expanded its range and can now be found across the entire USA, the northern parts of Mexico and most of Canada (Bendire, 1895; Scott, 1987; Rothstein, 1994; Ortega, 1998; Cruz *et al.*, 2000; Peterjohn *et al.*, 2000; Powell, 2017). This range expansion has been linked to the transformation of forests to agricultural land and increasing urbanization (Powell, 2017; CABI, 2018). Brown-headed Cowbirds breed from south-east Alaska through southern Canada and across the entire continental USA to central Mexico (Roof, 1997).

At the end of the 20th century, the species had an estimated global population of 20–40 million individuals (Fraga and Garcia, 2019). An estimated 77% of the population frequents the USA for some part of the year, with Brown-headed Cowbirds being especially common in the central parts of North America (Ortega, 1998; Peterjohn *et al.*, 2000; Powell, 2017). Mexico sees visits by around 31% of the population, with only 14% of the population frequenting Canada in a year (Powell, 2017).

13.4 Description

Brown-headed Cowbirds are sexually dimorphic (Fig. 13.2A, B) (Byerley, 2013; CABI, 2018). Males are slightly larger than females with lengths of 19–22 cm, body masses of 42–50 g and an average wingspan of 36 cm (CABI, 2018; Fraga and Garcia, 2019). Females are 16–20 cm in length with body masses of 38–45 g and have an average wingspan of 32–38 cm (Byerley, 2013; Fraga and Garcia, 2019).

The males are more brightly coloured with a glossy green-black appearance and a characteristic brown head and neck (Scott, 1987; Byerley, 2013; CABI, 2018). Females, in contrast, are a dull brown or grey colour, and may have a lightly coloured streaked breast (Scott, 1987; Byerley, 2013). Both sexes have black eyes and long pointed wings (male wing chord length: 10.4–11 cm; female wing chord length: 9.6–10.1 cm) (Roof, 1997; Byerley, 2013; CABI, 2018). Both sexes have a grey conical beak with a curved upper mandible; this curve is more pronounced in males (Byerley, 2013; CABI, 2018).

Juvenile Brown-headed Cowbirds are similar to females in their colouring but with a more prominent streaked breast (Fig. 13.2D)

*Corresponding author: wilsonamyleigh@gmail.com

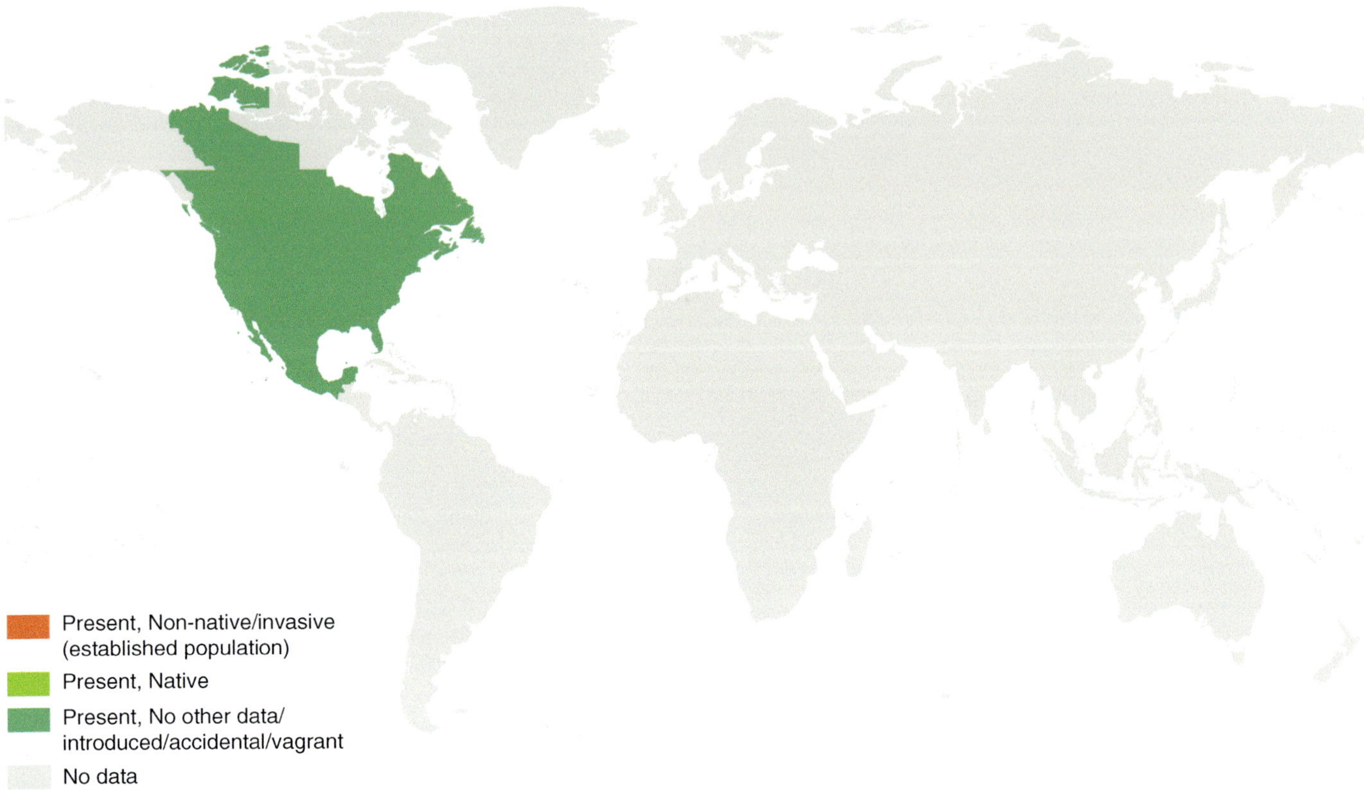

Fig. 13.1. Global distribution of the Brown-headed Cowbird (*Molothrus ater*) showing its natural (green) and invaded (red) ranges.

Fig. 13.2. (A) A male Brown-headed Cowbird. (©Photograph: R. Lazar.) (B) A female Brown-headed Cowbird. (©Photograph: M.M. Doucet.) (C) A male Brown-headed Cowbird moulting into adult plumage. (©Photograph: P. Wolf.) (D) A juvenile Brown-headed Cowbird. (©Photograph: Cephas. CC BY-SA 3.0; https://creativecommons.org/licenses/by-sa/3.0.) (E) A Brown-headed Cowbird egg (bottom left) in the nest of an Eastern Phoebe (*Sayornis phoebe*). (©Photograph: Gala Web Design. CC BY 3.0; https://creativecommons.org/licenses/by/3.0.) (F) A Brown-headed Cowbird chick (top) in the nest of a Savannah Sparrow (*Passerculus sandwichensis*). (©Photograph: K. Fleming. CC BY-SA 3.0; https://creativecommons.org/licenses/by-sa/3.0.)

(Scott, 1987). Young males when moulting into their adult plumage display a patchwork of buff, brown and black colouring (Fig. 13.2C) (Scott, 1987).

The call of Brown-headed Cowbirds consists of a combination of harsh rattles and squeaky gurgles and whistles (Scott, 1987).

13.5 Diet

Brown-headed Cowbirds forage in groups and feed primarily on the ground in open areas, with seeds from grasses, weeds and, to a lesser degree, crop grains forming most of their diet (Mayfield, 1965; Darley, 1968; Ehrlich *et al.*, 1988; Roof, 1997; Powell, 2017). Brown-headed Cowbirds are often found feeding near livestock, particularly large ungulates, as their feeding habits often flush out insects and create microhabitats that increase insect abundance (Goguen and Mathews, 2000). Insects such as grasshoppers and beetles make up the remainder (about 25%) of their diet (Ehrlich *et al.*, 1988; Powell, 2017). Female Brown-headed Cowbirds have also been observed to eat snail shells and occasionally eggs from parasitized nests (Powell, 2017). This behaviour is thought to be a result of their high calcium requirements due to the large numbers of eggs laid during their breeding season (Powell, 2017).

13.6 Introduction and Invasion Pathways

Clearing of forests for agricultural and grazing land and urbanization have enabled Brown-headed Cowbirds to rapidly expand their ranges (Roof, 1997; Georgia Museum of Natural History, 2000). Permanent feedlots, mowed grass and pastures enable cowbirds, which may have historically been associated with nomadic herds of bison, to spend the entire breeding season in the same place and even establish breeding territories (Rothstein and Robinson, 1998). The pre-Columbian distribution of cowbirds, however, remains poorly known; it is possible that they were much more widespread when Native American populations were much larger and maintained more agricultural openings.

13.7 Breeding Behaviour

Brown-headed Cowbirds are obligate brood parasites and have been observed to successfully parasitize over 150 host species, mostly songbirds (Payne, 1977; Mayfield, 1977; Lowther, 1993; Davies, 2000; Powell, 2017; CABI, 2018). Breeding occurs from mid-April until mid-July with peak breeding occurring in May (Georgia Museum of Natural History, 2000; Fraga and Garcia, 2019). Female cowbirds find nests by flushing out birds from their nests or by quietly monitoring for signs of nest building (Powell, 2017). Brown-headed Cowbirds lay white to greyish white eggs (1.8–2.5 cm long, 1.5–1.8 cm wide) with brown or grey spots (Fig. 13.2E) (Powell, 2017). In order to give their young an advantage, female Brown-headed Cowbirds are known to kick out an egg from some nests, and in some cases even the nestling, of the host species before laying their own when there is more than one egg (or nestling) in the nest already (Ehrlich *et al.*, 1988; Marvil and Cruz, 1989; CABI, 2018). Female Brown-headed Cowbirds have also occasionally been observed to eat a host egg (Ehrlich *et al.*, 1988).

Brown-headed Cowbirds lay one to two eggs in each nest and can lay up to 40 eggs in a breeding season (Jackson and Roby, 1992; Lowther, 1993; CABI, 2018). They are physiologically capable of laying as many as 77 eggs in a season (Holford and Roby, 1993). They have a shorter incubation period (10–13 days) than most of their hosts and are born naked except for a few tufts of down (Fig. 13.2F) (Ehrlich *et al.*, 1988; Marvil and Cruz, 1989; Powell, 2017; CABI, 2018). Cowbird chicks grow faster than most of their host species and fledge in 8–13 days (Ehrlich *et al.*, 1988; Marvil and Cruz, 1989; Powell, 2017; CABI, 2018).

13.8 Habitat

Brown-headed Cowbirds are found in nearly all open habitats when foraging (Friedmann, 1929; Mayfield, 1965; Dufty, 1982; Rothstein *et al.*, 1986; Fraga and Garcia, 2019). They usually forage on the ground and therefore prefer areas with short grass or bare ground, such as mowed or grazed grasslands, permanent pastures, feedlots and agricultural fields (Friedmann, 1929; Mayfield, 1965; Dufty, 1982; Rothstein *et al.*, 1986; Georgia Museum of Natural History, 2000; Powell, 2017; Fraga and Garcia, 2019). Typically, female cowbirds spend most of the mornings searching for nests, often in forest habitats where most species accept cowbird eggs, and then commute up to 15 km to open areas where they forage in the afternoons (Rothstein and Robinson, 1998; Goguen and Mathews 2000). They also search for nests in more open habitats where many host species have defences against brood parasitism such as egg ejection and abandonment of parasitized nests (Rothstein and Robinson, 1998).

13.9 Impacts

Brown-headed Cowbirds successfully parasitize over 150 host species, mostly songbirds (CABI, 2018). This parasitism is a major threat to songbird population numbers, particularly those species that have short breeding seasons and are unable to re-nest (CABI, 2018). In some highly fragmented forest regions, more than two-thirds of host nests contain cowbird eggs, and some host species average two to three cowbird eggs per nest (Robinson *et al.*, 1995). Parasitism, along with habitat loss, poses a major threat to endangered species with small population numbers such as the endangered Kirtland's Warbler (*Dendroica kirtlandii*) and Black-capped Vireo (*Vireo atricapilla*) (Griffith and Griffith, 2000; Powell, 2017).

13.10 Control

Manual trapping of adult Brown-headed Cowbirds has been found to be the most efficient way of killing large numbers of individuals in areas where endangered cowbird hosts are concentrated (Griffith and Griffith, 2000; Meuhter, 2003). Eggs are also physically removed from host nests (Meuhter, 2003). These methods, however, are impractical on larger spatial scales.

The best way to manage cowbirds is to reduce forest fragmentation, especially by minimizing openings within forests that create cowbird feeding habitats such as pastures, mowed grass and feedlots (Lebbin *et al.*, 2010). Large forest tracts with few openings are essentially free of cowbird parasitism; such tracts may act as source habitats for many hosts that produce almost no young in highly fragmented forests (Robinson *et al.*, 1995).

In the specific case of parasitism of Acadian Flycatchers (*Empidonax virescens*) by Brown-headed Cowbirds, it has been suggested that the removal of the invasive Amur honeysuckle (*Lonicera maackii*) may help to reduce brood parasitism on this species (Rodewald, 2011).

13.11 Uses

Brown-headed Cowbirds may help to control insect populations (Byerley, 2013).

13.12 Acknowledgements

Grateful thanks to Scott K. Robinson for constructive comments on a draft of this chapter.

13.13 References

Bendire, C.E. (1895) *Life Histories of North American Birds*, Vol. 2: *United States National Museum Special Bulletin*, Part 3. US Government Printing Office, Washington, DC.

Byerley, B. (2013) Animal Diversity Web (*Molothrus ater*). Available at: https://animaldiversity.org/accounts/Molothrus_ater/ (accessed 27 October 2019).

CABI (2018) *Molothrus ater* (Brown-headed Cowbird). In: *Invasive Species Compendium*. CAB International, Wallingford, UK. Available at: www.cabi.org/isc/datasheet/75200 (accessed 27 October 2019).

Cruz, A., Prather, J.W., Post, W. and Wiley, J.W. (2000) The spread of Shiny and Brown-headed cowbirds into the Florida region. In: Smith, J.N.M., Cook, T.L., Rothstein, S.I., Robinson, S K. and Sealy, S.G. (eds) *Ecology and Management of Cowbirds and Their Hosts: Studies in the Conservation of North American Passerine Birds*. University of Texas Press, Austin, Texas, pp. 47–57.

Darley, J.A. (1968) The social organization of breeding brown-headed cowbirds. PhD thesis, University of Western Ontario, London, Ontario.

Davies, N.B. (2000) *Cuckoos, Cowbirds and Other Cheats*. Poyser, London.

Dufty, A.M. (1982) Movements and activities of radio-tracked Brown-headed Cowbirds. *Auk* 99, 316–327.

Ehrlich, P.R., Dobkin, D.S. and Wheye, D. (1988) *The Birder's Handbook: A Field Guide to the Natural History of North American Birds*. Simon & Schuster, New York.

Fraga, R. and Garcia, E.F.J. (2019) Brown-headed Cowbird (*Molothrus ater*). In: del Hoyo, J., Elliott, A., Sargatal, J., Christie, D.A. and de Juana, E. (eds) *Handbook of the Birds of the World Alive*. Lynx Edicions, Barcelona, Spain. Available at: www.hbw.com/node/62295 (accessed 27 October 2019).

Friedmann, H. (1929) *The Cowbirds: A Study in the Biology of Social Parasitism*. Charles C. Thomas, Springfield, Illinois.

Georgia Museum of Natural History (2000) *Molothrus ater*. Georgia Department of Natural Resources, Atlanta, Georgia.

Goguen, C.B. and Mathews, N.E. (2000) Local gradients of cowbird abundance and parasitism relative to livestock grazing in a western landscape. *Conservation Biology* 14, 186–1869.

Griffith, J.T. and Griffith, J.C. (2000) Cowbird control and the endangered Least Bell's Vireo: a management success story. In: Smith, J.N.M., Cook, T.L., Rothstein, S.I., Robinson, S.K. and Sealy, S.G. (eds) *Ecology and Management of Cowbirds and Their Hosts*. University of Texas Press, Austin, Texas, pp. 342–356.

Holford, K.C. and Roby, D.D. (1993) Factors limiting fecundity of captive brown-headed cowbirds. *Condor* 95, 536–545.

Jackson, N.H. and Roby, D.D. (1992) Fecundity and egg-laying patterns of captive yearling brown-headed cowbirds. *Condor* 92, 585–589.

Lebbin, D.J., Parr, M.J. and Fenwick, G.H. (2010) *The American Bird Conservancy Guide to Bird Conservation*. University of Chicago Press, Chicago, Illinois.

Lowther, P.E. (1993) Brown-headed Cowbird (*Molothrus ater*). In: Poole, A.F. and Gill, F.B. (eds) *The Birds of North America*. Cornell Lab of Ornithology, Ithaca, New York.

Marvil, R.E. and Cruz, A. (1989) Impact of Brown-headed Cowbird parasitism on the reproductive success of the Solitary Vireo. *Auk* 106, 476–480.

Mayfield, H.F. (1965) The Brown-headed Cowbird with old and new hosts. *Living Bird* 4, 13–28.

Mayfield, H.F. (1977) Brown-headed cowbird: agent of extermination? *American Birds* 31, 107–113.

Meuhter, V. (2003) *Cowbirds and Conservation*. National Audubon Society, New York.

Ortega, C.P. (1998) *Cowbirds and Other Brood Parasites*. University of Arizona Press, Tucson, Arizona.

Payne, R.B. (1977) The ecology of brood parasitism in birds. *Annual Review of Ecology and Systematics* 8, 1–28.

Peterjohn, B.G., Sauer, J.R. and Schwarz, S. (2000) Temporal and geographic patterns in population trends of Brown-headed Cowbirds. In: Smith, J.N.M., Cook, T.L., Rothstein, S.I., Robinson, S.K. and Sealy, S.G. (eds) *Ecology and Management of cowbirds and Their Hosts*. University of Texas Press, Austin, Texas, pp. 21–34.

Powell, H. (2017) Brown-headed Cowbird: life history. The Cornell Lab of Ornithology, Ithaca, New York. Available at: www.allaboutbirds.org/guide/Brown-headed_Cowbird/lifehistory (accessed 27 October 2019).

Robinson, S.K., Thompson, F.R., Donovan, T.M., Whitehead, D.R. and Faaborg, J. (1995) Regional effects of forest fragmentation on the nesting success of migratory birds. *Science* 267, 1987–1990.

Rodewald, D. (2011) Spreading messages about invasives. *Diversity and Distributions* 18, 97–99.

Roof, J. (1997) Molothrus ater*: Brown-headed Cowbird*. University of Michigan, Ann Arbor, Michigan. Available at: https://animaldiversity.org/accounts/molothrus_ater/ (accessed 27 October 2019).

Rothstein, S.I. (1994) The cowbird's invasion of the Far West: history, causes and DADSF consequences experienced by host species. *Studies in Avian Biology* 15, 301–315.

Rothstein, S.I. and Robinson, S.K. (1998) Parasitic birds and their hosts. Oxford University Press, Oxford.

Rothstein, S.I., Yokel, D.A. and Fleischer, R.C. (1986) Social dominance, mating and spacing systems, female fecundity, and vocal dialects in captive and free ranging Brown-headed Cowbirds. *Current Ornithology* 3, 127–185.

Scott, S.L. (1987) *Field Guide to the Birds of North America*, 2nd edn. National Geographic Society, Washington, DC.

14 Rock Dove (*Columba livia* Gmelin, 1789)

Ndivhuwo Shivambu*, Cavin T. Shivambu and Colleen T. Downs
Centre for Excellence in Invasion Biology and Centre for Functional Biodiversity, School of Life Sciences, University of KwaZulu-Natal, Pietermaritzburg 3209, South Africa

Citation: Shivambu, N., Shivambu, C.T. and Downs, C.T. (2020) Rock dove (*Columba livia* Gmelin, 1789). In: Downs, C.T. and Hart, L.A. (eds) *Invasive Birds: Global Trends and Impacts*. CAB International, Wallingford, UK, pp. 109–117.

14.1 Common Names

Rock Dove, Rock Pigeon, Common Pigeon, Feral Pigeon (Rose *et al.*, 2006; Donegan, 2015).

14.2 Nomenclature

The Rock Dove (*Columba livia* Gmelin, 1789) belongs to the Columbidae family, with 12 subspecies, where some of the species are derived from Feral Pigeons (*Columba livia domestica*) (Gibbs *et al.*, 2000). Many of these subspecies are from different geographical areas (Fig. 14.1):

- *C. l. livia* occurs in Asia towards western Kazakhstan, Iran, Iraq, Georgia, Cyprus, Turkey, North Africa, Europe (western and southern parts) and the northern Caucasus.
- *C. l. atlantis* Bannerman, 1931, is found in the Azores and Cape Verde, and in Madeira. The populations of this subspecies are varied with uneven upperparts covering the black wing bars and are derived from Feral Pigeons.
- *C. l. canariensis* Bannerman, 1914, occurs in the Canary Islands and is usually smaller than other subspecies.
- *C. l. gymnocyclus* Gray, 1856, is native to Benin, Guinea, Ghana, Nigeria and Senegal. This subspecies is relatively smaller and darker than *C. l. livia* with a blackish colour on the head and rump. The underparts have white marks and the iridescence coloration of the nape extends on to the head.
- *C. l. targia* von Schweppenburg, 1916, occurs and breeds in the mountains of the eastern Sahara to Sudan. It is slightly smaller than the usual form, with a somewhat grey rump, and the back is a uniform grey colour together with the mantle, instead of white.
- *C. l. dakhlae* Meinertzhagen, 1928, occurs in central Egypt with a relatively smaller body size and is much lighter in colour than the other allopatric subspecies.
- *C. l. schimperi* Bonaparte, 1854, occurs in northern Sudan and in the southern Nile Delta. This subspecies closely resembles *C. l. targia* but has a distinctly lighter mantle.
- *C. l. palaestinae* Zedlitz, 1912, is native to Arabia and occurs from Syria to Sinai. It has darker plumage and is larger than *C. l. schimperi*.
- *C. l. gaddi* Zarodney & Looudoni, 1906, occurs and breeds in Azerbaijan and Iran, and east to Uzbekistan. It is larger and paler than *C. l. palaestinae*.
- *C. l. neglecta* Hume, 1873, is native in the eastern central Himalayas. Its body size is similar to the majority of the other subspecies, but it has darker plumage and a widespread rainbow-like shine on its neck.
- *C. l. intermedia* Strickland, 1844, is found within the range of *C. l. neglecta*, which is in India, south of the Himalayas and in Sri Lanka. It is comparable in appearance to *C. l. neglecta*, but its back is less contrasting and the bird is much darker.
- *C. l. nigricans* Buturlin, 1908, is native to northern China and Mongolia. It is mutable and is derived from Feral Pigeons (Gibbs *et al.*, 2000). In many countries, genetic studies linking these subspecies to their founder populations are lacking (Johnston *et al.*, 2006; Jacob *et al.*, 2016).

14.3 Distribution

The Rock Dove is one of the most widespread invasive species and has a large native range compared with other invasive bird species (Stern and Dickinson, 2010). Its native distribution range includes southern Europe, northern parts of Africa and large parts of the Middle East, and extends into southern Asia

*Corresponding author: ndivhuwomaligana@gmail.com

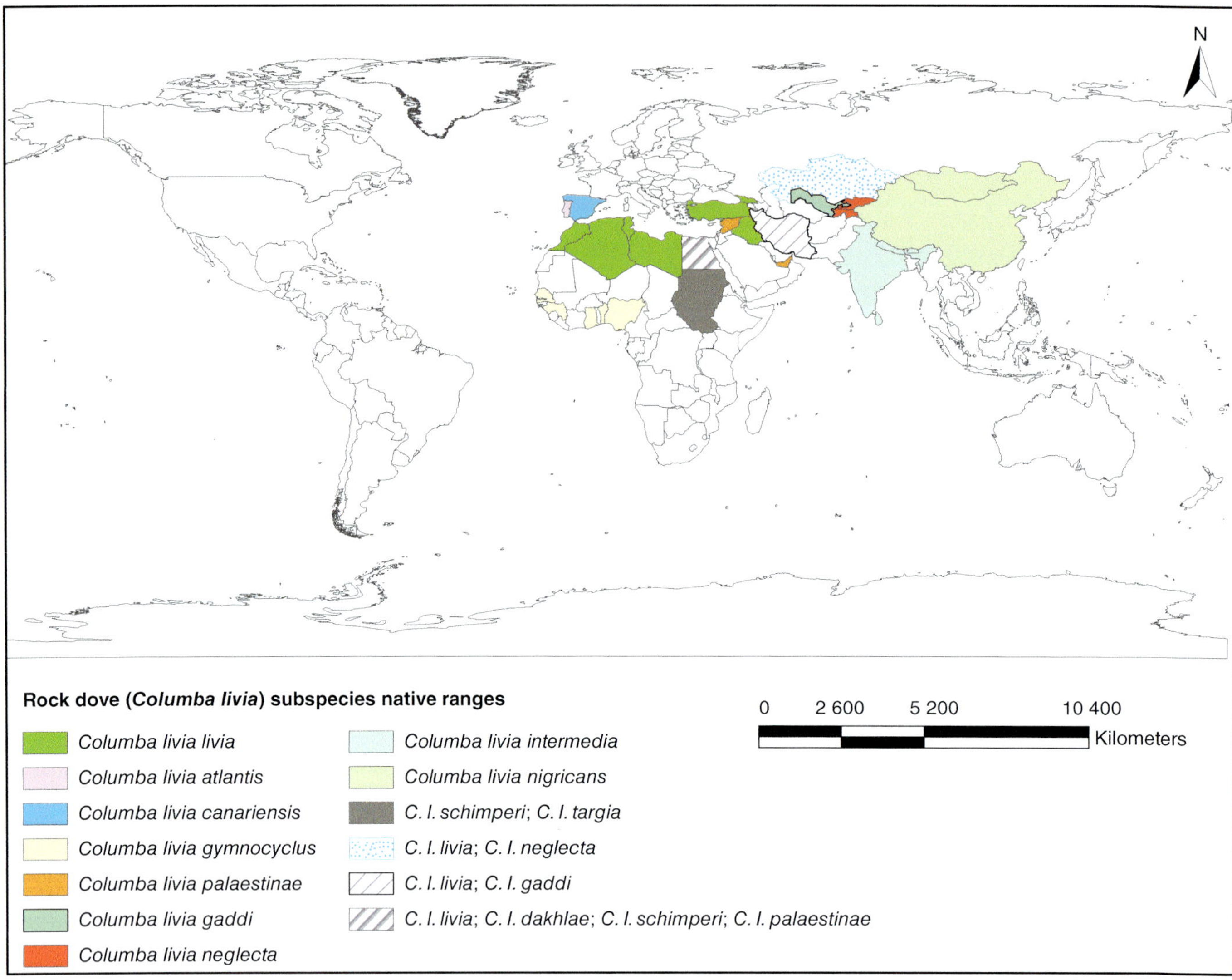

Fig. 14.1. The native distribution ranges of the 12 subspecies of Rock Dove (*Columba livia*) around the world. Each native range is represented by a colour and where more than one subspecies is listed, ranges overlap.

(Fig. 14.2) (Johnston, 1992; Stern and Dickinson, 2010). It is regarded as one of the world's worst invasive bird species as it occurs in approximately 64 countries outside its native range, mostly in urban and rural areas (Sacchi *et al.*, 2002; Rose *et al.*, 2006; Przybylska *et al.*, 2012; BirdLife International, 2016). The distribution of this species is still expanding, with unknown estimated populations in many places (Ferman *et al.*, 2010; Przybylska *et al.*, 2012; BirdLife International, 2016).

The Rock Dove's native range includes: Afghanistan, Albania, Algeria, Andorra, Armenia, Austria, Azerbaijan, Belarus, Belgium, Bhutan, Bosnia and Herzegovina, Bulgaria, Cape Verde, Chad, China, Croatia, Cyprus, Czech Republic, Egypt, Estonia, Ethiopia, Faroe Islands, Finland, France, Georgia, Germany, Greece, Hungary, Iceland, India, Iran, Iraq, Ireland, Israel, Israel, Italy, Jordan, Kazakhstan, Kyrgyzstan, Latvia, Lebanon, Libya, Lithuania, Luxembourg, Macedonia, Mali, Malta, Mauritania, Moldova, Monaco, Mongolia, Montenegro, Morocco, Myanmar, Nepal, Norway, Oman, Pakistan, Palestine, Poland, Portugal, Qatar, Romania, Russian Federation, San Marino, Saudi Arabia, Serbia, Slovakia, Slovenia, Spain, Sudan, Sweden, Switzerland, Syria, Tajikistan, Tunisia, Turkey, Turkmenistan, the UK, Ukraine, United Arab Emirates, Uzbekistan, Western Sahara and Yemen (BirdLife International, 2016; Global Invasive Species Database, 2019).

The Rock Dove's introduced range includes: American Samoa, Angola, Anguilla, Argentina, Australia, Austria, the Bahamas, Bangladesh, Barbados, Belgium, Belize, Bermuda, Bhutan, Bolivia, Botswana, Brazil, Brunei, Cambodia, Cameroon, Canada, the Cayman Islands, Chile, Colombia, Comoros, Costa Rica, Cuba, Czech Republic, Djibouti, the Dominican Republic, Ecuador, El Salvador, Fiji, French Guiana, French Polynesia, Gibraltar, Grenada, Guadeloupe, Guatemala, Guyana, Eswatini, Haiti, Honduras, Hong Kong, Hungary, India, Indonesia, Jamaica, Japan, Jersey, Kiribati, Lesotho, Liechtenstein, Madagascar, Malaysia, Maldives, Martinique, Mayotte, Mexico, Montserrat, the Netherlands, the Netherlands Antilles,

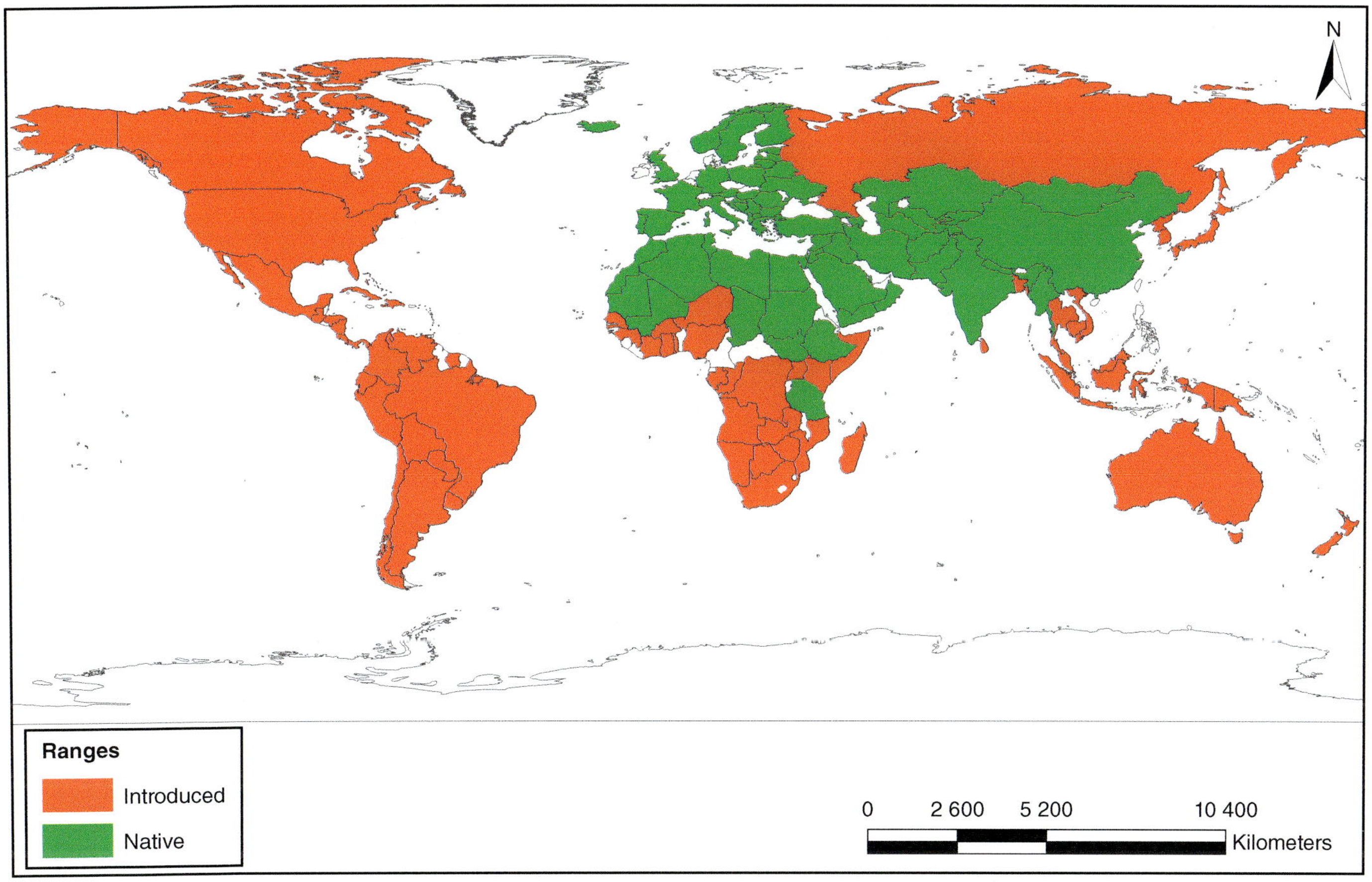

Fig. 14.2. Global distribution of the Rock Dove showing its native (green) and introduced (red) ranges.

New Zealand, Nicaragua, Norfolk Island (Pacific Ocean), Panama, Papua New Guinea, Paraguay, Peru, Portugal, Puerto Rico, Saint Helena, Saint Kitts and Nevis, Saint Lucia, Saint Pierre and Miquelon, Saint Vincent and the Grenadines, Samoa, Singapore, South Africa, Spain, Sri Lanka, Suriname, Switzerland, Taiwan, Thailand, Tonga, Trinidad and Tobago, Turks and Caicos Islands, the USA, Uruguay, Venezuela, Wallis and Futuna, and Zimbabwe (BirdLife International, 2016; Global Invasive Species Database, 2019).

14.4 Description

The Rock Dove has a range of subspecies, each with different morphometric and colour variations (Gibbs *et al.*, 2000; Stern and Dickinson, 2010). In general, their body length is 32–33 cm, with a 62–72 cm wingspan (Dean, 2005). Although the body mass of the domesticated Rock Dove can exceed the norm, the feral Rock Dove weighs between 238 and 400 g (Gibbs *et al.*, 2000; Dean, 2005). Generally, the head and neck are dark grey in colour, with colourful iridescence along the neck and chest area (Fig. 14.3A, B). The iris is orange-red, sometimes golden with an inner ring that is paler, and the eyelids are grey. The bill is brownish-black with a visible white powdery cere and the feet are a pink-red colour. Their wing chord length is 22.3–23.0 cm, bill length is 1.8–1.9 cm, tarsus length is 2.6–3.5 cm and the tail is around 9.6–11 cm long (Gibbs *et al.*, 2000; Dean, 2005). Both sexes are identical, but the female has less iridescence on her neck and breast (Gibbs *et al.*, 2000). The juvenile is identical to the adults, but its colour is cloudier, with wing bars less visible, and they lack iridescence on the neck (Dean, 2005). The most distinctive characteristics to identify pure Rock Doves are the white lower back and the two distinctive black bars on the pale grey wings (Gibbs *et al.*, 2000; Dean, 2005). Some hybrid Rock Doves have variable plumages, which makes species identification difficult (Le Roux *et al.*, 2012).

14.5 Diet

The Rock Dove is an opportunistic and generalist feeder (Spennemann and Watson, 2017). It is commensal with humans, particularly in cities, where it feeds on a variety of abundant anthropogenic foods (Bull, 1973; Sacchi *et al.*, 2002, Ferman *et al.*, 2010). These include discarded foods on the streets, in garbage and refuse bins, in parks and in shopping centres (e.g. white bread, deep-fried potato chips, meat), but they also feed on agricultural crops (e.g. maize, grits, barley, wheat, red and white grapes, sorghum, safflower, green peas, yellow peas, vetches), grass and weed seeds, and insects (Murton

Fig. 14.3. Rock Doves on Durban Beach, South Africa. (A) Front view. (B) Rear view. (©Photographs: H. Jordaan.)

and Westwood, 1966; Jokimäki and Suhonen, 1998; Spennemann and Watson, 2017). Humans have contributed to their survival in the cities, for example in most European cities the public feed Rock Doves and this behaviour has allowed weak and handicapped birds to survive (Haag-Wackernagel, 1995).

14.6 Introduction and Invasion Pathways

The Rock Dove is well established and has the largest avian invaded range across the world (Stern and Dickinson, 2010; BirdLife International, 2016; Global Invasive Species Database, 2019). The abundance of food on city streets, parks and agricultural lands, and ample breeding and roosting sites on man-made infrastructures has facilitated the Rock Dove's establishment and spread (Sacchi *et al.*, 2002; Buijs and van Wijnen, 2003; Rose *et al.*, 2006). Generally, where Rock Doves have established, their populations are high across the world. In Amsterdam, a total of 11,885 individuals were counted in 894 locations (Buijs and van Wijnen, 2003). In Europe, the total population is estimated to be between 17 million and 28 million birds (Ferman *et al.*, 2010).

The Rock Dove was introduced in many countries as a domesticated bird through the pet trade, and then became invasive through release and escape pathways (Rappole *et al.*, 2000; Burivalova *et al.*, 2017). Europeans also contributed to the early introductions of Rock Doves through transportation of domesticated animals to new locations (Robbins, 1995). In North America, European settlers in the 17th century brought them as a food source, but they escaped from captivity and formed feral populations (Johnson, 1998). This also occurred in other parts of the world (Eguchi and Amano, 2004). Rock Doves were used as messengers during war (Wendell, 1977), and the first country to do so at the beginning of 1891 was France (Snyders, 2015). Homing pigeons were also introduced to South Africa for communication during war between 1899 and 1902 (Lincoln, 1927). The pigeons would transfer messages from one part of the English army to another (Lincoln, 1927). The introduction of Rock Doves in some countries was through racing competitions, with some racing pigeons failing to return to their lofts after competitions (Robbins 1995). There is little information on their introductions to islands, but on the Galapagos Islands, it is reported that the total population of Rock Doves was produced from four captive individuals introduced in 1972 or 1973 (Harmon *et al.*, 1987).

Some Rock Dove introductions were through natural dispersal (Johnston and Janiga, 1995), but relatively few studies on their movement patterns have been conducted (Rose *et al.*, 2005). They are not migratory, but daily flight distances reported can be up to 50 km (Johnston and Janiga, 1995). In the Mediterranean (Sardinia), they flew up to 19 km between foraging and roosting sites (Baldaccini *et al.*, 2000). However, other studies have shown that Rock Doves do not fly far and that their movement is governed by food availability (Hetmański, 2013; Soldatini *et al.*, 2006).

14.7 Breeding Behaviour

The breeding biology of Rock Doves varies geographically (Dean, 2005; Kasprzak *et al.*, 2006). In many countries, Rock Doves breed all year, for example in the USA (Preble and Heppner, 1981; Johnson 1998), Switzerland (as long as food is available; Stock and Haag-Wackernagel, 2016), the UK (peaking between April and July) (Lees, 1946) and South Africa (Harris *et al.*, 2016). However, in Poland, they breed between January and May (Kasprzak *et al.*, 2006; Hetmański, 2013). The pair bonds may last for up to 12 years (Cramp, 1985). During courtship, the male leaves the roost with loud wing clapping and glides back in a long arc with wings held in a V-shape (Cramp, 1985). The male lands next to the female, bows and turns his chest aside, spreads his tail and points his bill down; if successful, copulation occurs (Cramp, 1985; Dean, 2005).

The Rock Dove nest is built by both sexes but mainly by the female (Cramp, 1985). It is built using sticks, twigs, feathers and anthropogenic materials such as wire and plastic (Fig. 14.4) (Cramp, 1985). Natural nests are usually built on the edge of the cliffs (Murton and Clarke, 1968; Cramp, 1985; Stern and Dickinson, 2010), while urban nest sites are typically on roofs and edges of buildings (Fenďa and Kicková, 2009; Dobeic *et al.*, 2011). They lay one or two eggs in a 48 h interval, which are subelliptical oval in shape (Fig. 14.4) (Cramp, 1985; Robertson, 1988). The eggs are white, weigh an average of 10 g and are 39 × 29 mm (Cramp, 1985; Robertson, 1988). They are incubated by both sexes for 16–19 days, with the male usually incubating during the early mornings for 1.5–2.5 h (Cramp, 1985; Łukasiewicz, 2014). The newly hatched chicks have a lead-grey bill with a pinkish-white tip (Cramp, 1985). Both adults care for the young, feeding them crop milk and proportionately more food as the chicks age (Cramp, 1985). The nestling period lasts between 35 and 37 days (Cramp, 1985; Stock and Haag-Wackernagel, 2016). Breeding success is reported to be approximately 37.4% (Stock and Haag-Wackernagel, 2016).

14.8 Habitat

The Rock Dove's native and invasive range is broad and covers a large extent of the world (Fig. 14.2), indicating that they can adapt and persist in both moderate and extreme climates, which can differ from their native range (Ferman *et al.*, 2010). This dove occurs in a wide range of habitats including urban areas (Fontana *et al.*, 2011), rural expanses (Garaffa *et al.*, 2009) and farmlands (Herzon *et al.*, 2008). The Rock Dove thrives in urban areas, especially in city centres and parks where there is an abundance of discarded anthropogenic food and tall buildings provide roosting and breeding sites (Giunchi *et al.*, 2008; Krimowa, 2012; Pike *et al.*, 2017; Spennemann and Watson, 2017). In its native range, it similarly occurs on farmlands (Dhindsa and Saini, 1994; Kale *et al.*, 2014) and urban areas (Michaelsen and Refvik, 2003; Kale *et al.*, 2012; Ali *et al.*, 2013).

Fig. 14.4. A typical Rock Dove nest with a clutch size of two. The photograph was taken at Verolavecchia, Italy. (©Photograph: Pierandrea Brichetti; https://www.hbw.com/ibc/photo/rock-dove-columba-livia/nest-containing-two-eggs).

14.9 Impacts

In the Rock Dove's native and introduced ranges such as Asia (Kale *et al.*, 2014), Europe (Hulme, 2007), Africa (Harris *et al.*, 2016) and America (Pimentel *et al.*, 2005), it is regarded as a common pest species. This is probably due to their defaecation on landmarks, sidewalks, statues and buildings (Fig. 14.5) (Sacchi *et al.*, 2002; Jerolmack, 2008; Silva and Medeiros, 2008). They negatively affect other species, for example the breeding success of White-tailed Tropicbirds *Phaethon lepturus* in Bermuda is compromised through nest competition (Dobson and Madeiros, 2008). They impact the economy negatively as they depredate crops (Jokimäki and Suhonen, 1998; Spennemann and Watson, 2017). For example, they cause damage to maize, seeds, legumes (soybeans and chickpeas) and sunflower crops (van Niekerk and van Ginkel, 2004; Giunchi *et al.*, 2012). In the USA alone, the damage associated with feral pigeons in urban areas amounts to around US$1.1 billion annually (Pimentel *et al.*, 2000).

The Rock Dove carries pathogens, diseases and parasites that are harmful to humans and other animals (Mushi *et al.*, 2000; Tietz Marques *et al.*, 2009; Gasparini *et al.*, 2011; Dolz *et al.*, 2013). A study by Al-Harbi (2003) in Libya indicated that

Fig. 14.5. Faecal contamination by Rock Doves in the city centre of Pietermaritzburg, KwaZulu-Natal Province, South Africa. (©Photographs: Ndivhuwo Shivambu and Cavin Shivambu.)

their faecal coliform bacteria contaminates artificial ponds and affects tilapia fish species such as *Oreochromis niloticus* and *Oreochromis aureus*. In the Galapagos Islands, Rock Doves carry the parasite *Trichomonas gallinae*, which causes a potentially fatal disease that affects native species such as the Galapagos Dove (*Zenaida galapagoensis*; Harmon *et al.*, 1987). In addition, they can also be a source of air- and waterborne contamination with *Enterocytozoon bieneusi* fungal spores, which affect human and domestic animal health (Graczyk *et al.*, 2007). These birds can harbour up to 60 different pathogens including *Salmonella enterica*, *Camplyobacter* spp., *Chlamydophila psittaci* and *Cryptococcus neoformans*, which have been transmitted to humans (Haag-Wackernagel and Moch, 2004). Additionally, Rock Doves are associated with ectoparasites such as bugs (e.g. *Cimex lectularius*), fleas (e.g. *Ceratophyllus gallinae*), flies (e.g. *Fannia canicularis*), mites (e.g. *Dermanyssus gallinae*) and ticks (e.g. *Argas reflexus*) that are known to infest humans (Haag-Wackernagel, 2005).

14.10 Control

Information on the breeding biology of this invasive species is crucial for effective management and eradication programmes (Phillips *et al.*, 2012). Physical and chemical methods have been used to control Rock Doves. Between 2001 and 2005, they were successfully removed from the Galapagos Islands (Galapagos National Park and the Charles Darwin Foundation, 2004; Phillips *et al.*, 2012). Several methods were used including shooting, poisoning with α-chloralose baits and trapping at nest sites (Brand *et al.*, 2003; Galapagos National Park and the Charles Darwin Foundation, 2004; Phillips *et al.*, 2012). In Bermuda, a strategy to control Rock Doves was implemented that included culling and the placement of artificial burrows to protect tropicbirds (Outerbridge, 2016). Contraceptives have also been effective; for example, in Ljubljana, Slovenia, numbers of Rock Doves were reduced by 49.1% using contraceptives (Dobeic, 2015). The numbers of Rock Doves have also successfully been reduced by removing the eggs and replacing them with fake eggs (Felder *et al.*, 2007; Jacquin *et al.*, 2010). Another method is to reduce the amount of food available in rural and urban areas, which has decreased the population of Rock Doves in Spain (Senar *et al.*, 2017). Rock Doves and their eggs are preyed on by many other species. They form a particularly important prey base for urban birds of prey such as Peregrine Falcons (*Falco peregrinus*; López-López *et al.*, 2009). In south-west England, 47% of the diet of Peregrine Falcons was composed of pigeons and doves, while in Porto Alegre, Brazil, it was 68.5% and in Santos it was 80.9% (Silva 1997; Drewitt and Dixon, 2008). In cities such as New York and London, trained hawks or falcons have been deployed to control Rock Doves (Felder *et al.*, 2007). In Liverpool, UK, robotic hawks were used to scare away roosting populations of Rock Doves in the city (Felder *et al.*, 2007). Consequently, in some areas, raptors are used for Rock Dove control (Felder *et al.*, 2007).

14.11 Uses

The Rock Dove is one of the most common pet birds worldwide, and breeding this species has become a popular hobby, with over 350 different breeds (Shapiro and Domyan, 2013). One of the breeds, the homing pigeon, is commonly kept by pigeon fanciers for racing competitions (de Herdt and Pasmans, 2009). In Japan and the Galapagos Islands, pigeons are kept as a food source (Eguchi and Amano, 2004; Phillips *et al.*, 2003). Pigeon meat (squab) is also served at restaurants (e.g. Chinese and French) and is delicious and also expensive (Darwati *et al.*, 2010; Cummins, 2018). Pigeons are also used in laboratories as models for biomedical research (Clayton and Emery, 2015). In cities, most people generally do not consider pigeons a problem as they enjoy watching and feeding them (Gompertz, 1957); however, attitudes are changing because of the negative impacts of pigeon excreta (Felder *et al.*, 2007; Dobeic *et al.*, 2011). Ancient Egyptians used these birds for ceremonial and culinary purposes, and in some parts of the world they have also been used as messengers during both war and peace times (Wendell, 1977; Shapiro and Domyan, 2013; Snyders, 2015).

14.12 Notes

The Rock Dove is regarded as one of the most intelligent bird species. A study by Scarf *et al.* (2011) revealed that they can distinguish stimuli differing in numerosity and their performance is indistinguishable from that demonstrated by monkeys. Another study showed that the Rock Dove can recognize and correctly classify individual heterospecifics based on facial information (Stephan *et al.*, 2012). Pigeons are also known for their homing ability, and return to their nests even when they have been released at unfamiliar locations (Walcott, 1996). Homing pigeons used in pigeon racing competitions have been recorded flying as far as 1800 km (Walcott, 1996).

14.14 References

Al-Harbi, A.H. (2003) Faecal coliforms in pond water, sediments and hybrid tilapia *Oreochromis niloticus* × *Oreochromis aureus* in Saudi Arabia. *Aquaculture Research* 34, 517–524.

Ali, S., Rakha, B.A., Hussain, I., Nadeem, M.S. and Rafique, M. (2013) Ecology of feral Pigeon (*Columba livia*) in urban areas of Rawalpindi/Islamabad, Pakistan. *Pakistan Journal of Zoology* 45, 1229–1234.

Baldaccini, N.E., Giunchi, D., Mongini, E. and Ragionieri, L. (2000) Foraging flights of wild rock doves (*Columba l. livia*): a spatio-temporal analysis. *Italian Journal of Zoology* 67, 371–377.

BirdLife International (2016) Rock Dove: *Columba livia*. The IUCN Red List of Threatened Species 2016. Available at: https://www.iucnredlist.org/species/22690066/86070297 (accessed 27 October 2019).

Brand, P.R., Snell, H.L. and Vargas, H. (2003) Feral rock doves in the Galápagos Islands: biological and economic threats. *Noticias de Galápagos* 62, 6–11.

Buijs, J.A. and van Wijnen, J.H. (2003) Survey of feral rock doves (*Columba livia*) in Amsterdam, a bird–human association. *Urban Ecosystems* 5, 235–241.

Bull, J. (1973) Exotic birds in the New York City area. *Wilson Bulletin* 85, 501–505.

Burivalova, Z., Lee, T.M., Hua, F., Lee, J.S.H., Prawiradilaga, D.M. and Wilcove, D.S. (2017) Understanding consumer preferences and demography in order to reduce the domestic trade in wild-caught birds. *Biological Conservation* 209, 423–431.

Clayton, N.S. and Emery, N.J. (2015) Avian models for human cognitive neuroscience: a proposal. *Neuron*, 86, 1330–1342.

Cramp, S. (1985) *The Birds of the Western Palearctic*, Vol. 4. Oxford University Press, Oxford.

Cummins, E. (2018) Americans used to eat Pigeon all the time – and it could be making a comeback. Available at: https://www.popsci.com/eat-pigeons-squab/ (accessed 22 March 2019).

Darwati, S., Martojo, H., Sumantri, C., Sihombing, D.T.H. and Mardiastuti, A. (2010) Productivity, repeatability of productive and reproductive traits of local Pigeon. *Journal of the Indonesian Tropical Animal Agriculture* 35, 268–274.

de Herdt, P. and Pasmans, F. (2009) Pigeons. In: Tully, T.N., Dorrestein, G.M. and Jones, A.K. (eds) *Handbook of Avian Medicine*, 2nd edn. W.B. Saunders, Edinburgh, UK, pp. 350–376.

Dean, W.R.J. (2005) Rock dove. In: Hockey, P.A.R., Dean, W.R.J. and Ryan, P. (eds) *Roberts Birds of Southern Africa*, 7th edn. Trustees of the John Voelker Bird Book Fund, Cape Town, pp. 276–277.

Dhindsa, M.S, and Saini, H.K (1994). Agricultural ornithology; an Indian perspective. *Journal of Biosciences* 19, 391–402.

Dobeic, M. (2015) Feral Pigeon (*Columba livia domestica* var. *urbana*) control in Ljubljana. PhD thesis, University of Ljubljana, Ljubljana, Slovenia.

Dobeic, M., Pintaric, Š., Vlahović, K. and Dovc, A. (2011) Feral Pigeon (*Columba livia*) population management in Ljubljana. *Veterinarski Arhiv* 81, 285–298.

Dobson, A.F. and Madeiros, J. (2008) Threats facing Bermuda's breeding seabirds: measures to assist future breeding success. In: *Proceedings of the 4th International Partners in Flight Conference: Tundra to Tropics*, McAllen, Texas, pp. 223–226.

Dolz, G., Solórzano-Morales, Á., Angelova, L., Tien, C., Fonseca, L. and Bonilla, M. (2013) *Chlamydia psittaci* genotype B in a Pigeon (*Columba livia*) inhabiting a public place in San José, Costa Rica. *Open Veterinary Journal* 3, 135–139.

Donegan, T.M. (2015) The Pigeon names *Columba livia*, *C. domestica* and *C. oenas* and their type specimens. *Bulletin of the British Ornithologists' Club* 136, 15–28.

Drewitt, E.J. and Dixon, N. (2008) Diet and prey selection of urban-dwelling Peregrine Falcons in southwest England. *British Birds* 101, 58–67.

Eguchi, K. and Amano, H.E. (2004) Spread of exotic birds in Japan. *Ornithological Science* 3, 3–11.

Felder, S., Casertano, M. and Roberts, A. (2007) Curbing the pigeon conundrum. *The New York Times*. Available at: www.nytimes.com/packages/pdf/nyregion/city_room/20071112_pigeonreport.pdf (accessed 27 October 2019).

Fenďa, P. and Kicková, G. (2009) Mites (Acari, Mesostigmata) in the nests of feral pigeons (*Columba livia* f. *domestica*) in Slovakia. *Folia Faunistica Slovaca* 14, 31–35.

Ferman, L.M., Peter, H.U. and Montalti, D. (2010) A study of feral Pigeon *Columba livia* var. in urban and suburban areas in the city of Jena, Germany. *Arxius de Miscellània Zoològica* 8, 1–8.

Fontana, C.S., Burger, M.I. and Magnusson, W.E. (2011) Bird diversity in a subtropical South-American City: effects of noise levels, arborisation and human population density. *Urban Ecosystems* 14, 341–360.

Galapagos National Park and the Charles Darwin Foundation (2004) Rock pigeons go flying: eradication of introduced rock pigeons. Available at: www.hear.org/galapagos/invasives/topics/management/vertebrates/projects/doves.htm (accessed 15 March 2019).

Garaffa, P.I., Filloy, J. and Bellocq, M.I. (2009) Bird community responses along urban–rural gradients: does the size of the urbanized area matter? *Landscape and Urban Planning* 90, 33–41.

Gasparini, J., Erin, N., Bertin, C., Jacquin, L., Vorimore, F., *et al.* (2011) Impact of urban environment and host phenotype on the epidemiology of Chlamydiaceae in feral Pigeons (*Columba livia*). *Environmental Microbiology* 13, 3186–3193.

Gibbs, D., Barnes, E. and Cox, J. (2000) Pigeons and doves. In: Gibbs, D., Barnes, E. and Cox, J. (eds) *A Guide to the Pigeons and Doves of the World*. Pica Press, London, pp. 176–179.

Giunchi, D., Albores-Barajas, Y.V., Baldaccini, N.E., Vanni, L. and Soldatini, C. (2008) Feral pigeons: problems, dynamics and control methods. *Integrated Pest Management and Pest Control – Current and Future Tactics* 10, 211–244.

Giunchi, D., Albores-Barajas, Y.V., Baldaccini, N.E., Vanni, L. and Soldatini, C., (2012) Feral pigeons: problems, dynamics and control methods. In: Soloneski, S. and Larramendy, M. (eds) *Integrated Pest Management and Pest Control – Current and Future Tactics*. Intech, Rijeka, Croatia, pp. 215–240.

Global Invasive Species Database (2019) Species profile: *Columba livia*. Available at: www.iucngisd.org/gisd/speciesname/Columba+livia (accessed 15 March 2019).

Gompertz, T. (1957) Some observations on the feral Pigeon in London (with addendum: two cases of polyneuritis in feral Pigeons. Derek Goodwin). *Bird Study* 4, 2–13.

Graczyk, T.K., Sunderland, D., Rule, A.M., da silva, A.J., Moura, I.N.S., *et al.* (2007) Urban feral Pigeons (*Columba livia*) as a source for air- and waterborne contamination with *Enterocytozoon bieneusi* spores. *Applied and Environmental Microbiology* 73, 4357–4358.

Haag-Wackernagel, D. (1995) Regulation of the street Pigeon in Basel. *Wildlife Society Bulletin* 23, 256–260.

Haag-Wackernagel, D. (2005) Parasites from feral Pigeons as a health hazard for humans. *Annals of Applied Biology* 147, 203–210.

Haag-Wackernagel, D. and Moch, H. (2004) Health hazards posed by feral Pigeons. *Journal of Infection* 48, 307–313.

Harmon, W.M., Clark, W.A., Hawbecker, A.C. and Stafford, M. (1987) *Trichomonas gallinae* in columbiform birds from the Galapagos Islands. *Journal of Wildlife Diseases* 23, 492–494.

Harris, E., de Crom, E.P., Labuschagne, J. and Wilson, A. (2016) Urban environment use by speckled (*Columba guinea*) and feral (*Columba livia*) Pigeons on the University of South Africa's Muckleneuk campus. *Applied Ecology and Environmental Research* 14, 399–419.

Herzon, I., Auninš, A., Elts, J. and Preikša, Z. (2008) Intensity of agricultural land-use and farmland birds in the Baltic States. *Agriculture, Ecosystems and Environment* 125, 93–100.

Hetmański, T. (2013) Dispersion asymmetry within a feral Pigeon *Columba livia* population. *Acta Ornithologica* 42, 23–31.

Hulme, P.E. (2007) Biological invasions in Europe: drivers, pressures, states, impacts and responses. *Biodiversity Under Threat* 25, 56–80.

Jacob, G., Prévot, A.C. and Baudry, E. (2016) Feral Pigeons (*Columba livia*) prefer genetically similar mates despite inbreeding depression. *PLoS One* 11, e0162451.

Jacquin, L., Cazelles, B., Prévot-Julliard, A.C., Leboucher, G. and Gasparini, J. (2010) Reproduction management affects breeding ecology and reproduction costs in feral urban pigeons (*Columba livia*). *Canadian Journal of Zoology* 88, 781–787.

Jerolmack, C. (2008) How Pigeons became rats: the cultural–spatial logic of problem animals. *Social Problems* 55, 72–94.

Johnson, R.F. (1998) Feral Pigeon. *Emporia State University* 45, 1–16.

Johnston, R.F. (1992) Geographic size variation in Rock Pigeons, *Columba livia*. *Italian Journal of Zoology* 59, 111–116.

Johnston, R.F. and Janiga, M. (1995) *Feral Pigeons*. Oxford University Press, New York.

Johnston, R.F., Siegel-Causey, D. and Johnson, S.G. (2006) European populations of the rock dove *Columba livia* and genotypic extinction. *American Midland Naturalist* 120, 1–10.

Jokimäki, J. and Suhonen, J. (1998) Distribution and habitat selection of wintering birds in urban environments. *Landscape and Urban Planning* 39, 253–263.

Kale, M., Dudhe, N., Kasambe, R., Chakane, S. and Bhattacharya, P. (2012) Impact of urbanization on avian population and its status in Maharashtra state, India. *International Journal of Applied Environmental Sciences* 7, 59–76.

Kale, M.A., Dudhe, N., Kasambe, R. and Bhattacharya, P. (2014) Crop depredation by birds in Deccan plateau, India. *International Journal of Biodiversity* 2014, 1–8.

Kasprzak, M., Hetmański, T. and Kulczykowska, E. (2006) Changes in hematological parameters in free-living Pigeons (*Columba livia* f. *urbana*) during the breeding cycle. *Journal of Ornithology* 47, 599–604.

Krimowa, S. (2012) Pigeons and people: resource ecology and human dimensions of urban wildlife. MSc thesis, Victoria University of Wellington, Wellington, New Zealand.

Le Roux, P., Kok, O. and Butler, H. (2012) Rock Pigeon × feral Pigeon hybrid? *South African Journal of Science and Technology* 26, 299–302.

Lees, J. (1946) All the year breeding of the rock dove. *British Birds* 39, 136–141.

Lincoln, F.C. (1927) The military use of the homing pigeon. *Wilson Bulletin* 39, 67–74.

López-López, P., Verdejo, J. and Barba, E. (2009) The role of Pigeon consumption in the population dynamics and breeding performance of a peregrine falcon (*Falco peregrinus*) population: conservation implications. *European Journal of Wildlife Research* 55, 125–132.

Łukasiewicz, M. (2014) Biology of embryo development in Pigeon *Columba livia* domesticus in conditions of artificial incubation. *Advances in Animal and Veterinary Sciences* 2, 401–406.

Michaelsen, T.C. and Refvik, S.I. (2003) Feral rock doves breeding on the coast of Norway. *Ornis Norvegica* 26, 119–124.

Murton, R. and Clarke, S. (1968) Breeding biology of rock doves. *British Birds* 61, 429–448.

Murton, R.K. and Westwood, N.J. (1966) The foods of the Rock Dove and Feral Pigeon. *Bird Study* 13, 130–146.

Mushi, E.Z., Binta, M.G., Chabo, R.G., Ndebele, R. and Panzirah, R. (2000) Parasites of domestic pigeons (*Columba livia domestica*) in Sebele, Gaborone, Botswana. *Journal of the South African Veterinary Association* 71, 249–250.

Outerbridge, M.E. (2016) Feral Pigeon (*Columba livia domestica*) management plan for Bermuda. Department of Environment and Natural Resources, Government of Bermuda, Bermuda, pp. 1–29.

Phillips, R.B., Cooke, B.D., Carrión, V. and Snell, H.L. (2012) Eradication of rock Pigeons, *Columba livia*, from the Galápagos Islands. *Biological Conservation* 147, 264–269.

Phillips, R.B., Snell, H.L. and Vargas, H. (2003) Feral Rock Dove in the Galapagos Islands: biological and economic threats. *Noticias de Galapagos* 62, 6–11.

Pike, M., Spennemann, D.H.R. and Watson, M.J. (2017) Building use by urban commensal avifauna in Melbourne central business district, Australia. *Emu* 117, 284–289.

Pimentel, D., Lach, L., Zuniga, R. and Morrison, D. (2000) Environmental and economic costs of nonindigenous species in the United States. *BioScience* 50, 53–66.

Pimentel, D., Zuniga, R. and Morrison, D. (2005) Update on the environmental and economic costs associated with alien-invasive species in the United States. *Ecological Economics* 52, 273–288.

Preble, D.E. and Heppner, F.H. (1981) Breeding success in an isolated population of rock doves. *Wilson Bulletin* 93, 357–362.

Przybylska, K., Haidt, A., Myczko, Ł., Ekner-Grzyb, A., Rosin, Z.M., *et al.* (2012) Local and landscape-level factors affecting the density and distribution of the feral Pigeon *Columba livia* var. *domestica* in an urban environment. *Acta Ornithologica* 47, 37–45.

Rappole, J.H., Derrickson, S.R. and Hubálek, Z. (2000) Migratory birds and spread of West Nile virus in the Western Hemisphere. *Emerging Infectious Diseases* 6, 319–328.

Robbins, C.S. (1995) Non-native birds. In: LaRoe, E.T., Farris, G.S., Puckett, C.E., Doran, P.D. and Mac, M.J. (eds) *Our Living Resources*. National Biological Service, Washington, DC, pp. 437–440.

Robertson, H.A. (1988) Relationships between body weight, egg weight, and clutch size in pigeons and doves (Aves: Columbiformes). *Journal of Zoology* 215, 217–229

Rose, E., Nagel, P. and Haag-Wackernagel, D. (2005) Suitability of using the global positioning system (GPS) for studying feral Pigeons *Columba livia* in the urban habitat. *Bird Study* 52, 145–152.

Rose, E., Nagel, P. and Haag-Wackernagel, D. (2006) Spatio-temporal use of the urban habitat by feral Pigeons (*Columba livia*). *Behavioral Ecology and Sociobiology* 60, 242–254.

Sacchi, R., Gentilli, A., Razzetti, E. and Barbieri, F. (2002) Effects of building features on density and flock distribution of feral Pigeons *Columba livia* var. *domestica* in an urban environment. *Canadian Journal of Zoology* 80, 48–54.

Scarf, D., Hayne, H. and Colombo, M. (2011) Pigeons on par with primates in numerical competence. *Science* 334, 1664–1664.

Senar, J.C., Montalvo, T., Pascual, J. and Peracho, V. (2017) Reducing the availability of food to control feral Pigeons: changes in population size and composition. *Pest Management Science* 73, 313–317.

Shapiro, M.D. and Domyan, E.T. (2013) Domestic Pigeons. *Current Biology* 23, 302–303.

Silva, R.S.E.(1997) Ecology and behavior of wintering *Falco peregrinus* (Falcoinformes: Falconidae) in Southeastern Brazil. *Ararajuba* 5, 203–208

Silva, A.M. and Medeiros, P.R. (2008) Pigeons (*Columba livia*) in an urban square: what can we learn from local people's knowledge and perception? *Gaia Scientia* 2, 32–40.

Snyders, H. (2015) 'More than just human heroes': the role of the Pigeon in the First World War. *Scientia Militaria: South African Journal of Military Studies* 43, 133–150.

Soldatini, C., Mainardi, D., Baldaccini, N.E. and Giunchi, D. (2006) A temporal analysis of the foraging flights of feral Pigeons (*Columba livia f. domestica*) from three Italian cities. *Italian Journal of Zoology* 73, 83–92.

Spennemann, D.H.R. and Watson, M.J. (2017) Dietary habits of urban Pigeons (*Columba livia*) and implications of excreta pH – a review. *European Journal of Ecology* 3, 27–41.
Stephan, C., Wilkinson, A. and Huber, L. (2012) Have we met before? Pigeons recognise familiar human faces. *Avian Biology Research* 5, 75–80.
Stern, C.A. and Dickinson, J.L. (2010) Pigeons. In: Michael, D.B. and Janice, M. (eds) *Encyclopedia of Animal Behavior*. Academic Press, London, pp. 723–730.
Stock, B. and Haag-Wackernagel, D. (2016) Food shortage affects reproduction of feral Pigeons *Columba livia* at rearing of nestlings. *Ibis* 158, 776–783.
Tietz Marques, S.M., Marinho de Cuadros, R., Jardim da Silva, C. and Baldo, M. (2009) Parasites of Pigeons (*Columba livia*) in urban areas of Lages, Southern Brazil. *Parasitología Latinoamericana* 62, 83–187.
van Niekerk, J.H. and van Ginkel, C.M. (2004) The feeding behaviour of Pigeons and doves on sown grain crops on the South African Highveld. *Ostrich* 75, 39–43.
Walcott, C. (1996) Pigeon homing: observations, experiments and confusions. *Journal of Experimental Biology* 199, 21–27.
Wendell, L. (1977) *The Pigeon*. Levi Publishing, Sumter, South Carolina.

15 Eurasian Collared-dove (*Streptopelia decaocto* Frivaldszky, 1838)

Cyril Eraud*
Office Français de la Biodiversité, Villiers-en-Bois, France

Citation: Eraud, C. (2020). Eurasian Collared-dove (*Streptopelia decaocto* Frivaldszky, 1838). In: Downs, C.T. and Hart, L.A. (eds) *Invasive Birds: Global Trends and Impacts*. CAB International, Wallingford, UK, pp. 118–131.

15.1 Common Names

Eurasian Collared-dove, Collared Dove, Collared Turtle Dove, Common Collared Dove, Indian/Eastern/Eurasian Ring Dove, Indian Ring Necked Dove.

15.2 Nomenclature

Three subspecies of Eurasian Collared-dove (*Streptopelia decaocto* Frivaldszky, 1838) were formerly considered: *S. d. decaocto*, *S. d. stoliczkae* and *S. d. xanthocyclus* (Roonwal, 1940). The last, from Myanmar and Yunnan, is now recognized as the species Burmese Collared-dove (*Streptopelia xanthocycla*; del Hoyo *et al.*, 2018; BirdLife International, 2018). *S. d. stoliczkae* is a synonym of *S. decaocto* (Romagosa, 2002).

15.3 Distribution

Until the 16th century, the Eurasian Collared-dove was confined to the Asian continent, east to the Levant: India, Sri Lanka and Myanmar (Fisher, 1953) and probably Afghanistan (Lever, 2005). Following natural expansion or human introductions, the current distribution spans four continents (Fig. 15.1). In Asia, it ranges from the Middle East (Turkey, Israel, Saudi Arabia, United Arab Emirates, Syria, Iran and Iraq) to Korean and north-eastern China (Fisher, 1953; Marchant, 1963; Ali and Akhtar, 2005; Hasan *et al.*, 2005; Amer, 2014). The species is also present in Japan where it was introduced in the 18th/19th centuries (Fisher, 1953; Lever, 2005). The southern limit in China remains undefined (Fujian region; Fisher, 1953). The northern Asian distribution range includes Kazakhstan, Mongolia and south-eastern Russia up to 50°N (50.50°N/115.11°E, Tasuchey; Goroshko, 2014). In Europe, the Eurasian Collared-dove is present in all countries from Greece, Italy and Spain in the south to Iceland, Norway (some occurrences up to 71°N; GBIF.org, 2020), Sweden and Russia (Mourmansk region; Nowak, 1989) in the north. In Eastern Europe, the limit is west from a line connecting the White and the Caspian Seas (Gibbs *et al.*, 2001). In Africa, the Eurasian Collared-dove has spread in all countries north of Sahara, from the Canary Islands to Egypt (Bergier *et al.*, 1999; Isenmann *et al.*, 2005, 2016; Moali and Isenmann, 2007; Bendjoudi *et al.*, 2015). The species was also recorded at several locations in Mauritania (Isenmann *et al.*, 2010) and is now well established on Cape Verde Islands (15°N; Hazevoet, 2012). In North America, the species is distributed over all the USA, including Alaska up to 65°N (Fairbanks) and the southern half of Canada. In the south, the Eurasian Collared-dove is established in Mexico (Almazán-Núñez, 2014; Blancas-Calva *et al.*, 2014), and some records are reported in Costa Rica and Panama. From southern Florida, the range extends to almost all of the Bahamian and Caribbean Islands, from Cuba to Trinidad and Tobago (Barré *et al.*, 1996). The species is probably expanding its range in South America, with some occurrences reported in Colombia (GBIF.org, 2020). The most southerly occurrences of free-ranging birds come from Lima, Peru, but their origin is unknown (Blancas-Calva and Blancas-Hernández, 2016).

The European population is estimated at 7,910,000–14,300,000 pairs or approximately 24 million–43 million individuals (BirdLife International, 2018). In other parts of the range, population sizes lack reliable estimates. Jennings (2010) estimated 2 million pairs in the Arabian Peninsula.

*Corresponding author: cyril.eraud@ofb.gouv.fr

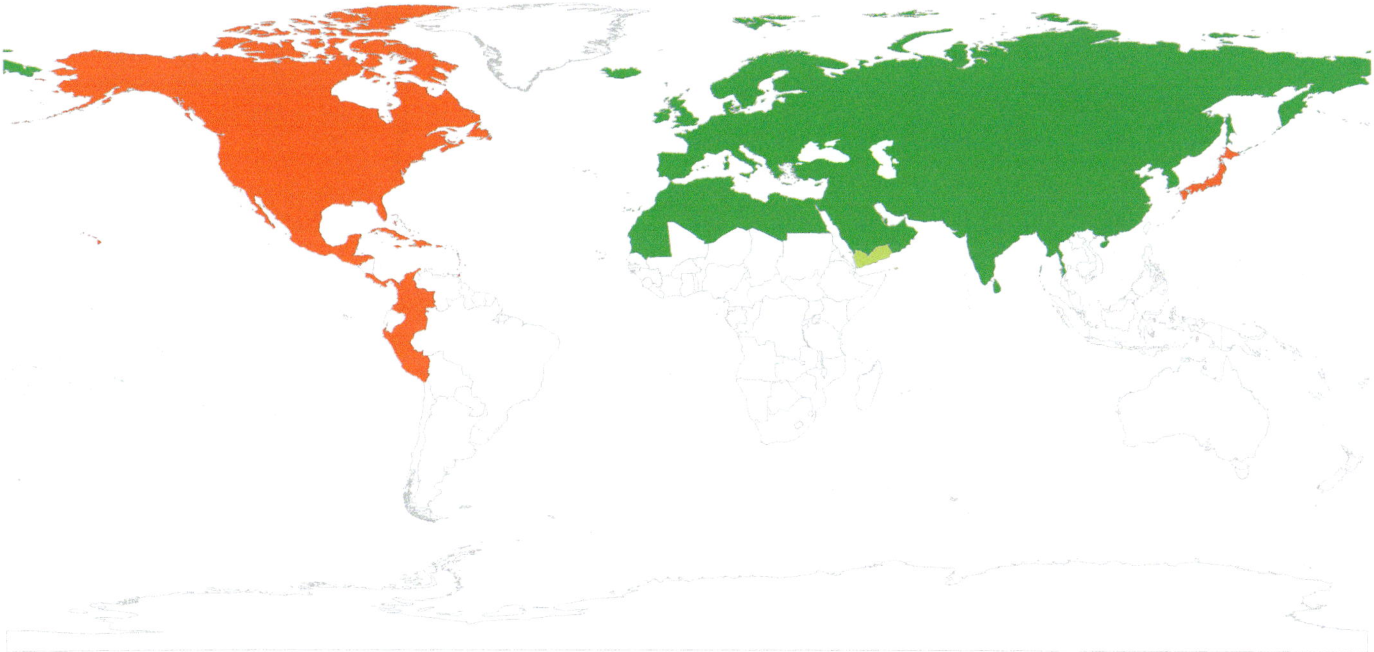

Fig. 15.1. Global distribution of the Eurasian Collared-dove showing natural (green) and invaded (red) distributions. Vagrant populations are shown in light green.

Fig. 15.2. An adult Eurasian Collared-dove. (© Photograph: François Sabathé.)

15.4 Description

The Eurasian Collared-dove is a medium-sized dove with pale buffy grey upperparts and grey pinkish underparts in adults (Fig. 15.2) (Gibbs *et al.*, 2001; Blasco Zumeta and Heinze, 2019). The wings are pale and the tail is grey with white edges and tips (Blasco Zumeta, 2019). From below, the blacker base of tail feathers contrasts clearly with the whiter tips. In juveniles, the upperparts and wing are pale with cream fringes. The legs are reddish/mauve in adults and dark grey or brownish-red in juveniles (depending on age). Adults have a narrow white/grey-bordered black half-collar on the neck. In juveniles, this collar is completed at about 2 months (Cramp, 1985). In adults, the chest is bright vinous-pink and the iris is red. In juveniles, the iris is brown. Sexual size dimorphism is reported across both native and invaded ranges, with the males being heavier and larger than the females (Table 15.1) but with a strong overlap, while plumage differences are also subtle. Consequently, sexing is difficult, even for birds in hands.

Across most of its current range, the species cannot be confused with most other dove species. In Europe, the Americas and the Caribbean, the Eurasian Collared-dove differs clearly from native dove species of similar size (e.g. European Turtle Doves (*Streptopelia turtur*), doves of the genus *Zenaida*). Confusion may occur with the domestic form of the African Collared-dove (*Streptopelia roseogrisea*), namely the Ringed Turtle Dove or Barbary Dove (*Streptopelia risoria*; Romagosa and McEneaney, 1999). The most reliable criteria include the undertail coloration pattern and male song characteristics (Romagosa and McEneaney 1999). In Africa and the Middle East, the species may now co-occur with several species of the genus *Streptopelia* with similar plumage features: the African Collared-dove (*S. roseogrisea*), Red-eyed Dove (*S. semitorquata*), Mourning Collared-dove (*S. decipiens*) and Vinaceous Dove (*S. vinacea*). However, the Eurasian Collared-dove can be readily distinguished from these species, except the first one (Gibbs *et al.*, 2001) as also described above. In Asia, the species can be confused with the Burmese Collared-dove (*Streptopelia xanthocycla*), which was formerly included in *S. decaocto* (*S. d. xanthocyclus*; Fisher, 1953). *S. xanthocyla* is darker with broad yellow (not white) orbital skin (del Hoyo *et al.*, 2018).

The Eurasian Collared-dove morphometric data suggest a size cline across the Eurasian portion of the species' range. Birds would be smaller and lighter in their ancestral range of distribution (India, Sri Lanka; Table 15.1) than in Europe (but see Kour 2016).

Table 15.1. Morphometrics of Eurasian Collared-doves. Data are means ± standard deviation or point estimates (range; number of samples). Note that some values may also include juveniles.

		Body weight (g)		Wing (mm)		Tail (mm)		
Continent	Country/season	Male	Female	Male	Female	Male	Female	Reference(s)
Asia	India/all year	152 ± 11.6 (115–184; *n* = 87)	146 ± 10.3 (113–176; *n* = 80)	173 ± 3.02 (169–176; *n* = 6)	–	–	–	Cramp (1985)
	India/all year	160 ± 1.9 (156–163; *n* = 37)	156 ± 1.84 (151–158; *n* = 28)	190 ± 5.99 (167–196; *n* = 37)	184 ± 9.57 (164–193; *n* = 28)	132 ± 3.61 (125–137; *n* = 37)	136 ± 8.98 (121–150; *n* = 28)	Kour (2016)
	Sri Lanka	–	–	165 ± 3.95 (159–171; *n* = 7)	161 ± 2.89 (158–164; *n* = 4)	–	–	Cramp (1985)
	India, Sri Lanka	–	–	165 ± 5.7 (152–178; *n* = 27)	162 ± 3.8 (155–168; *n* = 15)	–	–	Cramp (1985)
	Nepal/Bangladesh/ north-east India/eastern China, Korean, Japan	–	–	175 ± 4.0 (168–181; *n* = 15)	166 ± 4.8 (159–178; *n* = 13)	–	–	Cramp (1985)
	Central Asia (former *S. d. stolickzae*)	–	–	182 ± 7.3 (169–196; *n* = 16)	179 ± 8.1 (167–191; *n* = 9)	–	–	Cramp (1985)
	Iraq	–	–	165–180 (*n* = 11)		129–139 (*n* = 11)		Afrasiab *et al.* (2017)
	Middle East (Iraq, Israel, Afghanistan), north-east India	–	–	170 ± 5.6 (160–184; *n* = 28)	167 ± 4.4 (159–172; *n* = 20)	–	–	Cramp (1985)
Europe	Netherlands/ January	217 ± 6.46 (171–234; *n* = 22)	200 ± 15.9 (171–224; *n* = 13)	–	–	–	–	Cramp (1985)
	Netherlands/ March	206 ± 14.5 (173–243; *n* = 71)	197 ± 15.7 (162–231; *n* = 67)	–	–	–	–	Cramp (1985)
	Netherlands/all year	–	–	185 ± 4.33 (175–194; *n* = 85)	179 ± 4.08 (170–190; *n* = 83)	142 ± 4.53 (132–151; *n* = 73)	135 ± 4.87 (126–146; *n* = 79)	Cramp (1985)
	West Germany	197 (150–242; *n* = 103)	189 (125–229; *n* = 83)	–	–	–	–	Cramp (1985)
	Portugal/ May–July	194 ± 16.4 (175–225; *n* = 9)	198 ± 30 (165–260; *n* = 8)	–	–	–	–	Cramp (1985)

	Germany/ Balkans	–	–	177 ± 5.0(162–189; *n* = 104)	174 ± 5.4(160–188; *n* = 85)	–	–	Cramp (1985)
Africa	Algeria/all year?	197 ± 17.2 (171–224; *n* = 10)		–	–	–	–	Absi (2012)
	Algeria/ January–May	Urban: 181 ± 6.8 (*n* = 11); Suburban: 181 ± 0.9 (*n* = 11)		Urban: 165 ± 8.3 (*n* = 11); Suburban: 194 ± 7.9 (*n* = 11)		–	–	Belabed *et al.* (2013a)
	Algeria	195 ± 10.2(189–210; *n* = 4)	193 ± 16.3(170–220; *n* = 7)	–	–	–	–	Torki (2014)
America	USA, Florida	173 (*n* = 1)	160–180 (*n* = 3)	178 (*n* = 1)	167–180 (*n* = 3)	141 (*n* = 1)	134–138 (*n* = 3)	Smith and Kale (1986)
	Mexico/spring–summer	177	169	176	172	130	124	Salazar-Borunda *et al.* (2015)
	Mexico/ autumn–winter	167	150	171	161	134	130	Salazar-Borunda *et al.* (2015)
	Mexico/all year	109–205 (*n* = 40)		142–186 (*n* = 40)		104–149 (*n* = 40)		Salazar-Borunda *et al.* (2015)

There is insufficient information in America and North Africa to draw firm conclusions about geographical differences (Romagosa, 2002). In Eurasia, a coloration cline is also reported by Cramp (1985), with European birds being slightly paler than their Indian conspecifics. Afrasiab *et al.* (2017) reported an extreme (and unexplained) variation in coloration over the last 20 years in Baghdad (Iraq), where a quarter of the population has dark dorsal feathers with a nearly black breast and face.

15.5 Diet

Eurasian Collared-doves are mostly seed eaters throughout the year, feeding mainly on cultivated and weed seeds (see Cramp, 1985, for an extended list). The species may occasionally take berries (e.g. *Sambucus* spp.) and green parts of some plants, as well as seeds from deciduous trees (e.g. *Tilia*, *Acer* and *Viburnum* spp.). Consumption of invertebrates is anecdotal (molluscs, beetles, flies, aphids, lepidopteran larvae or termites; Rana, 1975; Cramp, 1985). Eurasian Collared-doves mostly take food items from the ground but also from elevated supplementary bird feeders, from standing plants (e.g. sunflowers) or directly from bushes/trees (berries). Like other columbids, the parents feed nestlings with crop milk in the first days following hatching and thereafter with a mix of seeds.

In Europe, North America, the Caribbean and North Africa, the Eurasian Collared-dove mostly feeds on food provided by human activities. Open chicken runs, barns with livestock, grain silos and docks/warehouses are common feeding sites throughout the year, where tens/hundreds of birds may congregate (Hudson, 1965, 1972; Coombs *et al.*, 1981; Barré *et al.*, 1996). In summer, the species gathers spilled grains from stubble or on roads. In spring and autumn, freshly sowed fields are used. Hence, in these regions, crop seeds form the bulk of the diet, supplemented by weed seeds. In some towns, it is not uncommon to see fearless individuals feeding on garbage in parks/gardens or on the outdoor floor of restaurants (e.g. feeding on bread). A wide range of seeds and plant items can be found in the stomach, depending on local availability. In Romania, a study (in August and October) based on 250 stomachs revealed that maize (*Zea mays*) represented 72.4% in weight, sunflower (*Helianthus annus*) 13.1% and wheat (*Triticum* spp.) 5.1%. The remainder included grapes (3.2%) and weed seeds (Cramp, 1985). In (formerly) Yugoslavia, wheat and maize were found dominating the diet. In Hungary, the diet was also reported to include maize, sunflower, millet (*Setaria* spp.) and various weed seeds (Cramp, 1985). In northern France, analysis of 481 stomach contents showed a dominance of crop seeds with important seasonal variation: mostly wheat in summer (69.9%), maize in autumn (73.9%) and a mix of cereal grains in winter (barley (*Hordeum* spp.), 46.4%; wheat, 41.8%; Sueur, 1999). In Algeria, analysis of ten stomach contents revealed the presence of crop seeds (*Triticum durum* and *Hordeum vulgare* comprising 24% of items) supplemented by weed seeds (*Carduus nutans*, *Stephanochilus omphalodes*, *Argyrolobium uniflorum*, *Fumaria officinalis*, *Convolvulus arvensis*), molluscs (16.3%) and green parts from *Atriplex* spp. (Absi, 2012). In North America (suburban areas of Florida), examination of eight stomachs found the presence of seeds from commercial bird feeder mixes (millet, milo, sunflower and maize; Romagosa, 2002).

In Iraq, the Eurasian Collared-dove was observed feeding in streets, gardens and yards, often close to domestic animals, or in fields (Marchant, 1963). Kour (2016) reported similar feeding tactics in detail in northern India (Jammu and Kashmir), where Eurasian Collared-doves gleaned spilled grains at rice shellers and grain stores or from harvested, sowed or ploughed fields. In winter months, the species feeds on wild seeds from both fields and uncultivated areas (e.g. roadsides, playgrounds) and by searching cow dung mounds (Kour, 2016). In village/urban areas, the species forages on feeders or waste food near households, market places or restaurants. Analysis of 65 stomach contents showed that cereal seeds (rice, maize, wheat, sorghum, lesser millet) predominated in the diet (52–61%), supplemented by weed seeds (18-20%), oilseeds (6–8%), pulses (Fabacea, 9–11%) and, minimally, invertebrates (1–3%) (Kour 2016). In Rajasthan, pearl millet (*Pennisetum glaucum*), wild indigo (*Tephrosia purpurea*), browntop millet (*Brachiaria ramosa*), *Citrullus vulgaris*, seeds of *Heliotropium strigosum* and termites occurred in the crops of nestlings (Rana, 1975).

15.6 Introduction and Invasion Pathways

About the Eurasian Collared-dove's expansion in Europe, James Fisher (1953) wrote, '...it has made one of the most remarkable range-changes to have been recorded, ever since man began to record the ranges and changes of birds.' At that time, Fisher did not suspect that only half of the way had been travelled. In less than 100 years, the species has significantly expanded its range throughout the world. This expansion results from two main independent events: (i) a colonization wave throughout Europe and North Africa, which started in the early 20th century from Balkan and Turkish free-ranging populations; and (ii) expansion across North/Central America and the Caribbean, initiated from various intentional introductions from the mid-1970s.

Historically, the Eurasian Collared-dove was restricted to India, Sri Lanka, Myanmar and neighbouring countries (Lever, 2005). The first mention at the southern edge of Europe was back in the 16th century. According to Stresemann (1950, cited in Fisher, 1953), the species was already present in Constantinople, Istanbul, by 1547. Its establishment in this region (Turkey, Balkan and Levant) over the following centuries is difficult to trace accurately and its origin – natural dispersion or introduction(s) – remains unclear. Following the detailed analysis of Fisher (1953), the distribution in south-eastern Europe would not have changed until the late 19th/early 20th century. Before 1912, its range covered the coastal region of Albania, Montenegro and part of the present-day Serbian coast, Macedonia, north-east Greece (Thrace region), northern Turkey (Marmara region), southern Bulgaria (as far north as Plovdiv) and probably (at the latest by 1920) the Black Sea Coast up to Dobruja, Romania. Presumably from Niš and Prokuplje, Serbia, the species reached Belgrade in 1912, which is thought to be the main starting point of the expansion wave in Europe (Fisher, 1953). In the early 1930s, the species occupied the Great Hungarian Plain (Monor, Berettyóújfalu and Paks), from where it has rapidly spread north-westward. The first breeding was reported from the Czech Republic in 1938, Austria and Poland in 1943, Ukraine in 1944, Germany in 1945, Italy in 1947, the Netherlands and Denmark in 1950, Sweden in 1951, Switzerland and France in 1952, Belgium, the UK and Norway in 1955, Estonia in 1957, Ireland in 1959 and Finland in 1966 (Cramp, 1985; Marja and Elts, 2018). Iceland was colonized by 1971. In newly colonized countries, the populations exhibited exponential growth. In the UK, the population has grown from four birds in 1955 to about 19,000 in 1964 (Hudson, 1965). In Sweden and the Netherlands, the population size was estimated at 1200 in 1961 and 1958, respectively.

South-western Europe was colonized by the Eurasian Collared-dove by the mid-1970s: the first breeding was reported in Spain in 1974, Corsica in 1975, Sardinia in 1977, Portugal in 1982 and Sicily in 1991 (Cramp, 1985; Snow and Perrins 1998). In Morocco, some (inaccurate) observations were reported as early as in the mid-1970s (Bergier *et al.*, 1999), but their number increased mostly from the first record of breeding birds in the city of Meknès in 1986 (Bergier *et al.*, 1999; Thévenot *et al.*, 2003). The species has spread rapidly in a south-westerly direction along the Atlantic coast. By the late 2000s, Eurasian Collared-doves occupied a wide coastal area lying along Tangier and Dakhla (23.7°N, 15.3°W) in the Western Sahara (Bergier *et al.*, 1999). In 1999, the species was observed along Mauritanian coasts (Nouackchott) and inland a few years later at Oued El Abiod in 2004 (17.21°N, 12.45°W; Isenmann *et al.*, 2010). At the extreme south-west, the species was first reported at Sal on Cape Verde Islands in 2006 and at São Nicolau in 2009 (Hazevoet, 2012). Expansion in Algeria has followed a similar pattern: from first observations at Annaba in 1994 (first breeding in 1996), the species has rapidly expanded along the Mediterranean coast towards Algiers, which was colonized in 2001 (Moali *et al.*, 2003). By the same date, occurrences had been reported in several western localities (e.g. Tlemcen and Beni Saf, Algeria), suggesting colonization from Morocco (Moali *et al.*, 2003). In parallel, the species has spread across the Algerian Sahara, reaching as far as Tamanrasset in 2007 (22.5°N; Moali and Isenmann, 2007), a distance travelled from Annaba of about 1500 km in 13 years. The species was known as early as 1963–1967 in Tibar, northern Tunisia (Moali and Isenmann, 2007), but the expansion was mainly initiated from the 1980s (Azafzaf *et al.*, 2015). In 2002, the species was observed further south at Douz (33.46°N; Moali and Isenmann, 2007) and up to Ghadamis in Libya in 2005 (Isenmann *et al.*, 2016). Some later observations near Tripoli in 2007 and Misrata in 2008 (Isenmann *et al.*, 2016) suggested that from Tunisia the Eurasian Collared-dove may have colonized western Libya. The same year, the species was also reported in the eastern part of the country (near Benghazi and further east in Derna in 2010 (Isenmann *et al.*, 2016). In the Middle East, detailed information on the colonization process is scarce: the species was reported in 1979 in Jordan and Egypt (at Suez, and subsequent spreading to Cairo; Cramp, 1985; Snow and Perrins 1998). The species was observed in Kuwait in 1963 (Vincent, 2008). In Bahrain and Qatar, the Eurasian Collared-dove was introduced in the 1960s following the release of imported caged birds from India and Iran (Lever, 2005). The occurrence in Arabia may have resulted from similar introductions and natural expansion (Lever, 2005).

The Eurasian Collared-dove colonization of North America is thought to have originated from the release in 1974 of around 50 birds of the nominate *S. d. decaocto* (Smith and Kale, 1986) by a pet breeder in Nassau, Bahamas (Smith, 1987; Romagosa and Labisky, 2000). Breeding was first recorded in the summer of 1975 in Nassau, followed by the establishment of a sustainable population and a rapid expanse throughout the Bahamas. By the mid-1980s, the Eurasian Collared-dove was common on several Bahamas islands (Andros, Abaco, Grand Bahama, Soanish Wells, Bilini and Berry Islands; Romagosa and McEneaney, 1999). Additional birds would also have been released on Abaco Island to decrease local hunting pressure on some native columbids (Smith, 1987). The colonization of southern Florida in the late 1970s/early 1980s from the Bahamas is the most likely hypothesis (Smith, 1987). The precise date of their arrival remains uncertain because of the confusion with Ringed Turtle Doves (Romagosa and Labisky, 2000). The first breeding was recorded in the spring of 1982 in Homestead, south of Miami (Smith and Kale, 1986). Smith (1987) was right when he suspected that 'the species will spread here in much the same fashion as in Europe'. From 1986, the species is recorded in historical bird surveys (Romagosa and Labisky, 2000; Hooten and Wikle, 2008) and from the early/ mid-1990s, the number of localities with occurrence of the species increased dramatically, as well as the extent of the species' range. Based on breeding bird surveys, the species was reported from Alabama in 1993, Georgia in 1995, Texas in 1996, Arkansas, Tennessee and South Carolina in 1999, Iowa and North Carolina in 2000, Kansas and Illinois in 2001, Oklahoma, Nebraska, New Mexico in 2002, and Colorado, Arizona and Kentucky in 2003. In 2007, the species was well established along the Gulf of Mexico, and occurrences were reported further east in California and Washington (Fujisaki *et al.*, 2010). By 2009, the species had been reported in Alaska and in several provinces of Canada (see https://ebird.org), with sightings in Saskatchewan as early as 2002 (Syroteuk, 2008). In Quebec, the species was first mentioned west of Ottawa at Ile du Grand-Calumet in 2004 (www.oiseauxqc.org; accessed 28 October 2019). While the expansion across North America was driven mainly by dispersing birds, several releases/translocations were also reported in the early to mid-1990s in California, Missouri, Illinois, Tennessee and Texas (Romagosa and McEneaney, 1999).

The Eurasian Collared-dove's progression southwards from the Mexico border and the Gulf of Mexico is less well documented. By the mid-2000s, the species was widely distributed across mexican states: e.g. it was reported in Baja California, Sinaloa, Jalisco, Hidalgo and far south in Veracruz, Oaxaca, Guerrero and Yucatán (Gómez de Silva, 2006a,b; Valencia-Herverth *et al.*, 2011; Chablé-Santos *et al.*, 2012; Tinajero and Rodríguez-Estrella, 2014). According to eBird records (eBird, 2020 see also GIB.org, 2020), the species was reported in Belize in 2005, in Costa Rica in 2008. By 2015–2016, reportings in Guatemala extended from Petén in the north to Escuintla on the Pacific Coast, with several records in Honduras. Recently (2018–2019), the species was sighted in Sonsonate and Morazán in El Salvador and in David and Colón in Panama. Local releases have probably accelerated the expansion of the species towards South America, creating hedge bridges far from the colonization front. For instance, the species was reported in 2005 in Valle del Cauca, Colombia (Estela, 2005). Further south, the species was reported in 2015 in Lima City, Peru (Blancas-Calva and Blancas-Hernández, 2016).

The colonization of the Lesser Antilles by the Eurasian Collared-dove was independent of its introduction in the Bahamas. About 20 birds were released in 1976 before Soufrière's eruption, in Saint-Claude, Guadeloupe, creating a sustainable population (Barré *et al.*, 1996), thought to be the source of the spread on neighbouring islands of Dominica (1987), Montserrat (1990), Martinique (1994), Nevis (1995) and Saint-Martin (2002) (Barré *et al.*, 1996; Francis, 1996; Levesque and Jaffard, 2003). Occurrences in other islands of the Antilles may also have resulted from local release. For instance, the species was recorded in Barbados in 1988, in Puerto Rico in 90s and in the Dominican Republic in 1997 (eBird, 2020; GBIF.org, 2020; see Luna *et al.*, 2018, for later mentions in Dominican Republic). Elsewhere in the world, several releases, or sightings of free-ranging individuals, were reported but without success in establishing sustainable populations (e.g. Australia and South Africa; Dean, 2000; Dyer *et al.*, 2017; GBIF.org, 2020).

The expansion wave of the Eurasian Collared-dove is thought to have resulted from two processes: the dispersal of birds creating head bridges ahead of the invasion front, and a subsequent filling of void areas as local populations grew (Hudson, 1972; Hengeveld, 1993; Romagosa and Labisky, 2000; Bendjoudi *et al.*, 2015; Bagi *et al.*, 2017). Subsequent fillings may last long after the species has successfully settled in a country (e.g. Eraud *et al.*, 2007).

Several Eurasian Collared-dove ring recoveries illustrate the capability of the species to disperse over long distance: a juvenile ringed in July 1961 and an adult ringed in December 1963 at Herford, Germany, were recovered around 900 km west in Perranporth in Cornwall, UK in June 1962 and 700 km west in Sunderland in the north-east of the UK in November 1964, respectively (Hudson, 1965). An adult banded in Wageningen in the Netherlands in February 1964 was recovered approximately 600 km north-west in Lytham St Annes in Lancashire, UK, in January 1965. Other long dispersal events have been reported in other directions: a nestling ringed in Holleben, Germany, in July 1951 was recovered dead in May 1952 600 km south-west in Juniville, France, and a bird ringed in January 1951 at Haldensleben, Germany, was recovered 600 km south in northern Italy 10 months later (Fisher, 1953). There is also a report of a bird ringed in Belgium and recovered in 1971 at Rabat, Morocco, which represents the first occurrence of this species in this country (Thévenot *et al.*, 2003).

The underlying causes of the Eurasian Collared-dove's rapid range expansion in Europe from the Balkans are unclear. Suggestions include mutation of a favourable gene (Mayr, 1951), climate change, and changes in agricultural practices and a subsequent increase in food supplies (Robertson, 1990; Bagi *et al.*, 2017). Other hypotheses imply some ecological/biological traits of the species, including changes in nesting behaviour (from nesting in buildings to in trees; Stresemann and Nowak, 1958), its adaptability to human urbanized and altered landscapes (Hudson, 1965; Bonter *et al.*, 2009; Fujisaki *et al.*, 2010; Bendjoudi *et al.*, 2015) and its potentially high reproductive output (Bagi *et al.*, 2017).

15.7 Breeding Behaviour

The Eurasian Collared-dove breeding phenology shows significant geographical differences, especially between historical and some newly colonized ranges (note that contrasts in habitat features and periods are largely confounded). The most striking differences concern the extent and the shape of egg-laying seasons (Table 15.2). In Rajasthan, north-western India, (Rana, 1975) and Iraq (Marchant, 1963), the egg-laying season starts mainly in April and lasts about 7 months. In the Indian state of Jammu and Kashmir, the egg-laying season starts earlier (March) and lasts 9 months (Kour, 2016). In Europe, North America and the Caribbean, the species can lay almost year-round, with a significant proportion of laying initiated as early as March (Raffaele *et al.*, 1998; Romagosa, 2002). In Algeria, laying is reported from February to autumn (Torki, 2014; Bendjoudi *et al.*, 2015). The distribution of clutches over the season in Europe and North Africa resembles a bell-shaped curve, with a laying peak in spring (46–61% in April–June). In Rajasthan, the egg-laying season shows two peaks, coinciding with seasonal fluctuation in the development of reproductive organs (Rana, 1975): a minor peak in spring (April) and a major one during the monsoon season (September). A different-shaped egg-laying season was reported in Jammu and Kashmir, but with a similar peak during the monsoon season (Kour, 2016). Climatic conditions and food availability are thought to be major determinants of the onset, extent and distributional form of the laying season (Rana, 1975; Robertson, 1990).

The Eurasian Collared-dove is monogamous. Pairs nest several times a year, with five broods on average in north-west Germany (Hofstetter, 1954, cited in Hudson, 1965) and three to five (up to nine) in the UK (Robertson 1990). Roonwal (1940) mentioned four to five broods for Indian birds, but according to Robertson (1990), more than three successful broods in a year would be unlikely (see also Kour, 2016). In Iraq, more than three broods would be rare (Marchant, 1963). Across native and invaded ranges, a typical clutch includes two eggs, but occasionally one, or up to three (Rana, 1975; Cramp, 1985; Romagosa, 2002; Kour, 2016): In eastern Germany, 97% of clutches included two eggs, 88.1% in the Czech Republic, 89% in the UK (Robertson, 1990), 100% in Iran (Hosseini-Moosavi *et al.*, 2013), 96% in Iraq (Marchant, 1963) and 84% in Algeria (Torki, 2014). In Rajasthan, a slightly lower value (78%) was reported (Rana, 1975). Invariably, eggs are white (Fig. 15.3). In two-egg clutches, the second egg is larger and heavier (Cramp, 1985; Robertson, 1990). Eggs tend to be larger and heavier in the European part of the range, probably in relation to differences in female body size/weight (Tables 15.1 and 15.3). Exceptions are the egg sizes and weights from Jammu and Kashmir (Table 15.3).

In the UK, incubation (by both parents) lasts 15 days ± 1.1 (mean ± SD, range 14–19; Robertson 1990) and mostly 14–16 days elsewhere in Europe (Gnielka, 1975, cited in

Table 15.2. Data on monthly distribution (%) of clutches (or nests occurrence) of Eurasian Collared-dove.

	Jan	Feb	Mar	Apr	May	Jun	Jul	Aug	Sep	Oct	Nov	Dec	No. nests	Reference(s)
India (Rajasthan)	0	0	0	22	8	0	5	25	36	3	0	0	59	Rana (1975)
India (Jammu and Kashmir)[a]	0	0	8	6	8	13	11	16	17	15	5	0	280	Based on 2010 and 2011 data from Kour (2016)
Iraq	0	0	2	26	29	11	13	15	5	0	0	0	62	Marchant (1963)
Continental Europe	1	2	11	22	18	13	18	11	6	1	0	0	1,989	Mean values from Kubik and Balat (1973); Pikula and Kubik (1978); Gnielka (1975); Saemann (1975)
Occidental Europe	<1	1	6	14	24	19	14	11	8	2	<1	<1	11,424	Mean values from Coombs *et al.* (1981); Robertson (1990); Hudson (1965); ONCFS Columbids scheme (Eraud and Boutin, 2008; Lormée, 2018)
Algeria[b]	0	8	19	19	19	22	11	3	0	0	0	0	37	Torki (2014)
Detailed studies in Europe														
Czech Republic	0	0	6	21	16	15	22	12	9	<1	0	1	302	Kubik and Balat (1973)[c]
Czech Republic	<1	1	10	27	18	9	18	11	5	0	0	0	384	Pikula and Kubik (1978)[c]
East Germany	0	2	15	25	20	14	13	7	4[d]		0	0	588	Gnielka (1975)[c]
East Germany	<1	4	12	14	17	15	17	13	7	0	0	0	715	Saemann (1975)[c]
Cambridgeshire	0	2	6	12	19	25	11	11	14	<1	0	0	170	Coombs *et al.* (1981)
Oxfordshire	0	<1	5	16	20	18	18	12	9	<1	0	0	142	Robertson (1990)
UK	0	0	4	12	32	14	15	9	6	5	2	1	95	Hudson (1965)
France[e]	<1	1	10	17	24	20	13	10	4	<0.1	<0.1	<0.1	11,017	ONCFS Columbids scheme (Eraud and Boutin, 2008, Lormée, 2018)

[a]Cumulated data over two seasons (2010–2011).
[b]Cumulated data over two seasons (2011–2012).
[c]Cited in Robertson (1990).
[d]For averaging, the value was distributed between September and October (2% in each).
[e]Cumulated data over 17 seasons (2001–2017; laying dates are back-calculated from nestling age at banding).

Fig. 15.3. An Eurasian Collared-dove egg and a freshly hatched chick. (©Photograph: OFB.)

Cramp, 1985) or the USA (Rose and Rose, 1999; Fielder *et al.*, 2012). A period of about 2 weeks was also reported in Iraq (Marchant, 1963) and in Jammu and Kashmir (14 ± 1.41; range 12–16; Kour, 2016). In Rajasthan, this period lasts longer in spring (range 18–19 days; Rana 1975). Hatching is asynchronous (Cramp, 1985), and fledging usually occurs 16–18 days later across most parts of the species range (Marchant, 1963; Cramp, 1985; Eraud *et al.*, 2009; Absi, 2012; Fielder *et al.*, 2012). In India, reported values range from 12 to 18 days in Rajasthan (Rana, 1975) and 14–18 days in Jammu and Kashmir (Kour, 2016). During this period, both parents feed the young by regurgitation and maintain sanitation (Cramp, 1985; Belabed *et al.*, 2013b; Kour, 2016). After leaving the nest, the young are still under parental care and reach independence at 30–40 days old (Cramp, 1985). Consecutive lays in the same nest are common (Cramp, 1985), and relaying may be prompt (a few days), sometimes when nestlings are still present in the nest (Robertson, 1990). The mean breeding success – defined as the percentage of chicks fledged from eggs laid – is highly variable among studies, ranging from around 31% to 97%, with no clear pattern between native and colonized ranges (Table 15.4). A survey conducted in Poland over 25 seasons (1974–1998) showed high year-to-year variability (20–41%; Górski and Antczak, 1999). In the UK, Poland and France, breeding success was shown as being lower early (Robertson, 1990; Górski and Antczak, 1999; Eraud and Boutin, 2008) and late (Górski and Antczak, 1999) in the season. In Rajasthan, breeding success was lower in dry periods (spring) than in the monsoon period (15.6% versus 39%; Rana, 1975). Mostly depredation, but also starvation, weather conditions, nest desertion, eggs/nestling falling or nestling death, were causes of failure (Rana, 1975; Coombs *et al.*, 1981; Robertson, 1990; Górski and Antczak, 1999; Eraud and Boutin, 2008; Kour, 2016). In Europe, corvids (*Corvus* spp., *Garrulus glandarius*, *Pica pica*) are the main sources of depredation (e.g. Górski and Antczak, 1999). In Rajasthan, the main avian predators are House Crows (*Corvus splendens*) and Indian Rollers (*Coracias benghalensis*) (Rana, 1975).

Eurasian Collared-dove nests are usually flimsy platforms of twigs, stems and roots, sometimes with other materials (e.g. feathers, wire, wool) (Rana, 1975; Cramp, 1985). Pairs often re-nest in the same nest (Ferguson-Lees, 1964; Robertson, 1990) and may occasionally renovate old nests from year to year (Rana, 1975) or nests from other species (Hudson, 1965). The height of nests ranges from 2 to 22 m and is highly variable, depending on local context. Across its whole range, the species nests in a wide variety of coniferous or deciduous trees and shrubs of native, fruit or ornamental species. Within its historical range, the species is reported rarely making use of man-made structures as nesting supports. In Jammu and Kashmir, none of 394 nests were found in buildings and only 5% used artificial supports such as electricity meters or TV antennae (Kour, 2016). In Rajasthan and Iraq, all nests (except one) were found in trees or bushes (Marchant, 1963; Rana, 1975). This nest-site preference contrasts with the wide use of artificial sites reported across recently invaded/colonized ranges. In the UK, the species was reported nesting on roofs, window sills, beams/rafters or machinery on farms or in outbuildings (Hudson, 1965;

Table 15.3. Size of Eurasian Collared-dove eggs across its distributional range. Data are means (SD/SE; sample size), where available.

Country	Length (mm)	Width (mm)	Mass (g)	Reference
India (Rajasthan)	26.82 (SE 0.40; n = 109)	20.75 (SE 0.38; n = 109)	7.5 (SE 0.25; n = 109)	Rana (1975)
India (Jammu and Kashmir)	29.91 (SD 1.84; n = 250)	24.09 (SD 2.01; n = 250)	11.87 (SD 1.40; n = 250)	Kour (2016)
Iran	30.86 (SD 0.93; n = 27)	23.35 (SD 0.84; n = 27)	8.64 (SD 0.69; n = 26)	Hosseini-Moosavi *et al.* (2013)
Algeria	–	–	8.9 (n = 4)	Torki (2014)
Algeria	31.24 (SD 0.89; n = 9)	23.96 (SD 0.42; n = 9)	–	Absi (2012)
Czech Republic	30.5 (n = 698)	23.6 (n = 698)	9.6 (n = 75)	Cramp (1985)
France (western)	30.64 (SD 1.43; n = 84)	23.63 (SD 0.89; n = 84)	9.37 (SD 0.97; n = 84)	Eraud (unpublished data from 2005)
England	31.28 (SD 1.29; n = 240)	24.11 (SD 0.69; n = 240)	10.0 (SD 0.8; n = 88)	Robertson (1990)

Table 15.4. Breeding success of Eurasian Collared-doves across its distributional range.

Country	Eggs laid	Hatched eggs (%)	Fledged chicks (%)	Overall success (%)	Reference
India (Rajasthan)	109	40	80	32	Rana (1975)
India (Jammu and Kashmir)	1 389	57	81	47	Kour (2016)
Iraq	77	64	55	35	Marchant (1963)
Iran	38	61	61	37	Hosseini-Moosavi *et al.* (2013)
Algeria	70	66	65	43	Torki (2014)
	26	–	–	39	Absi (2012)
Hungary	32	97	100	97	Tomasz (1955)[a]
Czech Republic	436	86	79	69	Kubik and Balat (1973)[a]
East Germany	1 359	–	–	28	Saemann (1975)[a]
Poland	15 764	–	–	31	Górski and Antczak (1999)
West Germany	175	–	–	54	Hofstetter (1954)[a]
UK	317	42	62	26	Coombs *et al.* (1981)[a]
	73	51	86	44	Coombs *et al.* (1981)[a]
	282	59	69	41	Robertson (1990)

[a]Cited in Robertson (1990).

Robertson, 1990). In Arkansas, 15 out of 20 nests were located on human-made structures (e.g. electrical substation, poles; Fielder *et al.*, 2012), and recently, similar behaviour was reported in Algeria and Morocco (Bougaham and Moulaï, 2013). In France, about two-thirds of nesting sites were artificial sites, and it was suggested that nesting in buildings might confer advantages in terms of breeding success (Eraud and Boutin, 2008).

15.8 Habitat

Given the large distribution of the Eurasian Collared-dove across the world, its habitat embraces a wide range of ecological and climatic contexts, from arid deserts to boreal zones, through temperate, insular and tropical contexts. One of the common features of newly colonized habitats is the high proximity of the species to human habitations and activities, contrary to what is observed in the native range. In India, the Eurasian Collared-dove mostly frequents open, cultivated and dry deciduous country with scatterings of trees and avoids city centres (Rana, 1975; Cramp, 1985). In Jammu and Kashmir, the species is observed in open areas or agricultural land with trees and scrubs, but also in suburban gardens and parks (Kour, 2016). In Europe and North Africa, the Eurasian Collared-dove frequents suburbs, urban parks, small towns and villages, as well as clusters of isolated houses, as long as nest sites and food resources are available; open countryside and the centre of large cities (but see Ferguson-Lees 1964) are usually less used (Hudson, 1965, 1972; Cramp, 1985; Merabet *et al.*, 2010; Bendjoudi *et al.*, 2015; Šálek, 2014; Marja and Elts, 2018). Similar habitat preferences are reported across the American and Caribbean range (Smith, 1987; Beckett *et al.*, 2007; Fujisaki *et al.*, 2010; Scheidt and Hurlbert, 2014; Johnston López *et al.*, 2015; Luna *et al.*, 2018). In both native and invaded ranges, the species avoids highly forested landscapes (Bonter *et al.*, 2009; Veech *et al.*, 2011; Scheidt and Hurlbert, 2014; Bresgunova *et al.*, 2017).

Across its worldwide range, the Eurasian Collared-dove is mostly sedentary and therefore tends to be absent (or scarce) from areas where temperatures are low and winter conditions harsh (e.g. Fujisaki *et al.*, 2010; Scheidt and Hurlbert, 2014). However, seasonal movements have been reported, leading to the desertion of some areas in winter, for example from the western Himalayas where the species breeds in spring at up to 2400 m (Cramp, 1985). In France, some passages of birds are also reported across the English Channel and along the Atlantic coast (Yeatman-Berthelot and Jarry, 1994; Dupuy *et al.*, 2018).

15.9 Impacts

The Eurasian Collared-dove outside its native range could outcompete native avifauna for resources (Beckett *et al.*, 2007), but consequences on native species remain equivocal. In southern Florida, the species was observed chasing other bird species from feeding stations (Romagosa and Labisky, 2000). Poling and Hayslette (2006) found a large degree of dietary overlap with the native mourning doves but with no behavioural dominance of the Eurasian Collared-dove, suggesting that the potential for negative effects of collared-doves on mourning dove populations may be lower than previously suspected (Poling and Hayslette, 2006). A case of piracy of a nest constructed and used by an American Robin (*Turdus migratorius*) was reported in Texas (Kasner and Pyeatt, 2016), but this behaviour is infrequent. In Europe, there are also anecdotal reports of the Eurasian Collared-dove chasing European Turtle Doves at feeding sites in the UK (Fletcher, 1979). While in Extremadura, Spain, the species has colonized rural habitats where it could interfere with the native Turtle Dove (Rocha-Camarero and Hidalgo, 2002), spatial segregation in other countries (e.g. France), probably prevents strong competition between these two species.

Another possible effect of increasing Eurasian Collared-dove populations is the risk of disease transmission to native species and livestock/poultry (Beckett *et al.*, 2007) through the carriage of protozoa, zoonotic bacteria or viruses. The species is a regular host of the protozoan parasite *Trichomonas gallinae*, with varying – although high – prevalence across its range (Iraq: 10%, Al-Bakry, 2009; Arizona, USA: 43%, Justice-Allen and Knox, 2014; UK: 86%, Lennon *et al.*, 2013; Germany: 66.7%, Quillfeldt *et al.*, 2018; southern Europe: 60%, Marx *et al.*, 2017) and across different lineages. In Malta, some individuals were found to be infected by a pathogenic lineage (i.e. A/B) known to cause mortality and subsequent population decline in several bird species (Marx *et al.*, 2017). Several zoonotic bacteria have also been isolated from the Eurasian Collared-dove. In the UK, the Eurasian Collared-dove is considered one of the species most affected by *Chlamydia psittaci* (Gough and Bevan, 1983; de Gruchy, 1983, cited in Colvile *et al.*, 2012) and a high prevalence (61%) was also reported in northern Italy (Donati *et al.*, 2015). In Iraq, a large spectrum of zoonotic bacteria was isolated from various tissues (*Salmonella* spp., *Shigella* spp., *Listeria monocytogenes*, *Brucella abortus*, *Campylobacter* spp.; Al-Jumaily *et al.*, 1989; Ahmed *et al.*, 2011). A similar spectrum was also isolated from eggshells, suggesting a potential role of shell dispersal by parents in spreading diseases (Al-Obaidi and Al-Shadeedi, 2016). The Eurasian Collared-dove is also thought to be a reservoir for some viruses that may cause economic damage, such as avian paramyxovirus serotype 1 (APMV-1 or Newcastle disease virus; Schuler *et al.*, 2012) and mosquito-borne viruses such as St Louis encephalitis virus and West Nile virus (Beckett *et al.*, 2007; Panella *et al.*, 2013) or pigeon circovirus (Kubícek and Taras, 2005). In Europe, AMPV-1 is endemic and two sublineages (4a and 4b) of a pigeon variant (pigeon paramyxomvirus serotype 1 (PPMV-1)) were isolated from Eurasian Collared-doves (Bonfante *et al.*, 2012). In the USA, PPMV-1 was recently associated with mortality events in both Eurasian Collared-doves and native columbids (Schuler *et al.*, 2012; Justice-Allen and Knox, 2014), but the role of the Eurasian Collared-dove in transmitting this disease to native birds has not been definitively clarified (Justice-Allen and Knox, 2014).

Economic damage caused by flocks of Eurasian Collared-doves also includes eating and fouling stored grains and food delivered to livestock or poultry (Hudson, 1965; Robertson, 1990). In northern India, Kour (2016) reported damage to newly sown crops (wheat) or pecking of mustard seeds from stacks left in the field after harvest. The need for fresh green food may also cause damage to vegetables grown in gardens (e.g. lettuce; Hudson, 1965). In the early 1960s, the calling of the species in the early morning was viewed as a source of annoyance by some boarding-house keepers (Hudson, 1965). However, it is not known if this is still the case.

15.10 Control

In Europe, the Eurasian Collared-dove is viewed as a native – not invasive – bird species, so hunting regulations follow prescriptions from the Birds Directive 2009/147/EC. The Eurasian Collared-dove is a legal hunted species in ten countries of the European Union (Birds Directive 2009/147/EC: Austria, Belgium, Czech Republic, Cyprus, Denmark, France (including French West Indies), Germany, Hungary, Romania, Slovakia) and in Switzerland, but is protected elsewhere. In the UK, legal protection was removed in 1977 (Aebischer 1995), and until recently, the species could be shot where it was viewed as a pest (Robertson, 1990). In the USA, where the Eurasian Collared-dove is considered an introduced species, hunting is encouraged year-round, with no limit in most states. It is probably not hunted in Morocco or Algeria, and is very likely to be exposed to a low hunting pressure in other North African countries. Hunting regulations elsewhere across the species' range are poorly known.

Hunting statistics for Eurasian Collared-doves in the USA are unknown, so it is unclear whether hunting is a significant means of regulation. However, the extraordinary progression of the species in North America testifies to the weak success of current hunting practices in curbing expansion of the species. In Europe, the number of birds shot is small relative to population sizes. Hirschfeld and Heyd (2005) and Hirschfeld and Attard (2017) estimated an annual bag size of fewer than 500,000 birds, most of them from France. In this country, bag size was estimated at 288,000–323,000 in the hunting season 1998/1999 and 176,000–202,000 in 2013/2014 (Lormée and Aubry, 2018). Current hunting practices in Europe probably have a weak impact on the overall dynamics of populations (see also Aebischer, 1995). If hunting can reduce populations in rural areas, urban/suburban populations probably remain unaffected (Romagosa, 2002).

In Europe and the USA, the Eurasian Collared-dove is predated by avian predators including corvids and raptors (e.g. hawks and owls), as well as by feral cats (*Felis catus*) (Robertson, 1990; Górski and Antczak, 1999; Romagosa, 2002; Eraud *et al.*, 2011; Bielefeld, 2014).

15.11 Uses

In the USA and Europe, the Eurasian Collared-dove is a game species. The species is also kept in captivity for ornamental purposes in numerous countries.

15.12 Notes

The very high association of the Eurasian Collared-dove with human activities outside its native range, both for breeding and foraging purposes, and its extended breeding season and related high number of broods, together with its high dispersal ability, make it particularly efficient in colonizing and establishing sustainable populations (Bagi *et al.*, 2017; Ingenloff *et al.*, 2017), especially in the context of land cover conversion into urban and agricultural areas (Romagosa, 2002). This overall dynamic makes the species particularly difficult to extirpate if nothing is done quickly after an intentional introduction.

15.13 Acknowledgements

Original data on breeding phenology in France was kindly provided by Hervé Lormée (OFB). Early versions were improved by comments from Alexandre Villers (OFB).

15.14 References

Absi, K. (2012) Nidification et reproduction des populations de tourterelles des bois, turque et maillées (*Streptopelia turtur, S. decaocto et S. senegalensis*) dans les oasis sud-est des Ziban. Masters thesis, Faculté des Sciences exactes et Sciences de la nature et de la Vie, Département des Sciences Agronomiques, Biskra, Algeria.

Aebischer, N.J. (1995) Investigating the effects of hunting on the survival of British pigeons and doves by analysis of ringing recoveries. *Journal of Applied Statistics* 22, 923–935.

Afrasiab, S., Al-Moussawi, A. and Mohammad, M.K. (2017) Color variation of *Streptopelia decaocto* (Aves, Columbidae) with some notes on endoparasites. *Bulletin of the Iraq Natural History Museum* 14, 267–273.

Ahmed, L.M., Al-Obaidi, F.A. and Al-Shadeedi, S.M. (2011) Prevalence of some zoonotic bacteria in wild birds in Kirkuk city. *Al-Anbar Journal of Veterinary Sciences* 4, 11–17.

Al-Bakry, H.S. (2009) Prevalence of avian trichomoniasis in different species of pigeons in Mosul. *Iraqi Journal of Veterinary Sciences* 23 105–109.

Ali, Z. and Akhtar, M. (2005) Bird surveys at wetlands in Punjab, Pakistan, with special reference to the present status of White-headed Duck *Oxyura leucocephala*. *Forktail* 21, 43–50.

Al-Jumaily, W.T., Al-Atar, M.A., Al-Tae, A.R., Mansour, A.D., Jiad, J.H. and Abdul-Latif, H. (1989) The incidence of salmonellae and serological evidence of Newcastle disease in some wild birds from Baghdad area. *Journal of Biological Science Research* 20, 213–219.

Almazán-Núñez, C. (2014).Nuevos registros de la paloma turca (*Streptopelia decaocto*) en el estado de Guerrero, México. *Acta Zoológica Mexicana* 30, 701–706.

Al-Obaidi, F.A. and Al-Shadeedi, S.M.J. (2016) Effect of season on some microbial counts contaminated egg shells of some native birds in Baghdad city. In: *Proceedings of the 4th International Conference of Genetic and Environment*. Association of Genetic and Environmental Resources Conservation (AGERC), Cairo, Egypt, pp. 748–754.

Amer, S.A.M. (2014) Preliminary molecular variability of some dove species inhabiting Saudi Arabia. *American Journal of BioScience* 2, 153.

Azafzaf, H., Feltrup-Azafzaf, C., Dlensi, H. and Isenmann, P. (2015) Nouvelles données sur l'Avifaune de Tunisie (2005–2014). *Alauda* 83, 7–28.

Bagi, Z., Kraus, R.H. and Kusza, S. (2017) A review of the invasive Eurasian Collared dove and possible research methods in the future. *Balkan Journal of Wildlife Research* 4, 1–10.

Barré, N., Feldmann, P., Tayalay, G., Roc, P., Anselme, M. and Smith, P.W. (1996) Status of the Eurasian Collared-dove (*Streptopelia decaocto*) in the French West Indies. *El Pitirre* 9, 2–3.

Beckett, S.M., Komar, N. and Doherty, P.F. Jr (2007) Population estimates for Eurasian Collared-dove in Northeastern Colorado. *Wilson Journal of Ornithology* 119, 471–475.

Belabed, A.I., Aouissi, H.A., Zediri, H., Djemadi, I., Driss, K., *et al.* (2013a) Effet de l'urbanisation sur le phénotype de la Tourterelle turque (*Streptopelia decaocto*) dans le Nord-Est Algérien. *Bulletin de l'Institut Scientifique, Section Sciences de la Vie* 35, 155–164.

Belabed, A.I., Djemadi, I., Zediri, H., Eraud, C. and Bouslama, Z. (2013b) Étude de l'investissement Parental Chez la Tourterelle Turque (*Streptopelia decaocto*) dans le Nord-est Algérien. *European Journal of Scientific Research* 94, 421–436.

Bendjoudi, D., Voisin, J.-F., Doumandji, S., Merabet, A., Benyounes, N. and Chenchouni, H. (2015) Rapid increase in numbers and change of land-use in two expanding Columbidae species (*Columba palumbus* and *Streptopelia decaocto*) in Algeria. *Avian Research* 6, 18.

Bergier, P., Franchimont, J. and Thévenot, M. (1999) Implantation et expansion géographique de deux espèces de columbidés au Maroc: aa Tourterelle turque *Streptopelia decaocto* et la Tourterelle Maillée *Streptopelia senegalensis*. *Alauda* 67, 23–36.

Bielefeld, R.R. (2014) Florida Burrowing Owl (*Athene cunicularia floridana*) preys on Eurasian Collared-Dove (*Streptopelia decaocto*). *Florida Field Naturalist* 42, 114–118.

BirdLife International (2018) Eurasian Collared-dove: *Streptopelia decaocto*. The IUCN Red List of Threatened Species 2018. Available at: www.iucnredlist.org/species/22727811/132178595 (accessed 27 October 2019).

Blancas-Calva, E. and Blancas-Hernández, J.C. (2016) Presencia de la paloma turca (*Streptopelia decaocto*) en la ciudad de Lima, Perú [Presence of the Eurasian Collared-Dove (*Streptopelia decaocto*) in Lima city, Peru]. *Huitzil, Revista Mexicana de Ornitología* 17, 111–114.

Blancas-Calva, E., Castro-Torreblanca, M. and Blancas-Hernández, J.C. (2014) Presencia de las palomas turca (*Streptopelia decaocto*) y africana de collar (*Streptopelia roseogrisea*) en el estado de Guerrero, México [Presence of the Eurasian Collared Dove (*Streptopelia decaocto*) and the African Collared Dove (*Streptopelia roseogrisea*) in the state Guerrero, Mexico]. *Huitzil, Revista Mexicana de Ornitología* 15, 10–16.

Blasco Zumeta, J. and Heinze, G.-M. (2019) Atlas de Identificación de las Aves de Aragón Identification/Atlas of Aragon's Birds. Available at: http://blascozumeta.com/introduction/ (accessed 27 October 2019).

Bonfante, F., Terregino, C., Heidari, A., Monne, I., Salviato, A., *et al.* (2012) Identification of APMV-1 associated with high mortality of collared doves (*Streptopelia decaocto*) in Italy. *Veterinary Record* 171, 327.

Bonter, D.N., Zuckerberg, B. and Dickinson, J.L. (2009) Invasive birds in a novel landscape: habitat associations and effects on established species. *Ecography* 33, 494–502.

Bougaham, A.F. and Moulaï, R. (2013) Première nidification de la Tourterelle turque *Streptopelia decaocto* (Frivaldszky, 1838) (Aves, Columbidae) sur des édifices humains en Algérie. *Bulletin de l'Institut Scientifique, Section Sciences de la Vie* 35, 151–153.

Bresgunova, O.O., Litinskiy, N.S., Litinskaya, Y.V. and Sinna, O.I. (2017) The distribution and numbers of the Collared Doves (*Streptopelia decaocto*) during the nesting period in Kharkiv City. *Branta* 20, 187–199.

Chablé-Santos, J., Gómez-Uc, E. and Hernández-Betancourt, S. (2012) Registros reproductivos de la paloma de collar (Streptopelia decaocto) en Yucatán, México [Reproductive records of the Eurasian Collared-Dove (*Streptopelia decaocto*) in Yucatan, Mexico]. *Huitzil, Revista Mexicana de Ornitología* 13, 1–5.

Colvile, K.M., Lawson, B., Pocknell, AM., Dagleish, M.P., John, S.K. and Cunningham, A.A. (2012) Chlamydiosis in British songbirds. *Veterinary Record* 171, 177.

Coombs, C.F.B., Isaacson, A.J., Murton, R.K., Thearle, R.J.P. and Westwood, N.J. (1981) Collared doves (*Streptopelia decaocto*) in urban habitats. *Journal of Applied Ecology* 18, 41–62.

Cramp, S. (ed.) (1985) *Handbook of the Birds of Europe, the Middle East and North Africa: The Birds of the Western Palearctic*, Vol. 4: *Terns to Woodpeckers*. Oxford University Press, Oxford.

Dean, W.R.J. (2000) Alien birds in southern Africa: what factors determine success? *South African Journal of Science* 96, 9–14.

de Gruchy, P.H. (1983) Chlamydiosis in collared doves. *Veterinary Record* 113, 327.

del Hoyo, J., Collar, N., Kirwan, G.M. and Garcia, E.F.J. (2018) Burmese Collared-dove (*Streptopelia xanthocycla*). In: del Hoyo, J., Elliot, A., Sargatal, J., Christie, D.A. and de Juana, E. (eds) *Handbook of the Birds of the World Alive*. Lynx Edicions, Barcelona, Spain.

Donati, M., Laroucau, K., Delogu, M., Vorimore, F., Aaziz, R., *et al.* (2015) *Chlamydia psittaci* in Eurasian Collared Doves (*Streptopelia decaocto*) in Italy. *Journal of Wildlife Diseases* 51, 214–217.

Dupuy, J., Quaintenne, G., Robin, F. and Couzi, L. (2018) Mise à jour du document de concept clé sur la période de reproduction et de migration prénuptiale des espèces d'oiseaux chassables dans l'UE : propositions de la LPO France. LPO, BirdLife France, Rochefort, France.

Dyer, E.E., Cassey, P., Redding, D.W., Collen, B., Franks, V., *et al.* (2017) The global distribution and drivers of alien bird species richness. *PLoS Biology* 15, e2000942.

eBird (2020) eBird: An online database of bird distribution and abundance [web application]. eBird, Cornell Lab of Ornithology, Ithaca, New York. Available: http://www.ebird.org (accessed 23 July 2020).

Eraud, C. and Boutin, J.-M. (2008) Tourterelle turque. In: *Tout le Gibier de France. Atlas de la Biodiversité de la Faune Sauvage, les 90 Espèces Chassables*. Hachette Pratique, Fédération Nationale des Chasseurs, Office National de la Chasse et de la Faune Sauvage, Paris, pp. 443–450.

Eraud, C., Boutin, J.-M., Roux, D. and Faivre, B. (2007) Spatial dynamics of an invasive bird species assessed using robust design occupancy analysis: the case of the Eurasian collared dove (*Streptopelia decaocto*) in France. *Journal of Biogeography* 34, 1077–1086.

Eraud, C., Jacquet, A. and Faivre, B. (2009) Survival cost of an immune soliciting in nature. *Evolution* 63, 1036–1043.

Eraud, C., Jacquet, A. and Legagneux, P. (2011) Post-fledging movements, home range, and survival of juvenile Eurasian collard-doves in western France. *The Condor* 113,150–158.

Estela, F. (2005) eBird Checklist: https://ebird.org/ebird/view/checklist/S25760413. eBird: An online database of bird distribution and abundance [web application]. eBird, Ithaca, New York. Available: http://www.ebird.org (accessed 23 July 2020).

Ferguson-Lees, I.J. (1964) Studies of less familiar birds 127. Collared Dove. *British Birds* 57, 170–175.

Fielder, J.M., Kannan, R., James, D.A. and Cunningham, J. (2012) Status, dispersal, and breeding biology of the exotic Eurasian Collared-Dove (*Streptopelia decaocto*) in Arkansas. *Journal of the Arkansas Academy of Science* 66, 55–61.

Fisher, J. (1953) The Collared Turtle Dove. *British Birds* 5, 153–182.

Fletcher, M.R. (1979) Aggression by Collared Dove to Turtle Doves. *British Birds* 72, 346.

Francis, J. (1996) The Eurasian Collared-Dove on Nevis. *El Pitirre* 9, 4.

Fujisaki, I., Pearlstine, E.V. and Mazzotti, F.J. (2010) The rapid spread of invasive Eurasian Collared Doves *Streptopelia decaocto* in the continental USA follows human-altered habitats. *Ibis* 152, 622–632.

Gbif.org (22 July 2020) GBIF Occurrence Download. https://doi.org/10.15468/dl.ufyysm

Gibbs, D., Barnes, E. and Cox, J. (2001) *Pigeons and Doves: a Guide to the Pigeons and Doves of the World*. Pica Press, Robertsbridge, UK.

Gnielka, R. (1975) Zur Brutbiologie der Türkentaube *Streptopelia decaocto*. *Ornithologische Mitteilungen* 27, 71–83.

Gómez de Silva, H. (2006a) Mexico. *North American Birds* 60, 290–295.

Gómez de Silva, H. (2006b) Mexico. *North American Birds* 60, 444–451.

Goroshko, O.A. (2014) The first breeding of Eurasian Collared Dove (*Streptopelia decaocto* Frivaldszky, 1838) in Baikal Region. *Baikalian Zoological Journal* 2, 103–104.

Górski, W. and Antczak, J. (1999) Breeding losses in an urban population of the Collared Dove *Streptopelia decaocto* in Slupsk, Poland. *Acta Ornithologica* 34, 191–198

Gough, R.E. and Bevan, B.J. (1983) Isolation and identification of *Chlamydia psittaci* from collared doves (*Streptopelia decaocto*). *Veterinary Record* 112, 552.

Hasan, A., Ali Khan, S. and Ahmad, S.I. (2005) Fish and birds in Keti Bundar, Shah Bundar and other parts of the Indus Delta. *Zoological Survey of Pakistan* 16, 33–39.

Hazevoet, C.J. (2012) Seventh report on birds from the Cape Verde Islands, including records of nine taxa new to the archipelago. *Zoologia Caboverdiana* 3, 1–28.

Hengeveld, R. (1993) What to do about the North American invasion by the Collared-Dove. *Journal of Field Ornithology* 64, 477–489.

Hirschfeld, A. and Attard, G. (2017) Bird hunting in Europe. An analysis of bag figures and their effect on the conservation of threatened species. *Berichte zum Vogelschutz* 53–54, 15–42.

Hirschfeld, A. and Heyd, A. (2005) Jagdbedingte Mortalität von Zugvögeln in Europa: Streckenzahlen und Forderungen aus Sicht des Vogel- und Tierschutzes. *Berichte zum Vogelschutz* 42, 47–74.

Hofstetter, F.-B. (1954) Untersuchungen an einer Population der Ttirkentaube. *Journal of Ornithology* 95, 348–410.

Hooten, M.B. and Wikle, C.K. (2008) A hierarchical Bayesian non-linear spatio-temporal model for the spread of invasive species with application to the Eurasian Collared-Dove. *Environmental and Ecological Statistics* 15, 59–70.

Hosseini-Moosavi, S.M., Behrouzi-Rad, B., Karimpour, R. and Nasab, S.M.A. (2013) Breeding Biology of Collared dove (*Streptopelia decaocto*) in Khuzestan Province, southwestern Iran. *Berkut* 22, 51–54.

Hudson, R. (1965) The spread of the Collared Dove in Britain and Ireland. *British Birds* 58, 105–139.

Hudson, R. (1972) Collared Doves in Britain and Ireland during 1965–1972. *British Birds* 65, 139–155.

Ingenloff, K., Hensz, C.M., Anamza, T., Barve, V., Campbell, L.P., *et al.* (2017) Predictable invasion dynamics in North American populations of the Eurasian collared dove *Streptopelia decaocto*. *Proceedings of the Royal Society B: Biological Sciences* 284, 20171157.

Isenmann, P., Gaultier, T., El Hili, A., Azafzaf, H., Dlensi, H. and Smart, M. (2005) *Oiseaux de Tunisie – Birds of Tunisia*. Société d'Etudes Ornithologiques de France, Paris.

Isenmann, P., Benmergui, M., Browne, P., Diam Ba, A., Hamallah Diagana, C., *et al.* (2010) *Oiseaux de Mauritanie – Birds of Mauritania*. Société d'Etudes Ornithologiques de France, Paris.

Isenmann, P., Hering, J., Brehme, S., Essghaier, M., Etayeb, K., *et al.* (2016) *Oiseaux de Libye – Birds of Libya*. Société d'Etudes Ornithologiques de France, Paris.

Jennings, M.C. (2010) *Atlas of the Breeding Birds of Arabia*. Senckenberg-Gesellschaft für Naturforschung, Frankfurt, Germany.

Johnston López, K., Macías Duarte, A. and Castillo Gámez, R.A. (2015) Urban birds in the Sonoran Desert: estimating population density from point counts. *Huitzil, Revista Mexicana de Ornitología* 16, 37–47.

Justice-Allen, A. and Knox, C. (2014) The prevalence of Pigeon Paramyxovirus 1 and *Trichomonas gallinae* in Band-tailed Pigeons (*Patagioenas fasciata*), Mourning Doves (*Zenaida macroura*), and White-winged Doves (*Zenaida asiatica*). Arizona Game and Fish Department, Phoenix, Arizona.

Kasner, A.C. and Pyeatt, D.N. (2016) Eurasian Collared-Dove (*Streptopelia decaocto*) usurps nest of American Robins (*Turdus migratorius*). *Wilson Journal of Ornithology* 128, 198–200.

Kour, D.N. (2016) Ecology of some species of Columbiformes from Jammu. PhD Thesis, University of Jammu, Jammu, India.

Kubícek, O. and Taras, L. (2005) Incidence of pigeon circovirus in Eurasian Collared-Dove (*Streptopelia decaocto*) detected by nested PCR. *Acta Veterinaria Brno* 74, 361–368.

Kubik, J. and Balat, F. (1973) Zur populationsdynamik der Türkentaube *Streptopelia decaocto* (Friv.) in Brno, CSSR. *Zoologicke Listy* 22, 59–72.

Lennon, R.J., Dunn, J.C., Stockdale, J.E., Goodman, S.J., Morris, A.J. and Hamer, K.C. (2013) Trichomonad parasite infection in four species of Columbidae in the UK. *Parasitology* 140, 1368–1376.

Lever, C. (2005) *Naturalised Birds of the World*. T. and A.D. Poyser, London.

Levesque, A. and Jaffard, M.-E. (2003) La Tourterelle turque en Guadeloupe: statut, répartition, propositions de gestion. Amazona-Guadeloupe. Available at : https://www.amazona-guadeloupe.com/_iserv/dlfiles/dl.php?ddl=rap-4.pdf (accessed 27 October 2019).

Lormée, H. (2018) Programme d'études Colombidés – bilan de la campagne 2017. Report of the Office National de la Chasse et de la Faune Sauvage, France.

Lormée, H. and Aubry, P. (2018) Estimation des tableaux de chasse de colombidés en France pour la saison 2013–2014. *Faune Sauvage* 318, 15–22.

Luna, Á., Romero-Vidal, P., Hiraldo, F. and Tella, J. (2018) Cities favour the recent establishment and current spread of the Eurasian collared dove *Streptopelia decaocto* (Frivaldszky, 1838) in Dominican Republic. *BioInvasions Records* 7, 95–99.

Marchant, S. (1963) The breeding of some Iraqi birds. *Ibis* 105, 516–557.

Marja, R. and Elts, J. (2018) Kaelus-turteltuvi (*Streptopelia decaocto*) elupaigakasutus, käitumine ja tegutsemiskohad 2017. aastal. *Hirundo* 31, 21–31.

Marx, M., Reiner, G., Willems, H., *et al.* (2017) High prevalence of *Trichomonas gallinae* in wild columbids across western and southern Europe. *Parasites and Vectors* 10, 242.

Mayr, E. (1951) Speciation in birds. In: Hörstadius, S. (ed.) *Proceedings of the 10th International Ornithological Congress*. Almqvist & Wiksell, Uppsala, Sweden, pp. 91–131.

Merabet, A., Doumandji, S. and Baziz, B. (2010) Expansion des Populations des Columbiformes au Sein des Oiseaux des Milieux Agricoles et Suburbains en Mitidja (Algérie). *European Journal of Scientific Research* 43, 113–126.

Moali, A. and Isenmann, P. (2007) La Tourterelle turque *Streptopelia decaocto* nicheuse en 2007 à Tamanrasset (Sahara Central, Algérie) et son expansion au Sahara. *Alauda* 75, 247–248.

Moali, A., Moali-Grine, N., Fellous, A. and Isenmann, P. (2003) Expansion spatiale de la Tourterelle turque *Streptopelia decaocto* et présence dans les parcs urbains du Pigeon ramier *Columba palumbus* en Algérie. *Alauda* 71, 371–374.

Nowak, E. (1989) Ausbreimng der Thrkentaube (*Streptopelia decaocto*) in der UdSSR: Umfrage 1988. *Journal of Ornithology* 130, 513–527.

Panella, N.A., Young, G. and Komar, N. (2013) Experimental infection of Eurasian collared-dove (*Streptopelia decaocto*) with West Nile virus. *Journal of Vector Ecology* 38, 210–214.

Pikula, J. and Kubik, V. (1978) Die Brutikologieder Türkentaube *Streptopelia decaocto* im milieuder Stadt Brno. *Acta scientiarum naturalium Academiae Scientiarum Bohemicae, Brno* 12, 1–40.

Poling, T.D. and Hayslette, S.E. (2006) Dietary overlap and foraging competition between Mourning Doves and Eurasian Collared-Doves. *Journal of Wildlife Management* 70, 998–1004.

Quillfeldt, P., Schumm, Y.R., Marek C., Mader, V., Fischer, D. and Marx, M. (2018) Prevalence and genotyping of *Trichomonas* infections in wild birds in central Germany. *PLoS One* 13, e0200798.

Raffaele, H.A., Wiley, J., Garrido, O.H., Keith, A. and Raffaele, J. (1998) *A Guide to the birds of the West Indies*. Princeton University Press, Princeton, New Jersey.

Rana, B.D. (1975) Breeding biology of the Indian ring dove in the Rajasthan Desert. *Auk* 92, 322–332.

Robertson, H.A. (1990) Breeding of Collared Doves *Streptopelia decaoctoin* rural Oxfordshire, England. *Bird Study* 37, 73–83.

Rocha-Camarero, G. and Hidalgo de Trucios, S. (2002) The spread of the Collared Dove *Streptopelia decaocto* in Europe: colonization patterns in the west of the Iberian Peninsula. *Bird Study* 49, 11–16.

Romagosa, C.M. and Labisky, R.F. (2000) The establishment and dispersal of the Eurasian Collared-Dove (*Streptopelia decaocto*) in Florida. *Journal of Field Ornithology* 71, 159–166.

Romagosa, C.M. (2002) Eurasian Collared-Dove (*Streptopelia decaocto*). In: *Birds of North America*. The Cornell Lab of Ornithology, Ithaca, New York. Available at: https://birdsna.org/Species-Account/bna/species/eucdov (accessed 27 October 2019).

Romagosa, C.M. and McEneaney, T. (1999) Eurasian Collared-Dove in North America and the Caribbean. *North American Birds* 53, 348–353.

Roonwal, M.L. (1940) On the subspecies of the Ring-dove, *Streptopelia decaocto* (Frivaldsky). *Records of the Indian Museum* 42, 437–452.

Rose, R.E. and Rose, M.C. (1999) Observations of nesting Eurasian Collared-Doves (*Streptopelia decaocto*) in Gulf Breeze, Florida. *Alabama Birdlife* 45, 1–3.

Saemann, D. (1975) Studien an einer Grossstadtpopulation der Tiirkentaube *Streptopelia decaocto* im Siiden der DDR. *Hercynia* 12, 361–388.

Salazar-Borunda, M.A., Martínez-Guerrero, J.H. and Pereda-Solís, M.E. (2015) Morphometrics and body condition index of Eurasian Collared-Dove and Mourning Dove in Durango, Mexico. *Open Journal of Ecology* 5, 33–38.

Šálek, M. (2014) Population density of Eurasian Collared Dove *Streptopelia decaocto* in different types of urban development. *Sluka* 10, 38–48.

Scheidt, S.N. and Hurlbert, A.H. (2014) Range expansion and population dynamics of an invasive species: the Eurasian Collared-Dove (*Streptopelia decaocto*). *PLoS One*, 9, e111510.

Schuler, K.L., Green, D.E., Justice-Allen, A.E., Jaffe, R., Cunningham, M., *et al.* (2012) Expansion of an exotic species and concomitant disease outbreaks: pigeon paramyxovirus in free-ranging Eurasian collared doves. *EcoHealth* 9, 163–170.

Smith, P.W. (1987) The Eurasian Collared-Dove arrives in the Americas. *American Birds* 41, 1371–1379.

Smith, P.W. and Kale, H.W. (1986) Eurasian Collared-Doves collected in Florida. *Florida Field Naturalist* 14, 104–107.

Snow, D.W. & Perrins, C.M. (1998) The Birds of the Western Palearctic Concise Edition Volume 1. Oxford University Press, Oxford.

Stresemann, E. and Nowak, E. (1958) Die ausbreitung der Türkentaube in Asien und Europa. *Journal of Ornithology* 99, 243–296.

Sueur, F. (1999) *La Tourterelle Turque*. Eveil Nature, Saint-Yrieix-sur-Charente, France.

Stresemann, E. (1950) Erste Bruten der Tiirkentaube in Deutschland. In: *Ornithologische Berichte* II, Heft 2, pp. 98–103.

Syroteuk, M. (2008) First occurrence of the Eurasian Collared-Dove (*Streptopelia decaocto*) in the west Kootenay Region of British Columbia. *Wildlife Afield* 5, 226–228.

Thévenot, M., Vernon, R. and Bergier, P. (2003) *The Birds of Morocco: An Annotated Check-list*. The Natural History Museum, Tring, UK.

Tinajero, R. and Rodríguez-Estrella, R. (2014) Incremento en la distribución y primer registro de anidación de la paloma de collar (*Streptopelia decaocto*) en la península de Baja California, México. *Revista Mexicana de Biodiversidad* 85, 898–909.

Tomasz, J. (1955) Contributions to the ecology of the Indian Ring Dove. *Aquila* 59–62, 101–143.

Torki, S. (2014) Reproduction des tourterelles dans la région des Ziban. Thesis, Université Kasdi Merbah, Faculté des Sciences de la Nature et de la Vie et Sciences de la Terre et de l'Univers, Ouargla, Algeria.

Valencia-Herverth, J., Valencia-Herverth, R., Mendiola-González, M.E., Sánchez-Cabrera, M. and Martínez-Morales, M.A. (2011) New and outstanding bird records to the State of Hidalgo, Mexico. *Acta Zoológica Mexicana* 27, 843–861.

Veech, J.A., Small, M.F. and Baccus, J.T. (2011) The effect of habitat on the range expansion of a native and an introduced bird species: habitat and range expansion. *Journal of Biogeography* 38, 69–77.

Vincent, P. (2008) *Saudi Arabia. An Environmental Overview*. Taylor & Francis, London.

Yeatman-Berthelot, D. and Jarry, G. (eds) (1994) *Nouvel Atlas des Oiseaux Nicheurs de France: 1985–1989*, 2nd edn. Revue et Corrigée, Société Ornithologique de France, Paris.

16 Chukar Partridge (*Alectoris chukar* Gray, 1830)

Ndivhuwo Shivambu*, Cavin T. Shivambu and Colleen T. Downs
Centre for Excellence in Invasion Biology and Centre for Functional Biodiversity, School of Life Sciences, University of KwaZulu-Natal, P/Bag X01, Pietermaritzburg 3209, South Africa

Citation: Shivambu, N., Shivambu, C.T. and Downs, C.T. (2020) Chukar Partridge (*Alectoris chukar* Gray, 1830). In: Downs, C.T. and Hart, L.A. (eds) *Invasive Birds: Global Trends and Impacts*. CAB International, Wallingford, UK, pp. 132–137.

16.1 Common Names

Chikone, Chickore, Chucor, Chukker, Chukar, Chukor, Chukru, Indian Chukor, Rock Partridge, Kabk, Kaukau, keklik, Nek-pa Zarkar (Christensen, 1970)

16.2 Nomenclature

The Chukar Partridge (*Alectoris chukar* Gray, 1830) belongs to the family Phasianidae (Barilani *et al.*, 2007). The variations in plumage within the widespread distributed groups of this species that have been designated as subspecies have often led to complications in identification of the Chukar Partridge (Hartert, 1925). There are currently 16 recognized subspecies of Chukar Partridge (McGowan and Kirwan, 2019):

- *A. c. chukar* Gray, 1830, is native to the eastern parts of Afghanistan to Nepal.
- *A. c. cypriotes* Hartert, 1917, is an island chukar, native to the south and eastern parts of Bulgaria, and southern parts of Crete, Cyprus, Syria and Rhodes.
- *A. c. dzungarica* Sushkin, 1927, is native to the north-western parts of Mongolia extending to Altai, Russia, and the eastern parts of Shannan, Tibet.
- *A. c. falki* Hartert, 1917, is native to the north-eastern part of Kazakhstan to the central-northern parts of Afghanistan extending to the Pamir Mountains and western China (West of Xinjiang).
- *A. c. kleini* Hartert, 1925, native to south-eastern parts of Bulgaria, north-eastern Greece and northern Turkey, and Caucasus.
- *A. c. koroviakovi* Zarudny, 1914, the Persian Chukar, is native to the eastern parts of Iran to the western parts of Pakistan.
- *A. c. kurdestanica* Meinertzhagen, 1923, the Kurdestan Chukar, is native to the southern parts of the Caucasus Mountains to south-eastern Turkey, including the northern parts of Syria, Iraq and Iran.
- *A. c. pallescens* Hume, 1873, the Northern Chukar, is native to north-eastern Afghanistan to Ladakh and western Tibet, south and east of the Pamir Mountains to north-western parts of India and west of the Tibetan Plateau.
- *A. c. pallida* Hume, 1873, is native to north-western China, west and south of Xinjiang.
- *A. c. potanini* Sushkin, 1927, is native to the western parts of Mongolia and the northern parts of China including northern Gansu and inner-central Mongolia.
- *A. c. pubescens* Swinhoe, 1871, is native to the inner parts of Mongolia to Liaoning towards China in the north-western parts of Sichuan and eastern Qinghai.
- *A. c. sinaica* Bonaparte, 1858, the Desert Chukar, is native to the northern parts of the Syrian Desert extending to the southern areas of the Sinai Peninsula and probably the north-western parts of Saudi Arabia.
- *A. c. subpallida* Zarudny, 1914, is native to Uzbekistan and the south-east of Kazakhstan and Tajikistan, particularly in Kyzylkum and Karakum mountains.
- *A. c. werae* Zarudny & Loudon, 1904, the Iranian Chukar, is native to the eastern parts of Iraq and south-western Iran.
- *A. c. shestoperovi* Sushkin, 1927, is native to the northern parts of Afghanistan and Turkmenistan.
- *A. c. fallax* Sushkin, 1927, is native to the north-western parts of China including the Tien Shannan (Tibet) mountains and the north-western parts of Xinjiang.

*Corresponding author: ndivhuwomaligana@gmail.com

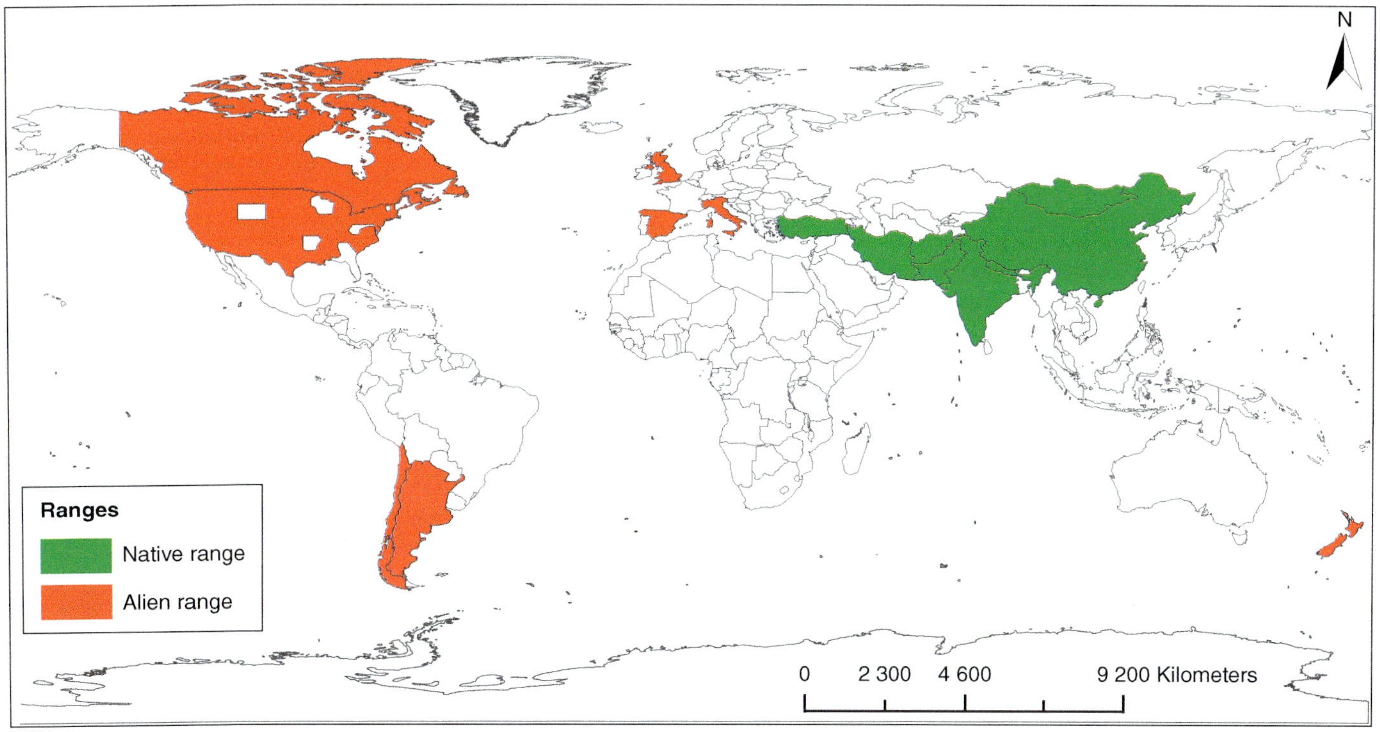

Fig. 16.1. Global distribution range the Chukar Partridge (*Alectoris chukar*) showing the native (green) and alien (red) ranges.

The Chukar Partridge has been considered to form part of a species complex with the European Partridge (*Perdix perdix*), the Rock Partridge (*Alectoris graeca*, also known as the Greek Partridge), Philby's Partridge (*Alectoris philbyi*) and Przevalski's Partridge (*Alectoris magna*) and was considered as a conspecific mainly with the Rock Partridge (Hartert, 1925; Christensen, 1970).

16.3 Distribution

The native range of the Chukar Partridge is Asia (Fig. 16.1) (Young, 1981; Khan *et al.*, 2017). The distribution in its native range stretches from the Aegean Sea through the western parts of Asia including Afghanistan, Assyria, Israel, Iran, Lebanon, Pakistan, Palestine, Turkey and India, to the inner areas of Western Himalaya to Nepal, Mongolia and China (Young, 1981; Kark, 2001; Madge and McGowan, 2002; Barbanera *et al.* 2007). The Chukar Partridge has invaded countries such as Argentine, Canada, Chile, Hawaii, Italy, New Zealand, Saint Helene, Spain and the UK (Long, 1981), and the USA (Alabama, Arizona, California, Colorado, Connecticut, Idaho, Illinois, Indiana, Iowa, Kansas, Kentucky, Louisiana, Massachusetts, Michigan, Minnesota, Mississippi, Missouri, Montana, Nebraska, Nevada, New Hampshire, New Mexico, New York, North Carolina, North Dakota, Ohio, Oklahoma, Oregon, Pennsylvania, Rhode Island, South Carolina, South Dakota, Tennessee, Texas, Utah, Virginia and Washington) (Christensen, 1970; Long, 1981; Lever, 1987; Sol *et al.*, 2012; Moulton *et al.*, 2015).

16.4 Description

The Chukar Partridge is a medium-sized, rounded species with a body length ranging from 32 to 35 cm. It has a grey breast, a light brown back and a black sharply defined gorget around the neck and on the forehead across the eyes (Fig. 16.2) (Christensen, 1970). The colour of skin around the eyes is coral red and is similar to the beak and feet. The sexes are similar, females with slightly smaller size with unspurred wings (Rasmussen and Anderton, 2005). The Chukar Partridge's tail consists of 14 feathers; the third primary feather is the longest while the first primary feather is at the same level as the fifth and sixth primary feathers (Blanford, 1898). The Chukar Partridge resembles the Rock Partridge, but the latter is dark brown on its back with a yellowish stain on the foreneck (Watson, 1962). The Chukar Partridge is distinguished from the Red-legged Partridge (*Alectoris rufa*) by the black collar that breaks into dark lines adjacent to the breast. The Chukar Partridge has a

Fig. 16.2. An adult Chukar Partridge at Weltvogelpark, Walsrode Bird Park, Germany. (©Photograph: Olaf Oliviero Riemer; licensed under Creative Commons Attribution-Share Alike 3.0 Unported.)

noisy call (chuck-chuck-chukar-chukar) from which its name is derived (Baker, 1928).

16.5 Diet

The Chukar Partridge feeds on a variety of food items, commonly seeds, insects, wild fruits and sometimes sand, which helps with the grinding process (Oates, 1898; Christensen, 1970). In the USA, they feed mostly on the seeds of grasses and weeds, particularly drooping brome (*Bromus tectorum*; Christensen, 1970; Johnsgard, 1973), while those in Pakistan and Kashmir prefer the seeds of *Eragrostis* spp. (Oakleaf and Robertson, 1971). The Desert Chuckar (*A. c. sinaica*) feeds on succulent vegetation to acquire its water needs and to save the energy it would otherwise need to search for water in the desert (Degen *et al.*, 1984). A total of 72 different food items were isolated from the gizzards of Chukar Partridges in Oregon, USA, including cheatgrass, wild onion (*Allium* spp.), seeds, leaves and shoots of grasses, subterranean bulbils of prairie starflower, and plant parts including leaves, buds, stems, capsules and bracts (Christensen, 1970; Walter and Reese, 2003). Arthropods such as ants, centipedes, bugs, beetles (including cocoons), butterflies and moths, caterpillars, grasshoppers, Jerusalem crickets, spiders and gull midges are also consumed by the Chukar Partridge (Christensen, 1970; Walter and Reese, 2003). Other food items reported include small rodents' faeces and wood (Christensen, 1970). The foraging behaviour of Chukar Partridges is mainly during mid-morning, extending through the afternoon (Christensen, 1970). On hot days, they spend most of their time in the shade, under the shrubs or canyons near water supplies (Christensen, 1970; Johnsgard, 1973).

The movement of the Chukar Partridge involves searching for food and water; for example, in the semi-arid Negev Desert, Israel, it was reported to move around farms as large as 20 ha feeding on sprouting sugar beets (Benjamini, 1980). Walter (2002) reported that the Chukar Partridge can fly at a radius of 280 m with a water source and food at the centre. The maximum distances travelled by the Chukar Partridge were recorded in New Mexico, which included 61 km in 7 months, 35 km in a year, and between 61 and 64 km in a year (Johnsgard, 1973). The movement of this species is limited by ambient temperature, whereby they move less during high temperatures to avoid water loss (Degen *et al.*, 1984; Cole *et al.*, 1995; Christensen, 1996; Khan *et al.*, 2017). The extent of movement of Chukar Partridges varies in hot, arid and semi-arid areas that have little water or widely dispersed food containing water such as onion bulbs (*Allium* spp.), bulbous blue grass bulbs (*Poa bulbosa*), and hawksbeard seed heads (*Crepis* spp.) (Christensen, 1970; Larsen *et al.*, 2010).

16.6 Introduction and Invasion Pathways

The Chukar Partridge has been introduced in many parts of world as a game bird (Long, 1981; Pyle and Pyle, 2009). In North America, the first population of Chukar Partridges was collected from Nepal and Afghanistan in 1893, and it was introduced as a game bird, which provided an income to government wildlife agencies through hunting (Gullion and Christensen, 1957; Christensen, 1996; Pyle and Pyle, 2009). Subsequently, the population became feral and occupied inaccessible areas such as barren slopes, rocky areas and extensive thickets (Gullion and Christensen, 1957; Cole *et al.*, 1995; Khan *et al.*, 2017). Since then, the feral population of Chukar Partridges has increased, and it is now reported as one of the worst invasive species in the USA, occupying 90% of the states (Fig. 16.1) (Christensen, 1970; Long, 1981; Lever, 1987; Sol *et al.*, 2012; Moulton *et al.*, 2015).

In other parts of the world, the Chukar Partridge has been introduced as part of the pet trade (Mishra and Fitzherbert, 2004; Mahmood *et al.*, 2011; Abbas *et al.*, 2015). This species is known to be charismatic and one of the most commonly traded caged pet species; for example, in Afghanistan, they are mainly sold as pets although some are used for bird fights (Mishra and Fitzherbert, 2004). In India, a number of pet store owners were reported to have smuggled Chukar and European or Grey Partridges without possession of licences (Mahmood *et al.*, 2011), while in Pakistan, the Chukar Partridge and the Ram Chukar or Snow Partridge (*Lerwa lerwa*) are also sold in large quantities (Abbas *et al.*, 2015). In California, USA, caged Chukar Partridges were released, or some escaped captivity and became feral (Hardy, 1973).

Feral populations of Chukar Partridge have established in parts of Argentina, Canada, California, Chile, Hawaii and New Zealand (Hardy, 1973; Long, 1981). In South Africa, they were introduced to Robben Island in 1964, but the numbers have been reported to be dropping (Hockey *et al.*, 2005; Picker and Griffiths, 2013). Dyer (1996) reported that several snakes on Robben Island prey on birds including the Chukar Partridge. In New South Wales, Australia, a breeding population was introduced but failed to persist and became extinct (Christidis and Boles, 2008). In Europe, this species was introduced in many areas to restock the level of partridge game such as the Red-legged Partridge (Blanco-Aguiar *et al.*, 2008).

16.7 Breeding Behaviour

The Chukar Partridge is regarded as a monogamous bird (Christensen, 1970; Mahmood *et al.*, 2019). It frequently makes a noisy call (chuck-chuck-chukar-chukar) with the frequency of the call ranging from 0.15 to 0.3 per min as a sign of courtship (Mahmood *et al.*, 2019). Once paired, the male Chukar Partridge exhibits aggressive behaviour and fights any male that enters its territory (Mackie and Buechner, 1963). The Chukar Partridge has been reported to breed in early February or mid-March in Nevada, USA (Christensen, 1970), while in its native range in Pakistan (e.g. Lower Dir District, Khyber Pakhtunkhwa), breeding occurs between February and March, with a breeding peak in April (Mahmood *et al.*, 2019). In Bulgaria, it breeds between March and June (Gruychev, 2016). In Pakistan, it has been reported that if the nest of the Chukar Partridge is destroyed or the eggs do not hatch, the bird will re-nest if climatic conditions permit (Biddulph, 1881). The nest resembles a dust bowl, made of dry grasses, stems and feathers, and is built on the ground, mainly under small trees or tall grasses such as *Atriplex* spp., with clutch sizes of between nine and 16 (Fig. 16.3) (Christensen, 1970; Hemonides, 2013; Mahmood *et al.*, 2019). The mean range rate of laying eggs is 1.3 eggs/day (Mackie and Buechner, 1963). The eggs are uniform and oblong in shape, yellow-white in colour with numerous speckles and brown dots, with the weight ranging from 16 to 25 g, and an average length and width of 42 mm and 32 mm, respectively (Mackie and Buechner, 1963; Hemonides, 2013). Only the female Chukar Partridge incubates the eggs for between 21 and 24 days (Mackie and Buechner, 1963; Hemonides, 2013). The male stays away from the nest but rejoins the family when the eggs are hatched (Goodwin, 1953; Mahmood *et al.*, 2019). After the eggs hatch, both adults take care of the brood until they are fully developed (Christensen, 1970). The young chicks feed on insects and other small animal parts, and adults consume more than 15% of the food by volume compared with the chicks (Johnsgard, 1973). The hatching success is generally high (75–85%), depending on the different nests and geographical area (Mahmood *et al.*, 2019).

Fig. 16.3. A typical active nest of Chukar Partridge with a clutch size of eight in the Lower Dir District, Khyber Pakhtunkhwa, Pakistan. (©Photograph: from Mahmood *et al.*, 2019.)

There are several factors such as food availability, photoperiod, temperature and water availability that determine whether breeding will occur or not (Christensen, 1970). In addition, climate change, including abnormally heavy rainfall and drought, can directly affect the success of the hatch and may also result in nest loss (Christensen, 1970).

16.8 Habitat

The Chukar Partridge generally prefers mountainous areas with barren slopes, rocky areas with concentrated thickets and shrubs, or dry and semi-arid environments with relatively low rainfall and some dense vegetation that is used mainly for roosting (Gullion and Christensen, 1957; Christensen, 1970; Cole *et al.*, 1995; Hemonides, 2013; Khan *et al.*, 2017; Mahmood *et al.*, 2019). In its native range, the habitat of the Chukar Partridge includes open areas with scattered shrubs, cultivated land such as agricultural areas, semi-desert and mountainous rocky areas with grasses (Young, 1981; Rawat, 1998; Khan *et al.*, 2017). In Israel, Jordan and the Himalayas, this species was found in areas of low altitude as low as 400 m above sea level (a.s.l.), whereas in more eastern areas it is found mainly at altitudes between 2000–4000 m a.s.l., except for Pakistan where they occur at altitudes of 600 m a.s.l. (Whistler, 1949; Rasmussen and Anderton, 2005). Stuart (1922) reported that this species is not found in areas with extreme humidity and rainfall. It can survive in areas with arid to semi-arid climates where water is generally available and ambient temperatures vary (Cole *et al.*, 1995; Christensen, 1996; Khan *et al.*, 2017).

16.9 Impacts

The introduction of the Chukar Partridge has been reported to negatively affect the genetic diversity of native Red-legged Partridges and Rock Partridges through hybridization (Barbanera *et al.*, 2005; Barilani, 2007; Blanco-Aguiar *et al.*, 2008). The Chukar Partridge is aggressive, and Harper *et al.* (1958) observed an adult killing a Gambel's Quail (*Callipepla gambelii*) at a pond in California, USA. In North America, the Chukar Partridge has been reported to aid in the dispersal and germination of invasive cheatgrass (*Bromus tectorum*) (Walter and Reese, 2003). Chukar Partridges are important agricultural pests and have been reported to cause damage to cereals, tomatoes, prickly pears, apples, potatoes and sunflowers (Christensen 1970; Moran and Keidar, 1993). This bird is also associated with avian diseases such as *Chlamydia* spp. that can be passed to humans (Erbeck and Nunn, 1999). Chukar Partridges also carry influenza A virus, which can be transmitted to humans and other mammals (Humberd *et al.*, 2006). In addition, viruses that can cause diseases to humans such as eastern viral encephalomyelitis and West Nile virus, and other parasites have been isolated from this species (Moulthrop and Gordy, 1960; Sychra, 2005; Wünschmann and Ziegler, 2006; Hemonides, 2013). West Nile virus could pose risk to humans and other domestic animals such as the Himalayan monal (*Lophophorus impeyanus*; Wünschmann and Ziegler, 2006). On the Island of

Maui, Hawaii, Chukar Partridges have been reported to play an important role in the ecosystem by aiding seed dispersal and germination of native species in degraded ecosystems (Cole *et al.*, 1995).

16.10 Control

Not much is known about the control of the Chukar Partridge. So far, only a biological method has been applied, where genetic screening was used to carefully identify local populations of Chukar Partridge that are more genetically pure than others (Barbanera *et al.*, 2007).

16.11 Uses

The Chukar Partridge is one of the popular game birds and is a valued pet bird in the Middle East (Mobini, 2012). It can be relatively easily bred in captivity, generally with high rates of productivity (Cowell, 2008; Scott *et al.*, 2009; Yilmaz and Tepeli, 2009). During the breeding season, these birds can be aggressive, and as a result of this behaviour, people usually keep them as fighting birds (Stuart, 1922). Humans also keep this species for meat and egg production (Ozek, 2006; Pourghanbari *et al.*, 2016).

16.12 Notes

In the Mediterranean and Middle East regions, a female Chukar Partridge is regarded as a symbol of passion and motherliness, while a male is considered a symbol of intelligence, courage and evil (Al-Barwari and Saeed, 2012). This bird is the national bird of Pakistan, and people in Punjab, India, consider it a symbol of intense and unrequited love (Ali *et al.*, 2017). The Chukar Partridge is said to always look at the moon, and for this reason, it is considered to be in love with the moon (Temple, 1884).

16.13 References

Abbas, F.I., Rooney, T.P., Mian, A., Bhatti, Z.I. and Haider, J. (2015) The distribution, population status, and wildlife product trade of Himalayan Musk Deer in Gilgit-Baltistan, Pakistan. *Journal of Bioresource Management* 2, 38–47.

Al-Barwari, S. and Saeed, I. (2012) Parasitosis of the Chukar Partridge, *Alectoris chukar* in North Iraq. *Turkey Journal of Parasitology* 36, 240–246.

Ali, H., Rehman, H.U. and Usman, K. (2017) Distribution, population, and habitat utilization of Alectoris chukar in district Malakand Khyber Pakhtunkhwa, Pakistan. *Journal of Entomology and Zoology Studies* 5, 964–966.

Baker, E.S. (1928) Fauna of British India-birds. *Indian Forester* 54, 183–184.

Barbanera, F., Negro, J. J., Di Giuseppe, G., Bertoncini, F., Cappelli, F. and Dini, F. (2005) Analysis of the genetic structure of red-legged partridge (*Alectoris rufa*, Galliformes) populations by means of mitochondrial DNA and RAPD markers: a study from central Italy. *Biological Conservation* 122, 275–287.

Barbanera, F., Guerrini, M., Hadjigerou, P., Panayides, P., Sokos, C., *et al.* (2007) Genetic insight into Mediterranean chukar (*Alectoris chukar*, Galliformes) populations inferred from mitochondrial DNA and RAPD markers. *Genetica* 131, 287–298.

Barilani, M., Bernard-Laurent, A., Mucci, N., Tabarroni, C., Kark, S., *et al.* (2007) Hybridisation with introduced chukars (*Alectoris chukar*) threatens the gene pool integrity of native rock (*A. graeca*) and red-legged (*A. rufa*) partridge populations. *Biological Conservation* 137, 57–69.

Benjamini, L. (1980) Feeding behavior of the chukar (*Alectoris chukar*) in sugar beet plantations. *Phytoparasitica* 8, 3–18.

Biddulph, J. (1881) The birds of Gilgit. *Stray Feathers* 9, 301–366.

Blanco-Aguiar, J.A., González-Jara, P., Ferrero, M.E., Sánchez-Barbudo, I., Virgós, E., *et al.* (2008) Assessment of game restocking contributions to anthropogenic hybridization: the case of the Iberian red-legged partridge. *Animal Conservation* 11, 535–545.

Blanford, W.T. (ed.) (1898) *The Fauna of British India: Including Ceylon and Burma*. Taylor & Francis, London.

Christensen, C. (1996) Chukar: *Alectoris chukar*. In: Poole, A and Gill, F, (eds) *The Birds of North America*. Academy of Natural Sciences of Philadelphia, Philadelphia, Pennsylvania, pp. 1–20.

Christensen, G.C. (1970) The Chukar Partridge: its introduction, life history, and management. Biological Bulletin No. 4, Nevada Department of Wildlife Reno, Nevada.

Christidis, L. and Boles, W.E. (2008) *Systematics and Taxonomy of Australian Birds*. CSIRO Publishing. Collingwood, Australia.

Cole, F., Loope, L., Medeiros, A., Raikes, J. and Wood, C. (1995) Conservation implications of introduced game birds in high elevation Hawaiian shrub land. *Conservation Biology* 9, 306–313.

Cowell, D. (2008) Chukar partridge (*Alectoris chukar*). Game Bird and Water Fowl website. Available at: www.gbwf.org/francolin/chukar1.php (accessed 27 October 2019).

Degen, A.A., Pinshow, B. and Shaw, P.J. (1984) Must desert Chukars (*Alectoris chukar sinaica*) drink water? Water influx and body mass changes in response to dietary water content. *Auk* 101, 47–52.

Dyer, B.M. (1996) Predation by snakes on seabirds at three South African islands. *South African Journal of Marine Science* 17, 309–313.

Erbeck, D.H. and Nunn, S.A. (1999) Chlamydiosis in pen-raised bobwhite quail (*Colinus virginianus*) and chukar partridge (*Alectoris chukar*) with high mortality. *Avian Diseases* 43,798–803.

Goodwin, D. (1953) Observations on voice and behaviour of the red-legged partridge (*Alectoris rufa*). *Ibis* 95, 581–614.

Gruychev, G., (2016) Declining populations of Chukar Partridge (*Alectoris chukar*) in Bulgaria. *Turkish Journal of Zoology* 40, 818–823.

Gullion, G.W. and Christensen, G.C. (1957) A review of the distribution of gallinaceous game birds in Nevada. *Condor* 59, 128–138.

Hardy, J.W. (1973) Feral exotic birds in southern California. *Wilson Bulletin* 85, 506–512.

Harper, H.T, Beverly, H.H. and William, D.B. (1958) The chukar partridge in California. *California Fish and Game* 44, 5–50.

Hartert, E. (1925) A new form of chukar partridge (*Alectoris graeca kleini*). *Novitates Zoologicae* 32, 137.

Hemonides, E. (2013) Breeding Chukar Partridges (*Alectoris chukar*) in Cyprus. Thesis, Department of Animal Breeding, University of Veterinary Medicine Budapest, Budapest. Available at: www.huveta.hu/bitstream/handle/10832/958/HemonidesThesis.pdf;jsessionid=A916A9CAE1686E29FE-9229B54A9131B4?sequence=1 (accessed 23 May 2019).

Hockey, P.A.R., Dean, W.R.J. and Ryan, P.G. (2005) *Roberts Birds of Southern Africa*, 7th edn. Trustees of the John Voelker Bird Book Fund, Cape Town.

Humberd, J., Guan, Y. and Webster, R.G. (2006) Comparison of the replication of influenza A virus in Chinese ring-necked pheasants and chukar partridges. *Journal of Virology* 80, 2151–2161.

Johnsgard, P.A. (1973) Chukar Partridge. *Ornithology Commons* 32, 489–501.

Kark, S. (2001) Shifts in bilateral asymmetry within a distribution range: the case of the chukar partridge. *Evolution* 55, 2088–2096.

Khan, Y.Q., Ahmad, M., Naeem, M., Khan, M., Ali, H., *et al.* (2017) Distribution, population, and habitat utilization of *Alectoris chukar* in district Malakand Khyber Pakhtunkhwa, Pakistan. *Journal of Entomology and Zoology Studies* 5, 964–966.

Larsen, R.T., Bissonette, J.A., Flinders, J.T., Hooten, M.B. and Wilson, T.L. (2010) Summer spatial patterning of chukars in relation to free water in western Utah. *Landscape Ecology* 25, 135–145.

Lever, C. (1987) *Naturalized Birds of the World*. Longman Scientific and Technical Publisher, Harlow, UK.

Long, J.L. (1981) *Introduced Birds of the World: The Worldwide History, Distribution and Influence of Birds Introduced to New Environments*. Agricultural Protection Board of Western Australia, Terrey Hills, Australia.

Mackie, R.J. and Buechner, H.K. (1963) The reproductive cycle of the chukar. *Journal of Wildlife Management* 27, 246–260.

Madge, S. and McGowan, P. (2002) *Pheasants, Partridges and Grouse*. A. & C. Black, London.

Mahmood, T., Shah, S.M.A., Rais, M. and Nadeem, M.S. (2011) An investigation of animal species trade at pet shops of Rawalpindi and Multan cities. *Journal of Animal and Plant Science* 21, 822–929.

Mahmood, T., Ahmad, I., Akrim, F., Hamid, A., Waseem, M., *et al.* (2019) Breeding Ecology of Chukor Partridge (*Alectoris chukar*) in Lower Dir District, Khyber Pakhtunkhwa, Pakistan. *Pakistan Journal of Zoology* 51, 265–271.

McGowan, P.J.K. and Kirwan, G.M. (2019) Chukar (*Alectoris chukar*). In: del Hoyo, J., Elliott, A., Sargatal, J., Christie, D.A. and de Juana, E. (eds) *Handbook of the Birds of the World Alive*. Lynx Editions, Barcelona, Spain.

Mishra, C. and Fitzherbert, A. (2004) War and wildlife: a post-conflict assessment of Afghanistan's Wakhan Corridor. *Oryx* 38, 102–105.

Mobini, B. (2012) Microscopic study of the gall bladder of the chukar partridge (*Alectoris chukar*). *Bulgarian Journal of Veterinary Medicine* 15, 73–78.

Moran, S. and Keidar, H. (1993) Checklist of vertebrate damage to agriculture in Israel. *Crop Protection* 12, 173–182.

Moulthrop, I.M. and Gordy, B.A. (1960) Eastern viral encephalomyelitis in chukar (*Alectoris graeca*). *Avian Diseases* 4, 380–383.

Moulton, M.P., Cropper, W.P. Jr and Broz, A.J. (2015) Inconsistencies among secondary sources of Chukar Partridge (*Alectoris chukar*) introductions to the United States. *PeerJ* 3, e1447.

Oakleaf, R.J. and Robertson, J.H. (1971) Fall food items utilized by Chukars in Kashmir, India. *Journal of Wildlife Management* 35, 395–397.

Oates, E.W. (1898) The red-legged Partridges. In: *A Manual of the Game Birds of India*, Part 1: *Land Birds*. A.J. Combridge, Mumbai, India, pp. 179–183.

Ozek, K. (2006) The optimum protein content in high-energy starter diet for Chukar Partridge (*Alertons chukar chukat*). *International Journal of Poultry Science* 5, 522–525.

Picker, M. and Griffiths, C. (2013) *Alien and Invasive Animals: A South African Perspective*. Penguin Random House, Cape Town, South Africa.

Pourghanbari, G.H., Nili, H., Habibi, H., Morovati, M., Salehi, E. and Sadoughifar, R. (2016) Response of chukar partridge performance and blood parameters to different dietary crude protein. *Advances in Bioresearch* 7, 162–169.

Pyle, R.L. and Pyle, P. (2009) *The Birds of the Hawaiian Islands: Occurrence, History, Distribution, and Status*. B.P. Bishop Museum, Honolulu, Hawaii.

Rasmussen, P.C. and Anderton, J.C. (2005) *Birds of South Asia: The Ripley Guide*, Vol. 2. Smithsonian Institution, Washington, DC, and Lynx Ediciones, Barcelona, Spain.

Rawat, G.S. (1998) Temperate and alpine grasslands of the Himalaya: ecology and conservation. *Parks* 8, 27–36.

Scott, P., Turner, A., Bibby, S. and Chamings, A. (2009) Structure and dynamics of Australia's commercial poultry and ratite industries. Department of Agriculture, Fisheries and Forestry. Moonee Ponds, Australia.

Sol, D., Maspons, J., Vall-Llosera, M., Bartomeus, I., García-Peña, G.E., *et al.* (2012) Unraveling the life history of successful invaders. *Science* 337, 580–583.

Stuart, B.E.C. (1922) The game birds of India, Burma and Ceylon, part 31. *Journal of the Bombay Natural History Society* 28, 306–312.

Sychra, O. (2005) Chewing lice (Phthiraptera: Amblycera, Ischnocera) from chukars (*Alectoris chukar*) from a pheasant farm in Jinacovice (Czech Republic). *Veterinary Medical Journal of Czech* 50, 213–218.

Temple, R.C. (1884) *The Legends of the Panjâb*, Vol. 2. Education Society's Press, Mumbai, India.

Walter, H. (2002) Natural history and ecology of the chukar (*Alectoris chukar*) in the northern Great Basin. *Great Basin Birds* 5, 28–37.

Walter, H. and Reese, K.P. (2003) Fall diet of chukars (*Alectoris chukar*) in eastern Oregon and discovery of ingested lead pellets. *Western North American Naturalist* 63, 402–405.

Watson, G.E. (1962) Three sibling species of *Alectoris* Partridge. *Ibis* 104, 353–367.

Whistler, H. (1949) Alectoris Gracea (Meisner). Chkor. In: *Popular Handbook of Indian Birds*, 4th edn. Gurney and Jackson, London. pp. 428–430.

Wünschmann, A. and Ziegler, A. (2006) West Nile virus-associated mortality events in domestic Chukar partridges (*Alectoris chukar*) and domestic Impeyan pheasants (*Lophophorus impeyanus*). *Avian Diseases* 50, 456–459.

Yilmaz, A. and Tepeli, C. (2009) Breeding performance of a captive Chukar partridge (*Alectoris chukar*) flock. *Journal of Animal and Veterinary Advances* 8, 1584–1588.

Young, J.A. (1981) Chukar partridge. *Rangelands* 3, 166–168.

17 Cattle Egret (*Bubulcus ibis* Linnaeus, 1758)

Jen Cele and Colleen T. Downs*

Centre for Excellence in Invasion Biology, School of Life Sciences, University of KwaZulu-Natal, Pietermaritzburg 3209, South Africa

Citation: Cele, J. and Downs, C.T. (2020). Cattle Egret (*Bubulcus ibis* Linnaeus, 1758) In: Downs, C.T. and Hart, L.A. (eds) *Invasive Birds: Global Trends and Impacts*. CAB International, Wallingford, UK, pp. 138–143.

17.1 Common Names

Cattle Egret, Western Cattle Egret, Eastern Cattle Egret, Buff-Backed Heron, Cow Crane, Cow Bird, Cow Heron, Elephant Bird, Tick Bird, Rhinoceros Egret, Hippopotamus Egret, Ilanda (Zulu/Xhosa), Koereiger (Dutch), Kuhreiher (German), Heron Garde-Boeuf (French), Kohager (Swedish), Depsulgabuey or Garrapatosa Garza De Ganado, Garza De Vaquera, Garcita de Ganado, Garcilla Garrapatera (Spanish), Busluisvoel (Afrikaans), Abu Qerdan (Arabic).

17.2 Nomenclature

Although debated for many years by taxonomists, Cattle Egrets (*Bubulcus ibis* Linnaeus, 1758) are widely accepted today as members of the family Ardeidae. They were originally described as *Ardea ibis* (Linnaeus, 1758), erroneously classified as an *ibis* species due to similarities with the African Sacred Ibis (*Threskiornis aethiopicus*) and were later classified in the genus *Bubulcus* by Bonaparte in 1855 because of their association with cattle (Jobling, 2010; Chittenden *et al.*, 2016). Three subspecies are recognized: *B. i. coromandus*, which occurs in Asia and Oceania; *B. i. seychellarum*, which is present in the Seychelles, and *B. i. ibis*, which is the most widely distributed occurring in Africa, southern Europe, west Asia and America (Ahmed, 2011; Congrains *et al.*, 2016).

17.3 Distribution

Cattle Egrets are cosmopolitan birds with a wide global distribution (Fig. 17.1). This is because of their successful breeding and adaptive nature (Moralez-Silva and Del Lama, 2014; Congrains *et al.*, 2016; Toloa *et al.*, 2017). They have undergone a dramatic range expansion since the late 1800s, spreading throughout Africa into southern Europe and beyond (Louw, 2005). Cattle Egrets are considered native to Africa, southern Europe (Portugal and Spain), the Asian tropics (a southward range from Japan to India) and northern Australia (Arendt, 1988; Moralez-Silva and Del Lama, 2014). Their current range spans continents because of decades of intra and intercontinental migrations (Arendt, 1988). Population estimates are at 1 million individuals and thus they are a species of least concern (BirdLife International, 2016).

As early as the late 1800s, sightings were reported in the New World, thought to be accidental visitors. However, a specimen in South America was formally recorded and verified years later. Emmet Blake, near Buxton, East Coast, British Guiana, found this singular bird in 1937 (Crosby, 1972; Telfair, 1983). Not long after, specimens were collected in Surinam and Venezuela, supporting the theory of a migration rather than an introduction by explorers of a captive bird (Crosby, 1972; Moralez-Silva and Del Lama, 2014). This was further supported by sightings of entire flocks, up to 105 individuals in Surinam in 1946 (Crosby, 1972). They were considered migrants passing through the region until the discovery of nesting colonies in Columbia, along with year-long sightings by 1958.

Sightings were made in Florida, USA, during the late 1940s but were not reported. In the spring of 1952 at Lake Okeechobee, Florida, a Cattle Egret was photographed and misidentified as a Snowy Egret (*Egretta thula*), while in that same season a specimen was caught in Wayland, Massachusetts

*Corresponding author: downs@ukzn.ac.za

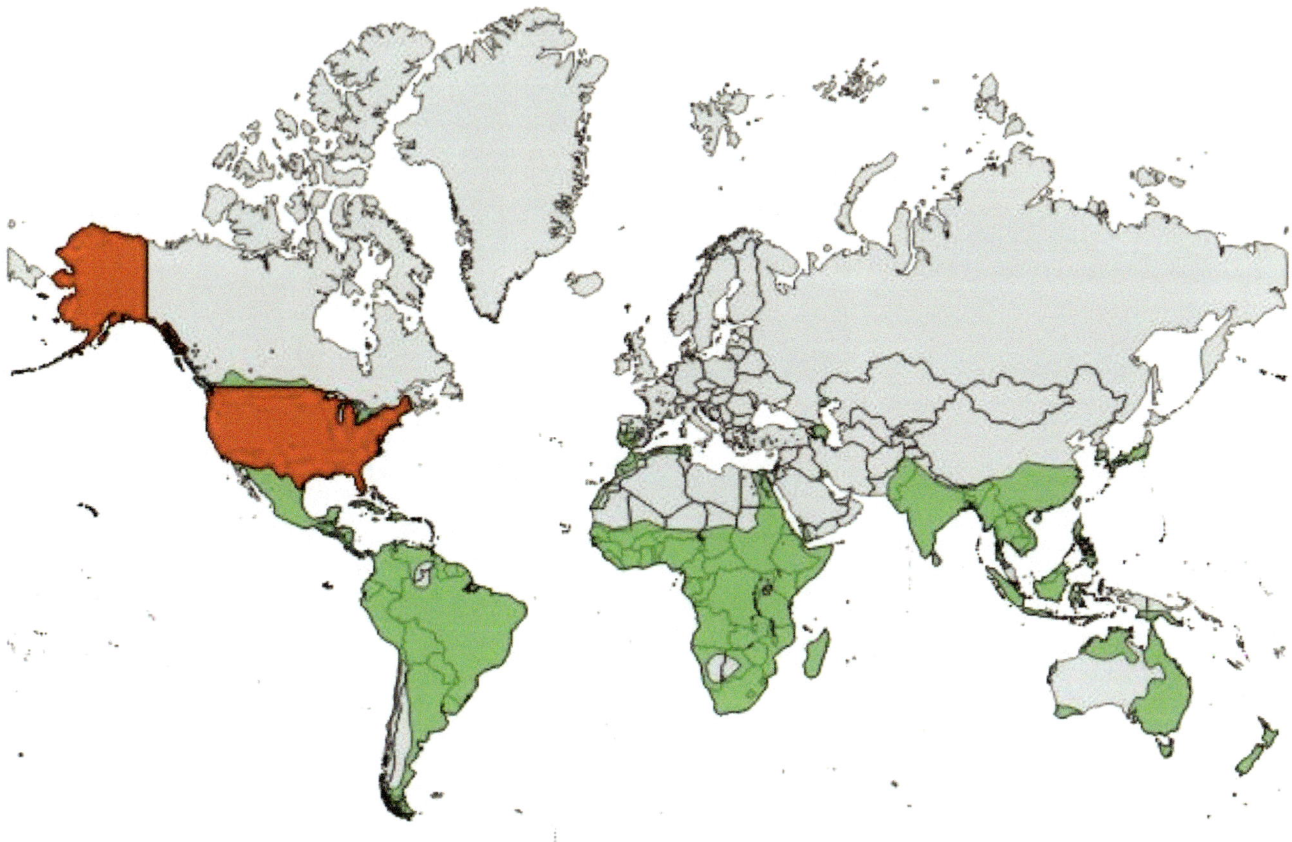

Fig. 17.1. Global distribution of the Cattle Egret (*Bubulcus ibis*) showing the natural (green) and invaded (red) ranges.

(Crosby, 1972). Their distribution along North America remained coastal in the early years of invasion into the USA (Crosby, 1972; Miller and Ryder, 1978; Telfair, 1983). By 1962, there was evidence of breeding in Missouri, in the mid-west. In 1977, nesting pairs were discovered as far inland as Pelican Island, Colorado (Miller and Ryder, 1978; Telfair, 1983).

17.4 Description

Cattle Egrets have little to no sexual dimorphism (Louw, 2005). They are medium-sized birds (46–56 cm in length) with a wingspan of up to 88–96 cm and heron-like characteristics (Fig. 17.2B). Adults weigh 360 g on average (340–390 g), but males can reach 512 g (Scott, 1987; Telfair, 2006). While most herons have long elongated necks, the Cattle Egret's is short and thick. They have a long sharp, yellow bill, which is orange in breeding adults. Their lore and iris are also yellow.

Cattle Egret chicks have white down feathers, a bushy crown and dark olive-green legs and a yellow bill, which curves downwards. Fledglings exhibit black legs and bill, becoming yellow in juveniles. Juveniles are nearly indistinguishable from adults except over breeding periods (Fig. 17.2A). Sexual maturity is reached as early as 2 years of age when their legs turn from a greyish-yellow to bright red during the breeding season. This is also when their typically white plumage becomes rufous around the breast, back and crown (Fig. 17.2C) (Kaufman, 1996; Telfair 2006).

Males at times have darker breeding plumage than females. Each subspecies experiences a slightly different extent of breeding plumage. The breeding plumage of *B. i. coromandus* is thick in texture, short and expresses a richer rust coloration covering the throat, which is widespread along the neck and expressed in the ear coverts. *B. i. ibis* has longer wispier breeding plumage with a buff pigmentation around the crown, chest, back and nape (Ahmed, 2011).

17.5 Diet

Although Cattle Egrets require proximity to a body of water for roosting and nesting, they mostly forage in terrestrial habitats (Congrains *et al.*, 2016). They are closely associated with mega-herbivores (Fig. 17.2D) (Rice, 1963; Siegfried, 1971a,b; Browder, 1973; Thompson *et al.*, 1982; Kopij, 1999; Telfair, 2006; Toloa *et al.*, 2017). In the presence of the African buffalo (*Syncerus caffer*), cattle (*Bos* spp.), zebra (*Equus quagga*) and even lawn mowers (Fig. 17.2E), their diet is mostly insectivorous consisting of Annelida (especially earthworms). Orthoptera (grasshoppers), and Lepidoptera (caterpillars) also are import in the diet (Siegfried, 1971a, b; Browder, 1973; Kopij, 1999). Their preference for Diptera (flies) is especially important for

Fig. 17.2. (A) A juvenile Cattle Egret. (B) A non-breeding adult. (C) Breeding adults with rufous plumes. (D, E) Foraging in association with large mammals (D) and on lawns (E). (F) A stick nest. (G) Large interspecies roosting flocks on reeds and fallen trees at the water's edge, with nesting birds in trees above. All images taken in South Africa. (©Photographs: L. Hart (A–C, E–G; H. Jordaan (D).)

fish farmers in India (Seedikkoya and Azeez, 2009) and cattle ranchers around the world (Siegfried, 1971a, b; Thompson *et al.*, 1982; Kopij, 1999). This commensal strategy allows low-energy expenditure while maximizing energy intake.

In the dry seasons, Cattle Egrets forage in the absence of mega-herbivores. During this time, they are opportunistic feeders with a broad diet (Blaker, 1967; Siegfried, 1971a, b; Browder, 1973; Kopij, 1999). Small birds, reptiles, molluscs, rodents and spiders have all been recorded as part of their diet (Siegfried, 1971a, b; Kopij, 1999).

In urban environments, they have been observed feeding on bush crickets under street lights (Toloa *et al.*, 2017), and often fly distances from roosts to forage in municipal refuse landfills (Seedikkoya *et al.*, 2007; Kuranchie *et al.*, 2013; Toloa *et al.*, 2017). They thrive in habitats like landfills or municipal solid-waste dumps. These habitats become suitable as they are covered with prey ranging from maggots to insects and smaller birds (Seedikkoya *et al.*, 2007; Kuranchie *et al.*, 2013).

17.6 Introduction and Invasion Pathways

The Cattle Egret is considered an invasive species by the International Union for Conservation of Nature's Invasive Species Specialist Group (Moralez-Silva and Del Lama, 2014).

Cattle Egrets have successfully invaded many regions of the world due to their highly adaptive nature and as a result of undergoing both intra- and intercontinental migrations. As migratory birds (Browder, 1973), they undergo long-distance migrations in a flock of other heronry birds (Arendt, 1988). Given their high adaptability, Cattle Egrets have successfully formed colonies in new territories throughout North America (Congrains *et al.*, 2016). The Cattle Egret is an invasive alien in the Galapagos Islands (Phillips *et al.*, 2012; Moralez-Silva and Del Lama, 2014) and in Hawaii (Stone and Anderson, 1988; Moralez-Silva and Del Lama, 2014). They have been dispersing since the 1940s from the Malay Archipelago to New Guinea and Australia. Covering approximately 1600 km annually, Australia's migratory Cattle Egrets flock to New Zealand (Ahmed, 2011).

Believed to have originated in eastern Africa, the Cattle Egret is now widely dispersed throughout the African continent (Arendt, 1988). Ringing data reveal that Cattle Egrets can undertake a flight distance of 3750 km from their natal site. A recovery was made in northern Central African Republic of a bird ringed in Johannesburg (Kopij, 2017), an interesting find as South African Cattle Egrets are thought to have originally migrated from Central Africa.

It is thought that Cattle Egrets migrated from Africa to South America, invading North America in an upward migration along the east coast and then across the Caribbean (Crosby, 1972; Browder, 1973). While some argue for a direct migration from Africa to North America, others find this unlikely due to the prevailing winds, despite a sighting aboard a trawler at sea, over 480,000 km off the coast of Newfoundland, Canada (Crosby, 1972). Based on meteorological projections for an 1877 discovery in Suriname and Guyana, it is possible that these early crossings originated from Senegal using tailwinds. Possible flight trajectories were simulated, proposing a 132 h direct flight from the shores of West Africa (Massa *et al.*, 2014).

Years of ringing Cattle Egrets have enabled birders to track their movements. Australian *B. i. comorandus* has been confirmed to migrate south to New Zealand in winter. A bird tagged in the coastal town of Ballina, Australia, was recorded near Rangiriri, New Zealand, in the winter of 1990. In April 1991, a bird from Shortland, Australia, was recovered at Tauroa Point in Northland, New Zealand, while another ringed in Shortland wintered in Rangiriri, New Zealand. It is reported that migrants are restricted to a localized range returning to or near the same roost each year (Maddock and Geering, 1994).

17.7 Breeding Behaviour

The Cattle Egret is a highly adaptable wetland-associated species (Arendt, 1988; Goutner *et al.*, 1991; Kopij, 2008). Cattle Egrets roost in mixed colonies with either terrestrial or aquatic birds (Louw, 2005; Kopij, 2008). It is thought that Cattle Egrets have higher rates of reproductive success when nesting in mixed colonies (Arendt, 1988; Belzer and Lombardi, 1989; Toloa *et al.*, 2017). Nesting typically occurs in a tree or bush overlooking water (Fig. 17.2G). Wetland habitats such as mangroves, reed beds, thicket and marshes are ideal for nesting colonies (Arendt, 1988).

In Uttar Pradesh, India, mesquite bushes (*Prosopis juliflora*) were found to host colonies of up to 436 Cattle Egret nests (Dwevedi *et al.*, 2015). Unlike other findings, this site is not in proximity to water and features a monospecific colony (Dwevedi *et al.*, 2015). In Kampala, Uganda, although approximately 13,000 individual Cattle Egrets were found roosting throughout the city, only 1500 nests were recorded for 2014 (Toloa *et al.*, 2017). These nesting birds are thought to either form subpopulations or breed twice a year. A bimodal breeding pattern was recorded, which coincides with local rainfall patterns. With the rainfall comes the arrival of grasshoppers and bush crickets (thought to be attracted by the city lights), thus creating a localized food source (Toloa *et al.*, 2017).

When ready to attract a mate, Cattle Egret males are very vocal (Blaker, 1967; McKilligan, 1990). They display together while establishing their territories. Display behaviours include neck extensions, partially raised plumes and an open bill. In the courtship of females, males can be heard giving a 'raa' call. Males display circle flights and then wingspread. If a male is yet to attract a female, aggression is increased and the displays become physical. Cattle Egrets have been observed undertaking aerial fights and bill jabbing. Due to the male's aggression, females encroach from behind. After a few hours of courtship, a 'thonk' call is heard as pairs are formed, along with back biting and bill clappering behaviours. This can last for up to 4 days (Blaker, 1967). Males that have been successful at forming a pair become territorial of their nests and chosen mate for the season (McKilligan, 1990).

Cattle Egret nests are generally an untidy platform of mainly dry sticks with some stems of weeds and/or reeds and occasionally lined with grass (Fig. 17.2F) (Blaker, 1967; Louw, 2005). Both Cattle Egret parents share responsibilities for incubating the eggs, as well as for feeding and guarding the chicks. Eggs are laid 2 days apart, followed by a 21–26-day incubation period (Blaker, 1967). Clutch sizes are typically three eggs but range from one to five, accounting for their relatively high breeding success (Kopij, 1999; Hilaluddin *et al.* 2003). If nesting fails, Cattle Egrets will reattempt nesting if the breeding season has not ended. Hatching 1–2 days apart, chicks become vocal and aggressive with age. Lack of food resources can cause sibling aggression, with the first two hatchlings having the highest chance of survival. This is mostly experienced in their native range, and this is probably the reason for asynchronistic hatching. Chicks are fully feathered within 21 days, allowing them the mobility to hop in and out of their nest. By day 30, they are independent fledglings (Blaker, 1967; Kopij, 1999).

Reproductive success is higher in the New World given the abundance of food provided by anthropogenic landscapes (Telfair, 1983). Third chicks have a 98% survival rate and fourth chicks a 97% chance of survival to fledging. Feeding restrictions are limited only by parental ability to deliver food to chicks rather than by availability (Telfair, 1983).

17.8 Habitat

Cattle Egret nesting and roosting sites are generally near water (Fig. 17.2G), but there are exceptions (Krebs *et al.*, 1994;

Parkes *et al*., 2012; Dwevedi *et al*., 2015), especially in urban areas (Louw, 2005; J. Cele and C.T. Downs, unpublished data). Foraging occurs in shallow marshes and at the edges of streams and rivers alongside other wading birds (Siegfried, 1971a,b). However, generally wetland habitats offer limited foraging potential (Krebs *et al*., 1994). Therefore, they naturally occur in grassland habitats in periods of high insect abundance where they forage, obtaining a protein-rich diet in the presence of large grazers or foraging unassisted in inundated plains (Fig. 17.2D) (Kopij, 1999). Recent conversions of forests to pastureland for cattle ranching and for cultivation has aided in the expansion of suitable foraging habitats (Arendt, 1988; Telfair, 2006; Ahmed, 2011). Anthropogenic foraging habitats include urban parks, wastewater treatment facilities, organic waste refuse dumps or landfill sites, golf courses, pastures and croplands (Louw, 2005; Toloa *et al*., 2017; J. Cele and C.T. Downs, unpublished data).

17.9 Impacts

American studies have noted competition among Cattle Egrets and native herons but not for food, as they differ in diet (Burger, 1978; Arendt, 1988). The aggressive Cattle Egret generally outcompetes for nesting space (Burger, 1978; Dami *et al*., 2006), as has been observed in Camargue, France, with Little Egrets (*Egretta garzetta*; Dami *et al*., 2006). Cattle Egrets have been observed chasing young Snowy Egrets in New Jersey, USA, resulting in higher reproductive success (Burger, 1978). They have been observed eating the chicks of native aquatic birds in Hawaii and the Fernando de Noronha archipelago, along with the endemic Noronha skink (*Euprepis atlanticus*; Moralez-Silva and Del Lama, 2014).

In urban areas, Cattle Egrets are regarded as a nuisance species (Parkes *et al*., 2012). Although quiet at foraging sites, they can be very noisy at roosts, and messy (J. Cele and C.T. Downs, unpublished data). Roosting trees near businesses or residences are subject to removal (Parkes *et al*., 2012). This view is not a globally shared one, as in some countries they act as a biocontrol. For example, in Kampala, Uganda, there are influxes of grasshoppers coinciding with rainfall. These grasshoppers often feed on cereal crops, a stable agricultural product. Additionally, without predation, grasshoppers and bush crickets would overfeed on grasses grazed by livestock. The presence of Cattle Egrets, which feed on grasshoppers, is welcomed by farmers who would otherwise lose revenue (Toloa *et al*., 2017).

In Indian fishing villages and organic waste refuse dump sites, Cattle Egrets can be found in large numbers feeding on maggots including the house fly (*Musca domestica*) and bluebottle fly (*Calliphora* spp.) (Seedikkoya *et al*., 2007; Seedikkoya and Azeez, 2009). These would otherwise pose a public health risk, as they are carriers of pathogens such as cholera and typhoid once metamorphosed (Seedikkoya *et al*., 2007). Cattle Egrets removed about 175–200 g maggots/day in a fishing village on the south-west coast of India where foraging took place for about 25 ha along Puthiyappa Beach (Seedikkoya and Azeez, 2009).

17.10 Control

Cattle Egrets are often territorial and aggressive (Blaker, 1967; Dusi and Dusi, 1968; Burger, 1978; Dami *et al*., 2006). They breed in large colonies, so predation efforts are made but are often unsuccessful (Blaker, 1967; Dusi and Dusi, 1968). In cases where predator numbers are high, entire heronries have lost young, and at times even adults are preyed on (Dusi and Dusi, 1968). Predators include snakes, birds of prey and mammalian small carnivores (Blaker, 1967; Dusi and Dusi, 1968). In areas where they have been introduced, roosting trees are removed to deter them, and they are sometimes trapped or shot.

17.11 Uses

In the Omi-aro and Labaka villages of Kwara State, Nigeria, the Cattle Egret is used for an annual ritual. They are taken from their nests along the riverbanks. The adult Cattle Egret is eaten as well as the eggs, leaving the feathers for use in the Iya Efun ceremony (Weliange *et al*., 2015). Nigeria is one of several countries with reported use for Cattle Egrets (Nikolaus, 2001; Williams *et al*., 2014).

17.12 Notes

As described earlier, Cattle Egrets have undergone both intra- and intercontinental migrations. They have successfully invaded many regions of the world due to their highly adaptive nature. They are gregarious creatures who migrate, roost and forage with other species and show behavioural plasticity to persist in new environments.

17.13 References

Ahmed, R. (2011) Subspecific identification and status of Cattle Egret. *Dutch Birding* 33, 294–304.

Arendt, W.J. (1988) Range expansion of the cattle egret (*Bubulcus ibis*) in the Greater Caribbean Basin. *Colonial Waterbirds* 11, 252–262.

Belzer, W. and Lombardi, J. (1989) Cattle egret symbiosis and heronry abandonment. *Colonial Waterbirds* 12, 115–117.

BirdLife International (2016) *Bubulcus ibis*. The Birdlife Red List of Threatened Species 2016. Available at: www.iucnredlist.org/species/22697109/86454050 (accessed 27 October 2019).

Blaker, D. (1967) Behaviour of the Cattle Egret *Ardeola ibis*. *Ostrich* 40, 75–129.

Browder, J.A. (1973) Long-distance movements of Cattle Egrets. *Bird-Banding* 44, 158–170.

Burger, J. (1978) Competition between Cattle Egrets and Native North American Herons, Egrets, and Ibises. *The Condor* 80, 15–23.

Chittenden, H., Davies, G. and Weiersbye, I. (2016) *Roberts Bird Guide*, 2nd edn. Jacana Media, Cape Town.

Congrains, C., Carvalho, A.F., Miranda, E.A., Cumming, G.S., Henry, D.A., *et al.* (2016) Genetic and paleomodelling evidence of the population expansion of the cattle egret *Bubulcus ibis* in Africa during the climatic oscillations of the Late Pleistocene. *Journal of Avian Biology* 47, 846–857.

Crosby, G.T. (1972) Spread of the Cattle Egret in the Western Hemisphere. *Bird-Banding* 43, 205–212

Dami, L., Bennetts, R. and Hafner, H. (2006) Do Cattle Egrets exclude Little Egrets from settling at higher quality sites within mixed-species colonies? *Waterbirds* 29, 154–162.

Dusi, J. and Dusi, R.T. (1968) Ecological factors contributing to nesting failure in a heron colony. *Wilson Bulletin* 80, 458–466.

Dwevedi, R., Kumar, A. and Mylswamy, M. (2015) A monospecific colony of Cattle Egret *Bubulcus ibis* in agricultural landscape of central Uttar Pradesh, India. *Indian Birds* 10, 149–150.

Goutner, V., Hans, J., Kazantzidis, S. and Naziridis, T. (1991) The occurrence of Cattle Egret in Greece. *Rivista Italiana di Ornitologia* 61, 107–112.

Hilaluddin, Shah, J. and Shawl, T. (2003) Nest site selection and breeding success by Cattle Egret and Little Egret in Amroha, Uttar Pradesh, India. *Waterbirds* 26, 444–448.

Jobling, J.A. (2010) *The Helm Dictionary of Scientific Bird Names*. Christopher Helm, London.

Kaufman, K. (1996) *Lives of North American birds*. Houghton Mifflin Company, Boston, Massachusetts.

Kopij, G. (1999) Food of the Cattle Egret (*Bubulcus ibis*) in South African grassland. *Vogelwarte* 40, 98–109.

Kopij, G. (2008) Range and population expansion of the Cattle Egret Bubulcusibis in Lesotho. *Ostrich* 79, 245–248.

Kopij, G. (2017) Migratory connectivity of South African Cattle Egrets (*Bubulcus Ibis, Ciconiiformes, Ardeidae*). *Zoological Journal* 96, 418–428.

Krebs, E.A., Riven-Ramsey, D. and Hunte, W. (1994) The colonization of Barbados by Cattle Egrets (Bubulcus ibis) 1956–1990. *Colonial Waterbirds* 17, 86–90.

Kuranchie, A., Kyerematen, R., Attuquayefio, K.D. and Lars H.H. (2013) Foraging activities, success and efficiency of Cattle Egrets (*Bubulcus ibis*) in three habitat types in the Greater Accra Region of Ghana. *Journal of Biological and Food Science Research* 2, 45–50.

Louw, K. (2005) Cattle Egret. In: Hockey, P.A.R., Dean, W.R.J. and Ryan, P. (eds) *Roberts Birds of Southern Africa*, 7th edn. Trustees of the John Voelcker Bird Book Fund, Cape Town, pp. 592–594.

Maddock, M. and Geering, D. (1994) Range expansion and migration of the Cattle Egret. *Ostrich* 65, 191–203.

Massa, C., Doyle, M. and Fortunato, R.C. (2014) On how Cattle Egret (*Bubulcus ibis*) spread to the Americas: meteorological tools to assess probable colonization trajectories. *International Journal of Biometeorology* 58, 1879–1891.

McKilligan, N. (1990) Promiscuity in the Cattle Egret (*Bubulcus ibis*). *Auk* 107, 334–341.

Miller G.C. and Ryder R.A. (1978) Cattle Egret in Colorado. *Western Birds* 10, 37–41.

Moralez-Silva, E. and Del Lama, S.N. (2014) Colonization of Brazil by the cattle egret (*Bubulcus ibis*) revealed by mitochondrial DNA. *NeoBiota* 21, 49–63.

Nikolaus, G. (2001) Bird exploitation for traditional medicine in Nigeria. *Malimbus* 23, 45–55.

Parkes, M.L., Mora, M.A. and Feagin, R.A. (2012) Using scale, cover type and GIS to evaluate nuisance Cattle Egret colony site selection. *Waterbirds* 35, 56–63.

Phillips, R.B., Wiedenfeld, D.A. and Snell, H.L. (2012) Current status of alien vertebrates in the Galapagos Islands: invasion history, distribution, and potential impacts. *Biological Invasions* 14, 461–480.

Rice, D.W. (1963) Birds associating with elephants and hippopotamuses. *Auk* 80, 196–197.

Scott, S.L. (1987) *Field Guide to the Birds of North America*. National Geographic Society, Washington, DC.

Seedikkoya, K. and Azeez, P.A. (2009) Cattle Egrets and dry fish business in coastal areas. *Nature Precedings*, 5 January 2009. Available at: http://precedings.nature.com/documents/2752/version/1/files/npre20092752-1.pdf (accessed 27 October 2019).

Seedikkoya, K., Azeez, P.A. and Shukkar, E.A.A. (2007) Cattle Egret as a biocontrol agent. *Zoos' Print Journal* 22, 2864–2866.

Siegfried, W.R. (1971a) Feeding activity of the Cattle Egret. *Ardea* 59, 38–46.

Siegfried, W.R. (1971b) The food of the Cattle Egret. *Journal of Applied Ecology* 8, 447–468

Stone, C.P. and Anderson, S.J. (1988) Introduced animals in Hawaii's natural areas. In: *Proceedings of the Thirteenth Vertebrate Pest Conference*. University of Nebraska, Lincoln, Nebraska. Available at: https://digitalcommons.unl.edu/cgi/viewcontent.cgi?article=1027&context=vpcthirteen (accessed 27 October 2019).

Telfair, R.C. II (1983) The cattle egret: a Texas focus and world view. Texas Agricultural Experimentation Station, Texas A&M University, College Station, Texas.

Telfair, R.C. II (2006) Cattle Egret (*Bubulcus ibis*). In: *The Birds of North America*. The Cornell Lab of Ornithology, Ithaca, New York. Available at: https://birdsna.org/Species-Account/bna/species/categr/introduction (accessed 27 October 2019).

Thompson, C.F., Lanyon, S.M. and Thompson, K.M. (1982) The influence of foraging benefits on association of cattle egrets (*Bubulcus ibis*) with cattle. *Oecologia* 52, 167–170.

Toloa, E., Kibuule, M., Blasberg, D. and Pomeroy, D. (2017) The increasingly urban status of the Cattle Egret *Bubulcus ibis* in Uganda, with some observations on its variable breeding seasons and associated species. *Scopus* 37, 14–23.

Weliange, W., Kolawole, R.A., Prasannajith, N.S., Afolabi, A.S. and Amaechi, E.C. (2015) Ethno-ornithological knowledge and uses of birds in Omi-Aro and Labaka villages, Kwara State, Nigeria. *Malimbus* 37, 41–54.

Williams, V.L., Cunningham, A.B., Kemp, A.C. and Bruyns, R.K. (2014) Risks to birds traded for African traditional medicine: a quantitative assessment. *PLoS One* 9, e105397.

18 Red Junglefowl (*Gallus gallus* Linnaeus, 1758), Grey Junglefowl (*Gallus sonneratii* Temmink, 1813) and Green Junglefowl (*Gallus varius* Shaw, 1798)

Preshnee Singh*, David A. Ehlers Smith and Colleen T. Downs
Centre for Excellence in Invasion Biology, School of Life Sciences, University of KwaZulu-Natal, Pietermaritzburg 3209, South Africa

Citation: Singh, P., Ehlers Smith, D.A., and Downs, C.T. (2020). Red Junglefowl (*Gallus gallus* Linnaeus, 1758), Grey Junglefowl (*Gallus sonneratii* Temmink, 1813) and Green Junglefowl (*Gallus varius* Shaw, 1798). In: Downs, C.T. and Hart, L.A. (eds) *Invasive Birds: Global Trends and Impacts*. CAB International, Wallingford, UK, pp. 144–148.

18.1 Common Names

Red Junglefowl: Domestic Fowl, Feral Chicken, Junglefowl, Red Jungle Fowl, Red Junglefowl, Red Jungle-fowl, Wild Junglefowl, Burmese Red Junglefowl (Avibase, 2018). Grey Junglefowl: Gray Junglefowl, Gray Jungle-fowl, Grey Junglefowl, Grey Jungle-fowl, Sonnerat's Junglefowl (Avibase, 2018). Green Junglefowl: Green Jungle Fowl, Green Jungle-fowl, Javan Junglefowl (Avibase, 2018).

18.2 Nomenclature

The Red Junglefowl (*Gallus gallus* Linnaeus, 1758) has five subspecies: *G. g. bankivai*, *G. g. gallus*, *G. g. jabouillei*, *G. g. murghi* and *G. g. spadiceus*. The Grey Junglefowl (*Gallus varius* Temmink, 1813) and Green Junglefowl (*Gallus sonneratii* Shaw, 1798) both have no subspecies (del Hoyo *et al.*, 2014).

18.3 Distribution

18.3.1 Native distribution

The natural range of Red Junglefowl is determined by the subspecies. *G. g. gallus* is distributed from northern Myanmar to eastern Thailand. *G. g. jabouillei* is endemic to Vietnam but is also found in southern China in Yunnan and Guangxi Provinces and on Hainan Island. *G. g. bankiva* is endemic to Indonesia and is found in southern Sumatra, Java and Bali. *G. g. murgha* is found in Nepal, Bangladesh and northern India. *G. g. spadiceus* is native to the Malay Peninsula, northern Sumatra and northern Myanmar into southern Yunnan, China. Thus, collectively the natural range of the Red Junglefowl is the Asian continent from China and India at the north-western extremity and Bali at the south-eastern extremity (Fig. 18.1) (Avibase, 2018).

The natural distribution of the Grey Junglefowl is concentrated around the Indian Peninsular but may be found in the states of Gujarat, Madhya Pradesh and southern Rajasthan (Fig. 18.1) (Avibase, 2018).

The natural distribution of the Green Junglefowl is on the Indonesian islands of Java, Bali, Lombok, Sumbawa, Flores and Alor (Fig. 18.1) (Avibase, 2018).

18.3.2 Hybridization, domestication and introduced distribution

The Red Junglefowl is considered to be one of the earliest domesticated and released species of animal in human history, with archaeological evidence suggesting that Red Junglefowl were transported into northern China approximately 8000 years ago (West and Zhou, 1989). Domesticated poultry then spread into western China and as far south as Indonesia by approximately 7400 years ago (Miao *et al.*, 2013), into the Pacific Islands by around 3000 years ago (Steadman *et al.*, 1990), were established in Europe by about 1600 years ago (Sykes, 2012) and spread as far as South America by approximately 700 years ago (Storey *et al.*, 2007; Cassey *et al.*, 2015). The species has long been considered – including by Charles Darwin – to be the precursor to domesticated poultry and the origin of invasive *Gallus* spp. globally. However, recent genetic evidence has indicated that several genetic and phenotypic markers originate from the Grey Junglefowl, including the yellow pigment in the legs and other anatomical features of all domesticated poultry

*Corresponding author: preshnee.singh@gmail.com

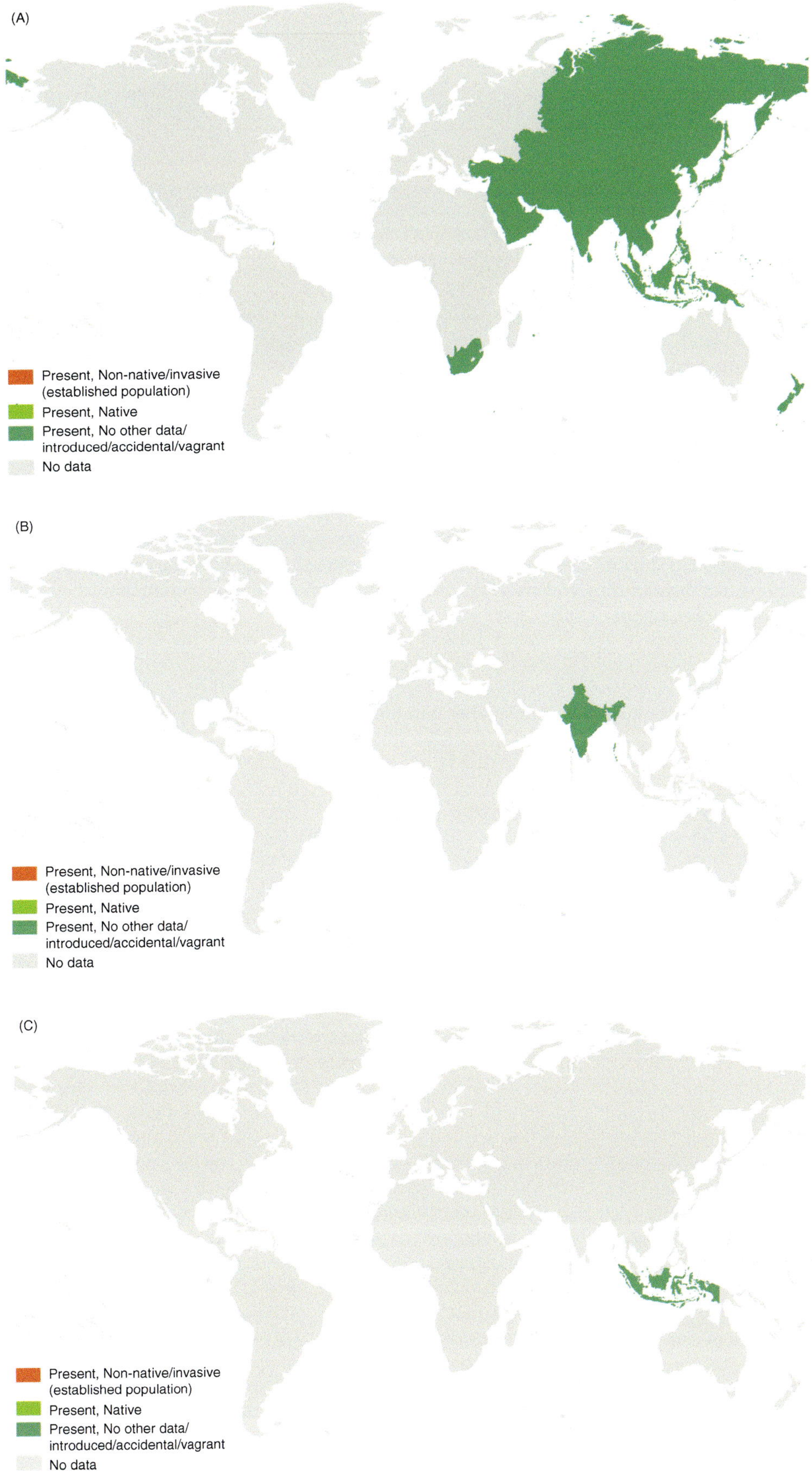

Fig. 18.1. Global distribution of the Red Junglefowl (*Gallus gallus*) (A), Grey Junglefowl (*Gallus sonneratii*) (B) and Green Junglefowl (*Gallus varius*) (C) showing the native (blue) and invasive (red) ranges.

(Eriksson *et al.*, 2008). Indeed, Red and Grey Junglefowl are known to hybridize within their limited sympatric distribution (del Hoyo *et al.*, 1994; Nishibori *et al.*, 2005; Sambandam *et al.*, 2015).

The Green Junglefowl has had limited invasive impact, despite its presence on the Global Invasive Species Database (www.iucngisd.org/gisd/, accessed 30 October 2019), with small non-native populations only known from the Cocos (Keeling) Islands, Australia, and Anguilla in the eastern Caribbean. Male Green Junglefowl are bred with Red Junglefowl or domestic chickens to produce an infertile offspring known in the Java region as a 'bekisar' (del Hoyo *et al.*, 1994). Many Green Junglefowl males are removed from local populations for this. This practice originated on Kangean Islands (north-east of Java) and has grown in popularity, with people on the north coast of Bali and East Java now engaging in this. These 'bekisar' are used in vocal competitions. From 1994, the 'bekisar' has been the provincial bird of East Java and has much cultural significance for the region (del Hoyo *et al.*, 1994).

Examination of morphology and behaviour show that domestic chickens are most closely related to the Red Junglefowl (Desta, 2019). The Red Junglefowl and Ceylon Junglefowl (*Gallus lafayettii*) are closely related, with the Grey and Green Junglefowl more distant, although there are features from both Grey and Green Junglefowl found in domestic chicken breeds (Desta, 2019). Thus, given the difficulty of disentangling the prehistoric domestication, hybridization and introduction pathways by Junglefowl species (Eriksson *et al.*, 2008), we treat the species found outside of its natural range as a single hybrid Junglefowl species.

The invasive range of the Junglefowl now includes the major landmasses of Australasia, including Australia, New Zealand, Vanuatu, New Caledonia, Fiji, Samoa and Hawaii. It is found on Réunion Island in the southern Indian Ocean, and the Bahamas, Trinidad and Tobago, and the Dominican Republic in the Caribbean. It was also introduced to the USA, Ecuador and Chile (Avibase, 2018). The domesticated chicken may be found in virtually every country on Earth.

Fig. 18.2. Adult male Green Junglefowl. (©Photograph Francis Smith).

18.4 Description

Red Junglefowl are sexually dimorphic. Males are around 65–70 cm long, weighing 672–1450 g. Females are around 42–46 cm long and weigh 485–1050 g (del Hoyo *et al.*, 1994). Males have a dark metallic green body and a long-arched tail, which sometimes reflects blue or purple (Fig. 18.2A) (Glenister, 1951; del Hoyo *et al.*, 1994). They have a white earlobe (del Hoyo *et al.*, 1994) with a red comb, face and wattles (Glenister, 1951). They have orange on top of the head, nape and hackles with a maroon back and an orange-red rump (Glenister, 1951). The wings are also dark metallic green but maroon at the shoulders with yellowish chestnut at the edges (Glenister, 1951). Females are brown vermiculated with black. The top of the head is chestnut, the face is reddish and the tail is brownish black (Glenister, 1951). The neck and upper mantle are yellowish-orange pencilled with black (Glenister, 1951). They lack wattles and have a small comb. The subspecies variations differ in coloration and the length and shape of the male hackles during breeding season. In *G. g. bankiva*, the hackles are redder and rounder and in *G. g. murgha* they are pointed with dark markings, while in *G. g. jabouillei* the males have smaller combs and hackles and no white earlobe while the females are generally darker (del Hoyo *et al.*, 1994).

Grey Junglefowl are sexually dimorphic. Males are around 70–80 cm long, weighing 790–1136 g and females are about 38 cm long, weighing 705–790 g (del Hoyo *et al.*, 1994). Males have a smaller comb and hackles, in comparison with the Red Junglefowl, with a distinguishable neck pattern of ochre spots on black (del Hoyo *et al.*, 1994). The female differs from other *Gallus* spp. with a white streaked breast and unbarred flight feathers (del Hoyo *et al.*, 1994).

Green Junglefowl are sexually dimorphic. Males are around 70 cm long and females about 40 cm. They weigh between 454 and 795 g (del Hoyo *et al.*, 1994). Males are unmistakable with a scaled appearance over dark green plumage that appears black from a distance (del Hoyo *et al.*, 1994). Their multi-coloured combs, light blue turning purple or red towards the top, and hackles are also distinctive (del Hoyo *et al.*, 1994; Avibase, 2018). They have bare facial skin showing scarlet red faces and some yellow facial skin extending below each ear delineating the plumed hackles from the gular lappet (del Hoyo *et al.*, 1994; Avibase, 2018). The wattle is bordered with blue edges and is yellow towards the throat (Avibase, 2018). The mantles are multi-coloured iridescent with blue at the base showing various shades of gold to bronze-green (Avibase, 2018). The throat is reflective and appears violet at the proximal edges and blue at the distal edges (Avibase, 2018). The lesser coverts are burnt orange and bronze-black at the centres, while the greater secondary coverts are ochre (Avibase, 2018). The breast and ventral regions are black (Avibase, 2018). Female is mostly brown with occasional green feathers and no comb. They differ from other *Gallus* spp. by having a combination of scaly upperparts, barred wings and no bold patterning on the breast (del Hoyo *et al.*, 1994).

Domesticated chickens have various breeds, which may share some of the features found with other *Gallus* spp., especially in hybrids with these species. Male Red and Grey

Junglefowl display eclipse plumage, which is a trait that has disappeared from domestic chickens (Desta, 2019).

18.5 Diet

Junglefowl are omnivorous with a diverse and opportunistic diet, with seasonal preferences most likely depending on availability (del Hoyo *et al.*, 1994; Avibase, 2018). They feed on seeds, shoots and invertebrates, which include insects and insect eggs from beetles to grasshoppers and termites, and are sometimes found feeding on cultivated crops in rice fields and on palm and coffee plantations (del Hoyo *et al.*, 1994; Avibase, 2018). They are also known to feed on figs, berries, leaves, fruit, worms and sometimes small reptiles (del Hoyo *et al.*, 1994).

18.6 Breeding Behaviour

Junglefowl are polygamous. The male Red Junglefowl have spurs on their lower legs behind and above the feet, which are used to fight off other males. During breeding, they display a behaviour known as 'tidbitting' to attract females, where males find food in the presence of a female and display by calling with bobbing and twitching head movements (del Hoyo *et al.*, 1994; Avibase, 2018). They then pick up and drop the food repeatedly until the female accepts. They breed from March–May, during the dry season, in India, although eggs have been found in January–October in different parts of the country (del Hoyo *et al.*, 1994). In Bangladesh, the season is from March–June and in the Malay Peninsula from December–May, peaking in January–February, and in China from February–May (del Hoyo *et al.*, 1994). They nest in dense secondary growth or bamboo forest under bushes or in bamboo clumps with clutches of five to six eggs, and have an incubation period of 18–20 days, incubated by the female (del Hoyo *et al.*, 1994).

Grey Junglefowl breed throughout the year with peaks in February and March (del Hoyo *et al.*, 1994). Regional peaks in breeding would include October–December in Western Nilgiris and March–August in Travancore (del Hoyo *et al.*, 1994). The clutch is usually four to five eggs, with clutches in the south suggested to be larger than those in the north (del Hoyo *et al.*, 1994). They nest in small hollows next to vegetation (del Hoyo *et al.*, 1994). Males have been recorded finding food for chicks (del Hoyo *et al.*, 1994).

Green Junglefowl live in groups of up to five, with females led by a dominant male. In breeding season, dominant males are challenged by males with no flocks. The displays involve clapping wings and crowing while fighting with their spurs. The breeding seasons for the Green Junglefowl are thought to be long and variable in Java with a season from June to November, and sometimes as early as March (del Hoyo *et al.*, 1994). Generally, the seasons are earlier in East Java compared with the west. They nest in ground depressions among dense vegetation on top of tree ferns (del Hoyo *et al.*, 1994). Clutches have three to four eggs, with a 21-day incubation period recorded in captivity (del Hoyo *et al.*, 1994). Chicks are a chocolate brown colour above with a brown neck and breast, and creamy white below (del Hoyo *et al.*, 1994).

Again, domesticated chickens may share some of these behaviours depending on the breed or the hybrids.

18.7 Habitat

Both Red and Green Junglefowl inhabit tropical and subtropical habitats, including mangroves, to altitudes up to 2000 m. They seem to prefer flat or gently sloping terrain and edge or secondary habitats to forest (del Hoyo *et al.*, 1994; Desta, 2019). The Green Junglefowl has also been known to inhabit lowland moist forest and shrub land, typically breeding along coastal areas (del Hoyo *et al.*, 1994). The Grey Junglefowl seems to prefer understorey vegetation and forests up to 1500 m above sea level (del Hoyo *et al.*, 1994). They have been spotted frequently in places with overgrown lantana (*Lantana camara*) (del Hoyo *et al.*, 1994). All Junglefowl are found in human-altered habitats such as rubber, oil, tea and coffee plantations (del Hoyo *et al.*, 1994; Desta, 2019).

18.8 Impacts

Although Junglefowl species have been found to feed on crops, there does not seem to be a major reported impact on agriculture (Peh, 2010). The Grey Junglefowl has been found to feed on *L. camara* berries and may be contributing to the spread of this invasive plant (del Hoyo *et al.*, 1994). Junglefowl seem to prefer human-altered habitats (Desta, 2019) and their closeness to people brings potential risk for avian diseases.

18.9 Control

There seem to be no recorded measures for control. Junglefowl are typically shot but this is mostly for food. There is potential to use avian birth control products as are used to control Feral Pigeons (*Columba livia*) and other bird invasives.

18.10 Uses

The Red Junglefowl is typically shot and used for food (del Hoyo *et al.*, 1994) while male Green Junglefowl are captured and kept as pets because of their call and appearance, or are bred with Red Junglefowl and domestic fowl to create hybrids. These hybrids are used in vocal competitions and have cultural significance in the Java region (del Hoyo *et al.*, 1994). Hybridization of the Red Junglefowl and domestic chickens produces fertile offspring and is suggested to create more disease-resistant breeds (Desta, 2019). Domestic chickens are typically bred to be larger and produce more than Junglefowl (Desta, 2019).

18.11 References

Avibase (2018) *Junglefowl*. Available at: https://avibase.bsc-eoc.org/ (accessed 29 October 2019).

Cassey, P., Vall-Llosera, M., Dyer, E. and Blackburn, T.M. (2015). The biogeography of avian invasions: history, accident and market trade. In: Canning-Clode, J. (ed.) *Biological Invasions in Changing Ecosystems*. De Gruyter Open, Warsaw, pp. 37–54.

del Hoyo, J., Collar, N.J., Christie, D.A., Elliott, A. and Fishpool, L.D.C. (2014) *Illustrated Checklist of the Birds of the World*, Vol. 1: *Non-passerines*. Lynx Edicions, Barcelona, Spain, and BirdLife International, Cambridge, UK.

del Hoyo, J., Elliott, A. and Sargatal, J. (1994) *Handbook of the Birds of the World*, Vol. 2: *New World Vultures to Guineafowl*. Lynx Edicons, Barcelona, Spain.

Desta, T. (2019) Phenotypic characteristic of junglefowl and chicken. *World's Poultry Science Journal* 75, 69–82.

Eriksson, J., Larson, G., Gunnarsson, U., Bed'hom, B., Tixier-Boichard, M., *et al.* (2008) Identification of the yellow skin gene reveals a hybrid origin of the domestic chicken. *PLoS Genetics* 4, e1000010.

Glenister, A.G. (1951) *The Birds of the Malay Peninsula Singapore and Penang*. Oxford University Press, London.

Miao, Y.W., Peng, M.-S., Wu, G.-S., Ouyang, Y.-N., Yang, Z.-Y., *et al.* (2013) Chicken domestication: an updated perspective based on mitochondrial genomes. *Heredity* 110, 277–282.

Nishibori, M., Shimogiri, T., Hayashi, T. and Yasue, H. (2005) Molecular evidence for hybridization of species in the genus *Gallus* except for *Gallus varius*. *Animal Genetics* 36, 367–375.

Peh, K.S.-H. (2010) Invasive species in Southeast Asia: the knowledge so far. *Biodiversity Conservation* 19, 1083–1099.

Sambandam, S., Fernandes, M., Thakur, M. and Kaul, R. (2015) Conservation of Red Junglefowl (*Gallus gallus*) in India. Technical Report, Wildlife Institute of India, Dehradun, India.

Steadman, D.W., Greiner, E.C. and Wood, C.S. (1990) Absence of blood parasites in indigenous and introduced birds from the Cook Islands, Pacific South. *Conservation Biology* 4, 398–404.

Storey, A.A., Ramírez, J.M., Quiroz, D., Burley, D.V., David, J., *et al.* (2007) Radiocarbon introduction and DNA evidence for a pre-Columbian to Chile of Polynesian chickens. *Proceedings of the National Academy of Sciences USA* 104, 10335–10339.

Sykes, N. (2012) A social perspective on the introduction of exotic animals: the case of the chicken. *World Archaeology* 44, 158–169.

West, B. and Zhou, B. (1989) Did chickens go north? New evidence for domestication. *World's Poultry Science Journal* 45, 205–218.

19 House Finch (*Haemorhous mexicanus* Müller, 1776)

Sarah E. Britton and Alexander V. Badyaev*
Department of Ecology and Evolutionary Biology, University of Arizona, Tucson, AZ 85721, USA

Citation: Britton, S.E. and Badyaev, A.V. (2020) House Finch (*Haemorhous mexicanus* Müller, 1776). In Downs, C.T. and Hart, L.A. (eds) *Invasive Birds: Global Trends and Impacts*. CAB International, Wallingford, UK, pp. 149–154.

19.1 Common Name

House Finch.

19.2 Distribution

Until the early 1930s, the native range of House Finches (*Haemorhous mexicanus* Müller, 1776) was mainly confined to the south-western USA (primarily Arizona and southern and coastal California) and north-western Mexico (Fig. 19.1). In 1939, fewer than 100 finches captured in Santa Barbara, California, were released from a pet store in Brooklyn, New York (Elliot and Arbib, 1953; Mundinger, 1975). From this small release, House Finches have greatly expanded their range in eastern North America without any further introductions. House Finches were also introduced to Hawaii in 1859 from the San Francisco Bay area, and since 1911, the species occurs on all major islands (Grinnell, 1911). All populations, native and recently established, are currently undergoing exceptional range expansion, such that the native western populations are expanding from south-western USA and populations expanding from the eastern introduction in New York have now converged in numerous locations. The current range in North America extends from Florida in the south to New Brunswick and Quebec in the north, and from Quebec City across Canada to Prince George, British Columbia. Along the west coast, populations are found from Oaxaca and Chiapas in southern Mexico to Whitehorse, Yukon, and to Anchorage, Alaska. Within these boundaries the species distribution is largely continuous (Fig. 19.1).

19.3 Description

The House Finch is small (19–22 g), grey-brown in colour with streaking below (Fig. 19.2). Males have carotenoid-based pigmentation ranging in colour from pale yellow to orange to purple occurring on the crown, back, eyebrow stripe, cheek, shoulder patch, rump and ventral plumage. Young females occasionally show faint reddish pigmentation on the rump but otherwise females lack red pigmentation. Both the biochemical composition of the carotenoid pigmentation and its plumage expression vary extensively within and among individuals and populations (Michener and Michener, 1931; Brush and Power, 1976; Hill, 1993; Badyaev and Young, 2004; Badyaev and Hill, 2002), reflecting remarkable versatility in biochemical conversion and accommodation of variable dietary carotenoids across the species' vast geographical and ecological ranges (Badyaev *et al.*, 2017; Badyaev and Morrison, 2018).

The House Finch shows extensive morphological variability among populations. The birds with the heaviest mass (mean ± sd) occur in the northern parts of its range, in Montana (males: 21.71 ± 1.38 g; females; 21.57 ± 1.40 g) and Michigan (males; 21.45 ± 1.24 g; females: 22.12 ± 1.46 g). The smallest birds occur in the most tropical regions, in Hawaii (males: 19.40 ± 1.36 g; females: 19.17 ± 1.32 g) and Mexico (males: 19.51 ± 1.26 g; females: 19.39 ± 1.22 g) (Badyaev and Hill, 2000). Birds tend to be heavier in winter than during the summer breeding season (e.g. in Wisconsin, winter body mass was 22.4 ± 0.1 g and breeding body mass was 21.4 ± 0.3 g; Hartup *et al.*, 2004).

Sexual dimorphism varies extensively across House Finch populations (Badyaev and Hill, 2000) and evolves rapidly within populations (Badyaev, 2009, 2010), with frequent reversals, but in most populations, males tend to have proportionally longer

*Corresponding author: abadyaev@email.arizona.edu

Fig. 19.1. Distribution of the House Finch (*Haemorhous mexicanus*) in North America.

wings, shorter bills, longer tails and smaller body mass than females. The birds with the longest bills occur in Mexico (males: 9.08 ± 0.32 mm; females: 8.99 ± 0.25 mm), while those with the shortest bills are in Hawaii (males: 8.20 ± 0.37 mm; females: 7.99 ± 0.40 mm). The birds with longest tails occur in New York (males: 67.10 ± 8.49 mm; females: 67.94 ± 5.55 mm), while those with the shortest tails are in Hawaii (males: 58.58 ± 2.63 mm) and in Alabama (females: 57.49 ± 2.68 mm). Birds with the longest wings occur in Hawaii (males: 80.98 ± 2.15 mm; females: 79.37 ± 1.50 mm) and Mexico (males: 80.74 ± 1.99 mm; females: 78.32 ± 2.04 mm), whereas birds with the proportionally shortest wings are in Montana (males: 79.30 ± 1.81 mm; females: 77.32 ± 2.06 mm) and Alabama (males: 79.72 ± 1.91 mm; females: 77.13 ± 2.02 mm). Birds with the longest tarsi occur in Mexico (males: 17.43 ± 0.55 mm; females: 17.73 ± 0.56 mm) and Alabama (males: 17.45 ± 1.19 mm), and birds with the shortest tarsi in California (males: 17.10 ± 0.53 mm) and Alabama (females: 16.80 ± 0.65 mm) (Badyaev and Hill, 2000).

19.4 Diet

Throughout the House Finch's range, common foods are seeds, buds and leaves, flowers, pollen and fruits. Nestlings' diet can include occasional insect larvae. In urban areas, they rely heavily on feeders. In these areas, a preference for sunflower seeds (Aldrich, 1982) exerts strong selection pressure on beak configurations and development, resulting in recurrent convergence to an 'urban' beak morphology during species dispersal (Badyaev, 2014) and associated changes in song characteristics and population structure (Badyaev *et al.*, 2008b).

In California, most commonly consumed plant seeds include Napa thistle (*Centaurea melitensis*), black mustard (*Brassica nigra*), wild mustard (*Raphanus sativus*), amaranth (*Amaranthus* spp.), alfilaria (*Erodium* spp.), knotweed (*Polygonum aviculare*) and turkey mullein (*Eremocarpus setigerus*), as well as commercial fruits in the summer (Beal, 1907). In the south-west Arizona deserts, the species' diet includes cactus and grass seeds, as well as fruits and flowers of prickly

Fig. 19.2. Adult male (left) and female (right) House Finches. (©Photograph: Alex Badyaev: www.tenbestphotos.com.)

pear (*Opuntia* spp.), ocotillo (*Fouquieria splendens*), creosote bush (*Larrea tridentata*), ironwood (*Olneya tesota*), fishhook cacti (*Mammillaria microcarpa*) and pollen and fruits of saguaro (*Cereus giganteus*) (Badyaev *et al.*, 2012). In the east, a common fruit staple is mulberry (*Morus* spp.). In Hawaii, their diet includes ironwood (*Casuarina equisetefolia*), Formosan koa (*Acacia confusa*), pink tecoma (*Tabebuia pentaphylla*), broad-leaved plantain (*Plantago major*), pear thistle (*Cirsium vulgare*) and fruits including banyan (*Ficus* spp.), guava (*Psidium guajava*) and rose apple (*Eugenia* spp.) (Hirai, 1975).

19.5 Introductions and Invasion Pathways

The expansions of the ancestral House Finch populations in the south-western USA (to the north and east) and of the introduced population in the north-eastern USA (to the north, west and south) occurred at remarkably similar rates and times, despite the lack of connectivity between these areas during the early stages of range expansion (Badyaev *et al.*, 2012). In the well-studied sequence of House Finch population establishment in Montana, the dispersal occurred in a stepping-stone manner, where populations persisted for five to ten generations before a group of birds undertook a long-distance dispersal leap, establishing a new population (Badyaev and Walsh, 2014), a pattern concordant with that predicted by the Allee effect (Veit and Lewis, 1996).

In the 1950s, House Finch native populations began to expand east and north through Oklahoma, Kansas and Nebraska, moving northwards on both sides of the Continental Divide. The western branch of this expansion reached western Montana in 1969, while expansion east of the Continental Divide reached south-eastern Montana in the mid-1950s and central Montana in the early 1980s (Badyaev *et al.*, 2012). Expansion reached Alberta and Saskatchewan in the 1990s (Koes and Taylor, 1993) and Alaska in 1996. Southward expansion of native populations reached southern Mexico in 1980 (Peterson and Chalif, 1989).

Over a similar time frame, expansion occurred from the introduced House Finches in New York. By the 1970s, populations existed in Vermont, New Hampshire and Maine (Badyaev *et al.*, 2012). The first records in Ontario were in 1976 (Goodwin, 1976), and the birds were well established in Quebec by the 1990s (Aubry and Yank, 1993). In the 1970s, expansion occurred southwards to the Carolinas and Georgia, reaching Florida by the early 1990s and the Everglades by 2005 (Pranty, 2005).

Since the late 1990s, native western and introduced eastern House Finch populations have formed a largely contiguous range (Badyaev *et al.*, 2012), although the two invasions are commonly separated by local geographic barriers. For example, in central Montana, eastern and western branches of recently established populations are largely separated by the Continental Divide, although exchange of dispersing birds between the branches is common.

Current House Finch range expansion, especially in western North America, probably represents recolonization of the ancestral range, as fossil records indicate the presence of the species in Gunnison Basin, Colorado, from 9000 years ago (Walker, 2001) and in New Mexico from the late Pleistocene (Howard and Miller, 1933).

19.6 Breeding Behaviour

The House Finch is socially monogamous with only 4–9% of offspring resulting from extra-pair mating (Oh and Badyaev, 2006; Lindstedt *et al.*, 2007). Pair formation begins in winter flocks and continues throughout the breeding season (Thompson, 1960). Gonadal cycles are controlled by photoperiodicity, although spermatogenesis is not sensitive to photoperiod (Hamner, 1966), and ambient temperature does not affect reproductive preparations (Watts *et al.*, 2018). In females, breeding experience modulates sensitivity to photoperiod (Salvante *et al.*, 2013), potentially explaining the experience-related adjustment of breeding phenology to local conditions during range expansion (Badyaev and Oh, 2008).

House Finch nest building is from early February in southern populations (e.g. Arizona, Alabama) to mid-March–April in the north (Montana). Both sexes participate in nest site prospecting, but only the female builds. Nest sites are off the ground and always include a solid base with an overhanging structure but otherwise are highly variable and include pine and spruce trees (including at the top of the canopy), buildings with ivy, pandanus and palms (Hawaii), cholla cacti (Arizona desert), olives, junipers, man-made structures, hanging flower pots and abandoned farm equipment (Badyaev *et al.*, 2012).

House Finch clutch size ranges from two to seven eggs and commonly declines as the season progresses. For example, in Arizona's urban population, first clutches are 3.9 ± 0.8 eggs, second clutches are 3.6 ± 0.9 eggs, third ones are 3.3 ± 0.6 eggs and fourth ones are 2.8 ± 0.4 eggs (Badyaev *et al.*, 2012). The largest clutches are found in the northern populations (western Montana: 4.6 ± 0.03 eggs; Ontario: 4.5 ± 0.63 eggs; Peck and James, 1987) central California (4.4 ± 0.58 eggs; Evenden, 1957) and north-eastern USA (4.5 ± 0.86 eggs; Wootton, 1986). Smaller clutches are found in southern regions such as the Arizona desert (3.5 ± 0.4 eggs; Badyaev *et al.*, 2012), and Hawaii Island (3.9 ± 0.7 eggs; van Riper, 1976). Typically, the House Finch has at least two broods per season, but four and five broods per season occur during the early stages of population establishment in northern populations (Badyaev *et al.*, 2012).

House Finch eggs are pale blue with dark speckling (Harrison, 1978) and egg measurements and structure (shell thickness and porosity) vary geographically (Badyaev *et al.*, 2006c). In humid climates, the eggs are larger and eggshells are thicker with lower pore density than in dry climates. For example, Alabama populations have thicker eggshells (0.08–0.10 mm) and lower pore density (19.92 ± 3.94 pores/cm^2) than those in Arizona (0.06–0.08 mm, 39.5 ± 7.43 pores/cm^2) or Montana (41.30 ± 10.42 pores/cm^2) (Stein and Badyaev, 2011). Females lay one egg/day and the onset of incubation depends heavily on female experience with local environmental conditions. Immigrant females in established populations or females during the initial population establishment tend to incubate from the first egg, whereas older or local females always incubate when the ambient temperature is outside of egg tolerance range, but otherwise incubate from the last egg (Badyaev *et al.*, 2003a; Badyaev and Oh, 2008).

In south-west Arizona, the House Finch nest temperature over the course of incubation averaged 32.03 ± 3.21°C, and during off-bouts, the nest temperature cooled to an average of 19.54 ± 5.66°C early in the season and reached a mean high temperature of 37.91 ± 2.74°C late in the season. The mean off-bout length during incubation in the south-west Arizona population is 23.83 ± 8.50 min (Stein *et al.*, 2009). Incubation lasts 13–14 days (Thompson, 1960).

Environmentally induced onset of incubation and its hormonal basis create pronounced gradients of hormones, nutrients and antioxidants during oogenesis that, in turn, result in sex-biased allocation of these substances in sequentially ovulating oocytes (Badyaev *et al.*, 2005, 2006a, 2008a). Sex-biased laying order and sex-specific allocation of substances interact with flexible incubation periods to produce highly variable periods of growth for male and female nestlings and associated variation in sexual size dimorphism (Badyaev *et al.*, 2003b). By increasing within-population morphological variation and shifting morphological distribution, maternally induced dimorphism reduces offspring mortality in newly established populations by 20–30% and ultimately enables initial persistence of populations under novel environmental conditions (Badyaev *et al.*, 2002). This strategy is likely a co-option of mechanisms that underlie maternally induced sex-specific modifications of offspring growth in the native population of this species (Badyaev *et al.*, 2006b).

19.7 Habitat

The House Finch's native habitat is open, dry desert landscapes including desert grassland, desert shrubland (especially areas with cholla cacti), chaparral, oak savannah, juniper-oak woodland, riparian areas, open coniferous forest and subalpine bushes at elevations up to 3,500 m above sea level (Fig. 19.3). The House Finch is also common in urban areas. In the eastern USA, the House Finch is primarily associated with humans and is common in suburban areas with lawns and wooded areas but is also found in urban areas. In the northern part of the eastern range, breeding habitats are often associated with coniferous trees (Fernández-Juricic *et al.*, 2005), but in both the east and west, the species avoids dense wooded areas (White *et al.*, 1996). In Hawaii, the House Finch is also closely associated with human settlements including agricultural areas, and they also occur in grasslands and native forests, but avoid dense rainforest (Berger, 1981).

19.8 Impacts

The House Finch can cause damage to commercial fruits, especially in Hawaii, southern California (Woods, 1968) and north-western Montana. They are also known to cause damage to grain (Berger, 1981).

19.9 Control

The most common natural predators of the House Finch include Cooper's Hawks (*Accipiter cooperii*), Sharp-shinned Hawks (*Accipiter striatus*), Northern Goshawks (*Accipiter gentilis*), Northern Harriers (*Circus cyaneus*), American Kestrels (*Falco sparverius*), Greater Roadrunners (*Geococcyx californianus*), mammals such as domestic cats (*Felis catus*), and snakes. The most common nest predators are Blue Jays (*Cyanocitta cristata*), Common Grackles (*Quiscalus quiscula*), Common Crows (*Corvus brachyrhynchos*), Steller's Jays (*Cyanocitta stelleri*), Clark's Nutcrackers (*Nucifraga columbiana*), Black-billed Magpies (*Pica hudsonia*) and Red-winged Blackbirds (*Agelaius phoeniceus*) in the north and Cactus Wrens (*Campylorhynchus brunneicapillus*), Gila Woodpeckers (*Melanerpes uropygialis*) and Curve-billed Thrashers (*Toxostoma curvirostre*) in the south.

Fig. 19.3. Adult male (left) and female (right) at the nest in native habitat. (©Photograph: Alex Badyaev: www.tenbestphotos.com.)

Mammals that predate nests include domestic cats, various squirrels and rats (Badyaev *et al.*, 2012). In Hawaii, rats are the primary nest predator (Hirai, 1975).

A major cause of death is mycoplasmal conjunctivitis caused by *Mycoplasma gallisepticum* (Cookson and Shivaprasad, 1994; Hochachka and Dhondt, 2000). First reported in 1994 in the mid-Atlantic States, the disease accounted for a 60% decline in eastern populations in the first three years (Dhondt *et al.*, 1998). The first record of the disease in recently established western populations was in 2002 (Duckworth *et al.*, 2003). In 2009, the disease was reported in south-western Arizona, resulting in up to 70% mortality (Badyaev *et al.*, 2012). Other major sources of mortality are winter ambient temperatures and snow cover in the north, and nest mite infestation in the south.

19.10 Notes

Many reasons have been proposed to explain the exceptional colonization ability of this species, including the unusually flexible growth trajectories that can produce a diverse array of morphologies (Badyaev and Martin, 2000), environmentally induced maternal strategies that enable initial survival of populations (Badyaev *et al.*, 2005), rapid acclimation abilities (Dawson *et al.*, 1983), superior competitive abilities over birds of similar niches (Wootton, 1987; Cooper *et al.*, 2007) and remarkably few dependencies on other species in the diverse ecosystems that the species now occupies. Nevertheless, the extent to which these factors are causes or consequences of the unusually wide ecological distribution is unclear.

19.11 References

Aldrich, J. (1982) Rapid evolution in the House Finch (*Carpodacus mexicanus*). *Journal of the Yamashina Institute for Ornithology* 14, 179–186.

Aubry, Y. and Yank, R. (1993) Quebec Region. *American Birds* 47, 393.

Badyaev, A.V. (2009) Evolutionary significance of phenotypic accommodation in novel environments: an empirical test of the Baldwin effect. *Philosophical Transactions of the Royal Society B: Biological Sciences* 364, 1125–1141.

Badyaev, A.V. (2010) The beak of the other finch: coevolution of genetic covariance structure and developmental modularity during adaptive evolution. *Philosophical Transactions of the Royal Society B: Biological Sciences* 365, 1111–1126.

Badyaev, A.V. (2014) Reconciling innovation and adaptation during recurrent colonizations: molecular, genetic, and developmental bases. In: Gil, D. and Brumm, H. (eds) *Avian Urban Ecology: Behavioural and Physiological Adaptations*. Oxford University Press, Oxford, pp. 155–169.

Badyaev, A.V. and Hill, G.E. (2000) The evolution of sexual dimorphism in the House Finch: I. Population divergence in morphological covariance structure. *Evolution* 54, 1784–1794.

Badyaev, A.V. and Hill, G.E. (2002) Parental care as a conditional strategy: distinct reproductive tactics associated with elaboration of plumage ornamentation in the House Finch. *Behavioral Ecology* 13, 591–597.

Badyaev, A.V. and Martin, T.E. (2000) Individual variation in growth trajectories: phenotypic and genetic correlations in ontogeny of the House Finch (*Carpodacus mexicanus*). *Journal of Evolutionary Biology* 13, 290–301.

Badyaev, A.V. and Morrison, E.S. (2018) Emergent buffering balances evolvability and robustness in the evolution of phenotypic flexibility. *Evolution* 72, 647–662.

Badyaev, A.V. and Oh, K.P. (2008) Environmental induction and phenotypic retention of adaptive maternal effects. *BMC Evolutionary Biology* 8, e3.

Badyaev, A.V. and Walsh, J.B. (2014) Epigenetic processes and genetic architecture in character origination and evolution. In: Charmantier, A., Garant, D. and Kruuk, L.E.B. (eds) *Quantitative Genetics in the Wild*. Oxford University Press, Oxford, pp. 177–189.

Badyaev, A.V. and Young, R.L. (2004) Complexity and integration in sexual ornamentation: an example with carotenoid and melanin plumage pigmentation. *Journal of Evolutionary Biology* 17, 1317–1327.

Badyaev, A.V., Hill, G.E., Beck, M.L., Dervan, A.A., Duckworth, R.A., *et al.* (2002) Sex-biased hatching order and adaptive population divergence in a passerine bird. *Science* 295, 316–318.

Badyaev, A.V., Beck, M.L., Hill, G.E. and Whittingham, L.A. (2003a) The evolution of sexual size dimorphism in the House Finch: V. Maternal effects. *Evolution* 57, 384–396.

Badyaev, A.V., Hill, G.E. and Beck, M.L. (2003b) Interaction between maternal effects: Onset of incubation is related to offspring sex in a passerine bird. *Oecologia* 135, 386–390.

Badyaev, A.V., Schwabl, H., Young, R.L., Duckworth, R.A., Navara, K. and Parlow, A.F. (2005) Adaptive sex differences in growth of pre-ovulation oocytes in a passerine bird. *Proceedings of the Royal Society Biological Sciences Series B* 272, 2165–2172.

Badyaev, A.V., Acevedo Seaman, D., Navara, K.J., Hill, G E. and Mendonça, M.T. (2006a) Evolution of sex-biased maternal effects in birds: III. Adjustment of ovulation order can enable sex-specific allocation of hormones, carotenoids, and vitamins. *Journal of Evolutionary Biology* 19, 1044–1057.

Badyaev, A.V., Hamstra, T.L., Oh, K.P. and Acevedo Seaman, D. (2006b) Sex-biased maternal effects reduce ectoparasite-induced mortality in a passerine bird. *Proceedings of the National Academy of Sciences USA* 103, 14406–14411.

Badyaev, A.V., Oh, K.P. and Mui, R. (2006c) Evolution of sex-biased maternal effects in birds: II. Contrasting sex-specific oocyte competition in native and recently established populations. *Journal of Evolutionary Biology* 19, 909–921.

Badyaev, A.V., Young, R.L., Hill, G.E. and Duckworth, R.A. (2008a) Evolution of sex-biased maternal effects in birds: IV. Intra-ovarian growth dynamics can link sex-determination and sex-specific acquisition of resources. *Journal of Evolutionary Biology* 21, 449–460.

Badyaev, A.V., Young, R.L., Oh, K.P. and Addison, C. (2008b) Evolution on a local scale: developmental, functional, and genetic bases of divergence in bill form and associated changes in song structure between adjacent habitats. *Evolution* 62, 1951–1964.

Badyaev, A.V., Belloni, V. and Hill, G.E. (2012) House Finch (*Haemorhous mexicanus*). In: Poole, A. (ed.) *The Birds of North America*. The Cornell Lab of Ornithology, Ithaca. New York. Available at: https://birdsna.org/Species-Account/bna/species/houfin (accessed 29 October 2019).

Badyaev, A.V., Potticary, A.L. and Morrison, E.S. (2017) Most colorful example of genetic assimilation? Exploring the evolutionary destiny of recurrent phenotypic accommodation. *American Naturalist* 190, 266–280.

Beal, E.E.L. (1907) Birds of California in relation to the fruit industry, Part 1. *Biological Survey Bulletin* 30, 34.

Berger, A.J. (1981) *Hawaiin Birdlife*, 2nd edn. University of Hawaii Press, Honolulu, Hawaii.

Brush, A.H. and Power, D.M. (1976) House Finch pigmentation: carotenoid metabolism and the effect of diet. *Auk* 93, 725–739.

Cookson, K.C. and Shivaprasad, H.L. (1994) *Mycoplasma gallisepticum* infection in chukar partridges, pheasants, and peafowl. *Avian Diseases* 38, 914–921.

Cooper, C.B., Hochachka, W.M. and Dhondt, A.A. (2007) Contrasting natural experiments confirm competition between house finches and house sparrows. *Ecology* 88, 864–870.

Dawson, W.R., Marsh, R.L., Buttemer, W.A. and Carey, C. (1983) Seasonal and geographic-variation of cold resistance in House Finches *Carpodacus mexicanus*. *Physiological Zoology* 56, 353–369.

Dhondt, A.A., Tessaglia, D.L. and Slothower, R.L. (1998) Epidemic mycoplasmal conjunctivitis in House Finches from eastern North America. *Journal of Wildlife Diseases* 34, 265–280.

Duckworth, R.A., Badyaev, A.V., Farmer, K.L., Hill, G.E. and Roberts, S.R. (2003) First case of mycoplasmosis in the native range of the House Finch (*Carpodacus mexicanus*). *Auk* 120, 528–530.

Elliot, J.J. and Arbib, R.S.J. (1953) Origin and status of the House Finch in the eastern United States. *Auk* 70, 31–37.

Evenden, F.G. (1957) Observations on nesting behavior of the house finch. *Condor* 59, 112–117.

Fernández-Juricic, E., Poston, R., de Collibus, K., Morgan, T., Bastain, B., *et al.* (2005) Microhabitat selection and singing behavior patterns of male House Finches (*Carpodacus mexicanus*) in urban parks in a heavily urbanized landscape in the Western U.S. *Urban Habitats* 3, 49–69.

Goodwin, C.E. (1976) The spring migration. Ontario Region. *American Birds* 30, 832–836.

Grinnell, J. (1911) The linnet of the Hawaiian Islands: a problem in speciation. *University of California Publications in Zoology* 7, 79–95.

Hamner, W.M. (1966) Photoperiodic control of annual testicular cycle in House Finch *Carpodacus Mexicanus*. *General and Comparative Endocrinology* 7, 224–233.

Harrison, C.J.O. (1978) *A Field Guide to the Nests, Eggs and Nestlings of North American Birds*. Collins, Toronto, Canada.

Hartup, B.K., Stott-Messick, B.A., Guzy, M. and Ley, D.H. (2004) Health survey of House Finches (*Carpodacus mexicanus*) from Wisconsin. *Avian Diseases* 48, 84–90.

Hill, G.E. (1993) Geographic variation in the carotenoid plumage pigmentation of male House Finches (*Carpodacus mexicanus*). *Biological Journal of the Linnean Society* 49, 63–89.

Hirai, L.T. (1975) The Hawaiian House Finch. *Elepaio* 36, 1–5.

Hochachka, W.M. and Dhondt, A.A. (2000) Density-dependent decline of host abundance resulting from a new infectious disease. *Proceedings of the National Academy of Sciences USA* 97, 5303–5306.

Howard, H. and Miller, A.H. (1933) Bird remains from cave deposits in New Mexico. *Condor* 35, 15–18.

Koes, R.F. and Taylor, P. (1993) Prairie Provinces region. *American Birds* 47, 269–270, 423–425.

Lindstedt, E.R., Oh, K.P. and Badyaev, A.V. (2007) Ecological, social, and genetic contingency of extrapair behavior in a socially monogamous bird. *Journal of Avian Biology* 38, 214–238.

Michener, H. and Michener, J.R. (1931) Variation in color of male House Finches. *Condor* 33, 12–19.

Mundinger, P.C. (1975) Song dialects and colonization in the House Finch. *Condor* 77, 407–422.

Oh, K.P. and Badyaev, A.V. (2006) Adaptive genetic complementarity coexists with selection for elaborate sexual traits in mate choice of a passerine bird. *Proceedings of the Royal Society of London B: Biological Sciences* 273, 1913–1919.

Peck, G.K. and James, R.D. (1987) *Breeding Birds of Ontario: Nidiology and Distribution*. Toronto, Miscellaneous Publications of the Royal Ontario Museum, Ontario, Canada.

Peterson, R.T. and Chalif, E.L. (1989) *Aves de México: Guía de campo [A Field Guide to Mexican Birds and Adjacent Central America]*. Editorial Diana, Mexico City, Mexico.

Pranty, B. (2005) Florida. *North American Birds* 59, 62–65.

Salvante, K.G., Dawson, A., Aldredge, R.A., Sharp, P.J. and Sockman, K.W. (2013) Prior experience with photostimulation enhances photo-induced reproductive response in female House Finches. *Journal of Biological Rhythms* 28, 38–50.

Stein, L.R. and Badyaev, A.V. (2011) Evolution of eggshell structure during rapid range expansion in a passerine bird. *Functional Ecology* 25, 1215–1222.

Stein, L.R., Oh, K.P. and Badyaev, A.V. (2009) Fitness consequences of male provisioning of incubating females in a desert passerine bird. *Journal of Ornithology* 151, 227–233.

Thompson, W.L. (1960) Agonistic behavior in the House Finch, II: factors in aggressiveness and sociality. *Condor* 62, 378–402.

van Riper, C. III. (1976) Aspects of House Finch breeding biology in Hawaii. *Condor* 78, 224–229.

Veit, R.R. and Lewis, M.A. (1996) Dispersal, population growth, and the Allee effect: dynamics of the House Finch invasion of eastern North America. *American Naturalist* 148, 255–274.

Walker, D.N. (2001) Faunal remains from the Chance Gulch site (5GN817), 2000–2002 excavations. PhD thesis, Western State College, Colorado.

Watts, H.E., Jimenez, D., Pacheco, V. and Vilgalys, T.P. (2018) Effects of temperature on the timing of breeding and molt transitions in House Finches. *Journal of Experimental Biology* 221, jeb185058.

White, D.H., Kepler, C.B., Hattfield, J.S., Sykes Jr., P.W. and Seginak, J.T. (1996) Habitat associations of birds in the Georgia Piedmont during winter. *Journal of Field Ornithology* 67, 159–166.

Woods, R.S. (1968) *Carpodacus mexicanus frontalis* (Say). House Finch. *United States National Museum Bulletin* 188, 290–314.

Wootton, J.T. (1986) Clutch-size differences in western and introduced eastern populations of House Finches: patterns and hypotheses. *Wilson Bulletin* 98, 459–462.

Wootton, J.T. (1987) Interspecific competition between introduced House Finch populations and two associated passerine species. *Oecologia* 71, 325–331.

20 Common Waxbill (*Estrilda astrild* Linnaeus, 1758)

Joana Ribeiro[1,2]*, Neftalí Sillero[3], Ricardo Jorge Lopes[1], Martin J.P. Sullivan[4], Joana Santana[1,2], César Capinha[1,2] and Luís Reino[1,2]

[1]CIBIO/InBio, Centro de Investigação em Biodiversidade e Recursos Genéticos, Laboratório Associado, Universidade do Porto, Campus Agrário de Vairão, 4485-661 Vairão, Portugal; [2]CIBIO/InBio, Centro de Investigação em Biodiversidade e Recursos Genéticos, Laboratório Associado, Instituto Superior de Agronomia, Universidade de Lisboa, Tapada da Ajuda, 1349-017 Lisbon, Portugal; [3]Centro de Investigação em Ciências Geo-Espaciais (CICGE), Faculdade de Ciências da Universidade do Porto (FCUP), Alameda do Monte da Virgem, 4430-146 Vila Nova de Gaia, Portugal; [4]School of Geography, University of Leeds, Leeds LS2 9JT, UK

Citation: Ribeiro, J., Sillero, N., Lopes, R., Sullivan, M., Santana, J., Capinha, C. and Reino, L. (2020) Common Waxbill (*Estrilda astrild* Linnaeus, 1758). In: Downs, C.T. and Hart, L.A. (eds) *Invasive Birds: Global Trends and Impacts*. CAB International, Wallingford, UK, pp. 155–158.

20.1 Common Names

Common Waxbill, St Helena Waxbill, Rooibeksysie (Afrikaans, South Africa).

20.2 Distribution

The Common Waxbill (*Estrilda astrild* Linnaeus, 1758) is a passerine native to the Afro-tropical region, with an extensive sub-Saharan distribution (Nuttall, 2005; Payne, 2010) (Fig. 20.1). It is also one of the most successful alien bird species globally, sustaining naturalized populations in Europe (Portugal and Spain), South and North America (Brazil, USA) and on a number of islands (e.g. Azores, Reunion, Oahu (Hawaii), Society Islands (Tahiti), New Caledonia, Amirante, parts of the Seychelles Archipelago, Mauritius, Rodrigues, Ascension, St Helena, São Tomé and Príncipe, Bermuda, Trinidad, Cape Verde and Canary Islands) (Xavier, 1968; Reino and Silva, 1998; Stiels *et al.*, 2011).

20.3 Description

The Common Waxbill is a small, grey-brown-coloured finch, distinguished by its red conical bill and face patch (Fig. 20.2). This finch's bill looks as if it has been dipped in red wax, explaining its common name. The cheeks, throat and belly are whitish-grey, while the rest of the plumage is finely barred, and the underside has a dusting of red. Adult Common Waxbills have a wingspan of 12–14 cm and a length of around 11.5 cm. They weigh approximately 8.9 g. The species has a relatively long, slender tail and rounded wings. Females are paler overall, with less red along the belly. The plumage of juveniles is duller than that of the adults, with little red on the underbelly and none on the bill. Nestlings have conspicuous white gape flanges along the edges of their mouths (Burton and Burton, 2002; Schuetz, 2005b).

20.4 Diet

The Common Waxbill is a granivorous bird, foraging in low vegetation. Due to its small size, it is extremely agile (Dunning, 2008), and is often observed taking small seeds upside down or in the bent stems of fragile plants (Clement *et al.*, 1993; Matias, 2002).

20.5 Introduction and Invasion Pathways

The Common Waxbill is a passerine bird of the family Estrildidae, often described as the 'single most important avicultural family', disputing this status only with the family Psittacidae. Commonly kept as caged birds around the world, waxbills have been used at least since the 19th century in Europe and Brazil (Sick, 1997; Cardoso and Reino, 2018), being transported by humans around the world ever since (Stiels *et al.*, 2011). Although a melodious and musical song is

*Corresponding author: joanateixeiraribeiro@gmail.com

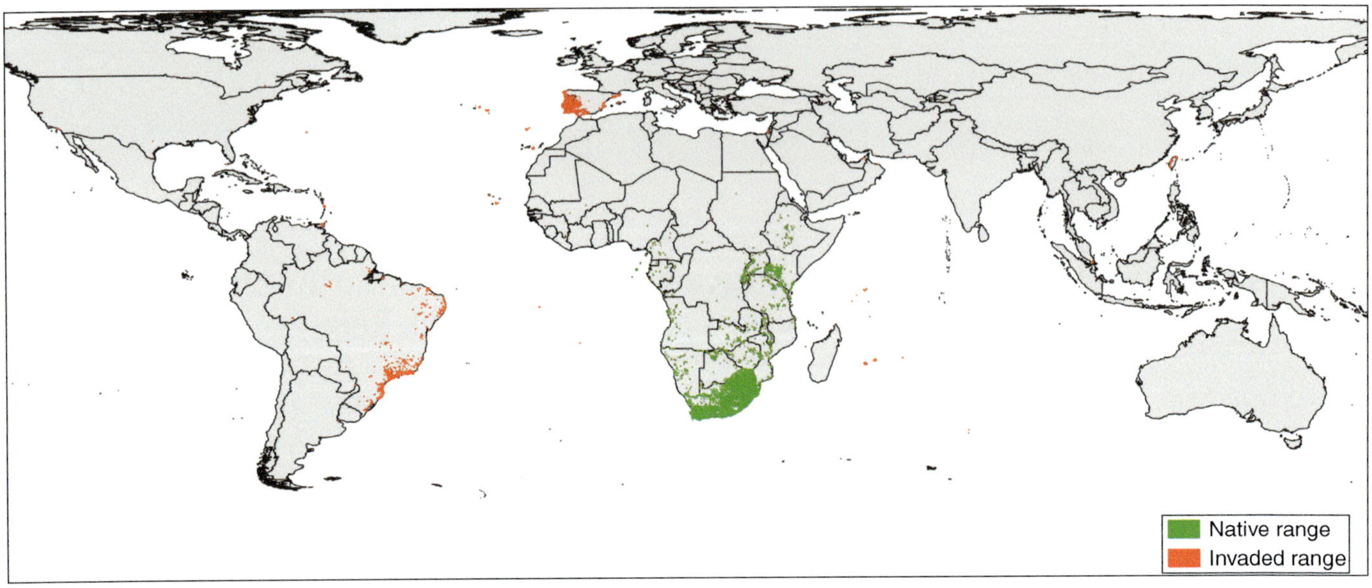

Fig. 20.1. Global distribution of the Common Waxbill (*Estrilda astrild*) showing the native (green) and invaded (red) ranges. Arrows indicate insular established invasive populations of Common Waxbills.

Fig. 20.2. Adult male Common Waxbill. (© Photograph: Gonçalo Cardoso.)

lacking, its popularity may be explained by: (i) its colourful plumage, easy management and suitability for keeping in cages or aviaries; (ii) their easy transportation and resilience to relatively long periods in cages, while in transit; (iii) the reasonable price and abundance in pet shops; and (iv) their suitability for novice breeders, requiring little effort.

20.6 Breeding Behaviour

While not territorial, Common Waxbills are highly gregarious, nesting scattered in loose colonies, with each pair building a domed nest and providing biparental care to nestlings (Clement *et al.*, 1993; Payne, 2010). Like all finches of the genus *Estrilda*, Common Waxbills are monogamous. Their song and display are both important aspects of courtship, and pair formation usually begins with a 'curtsy' and song exchange between the two prospective mates. Allopreening occurs frequently between the mates. During the nest building and solicitation period, both males and females may participate in stem displays to their mates – a form of display during which a stem is held in the beak. The male sings in an irregular pattern during this display, while the female remains silent. After pairing off, they separate from the larger flock to breed. Occasionally, they can be found in small territories adjacent to other pairs (Schuetz, 2005a). Males will also display to other females in the flock (those that are not his mate), but this display does not begin with a curtsy and is a type of 'fluffed singing'. Females almost always flee when they receive these displays, as they are advances from males with whom they are not mated. Males will attempt to mate with females that do not flee from his advances (Kunkel, 1968).

The time of breeding is highly variable across the Common Waxbill's native range (Payne, 2010). However, the breeding season usually takes place in midsummer in the northern and southern hemispheres, respectively. For example, in southern Africa, breeding occurs between September and January (Nuttall, 2005).

The Common Waxbill nest is a woven, spherical mass of grasses with a narrow entrance (Fig. 20.3). Nests are generally on or close to the ground, hidden in grassy vegetation. Nests are mostly built by females with the assistance of males in decorating and lining the inside with feathers. Both parents place animal scats in the nest throughout the nesting period to divert predators. A separate 'cock's nest' is built on top of the main nest, supposedly to serve as a resting place for the parent who is not

Fig. 20.3. An *Estrilda astrild* nest, with the main (low) and secondary entrance (up). (© Photograph: Cristiana Marques.)

incubating (Kunkel, 1968; Schuetz, 2005b). Common Waxbills have a clutch size of four to six eggs and may raise several broods a year. Incubation lasts 11–12 days, and fledging takes 17–21 days. Both male and female incubate and feed the helpless, altricial young. Common Waxbill juveniles reach reproductive maturity at 6–12 months of age (Burton and Burton, 2002).

20.7 Habitat

Common Waxbills are associated mostly with open habitats, often close to water bodies (Clement *et al.*, 1993; Payne, 2010). They may also be found in open mesic habitats, such as anthropogenic landscapes (Reino and Silva, 1998). Common Waxbills are sensitive to low temperatures (<15°C) and seem to suffer in particular from cold and wet weather conditions (Steinbacher and Wolters, 1965; Nicolai and Steinbacher, 2007), which may explain their ability to adapt to patchy landscapes and temperate or Mediterranean climates, with relatively mild winters and summers.

20.8 Impacts

The ecological role of the Common Waxbill varies according to its location. In Cape Verde and the Seychelles, invasive Common Waxbill populations have been shown to have negative effects on the crops they consume (Silva *et al.*, 2002). However, in the Iberian Peninsula, waxbills occupy a niche marginal to the ecological space occupied by native passerines, and thus appear not to compete strongly with natives (Batalha *et al.*, 2013). In the newly invaded Brazilian territories, Common Waxbills are reported to feed mostly on introduced grass species, which are rarely consumed by native Brazilian bird species (Silva *et al.*, 2018). Despite often being considered an invasive species, there are no reports of Common Waxbills displacing native species. However, as granivores, they may play a role in seed dispersal of the plants they feed on, which may be relevant if the consumed species are invasive.

20.9 Control

Common Waxbill young may suffer from an increased risk of predation due to their nests being positioned so close to the ground (Schuetz, 2005a). For example, in their native range, mice and snakes are reported as common predators targeting Common Waxbill eggs and young. However, predation on waxbills in invaded ranges has not been reported.

20.10 Uses

Common Waxbills have been kept as caged birds for human enjoyment since the 19th century.

20.11 Notes

Across its native range, the time of breeding for Common Waxbills is highly variable, often considered prolonged throughout the year (Payne, 2010; Sanz-Aguilar *et al.*, 2015). This opportunistic breeding strategy allows the Common Waxbill to capitalize on favourable conditions for reproduction whenever they occur. In addition, the ability of waxbill flocks to make vagrant movements in search of suitable habitat is equally regarded as important for their establishment and invasion success in many of their non-native areas (Payne, 2010). Another key to their success may be their release from parasite loads in non-native ranges. Lopes *et al.* (2018) reported a relatively low prevalence of parasites in Common Waxbills established in Portugal, contrasting with significantly higher prevalence in native regions.

20.12 References

Batalha, H.R., Ramos, J.A. and Cardoso, G.C. (2013) A successful avian invasion occupies a marginal ecological niche. *Acta Oecologica* 49, 92–98.

Burton, M. and Burton, R. (2002) *The Wildlife Encyclopedia*. Marshall Cavendish Corporation, Tarrtown, New York.

Cardoso, G.C. and Reino, L. (2018) Ecologically benign invasions: the invasion and adaptation of common waxbills (*Estrilda astrild*) in Iberia. In: Queiroz, A.I. and Pooley, S. (eds) *Histories of Bioinvasions in the Mediterranean*. Springer, Cham, Switzerland, pp. 149–169.

Clement, P., Harris, A. and Davis, J. (1993) *Finches and Sparrows: An Identification Guide*. Princeton University Press, Princeton, New Jersey.

Dunning, J. (2008) *CRC Handbook of Avian Body Masses*. CRC Press, Boca Raton, Florida.

Kunkel, P. (1968) Displays facilitating sociability in waxbills of the Genera *Estrilda* and *Lagonosticta* (Fam. Estrildidae). *Behaviour* 29, 237–260.

Lopes, R.J., Correia, J., Batalha, H. and Cardoso, G.C. (2018) Haemosporidian parasites missed the boat during the introduction of common waxbills (*Estrilda astrild*) in Iberia. *Parasitology* 145, 1493–1498.

Matias, R. (2002) *Aves Exóticas que Nidificam em Portugal Continental*. Institudo da Conserva da Natureza, Sociedade Portugesa para o Estudo das Aves, Lisbon.

Nicolai, J. and Steinbacher, J. (2007) *Handbuch der Vogelpflege. Prachtfinken. Afrika*, 3rd edn. Eugen Ulmer, Stuttgart, Germany.

Nuttall, R.J. (2005) Common Waxbill. In: Hockey, P.A.R., Dean, W.R.J. and Ryan, P.G. (eds) *Roberts Birds of Southern Africa*, 7th edn. Trustees of the John Voelcker Bird Book Fund, Cape Town, pp. 1051–1052.

Payne, R. (2010) Family Estrildidae (Waxbills). In: del Hoyo J., Elliott, A. and Christie D.A. (eds) *Handbook of the Birds of the World*, Vol. 15: *Weavers to New World Warblers*. Lynx Edicions, Barcelona, Spain, pp. 234–377.

Reino, L.M. and Silva, T. (1998) The distribution and expansion of the Common Waxbill in the Peninsula Iberica. *Biological Conservation and Fauna* 102, 163–167.

Sanz-Aguilar, A., Carrete, M., Edelaar, P., Potti, J. and Tella, J.L. (2015) The empty temporal niche: breeding phenology differs between coexisting native and invasive birds. *Biological Invasions* 17, 3275–3288.

Schuetz, J.G. (2005a) Common waxbills use carnivore scat to reduce the risk of nest predation. *Behavioral Ecology* 16, 133–137.

Schuetz, J.G. (2005b) Reduced growth but not survival of chicks with altered gape patterns: implications for the evolution of nestling similarity in a parasitic finch. *Animal Behaviour* 70, 839–848.

Sick, H. (1997) *Ornitologia Brasileira*. Nova Fronteira, Rio de Janeiro, Brazil.

Silva, J.M.C., Souza, M.A., Ribeiro, V. and Machado, R.B. (2018) Niche expansion of the common waxbill (*Estrilda astrild*) in its non-native range in Brazil. *Biological Invasions* 20, 2635–2646.

Silva, T., Reino, L.M. and Borralho, R. (2002) A model for range expansion of an introduced species: the common waxbill *Estrilda astrild* in Portugal. *Diversity and Distributions* 8, 319–326.

Steinbacher, J. and Wolters, H.E. (1965) *Vogel in Kafig und Voliere. Ein Handbuch fur Vogelliebhaber. Prachtfinken*, Vol. 1: *Astrilde*, 2nd edn. Hans Limberg, Aachen, Germany.

Stiels, D., Schidelko, K., Engler, J.O., van den Elzen, R. and Rödder, D. (2011) Predicting the potential distribution of the invasive Common Waxbill *Estrilda astrild* (Passeriformes: Estrildidae). *Journal of Ornithology* 152, 769–780.

Xavier, A. (1968). Bicos de Lacre em Óbidos. *Cyanopica* 1, 77–78.

21 Scaly-breasted Munia (*Lonchura punctulata* Linnaeus 1758)

Jason M. Gleditsch[1]* and Daniel M. Brooks[2]

[1]*Deptartment of Natural Resources and Environmental Science, University of Illinois, Champaign-Urbana, Urbana, IL 61801, USA;* [2]*Houston Museum of Natural Science, Department of Vertebrate Zoology, Houston, TX 77019, USA*

Citation: Gleditsch, J.M. and Brooks, D.M. (2020) Scaly-breasted Munia (*Lonchura punctulata* Linnaeus 1758). In: Downs, C.T. and Hart, L.A. (eds) *Invasive Birds: Global Trends and Impacts*. CAB International, Wallingford, UK, pp. 159–162.

21.1 Common Names

Nutmeg Finch, Nutmeg Mannikin, Nutmeg Munia, Ricebird, Scaly-breasted Manakin, Scaly-breasted Munia, Spice Finch, Spice Munia, Spotted Mannikin, Spotted Munia.

21.2 Distribution

There have been 12 subspecies of the Scaly-breasted Munia (*Lonchura punctulata* Linnaeus 1758) described that are native to northern Pakistan, India, southeast Asia, and the multiple islands and archipelagos of the region, including the Philippines, Polynesia, Thailand and Sri Lanka (Fig. 21.1). Specific distributions of the subspecies are as follows (Lepage *et al.*, 2014):

- *L. p. punctulata*: Nepal to Sikkim, India and Sri Lanka.
- *L. p. nisoria*: Java, Bali, Lombok and Sumbawa.
- *L. p. holmesi*: south-east Borneo (Kalimantan).
- *L. p. fretensis*: southern Thailand and Malay Peninsula to Sumatra and adjacent islands.
- *L. p. baweana*: Bawean Island.
- *L. p. subundulata*: north-east India (Assam) to Bhutan and western Myanmar.
- *L. p. topela*: southern China to northern Thailand, Indochina, Hainan and Taiwan.
- *L. p. yunnanensis*: north-east Myanmar and south-west China.
- *L. p. cabanisi*: the Philippines (Luzon, Mindoro, Panay, Cebu, Calauit and Palawan).
- L. p. particeps: Sulawesi.
- *L. p. sumbae*: Sumba (Lesser Sundas).
- *L. p. blasii*: Flores, Timor, Tanimbar Islands and adjacent Lesser Sundas.

The Scaly-breasted Munia has also been introduced to Mauritius, Réunion, French Polynesia, Guadeloupe, Hispaniola, Hawaii, Australia, Japan, Puerto Rico, Cuba, United Arab Emirates Portugal, Yucatán Peninsula, Cayman Islands and the USA (California, Texas, Florida and Mississippi; Fig. 21.1) (BirdLife International, 2019).

21.3 Description

The Scaly-breasted Munia is about 12 cm long and weighs 12–16 g (Payne, 2010). Adult birds have a dark brown head with lighter brown wings, tail and back (Fig. 21.2). The contour feathers on the breast and underside of adult birds are whitish grey with brown edging, creating a scalloping pattern on the underside of the bird. The undertail coverts do not have this pattern and appear whitish grey. Feathers on the head, back and wings have barely noticeable barring and pale shaft streaks (Payne, 2010). On the tail, the central rectrices have a golden tinge (Payne, 2010). The bills are blackish with a paler blueish grey at the base of the lower mandible. The only differences between the sexes are darker markings on the underside and a darker throat in males (Rasmussen and Anderton, 2005). However, there can be variations in plumage colour and size across the different populations in their native range (Payne, 2010).

Juveniles are lighter in colour having uniform pale-brown upperparts and lacking the dark head found in adults. Scaly-breasted Munia juveniles can often be confused with juveniles

*Corresponding author: jason.gleditsch@temple.edu

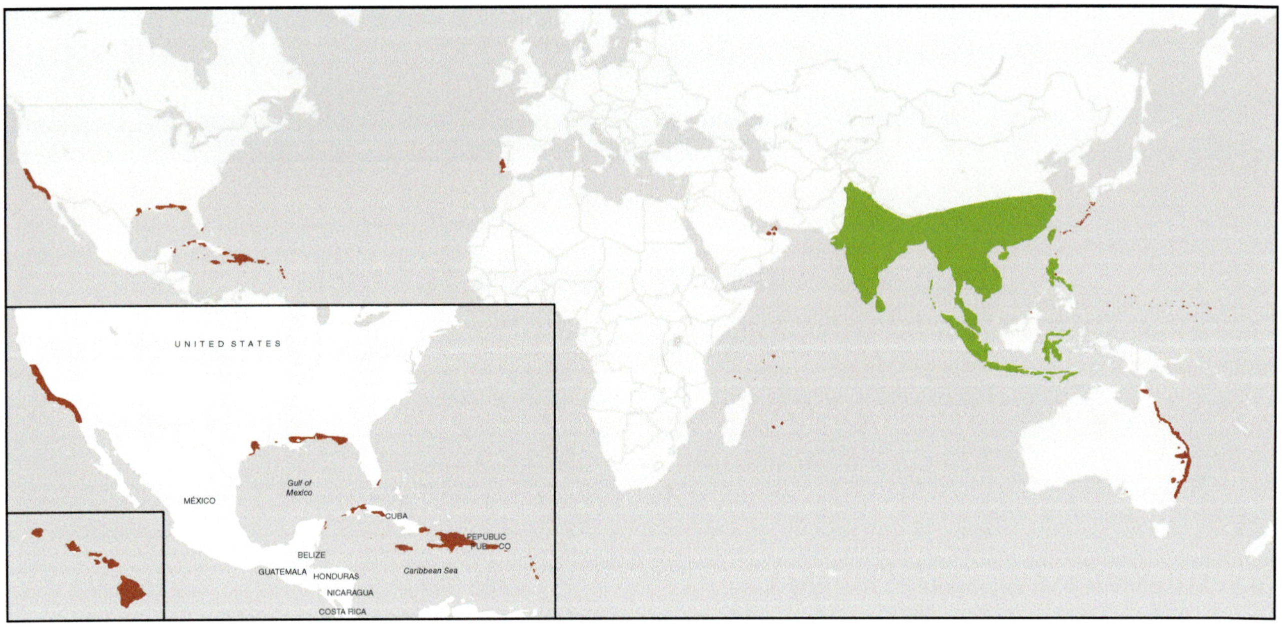

Fig. 21.1. Global distribution of the Scaly-breasted Munia (*Lonchura punctulata*) showing the native (green) and introduced (red) ranges; the inset shows North America. (Data from Sullivan *et al.*, 2009; Dyer *et al.*, 2017; Birdlife International, 2019).

Fig. 21.2. An adult Scaly-breasted Munia. (©Photograph: Doug Fincher.)

of other munia species such as the Tricolored Munia (*Lonchura malacca*) and Black-throated Munia (*Lonchura kelaarti*) (Restall, 1997; Rasmussen and Anderton, 2005).

The Scaly-breasted Munia song starts with repeated quiet notes followed by a series of whistles and churrs and ending with a slurred, longer 'weee'. They have two different contact calls. One contact call is a softer repeated 'tit-ti, tit-ti', while the other is a louder 'kit-teee, kit-teee'. The louder call is disyllabic, with the tone differing between sexes. The flight call is a rising 'puip' or 'sieuw', often heard in flocks (song characteristics adapted from Payne, 2010).

21.4 Diet

The Scaly-breasted Munia feeds mainly on seeds from grasses, crops and casuarinas (Ali and Ripley, 1987). Like some other munias, they may also feed on filamentous algae in shallow water (Wells, 2007; Payne 2010). In addition, Scaly-breasted Munias have been observed eating fleshy fruit such as those of *Lantana* spp. (Payne, 2010). Some insects and household scraps are also taken for food (Payne, 2010).

Invasive populations of Scaly-breasted Munias in Texas consume a variety of grass seeds but are common at feeders in urban areas, comprising 67% of all foraging observations in that region (Conn *et al.*, 2017).

21.5 Introduction and Invasion Pathways

Scaly-breasted Munias are a common cage bird and are used for religious ceremonies. It is thought that the introduced populations in the USA (California, Texas and Florida) were the result of escaped pets or were released intentionally as a part of ceremonies and rituals (Garrett and Garrett, 2016; Conn *et al.*, 2017). In Hawaii, the birds seem to have been introduced by M. Hillebrand in 1866 after an expedition to Asia (Meier, 2005; Pyle and Pyle, 2017). After the initial introduction to Honolulu, it is believed that they then dispersed throughout the other islands from this initial introduced population, as flocks have been observed at sea (Pyle and Pyle, 2017). It is thought that the ecological plasticity of the Scaly-breasted Munia is why it is so successful throughout its introduced range. In addition, the frequency at feeders in urban areas could be another reason why they are so widespread.

21.6 Breeding Behaviour

Scaly-breasted Munia breeding occurs throughout the year, with peaks during various months depending on the region. In India breeding occurs mainly in May–September, in Singapore they breed from February to September; in Borneo and Flores from March to July, and in Timor from May to June. In the Malay Peninsula, breeding occurs in all months except November, with the last brood usually fledging in October. Breeding occurs in almost all months in the Philippines, Thailand and Indonesia (breeding periods adapted from Voous, 1950; Payne, 2010).

Scaly-breasted Munia males attract females by holding grass in the bill and carrying it around (Payne, 2010). The male drops the grass when singing, during which he pivots his body and swings his head from side to side (Payne, 2010). The male then perches upright, fluffs its feathers and bobs, stretching and bending its legs (Payne, 2010).

In India, the Scaly-breasted Munia is known to nest in grass averaging 2.23 ± 0.64 m from the ground (Gokula, 2001). However, nests are rarely found below 4 m and can be found at up to 13 m concealed in trees, ferns, palms, creepers and epiphytes including orchids (Payne, 2010). Their nests are loosely constructed balls made of grass heads, strips of leaves and twigs, with a lower opening on the side (Payne, 2010), constructed by weaving the grass and leaves together. Inside, the nests are lined with fine grass. One nest in India took 6 days to finish and for egglaying to initiate (Lamba, 1974). Occasionally, old nests are used. Nests are often in colonial groups but can also be solitary in some areas. When nests are found in colonies, there can be ten or more nests in a single tree (Payne, 2010).

Clutches of three to six eggs are laid and incubated for 14–15 days (Gokula, 2001; Payne, 2010). After hatching, the nestlings stay in the nest for 18–19 days (Payne, 2010). There have been clutches larger than six, but these are thought to be due to multiple females laying eggs in a single nest. Multiple females laying in the same nest has been observed in the Malay Peninsula. In addition, some nests can contain young of various ages, providing support for the notion that multiple females can use the same nest. Up to five young can fledge from a single nest (Payne, 2010). Scaly-breasted Munias can have up to four clutches per season but usually have two (Restall, 1997; Smithson, 1997).

In the introduced range, less is known about the breeding behaviour of Scaly-breasted Munias. In Hispaniola, breeding was from June to October (Payne, 2010), in Texas it was from April to September (Conn *et al.*, 2017), and in California it was from February to November (Smithson, 1997). In California, nests are most commonly found in introduced pines at an average height of 5.6 m, and in Texas nests are found in various plant species 1.8–4.0 m from the ground (Conn *et al.*, 2017).

In California, Scaly-breasted Munias are hosts for the introduced nest parasite the Pin-tailed Whydah (*Vidua macroura*), and researchers believe that the spread of the Pin-tailed Whydah is directly related to the occurrence of Scaly-breasted Munia (Garrett and Garrett, 2016). Additionally, there is a record of parasitism of a Scaly-breasted Munia nest by a Shiny Cowbird (*Molothrus bonariensis*) in Puerto Rico (Pérez-Rivera, 1986).

21.7 Habitat

Scaly-breasted Munias are found in a range of habitats including grasslands with bushes, trees and scrub. They can also be found in scrubby mangroves, coastal landfill and anthropogenic habitats such as gardens and cultivated lands (Payne, 2010). In India, they are especially common in paddy fields where they are considered a minor pest due to their feeding on grain. While they are found mainly in the plains, they are also in the foothills of the Himalayas and can be seen at altitudes up to 2500 m above sea level (a.s.l.). On the Myanmar-Chinese border, they have been observed at up to 3000 m a.s.l. (Payne, 2010). Typically, they are found close to water and grassy habitats. For instance, in Pakistan, they can only be found from Swat in the west to Lahore but are absent from the desert regions (Abbass *et al.*, 2010).

Outside their native range, Scaly-breasted Munias occupy habitats that have similar characteristics to their native range, with only a few areas such as south-eastern Australia and the Persian Gulf exhibiting habitats that are less like that of the native range (Stiels *et al.*, 2015). In Texas, they prefer weedy fields and detention ponds, and are commonly found in residential gardens; more than half of all observations were in large parks (Brooks and Page, 2012; Conn *et al.*, 2017). In Hawaii, they are most commonly found in lowland fields, open woodlands and parks. In addition, on the island of Oahu, they are more commonly observed in lowland forest with sparse canopies than intact forests (J.M. Gleditsch *et al.*, unpublished data). Records indicate that they have been found at up to 2500 m a.s.l. on Mauna Kea and are only occasionally found in openings to native forest (Pyle and Pyle, 2017). In Japan, they are found in riparian reed bed habitats (Eguchi and Amano, 2004).

21.8 Impacts

In many areas, Scaly-breasted Munias are regarded as an agricultural pest, feeding in large flocks on cultivated cereals such as rice. In Australia, they are believed to be a competitor with native estrildids, but in other areas of the introduced range, competition with native species has not been observed or reported (Conn *et al.*, 2017). Additionally, there may be some concern about this species and the spread of parasites and diseases, given their abundance at feeders and their propensity to form mixed-species flocks. The first report of *Sternostoma tracheacolum*, a respiratory parasite, in Hawaii was found in other estrildid finches in an atypical locale (Smith, 1973). However, Scaly-breasted Munias are not susceptible to avian malaria (*Plasmodium relictum*) in Hawaii (Atkinson *et al.*, 1995), suggesting that this concern may be specific to disease and location.

21.9 Control

Trapping of Scaly-breasted Munias has occurred in many areas, including parts of their native range, to control their impact on agriculture. In 1934, tens of thousands were trapped to minimize their impact on rice production in Hawaii, with limited success (Bryan, 1937).

21.10 Uses

The Scaly-breasted Munia is a common bird used in Buddhist Feng Sheng ceremonies, making up 35% of birds sold in Phnom Penh, Cambodia (Gilbert *et al.*, 2012). Additionally, Scaly-breasted Munias are relatively inexpensive cage birds, making them popular in the pet bird trade, or are purchased for release during religious ceremonies (Conn *et al.*, 2017).

21.11 Notes

The Scaly-breasted Munia is a model species for foraging and physiological studies due to its flocking behaviour and the ease of keeping them in captivity. Most notably they were used by Giraldeau *et al.* (1994) to test the producer–scrounger game, and later by others for understanding mixed-species flocks (e.g. Beauchamp and Livoreil, 1997; Giraldeau and Beauchamp, 1999; Rieucau and Giraldeau, 2008).

21.12 References

Abbass, D., Rais, M., Ghalib, S.A. and Khan, M.Z. (2010) First Record of Spotted Munia (*Lonchura punctulata*) from Karachi. *Pakistan Journal Zoology* 42, 503–505.

Ali, S. and Ripley, S.D. (1987) *Handbook of the birds of India and Pakistan*, Vol. 5: *Larks to Grey Hypocolius*, 2nd edn. Oxford University Press, Bombay, India.

Atkinson, C.T., Woods, K.L., Dusek, R.J., Sileo, L.S. and Iko, W.M. (1995) Wildlife disease and conservation in Hawaii: pathogenicity of avian malaria (*Plasmodium relictum*) in experimentally infected Iiwi (*Vestiaria coccinea*). *Parasitology* 111, S59–S69.

Beauchamp, G. and Livoreil, B. (1997) The effect of group size on vigilance and feeding rate in spice finches (*Lonchura punctulata*). *Canadian Journal of Zoology* 75, 1526-1531.

BirdLife International (2019) Species factsheet: *Lonchura punctulata*. Available at: http://datazone.birdlife.org/species/factsheet/scaly-breasted-munia-lonchura-punctulata (accessed 29 October 2019).

Brooks, D.M. and Page, G. (2012) High diversity of invasive passerids at a park in southeast Texas. *Texas Ornithological Society* 45, 23.

Bryan, E.H. Jr (1937) *Birds of Oahu. Paradise of the Pacific* 49(1), 22; (2), 24–25; (4), 24, (5) 22–31; (6), 20.

Conn, A., Lazenby, L. and Brooks D.M. (2017) Ecology, behavior and reproduction of an invasive population of Scaly-breasted Munias (*Lonchura punctulata*) in Houston, Texas. In: Brooks, D.M. (ed.) *Half a Century of Ornithology in Texas: The Legacy of Dr. Keith Arnold*. Miscellaneous Publications, Houston Museum of Natural Sciences, Houston, Texas, pp. 40–53.

Dyer, E.E., Redding, D.W. and Blackburn, T.M. (2017) The global avian invasions atlas, a database of alien bird distributions worldwide. *Scientific Data* 4, 170041.

Eguchi, K. and Amano, H.E. (2004) Spread of exotic birds in Japan. *Ornithological Science* 3, 3–11.

Garrett, J.F. and Garrett, K.L. (2016) The Pin-tailed Whydah as a brood parasite of the Scaly-breasted Munia in southern California. *Western Birds* 47, 314–320.

Gilbert, M., Sokha, C., Joyner, P.H., Thomson, R.L. and Poole, C. (2012) Characterizing the trade of wild birds for merit release in Phnom Penh, Cambodia and associated risks to health and ecology. *Biological Conservation* 153, 10–16.

Giraldeau, L.A. and Beauchamp, G. (1999) Food exploitation: searching for the optimal joining policy. *Trends in Ecology & Evolution* 14, 102–106.

Giraldeau, L.A., Soos, C. and Beauchamp, G. (1994) A test of the producer-scrounger foraging game in captive flocks of spice finches, *Lonchura punctulata*. *Behavioral Ecology and Sociobiology* 34, 251–256.

Gokula, V. (2001) Nesting ecology of the Spotted Munia *Lonchura punctulata* in Mudumalai Wildlife Sanctuary (South India). *Acta Ornithologica* 36, 1–5.

Lamba, B.S. (1974) Nest construction technique of the Spotted Munia, *Lonchura punctulata*. *Journal of Bombay Natural History Society* 71, 613–616.

Lepage, D., Vaidya, G. and Guralnick, R. (2014) Avibase – a database system for managing and organizing taxonomic concepts. *ZooKeys* 420, 117. Available at https://avibase.bsc-eoc.org/species.jsp?avibaseid=6D4F72B157D99627 (accessed 21 January 2019).

Meier, U.H. (2005) *Hawai'i's Pioneer Botanist Dr. William Hillebrand: His Life and Letters*. B.P. Bishop Museum Press, Honolulu, Hawaii.

Payne, R.B. (2010) Family Estrildidae (Waxbills). In: del Hoyo, J., Elliot, A. and Christie, D.A. (eds) *Handbook of the Birds of the World*, Vol. 15: *Weavers to New World Warblers*. Lynx Edicions. Barcelona, Spain, pp. 234–377.

Pérez-Rivera, R.A. (1986) Parasitism by the Shiny Cowbird in the interior parts of Puerto Rico. *Journal of Field Ornithology* 57, 99–104.

Pyle, R.L., and Pyle, P. (2017) *The Birds of the Hawaiian Islands: Occurrence, History, Distribution, and Status*. B.P. Bishop Museum, Honolulu, Hawaii. Available at http://hbs.bishopmuseum.org/birds/rlp-monograph/pdfs/07-ZOST-ESTR/NUMA.pdf (accessed 21 January 2019).

Rasmussen, P.C. and Anderton, J.C. (2005) *Birds of South Asia: The Ripley Guide*. Lynx Edicions, Barcelona, Spain, and Smithsonian Institution, Washington, DC.

Restall, R. (1997) *Munias and Mannikins*. Yale University Press, New Haven, Connecticut.

Rieucau, G. and Giraldeau, L.A. (2008) Group size effect caused by food competition in nutmeg mannikins (*Lonchura punctulata*). *Behavioral Ecology* 20, 421–425.

Smith, H.E. (1973) A rhinonyssid mite in atypical loci of Estrildine finches in Hawaii. *Journal of Parasitology* 59, 1148–1148.

Smithson, W.S. (1997) Breeding biology and habitat use of the Orange Bishop (*Euplectes franciscanus*) and Nutmeg Mannikin (*Lonchura punctulata*) in southern California: 1997 breeding season report. Unpublished report, California Department of Food and Agriculture, Sacramento, California.

Stiels, D., Gaißer, B., Schidelko, K., Engler, J.O. and Rödder, D. (2015) Niche shift in four non-native estrildid finches and implications for species distribution models. *Ibis* 157, 75–90.

Sullivan, B.L., Wood, C.L., Iliff, M.J., Bonney, R.E., Fink, D. and Kelling, S. (2009) eBird: a citizen-based bird observation network in the biological sciences. *Biological Conservation* 142, 2282–2292.

Voous, K.H. (1950) The breeding seasons of birds in Indonesia. *Ibis* 92, 279–287.

Wells, D.R. (2007) *The Birds of the Thai-Malay Peninsular*, Vol. II: *Passerines*. Christopher Helm, London.

22 Northern Red Bishop (*Euplectes franciscanus* Isert 1789)

Jason M. Gleditsch[1]*, Katherine G. Ohman[2], Chloe I. Marcheli[3], Alyssa R. Conn[2], Amy E. Dunham[3] and Daniel M. Brooks[2]

[1]*Department of Natural Resources and Environmental Science, University of Illinois, Champaign-Urbana, Urbana, IL 61801, USA;* [2]*Houston Museum of Natural Science, Department of Vertebrate Zoology, Houston, TX 77019, USA;* [3]*Department of BioSciences, Rice University, Houston, TX 77005, USA*

Citation: Gleditsch, J.M., Ohman, K., Marcheli, C., Conn, A., Dunham, A.E. and Brooks, D.M. (2020) Northern Red Bishop (*Euplectes franciscanus* Isert 1789). In: Downs, C.T. and Hart, L.A. (eds) *Invasive Birds: Global Trends and Impacts*. CAB International, Wallingford, UK, pp. 163–169.

22.1 Common Names

Franciscan Bishop, Orange Bishop, Northern Red Bishop, West Nile Red Bishop.

22.2 Distribution

The native range of the Northern Red Bishop (*Euplectes franciscanus* Isert 1789) is confined to sub-Saharan Africa (Fig. 22.1), reaching south into central Africa (BirdLife International, 2019), specifically, southern Mauritania south to northern Liberia, east to Eritrea, Ethiopia, north-western and southern Somalia, north-eastern Democratic Republic of the Congo, Uganda and Kenya (Lepage *et al.*, 2014). There are only two described subspecies: *E. f. franciscanus* in the western range and *E. f. pusillus* in the eastern range (Lepage *et al.*, 2014). However, the geographical delineation of the subspecies is thought to be unwarranted, making this species monotypic (Craig, 2010). The Northern Red Bishop is often considered a conspecific of the Southern Red Bishop (*Euplectes orix*), which is found in southern Africa, but this is not supported by molecular data (Craig, 2010). The similarities between these species may contribute to sightings of the Northern Red Bishop outside the native range through misidentification.

The Northern Red Bishop was introduced and established in Jamaica, Puerto Rico, Bermuda, Guadeloupe, Martinique and the USA (Florida, Texas, California, Arizona) (Fig. 22.1) (BirdLife International, 2019). Although they were initially introduced to Oahu, Hawaii, in 1965, they did not establish, with only occasional sightings since then around the Pearl Harbor region (Pyle and Pyle, 2017).

Much of the information in this chapter stems from data analysed from the state of Texas, USA, that were collected as part of a Citizen Science programme (Texas Invasive Bird Project: www.hmns.org/invasivebirds, accessed 30 October 2019), with methods similar to those of Conn *et al.* (2017) with dates spanning a decade (June 2008–July 2017). Of 216 sightings of Northern Red Bishops in Texas, 184 (85%) occurred in Houston and the surrounding area (Fig. 22.2). The majority occurred towards west Houston, with the highest concentrations at Addicks Reservoir/Bear Creek Park and Arthur Storey Park (Fig. 22.2).

22.3 Description

The Northern Red Bishop is around 11 cm long and weighs 12–22 g. Adult males during breeding season are all orange-red except for a black crown, mask, belly and flanks (Fig. 22.3). The wings (coverts and flight feathers), tail and thighs are brown. The tail is typically covered by red to orange uppertail and undertail coverts. The bill is thick, conical and black, and the legs are pale brown to flesh-coloured. The adult males during the non-breeding season are brown with dark streaks in the centre and paler edges of the feathers and are indistinguishable from females except in size, with females being smaller. Females also have a yellowish supercilium

*Corresponding author: jason.gleditsch@temple.edu

Fig. 22.1. Global distribution of the Northern Red Bishop (*Euplectes franciscanus*) showing the native (green) and introduced (red) ranges. (Data from Dyer *et al.*, 2017; Sullivan *et al.*, 2009).

and brown to buff lores (Fig. 22.3). The throat, belly and undertail coverts of females are white and the breast is buffy with light streaking. Female bills are brown to flesh-coloured. Juveniles are similar to females but have wider buffy feather edges until post-juvenile moult. Males do not moult into their breeding plumage until their second year.

The Northern Red Bishop's song consists of thin, squeaky notes followed by 'zee-zee-zee' buzzing and sizzling sounds. When flying, they make a rattling call and give a high-pitched 'tsip' as a contact call. Their alarm call is a harsh 'chak' sound (description information adapted from Craig, 2010).

22.4 Diet

Northern Red Bishops feed primarily on small grass seeds and some insects (Craig, 2010). They mainly consume seeds from grasses but also some crop seeds. Millet and sorghum (*Sorghum bicolor*) are preferred (Craig, 2010). Rice has been found to be a major component of their diet in regions of its native range, leading the Northern Red Bishop to be labelled as an agricultural pest. They forage mostly on the ground but will take insects in flight (Craig, 2010). During the non-breeding season, Northern Red Bishops often form mix-species flocks with other seed-eaters such as canaries (*Serinus* spp.), estrildid finches and congeners (Craig, 2010; Brooks and Page, 2012).

In the introduced range, Northern Red Bishops are commonly found at seed feeders, especially in the winter and early spring when seeds are less abundant (Garrett, 1998; Brooks and Page, 2012). In California, they have been observed feeding on emergent aquatic vegetation (e.g. *Polygonum* spp.; Garrett, 1998). They have been observed in mixed-species foraging flocks in both California and Texas (Garrett, 1998; Brooks and Page, 2012). The largest flocks of Northern Red Bishops in Houston occur between May and June, when flock sizes peak, averaging 19 birds in June (Fig. 22.4). Absolute peak flock sizes ranged from 40 individuals (November 2011) in a residential area to 70 (August 2009).

Foraging is the most frequent behaviour of Northern Red Bishops in Texas, accounting for 38% of all behaviour, followed by perching/resting at 26% and other activities (Fig. 22.5).

22.5 Introduction and Invasion Pathways

The Northern Red Bishop is a popular cage bird in the international pet trade and escaped and/or released birds are the

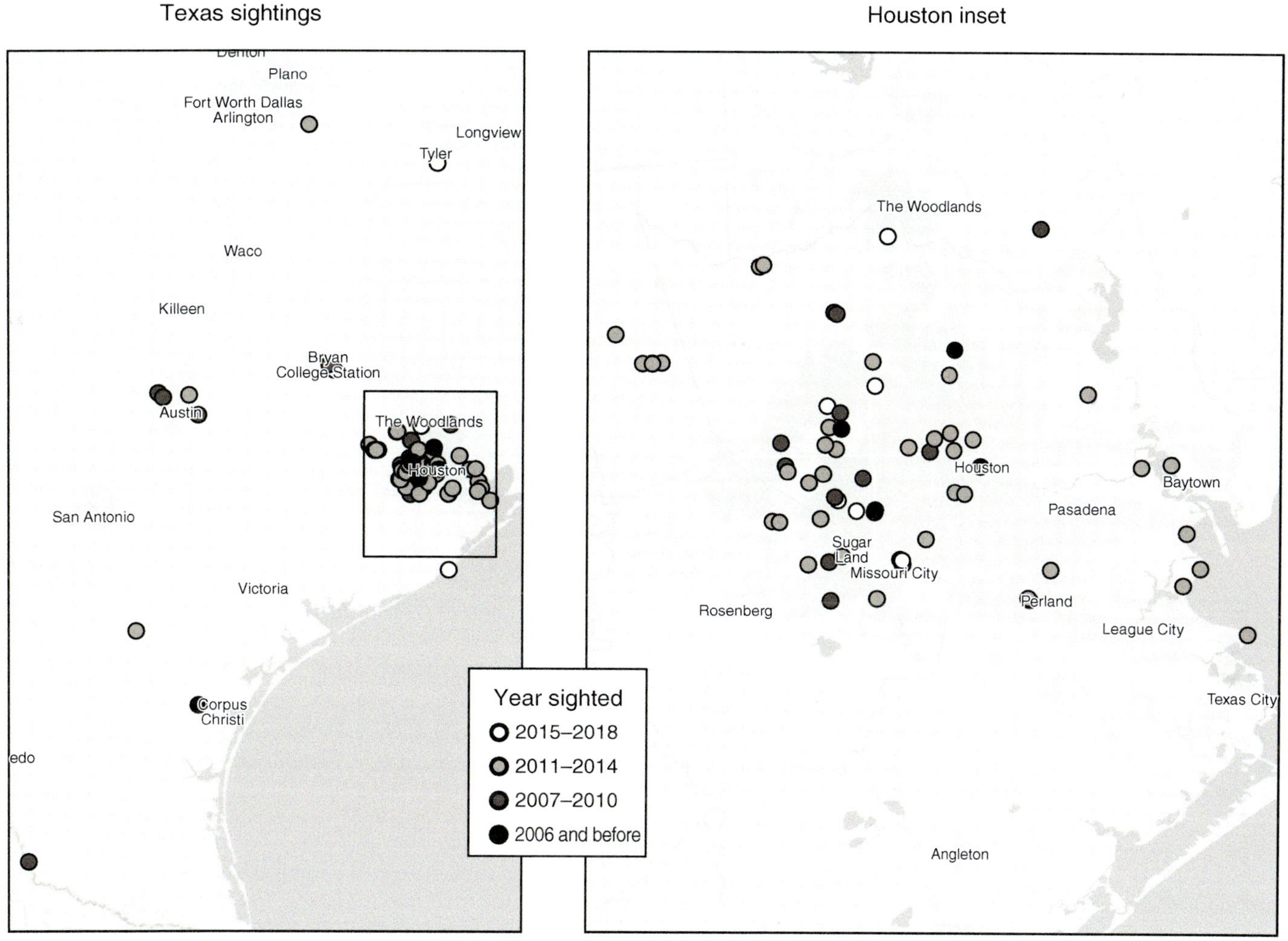

Fig. 22.2. Distribution of Northern Red Bishops in Texas, USA, and the city of Houston (inset), June 2008–July 2017.

Fig. 22.3. Male Northern Red Bishop in breeding plumage (left) and female (right). (©Photograph: Mark Lear.)

likely introduction pathway, whether as escaped pets or intentionally released as part of ceremonies and rituals. The first recorded sightings of Northern Red Bishops outside the native range were in Hawaii in 1965 (Pyle and Pyle, 2017) but this

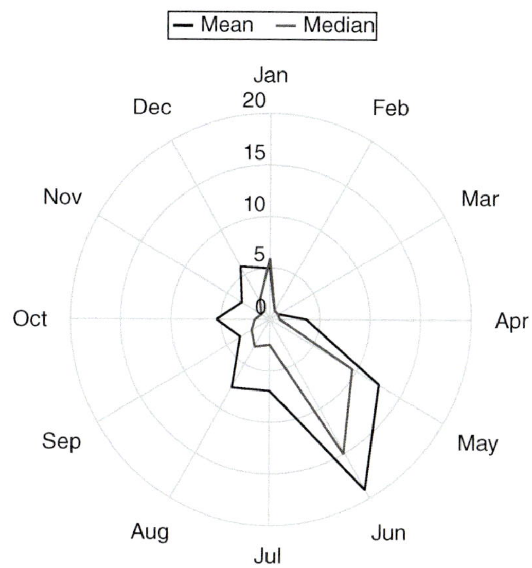

Fig. 22.4. Northern Red Bishop mean flock size by month in Texas, USA, June 2008–July 2017.

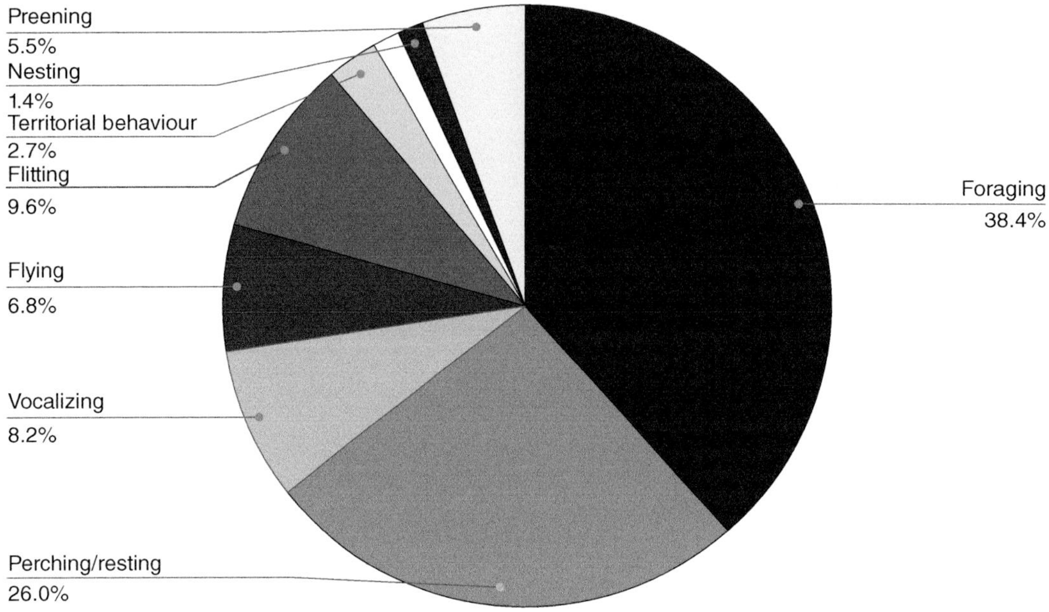

Fig. 22.5. Northern Red Bishop frequency of observed activity patterns in Texas, June 2008–July 2017.

population was unsuccessful at establishing. Later reports of Northern Red Bishops were recorded in Puerto Rico in 1972 (Long, 1981) and southern California in the 1970s (Garrett, 1998), where breeding populations have established. An established population was also reported in Arizona in 1998 (Dunn and Alderfer, 2008). They were not reported in Texas until 2002 and have been suspected of breeding since 2005 (Lockwood and Freeman, 2014). The population in Houston appears to be expanding over time (Fig. 22.2).

22.6 Breeding Behaviour

In their native range, Northern Red Bishops breed during various periods between February and November depending on location. For instance, in the most western part of the range (Gambia and Senegal), they breed from August to November, and the initiation and duration of the breeding season moves earlier in the year and longer along an eastward cline. The furthest eastern populations in Ethiopia breed in May–September. There are a few exceptions to this trend, with populations in Niger breeding in October, populations in Sudan breeding from February to March, and August to November, and populations in Somalia breeding in April and July (native range breeding periods adapted from Craig, 2010). In central portions of their range, the initiation of the breeding period appears to coincide with the end of the wet season, and the peak of breeding occurs during the start of the dry season (Cox *et al.*, 2013).

Northern Red Bishop males are territorial and are thought to be similar in behaviour to the Southern Red Bishop, which often defend their territories through direct confrontations (Craig, 1974). However, territories may be clustered (Craig, 2010). When females are observed in their territories, the male will display by 'bumble-flying' – flights within the territory where the male puffs its feathers and beats its wings slowly (Delacour and Edmond-Blanc, 1933). During these flights, the wings may make distinctive sounds (von Boetticher, 1952). The males of Southern Red Bishops will court any conspecific with brown plumage, regardless of whether it is an immature bird or female (Craig, 1974), and since the courtship of both species is so similar, this may also be the case for the Northern Red Bishop. Indeed, captive males have attempted to copulate with immature males (Craig, 1974). The Northern Red Bishop is polygynous and can have up to five females in a harem (Craig, 2010).

Northern Red Bishop nests are globular with a side entrance woven from coarse grass strips by the male. The nest is then lined by the female with grass flower heads (Craig, 2010). The nests are typically placed in grass, weeds and crops 1–2 m from the ground, and occasionally in bamboo, shrubs and small trees 3–6 m from the ground (Craig, 2010). The density of nests in millet is reported to be ten times greater than nest densities found in wild grasses (Craig, 2010). Clutches range from two to four blue eggs and are incubated only by the female for a period of 13–14 days (recorded in captivity; Craig, 2010). Nestlings are also only cared for by the female and typically fledge after 14–16 days.

In the introduced range, Northern Red Bishops breed in August–November in California, with nests averaging 1.1 m high (Garrett, 1998), and may be tied to the seeding *Echinochloa* spp. grasses and other important food resources (Smithson, 1997). Females possibly only produce one clutch of two to three eggs per year in California (Smithson, 1997). In the West Indies, they breed from March to November in grassy edges of sugarcane fields (Craig, 2010). In Texas, the male breeding plumage was observed with greatest frequency during summer and autumn (June–November; Fig. 22.6). While there was a single sighting of partially coloured plumage during winter and spring, the majority were in the autumn (Fig. 22.6). Full-coloured breeding plumage was only observed during the summer and autumn, with more sightings in the summer (Fig. 22.6).

Instances of territorial behaviour, reproductive behaviour and interspecific interactions were also observed during late summer (July–September) when most breeding plumage males were observed (Fig. 22.7).

22.7 Habitat

In their native range, Northern Red Bishops are commonly found in tall, open grasslands that can have some shrub encroachment (Craig, 2010). They are also found in tall crops and margins of agricultural lands. Some agriculture that they have been associated with includes rice in Mali, sugarcane in Mauritania, millet in Sudan and maize in Ethiopia (Craig, 2010). In the eastern portion of their native range, they can be found at elevations of between 600 and 1800 m above sea level (a.s.l.) (Stevenson and Fanshawe, 2004), and in Ethiopia they can be observed at up to 2000 m a.sl. but are also found in lowlands and coastal regions (Craig, 2010).

Northern Red Bishops were never observed in forest habitat during a study in Puerto Rico (Irizarry, 2012), and are thought to be associated mainly with the grassy margins of sugarcane fields in the West Indies (Craig, 2010). In California and Arizona, they inhabit weedy areas, river channels, flood basins and river bottoms (Garrett, 1998; Dunn and Alderfer, 2008). They are especially associated with *Echinochloa* spp.

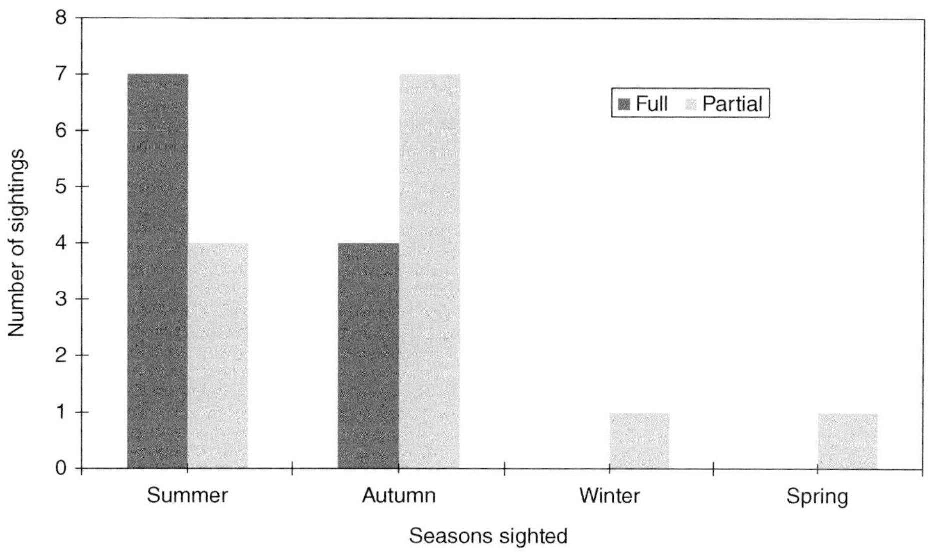

Fig. 22.6. Northern Red Bishop male breeding plumage sightings by season in Texas, June 2008–July 2017.

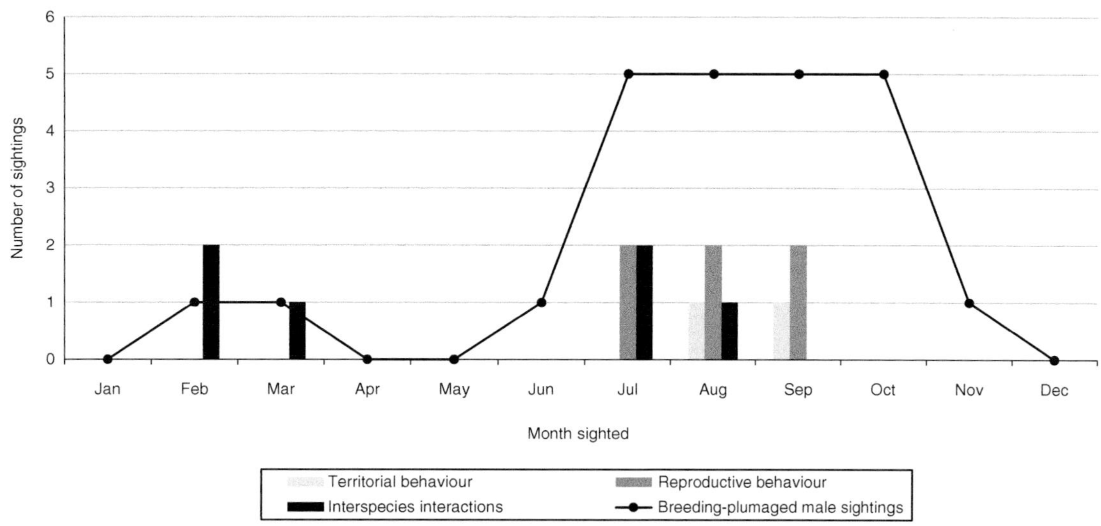

Fig. 22.7. Northern Red Bishop plumage synchronization with behaviour in Texas (June 2008 – July 2017).

grasses in California (Smithson, 1997). In Texas, they have been observed around urban and suburban habitats, including mowed fields, grassy edges of woods and golf courses, and ponds and river banks with high densities of reeds (Brooks and Page, 2012). In Texas, they are now moving away from their preferred habitat with masting seed heads of approximately 3 m in height to supplementary feeders (Brooks and Page, 2012).

In terms of microhabitat selection in Texas, of the 20 identified types of plants used for perching by the Northern Red Bishop, 35% (n = 7) were native to Texas, 45% (n = 9) were exotic species that occur outside of the native range of the Northern Red Bishop (Table 22.1), and 20% (n = 4) were indeterminate. Wild sunflower (*Helianthus* spp.; n = 4) and cattails (*Typha* spp.; n = 4) were the most frequent types of identified plant used for perching. Feeder perches (n = 5; Table 22.1) were the most used of the abiotic perch types observed. Of the 69 total biotic and abiotic perches, average reported perch height was 2.28 m off the ground. High perches were generally preferred (low perches $\leq$1.5 m: n = 23; high perches >1.5 m: n = 35) by Northern Red Bishops.

22.8 Impacts

In some areas of their native range, Northern Red Bishops are considered agricultural pests due to their foraging on and breeding in crops (e.g. Somalia) (Craig, 2010). In Texas, introduced Northern Red Bishops were found to be non-aggressive towards other species in ten interspecific interactions (Table 22.2). Two of the species – the Orange-cheeked Waxbill (*Estrilda melpoda*) and House Sparrow (*Passer domesticus*) – were also introduced (Table 22.2).

Table 22.1. Northern Red Bishop perch use in Texas.

Latin Name	Plant/Object	Origin	Reports	Height (m)	<=1.5 m	>1.5 m	No data
Anisacanthus wrightii	Flame Acanthus	N	1				X
Rhododendron sp.	Azalea	N/EO	1				X
Callistemon sp.	Bottlebrush	EO	1	4		X	
Verbena brasiliensis	Brazilian Vervain	EO	1	1.3	X		
Heterotheca subaxillaris	Camphorweed	N	1	1.3	X		
Typha sp.	Cattails	EO	4	2.65		X	
Triadica sebifera	Chinese tallow	EO	1	1	X		
Trifolium sp.	Clover	N/EO	1		X		
Lagerstroemia sp.	Crape Myrtle	EO	3	3		X	
Cupressus sp.	Cypress	EO	1	1	X		
Chilopsis linearis	Desert willow	N	1				X
Linum sp.	Flax	N/EO					X
Celtis occidentalis	Hackberry	N	2	3.6		X	
Hamelia patens	Hummingbird bush	EO	1				X
Ilex sp.	Holly Bush	N/EO	2	3		X	
Quercus sp.	Oak	N	3	3.5		X	
Nerium oleander	Oleander	EO	2				X
Quercus palustris	Pin Oak	EO	1	2		X	
	Unidentified grass		16	1.01	X		
	Unidentified rose		1				X
	Unidentified shrub		2	3.1		X	
	Unidentified tree		5	1.56		X	
Helianthus sp.	Wild sunflower	N	4	1.57		X	
Salix sp.	Willow	N	2	3		X	
	Abiotic Perches						
	brush pile		1	1.5	X		
	electrical wires		1				X
	feeder		5	2.03		X	
	fence		2				X
	metal post		1				X
	purple martin birdhouse		1	4.5		X	
	shepards hook		1	1.1	X		

N = Native Texas Plant
EO = Exotic plant whose native range lies outside the native distribution of the northern red bishop

Table 22.2. Northern Red Bishop interspecific foraging behaviour in Texas, USA.

Common name	Latin name	No. passive observations	Introduced
House Sparrow	*Passer domesticus*	2	✓
Orange-cheeked Waxbill	*Estrilda melpoda*	1	✓
Indigo Bunting	*Passerina cyanea*	2	
Painted Bunting	*Passerina ciris*	2	
Northern Cardinal	*Cardinalis cardinalis*	1	
Unidentified species		2	
Total	**5 species**	**10**	

22.9 Control

No control initiatives have been reported for Northern Red Bishops.

22.10 Uses

The Northern Red Bishop is sold in the world pet trade as a cage bird (BirdLife International, 2019).

22.11 References

BirdLife International (2019) *Euplectes franciscanus*. Available at: http://datazone.birdlife.org/species/factsheet/northern-red-bishop-euplectes-franciscanus (accessed 21 January 2019).

Brooks, D.M. and Page, G. (2012) High diversity of invasive passerids at a park in southeast Texas. *Texas Ornithological Society* 45, 23.

Conn, A., Lazenby, L. and Brooks D.M. (2017) Ecology, behavior and reproduction of an invasive population of Scaly-breasted Munias (*Lonchura punctulata*) in Houston, Texas. In: Brooks, D.M. (ed.) *Half a Century of Ornithology in Texas: The Legacy of Dr. Keith Arnold*. Miscellaneous Publications, Houston Museum of Natural Sciences, Houston, Texas, pp. 40–53.

Cox, D.T., Brandt, M.J., McGregor, R., Ottosson, U., Stevens, M.C. and Cresswell, W. (2013) The seasonality of breeding in savannah birds of West Africa assessed from brood patch and juvenile occurrence. *Journal of Ornithology* 154, 671–683.

Craig, A.J.F.K. (1974) Reproductive behaviour of the male Red Bishop Bird. *Ostrich* 45, 149–160.

Craig, A.J.F.K. (2010) Family Ploceidae (Weavers). In: del Hoyo, J., Elliot, A. and Christie, D.A. (eds) *Handbook of the Birds of the World*, Vol. 15: *Weavers to New World Warblers*. Lynx Edicions, Barcelona, Spain, pp. 74–197.

Delacour, J. and Edmond-Blanc, F. (1933) Monographie des Veuves (revision des genres *Euplectes* et Vidua). *L'Oiseau* 3, 519–562, 687–726.

Dunn, J.L. and Alderfer, J.K. (2008) *National Geographic Field Guide to the Birds of Western North America*. National Geographic Books, Washington, DC.

Dyer, E.E., Redding, D.W. and Blackburn, T.M. (2017) The global avian invasions atlas, a database of alien bird distributions worldwide. *Scientific Data* 4, 170041.

Garrett, K.L. (1998) Population trends and ecological attributes of introduced parrots, doves, and finches in California. In: Baker, R.O. and Crabb, A.C. (eds) *Proceedings of the 18th Vertebrate Pest Conference*. University of California, Davis, pp. 46–54.

Irizarry, J.I. (2012) Patch dynamics and permeability of fragmented habitat in southwestern Puerto Rico. MSc thesis, North Carolina State University, Raleigh, North Carolina.

Lepage, D., Vaidya, G. and Guralnick, R. (2014) Avibase – a database system for managing and organizing taxonomic concepts. *ZooKeys* 420, 117.

Lockwood, M.W. and Freeman, B. (2014) *The Texas Ornithological Society Handbook of Texas Birds*, 2nd edn. Texas A&M Press, College Station, Texas.

Long, J.L. (1981) *Introduced Birds of the World*. Reed, Wellington.

Pyle, R.L. and Pyle, P. (2017) *The Birds of the Hawaiian Islands: Occurrence, History, Distribution, and Status*. B.P. Bishop Museum, Honolulu, Hawaii. Available at: http://hbs.bishopmuseum.org/birds/rlp-monograph/NonEstablished.htm (accessed 21 January 2019).

Smithson, W.S. (1997) Breeding biology and habitat use of the Orange Bishop (*Euplectes franciscanus*) and Nutmeg Mannikin (*Lonchura punctulata*) in southern California: 1997 breeding season report. Unpublished report, California Department of Food and Agriculture, Sacramento, California.

Stevenson, T. and Fanshawe, J. (2004) *Birds of East Africa: Kenya, Tanzania, Uganda, Rwanda, Burundi*. A. & C. Black, London.

Sullivan, B.L., Wood, C.L., Iliff, M.J., Bonney, R.E., Fink, D. and Kelling, S. (2009) eBird: a citizen-based bird observation network in the biological sciences. *Biological Conservation* 142, 2282–2292.

von Boetticher, H. (1952) *Die Widahvögel und Witwen*. Geest & Portig, Leipzig, Germany.

23 Warbling White-eye (*Zosterops japonicus* Temminck & Schlegel 1845)

Matthew J. Burnett* and Colleen T. Downs
Centre for Excellence in Invasion Biology, School of Life Sciences, University of KwaZulu-Natal, Pietermaritzburg 3209, South Africa

Citation: Burnett, M.J. and Downs, C.T. (2020) Warbling White-eye (*Zosterops japonicus* Temminck and Schlegel 1845). In: Downs, C.T. and Hart, L.A. (eds) *Invasive Birds: Global Trends and Impacts*. CAB International, Wallingford, UK, pp. 170–174.

23.1 Common Names

Warbling White-eye, Japanese White-eye, Mountain White-eye, Daito White-eye, Oriental White-eye, Everett's White-eye, Enggano White-eye, Swinhoe's White-eye, Hume's White-eye, Sangkar White-eye, Ashy-Bellied White-eye, Javan White-eye, Mejiro (Japanese).

23.2 Nomenclature

It should be noted that White-eyes (*Zosterops* spp.) from the eastern hemisphere are regarded as having the fastest diversification rates among birds, which gave rise to naming the family as the 'great speciators' by Cornetti *et al.* (2015). The Warbling White-eye (*Zosterops japonicus* Temminck & Schlegel 1845) has recently been rearranged along with two other white-eye species from the eastern hemisphere: the Oriental White-eye (*Z. palpebrosus*) and Mountain White-eye (*Z. montanus*) (IOC, 2019; Lim *et al.*, 2019). These species, originally set apart by plumaged features, have now been shown to have phylogenetic variability and have undergone a rearrangement into five species: the Oriental White-eye (*Z. palpebrosus*), Swinhoe's White-eye (*Z. simplex* – *Z. japonicus* and *Z. palpebrosus* variations merged), Javan White-eye (*Z. melanurus* – variation split off from *Z. palpebrosus*), Citronella White-eye (*Z. citronella* – variation split off from *Z. palpebrosus*) and Warbling White-eye (*Z. japonicus* – *Z. montanus* merged into the *Z. japonicus* group) (Fig. 23.1) (Lim *et al.*, 2019). There were previously eight subspecies of *Z. japonicus*, and the reclassification and the merger with *Z. montanus*, 15 are now recognized: *Z. j. japonicus*, *Z. j. loochooensis*, *Z. j. diatoensis*, *Z. j. stejnegeri*, *Z. j. alani*, *Z. j. insularis*, *Z. j. montanus*, *Z. j. whiteheadi*, *Z. j. halconensis*, *Z. j. parkesi*, *Z. j. pectoralis*, *Z. j. diuatae*, *Z. j. vulcani*, *Z. j. difficilis* and *Z. j. obstinatus* (Lim *et al.*, 2019; IOC, 2019).

23.3 Distribution

The natural distribution of the Warbling White-eye extends from Japan and east Asia to the Greater Sundas, Lesser Sundas and Philippines, Borneo and Indonesia, as recent merges in species variations have suggested (Fig. 23.2) (Lim *et al.*, 2019). Previously, distribution ranges included parts of mainland China and Thailand (Lim *et al.*, 2019), which is now the distribution for the suggested new species *Z. simplex*. The Warbling White-eye (*Z. japonicus*) was introduced into Korea and on to the islands of Hawaii where it has become invasive, with the potential to invade other parts of the Central Pacific and Australasia (Scott *et al.*, 1986).

23.4 Description

As with all white-eyes, the Warbling White-eye has a diagnostic bold, white eye-ring with an overall greenish appearance and shows no sexual dimorphism (Fig. 23.3) (Lekagul and Round, 1991). The throat and undertail coverts are a pale lemon yellow, with the upper-parts of the bird being olive-green with less yellow (Smythies, 1953). The feet and legs are black, with a black slightly down-curved bill that extends from a narrow yellow band on the forehead (Smythies, 1953; King and Dickinson, 1991). Juvenile birds are similar to adults; however, they have a fainter eye-ring (Robson, 2015). They are small songbirds, 10–12 cm in body length and weighing

*Corresponding author: matthewburnett014@gmail.com

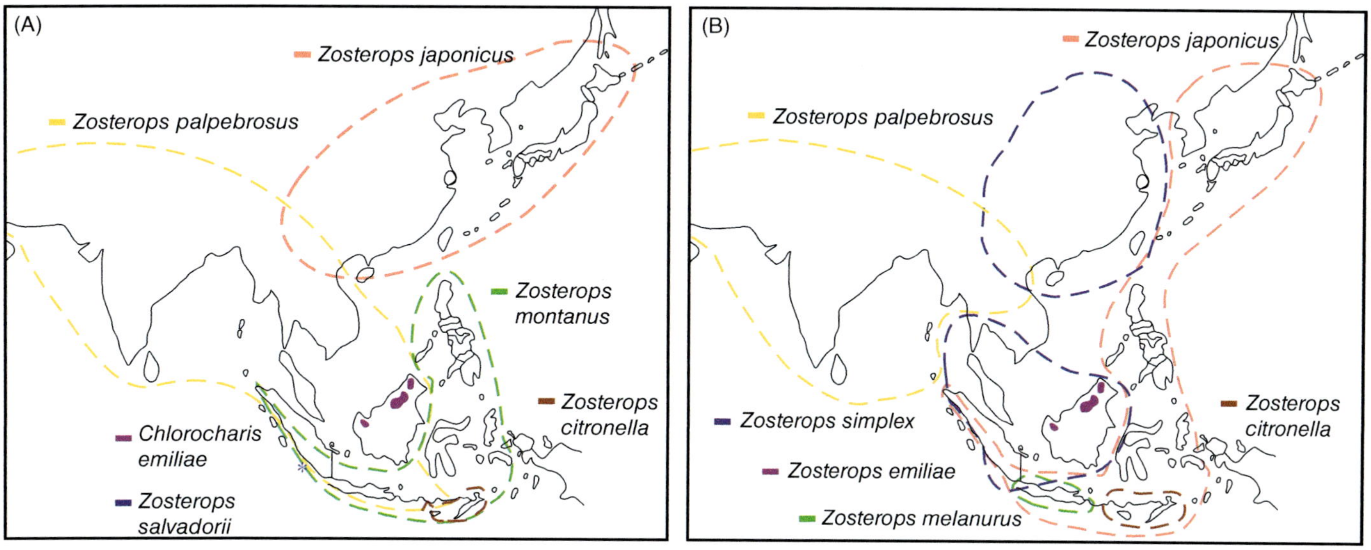

Fig. 23.1. (A) Natural distribution range of the three known *Zosterops* spp. affecting the reclassification of *Z. japonicus* (*Z. palpebrosus*, *Z. japonicus*, *Z. montanus*), with other species depicted that underwent similar reclassification. (B) ranges of the new suggested species with *Chlocharis emiliae* and *Z. salvadorii* being classified under *Z. emiliae* and still separated from *Z. japonicus*. (From Lim *et al.*, 2019.)

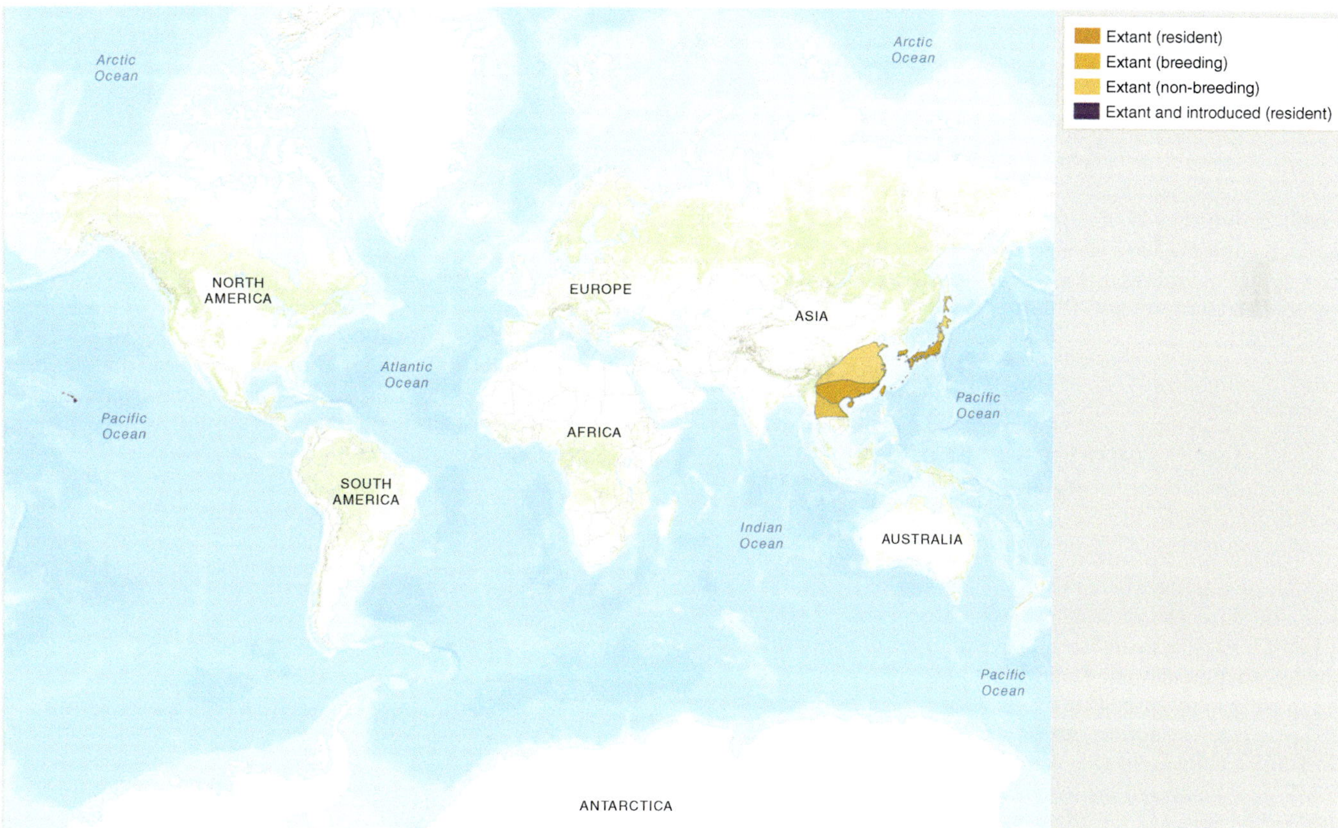

Fig. 23.2. Global distribution of the Warbling White-eye (*Zosterops japonicus*) and its introduced distribution depicted in the Central Pacific Islands including Hawaii (IUCN 2020). The distribution of the new arrangement is slightly different here compared with previous analysis, and because of the recent rearrangement, few databases have been updated (see Lim *et al.*, 2019 and Fig. 23.1).

Fig. 23.3. An adult Warbling White-eye. (©Photograph: James Eaton, Birdtour Asia.)

9.75–12.75 g (Brazil, 2018). Tables of morphological variations can be found in Lim *et al.* (2019). They often occur in flocks of five to 20 birds and are noisy with a simple high-pitched call (Pratt *et al.*, 1987). They are active and acrobatic by nature as they glean leaves and twigs in the canopy of trees and shrubs (Robson, 2015).

23.5 Diet

Warbling White-eyes are predominantly insectivorous, their stomach contents have shown them to be opportunistic with a diverse diet including, insect larvae, tiny fruit and nectar from flowers (MacKinnon and Phillipps, 2000; Gruner, 2004; Wada *et al.*, 2012).

23.6 Introduction and Invasion Pathways

All Asian white-eye species are regularly captured from wild populations across their range for the pet trade as a sought-after songbird (Scott *et al.*, 1986; Eaton *et al.*, 2015; Lewis 2017). This is jeopardizing wild populations of other white-eye species, particularly the Javan White-eye, which is regarded as the most heavily traded bird species in the world and is being threatened to extinction through trade (Lim *et al.*, 2019). The pet trade has brought the Warbling White-eye to many of the South and Central Pacific Islands (Kawakami and Higuchi, 2003). The invasive populations on the Hawaiian Islands are said to have come originally from the Warbling White-eye species group (*Z. japonicus*), introduced in 1929 as a pest control for crops on the islands. They have since invaded the surrounding islands in the Central Pacific through natural dispersion (Ely, 1971; Shallenberger, 1978).

23.7 Breeding Behaviour

Guest (1973) gave a detailed record of the breeding biology of Warbling White-eyes within an urban environment. The breeding season ranges from February to December, with a heightened period in July and August, where they form monogamous pairs and are highly territorial when nesting. Pairs are said to form during the non-breeding season where flocks of 20 can be seen together (Robson, 2015). They will lay two to five (average 3.1) pale blue eggs that take around 11 days to incubate, and both sexes share parental responsibility (Guest, 1973). The fledging period is approximately 20 days. Their nests are found at various heights above the ground in trees neatly nestled into a fork of a branch resembling a woven basket (average 56 mm in diameter, 41 mm deep) (Guest, 1973). Nesting material is diverse composing of grass, string, tin foil, leaves, moss and spider webs/cocoons. Nesting pairs hold on average an area of 6.5 ha, which is variable based on the availability of food to them; this could be smaller in natural areas where food is in higher abundance. Nesting pairs are tolerable to disturbances around the nest and unphased by the presence of other bird species nesting in proximity. The success rate from egg to fledging is high for such a small, altricial, tropical bird at 58.6% (Guest, 1973). Failure of nesting can be attributed to bad weather, poor nest construction, and interference by animals and humans. Furthermore, the adaptability of bettering a nest location after interference has been shown for this species (Guest, 1973; Horie and Takagi, 2012). Tables of nest measurements can be found in Guest (1973).

23.8 Habitat

Warbling White-eyes are habitat generalists. They occupy a wide range of tropical woody habitats and even persist in vegetated urban environments (Guest, 1973; Robson, 2015). Within their natural range, they occupy deciduous forest through to tropical rain forests but prefer open wooded habitats (Kurosawa and Askins, 2003). They frequent tree tops and canopies of rain forests, extending into secondary forest and shrubs and even coastal mangroves (Robson, 2015). They occur over a wide range of altitudes extending from sea level up to the treeline between 1525 and 1820 m in elevation, provided there is enough woody structure available to them (Robson, 2015).

23.9 Impacts of Species

23.9.1 Economic impact

Ironically, the Warbling White-eye, originally introduced on to the Hawaiian Islands to control insect pests on crops is now regarded as a crop pest itself, receiving the second highest number of complaints from farmers and affecting a wide range of crops from fruits to orchids (Koopman and Pitt, 2007).

23.9.2 Interspecific relationships with native species

The Warbling White-eye is considered to be the most abundant bird species in Hawaii and is regarded as a habitat generalist (Scott *et al.*, 1986; Boelman *et al.*, 2007). As a consequence, the Warbling White-eye competes directly for resources with native birds with similar niche guilds, many of which are faced with extinction, such as the Maui Creeper (*Paroreomyza montana*), the Hawaiian Amakihi (*Hemignathus virens*), the Hawaii Creeper (*Loxops mana*) and the Hawaii Akepa (*Loxops coccineus*) (Gruner, 2004; Samuel *et al.*, 2011; Rozek *et al.*, 2017) In some studies, this has been shown to negatively affect the juvenile growth rates of these native species and thus outcompete them (Freed and Cann, 2009; Rozek *et al.*, 2017). Interspecific competition was shown to negatively affect native birds because of limited resources rather than the Warbling White-eye negatively affecting their prey species (insects in this case) (Gruner, 2004). Their high abundance on the islands can be attributed to being habitat generalist and that the males improve the nest site selection over time to avoid nest predation (Horie and Takagi, 2012).

Warbling White-eyes are not specific to feeding off native plants and aid in the spread of the alien invasive plants that they frequent, threatening the native habitat suitable for native birds (Boelman *et al.*, 2007; Wu, 2012). Warbling White-eyes are found to be the most common visitor to the alien invasive fire-tree (*Myrica faya*) when in fruit (LaRosa *et al.*, 1985). Over one-third of faecal matter from Warbling White-eyes contains fertile seed from the fire-trees (Smathers and Gardner, 1979; LaRosa *et al.*, 1985; Woodward *et al.*, 1990). One positive interaction is that Warbling White-eyes, along with some other invaders, are replacing the endemic avifauna pollinators, ensuring the survival of certain endangered plants species (e.g. *Metrosideros polymorpha*) that have lost their avi-pollinators through extinctions (Chimera and Drake, 2010). Furthermore, on some islands, evidence suggests that the Warbling White-eye teaches native birds to feed off foreign food as part of their diet, as was the case for the Ogasawara Islands Honeyeater (*Apalopteron familiare*; Kawakami and Higuchi, 2003).

23.9.3 Spread of diseases to native species (avian malaria and pox)

Along with the introduction of non-native birds came avian-borne diseases to islands not adequately adapted to these diseases. The Warbling White-eye serves as a vector aiding in the spread of these diseases into isolated populations of native species, due to their high dispersal nature (Warner, 1968; Atkinson *et al.*, 2014). This is primarily in the case of avian malaria and pox (van Riper and Hansen, 2002; Woodworth *et al.*, 2005; Samuel *et al.*, 2011).

Overall, the Warbling White-eyes, because of their abundance on Hawaii and their nature as a habitat generalist, have negative effects on the islands' endemic species (Mountainspring and Scott, 1985; Gruner 2004). In a study by Baker *et al.* (2014), the Warbling White-eye was one of ten cases where non-native birds have been shown to drive extinction of native species, in this case due to direct competition with food resources. These effects, however, cannot be taken in isolation and are compounded by other factors such as habitat loss, invasion of mammalian species (rats) and other factors associated with island ecology (Mack *et al.*, 2000). In contrast, Lewis (2017), on the topic of naturalizing a non-native species, argued that, due to their high abundance, their presence in every habitat and their diverse food supply, the Warbling White-eye may be filling a role that might otherwise cause the loss of something not anticipated.

23.10 Control

The combination of local plants and native birds buffers the expansion of invasive plants and Warbling White-eyes, and thus intact native habitats are less susceptible to invasion by Warbling White-eyes (Boelman *et al.*, 2007). The idea behind ecological restoration through invasive eradication and minimization can limit the impact of Warbling White-eyes (Mack *et al.*, 2000; Boelman *et al.*, 2007). Predation on nests and individual birds is unlikely, unless in the presence of domestic cats (*Felis catus*), rodents and other small predatory mammals. Studies showed that native birds were more likely to be preyed on than Warbling White-eyes, as these birds have been known to adapt to predator avoidance techniques to increase their survival (Guest, 1973; Horie and Takagi, 2012). Bird trapping is being developed for the eradication of pest birds in Hawaii, where the Warbling White-eye is considered one of them (Koopman and Pitt, 2007).

23.11 Uses

Primarily, the Warbling White-eye makes an attractive caged bird as it tames well because of its gregarious nature, it has a melodious call and is a small bird (Lim *et al.*, 2019). The introduction to Hawaii, however, was because of their flocking and vigorous feeding nature that initially saw the species as a biological control for insects on the Hawaiian Islands. Its invasiveness has since then made the species unsuitable as a biological control agent (Scott *et al.*, 1986; Koopman and Pitt, 2007).

23.12 Notes

In the 1960s, the Warbling White-eye was sighted between Oahu and Johnston Atoll and on Johnston Atoll itself. These sightings were attributed as having come from the established population of Warbling White-eyes on the Hawaiian Islands (Ely, 1971). This is a remarkable distance for a bird of its size (400–1000 km depending on which part of the Hawaiian Islands), and through island hopping, it has occurred on other islands in the Central Pacific. Another interesting study on the species determined how the Warbling White-eye assisted in the dispersal of a snail species (*Tornatellides boeningi*) by ingesting these snails; the study by Wada *et al.* (2012) showed that around 14.3% of ingested snails passed through the digestive system unharmed.

23.13 References

Atkinson, C.T., Utzurrum, R.B., Lapointe, D.A., Camp, R.J., Crampton, L.H., *et al.* (2014) Changing climate and the altitudinal range of avian malaria in the Hawaiian Islands – an ongoing conservation crisis on the island of Kaua'i. *Global Change Biology* 20, 2426–2436.

Baker, J., Harvey, K.J. and French, K. (2014) Threats from introduced birds to native birds. *Emu-Austral Ornithology* 114, 1–12.

Boelman, N.T., Asner, G.P., Hart, P.J. and Martin, R.E. (2007) Multi-trophic invasion resistance in Hawaii: bioacoustics, field surveys, and airborne remote sensing. *Ecological Applications* 17, 2137–2144.

Brazil, M. (2018) *Birds of Japan*. Bloomsbury Publishing, London.

Chimera, C.G. and Drake, D.R. (2010) Patterns of seed dispersal and dispersal failure in a Hawaiian dry forest having only introduced birds. *Biotropica* 42, 493–502.

Cornetti, L., Valente, L.M., Dunning, L.T., Quan, X., Black, R.A., *et al.* (2015) The genome of the "great speciator" provides insights into bird diversification. *Genome Biology and Evolution* 7, 2680–2691.

Eaton, J.A., Shepherd, C.R., Rheindt, F.E., Harris, J.B.C., van Balen, S., *et al.*(2015) Trade-driven extinctions and near-extinctions of avian taxa in Sundaic Indonesia. *Forktail* 31, 1–12.

Ely, C.A. (1971) Pelagic observations of the Japanese White-eye in the central Pacific. *Condor* 73, 122–123.

Freed, L.A. and Cann, R.L. (2009) Negative effects of an introduced bird species on growth and survival in a native bird community. *Current Biology* 19, 1736–1740.

Gruner, D.S. (2004) Attenuation of top-down and bottom-up forces in a complex terrestrial community. *Ecology* 85, 3010–3022.

Guest, S.J. (1973) A reproductive biology and natural history of the Japanese White-eye (*Zosterops japonica japonica*) in urban Oahu. International Biological Program Technical Report, No. 29. Island Ecosystems IRP, US International Biological Program. Available at: http://hdl.handle.net/10125/25976 (accessed 29 October 2019).

Horie, S. and Takagi, M. (2012) Nest positioning by male Daito White-eyes *Zosterops japonicus daitoensis* improves with age to reduce nest predation risk. *Ibis* 154, 285–295.

IOC (2019) IOC World Bird List, version 9.2. Available at: www.worldbirdnames.org/updates/species-updates/ (accessed 28 February 2019).

IUCN (2020) Available at: https://www.iucnredlist.org/species/155158005/155636070#geographic-range (accessed 14 August 2020).

Kawakami, K. and Higuchi, H. (2003) Interspecific learning by the Ogasawara Islands Honeyeater *Apalopteron familiare* from the Japanese White-eye *Zosterops japonicus* on Hahajima, the Bonin Islands, southern Japan. *Ornithological Science* 2, 132–134.

King, B.F. and Dickinson, E.C. (1991) *Field guide to the birds of South-East Asia*. Collins, London.

Koopman, M.E. and Pitt, W.C. (2007) Crop diversification leads to diverse bird problems in Hawaiian agriculture. *Human-Wildlife Conflicts* 1, 235–243.

Kurosawa, R. and Askins, R.A. (2003) Effects of habitat fragmentation on birds in deciduous forests in Japan. *Conservation Biology* 17, 695–707.

LaRosa, A.M., Smith, C.W. and Gardner, D.E. (1985) Role of alien and native birds in the dissemination of firetree (Myricafaya Ait.-Myriacaceae) and associated plants in Hawaii. *Pacific Science* 39, 372–378.

Lekagul, B. and Round, P. D. (1991) *A guide to the birds of Thailand*. Saha Karn Bhact, Bangkok.

Lewis, D. (2017) *Belonging on an Island: Birds, Extinction, and Evolution in Hawai'i*. Yale University Press, New Haven, Connecticut.

Lim, B.T., Sadanandan, K.R., Dingle, C., Leung, Y.Y., Prawiradilaga, D.M., *et al.* (2019) Molecular evidence suggests radical revision of species limits in the great speciator white-eye genus *Zosterops*. *Journal of Ornithology* 160, 1–16.

Mack, R.N., Simberloff, D., Lonsdale, W.M., Evans, H., Clout, M. and Bazzaz, F.A. (2000) Biotic invasions: causes, epidemiology, global consequences, and control. *Ecological Applications* 10, 689–710.

MacKinnon, J. and Phillipps, K. (2000) *A Field Guide to the Birds of China*. Oxford University Press, Oxford.

Mountainspring, S. and Scott, J.M. (1985) Interspecific competition among Hawaiian forest birds. *Ecological Monographs* 55, 219–239.

Pratt, H.D., Bruner, P.L. and Berrett, D.G. (1987) *A Field Guide to the Birds of Hawaii and the Tropical Pacific*. Princeton University Press, Princeton, New Jersey.

Robson, C. (2015) *Birds of South-East Asia: Concise Edition*. Bloomsbury Publishing, London.

Rozek, J.C., Camp, R.J. and Reed, J.M. (2017) No evidence of critical slowing down in two endangered Hawaiian honeycreepers. *PLoS One* 12, e0187518.

Samuel, M.D., Hobbelen, P.H., DeCastro, F., Ahumada, J.A., LaPointe, D.A., *et al.* (2011) The dynamics, transmission, and population impacts of avian malaria in native Hawaiian birds: a modeling approach. *Ecological Applications* 21, 2960–2973.

Scott, J.M., Mountainspring, S., Ramsey, F.L. and Kepler, C.B. (1986) Forest bird communities of the Hawaiian Islands: their dynamics, ecology, and conservation. *Studies in Avian Biology* 9, 1–431.

Shallenberger, R.J. (1978) *Hawaii's Birds*. Hawaii Audubon Society, Honolulu, Hawaii.

Smathers, G.A. and Gardner, D.E. (1979) Stand analysis of an invading firetree (*Myrica faya* Aiton) population, Hawaii. *Pacific Science* 33, 239–255.

Smythies, B.E. (1953) *The Birds of Burma*. Oliver and Boyd, London.

van Riper, S.G. and Hansen, W.R. (2002) Epizootiology and effect of avian pox on Hawaiian forest birds. *Auk* 119, 929–942.

Wada, S., Kawakami, K. and Chiba, S. (2012) Snails can survive passage through a bird's digestive system. *Journal of Biogeography* 39, 69–73.

Warner, R.E. (1968. The role of introduced diseases in the extinction of the endemic Hawaiian avifauna. *The Condor* 70, 101–120.

Woodward, S.A., Vitousek, P.M., Matson, K., Hughes, F., Benvenuto, K. and Matson, P.A. (1990) Use of the exotic tree *Myrica faya* by native and exotic birds in Hawai'i Volcanoes National Park. *Pacific Science* 44, 88–93.

Woodworth, B.L., Atkinson, C.T., LaPointe, D.A., Hart, P.J., Spiegel, C.S., *et al.* (2005) Host population persistence in the face of introduced vector-borne diseases: Hawaii amakihi and avian malaria. *Proceedings of the National Academy of Sciences USA* 102, 1531–1536.

Wu, J.X. (2012) Movement patterns of the Omao and Japanese White-eye and implications for seed dispersal in Hawaii. PhD thesis, University of Hawaii at Hilo, Hilo, Hawaii.

24 House Crow (*Corvus splendens* Vieillot, 1817)

Cavin T. Shivambu*, Ndivhuwo Shivambu and Colleen T. Downs

Centre for Excellence in Invasion Biology and Centre for Functional Biodiversity School of Life Sciences, University of KwaZulu-Natal, P/Bag X01, Pietermaritzburg 3209, South Africa

Citation: Shivambu, C.T., Shivambu, N. and Downs, C.T. (2020) House Crows (*Corvus splendens* Vieillot, 1817). In: Downs, C.T. and Hart, L.A. (eds) *Invasive Birds: Global Trends and Impacts*. CAB International, Wallingford, UK, pp. 175–182.

24.1 Common Names

House Crow, Indian Grey-necked, Ceylon Crow, Colombo Crow.

24.2 Nomenclature

The House Crow (*Corvus splendens* Vieillot, 1817) belongs to the family Corvidae with five subspecies: *C. s. splendens*, *C. s. zugmayeri*, *C. s. protegatus*, *C. s. maledivicus* and *C. s. insolens* (Dean, 2005). In many countries, this species is named *C. splendens* because of the lack of genetic studies to determine which subspecies individuals belong to (Madge and Burn, 1994; Ottens, 2003).

24.3 Distribution

The House Crow's native range includes southern Iran, Pakistan, Burma, Nepal, Sri Lanka, Iran, Afghanistan, Thailand and extreme south-western China (Fig. 24.1) (Meininger *et al.*, 1980; Ali, 2002; Nyári *et al.*, 2006; Ryall, 2010). This species is regarded as one of the world's worst invasive species, established in approximately 25 countries (Suliman *et al.*, 2011). It is reported to be invasive in countries such as Hong Kong (Ryall, 2002), Singapore (Brook *et al.*, 2003), Ismailia Governorate, Egypt (Kamel, 2014), Tanzania (Shimba and Jonah, 2016) and South Africa (Nxele and Shivambu, 2018). In its native Pakistan (Khan, 2003) and India (Dhindsa *et al.*, 1991), it is regarded as a native invasive pest species.

The House Crow invaded range includes the following (Ryall, 2002; Brook *et al.*, 2003; Ottens and Ryall, 2003; Ottens, 2003; Suliman *et al.*, 2011; Kamel, 2014; Shimba and Jonah, 2016; Nxele and Shivambu, 2018):

- *Asia:* Bahrain, Chagos Archipelago, Hong Kong, Java, Sumatra, Israel, Kyushu, Jordan, Korea Republic, Kuwait, Peninsula Malaysia, Sabah, Oman, Qatar, Saudi Arabia, Singapore, Taiwan, Turkey, United Arab Emirates, Vietnam, Yemen.
- *Africa:* Benin, Djibouti, Egypt, Eritrea, Kenya, Madagascar, Mauritius, Morocco, Mozambique, Namibia, Seychelles, Somalia, South Africa, Sudan and Tanzania.
- *North America:* California, Florida, Louisiana, New Jersey and South Carolina.
- *Central America and Caribbean*: Barbados and Cuba.
- *South America:* Chile.
- *Europe:* Belgium, Cyprus, Denmark, France, Gibraltar, Hungary, Ireland, the Netherlands, Poland, Spain, the UK.
- *Oceania:* Australia.

Figure 24.1 indicates how well the House Crow has established outside its native range and how it is able to adapt and persist in different areas of the globe with varying climatic conditions. Its expansion varies in the different countries invaded.

24.4 Description

The House Crow is a medium-sized bird with a length of 43 cm and weighs about 245–300 g. Adults are shiny black, with the mantle and breast sooty grey to brown-grey (Fig. 24.2). The bill is black, and the eyes are brown. The legs and feet are black. Juveniles are dull, brownish black, and paler grey than adults on breast and mantle. Males and females are alike, although males are generally larger (Madge and Burn, 1994; Dean, 2005).

*Corresponding author: shivambucavin@gmail.com

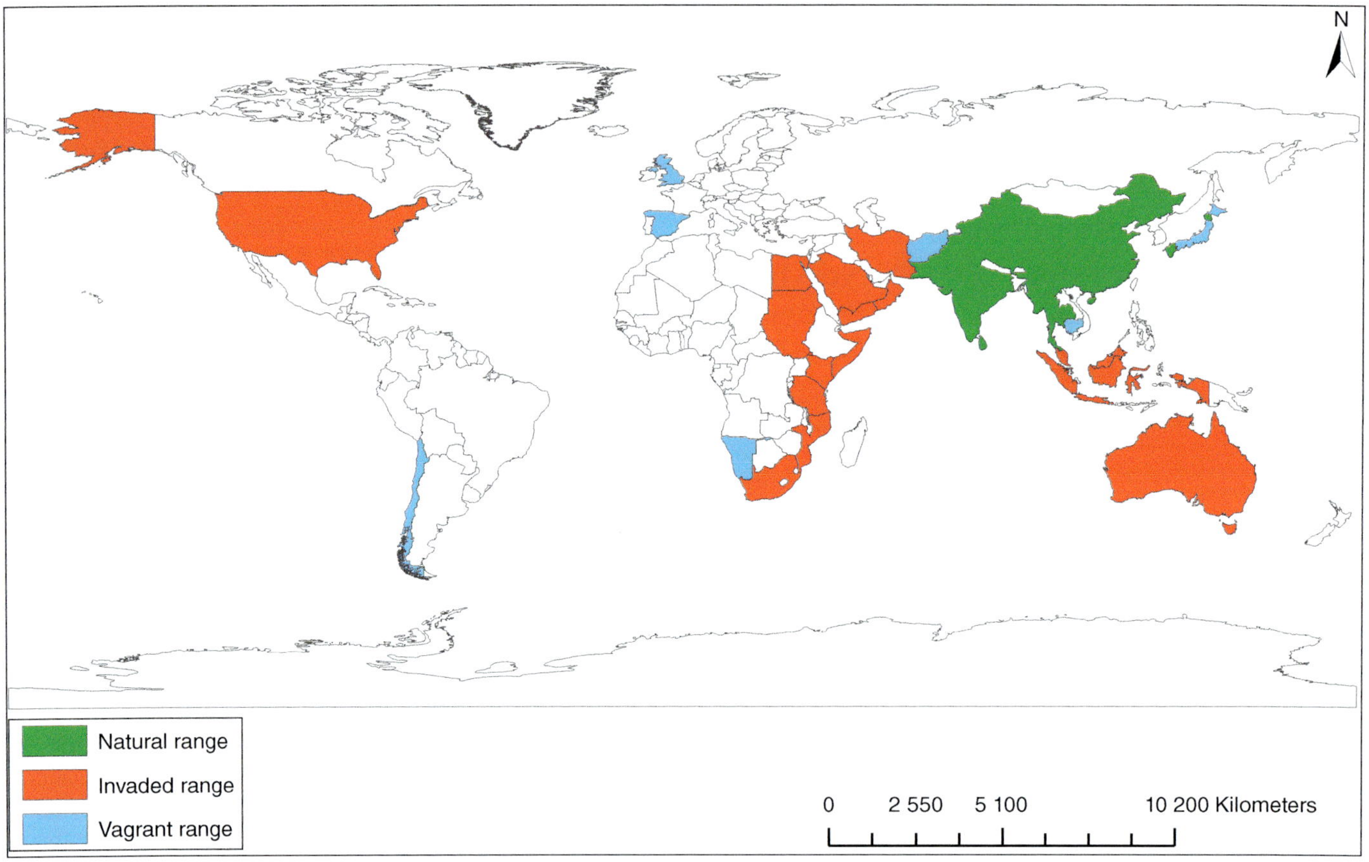

Fig. 24.1. Global distribution of the House Crow (*Corvus splendens*) showing its natural (green), invaded (red) and vagrant (blue) ranges.

Fig. 24.2. An adult House Crow flying over Wilson's Wharf, Durban, South Africa. (©Photograph: David Allan.)

24.5 Diet

House Crows feed on a variety of foods including amphibians, insects, small mammals, reptiles, molluscs, fruits and vegetables from gardens and farms, and garbage including human faeces and organic and inorganic materials (Suliman *et al.*, 2011; Koul and Sahi, 2013; Ahmed, 2014). The House Crow can also kill and prey on small livestock and poultry (Fraser *et al.*, 2015). It also feeds on anthropogenic waste food found in urban areas and has been reported to steal food from the public in Tanzania (Suliman *et al.*, 2011; Shimba and Jonah, 2016). In Malaysia, a total of 227 food items were identified in the stomach contents of urban House Crows, showing that this species is an opportunistic and generalist feeder (Wilson *et al.*, 2015).

24.6 Introduction and Invasion Pathways

The House Crow is one of the species that has established well around the world (Ryall, 2002; Brook *et al.*, 2003; Suliman *et al.*, 2011). It was introduced unintentionally to various countries, often on ships as a 'hitchhiker' to countries such as Tanzania (Shimba and Jonah, 2016), Socotra Island (Yemen) in 1995 and Malaysia (Suliman *et al.* 2011). It was intentionally introduced to some countries to clean garbage, and to control caterpillars, and ticks on livestock (Brook *et al.*, 2003; Fraser *et al.*, 2015). In Zanzibar, they were reported to be introduced in the 1890s from where they spread to Kenya in 1947, especially along the coastal areas between Mombasa and Malindi, as well as hitchhiking on ships travelling from Asia to East Africa (Ryall and Reid, 1987; Ryall, 1992).

House Crows are found in large numbers, especially in populated areas along the coast and in urban parks, cities and towns where there is anthropogenic waste food (Brook *et al.*, 2003; Wilson *et al.*, 2015). Although some House Crows are abundant inland in some countries (Wilson *et al.*, 2015; Shimba and Jonah, 2016), generally areas along the coast have more House Crows, as these areas are generally ports of entry where crows are introduced via ships (Suliman *et al.*, 2011; Wilson *et al.*, 2015). Wilson *et al.* (2015) reported that House Crows prefer contaminated areas because of the abundance of anthropogenic food littering, garbage and poor sanitation. A survey by Shimba (2011) in Dar es Salaam city, Tanzania, estimated that there were 800,000 House Crows in areas where there was poor sanitation. This species to date is continuing to spread because it is an opportunistic species with catholic feeding habits, and it can survive in a range of environmental conditions. For example, Ryall (2003) reported a House Crow breeding colony with winter ambient temperatures below −8°C.

The movement patterns and behaviour of House Crows have been relatively understudied in both their native and invasive ranges. However, Lim and Sodhi (2009) used radiotracking and found that House Crows travel up to 3.5 km from their roost sites. Observation data showed that the House Crow could move an average of 136 m in altitude, annually moving from 2000 to 4200 m above sea level in Mustang District, Nepal (Acharya and Ghimirey, 2013). The movement and range expansion of this species in Nepal was influenced by climate change (Acharya and Ghimirey, 2013). In its invasive range, the House Crow is generally resident and does not travel long distances (Dean, 2005; Lim and Sodhi, 2009; Acharya and Ghimirey, 2013).

24.7 Breeding Behaviour

House Crows have different breeding seasons across the globe. For example, in Singapore, they have been reported to breed throughout the year (Brook *et al.*, 2003), in South Africa between October and January (Allan and Davies, 2005; Dean, 2005), on Kharg Island from late May to early September (Behrouzi-Rad, 2010), in Pakistan from June to September (Awais *et al.*, 2015) and in Tanzania from late March to early May (Shimba and Jonah, 2016). The breeding biology of this species in other countries is still lacking, and information on breeding biology could be important in management and eradication programmes. The pairing of crows is through rubbing of bills and allopreening, which helps to strengthen the bond between the pairs, and this is occasionally followed by rapid copulating and allofeeding probably by both sexes (Dean, 2005). The allopreening and allobreeding take place in trees.

House Crow nests are built by the female only, but materials are provided by both sexes. The nests are generally made up of a large bowl of sticks with wire added in some areas (Fig. 24.3A). The inside of the nest is generally lined with soft grass, other plant materials and animal fibres (Fig. 24.3B). This type of nest is considered ‘long-lasting’ and it generally has an outside diameter of around 25–30 cm, depth of 7–10 cm, and a cup diameter of 12–15 cm (Ryall, 1990). The nest is placed at an average height of 6.8 m in the branches of trees that form a fork shape or in the outer bushy branches. Some House Crow nests are built on anthropogenic structures such as electricity pylons, with the height of nests varing from 15 to 49 m in India (Ali *et al.*, 2013). On Kharg Island, they nest on gas and oil pipes, window ledges, telephone and electricity poles, and television antennae (Ryall, 1990; Behrouzi-Rad, 2010).

The House Crow clutch size is two to five eggs with the potential of two or three eggs being hatched successfully (Allan and Davies, 2005; Dean, 2005; Awais *et al.*, 2015; Shimba and Jonah, 2016). The eggs are oval and pale bluish green or greyish green, and speckled with red-brown and grey, or brown dots markings (Tarboton, 2001). Their size is 37.0 × 26.9 cm, weighing around 13.1 g (Dean, 2005). The incubation period is 16–17 days and the eggs are incubated by both sexes but mainly by females at night (Snow *et al.*, 1997; Dean, 2005).

House Crow newly hatched chicks are fleshy coloured with the internal organs visible and are born with eyes closed. The nestling period is between 21 and 28 days, and the chick is fed by both adults (Maclean, 1993). In terms of breeding success, there is generally a 74% chance of House Crow eggs being hatched and a 54% chance that chicks will fledge successfully (Snow *et al.*, 1997).

24.8 Habitat

The native habitat of the House Crow is broad (Suliman *et al.*, 2011). It appears to thrive with anthropogenic land-use change

Fig. 24.3. (A) A typical House Crow nest with a clutch size of four. (B) A nest with two newly hatched House Crow chicks with internal viscera visible. Photos were taken at Merebank, Durban, South Africa. (©Photographs: David Allan.)

and inhabits agricultural and urban areas (Suliman *et al.*, 2011; Wilson *et al.*, 2015; Shimba and Jonah, 2016). Consequently, in its invaded range, it is similarly found in a range of habitat types but often in urban areas or on farmlands.

24.9 Impacts

The House Crow is regarded as one the world's worst invasive pest species. It has been reported to negatively affect the number of native birds, small vertebrates and invertebrates (Archer, 2001; Kamel, 2014) and to cause a decline in native birds by predation and competition (Ryall, 2010; Shimba and Jonah, 2016). For example, in Kenya, the House Crow has been reported to attack the nests of small native birds and Ploceidae weavers (Yap and Sodhi, 2004). In southern Africa, it destroys the eggs and kills the nestlings of birds such as Black-headed Herons (*Ardea melanocephala*), African Sacred Ibises (*Threskiornis aethiopicus*) and African Spoonbills (*Platalea alba*) (Dean, 2005).

The House Crow is also considered an agricultural pest; it damages orchards and destroys crops that are important to humans such as maize, cereal, groundnuts and wheat (Suliman *et al.*, 2011; Akram *et al.*, 2013; Kamel, 2014; Khan *et al.*, 2015). In addition, significant damage has been reported to crops, with 81% losses on maize, 55% on wheat and 55% on almonds (Dhindsa *et al.*, 1991; Reddy, 1998; Khan *et al.*, 2007).

Humans are also the victims of House Crows as this bird has aggressive behaviours. They have been reported to pose a risk of bird strike to airplanes, to snatch food from people and to attack people walking in parks and streets (Ryall, 1992; Suliman *et al.*, 2011; Shimba and Jonah, 2016). House Crows may also pose a health risk, as their pathogens can affect both humans and domestic animals (Suliman *et al.*, 2011). Bacterium species that can cause diseases in humans such as *Campylobacter* spp., *Escherichia coli* and *Salmonella* spp. are also associated with these birds (Roy *et al.*, 1998; Meier and Ryall, 2008). In villages, towns and cities where they are abundant, House Crows can make disturbing noise pollution and also contaminate roofs, walls, buildings and water with their excreta (Xi, 2009; Ahmed, 2014; Shimba and Jonah, 2016).

Although the House Crow is considered one of the worst invasive pest species, it has been used as a biological control for insect pests in agricultural sectors. For example, they have been used to control caterpillars in Malaysia, and ticks on livestock in Oman (Brook *et al.*, 2003; Fraser *et al.*, 2015). In Zanzibar, Tanzania, House Crows were introduced to reduce anthropogenic waste refuse as they are regarded as garbage scavengers (Brook *et al.*, 2003; Fraser *et al.*, 2015; Wilson *et al.*, 2015).

24.10 Control

Two types of control measure have been used to successfully control House Crows in areas of their invaded ranges, physical or mechanical control, and chemical control. The physical method was found to be useful in controlling House Crows, as evidenced by a successful eradication in Socotra, Yemen (Suleiman and Taleb, 2010; Suleiman *et al.*, 2010). Here the eradication of House Crows was achieved by collecting and killing of young birds, followed by shooting of adults. Shooting has also been shown to be successful for House Crow eradication in Australia where House Crows were shot at ports of entry, and this method has prevented House Crows from establishing in the country (Ryall, 2002, 2010). In Mauritius, poisoning using a chemical called Starlicide was used to successfully reduce the number of House Crows (Puttoo and Archer, 2004). Another successful poisoning method used in Mauritius was with an α-chloralose poison, which reduced the number of House Crows (Feare and Mungroo, 1990). No biological methods have been used to control House Crows, but in Israel, it has been reported that the Great Spotted Cuckoo (*Clamator glandarius*) is a natural enemy of the House Crow through brood parasitism (Yosef, 2002).

24.10.1 Climate matching as part of risk assessment

Risk-assessment schemes are evidence-based tools that aim to identify introduced species with a high likelihood of becoming problematic (Andreu and Vilà, 2010). One of the first formal risk-assessment screening tools was developed in Australia to prevent the introduction of weed species with invasion potential (Pheloung *et al.*, 1999). Recently, risk assessment has been adopted in many countries to predict the invasion potential of different taxa (Downey *et al.*, 2010; Kumschick and Richardson, 2013). Several approaches are used for the assessment of risks, including trait-scoring approaches, statistical approaches and rapid screening (Keller and Kumschick, 2017). Trait scoring is based on the traits that might be related to introduced species becoming invasive or causing harm to native species. Statistical approaches make use of trait lists, which are scored based on their invasiveness. Rapid screening uses climate matching, species distribution modelling and information on whether the species has a history of invasion elsewhere (Keller and Kumschick, 2017). Climate matching and/or species distribution modelling has been widely used as a rapid screening tool to determine the potential environmental suitability of a species to new environments (Thuiller *et al.*, 2005; Keller and Kumschick, 2017).

The use of risk assessment to assess the invasion potential for introduced animal species is not common around the world. We have used climate matching to understand the areas at risk of invasion should House Crows be introduced. This was done by using a list of global climate zones generated by Metzger *et al.* (2013) that contain native and exotic range distribution records for House Crows. The distribution records obtained from the literature and Global Biodiversity Information Facility (www.gbif.org/, accessed 29 October 2019) were prepared and cleaned with ArcGIS version 10.4.1 and Biogeo in R (Robertson *et al.*, 2016).

Native climate matching

The projected climate match from native to exotic ranges showed that the House Crow is well suited to a large range of climates across the globe (Fig. 24.4). The southern hemisphere has a large climatic suitability for House Crows to survive; this includes western and eastern coastal areas of North America, a large portion of northern regions of South

America, southern parts of Asia and the large northern region of Australia and Africa (with a low suitability in South Africa) (Fig. 24.4). The climatic suitability also matched some uninvaded countries and islands, and as a result, these uninvaded areas may become at risk of invasion by the House Crow if it was introduced.

Exotic climate matching

The climate from areas where House Crows are exotic and invasive matches similar climates in its native range. The climate suitability in North America is large in the western coastal areas, while South America, Africa, Australia and the southern parts of Asia have high suitability (Fig. 24.5). The generated

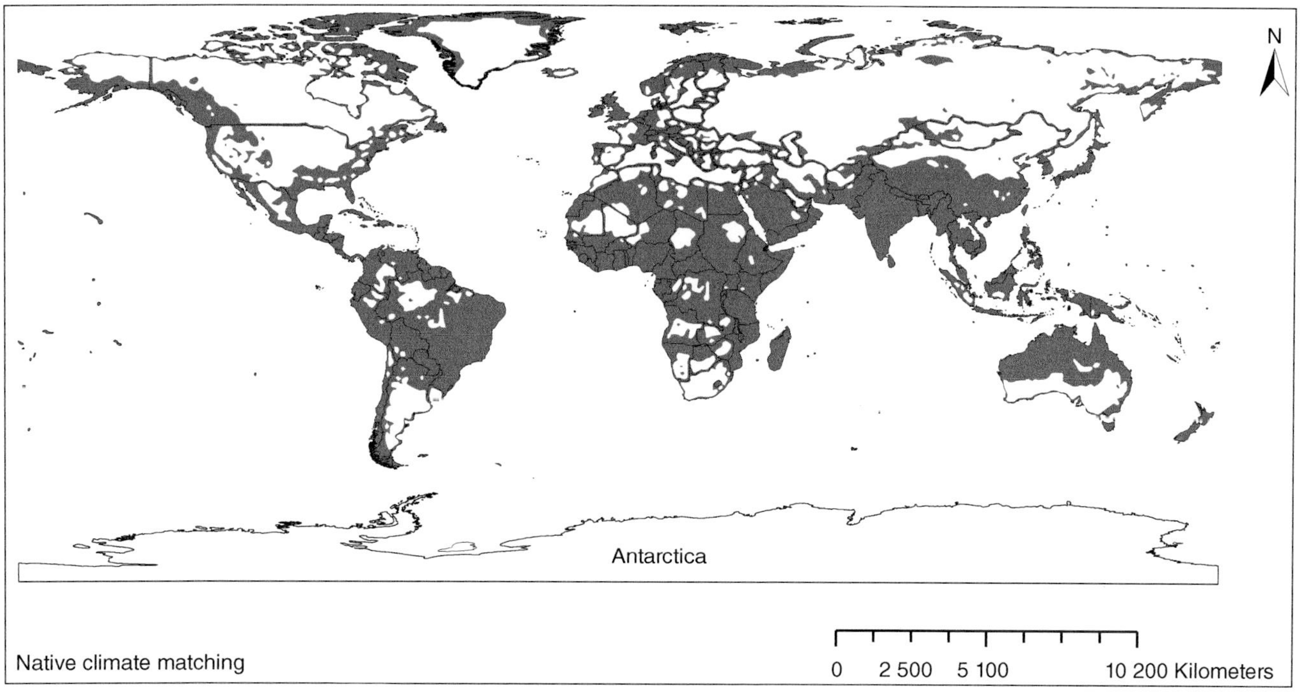

Fig. 24.4. The world climatic maps of Metzger *et al.* (2013) showing the global climatic classification of House Crows matched from its two native ranges (India and Pakistan) against world climates.

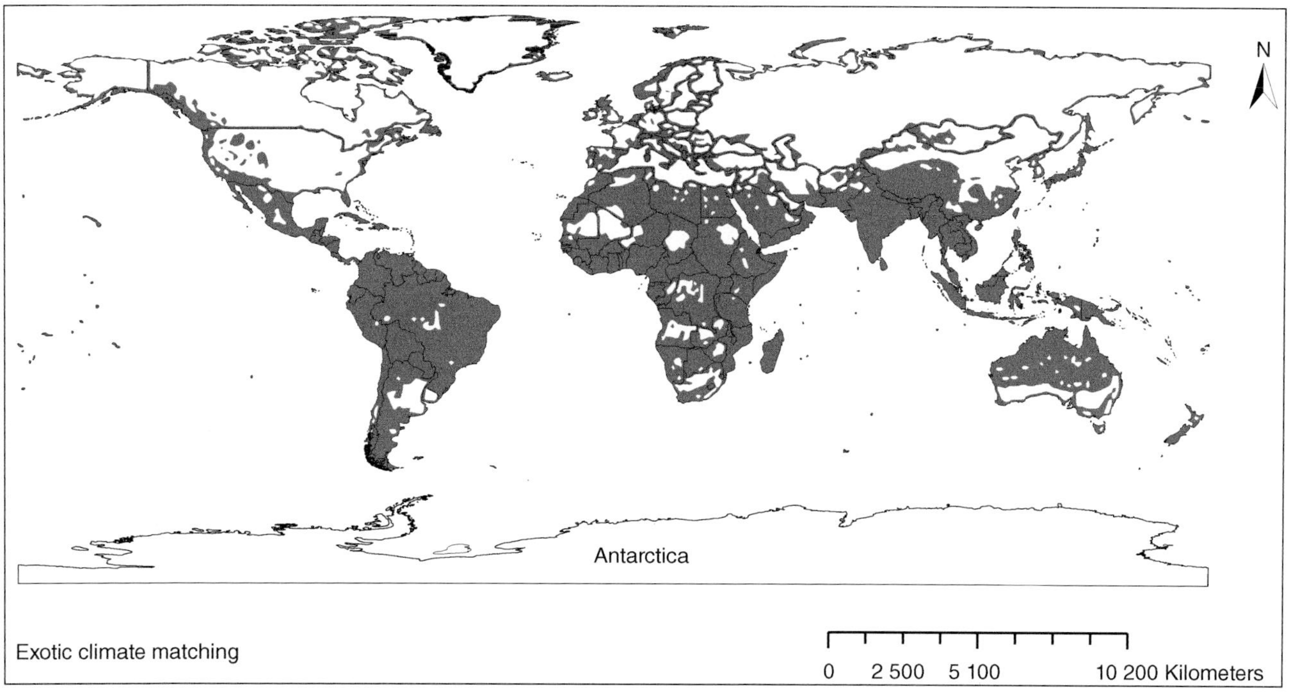

Fig. 24.5. The world climatic maps of Metzger *et al.* (2013) showing the global climatic classification of House Crows matched from its four exotic ranges (Kenya, Egypt, Singapore, and Malaysia) and where it is reported to be invasive in terms of world climates.

climate matching maps (native and exotic) showed that a large part of the world provides a suitable climate for the House Crow, indicating that this species is a high risk, and needs to be managed and eradicated in areas where it is known to occur as an invasive.

The ranges of climates suitable for House Crows are described as extremely hot and moist, warm temperature and mesic, hot and mesic, hot and dry, extremely hot and xeric, and extremely cold and mesic (Table 24.1). This indicates that House Crows can adapt to cold, hot and warm climatic

Table 24.1. A full description of climate zones that contributed to the suitability of House Crows during climate matching from native and exotic ranges across the world. Ranges with an asterisk are native ranges and those without are exotic ranges.

Climate code full name	Climate code	Range contributed
Extremely cold and wet	C3	India*
	D2	India*
	D3	India*
Cold and wet	E4	Malaysia
Extremely cold and mesic	F4	India*
	F13	India*
	F15	India*
Cold and mesic	G11	India*
	G8	India*
Cool temperature and dry	H8	Egypt
Cool temperature and moist	J3	India*
Cold and mesic	J4	India*
Warm temperature and mesic	K11	Pakistan*
	K12	India*, Malaysia
	K2	India*, Pakistan*
	K7	India*
	K9	Egypt
Hot and mesic	M1	India*, Malaysia
	M2	India*, Pakistan*
	M3	Kenya, Pakistan*
	M4	India*
	M5	India*, Pakistan*, Malaysia
	M6	India*, Kenya, Pakistan*
	M7	India*, Kenya
	M8	India*, Malaysia
Hot and dry	N10	Egypt
	N3	India*
	N5	Pakistan*
	N8	India*
Hot and arid	O1	Pakistan*, Egypt
	O2	Pakistan*, Egypt
	O3	Kenya, Egypt
Extremely hot and arid	P1	India*, Pakistan*
	P2	India*, Pakistan*, Egypt
Extremely hot and xeric	Q1	India*
	Q2	India*, Egypt
	Q3	India*
	Q4	India*
Extremely hot and moist	R1	India*, Kenya
	R2	India*, Kenya
	R3	India*, Malaysia
	R5	India*, Kenya
	R6	India*, Malaysia
	R7	India*
	R8	Singapore, Malaysia
	R9	India*, Kenya
	R10	Singapore, Malaysia

conditions. As a result, the House Crow has a high potential risk of invasion globally.

24.11 Uses

The House Crow is useful in tidying up anthropogenic waste refuse and has been used as a biological control agent for agricultural pests, ticks on livestock and to control caterpillars (Brook *et al.*, 2003; Fraser *et al.*, 2015; Wilson *et al.*, 2015). This bird also has some important social benefits in some human cultures. For example, the Jirels people of Central Nepal use the flesh of House Crows for treatment of infertility in men, hatchlings are used for fever and the flesh of the hatchlings is used for body aches (Bagde and Jain, 2015). In Hindu religions, the House Crow is used for the religious ritual 'pinda pradhana' where cooked rice balls are offered to House Crows, and this bird is also associated with the souls of ancestors (Kandari *et al.*, 2014). In Sri Lanka, the Sinhalese people perceive the calls of the House Crow as an indication of misfortune (Madduma Bandara, 2009).

24.12 Notes

The House Crow is a highly social bird that is known to live in close association with humans (Fraser *et al.*, 2015). They are generally not afraid of people and are known to disturb tourists with their loud noise and aggressive attacks when attempting to steal food (Kamel, 2014). House Crows are considered one of the most intelligent bird species and are known as a food-storing bird (Emery and Clayton, 2004; Srivastava *et al.*, 2016). House Crows are successful invaders because they have few natural enemies and are known to scare off their predators by mobbing up on them (Ryall, 1992; Chongomwa, 2011). These birds compete with native birds for nests and food and their aggressive behaviour has displaced other birds from their feeding sites (Chongomwa, 2011). For example, the Pied Crow *Corvus albus* has been displaced by House Crows in Mombasa and Kenya (Ryall, 1992). House Crows have a high potential rate to reproduce, they are generalist feeders and they can thrive in highly modified human environments (Brook *et al.*, 2003) which facilitates their invasion potential.

24.13 References

Acharya, R. and Ghimirey, Y. (2013) House Crow (*Corvus splendens*) heading upwards in Nepal: possible climate change impact influencing its movement. *Ibisbill* 2, 181–187.

Ahmed, A.M.K. (2014) Potential impacts of invasive House Crows (*Corvus splendens*) bird species in Ismailia Governorate, Egypt: ecology, control and risk management. *Journal of Life Sciences and Technologies* 2, 86–89.

Akram, N., Khan, H.A. and Javed, M. (2013) Inhibiting the House Crow (*Corvus splendens*) damage on maize growth stages with reflecting ribbons in farmland. *Journal of Animal and Plant Sciences* 23, 182–189.

Ali, A.M.S., Kumar, S.R. and Arun, P. (2013) House Crow *Corvus splendens* nesting on pylons, Kutch district, Gujarat, India. *Forktail* 29, 148–150.

Ali, S. (2002) *The Book of Indian Birds*, 13th edn. Oxford University Press, New Dehli, India.

Allan, D.G. and Davies, G.B. (2005) Breeding biology of House Crows (*Corvus splendens*) in Durban, South Africa. *Ostrich* 76, 21–32.

Andreu, J. and Vilà, M. (2010) Risk analysis of potential invasive plants in Spain. *Journal for Nature Conservation* 18, 34–44.

Archer, A.L. (2001) Control of the Indian House Crow *Corvus splendens* in eastern Africa. *Ostrich* 72, Suppl. 15, 147–152.

Awais, M., Ahmed, S., Mahmood, S., Mahmood, T. and Alvi, H. (2015) Breeding biology of the House Crow *Corvus splendens* at Hazara. *Podoces* 10, 1–7.

Bagde, N. and Jain, S. (2015) Study of traditional man–animal relationship in Chhindwara district of Madhya Pradesh, India. *Journal of Global Biosciences* 4, 1456–1463.

Behrouzi-Rad, B. (2010) Population estimation and breeding biology of the House Crow *Corvus splendens* on Kharg Island, Persian Gulf. *Podoces* 5, 87–94.

Brook, B.W., Sodhi, N.S., Soh, M.C.K. and Lim, H.C. (2003) Abundance and projected control of invasive House Crows in Singapore. *Journal of Wildlife Management* 67, 808–817.

Chongomwa, M.M. (2011) Mapping locations of nesting sites of the Indian House Crow in Mombasa. *Journal of Geography and Regional Planning* 4, 87–97.

Dean, W.R.J. (2005) House Crow. In: Hockey, P.A.R., Dean, W.R.J. and Ryan, P. (eds) *Roberts Birds of Southern Africa*, 7th edn. Trustees of the John Voelker Bird Book Fund, Cape Town, pp.721–722.

Dhindsa, K.K., Ghazi, N.K. and Tribasti, P. (1991) Damage proportions to sprouting sunflowers. *Tropical Pest Management* 37, 1607–1612.

Downey, P.O., Johnson, S.B., Virtue, J.G. and Williams, P.A. (2010) Assessing risk across the spectrum of weed management. *Plant Sciences Reviews* 2010, 203–217.

Emery, N.J. and Clayton, N.S. (2004) The mentality of crows: convergent evolution of intelligence in corvids and apes. *Science* 306, 19.0–19.7.

Feare, C.J. and Mungroo, Y. (1990) The status and management of the House Crow *Corvus splendens* (Vieillot) in Mauritius. *Biological Conservation* 51, 63–70.

Fraser, D.L., Aguilar, G., Nagle, W., Galbraith, M. and Ryall, C. (2015) The house crow (*Corvus splendens*): a threat to New Zealand? *ISPRS International Journal of Geo-Information* 4, 725–740.

Kamel, A.M. (2014) Potential impacts of invasive House Crows (*Corvus splendens*) bird species in Ismailia Governorate, Egypt: ecology, control and risk management. *Journal of Life Sciences and Technologies* 2, 86–89.

Kandari, L.S., Bisht, V.K., Bhardwaj, M. and Thakur, A.K. (2014) Conservation and management of sacred groves, myths, and beliefs of tribal communities: a case study from North India. *Environmental Systems Research* 3, 1–10.

Keller, R.P. and Kumschick, S. (2017) Promise and challenges of risk assessment as an approach for preventing the arrival of harmful alien species. *Bothalia* 47, 1–8.

Khan, H.A. (2003) Damage patterns of House Crow (*Corvus splendens*) on some food crops in Faisalabad. *Pakistan Journal of Biological Sciences* 6, 188–190.

Khan, H.A, Jabeen, G. and Anwar, N. (2007) Foraging rhythms of House Crow (*Corvus splendens*) for 30 and 60-minute durations on some crops in an agro-ecosystem in Faisalabad. *Pakistani Journal of Agricultural Science* 44, 283–288.

Khan, H.A., Javed, M. and Zeeshan, M. (2015) Damage assessment and management strategies for House Crow (*Corvus splendens* L) on the seedling stages of maize and wheat in an irrigated agricultural farmland of Punjab, Pakistan. *Journal of Entomology and Zoology Studies* 3, 151–155.

Koul, S. and Sahi, D.N. (2013) Feeding ecology of House Crow (*Corvus splendens*) in open agricultural fields in Jammu (J&K), India. *International Research Journal of Environment Sciences* 2, 85–87.

Kumschick, S. and Richardson, D.M. (2013) Species-based risk assessments for biological invasions: advances and challenges. *Diversity and Distributions* 19, 1095–1105.

Lim, H.C. and Sodhi, N.S. (2009) Space use and habitat selection of House Crows in a tropical urban environment: a radio-tracking study. *Raffles Bulletin of Zoology* 57, 561–568.

Maclean, G.L. (1993) *Roberts Birds of Southern Africa*, 6th edn. John Voelker Bird Book Fund, Cape Town.

Madduma Bandara, C. (2009) Exploring the link between culture and biodiversity in Sri Lanka. *Environmental Journal for the Global Community* 4, 1–23.

Madge, S. and Burn, H. (1994) *Crows and Jays: A Guide to the Crows, Jays and Magpies of the World*. Christopher Helm, London.

Meier, G.G. and Ryall, C. (2008) House Crow in the Middle East. *Wildlife Middle East News* 3, 8237–1990.

Meininger, P.L., Mullie, W.C. and Bruun, B. (1980) The spread of the House Crow, *Corvus splendens*, with special reference to the occurrence in Egypt. *Le Gerfaut* 70, 245–250.

Metzger, M.J., Bunce, R.G.H., Jongman, R.H.G., Sayre, R., Trabucco, A. and Zomer, R. (2013) A high-resolution bioclimate map of the world: a unifying framework for global biodiversity research and monitoring. *Global Ecology and Biogeography* 22, 630–638.

Nxele, B. and Shivambu, C. (2018) House Crow (*Corvus splendens*) eradication measures from eThekwini Municipality, KwaZulu-Natal, South Africa. *Journal of Biodiversity Management & Forestry* 7, 1–5.

Nyári, Á., Ryall, C. and Townsend Peterson, A. (2006) Global invasive potential of the House Crow *Corvus splendens* based on ecological niche modelling. *Journal of Avian Biology* 37, 306–311.

Ottens, G. (2003) Background and development of the Dutch population House Crows *Corvus splendens*. *Limosa* 76, 69–74.

Ottens, G. and Ryall, C. (2003) House Crows in the Netherlands and Europe. *Dutch Birding* 25, 312–319.

Pheloung, P.C., Williams, P.A. and Halloy, S.R. (1999) A weed risk assessment model for use as a biosecurity tool evaluating plant introductions. *Journal of Environmental Management* 57, 239–251.

Puttoo, M. and Archer, T. (2004) Control and/or eradication of indian crows (*Corvus splendens*) in Mauritius. *Agricultural and Sugar Review of Mauritius* 83, 209–399.

Reddy, V.R. (1998) Bird damage to maize crop on the student's research farm at Rajendranagar, Hyderabad, Andhra Pradesh. *Pavo* 36, 77–78.

Robertson, M.P., Visser, V. and Hui, C. (2016) Biogeo: an R package for assessing and improving data quality of occurrence record datasets. *Ecography* 39, 394–401.

Roy, P., Venugopalan, A.T. and Manvell, R. (1998) Isolation of Newcastle disease virus from an Indian house crow. *Tropical Animal Health and Production* 30, 177–178.

Ryall C. and Reid, C. (1987) The Indian House Crow in Mombasa. *Swara* 10, 9–12.

Ryall, C. (1990) Notes on nest construction by the Indian House Crow, *Corvus splendens* and other aspects of its breeding biology in Mombasa, Kenya. *Scopus* 14, 14–16.

Ryall, C. (1992) Predation and harassment of native bird species by the Indian House Crow *Corvus splendens* in Mombasa, Kenya. *Scopus* 16, 1–8.

Ryall, C. (2002) Further records of range extension in the House Crow *Corvus splendens*. *Bulletin of British Ornithologists Club* 122, 231–240.

Ryall, C. (2003) Notes on ecology and behaviour of House Crows in Hoek van Holland. *Dutch Bird* 25, 167–172.

Ryall, C. (2010) Further records and updates of range extension in House Crow *Corvus splendens*. *Bulletin of the British Ornithologists' Club* 130, 246–254.

Shimba, M. (2011) Nest success of the Indian House Crow in Dar es Salaam. Undergraduate thesis, University of Dar es Salaam, Dar es Salaam, Tanzania.

Shimba, M.J. and Jonah, F.E. (2016) Nest success of the Indian House Crow *Corvus splendens*: an urban invasive bird species in Dar es Salaam, Tanzania. *Ostrich* 88, 27–31.

Snow, D.W., Perrins, C.M., Hillcoat, B., Gillmor, R. and Roselaar, C.S. (1997) *The Birds of the Western Palearctic*, concise edn. Oxford University Press, Amsterdam.

Srivastava, U.C., Singh, D., Kumar, P. and Singh, S. (2016) Neuronal diversity and their spine density in the hippocampal complex of the House Crow (*Corvus splendens*), a food-storing bird. *Canadian Journal of Zoology* 94, 541–553.

Suleiman, A., Meier, G. and Haverson, P. (2010) Confirmed eradication of the House Crow from Socotra Island, Republic of Yemen. *Wildlife Middle East* 4, 1–12.

Suleiman, A.S. and Taleb, N. (2010) Eradication of the house crow *Corvus splendens* on Socotra, Yemen. *Sandgrouse* 32, 136–140.

Suliman, A.S., Meier, G.G. and Haverson, P.J. (2011) Eradication of the House Crow from Socotra Island, Yemen. In: Veitch, C.R., Clout, M.N. and Towns, D.R. (eds) *Island Invasives: Eradication and Management*. IUCN, Gland, Switzerland, pp. 361–363.

Tarboton, W. (2001) *Guide to the Nests and Eggs of Southern African Birds*. Struik Publishers, South Africa.

Thuiller, W., Richardson, D.M., Pyšek, P., Midgley, G.F., Hughes, G.O. and Rouget, M. (2005) Niche-based modelling as a tool for predicting the risk of alien plant invasions at a global scale. *Global Change Biology* 11, 2234–2250.

Wilson, R.F., Sarim, D. and Rahman, S. (2015) Factors influencing the distribution of the invasive house crow (*Corvus splendens*) in rural and urban landscapes. *Urban Ecosystems* 18, 1389–1400.

Xi, N.Z. (2009) House Crows: potential Invasive Alien Species in Hong Kong. *Science in Hong Kong* 6, 22–24.

Yap, C.A.M. and Sodhi, N.S. (2004) Southeast Asian invasive birds: ecology, impact and management. *Ornithological Science* 3, 57–67.

Yosef, R. (2002) Second breeding record of Great Spotted Cuckoo *Clamator glandarius* in Eilat. *Sandgrouse* 24, 142–143.

25 Australian Magpie (*Gymnorhina tibicen* Latham, 1802)

Dai K.J. Morgan[1]*, John G. Innes[2] and Joseph R. Waas[3]

[1]*Applied and Environmental Sciences, NorthTec, Whangarei 0110, New Zealand;* [2]*Manaaki Whenua Landcare Research, Hamilton 3240, New Zealand;* [3]*School of Science, University of Waikato, Hamilton 3240, New Zealand*

Citation: Morgan, D.K.J., Innes, J.G. and Waas, J.R. (2020) Australian Magpie (*Gymnorhina tibicen* Latham, 1802). In: Downs, C.T. and Hart, L.A. (eds) *Invasive Birds: Global Trends and Impacts*. CAB International, Wallingford, UK, pp. 183–193.

25.1 Common Names

Magpie, Australian Magpie, White-backed Magpie, Bell Magpie, Black-backed Magpie, Western Magpie, Makipae (Māori).

25.2 Distribution

The Australian Magpie (*Gymnorhina tibicen* Latham, 1802) is native to Australia and New Guinea. They have an almost ubiquitous distribution across Australia with the exception of the northern tips of the Top End and Cape York, the Gibson and Great Sandy Deserts, and the western side of Tasmania (Schodde and Mason, 1999) (Fig. 25.1). In Australia, eight subspecies have been identified (Schodde and Mason, 1999) that can differ considerably in size, plumage markings and geographical location (Table 25.1); however, interbreeding occurs where subspecies meet, and intermediate varieties are common (Schodde and Mason, 1999). In addition to the Australian subspecies, *G. t. papuana* naturally occurs in New Guinea in the Trans-Fly region from Princess Marianne Strait to Dogwa on the Oriomo River (Higgins *et al.*, 2006; Beehler and Pratt, 2016).

The Australian Magpie was introduced successfully to New Zealand and Fiji (Fig. 25.1) and unsuccessfully introduced to the Solomon Islands and Sri Lanka (Higgins *et al.*, 2006). In New Zealand, Australian Magpie subspecies are not recognized (Gill *et al.*, 2010) and, as such, recent reports of their distribution are only at the species level (e.g. Heather and Robertson, 1996; Robertson *et al.*, 2007). However, most of the magpies introduced to New Zealand were collected from Victoria or Tasmania and were of the white-backed variety (most likely *G. t. tyrannica* and *G. t. hypoleuca*) (McIlroy, 1968). Accordingly, historical reports indicate that White-backed Magpies are the most widely distributed variety and occupy most suitable habitats (see section 25.7) across the North Island (Heather and Robertson, 1996; Robertson *et al.*, 2007). In the South Island, they are also widely distributed, although uncommon in the north-western regions (Nelson and Marlborough); they are also rare on Stewart Island in the far south (Heather and Robertson, 1996; Robertson *et al.*, 2007). In addition, they do not occur in the alpine regions of the Southern Alps or in large tracts of continuous native forest (e.g. Fiordland) (Heather and Robertson, 1996; Robertson *et al.*, 2007).

The black-backed form is relatively abundant in localized areas, such as the lower North Island (Hawkes Bay, Central Plateau and parts of Taranaki) and between the Ashley and Conway Rivers on the east coast of the South Island (McCaskill, 1945; McIlroy, 1968). Some authors report that Black-backed Australian Magpies and intermediates occur outside these areas (McCaskill, 1945; McIlroy, 1968; O'Donnell, 2002; Morgan, 2008). None of the historical literature mention the release of the black-backed form. It is likely that some birds collected from the Victoria region were *G. t. tibicen* (McIlroy, 1968).

In the Fijian Islands, Australian Magpies are mainly restricted to the island of Tavenui but are occasionally reported on the nearby Vanua Levu and Cicia (Fig. 25.1) (Clunie *et al.*, 1978; Clunie, 2007). Both white- and black-backed varieties are present in the Fijian population, and intermediates occur (Clunie *et al.*, 1978).

25.3 Description

Unless otherwise stated, this description of the Australian Magpie is largely based on information presented in Heather

*Corresponding author: dmorgan@northtec.ac.nz

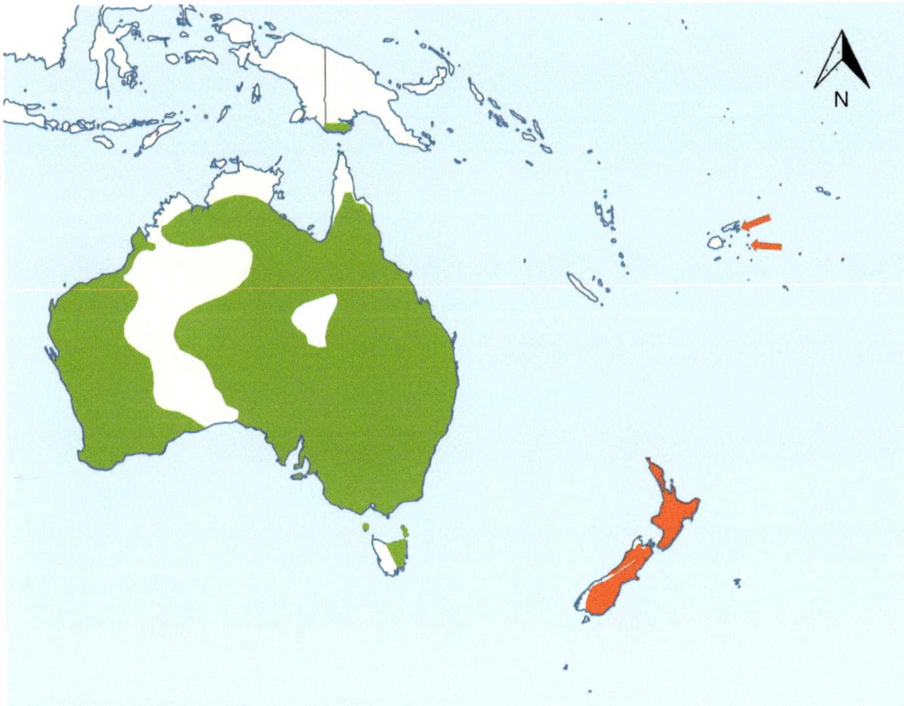

Fig. 25.1. Distribution of the Australian Magpie (*Gymnorhina tibicen*) showing introduced (red) and natural (green) ranges. Areas where Australian Magpies have been unsuccessfully introduced are not included (see text). The red arrows indicate the Fijian Islands of Vanua Levu (upper arrow) and Tavenui, and of Cicia (lower arrow). (Data from Schodde and Mason (1999), Higgins *et al.* (2006), Clunie (2007) and Robertson *et al.* (2007)).

and Robertson (1996), Kaplan (2004) and Higgins *et al.* (2006). Nine subspecies of Australian Magpie have been recognized and variations in several morphological characteristics have been identified (morphological data for subspecies residing in Australia and New Zealand are presented in Table 25.1). In addition, hybridization has been recorded in areas where different subspecies occur sympatrically (McIlroy, 1968; Kallioinen *et al.*, 1995; Morgan, 2008), adding further variation. Despite this, magpie subspecies can be broadly described as one of three forms: 'white-backed', 'black-blacked' or 'western' (Table 25.1).

Adult Australian Magpies have a mean height of approximately 410 mm and on average weigh 220–350 g; males are generally larger than females (Table 25.1) (Higgins *et al.*, 2006). For all forms, the head, scapular, wings, ventral region and thigh plumage are black, and they also have black legs and feet (Fig. 25.2). In adult males, the black plumage is very dark and often has a bluish sheen, while in females it is usually a duller shade of black. The bill of adult birds is bluish-white and darkens to a black tip (Fig. 25.2) which generally tapers to a straight point; however, in some individuals the upper mandible can slightly extend past the lower mandible to form a hook shape (McIlroy, 1968).

In the white-backed form, the nape, hind neck, mantle, wing coverts, and upper- and under-tail coverts are white (Fig. 25.2). For males, the mantle region is usually chalky-white and there is a crisp transition between the white and black plumage. In contrast, female napes and mantles can have varying amounts of grey mottling. The black-backed forms are very similar to white-backed forms except that the mantle and scapular plumage are a very dark black for males, and a slightly duller shade of black for females. In addition, females have a milky- or dull-grey mantle. For the western form, male plumage closely resembles that of the male white-backed form, as their backs are uniformly white. In contrast, western females are more similar to black-backed females, as they exhibit a black mantle with white scaling.

Morphological variations in the New Guinean subspecies *G. t. papuana* are poorly known, but they appear to exhibit traits that resemble several of the other subspecies (Black, 1986). For example, their dorsal plumage is similar to the western form *G. t. dorsalis*, but they have longer bills and shorter wings, which are more closely aligned to *G. t. longirostris* and *G. t. eylandtensis*, respectively (Black, 1986; Higgins *et al.*, 2006).

Juvenile and immature Australian Magpies have similar plumage markings to their parents, except that the black plumage is duller than adults, and often tinged with brown coloration. Any white plumage is usually heavily mottled with grey.

25.4 Diet

Australian Magpies predominantly consume invertebrate prey located on or just under the ground surface (McIlroy, 1968; Vestjens and Carrick, 1974; Veltman and Hickson, 1989; Kaplan, 2004) using visual, olfactory or vibratory stimuli (Floyd and Woodland, 1981). Vestjens and Carrick (1974) identified and enumerated prey items from 1319 stomach samples

Table 25.1. Distribution and morphological variation of Australian Magpie subspecies in Australia and New Zealand[a]. Unless otherwise stated, measurements are presented for subspecies residing in Australia; 'NZ' indicates that data were collected from subspecies residing in New Zealand. (Data from McIlroy, 1968; Schodde and Mason, 1999; Kaplan, 2004.)

				Bill length (mm)		Wing length (mm)	
Subspecies	Distribution	Form	Relative size	Male	Female	Male	Female
G. t. eylandtensis	Northern Territory, north-west Queensland	Black-backed	Small	56–62	51–57	230–255	225–245
G. t. terraereginae	Queensland, inland New South Wales, northern Victoria, western South Australia	Black-backed	Small-medium	48–58	47–53	245–265	235–255
G. t. tibicen	South-east Queensland, eastern New South Wales, New Zealand	Black-backed	Large	48–55 NZ: 49–55	45–50 NZ: 45–54	260–385 NZ: 256–282	255–270 NZ: 235–282
G. t. tyrannica	Victoria, south-east South Australia, New Zealand	White-backed	Very large	52–57	47–53	270–290	260–280
G. t. hypoleuca	Tasmania, New Zealand	White-backed	Small	43–47 NZ: 47–59	38–43 NZ: 45–53	248–258 NZ: 251–296	235–245 NZ: 230–276
G. t. telonocua	Southern South Australia	White-backed	Medium	50–56	45–50	255–265	245–255
G. t. dorsalis	South-western West Australia	Western	Medium	56–60	48–54	258–270	240–255
G. t. longirostris	North-western West Australia	Black-backed	Medium	60–65	55–60	245–260	235–250

[a]Most releases of White-backed Magpies into New Zealand comprised birds collected from Tasmania (McIlroy 1968); therefore, data for the white-backed form are only presented for *G. t. hypoleuca*.

Fig. 25.2. An adult male Australian Magpie of the white-backed form. (©Photograph: Shane Gibson.)

from Black-backed Magpies at a site close to Canberra, Australia. Annelids (i.e. Oligochaeta earthworms), arachnids (Araneida spider adults) and insects were the most observed prey classes. Of the insects, ants (of several genera) were most consumed, being found in 70% of samples. Scarab beetles (including larvae), weevils, carabid beetles, grasshoppers and moth larvae were observed in more than 30% of stomachs.

In New Zealand, results reported by Veltman and Hickson (1989) and McIlroy (1968) were similar to the Australian study in terms of invertebrate prey items taken by Australian Magpies; however, the relative importance of prey groups varied greatly.

Variability in the relative importance of prey groups may reflect a functional response to the availability of vertebrate prey (Holling, 1965). Veltman and Hickson (1989) assessed prey remains from faecal samples (in addition to using other techniques) to quantify the relative abundance of different invertebrates in the diet of magpies; most prey were taken in the same relative proportions by both flock and territorial magpies (Veltman and Hickson, 1989).

Australian Magpies can also consume appreciable amounts of plant material, but there is considerable variation across diet studies in this regard. For example, Veltman and Hickson (1989) found no evidence of green leaf material in faecal samples, and Vestjens and Carrick (1974) only detected very small amounts (considered to have been consumed accidentally when invertebrates were taken). In contrast, clover leaf and grass blades were found in 41% and 80% of gizzards, respectively, by McIlroy (1968); the amount of plant material in gizzards appeared to be negatively correlated with the quantity of invertebrates, indicating that plant material was only taken if invertebrate prey were not available (McIlroy, 1968).

Being opportunistic foragers, Australian Magpies have also occasionally been recorded consuming a wide range of vertebrate prey, such as bird eggs and nestlings, lizards, mice (*Mus musculus*) and frogs (McIlroy, 1968; Vestjens and Carrick, 1974; Veltman and Hickson, 1989). Australian Magpies are also regular visitors to supplementary bird feeders (e.g. Rollinson *et al.*, 2003); however, one experimental study demonstrated that they do not become dependent on supplementary foods and that these foods were rarely fed to nestlings (O'Leary and Jones, 2006).

25.5 Introduction and Invasion Pathways

Australian Magpies have been introduced to several Australian islands (e.g. Kangaroo Island, Rottnest Island; Higgins *et al.*, 2006); however, here we will only describe those introductions outside Australia.

25.5.1 New Zealand

Most Australian Magpie introductions occurred between 1864 and 1874; however, smaller-scale introductions continued through to the early 1900s (Drummond, 1907; Thomson, 1922; Oliver, 1955). Initially, liberations were conducted by the Canterbury and Otago Acclimatisation Societies. By 1871, Canterbury had 86 birds released at several sites; in other liberations in the region, however, numbers were not recorded (but all were from Victorian or Tasmanian stock) (Drummond, 1907; Thomson, 1922; McIlroy, 1968). Liberations appeared to have been very successful, which may partly have been due to magpies being protected until 1951 under New Zealand law (McIlroy, 1968), and they quickly established in Canterbury (Thomson, 1922; McCaskill, 1945).

In Otago, at least five Australian Magpie liberations of a total of 111 birds had been completed by 1869 (McCaskill, 1945), and initial reports suggested that the populations were doing well (breeding was observed in several areas) (Thomson, 1922). However, for reasons unknown, the Otago populations had disappeared by the 1920s; the extirpation may have occurred because birds were 'shot or taken by boys' (Thomson, 1922). Reinvasion occurred rapidly, however, and by the 1940s, they were relatively common across much of this province (McCaskill, 1945).

Liberations of Australian Magpies to North Island sites were not well documented. The Auckland Acclimatisation Society introduced ten birds in 1867 and a further bird in 1870 (Thomson, 1922); however, there were many other liberations where numbers were not recorded. For example, Drummond (1907) reported numerous liberations across the North Island up until 1902. In addition, the Wellington Acclimatisation Society introduced 260 birds in 1874 (Thomson, 1922).

Australian Magpies have successfully spread from their original release points and are now common or abundant in most areas where suitable habitat exists (see section 25.7) (McCaskill, 1945; McIlroy, 1968; Bull *et al.*, 1985; Heather and Robertson, 1996; Robertson *et al.*, 2007).

25.5.2 Fiji

Relatively little information on Australian Magpie liberations in Fiji has been published. Clunie *et al.* (1978) reported that the first introductions were to Tavenui Island in 1883. Subsequent releases were conducted in the early 1900s, and again in the 1930s and 1940s (Clunie *et al.*, 1978; Clunie, 2007; Kaplan and Rogers, 2013); however, subsequent releases during the 1900s appear to have been of single pairs (Clunie *et al.*, 1978). Both *G. t. tibicen* and *G. t. hypoleuca* were liberated; the two varieties interbreed and hybrids are common (Clunie, 2007).

Published reports suggest that Australian Magpies were only introduced to Tavenui Island, where they are still present (Clunie *et al.*, 1978); however, they have also established on Cicia Island (Clunie, 2007) although it is not clear whether they were intentionally liberated there or were self-introduced. Vagrants are occasionally seen on Vanua Levu Island (Clunie *et al.*, 1978).

25.5.3 Solomon Islands

The exact date when Australian Magpies (probably *G. t. tibicen*; Cain and Galbraith, 1956) were introduced to Guadalcanal Island is unknown. Sporadic observations of small numbers were reported from the 1940s to the mid-1950s between the Teneru River and Honiara (Cain and Galbraith, 1956), but the population appears to have failed as they no longer occur there (Dutson, 2011).

25.5.4 Sri Lanka

In 1905, nine birds (*G. t. tibicen*) were introduced to Nuwara Eliya (Long, 1981); however, no other details on this liberation are known. The introduction was unsuccessful, and magpies are not included in recent Sri Lankan bird lists (e.g. Lepage, 2018).

25.6 Breeding Behaviour

25.6.1 Social systems

Australian Magpies can have complex social systems, with several types of groupings identified (Carrick, 1972), the most common being: (i) magpies that defend territories year-round; and (ii) magpies that occur in semi-nomadic, non-territorial flocks. Carrick (1972) identified four different categories of territorial group, based largely on the quality of occupied habitat. 'Permanent' territories are the highest-quality territories (with respect to reproductive benefits) and are located where the nesting and foraging habitat supports adults and the production of offspring. 'Marginal' groups defend territories that have either poor foraging opportunities or few trees suitable for nesting. As a result, productivity is considerably lower than that experienced by birds defending permanent territories. 'Mobile' groups have separate foraging and nesting territories, and breeding attempts always fail. Finally, magpies from 'open' groups defend foraging areas containing no nesting trees, and breeding does not occur.

Birds that are part of non-territorial flocks can be of either sex and of any age; however, they generally consist of immature birds that have been evicted from natal territories. Adult birds that have either been displaced from a breeding territory or have lost a partner may also join these flocks (Carrick, 1972; Veltman, 1989). While non-territorial flocks can contain adult birds, no breeding takes place; nevertheless, social bonds can form between flocking adults, sometimes leading to the acquisition of a breeding territory outside the flock (Carrick, 1972; Veltman, 1989).

Australian Magpie breeding territories can be defended by a monogamous pair or by up to 26 birds (Robinson, 1956; Carrick, 1972; Veltman, 1989; Hughes and Mather, 1991). In Australia, geographical variation in the number of members that defend breeding territories has regularly been reported. For example, territories that are defended by pairs are found more commonly in the north-east regions, with multi-member territories becoming more frequent in the south-east (Carrick, 1972; Shurcliff and Shurcliff, 1974; Hughes *et al.*, 1996). In contrast, territories are defended by significantly more birds in south-western populations (Baker *et al.*, 2000). Reasons for geographical variation in territorial group size are not fully understood (Kaplan, 2004).

In New Zealand, up to ten birds have been observed defending breeding territories (Heather and Robertson, 1996); however, Veltman (1984) reported that most of the territories in her study were defended by pairs or trios. Kaplan and Rogers (2013) reported that the number of magpies defending territories in their Fijian study was between three and five.

The size of Australian Magpie breeding territories appears to be highly variable with territory sizes of 1.5–24.3 ha reported (Carrick, 1972; Shurcliff and Shurcliff, 1974; Hughes *et al.*, 1983; Veltman, 1984; Hughes *et al.*, 1996; Cox and Bauer, 1997; Rollinson, 2004). Territory size appears to depend on both resource quality and the number of group members. For example, Rollinson (2004) found that territories were significantly smaller in areas with a high proportion of short grass and more terrestrial invertebrate prey. In contrast, Shurcliff and Shurcliff (1974) found that territory size was positively correlated with the number of birds in the group, but the average area per magpie stayed approximately the same as group size increased. It is important to note, however, that other studies have found no correlation between territory size and the number of group members (Carrick, 1963; Veltman, 1989).

25.6.2 Breeding biology

The Australian Magpie has an extended breeding season, often occurring from June to January (early Austral winter–mid-summer); however, the initiation of breeding varies greatly between areas and is affected by both climate and resource availability. For example, in the tropical and subtropical northern areas of Australia, which are affected by a summer wet season, breeding typically occurs between June and September when it is drier (Kaplan, 2004). In contrast, in more temperate areas of Australia, breeding generally occurs later (August–November) when average temperatures are warmer (Carrick, 1972).

Considerable variation in the timing of Australian Magpie breeding can occur within regions, as laying dates for magpies nesting in suburban Brisbane, Australia, were significantly earlier compared with those of magpies living in adjacent rural areas (Rollinson and Jones, 2003). Early breeding in suburban populations may be explained by warmer temperatures, the watering of lawns (which increases the abundance of invertebrate populations) or the provision of food by people (Rollinson and Jones, 2003). In New Zealand, the breeding season typically extends from July to November, with nest building beginning in June (McCaskill, 1945; McIlroy, 1968; Veltman, 1984; Heather and Robertson, 1996). In Fiji, breeding has been reported to occur between August and November (Clunie, 2007).

Australian Magpies build open-cup nests that often have distinctive layering (McCaskill, 1945; McIlroy, 1968; Kaplan, 2004). The outer layer is made up of larger sticks or branches that result in a large semi-circular bowl (approximately 30 cm in diameter and 20 cm deep), while the inner layer is dominated by finer material, such as grass, smaller twigs and wool (McCaskill, 1945; McIlroy, 1968; Kaplan, 2004). The materials used in the construction of the inner layer are much finer and are woven together more closely. This results in the inner nest being well insulated from the wind (Kaplan, 2004). Typically, the internal dimensions of the nest are approximately 17 cm in diameter and 10 cm deep (McIlroy, 1968; Higgins, *et al.*, 2006). A variety of different plant species have been recorded in the composition of nests; in areas near humans, items such as wire, frayed rope, sacking material and cotton wool have been reported (e.g. McIlroy, 1968; Kaplan, 2004). Nest site selection and construction are only undertaken by females (Carrick, 1972).

Nests are often located in forks of the exposed upper reaches of tall trees and are therefore generally very conspicuous; nevertheless, this may confer fitness benefits through enhanced early predator detection (Kaplan, 2004). The height that nests are located is heavily influenced by available vegetation in the breeding area and the location of suitable branches in the trees that can securely support and anchor the nest (Kaplan, 2004). For example, McIlroy (1968) reported that nests in common broom (*Cytisus scoparius*) were less than 2 m from the ground, those in willow (*Salix* spp.) up to 12 m and those in macrocarpa (*Cupressus macrocarpa*) and pine (*Pinus* spp.) up to 18 m. Kaplan (2004) reported that nest heights in mature gum trees (*Eucalyptus* spp.) tended to be relatively high (15 m), as the smaller branches that can be used to anchor nests are only found in the upper reaches.

On average, three or four eggs (range one to six) are laid per clutch (Carrick, 1963, 1972; Veltman, 1984). The interval between consecutive eggs laid within a clutch is not known, but is thought to be about 24 h (Higgins *et al.*, 2006); however, reports of 2 and 3 days between consecutive eggs in a clutch exist (McIlroy, 1968). Eggs (approximately 40 × 27 mm) vary in coloration between pale blue and green, with olive, reddish or purple tinges (McIlroy, 1968; Heather and Robertson, 1996; Kaplan, 2004). Females alone incubate the clutch for 19–24 days (McCaskill, 1945; Carrick, 1972; Veltman, 1989; Rollinson, and Jones, 2003; Kaplan, 2004); the length of incubation does not appear to be affected by external factors (Rollinson and Jones, 2003). During incubation, the female can be fed by her mate or she may leave the nest unattended for short stints to forage (Kaplan, 2004).

Historical reports suggest that eggs hatch asynchronously at approximately 24 h intervals (e.g. McCaskill, 1945; Hobbs, 1972); however, more recent research indicates that hatching is synchronous (Kaplan, 2004). Hatched chicks are altricial and featherless; their eyes do not open during their first 10 days or so (Kaplan, 2004). Brooding is done by the female, although males may brood for short periods (i.e. while the female forages) (Higgins *et al.*, 2006).

There may be considerable variation in which members of a territorial group of Australian Magpies feed the nestlings. For example, in some populations, only the breeding male and female have been observed feeding nestlings, despite the presence of extra birds within the territory (Veltman, 1989); however, nestlings have been observed being fed by auxiliary group members in other studies (Brown and Farabaugh, 1991; Hughes *et al.*, 1996) that may or may not be related to the nestlings (Finn and Hughes, 2001).

The time to fledging of Australian Magpies is 28–38 days (McCaskill, 1945; Heather and Robertson, 1996; Rollinson and Jones, 2003; Kaplan, 2004). Rollinson and Jones (2003) suggest that the availability of food may affect the time taken to fledge, as they observed significantly shorter fledging times in suburban territories (where invertebrate prey was in greater abundance) compared with rural territories.

After Australian Magpies nestlings fledge, they are heavily dependent on other members of the territorial group, as they are not able to forage independently and are poor flyers, especially in the first two weeks (Kaplan, 2004). During this period, they continue to be fed by parents or other members of the territorial group (Carrick, 1972; Veltman, 1989; Hughes *et al.*, 1996; Finn and Hughes, 2001). The age when fledglings start foraging independently can vary greatly, and durations as short as 20 days or as long as 2–3 months have been reported (Higgins *et al.*, 2006); however, some fledglings continue to beg for food for 8–9 months (Higgins *et al.*, 2006).

If Australian Magpie juveniles disperse from the natal territory, the event generally occurs before the start of the following breeding season (Veltman and Carrick, 1990); however, once again, there is considerable variation in the timing of dispersal. For example, when the mean rate of extra-group paternity was comparatively low (44%) in a population, more than half of the juveniles dispersed from the natal territory; however, when extra-group paternity was high (82%), dispersal was almost zero (Durrant and Hughes, 2005). Within populations, there are also gender differences in dispersal, as juvenile males in Carrick's (1972) study were much more likely to leave the natal territory than females.

It is difficult to compare Australian Magpie breeding performance between different studies because reproductive rates are presented in different formats, for example as fledglings per territory (e.g. Hughes *et al.*, 1996), juveniles per adult female (e.g. Carrick, 1972) or juveniles per breeding female (e.g. Veltman, 1989). What is clear is that reproductive success is heavily influenced by territory quality. In a 12-year study by Carrick (1972) in open savannah woodland habitat in Australia, permanent territories produced the most juveniles (on average 0.56 per adult female), while only 0.09 juveniles per adult female came from marginal territories. Magpies from mobile territories always failed to produce juveniles, and no breeding occurred in open territories (Carrick, 1972).

Australian Magpie reproductive success has also been shown to vary greatly between breeding seasons within the same population. For example, there was an approximately 25% difference in the number of fledglings per breeding pair over two consecutive breeding seasons in one Australian study (Rollinson and Jones, 2003). This difference was thought to be due to climatic variation; in the year with the lower fledging rate, there was a significant drought, which was thought to have reduced the availability of prey (Rollinson and Jones, 2003). Other factors such as predation pressure and anthropomorphic factors (e.g. juveniles being struck by vehicles) also appear to affect reproductive performance; however, the relative importance of each mechanism is often unknown, as determining the exact cause of a failure is difficult (Rollinson and Jones, 2003).

In New Zealand, the only study that has measured Australian Magpie reproductive performance over several consecutive seasons reported an average rate of 1.3 juveniles per breeding female (Veltman, 1989). In this study, there was no difference in the reproductive rate of breeding females from territories defended by pairs compared with those defended by groups of birds; however, when productivity was compared on a per capita basis, the reproductive rate was significantly higher in territories defended by pairs (Veltman, 1989).

There are no published data on reproductive rates of magpies within their Fijian range.

25.7 Habitat

In both their native and introduced ranges, Australian Magpies are commonly associated with open habitats that have a high proportion of low ground cover for hunting invertebrate prey, and a moderate to low density of tall trees for nesting and roosting (Robinson, 1956; Carrick, 1963, 1972; Brown and Veltman, 1987; Heather and Robertson, 1996; Schodde and Mason, 1999). However, magpies can be highly adaptable and have been recorded in a wide range of natural and modified landscapes where these features occur (e.g. McIlroy, 1968; Jones, 2002; Morgan *et al.*, 2012; Kaplan and Rogers, 2013).

Historically, the most suitable habitats for magpies in Australia were semi-arid grasslands, savannah or woodland habitats, found in inland regions of the continent, as coastal areas were mainly covered in closed forest (Australian Surveying and Land Information Group, 1990). However, since European colonization, approximately 43% of forest (regarded as poor magpie habitat; Carrick, 1972) has been lost (Noble *et al.*, 1996), improving conditions for magpies. In addition, closed woodland and low and tall shrubland have all been reduced drastically (by 32, 76 and 37%, respectively; Noble *et al.*, 1996). Much of this vegetation was cleared for agricultural or urban development, creating suitable habitat (albeit highly modified) that has been exploited by magpies (Jones, 2002; Barrett *et al.*, 2003). Accordingly, in Australia, magpies are very common in developed open areas such as reserves, amenity parks and school grounds (Rollinson and Jones, 2003; O'Leary and Jones, 2006). They also inhabit residential gardens and backyards across Australia, foraging on natural prey items and anthropomorphic food sources (Rollinson *et al.*, 2003; Ishigame *et al.*, 2006; O'Leary and Jones, 2006).

In contrast, Australian Magpies in New Zealand are relatively uncommon in urban areas (e.g. van Heezik *et al.*, 2008; Fitzgerald and Innes, 2013), despite these areas sharing the same characteristics as Australian equivalents. In New Zealand, magpies are most frequently found in farmland with scattered trees, shelterbelt trees or adjacent forest (McIlroy, 1968; Veltman, 1989; Heather and Robertson, 1996).

There is little published literature describing habitat preferences of magpies in Fiji or New Guinea. In Fiji, they have been recorded on coconut plantations, in coastal areas and in wooded areas (Kaplan and Rogers, 2013), while in New Guinea they inhabit savannah and woodland habitat and rainforest edges (Toon *et al.*, 2017), all areas that have open areas of low-growing vegetation and tall trees.

25.8 Impacts

25.8.1 Positive impacts

Australian Magpies were intentionally introduced to Fiji and New Zealand as biocontrol agents (Thomson, 1922; McCaskill, 1945; McIlroy, 1968; Clunie, 2007). In Fiji, they were specifically introduced to control the coconut stick insect (*Graeffea* spp.), which causes considerable damage to coconut crops (Paine, 1968). Unfortunately, they had little impact on these pests, as the insects spend the majority of their life cycle in the coconut palm canopy (Paine, 1968) and magpies forage predominantly on the ground (Vestjens and Carrick, 1974; Brown and Veltman, 1987; Veltman and Hickson, 1989).

Australian Magpies were released in New Zealand to help control pasture pest invertebrates; while anecdotal evidence suggests that they may have been effective in this function, no empirical studies exist to support the claim (McIlroy, 1968; McDowall, 1994). Conversely, there is also anecdotal evidence that invertebrate prey abundances in some areas are actually higher where there are magpies, which may be the result of other predators being driven from these areas by magpies (McCaskill, 1945). While this hypothesis has not been tested, rural bird species were shown to avoid being in close proximity to territorial magpies in one New Zealand study (Morgan *et al.*, 2006a; but also see Borowske *et al.*, 2012).

In New Zealand, avian prey species nesting close to magpies may have higher fledging success through a reduction in predation from the Swamp Harrier or Kāhu (*Circus approximans*) (e.g. McIlroy, 1968). Swamp Harriers appear to be regarded as a serious threat to magpies and have been attacked more frequently than any other bird species in New Zealand (Morgan *et al.*, 2005, 2006a). Furthermore, Morgan *et al.* (2006b) reported that Swamp Harriers were an important predator of bird nests, predominantly the Eurasian Blackbird (*Turdus merula*) and Song Thrush (*Turdus philomelos*), in rural New Zealand and accounted for 36% of recorded predation events. They noted, however, that the study was conducted in areas where Australian Magpies were common and not controlled. It is not known whether nesting success of bird species decreases in areas where magpie populations have been reduced.

25.8.2 Negative impacts

Australian Magpies attack a wide range of species in both Australia and New Zealand, including birds, humans, sheep (*Ovis aries*), cats (*Felis catus*), dogs (*Canis familiaris*), possums (*Trichosurus vulpecula*), rabbits (*Oryctolagus cuniculus*), hedgehogs (*Erinaceus europaeus*) and horses (*Equus ferus caballus*) (Chisholm, 1910; McCaskill, 1945; Hall, 1962; McIlroy, 1968; Paton, 1977; Ashton, 1986; Barr, 1986; Cox and Bauer, 1997; Jones, 2002; Morgan *et al.*, 2005); however, the two groups that appear to be targeted most often by magpies are diverse terrestrial birds (e.g. Morgan *et al.*, 2005) and humans (e.g. Jones, 2002).

25.8.3 Impacts on humans

The magpie–human conflict is significant across the native range of Australian Magpies, and a considerable amount of research has been conducted into understanding the interactions and reducing attack probabilities (e.g. Jones *et al.*, 1980; Jones and Thomas, 1998, 1999; Cilento and Jones, 1999; Jones, 2002; Warne and Jones, 2003). However, quantifying the impact that magpies have on people in Australia is outside the scope of this chapter. Instead, we will focus on the potential impact magpies have on humans within their introduced ranges.

We are not aware of any published reports of Australian Magpies attacking people within their Fijian range. In New Zealand, many such reports exist (McCaskill, 1945; McIlroy, 1968); however, there is little evidence that they are having a significant impact on the social or physical well-being of people. For example, while several local government agencies in New Zealand include Australian Magpies within their regional pest management strategies, they are only considered 'of nuisance' to people during the breeding season due to occasional swooping or attacks (e.g. Auckland Regional Council, 2007). In addition, only 113 compensation claims relating to injuries received from magpie attacks were made between 2013 and 2017 in New Zealand (Analytics and Reporting, Accident Compensation Corporation NZ, 2018, unpublished data). The minor effect that Australian Magpies appear to have on people in New Zealand may partly be due to them being relatively uncommon within urban areas (Day, 1995; Fitzgerald and Innes, 2013) where most (86%) of the human population resides (see archive.stats.govt.nz/browse_for_stats/population/Migration/internal-migration/urban-rural-migration.aspx; accessed 28 January 2019). Accordingly, there are fewer opportunities for interactions between magpies and people in New Zealand.

25.8.4 Impacts on other birds

We are not aware of any published reports of Australian Magpies having any negative impacts on birds within their Fijian range. Indeed, magpies in Fiji appear to be considerably less aggressive and are described as 'shy and flighty, often hiding amongst shrubs and palms' (Kaplan and Rogers, 2013). In New Zealand, there are numerous reports of Australian Magpies attacking and even killing other birds (e.g. Morgan *et al.*, 2005). As such, concerns have been raised about the possibility they are having negative impacts on the abundance and/or distribution of native birds (e.g. McCaskill, 1945; Barrington, 1995; McKay, 1997).

While Australian Magpies clearly attack many other bird species, the proximate reason(s) and adaptive benefit(s) for this aggressive behaviour is not clear. For example, in New Zealand, 45 bird species have been observed being attacked by magpies but many of these reported attacks are on birds that are neither predators of, nor have resource overlap with, magpies (Morgan *et al.*, 2005). Furthermore, magpies do not appear to consider other birds as prey, as they rarely kill adults (Morgan *et al.*, 2006a), nestlings or eggs (Boulton and Cassey, 2006; Morgan *et al.*, 2006b, 2011).

Several studies have attempted to quantify the impact that Australian Magpies have on birds in New Zealand across different landscape scales. Innes *et al.* (2012) reduced magpie populations over five large (approximately 900 ha) rural blocks over a 3-year period and measured changes in the abundance of other birds (in relation to adjacent areas where magpies were not controlled). Small increases in only one native species, the New Zealand Pigeon or Kererū (*Hemiphaga novaeseelandiae*) and five exotic species were observed (Innes *et al.*, 2012). Interestingly, a small decrease in the abundance of the native New Zealand Fantail or Pīwakawaka (*Rhipidura fuliginosa*) was also observed in magpie removal areas (Innes *et al.*, 2012). Innes *et al.* (2012) concluded that there was little evidence that magpies are serious threats to other birds over landscape scales; however, the impact they may have on other birds at smaller scales could not be determined from their experiment (i.e. magpies could be excluding other birds from localized areas of high resource availability). Such small-scale displacements may not have been detected in the large-scale magpie removal experiment (Innes *et al.*, 2012).

To determine whether Australian Magpies influence the local movements of other birds in rural landscapes, Morgan *et al.* (2006a) compared the number of each species of bird that flew or foraged within 50 m of territorial magpie groups and non-territorial magpie flocks with those in adjacent 'magpie-free' areas over an entire year. In comparison with the magpie-free areas, significantly fewer birds of a range of species foraged close to both territorial and flocking magpies (Morgan *et al.*, 2006a). Furthermore, fewer birds flew close (less than 50 m) to territorial magpie groups (Morgan *et al.*, 2006a). For generalist resource consumers, such as insectivores and granivores (which include many of the exotic birds in rural New Zealand), being displaced a few hundred metres may not incur a significant cost, as these resources are fairly ubiquitous over the landscape; however, for specialist resource consumers that may exploit highly localized food sources, being displaced from a high-quality patch may incur a considerable fitness cost. For example, if nectar- or fruit-feeding specialists, such as the native Tui (*Prosthemadera novaeseelandiae*) or New Zealand Pigeon, are displaced from high-quality patches, such as forest fragments that are relatively rare in rural New Zealand, they may have to fly long distances to find alternate food sources. It is important to note that the majority of bird species monitored by Morgan *et al.* (2006a) were exotic granivores or ground-feeding insectivores, such as the Common Starling (*Sturnus vulgaris*), Blackbird and Yellowhammer (*Emberiza citronella*); native species were poorly represented in this study, as most of the areas where fieldwork was conducted did not have optimal habitat for native birds (e.g. forest fragments).

Accordingly, to test the prediction that removing Australian Magpies from areas of high resource availability increases visitations from native birds, magpie populations were significantly reduced around 11 forest fragments and rural gardens containing nectar- or fruit-producing trees over a 6-week period when food was abundant (Morgan *et al.*, 2012). The response of native birds to magpie removal was compared with other forest fragments and gardens where no removal was conducted. Significantly reducing the relative abundance of magpies had little effect on the relative abundance of any of the native birds that were monitored within treatment sites (Morgan *et al.*, 2012). This suggested that magpies are not having a significant impact on these native birds at this scale.

In summary, there is no evidence that Australian Magpies are having negative impacts in Fiji. In New Zealand, the perception that magpies are having negative impacts on the abundance or distribution of native birds appears to be based largely on anecdotal observations of interspecific aggressive acts that occur infrequently. The fact that magpies exhibit heightened levels of aggression towards other animals, especially birds, is not in dispute; however, their conspicuous behaviours and appearance, and habitat preferences may result in attacks being more likely to be observed by people. Accordingly, while magpies may occasionally chase and even kill other birds (Morgan *et al.*, 2005), this does not appear to have an impact at a population level (Morgan *et al.*, 2006a, 2012; Innes *et al.*, 2012).

25.9 Control

Information on control methods for Australian Magpies within their introduced range is only available from New Zealand sources, as they do not appear to be regarded as a pest species within Fiji, and populations are not controlled there. Within New Zealand, no proven 'best practice' magpie control regimes exist, as few large control regimes have been conducted. There are, however, several methods that are used regularly to reduce magpie populations. The following information is derived primarily from the website of the Waikato Regional Council (www.waikatoregion.govt.nz/services/regional-services/plant-and-animal-pests/animal-pests/magpies/, accessed 28 January 2019), a territorial local authority in the Waikato, North Island, New Zealand.

25.9.1 Trapping

Australian Magpies can easily be caught using Larsen-style live-capture traps. These traps have multiple compartments that are each held open by a false perch, which breaks after the bird has entered, closing the door and trapping the individual. Captured birds then need to be euthanized by the trapper. New Zealand legislation stipulates that traps must be checked within 12 h of sunrise the day after they are set. Australian Magpies can be lured into the trap by: (i) baiting with food (e.g. animal fat, pet food, egg); (ii) positioning a mirror at the back of the traps, or (iii) using a live conspecific 'call bird' in a cage adjacent to the open trap compartments (Morgan *et al.*, 2007). The call bird potentially simulates a territorial intrusion (Morgan *et al.*, 2007) and target magpies inadvertently walk into the trap while investigating the 'intruder'. The use of live-capture traps appears to be the most commonly used magpie control method in New Zealand (Innes *et al.*, 2012; Morgan *et al.*, 2012), and there are several commercially available varieties, or they can be home-built. Furthermore, a large number of traps can be deployed over a broad area and still be maintained by a relatively small number of people, making this method an efficient way to control magpies on a landscape scale (e.g. Innes *et al.*, 2012). This method also has the advantage of reducing non-target species by-catch, as non-target animals are not harmed and can be released during trap checks.

25.9.2 Shooting

Using a low-calibre rifle or a high-powered air rifle to shoot Australian Magpies may be effective at removing small numbers of birds but is unlikely to have a significant impact on a population; however, no published studies have measured the efficacy of this method when used in isolation. Magpies have been shown to have long memories (Kaplan, 2019), and it is likely that they will become wary of shooters if they have been shot at and survive. To increase success rates, it is recommended that model magpie decoys are deployed, and distress calls are broadcast from a speaker close to the decoy to draw birds closer to the shooter. The shooter should also be camouflaged or make use of natural cover, and attach a noise moderator to the firearm to reduce conspicuousness.

25.9.3 Poisoning

The use of the narcotic poison α-chloralose can be employed to potentially control large numbers of Australian Magpies. Under New Zealand regulations, an operator does not need to hold a poisons licence when α-chorolose is used in concentrations of 2.5% or less. It is recommended that concentrations that do not exceed this threshold are used when managing magpies, as this will anaesthetise them rather than cause mortality; comatose birds can then be destroyed humanely. Non-target species should be collected and kept warm until they revive and can subsequently be released; however, there is a risk that birds smaller than magpies may die when consuming bait containing 2.5% α-chloralose (New Zealand Landcare Trust, 2016). Bread is often used as the vector to deliver α-chloralose, and it is essential that extensive pre-feeding is done leading up to the operation in order for the magpies to become familiar with a potentially novel food type. Pre-feeding also gives the operator an opportunity to estimate the size of the magpie population in the vicinity, which allows an appropriate amount of toxic bait to be prepared.

25.9.4 Efficacy of magpie control methods

Only two studies have measured the response of Australian Magpie populations to magpie control (Innes *et al.*, 2012; Morgan *et al.*, 2012), so it is difficult to draw conclusions on the efficacy of the control methods outlined above. Furthermore, both studies predominantly or exclusively used trapping methods to remove magpies. Despite the small sample size, both studies showed that magpie populations can be significantly reduced during control operations. For example, Innes *et al.* (2012) reported that the relative abundance of magpies declined by, on average, 76% across five 900 ha blocks in rural areas over a 3-year control programme; this study also employed non-trapping methods (shooting and using α-chloralose). Morgan *et al.* (2012) significantly reduced the relative abundance of magpies by 52% around several forest fragments and rural gardens during a 6-week period using only trapping. Both of these studies (also see Morgan *et al.*, 2006a) implemented magpie removal operations in the winter/spring period, which is during the magpie breeding season (Heather and Robertson, 1996) and when territorial magpies are generally most aggressive (Cilento and Jones, 1999). Therefore, the capture rates during these studies may have been higher compared with other times of the year due to seasonal effects. Further research is needed to determine the efficacy of each removal method, and how capture rates change seasonally.

25.10 Uses

There are no known uses of Australian Magpies.

25.11 Acknowledgements

D.K.J.M. would like to acknowledge the NorthTec Research Fund. J.G.I. would like to acknowledge the Strategic Science Investment Funding for Crown Research Institutes from the New Zealand Ministry of Business, Innovation and Employment.

25.12 References

Ashton, C. (1986) Predation of young bee-eaters by a magpie. *Bird Observer* 656, 96.

Auckland Regional Council (2007) Auckland regional pest management strategy: 2007–2012. Auckand Regional Council, Auckland, New Zealand.

Australian Surveying and Land Information Group (1990) *Atlas of Australian Resources: Land Resources*, Vol. 6. Commonwealth Government Printer, Canberra.

Baker, A.M., Mather, P.B. and Hughes, J.M. (2000) Population genetic structure of Australian magpies: evidence for regional differences in juvenile dispersal behaviour. *Heredity* 85, 167–176.

Barr, J. (1986) Magpie attack on sparrow. *Canberra Bird Notes* 11, 133.

Barrett, G., Silcocks, A., Barry, S., Cunningham, R. and Poulter, R. (2003) *The New Atlas of Australian Birds*. Birds Australia, Victoria, Australia.

Barrington, J. (1995) Magpies – a black and white issue. *Forest and Bird Magazine* 275, 4.

Beehler, B.M. and Pratt, T.K. (2016) *Birds of Papua New Guinea: Distribution, Taxonomy and Systematics*. Princeton University Press, New Jersey.

Black, A. (1986) The taxonomic affinity of the New Guinean magpie *Gymnorhina tibicen papuana*. *Emu* 86, 65–70.

Borowske, A., Morgan, D.K.J. and Waas, J.R. (2012) Do heterospecific size and demeanour influence visitation behaviour of birds at urban foraging patches? *Journal of Ethology* 30, 75–82.

Boulton, R.L. and Cassey, P. (2006) An inexpensive method for identifying predators of passerine nests using tethered artificial eggs. *New Zealand Journal of Ecology* 30, 377–385.

Brown, E.D. and Farabaugh, S.M. (1991) Song sharing in a group-living song bird, the Australian magpie, *Gymnorhina tibicen*. Part III. Sex-specificity and individual specificity of vocal parts in communal chorus and duet songs. *Behaviour* 118, 244–274.

Brown, E.D., and Veltman, C.J. (1987) Ethogram of the Australian magpie (*Gymnorhina tibicen*) in comparison to other Cracticidae and *Corvus* species. *Ethology* 76, 309–333.

Bull, P.C., Gaze, P.D. and Robertson, C.J.R. (1985) *The Atlas of Bird Distribution in New Zealand*. Ornithological Society of New Zealand, Wellington.

Cain, A.J. and Galbraith, I.C.J. (1956) Field notes on birds of the eastern Solomon Islands. *Ibis* 98, 262–295.

Carrick, R. (1963) Ecological significance of territory in the Australian magpie, *Gymnorhina tibicen*. In: *Proceedings of the 13th International Ornithological Congress*, II. American Ornithologists Union, Washington, DC, pp. 740–753.

Carrick, R. (1972) Population ecology of the Australian black-backed magpie, royal penguin, and silver gull. In: *Population Ecology of Migratory Birds – A Symposium*. Wildlife Research Report, No. 2. US Department of the Interior, Washington, DC, pp. 41–99.

Chisholm, A.H. (1910) The magpie defamed. *Emu* 9, 246.

Cilento, N.J. and Jones, D.N. (1999) Aggression by Australian magpies *Gymnorhina tibicen* toward human intruders. *Emu* 99, 85–90.

Clunie, F. (2007) *Birds of the Fiji Bush*, 3rd edn. Fiji Museum, Suva, Fiji.

Clunie, F., Kinsky, F.C. and Jenkins, J.A.F. (1978) New bird records from the Fiji Archipelago. *Notornis* 25, 118–127.

Cox, S.J. and Bauer, J.J. (1997) Species interactions between the white-winged chough and Australian magpie in a fragmented landscape. *Pacific Conservation Biology* 3, 289–294.

Day, T.D. (1995) Bird species composition and abundance in relation to native plants in urban gardens, Hamilton, New Zealand. *Notornis* 42, 175–186.

Drummond, J. (1907) Dates on which introduced birds have been liberated, or have appeared, in different districts of New Zealand. *Transactions and Proceedings of the New Zealand Institute* 39, 503–508.

Durrant, K.L. and Hughes, J.M. (2005) Differing rates of extra-group paternity between two populations of the Australian magpie (*Gymnorhina tibicen*). *Behavioral Ecology and Sociobiology* 57, 536–545.

Dutson, G. (2011) *Birds of Melanesia – Bismarcks, Solomons, Vanuatu and New Caledonia*. Christopher Helm, London.

Finn, P.G. and Hughes, J.M. (2001) Helping behaviour in Australian magpies, *Gymnorhina tibicen*. *Emu* 101, 57–63.

Fitzgerald, N. and Innes, J. (2013) Hamilton City biennial bird counts: 2004–2012. Landcare Research Contract Report, LC 1484. Prepared for the Hamilton City Council, Hamilton, New Zealand.

Floyd, R.R. and Woodland, D.J. (1981) Localization of soil dwelling scarab larvae by the black-backed magpie, *Gymnorhina tibicen* (Latham). *Animal Behaviour* 29, 510–517.

Gill, B. J., Bell, B.D., Chambers, G.K., Medway, D.G., Palma, R.L., *et al.* (2010) *Checklist of the Birds of New Zealand, Norfolk and Macquarie Islands, and the Ross Dependency, Antarctica*, 4th edn. Te Papa Press, Wellington.

Hall, A.G. (1962) Black shag forced down by Australian magpies. *Notornis* 10, 44.

Heather, B.D. and Robertson, H.A. (1996) *A Field Guide to the Birds of New Zealand*. Viking, Auckland, New Zealand.

Higgins, P.J., Peter, J.M. and Cowling, S.J. (eds) (2006) *Handbook of Australian, New Zealand and Antarctic Birds*, Vol. 7a: *Boatbill to Starlings*. Oxford University Press, Melbourne, Australia.

Hobbs, J.N. (1972) Magpies defending nest after departure of young. *Australian Field Ornithology* 4, 131–132.

Holling, C.S. (1965) The functional response of predators to prey density and its role in mimicry and population regulation. *Memoirs of the Entomological Society of Canada* 97, 5–60.

Hughes, J.M. and Mather, P.B. (1991) Variation in the size of territorial groups in the Australian magpie, *Gymnorhina tibicen*. *Proceedings of the Royal Society of Queensland* 101, 13–19.

Hughes, J.M., Hesp, J.D.E., Kallioinen, R., Kempster, M., Lange, C.L., *et al.* (1996) Differences in social behaviour between populations of the Australian magpie *Gymnorhina tibicen*. *Emu* 96, 65–70.

Hughes, J.M., Pearce, B.J. and Vockenson, K. (1983) Territories of the Australian magpie (*Gymnorhina tibicen*) in south-east Queensland. *Emu* 83, 108–111.

Innes, J., Spurr, E.B., Arnold, G.C., Morgan, D., Waas, J.R. and Watts, C. (2012) Using 5-minute counts to study magpie impacts on other birds in New Zealand. *New Zealand Journal of Ecology* 36, 324–332.

Ishigame, G., Baxter, G.S. and Lisle, A.T. (2006) Effects of artificial foods on the blood chemistry of the Australian magpie. *Austral Ecology* 31, 199–207.

Jones, D. (2002) *Magpie Alert: Learning to Live with a Wild Neighbour*. University of New South Wales Press, Sydney, Australia.

Jones, D.N. and Thomas, L.K. (1998) Managing to live with Brisbane's wildlife: magpies and the management of positive and negative interactions. *Proceedings of the Royal Society of Queensland* 107, 45–49.

Jones, D.N. and Thomas, L.K. (1999) Attacks on humans by Australian magpies: management of an extreme suburban human–wildlife conflict. *Wildlife Society Bulletin* 27, 473–478.

Jones, L., Darroch, R.K., Gilding, J. and Bennett, D.H. (1980) A review of seasonal and ecological factors in Australian magpie *Gymnorhina tibicen* attacks on people. *Bird Behaviour* 2, 113–117.

Kallioinen, R.U.O., Hughes, J.M. and Mather, P.B. (1995) Significance of back colour in territorial interactions in the Australian magpie. *Australian Journal of Zoology* 43, 665–673.

Kaplan, G. (2004) *Australian Magpie: Biology and Behaviour of an Unusual Songbird*. CSIRO Publishing, Collingwood, Australia.

Kaplan, G. (2019) *Australian Magpie: Biology and Behaviour of an Unusual Songbird*. Second Edition. CSIRO Publishing, Australia.

Kaplan, G. and Rogers, L.J. (2013) Stability of referential signalling across time and locations: testing alarm calls of Australian magpies (*Gymnorhina tibicen*) in urban and rural Australia and in Fiji. *PeerJ* 1, e112.

Lepage, D. (2018) Avibase – Bird Checklists of the World: Sri Lanka. Available at: https://avibase.bsc-eoc.org/checklist.jsp?region=LK (accessed 30 October 2019).

Long, J.T. (1981) *Introduced Birds of the World: Worldwide History, Distribution and Influence of Birds Introduced to New Environments*. David & Charles, London.

McCaskill, L.W. (1945) Preliminary report on the present position of the Australian magpies (*Gymnorhina hypoleuca* and *G. tibicen*) in New Zealand. *New Zealand Bird Notes* 1, 86–104.

McDowall, R.M. (1994) *Gamekeepers for the Nation: The Story of New Zealand's Acclimatisation Societies, 1861–1990*. Canterbury University Press, Christchurch, New Zealand.

McIlroy, J.C. (1968) The biology of magpies (*Gymnorhina* spp.) in New Zealand. MAgSc thesis, Lincoln College, Lincoln, New Zealand.

McKay, D. (1997) Quardle ardle oodle: perceived value and appropriate controls of the Australian bell magpie (*Gymnorhina* spp.) in South Island, New Zealand. MSc thesis, Lincoln University, Lincoln, New Zealand.

Morgan, D. (2008) The proportion of white-backed and black-backed Australian magpies (*Gymnorhina tibicen*) in a Waikato population. *Notornis* 55, 155–157.

Morgan, D., Waas, J.R. and Innes, J. (2005) Magpie interactions with other birds in New Zealand: results from a literature review and public survey. *Notornis* 52, 61–74.

Morgan, D., Waas, J.R. and Innes, J. (2006a) Do territorial and non-breeding Australian magpies influence the local movements of rural birds in New Zealand? *Ibis* 148, 330–342.

Morgan, D., Waas, J.R. and Innes, J. (2006b) The relative importance of Australian magpies (*Gymnorhina tibicen*) as nest predators of rural birds in New Zealand. *New Zealand Journal of Zoology* 33, 17–29.

Morgan, D.K.J., Waas, J.R. and Innes, J. (2007) Can redirected aggression explain interspecific attacks by Australian magpies on other birds? *Behaviour* 144, 767–786.

Morgan, D.K.J., Waas, J.R., Innes, J. and Fitzgerald, N. (2011) Identification of nest predators using continuous time-lapse recording in a New Zealand city. *New Zealand Journal of Zoology* 38, 343–347.

Morgan, D.K.J., Waas, J.R., Innes, J. and Arnold, G. (2012) Native bird abundance after Australian magpie (*Gymnorhina tibicen*) removal from localised areas of high resource availability. *New Zealand Journal of Ecology* 36, 333–339.

New Zealand Landcare Trust (2016) *Northland Pest Control Guidelines, 2016*. New Zealand Landcare Trust, Hamilton, New Zealand.

Noble, I., Barson, M., Dumsday, R., Friedel, M., Hacher, R., *et al.* (1996) Land resources. In: Taylor, R. (ed.) *Australia: State of the Environment*. CSIRO Publishing, Collingwood, Australia, pp. 6.1–6.55.

O'Donnell, C.F.J. (2002) Classified summarised notes, South Island and outlying islands, 1 July 1999 to 30 June 2000. *Notornis* 49, 111–114.

O'Leary, R. and Jones, D.N. (2006) The use of supplementary foods by Australian magpies *Gymnorhina tibicen*: implications for wildlife feeding in suburban environments. *Austral Ecology* 31, 208–216.

Oliver, W.R.B. (1955) *New Zealand Birds*. Coulls Somerville Wilkie, Wellington.

Paine, R.W. (1968) Investigations for the biological control in Fiji of the coconut stick-insect *Graeffea crouanii* (Le Guillou). *Bulletin of Entomological Research* 57, 567–604.

Paton, D C. (1977) Magpies attacking blackbirds. *South Australian Ornithologist* 27, 185.

Robertson, C.J.R., Hyvönen, P., Fraser, M.J. and Pickard, C.R. (2007) *Atlas of Bird Distribution in New Zealand 1999–2004*. Ornithological Society of New Zealand, Wellington.

Robinson, A. (1956) The annual reproductive cycle of the magpie, *Gymnorhina dorsalis* Campbell. *Emu* 56, 235–336.

Rollinson, D.J. (2004) Synanthropy of the Australian magpie: a comparison of populations in rural and suburban areas of southeast Queensland, Australia. PhD thesis, Griffith University, Brisbane, Australia.

Rollinson, D.J. and Jones, D.N. (2003) Variation in breeding parameters of the Australian magpie *Gymnorhina tibicen* in suburban and rural environments. *Urban Environments* 6, 257–269.

Rollinson, D.J., O'Leary, R. and Jones, D.N. (2003) The practice of wildlife feeding in suburban Brisbane. *Corella* 27, 52–58.

Schodde, R. and Mason, I.J. (1999) *The Directory of Australian Birds*. CSIRO Publishing, Collingwood, Australia.

Shurcliff, A. and Shurcliff, K. (1974) Territory in the Australian magpie (*Gymnorhina tibicen*): an analysis of its size and change. *South Australian Ornithologist* 26, 127–132.

Thomson, G.M. (1922) *The Naturalisation of Animals and Plants in New Zealand*. University Press, Cambridge.

Toon, A., Drew, A., Mason, I.J., Hughes, J.M. and Joseph, L. (2017) Relationships of the New Guinean subspecies, *Gymnorhina tibicen papuana*, of the Australian magpie: an assessment from DNA sequence data. *Emu* 117, 305–315.

van Heezik, Y., Smyth, A. and Mathieu, R. (2008) Diversity of native and exotic birds across an urban gradient in a New Zealand city. *Landscape and Urban Planning* 87, 223–232.

Veltman, C.J. (1984) The social system and reproduction in a New Zealand magpie population, and a test of a cooperative breeding hypothesis. PhD thesis, Massey University, Palmerston North, New Zealand.

Veltman, C.J. (1989) Flock, pair and group living lifestyles without cooperative breeding by Australian magpies *Gymnorhina tibicen*. *Ibis* 131, 601–608.

Veltman, C.J. and Carrick, R. (1990) Male-biased dispersal in Australian magpies. *Animal Behaviour* 40, 190–192.

Veltman, C.J. and Hickson, R.E. (1989) Predation by Australian magpies (*Gymnorhina tibicen*) on pasture invertebrates: are non-territorial birds less successful? *Australian Journal of Ecology* 14, 319–326.

Vestjens, W.J.M. and Carrick, R. (1974) Food of the black-backed magpie, *Gymnorhina t. tibicen*, at Canberra. *Australian Wildlife Research* 1, 71–83.

Warne, R.M. and Jones, D.N. (2003) Evidence of target specificity in attacks by Australian magpies on humans. *Wildlife Research* 30, 265–267.

26 Mallard (*Anas platyrhynchos* Linnaeus, 1758)

Matthieu Guillemain[1]*, Pär Söderquist[2], Jocelyn Champagnon[3] and Johan Elmberg[2]

[1]*Office Français de la Biodiversité, Unité Avifaune Migratrice, La Tour du Valat, Le Sambuc, 13200 Arles, France;* [2]*Department of Environmental Science and Bioscience, Kristianstad University, Kristianstad, Sweden;* [3]*Tour du Valat, Research Institute for the Conservation of Mediterranean Wetlands, Le Sambuc, 13200 Arles, France*

Citation: Guillemain, M., Söderquist, P., Champagnon, J. and Elmberg, J. (2020) Mallard (*Anas platyrhynchos* Linnaeus, 1758). In: Downs, C.T. and Hart, L.A. (eds) *Invasive Birds: Global Trends and Impacts*. CAB International, Wallingford, UK, pp. 194–199.

26.1 Common Names

Mallard, Northern Mallard.

26.2 Nomenclature

The Mallard (*Anas platyrhynchos*) complex encompasses very closely related species (considered subspecies by some authors) including the American Black Duck (*A. rubripes*), Mexican Duck (*A. diazi*), Mottled Duck (*A. fulvigula*), Laysan Duck (*A. laysanensis*), Hawaiian Duck (*A. wyvilliana*), Indian Spot-billed Duck (*A. poecilorhyncha*), Eastern Spot-billed Duck (*A. zonorhyncha*), Meller's Duck (*A. melleri*), Yellow-billed Duck (*A. undulata*) and Philippine Duck (*A. luzonica*). Consistently considered as Mallard subspecies are the birds found in Greenland (*A. p. conboschas*) and an extinct population in the western Pacific Islands (*A. p. oustaleti*) (Del Hoyo *et al.*, 1992; Baldassarre, 2014; Reeber, 2015).

26.3 Distribution

The Mallard is the most common and widespread of all ducks (Fig. 26.1), occurring in virtually all wetland types of the northern hemisphere south to the Tropic of Cancer, with strongholds in North America and Europe (Young, 2005). The estimated worldwide Mallard population size is approximately 20 million individuals, of which around 9 million occur in North America and 7.5 million in Europe (Wetlands International, 2018). Outside its natural range, deliberate introductions have given rise to significant feral populations in southern Australia, New Zealand, Hawaiian Islands, Bermuda, South Africa, Madagascar, Mauritius, Falkland Islands, Cayman Islands, New Caledonia and the United Arab Emirates, as well as on a very large number of other islands well outside the natural range (e.g. in Oceania) (Palmer, 1976; Young, 2005, 2013; Reeber, 2015). Beside such established and self-sustaining populations, Mallards are found as ornamental birds in city parks worldwide (Baldassarre, 2014). In addition to such introductions outside the species range, hand-reared Mallards are released in very large numbers within the original geographical range of the species for hunting purposes (e.g. approximately 3 million are released annually in Europe; Champagnon, 2011; Söderquist, 2015). Finally, there is a captive population, maintained mainly for food consumption, estimated at some 681 million individuals worldwide (Tanabe, 1995).

26.4 Description

The Mallard is the largest of all dabbling ducks, with a mean body mass slightly over 1 kg in adult males (Young, 2005). Sexual dimorphism of the plumage is very conspicuous; breeding males have a yellowish bill, green metallic head, narrow white collar, brown breast, vermiculated grey wings and flanks, black tail and undertail coverts, except for a white line on the sides of the tail (Fig. 26.2). Some tail feathers are curled

*Corresponding author: matthieu.guillemain@ofb.gouv.fr

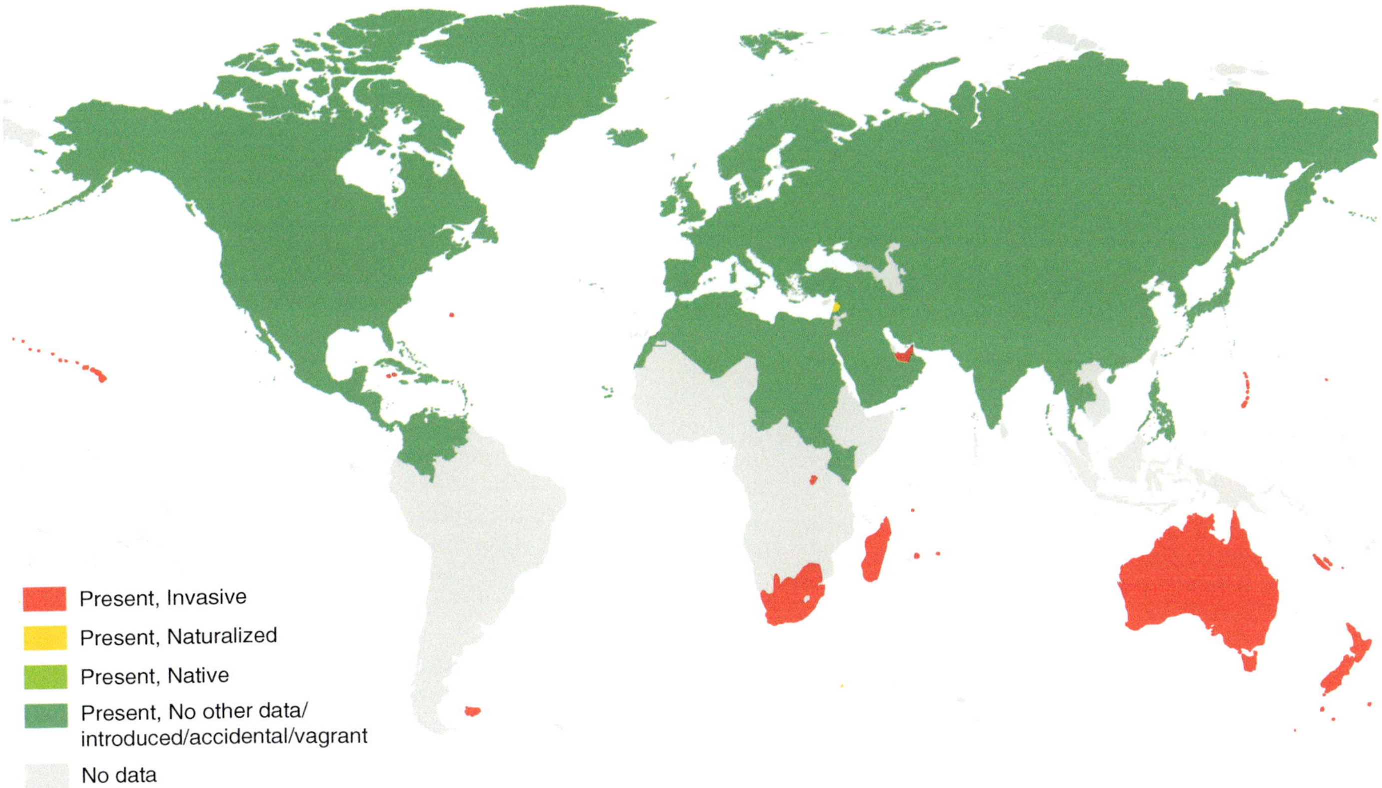

Fig. 26.1. Global distribution of the Mallard (*Anas platyrhynchos*).

Fig. 26.2. An adult male Mallard Duck with a female at the rear. (©Photograph: Maurice Benmergui.)

Fig. 26.3. A female Mallard with young. (©Photograph: Maurice Benmergui.)

upwards. After the breeding season, males moult into an eclipse plumage for a couple of months, when they have a camouflaged mottled brown appearance, much like what females have throughout the year (Figs 26.3 and 26.4). At that time, bill colour is the best character to tell the sexes apart: males have a plain greyish/yellowish bill, while females have an orange bill mottled black or brown. The legs are orange in both sexes. Males and females both have metallic-blue greater secondary coverts, forming a conspicuous blue wing bar in flight and often partly visible when the wing is folded in sitting and swimming birds. Age (differentiation between birds in their first year versus older) is difficult to tell in the field in fledged birds but can easily be determined with the bird in the hand, especially by examination of the black spot at the tip of the greater secondary coverts (Mouronval, 2016). The Mallard has been domesticated for over 2500 years, leading to selection of many breeds differing in size and colour: for example, the Rouen breed has the general appearance of a wild Mallard but can weigh up to 5 kg, while the Pekin Duck has a Mallard silhouette but sports an entirely white plumage (Kear, 1990). It is common that aberrant or white plumage appears among individuals of feral populations.

Fig. 26.4. (A, B) Mallards showing typical 'dabbling' feeding behaviour. Note that both Mallards are males moulting from eclipse to breeding plumage. (©Photographs: Maurice Benmergui.)

26.5 Diet

Mallards are mostly seed-eaters during the non-breeding season but turn to a protein-rich invertebrate diet during spring migration and the breeding season (Dessborn *et al.*, 2011). Depending on season and local conditions, a wide range of seeds and invertebrate species can be found in the Mallard diet, ranging from minute *Carex* spp. seeds to acorns, and including crop seeds on agricultural land (reviewed by Dessborn *et al.*, 2011). Invertebrate prey most often comprises molluscs, crustaceans and insects (both adults and larvae). Again, the size of regular prey can be extremely variable, from bryozoan statoblasts approximately 1 mm in size to crustaceans and insects over 50 mm in length (reviewed by Cramp and Simmons, 1977; see also Mouronval *et al.*, 2007). More exceptional Mallard prey includes small fish and amphibians, especially tadpoles (reviewed by Baldassarre, 2014). In urban environments, the Mallard is frequently a synanthrope, relying heavily on food provisioned by humans, especially bread and grain. Similarly, survival of hand-reared individuals released into wild populations often relies on provisioning of supplementary food (grain) by humans, because of the farmed ducks' inferior ability to forage efficiently on natural food sources compared with their wild conspecifics (Champagnon *et al.*, 2010, 2012; Söderquist *et al.*, 2014).

26.6 Introduction and Invasion Pathways

Mallards have mostly been introduced (outside the original geographical range) or released (within it) for ornamental or hunting purposes, although introductions of captive stock as farmed animals for human food consumption during the 19th century have also been reported (e.g. Hawaiian Islands; Pyle and Pyle, 2017). The conspicuous plumage of males and the tameness of the species make it a favourite in city parks where it readily settles and remains easily observed by the public. The ease with which it is bred in captivity explains why it can be released in such large numbers for hunting purposes. Escapees from small breeding farms are also frequent.

In Australia and New Zealand, Mallards were first introduced for hunting purposes in the 1860s from England, although with limited success until more intensive captive breeding and more massive imports from North America occurred in the 1930s (Frith, 1967; Falla *et al.*, 1979; Blakers *et al.*, 1984; see Dyer and Williams, 2010, and Guay *et al.*, 2015, for detailed accounts of Mallard introductions to New Zealand). The number and distribution of Mallards in the wild in Australia remain limited to this date, with birds being present more frequently in or close to larger cities. Conversely, the Mallard has become very abundant in New Zealand, and has long been the most numerous of ducks in the hunting bag in some districts (Frith, 1967).

Mallards were first introduced in South Africa during the 1940s where they flourished (Stafford, 2010). On some islands, Mallard introduction is a more recent event, such as in 1960 in Bermuda (Bermuda Audubon Society, 2018) and in 1983 in the Cayman Islands (Bradley and Rey-Millet, 1985). In some countries, the species was released even more recently: in 1995 in Saudi Arabia and in 2004 in Lebanon, probably for ornamental purposes (Banks *et al.*, 2008). In most cases, the size and trend in numbers of these introduced Mallard populations are unknown, but they can reach huge numbers, such as in New Zealand where the current population size is unknown but the estimated annual bag of Mallard and Mallard-like hybrids is 500,000 individuals (Williams, 2013).

Released Mallards often do not show migratory capacity similar to their wild conspecifics, with shorter migration distances in released individuals (Söderquist *et al.*, 2013), or even sedentary behaviour at lower latitudes (Champagnon, 2011; Guillemain *et al.*, 2015). Released Mallards are, however, capable of moving hundreds of kilometres when pushed by adverse weather conditions in northern areas, but such migration is usually later and over shorter distances than in wild conspecifics (Söderquist *et al.*, 2013). It is unknown whether these birds orient independently or merely follow the behaviour of wild conspecifics in the same flock.

As far as post-release range expansion outside the natural range is concerned, in temperate as well as in tropical regions where the species has been introduced, it is likely that it would mainly spread by gradual geographical expansion induced by local dispersal of individuals escaping high-bird-density areas. Some ringed feral individuals have, however, been shown to disperse over long distances (e.g. 2400 km between New Zealand and Australia; Tracey *et al.*, 2008).

26.7 Breeding Behaviour

One of the reasons for the great popularity of the Mallard with humans is its great fecundity. This trait was a main factor behind the domestication of the species by the Romans and Malays as early as 2500 years ago (Kear, 1990). It also explains why releases of hand-reared ducks for hunting purposes only concern Mallards and no other hunted species of ducks, which may be more difficult to keep and breed in captivity. Mallards are socially monogamous, i.e. they form pairs long before nesting and remain paired until incubation (Oring and Sayler, 1992). Extra-pair copulations forced by males are, however, frequent, as in most duck species (McKinney *et al.*, 1983; Baldassarre, 2014). Mallards generally have an extended breeding season, laying eggs from as early as February to as late as mid-June under natural conditions in the northern hemisphere (Cramp and Simmons, 1977; Young, 2005). Feral birds in urban areas can produce ducklings virtually throughout the year (Reeber, 2015). The nest is built on the ground by the female using on-site vegetation and its own feathers and down. Clutch size averages nine eggs (regular range seven to 13), with one egg laid per day and incubation lasting 27–28 days (Cramp and Simmons, 1977; Young, 2005). Only the female incubates the eggs and takes care of the ducklings, which soon leave the nest. During their first breeding season, females have lower nesting propensity than older females, probably related to inferior body condition (Devries *et al.*, 2008). The mean hatching rate in Mallards is highly variable, from around 35% to 70% under normal conditions (Cramp and Simmons, 1977; Young, 2005), but as low as 10–15% in some North American regions where surrounding habitats cause elevated predation rates (e.g. Prairie Pothole Region; Baldassarre, 2014). The number of fledged young per female (usually taking around 50–60 days) ranges from 3.5 to seven (Cramp and Simmons, 1977; Young, 2005). Heavy rainfall, cold temperatures, food limitation and natural predators are the main causes of breeding failure in this species (reviewed by Stafford and Pearse, 2007). Consequently, the release of such limiting factors in urban areas generally leads to introduced Mallards having a relatively high reproductive output.

The limited genetic distance among dabbling ducks (Sun *et al.*, 2017) and the aggressive mating system of these species both contribute to a high frequency of hybridization: hybrids between Mallards and no less than 62 other species have been recorded (mostly with dabbling ducks but also diving ducks, seaducks and even some goose species; Reeber, 2015).

26.8 Habitat

The Mallard is a ubiquitous species, being capable of using virtually all types of wetland both within its native range (e.g. in arctic tundra) and in the regions where it has been introduced (Cramp and Simmons, 1977; Baldassarre, 2014).

26.9 Impacts

The Mallard can provide valuable benefits to humans, through ecosystem services also provided by other duck species, such as seed and invertebrate dispersal, reduction of weed seeds in croplands and contribution to plant decomposition (including straw disposal) (reviewed by Green and Elmberg, 2014).

However, the Mallard has been listed among the most problematic alien species worldwide owing to its negative impacts on local wild duck populations (Global Invasive Species Database, 2015). Because of their large size and the fact that they are sometimes released in large numbers, Mallards may outcompete local birds through competition for food or space, especially when such populations are already in a poor conservation status due to other threats such as habitat change or introduced alien predators (e.g. Meller's Duck in Madagascar; Young and Rhymer, 1998; see also Owen *et al.*, 2006). Although the Mallard is a valuable ecosystem service provider when it comes to dispersal of plants, seeds and animal propagules in natural environments, it may also enhance the spread of unwanted alien plants and animals by the same means (Brochet *et al.*, 2009).

The main threat imposed by Mallard introductions, however, is through hybridization with local populations of other ducks. The very aggressive mating behaviour of the Mallard, combined with its great fecundity and genetic proximity to many other dabbling ducks, frequently leads to cross-breeds, with the offspring of such mixed-species pairs being fertile. Nine species of duck worldwide are threatened by hybridization with the Mallard: the American Black Duck, Mottled Duck and Mexican Duck in North America, the Hawaiian Duck (or Koloa) in Hawaii, the Grey Duck or Pacific Black Duck (*Anas superciliosa*) in New Zealand and Australia, Meller's Duck in Madagascar, and the Yellow-billed Duck and African Black Duck (*Anas sparsa*) in South Africa (Rhymer, 2006; Fox, 2009; Global Invasive Species Database, 2015).

It is worth noting the significant range in spatial effects of Mallards when it comes to competition and hybridization with indigenous duck species. Mallards are obviously a threat to small endemic duck populations on islands (e.g. the Hawaiian Duck; Pyle and Pyle, 2017), but are also a cause of major concern for more abundant and widespread continental species, such as the American Black Duck (Jarrett, 2005) and Mottled Duck (Callaghan, 2005) in North America, and the Yellow-billed Duck in South Africa (Young, 2005). In New Zealand, less than 5% of Grey Ducks are of pure genetic strain, with the remaining 95% being hybridized to some degree with Mallards, while the Mexican Duck is no longer considered a distinct species due to genetic mixing with Mallards (Global Invasive Species Database, 2015).

Within the native range of the species, the massive releases of hand-reared Mallards for hunting purposes also threaten the genetic integrity of local wild birds through genetic pollution from introgression of farmed Mallard genotypes (Söderquist *et al.*, 2017). This may lead to loss of local adaptations in wild populations, but also to introgression leading to the affected wild population becoming morphologically more like tame ducks in the long term (Söderquist *et al.*, 2017). The very large numbers of birds in some such farms, and the massive simultaneous releases of individuals into the wild, also cause epidemiological concern, as some diseases may spread within and between the dense duck stocks in farms before they are released and come into contact with wild conspecifics (see Vittecoq *et al.*, 2012 for the risk of avian influenza virus spread; and Fox, 2009, for duck virus enteritis). A high density of Mallards may also increase algal blooms, deoxygenation and loss of aquatic plants in wetlands, which may all be associated with increased risk of botulism outbreaks (Global Invasive Species Database, 2015).

As most released hand-reared Mallards are phenotypically similar to wild Mallards, another problem in native countries where large-scale releases occur is that this activity generally reduces the possibility of gauging the wild populations and the hunting bag. If released birds are not marked and identified from wild Mallards, it is difficult or impossible to reliably assess the abundance of the truly wild population and the proportion of wild birds in the hunting bag. This is obviously a problem for conservation and long-term sustainable harvest programmes, especially under the paradigm of adaptive harvest management.

26.10 Control

In Europe, the main means of control of released Mallards is hunting, leading to very large bags mainly comprising hand-reared individuals (e.g. 1.2 million shot in France during the 2013/2014 hunting season, the majority of which were considered released individuals; Guillemain *et al.*, 2016). In addition, in the absence of hunting, Mallards produced on duck farms seem to be poorly adapted to the wild environment, particularly owing to deficient foraging and digestion abilities, so that their survival rate is generally very low (4% annual survival rate in a hunting-free area in Camargue, southern France; Champagnon *et al.*, 2012).

A control experiment was carried out on Lord Howe Island in New South Wales, Australia, where hybridization with Mallard has led to the almost complete disappearance of native Pacific Black Ducks. Mallards and Mallard-looking hybrids were shot, trapped and hand captured, leading to a significant reduction in their numbers and concluding that complete eradication of the hybrids on the island was feasible (Tracey *et al.*, 2008). The situation may, of course, be much different in larger regions where more numerous Mallard populations and hybrids are present.

Control of Mallard populations is also undertaken in Hawaii, Florida and South Africa (in South Africa, it is classified as a 'category 2 invader' by the National Environmental Management: Biodiversity Act of 2004). The control plan in place in several provinces of the Republic of South Africa makes it illegal to import, trade or breed Mallards (Banks *et al.*, 2008). Control of Mallards in South Africa (Stafford, 2010) is mostly by trapping followed by euthanasia (Biosecurity Department of Environmental Affairs, 2015), although poisoning has also been described (Hart and Downs, 2015), for example by α-chloralose in Tasmania (after Tracey *et al.*, 2008).

26.11 Uses

The Mallard has been introduced outside its native range, or native Mallard populations have been supplemented with farmed individuals, mostly as a source of food ('provisioning ecosystem services') or game or for aesthetic pleasure ('cultural ecosystem services'; Green and Elmberg, 2014). These last two objectives have been fulfilled by the repeated introduction and release programmes worldwide: most people can enjoy viewing Mallards in parks and gardens, and released Mallards contribute to sustaining hunters' bags. It is not easy to assess how many of the 1.2 million Mallards shot annually in France are of captive origin (Guillemain *et al.*, 2016), but local genetic analyses in Camargue suggest that as much as 76% were released birds, 15% hybrids between wild and farmed birds and only 9% pure wild individuals (Champagnon *et al.*, 2013).

26.12 Notes

The very high adaptability of the Mallard to a wide range of habitats, for both foraging and breeding purposes, has allowed the species to settle easily and establish flourishing populations in many areas. Its aggressive breeding strategies, leading to high fertility rates, will continue to cause a threat to many other native duck populations through hybridization.

26.13 References

Baldassarre, G. (2014) *Ducks, geese and swans of North America*, Vol. 1. John Hopkins University Press, Baltimore, Maryland.

Banks, A.N., Wright, L.J., Maclean, I.M.D., Hann, C. and Rehfisch, M.M. (2008) Review of the status of introduced non-native waterbird species in the area of the African-Eurasian Waterbird Agreement: 2007 Update. BTO Research Report No. 489, Thetford, UK.

Bermuda Audubon Society (2018) Mallard. Available at: www.audubon.bm/birding/bermuda-birds/123-mallard (accessed 6 August 2018).

Biodiversity Department of Environmental Affairs (2015) Monitoring, control and eradication plans. Guidelines for species listed as invasive in terms of section 70 of National Environmental Management: Biodiversity Act, 2004 (Act No. 10 of 2004) (NEMBA) and as required by section 76 of this act. Biosecurity Department of Environmental Affairs, Cape Town.

Blakers, M., Davies, S.J.J.F. and Reilly, P.N. (1984) *The Atlas of Australian Birds*. Melbourne University Press, Melbourne, Australia.

Bradley, P.E. and Rey-Millet, Y.J. (1985) *Birds of the Cayman Islands*. P.E. Bradley, Cayman Islands.

Brochet, A.L., Guillemain, M., Fritz, H., Gauthier-Clerc, M. and Green, A.J. (2009) The role of migratory ducks in the long-distance dispersal of native plants and the spread of exotic plants in Europe. *Ecography* 32, 919–928.

Callaghan, D. (2005) Mottled Duck *Anas fulvigula*. In: Kear, J. (ed.) *Ducks, Geese and Swans*, Vol. 2. Oxford University Press, Oxford, pp. 517–520.

Champagnon, J. (2011) Conséquence des introductions d'individus dans les populations exploitées: l'exemple du Canard colvert *Anas platyrhynchos*. PhD thesis, University of Montpellier, Montpellier, France.

Champagnon, J., Guillemain, M., Elmberg, J., Folkesson, K. and Gauthier-Clerc, M. (2010) Changes in Mallard *Anas platyrhynchos* bill morphology after 30 years of supplemental stocking. *Bird Study* 57, 344–351.

Champagnon, J., Guillemain, M., Elmberg, J., Massez, G., Cavallo, F. and Gauthier-Clerc, M. (2012) Low survival after release into the wild: assessing "the burden of captivity" on Mallard physiology and behaviour. *European Journal of Wildlife Research* 58, 255–267.

Champagnon, J., Crochet, P.A., Kreisinger, J., Cížková, D., Gauthier-Clerc, M., *et al.* (2013) Assessing the genetic impact of massive restocking on wild mallard. *Animal Conservation* 16, 295–305.

Cramp, S. and Simmons. K.E.L. (1977) *The Birds of the Western Palearctic*, Vol. 1; *Ostrich to Ducks*. Oxford University Press, Oxford.

Del Hoyo, J., Elliott, A. and Sargatal, J. (1992) *Handbook of the Birds of the World*, Vol. 1. Lynx Edicions, Barcelona, Spain.

Dessborn, L., Brochet, A.L., Elmberg, J., Legagneux, P., Gauthier-Clerc, M. and Guillemain, M. (2011) Geographical and temporal patterns in the diet of pintail *Anas acuta*, wigeon *Anas penelope*, mallard *Anas platyrhynchos* and teal *Anas crecca* in the Western Palearctic. *European Journal of Wildlife Research* 57, 1119–1129.

Devries, J. H., Brookes, R.W.., Howerter, D.W. and Anderson, M.G. (2008) Effects of spring body condition and age on reproduction in mallards (*Anas platyrhynchos*). *Auk* 125, 618–628.

Dyer, J. and Williams, M. (2010) An introduction most determined: Mallard (*Anas platyrhynchos*) to New Zealand. *Notornis* 57, 178–195.

Falla, R.A., Sibson, R.B. and Turbott, E.G. (1979) *The New Guide to the Birds of New Zealand and Outlying Islands*. Collins, Auckland, New Zealand, and London.

Fox, A.D. (2009) What makes a good alien? Dealing with the problems of non-native wildfowl. *British Birds* 102, 660–679.

Frith, H.J. (1967) *Waterfowl in Australia*. Angus and Robertson, Sydney, Australia.

Global Invasive Species Database (GISD) (2015) *Anas platyrhynchos*. Available at: www.iucngisd.org/gisd/species.php?sc=1241 (accessed 6 August 2018).

Green, A.J. and Elmberg, J. (2014) Ecosystem services provided by waterbirds. *Biological Reviews* 89, 105–122.

Guay, P.J., Williams, M. and Robinson, R.W. (2015) Lingering genetic evidence of North American mallards (*Anas platyrhynchos*) introduced to New Zealand. *New Zealand Journal of Ecology* 39, 103–109.

Guillemain, M., Champagnon, J., Massez, G., Pernollet, C.A., George, T., *et al.* (2015) Becoming more sedentary? Changes in recovery positions of Mallard *Anas platyrhynchos* ringed in the Camargue, France, over the last 50 years. *Wildfowl* 65, 51–63.

Guillemain, M., Aubry, P., Folliot, B. and Caizergues, A. (2016) Duck hunting bag estimates for the 2013/14 season in France. *Wildfowl* 66, 126–141.

Hart, L. and Downs, C. (2015) Winged invaders: bird introductions. *Quest* 11, 38–41.

Jarrett, N.G. (2005) American Black Duck *Anas rubripes*. In: Kear, J. (ed.) *Ducks, Geese and Swans*, Vol. 2. Oxford University Press, Oxford, pp. 509–513.

Kear, J. (1990) *Man and Wildfowl*. T. and A.D. Poyser, London.

McKinney, F., Derrickson, S.R. and Mineau, P. (1983) Forced copulation in waterfowl. *Behaviour* 86, 250–294.

Mouronval, J.B. (2016) *Guide to the Sex and Age of European Ducks*. Office National de la Chasse et de la Faune Sauvage, Paris, France.

Mouronval, J.B., Guillemain, M., Canny, A. and Poirier, F. (2007) Diet of non-breeding waterfowl *Anatidae* and Coot *Fulica atra* on the Perthois gravel pits, northeast France. *Wildfowl* 57, 68–97.

Oring, L.W. and Sayler, R.D. (1992) The mating system of waterfowl. In: Batt, B.D.J., Afton, A.D., Anderson, M.G., Ankney, C.D., Johnson, D.H., *et al.* (eds) *Ecology and Management of Breeding Waterfowl*. University of Minnesota Press, Minneapolis, pp. 128–189.

Owen, M., Callaghan, D. and Kirby, J. (2006) Guidelines on avoidance of introductions of non-native waterbird species. AEWA Technical Series No. 10, Bonn, Germany. Available at: www.unep-aewa.org/sites/default/files/publication/ts12_guidelines_non-native-species_complete_0.pdf (accessed 30 October 2019).

Palmer, R.S. (ed.) (1976) *Handbook of North American Birds*, Vol. 2: *Waterfowl*. Yale University Press, New Haven, Connecticut, and London.

Pyle, R.L. and Pyle, P. (2017) *The Birds of the Hawaiian Islands: Occurrence, History, Distribution, and Status*. Version 2. B.P. Bishop Museum, Honolulu, Hawaï. Available at: http://hbs.bishopmuseum.org/birds/rlp-monograph/ (accessed 6 August 2018)

Reeber, S. (2015) *Wildfowl of Europe, Asia and North America*. Christopher Helm, London.

Rhymer, J.M. (2006) Extinction by hybridization and introgression in anatine ducks. *Acta Zoologica Sinica* 52, 83–109.

Söderquist, P. (2015) Large-scale releases of native species: the Mallard as a predictive model system. PhD thesis, Swedish University of Agricultural Sciences, Umea, Sweden.

Söderquist, P., Gunnarsson, G. and Elmberg, J. (2013) Longevity and migration distance differ between wild and hand-reared mallards *Anas platyrhynchos* in Northern Europe. *European Journal of Wildlife Research* 59, 159–166.

Söderquist, P., Norrström, J., Elmberg, J., Guillemain, M. and Gunnarsson, G. (2014) Wild Mallards have more "Goose-like" bills than their ancestors: a case of anthropogenic influence? *PLoS One* 9, e115143.

Söderquist, P., Elmberg, J., Gunnarsson, G., Thulin, C.G., Champagnon, J., *et al.* (2017) Admixture between released and wild game birds: a changing genetic landscape in European mallards (*Anas platyrhynchos*). *European Journal of Wildlife Research* 63, 98.

Stafford, J.D. and Pearse, A.T. (2007) Survival of radio-marked mallard ducklings in South Dakota. *Wilson Journal of Ornithology* 119, 585–591.

Stafford, L. (2010) Mallard strategy for South Africa. National Problem Bird Forum/CAPE Invasive Animal Working Group Report, Cape Town.

Sun, Z., Pan, T., Hu, C., Sun, L., Ding, H., *et al.* (2017) Rapid and recent diversification patterns in Anseriformes birds inferred from molecular phylogeny and diversification analyses. *PLoS One* 12, e0184529.

Tanabe, Y. (1995) History and phylogeny of Japanese native animals and strategies for their effective use. In: *The Third Ministry of Agriculture, Forestry and Fisheries, Japan (MAFF) International Workshop on Genetic Resources. Animal Genetic Resources: Efficient Conservation and Effective Use*. Agriculture, Forestry and Fisheries Research Council, Japan, pp. 17–36.

Tracey, J., Lukins, B. and Haselden, C. (2008) *Lord Howe Island Ducks: Abundance, Impacts and Management Options*. Invasive Animals Cooperative Research Centre, University of Canberra, Canberra.

Vittecoq, M., Grandhomme, V., Champagnon, J., Guillemain, M., Crescenzo-Chaigne, B., *et al.* (2012) High influenza A virus infection rates in mallards bred for hunting in the Camargue, South of France. *PLoS One* 7, e43974.

Wetlands International (2018) Waterbird population estimates. Available at: http://wpe.wetlands.org (accessed 18 July 2018).

Williams, M.J. (2013) Mallard. In: Miskelly, C.M. (ed.) *New Zealand Birds Online*. Available at: www.nzbirdsonline.org.nz/species/mallard (accessed 6 August 2018).

Young, G. (2005) Northern Mallard *Anas platyrhynchos*. In: Kear, J. (ed.) *Ducks, Geese and Swans*, Vol. 2. Oxford University Press, Oxford, 513–517.

Young, H.G. (2013) Mallard *Anas platyrhynchos* Canard Colvert. In: Safford, R.J. and Hawkins, A.F.A. (eds) *The birds of Africa*, Vol. VIII: *The Malagasy Region*. Christopher Helm, London, pp. 252–253.

Young, H.G. and Rhymer, J.M. (1998) Meller's duck: a threatened species receives recognition at last. *Biodiversity and Conservation* 7, 1313–1323.

27 Ruddy Duck (*Oxyura jamaicensis* Gmelin, 1789)

Carlos Gutiérrez-Expósito[1]*, Claire A. Pernollet[2], Tim Adriaens[3] and Iain Henderson[4]

[1]Estación Biológica de Doñana (CSIC), Americo Vespucio 26, Isla de la Cartuja, Sevilla 41032, Spain; [2]Conservatoire d'espaces naturels de Provence-Alpes-Côte d'Azur, 4 avenue Marcel Pagnol, 13100 Aix-en-Provence, France (CEN PACA); [3]Research Institute for Nature and Forest (INBO), Havenlaan 88 bus 73, B-1000 Brussels, Belgium; [4]National Wildlife Management Centre, Animal and Plant Health Agency, Sand Hutton, York YO41 1LZ, UK

Citation: Gutiérrez-Expósito, C., Pernollet, C.A., Adriaens, T. and Henderson, I. (2020) Ruddy Duck (*Oxyura jamaicensis* Gmelin, 1789). In: Downs, C.T. and Hart, L.A. (eds) *Invasive Birds: Global Trends and Impacts*. CAB International, Wallingford, UK, pp. 200-205.

27.1 Common Names

Ruddy Duck (English), Érismature Rousse (French), Malvasía Canela (Spanish), Rosse Stekelstaart (Dutch).

27.2 Nomenclature

The Ruddy Duck (*Oxyura jamaicensis* Gmelin, 1789) is a monotypic Nearctic species that until recently was considered a conspecific of the South American Andean Duck (*Oxyura ferruginea*) (del Hoyo and Collar, 2014).

27.3 Distribution

The Ruddy Duck breeds in North America, with migratory populations in western Canada and much of the USA, but it is resident in the south-western USA, Mexico and the West Indies (Fig. 27.1) (Kear, 2005). The distances travelled between the nesting and wintering sites can reach 1500 km; cases of long-distance migration up to 3500 km have also been documented (Johnsgard and Carbonell, 1996). It was introduced to the UK in the 1940s, and the first breeding in the wild was reported in 1960 (Hudson, 1976). Subsequently, it has bred in the UK, Iceland, Ireland, France, Belgium, the Netherlands, Germany and Morocco (Muñoz-Fuentes *et al.*, 2006). Between 1965 and 2009, it was sighted in 23 countries of the Western Palaearctic (Robertson *et al.*, 2015). Thanks to management efforts, the exotic population is extirpated in Spain and almost totally so in the UK and Belgium, but regular breeding and self-sustaining populations still occur in France and the Netherlands. Currently around 400 individuals are found in six to eight Western European countries (Robertson *et al.*, 2015).

27.4 Description

Similar to other stiff-tailed ducks, Ruddy Ducks are relatively small and compact diving ducks with a relatively long tail that is often raised upwards (tail of 6–8 cm in length). There is a strong sexual and phenological dimorphism, with males being slightly larger than females. Breeding males have overall reddish-brown body plumage with a strong black-and-white head pattern and a striking turquoise-blue concave (not swollen) bill, while the winter plumage is greyish-brown with a black bill (Fig. 27.2). In both plumages, the pure white cheeks and undertail coverts stand out. Females are greyish-brown with a dark cap and grey cheeks crossed by a horizontal line starting in the bill gape. The iris and legs are dark in both sexes. Males in their first winters are similar to adult females until the winter moult, when males start to develop the white cheeks. There is also a

*Corresponding author: carlines@ebd.csic.es

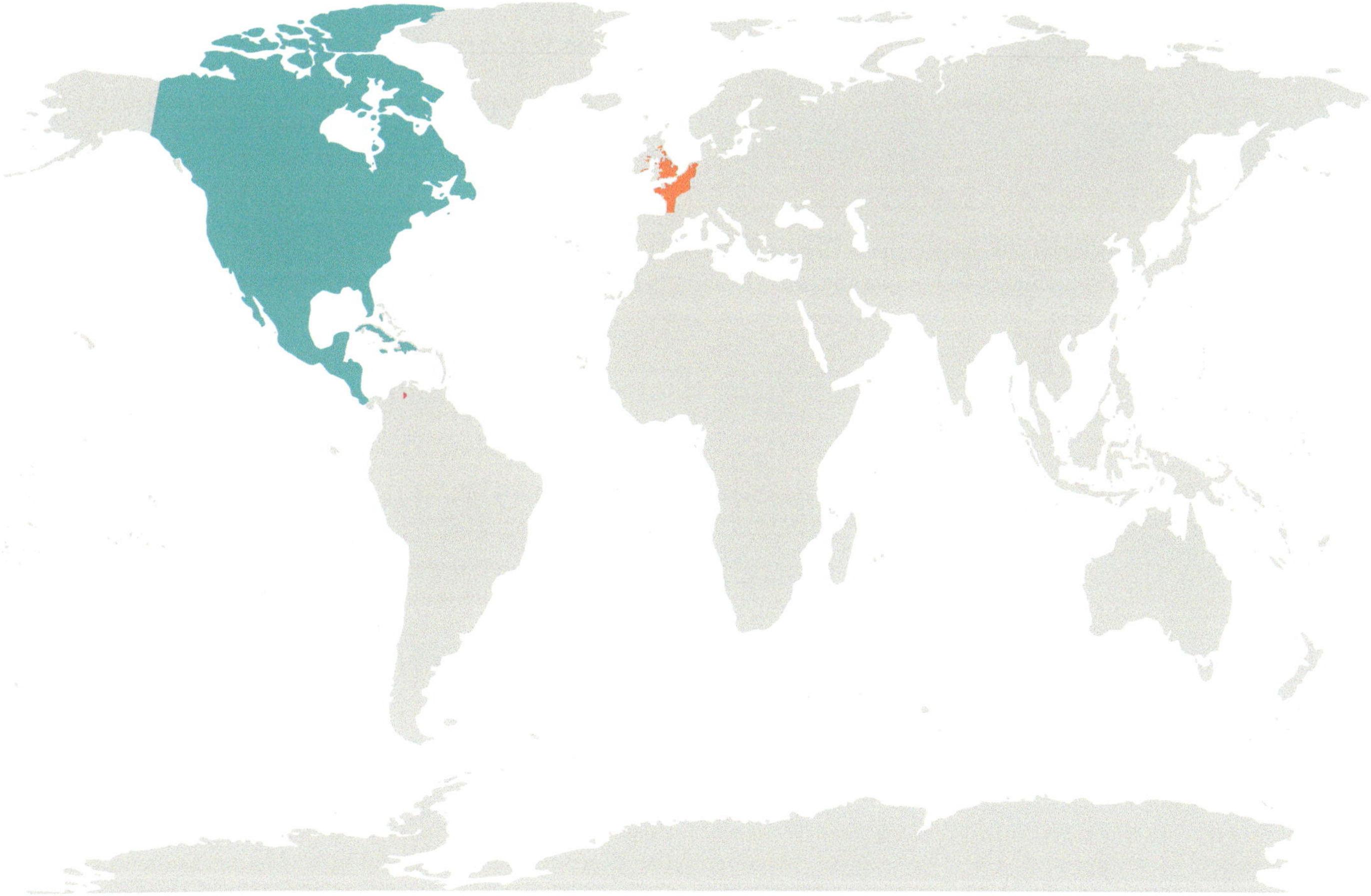

Fig. 27.1. Global distribution of the Ruddy Duck (*Oxyura jamaicensis*) showing its natural (blue) and invaded (red) ranges. The invaded range depicts only areas where the species is established (i.e. regular breeding).

size dimorphism, with males being slightly larger than females (body mass 539–794 g for males and 310–650 g for females). The mean male body length is 397 mm while that of females is 384 mm, and wing length is 154 mm and 148 mm for males and females, respectively (Brua, 2002).

Hybrids with the White-headed Duck (*Oxyura leucocephala*) have a great variation, mostly among males, but in general they are similar to this species but with some Ruddy Duck traits. Body feathers are usually greyer, lacking the brownish tones of White-headed Ducks, and the undertail coverts are almost pure white and not creamy or brownish. Although the bill profile is not as concave as that of the Ruddy Duck, hybrids never have the swollen aspect of the White-headed Duck (Fig. 27.3). The cheek stripe is wide and curved upright in the White-headed Duck and thin and straight in the Ruddy Duck, while hybrids have a mixed profile. The shape of the line between the bill base and head feathering is probably the best clue to correctly identify a hybrid, being S-shaped in the White-headed Duck and C-shaped in the Ruddy Duck, with the profile in hybrids closer to the former. In any case, a detailed observation of several characteristics is necessary for correct identification as it depends on the level of respective introgression of the two species (Urdiales and Pereira, 1993; Sáenz de Buruaga *et al.*, 2003).

Fig. 27.2. An adult male Ruddy Duck. (©Photograph: Dorian Anderson.)

27.5 Diet

The Ruddy Duck's diet is based primarily on small invertebrates captured from the muddy bottom of ponds and lakes during extended dives. Their main prey items are larval and

Fig. 27.3. Comparison of colour, bill shape, cheek stripe and bill-base profile among adult female *Oxyura* spp. ducks: White-headed Duck (left), hybrid (centre) and Ruddy Duck (right). Specimens held at the Doñana Biological Station (CSIC; Seville, Spain) scientific collection. (©Photograph: Carlos Gutiérrez-Expósito.)

pupal midges of the family Chironomidae, mainly of the genus *Chironomus* (Siegfried, 1973). In Spain, no differences in the diet were found among White-headed Ducks, Ruddy Ducks and their hybrids. Although based mainly on benthic chironomid larvae, up to 27 invertebrate families were identified. Green plant material and seeds were also detected (Sánchez *et al.*, 2000).

27.6 Introduction and Invasion Pathways

In 1948, seven Ruddy Ducks (four males and three females) were imported from the USA to the Wildfowl and Wetlands Trust facilities in Slimbridge, UK, for captive breeding (Hughes *et al.*, 1999). Up to 90 descendants were estimated to have escaped between 1953 and 1973 from this collection (Hudson, 1976), these being the founders of the British population whose first breeding event in the wild was confirmed in 1960. By 1980, up to 1500 wild birds were estimated in the UK (Muñoz-Fuentes *et al.*, 2006). A 1437% increase in the breeding range occurred between the two British Trust for Ornithology Bird Atlas surveys in 1968–1972 (Sharrock, 1976) and 1988–1991 (Gibbons *et al.*, 1993). With a yearly estimated growth rate of 15%, by 1998 the estimated population had grown to 4000 birds (Hughes *et al.*, 1999). Peak numbers in the UK probably occurred in the winter of 1999/2000 when the population was estimated to have reached around 6000 birds (Kershaw and Hughes, 2002). Research into control methods between 1999 and 2004 caused a slight decline in numbers, which accelerated when the UK Ruddy Duck Eradication programme (ERDUK) began in 2005.

The first sighting of the Ruddy Duck on the European mainland was reported in 1965 in Sweden, and by 1973, the species was observed for the first time in the Netherlands and Ireland, and then in France in 1974 (Hughes *et al.*, 1999). Between 1979 and 1994, Ruddy Ducks were sighted in 11 more European countries, as well as in Israel, Turkey and Morocco (Hughes *et al.*, 1999). By the end of the winter of 2009/10, when control measures had already started in the UK, France and Spain, Ruddy Ducks had been observed in 23 countries (Cranswick and Hall, 2010), but also in Algeria (Samraoui and Moussa, 2001) and the Balearic Islands (Muntaner, 2001). Significant numbers were found in four of these: 250 in the UK, 15 in Belgium, 220 in France and 60 in the Netherlands, and for the first time, numbers in continental Europe exceeded those in the UK (Cranswick and Hall, 2010). As a result of the ERDUK Life project in the UK and control measures taken in France, Belgium, Spain and the Netherlands, relatively significant numbers of Ruddy Ducks are currently present only in the Netherlands (60 in winter 2015/16) and France (200 in winter 2017/18) (see section 27.10).

Breeding of the Ruddy Duck has occurred in the UK, Ireland, Iceland, France, Belgium, Germany and the Netherlands. While there are records of pure Ruddy Duck breeding pairs in Spain and Morocco, most breeding events in these countries were hybridization cases with the White-headed Duck (see section 27.9).

27.7 Breeding Behaviour

The Ruddy Duck male courtship display consists mainly of 'bubbling'. After holding the tail, head and a couple of feather horns erect, the male beats his breast feathers with his bill, which forces any air out, thus forming bubbles on the water surface. As the sex ratio is highly biased towards males, these displays are evident in a sort of water-based lek until pair formation. Although monogamy appears to be the norm, pair bonds usually only last until the beginning of incubation (Brua, 2002). Females are highly philopatric to their nesting site (Siegfried, 1976; Johnsgard and Carbonell, 1996). Ruddy Ducks are reported to live for up to 13 years in the wild (Clapp *et al.*, 1982). Birds are sexually mature after 2 years (sometimes 1 year) (Brua 2002). Breeding starts in April–May and can last till October. Females usually lay clutches of six to eight (range of three to ten) (Siegfried, 1976; Johnsgard and Carbonell, 1996) relatively large eggs (62.3 × 45.6 mm) (Pelayo, 2001), which, after an incubation period of approximately 24 days, produce four to six very precocial ducklings, which stay with the female for 3–4 weeks. Although use of abandoned platforms of other ducks, coots or grebes has been reported, usually a floating nest is built with or on dead vegetation (e.g. *Typha*, *Scirpus* and *Carex* spp.; Siegfried, 1976) and is often well hidden in the vegetation (Brua, 2002). Parasitism of other waterfowl such as conspecifics (Reichart, 2008), other duck species, grebes, rails and bitterns (Eastman, 1999; Ehrlich *et al.*, 1988) has also been reported. Joyner (1975) found that 7.7% of duck nests in a Utah study were parasitized interspecifically by Ruddy Ducks. Such behaviour may well occur in the European population, but conclusive evidence is lacking. In North America, 34–73% of nests produce at least one duckling, with an average success rate of 55–60%, according to the authors. The egg hatching rate is estimated at 50–70% (Kear, 2005; Johnsgard and Carbonell, 1996).

27.8 Habitat

Ruddy Duck breeding occurs mainly on relatively small, scattered, open-surface freshwater wetlands (e.g. swamps, lakes, pools and marshes) with extensive emergent vegetation, while moulting, migration and wintering areas are usually larger permanent ponds, lakes and reservoirs of fresh waters where birds gather (Smith *et al.*, 2005). The American population exploits mainly brackish environments (e.g. estuarine, lagoons) during wintering (Johnsgard and Carbonell, 1996). In Spain, Ruddy Ducks have also been found at fish farms and sewage ponds at wastewater treatment plants, which are also widely used by White-headed Ducks outside the breeding season (authors' personal observations). Ruddy Ducks in Europe seem to be looking for eutrophic–hypertrophic habitats rich in organic matter and therefore favourable to the development of benthic maroinvertebrates. They use breeding habitats similar to those selected by the Common Pochard (*Aythya ferina*), Tufted Duck (*Aythya fuligula*) and Little Grebe (*Tachybaptus ruficollis*) (Kear, 2005).

27.9 Impacts

The main impact of Ruddy Ducks in Europe is the hybridization with the endangered White-headed Duck (Hall, 2016). Male Ruddy Ducks can join the White-headed Duck breeding males' parties, behaving much more aggressively and hoarding the visiting females. Cross-breeding has been shown to occur in the wild, and second-generation hybrids have been reported in both captivity and the wild (Hughes *et al.*, 1999; Muñoz-Fuentes *et al.*, 2007). In the western Mediterranean, White-headed Ducks are regularly present only in Morocco and Spain. Numbers have increased since 1977 when the entire Spanish population was reduced to 22 birds at Laguna de Zóñar (Córdoba, Spain). Protection of breeding and wintering sites and a ban on the hunting of White-headed Ducks from that date allowed the population to recover, and it is now stable at around 2000 birds. Ruddy Ducks are known to be much more aggressive in the breeding leks than its endangered autochthon counterpart. Together with the fact that hybrids can be fertile, and the strong genetic bottleneck suffered by the Spanish White-headed Duck population in the 1970s, hybridization with Ruddy Duck has become the main threat to the survival of the White-headed Duck in Europe. The first Ruddy Duck sighting in Spain occurred in 1983 and the first hybrids were reported by Spanish ornithologists in 1991 when two female White-headed Ducks produced at least nine hybrid chicks at El Hondo reservoir (Alicante, Spain). During the 1990s, hybridization was reported widely at this site and others in southern Spain (69 on at least 23 sites, with 68 culled), but the last hybridization event took place in 2004 at Puerto de Santa Maria (Cádiz, Spain) and no Spanish hybrids have been reported since 2007 when the last example was shot in Utrera (Seville, Spain) (Gutiérrez-Expósito *et al.*, 2015). Between 1996 and 2016, 50 observations of hybrids were made in Morocco. In France, two hybrid females were recorded in 2003, followed by one male in 2006. In at least five countries (France, Spain Italy, Morocco and Switzerland), the extent of occurrence of the two species overlaps partially or totally during breeding, migration or overwintering periods (Mouronval *et al.*, 2015).

27.10 Control

Once the risk of hybridization was identified as the major threat to the survival of the endangered Spanish population of White-headed Duck, an assessment of the feasibility of the total eradication of Ruddy Ducks in the UK was carried out by the Central Science Laboratory, an Executive Agency of the UK Department for Environment, Food and Rural Affairs (Defra). Regional control trials were conducted in three representative areas (Anglesey in north-west Wales, the West Midlands of England and Fife in eastern Scotland) between 1999 and 2002, by testing shooting, trapping and egg oiling as control methods. After culling 2600 birds, firearms were shown to be very effective, while trapping and egg oiling were relatively expensive and ineffective options (Hughes *et al.*, 2006). Based on the data from these regional control trials, it was estimated that in between 3 and 5 years, it would be possible to reduce the Ruddy Duck population by 97% if 14 or 15 officers could achieve an annual population reduction of between 65% and 70% (Smith *et al.*, 2005). With this background, and partly financed by the European Life-Nature Programme, project LIFE05 NAT/UK/000142 ERDUK (Eradication of Ruddy Ducks in the UK to protect the White-headed Duck) began in 2005, when the Ruddy Duck population was estimated to be around 5000 individuals (winter count). The project was highly successful, and by the winter of 2008/09, up to 6200 birds had been culled and a significant decline in the population had occurred, reducing the winter count to barely 1000 birds (Henderson, 2009). By the end of the project (March 2011), only around 100 Ruddy Ducks were left in the UK. Since then, the Defra has continued to finance an ongoing eradication programme (implemented by the Animal and Plant Health Agency, which had merged with parts of the Central Science Laboratory), which has seen numbers fall to around 20 individuals in 2018. Currently, Ruddy Ducks are believed to be present only in Northern Ireland, around Manchester in north-west England, in West London and in Kent. The effects of the ERDUK project were immediately noted in other countries: Ruddy Ducks disappeared from Ireland and Iceland, and the number of birds sighted in Spain declined dramatically. Unfortunately, by the beginning of this project, Ruddy Ducks were effectively established on the continent, and self-sustained populations were to be found in France, Belgium and the Netherlands. In France, control measures began as early as 1997, and in Belgium in 2009. Although significant efforts have been made in France, with the number of birds culled increasing in recent years (over 2000 eliminated by 2017), the Ruddy Duck population has been stable at around 200 individuals. A new Life *Oxyura* project (LIFE17 NAT/FR/000542) was approved, starting in autumn 2018 and running until 2023, which aims to totally eradicate Ruddy Ducks in France by contracting four persons to work exclusively on bird searching and culling. As a first result, in 2018, 177 Ruddy Duck were culled, of which 60% were adults, at Grand-Lieu Lake, Mayenne, Maine-et-Loire and other sites. At least 40 birds still remain on Grand-Lieu Lake and the search efforts to locate remaining birds will be intensified in coming years (J.-F. Maillard, personal communicaton).

In Belgium, before 2009, a few breeding pairs were locally dealt with by site managers. The species first bred successfully in 2008 in the Antwerp harbour area, with birds presumably originating from spill-over of a Dutch breeding area across the border (Spanoghe *et al.*, 2010). After this, an early-warning and rapid-response system was set up, making use of citizen science data. This resulted in about 80 birds being culled between 2009 and 2018. Some of these birds were wing clipped, indicating that escape is likely to be a contributing pathway for Ruddy Ducks in the wild. Actions are coordinated regionally and supported by a steering committee, which contains regional authorities and interest groups including representatives of bird keeper and bird protection organizations. Currently, only a small number of Ruddy Ducks (fewer than 10) are recorded yearly in wintering waterbird censuses, mostly in Flanders near the Dutch border. Although breeding occurs in some years, breeding pairs and any juveniles are culled, thus preventing any increase in numbers. In the Netherlands, Ruddy Ducks appeared for the first time in 1973 and have increased since the end of the 1990s to a reported maximum of 97 birds in the winter of 2005/06. After this, the number of wintering Ruddy Ducks dropped to an estimated 60–80 birds in 2013–2015 (Slaterus, 2016). The most recent country-wide census confirmed the presence of 81 birds in the winter of 2018, with birds being concentrated in three stronghold areas (Sovon Vogelonderzoek Nederland, 2019). This population appears to have been stable for several years, but due to several legal and practical constraints, limited action has been undertaken to reduce numbers: between 2013 and 2017 only 14 birds were eliminated (J. Cronau, personal communication).

In Spain, where Ruddy Duck hybridization occurred widely in the 1980s and 1990s, eradication attempts began relatively early, as Spanish authorities were aware of the very serious problem of genetic introgression that the arrival of this aggressive species posed for a genetically depleted White-headed Duck population that was slowly recovering after being on the verge of extinction in the late 1970s. Between 1984 and the end of the 20th century, the responsibility for eliminating Ruddy Ducks lay mostly with the Spanish regional governments and the management teams of some protected areas (e.g. El Hondo Natural Park, Doñana National Park). By the end of 1999, no fewer than 52 Ruddy Ducks and 45 hybrids had been culled. From 2000 onwards, a part-time specific team was contracted by the Spanish Government. In addition to White-headed Duck population monitoring by local observers, this team has the task of monitoring the whole White-headed Duck population in search of Ruddy Ducks and hybrids three times each year, and also acts as a rapid-response elimination force. Since then, 98 Ruddy Ducks have been eliminated but only 12 hybrids (Gutiérrez-Expósito *et al.*, 2015). Just a single hybridization event was reported in this period (whose descendants were immediately culled), and the other hybrids were believed to be survivors of the past decade or immigrants from Morocco, where a tiny Ruddy and White-headed Duck population that freely hybridized was reported (Thévenot *et al.*, 2003). Since 2011, no Ruddy Ducks or hybrids have been shot in Spain or Morocco, and sightings are very rare apart from a few short-staying vagrants (Gutiérrez-Expósito *et al.*, 2015).

In Europe, an international action plan to eradicate the Ruddy Duck from Europe has been adopted, which received support from the Council of Europe through the Bern Convention (Hall, 2016), but the approaches and intensity of control have varied widely among countries (Cranswick and Hall, 2010). Nevertheless, the numbers in 2013 were less than 7% of the 2000 population, making the objective of eradicating the Ruddy Duck from Europe achievable (Robertson *et al.*, 2015). The Ruddy Duck is currently listed as a species of European Union concern under Regulation (EU) 1143/2014 on invasive alien species. Therefore, since 3 August 2016, a ban on trade and restrictions on keeping, importing, selling, breeding and growing the Ruddy Duck has been implemented. Member States are also required to take measures for early detection and rapid eradication of this species.

27.11 Uses

No specific uses of the Ruddy Duck have been reported, apart from their use as ornamental species in zoological collections.

27.12 References

Brua, R.B. (2002) Ruddy Duck: *Oxyura jamaicensis*. In: Poole, A.F. and Gill, F.B. (eds) *The Birds of North America*. The Cornell Lab of Ornithology, Ithaca, New York. Available at: https://birdsna.org/Species-Account/bna/species/rudduc/introduction (accessed 30 October 2019).

Clapp, R.B., Morgan-Jacobs, D. and Banks, R.C. (1982) *Marine Birds of the Southeastern United States and Gulf of Mexico*. Part II: *Anseriformes*. US Fish and Wildlife Service, Office of Biological Services, Washington, DC.

Cranswick, P.A. and Hall, C. (2010) Eradication of the Ruddy Duck *Oxyura jamaicensis* in the Western Palaearctic: a review of progress and a revised action plan 2010–2015. WWT report to the Bern Convention, Slimbridge, UK.

Del Hoyo, J. and Collar, N.J. (2014) *International Illustrated Checklist of the Birds of the World*. Vol. 1: *Non-passerines*. Lynx Edicions, Barcelona, Spain.

Eastman, J.A. (1999) *Birds of Lake, Pond, and Marsh: Water and Wetland Birds of Eastern North America*. Stackpole Books, Mechanicsburg, Pennsylvania.

Ehrlich, P.R., Dobkin, D.S. and Wheye, D. (1988) *The Birder's Handbook*. Simon & Schuster, New York.

Gibbons, D.W., Reid, J.B. and Chapman, R.A. (eds) (1993) *The New Atlas of Breeding Birds in Britain and Ireland: 1988–1991*. T. and A.D. Poyser, London.

Gutiérrez-Expósito, C., Mouronval, J.-P., Vandegehuchte, M., Steeghs, M. and Henderson, I. (2015) The stretch towards eradication of Ruddy Ducks in Europe. An update to 2015. In: *10th European Vertebrate Pest Management Congress*, 21–25 September 2015, Seville, Spain.

Hall, C. (2016) A review of the progress against the Action Plan for Eradication of the Ruddy Duck *Oxyura jamaicensis* in the Western Palearctic, 2011–2015. WWT report to the Bern Convention, Slimbridge, UK.

Henderson, I. (2009) Progress of the UK Ruddy Duck eradication program. *British Birds* 102, 680–690.

Hudson, R. (1976) Ruddy Ducks in Britain. *British Birds* 69, 132–143.

Hughes, B., Criado, J., Delany, S., Gallo-Orsi, U., Green, A.J., *et al.* (1999) The status of the North American Ruddy Duck *Oxyura jamaicensis* in the Western Palearctic: towards an action plan for eradication. Report by the Wildfowl and Wetlands Trust to the Council of Europe, Strasbourg, France.

Hughes, B., Henderson, I. and Robertson, P. (2006) Conservation of the globally threatened white-headed duck, *Oxyura leucocephala*, in the face of hybridization with the North American ruddy duck, *Oxyura jamaicensis*: results of a control trial. *Acta Zoologica Sinica* 52, 576–578.

Johnsgard, P.A. and Carbonell, M. (1996) *Ruddy Ducks and Other Stifftails. Their Behaviour and Biology*. Animal Natural History Series. University of Oklahoma Press, Norman, Oklahoma.

Joyner, D.E. (1975) Nest parasitism and brood-related behaviour of the Ruddy Duck (*Oxyura jamaicensis*). PhD thesis, University of Nebraska, Lincoln, Nebraska.

Kear, J. (2005) *Ducks, Geese and Swans*, Vol. 1. Oxford University Press, Oxford.

Kershaw, M. and Hughes, B. (2002) The winter status and distribution of ruddy ducks in the UK 1966/67–1999/2000. WWT Wetlands Advisory Service report to the Central Science Laboratory. Department for Environment, Food and Rural Affairs, London.

Mouronval, J.-B, Maillard, J.-F. and Cugnasse, J.-M. (2015) Plan national de lutte contre l'érismature rousse (*Oxyura jamaicensis*) 2015–2025. Le ministère de l'Ecologie, du Développement durable et de l'Energie, Office National de la Chasse et de la Faune Sauvage, Paris.

Muñoz-Fuentes, V., Green, A.J., Sorenson, M.D., Negro, J.J. and Vilà, C. (2006) The ruddy duck *Oxyura jamaicensis* in Europe: natural colonization or human introduction? *Molecular Ecology* 15, 1441–1453.

Muñoz-Fuentes, V., Green, A.J., Negro, J.J. and Sorenson, M.D. (2007) Hybridization between white-headed ducks and introduced ruddy ducks in Spain. *Molecular Ecology* 16, 629–638.

Muntaner, J. (2001) Primera cita de malvasía canela *Oxyura jamaicensis* en las Baleares y problemas de conservación de la malvasía cabeciblanca *Oxyura leucocephala*. *Anuari Ornitològic de les Balears* 16, 41–46.

Pelayo, J.T. (2001) Correlates and consequences of egg size variation in wild Ruddy Ducks (*Oxyura jamaicensis*). MSc thesis, University of Saskatchewan, Saskatoon, Canada.

Reichart, L.M. (2008) Conspecific brood parasitism in ruddy ducks (*Oxyura jamaicensis*). PhD thesis, Washington State University, Pullman, Washington, DC.

Robertson, P.A., Adriaens, T., Caizergues, A., Cranswick, P.A., Devos, K., *et al.* (2015) Towards the European eradication of the North American ruddy duck. *Biological Invasions* 17, 9–12.

Sáenz de Buruaga M., Andrés, T., Bravo, M.A., Calzada, J., Garrido, H., *et al.* (2003) Reconocimiento de híbridos de *Oxyura leucocephala* × *Oxyura jamaicensis*. In: Capdevilla-Argüelles, L., Zilletti, B. and Pérez-Hidalgo, N. (eds) *Contribuciones al Conocimiento de les Especies Exóticas Invasoras*. Grupo Especies Invasoras. G.E.I. Serie Técnica No. 1, p. 235.

Samraoui, B. and Moussa, H. (2001) Premiére observation de l'Erismature Rousse *Oxyura jamaicensis* en Algérie. *Alauda* 69, 396.

Sánchez, M.I., Green, A.J. and Dolz, J.C. (2000) The diets of the White-headed Duck *Oxyura leucocephala*, Ruddy Duck *O. jamaicensis* and their hybrids from Spain. *Bird Study* 47, 275–284.

Sharrock, J.T.R. (1976) *The Atlas of Breeding Birds in Britain and Ireland*. British Trust for Ornithology, Tring, UK.

Siegfried, W.R. (1973) Summer food and feeding of the ruddy duck in Manitoba. *Canadian Journal of Ornithology* 51, 1293–1297.

Siegfried, W.R. (1976) Breeding biology and parasitism in the Ruddy Duck. *Wilson Bulletin* 88, 566–574.

Slaterus, R. (2016) Actueel voorkomen van Rosse Stekelstaarten in Nederland, situatie februari 2016. Sovon-rapport 2016/23. Sovon Vogelonderzoek Nederland, Nijmegen, The Netherlands.

Smith, G.C., Henderson, I. and Robertson, P.A. (2005) A model of ruddy duck *Oxyura jamaicensis* eradication for the UK. *Journal of Applied Ecology* 42, 546–555.

Sovon Vogelonderzoek Nederland (2019) Hoeveel rosse stekelstaarten overwinteren er momenteel in Nederland? *Kijk op Exoten* 26, 12–13.

Spanoghe, G., Faveyts, W. and Vermeersch, G. (2010) Broedende Rosse Stekelstaarten *Oxyura jamaicensis* in Vlaanderen: een aanwinst? *Natuur.oriolus* 76, 1–7.

Thévenot, M., Vernon, R. and Bergier, P. (2003) *The Birds of Morocco*. British Ornithologists Union, British Ornithologists Club, The Natural History Museum, Tring, UK.

Urdiales, C. and Pereira, P. (1993) Claves de identificación de *O. jamaicensis*, *O. leucocephala* y sus híbridos. Instituto para la Conservación de la Naturaleza, Madrid.

28 Egyptian Goose (*Alopochen aegyptiaca* Linnaeus, 1766)

Frank Huysentruyt[1]*, Corey T. Callaghan[2], Diederik Strubbe[3], Katherine Winston[4], Tim Adriaens[1] and Daniel M. Brooks[4]

[1]Research Institute for Nature and Forest (INBO), Havenlaan 88 bus 73, B-1000 Brussels, Belgium; [2]School of Biological, Earth and Environmental Sciences, UNSW Sydney, Sydney, NSW 2052, Australia; [3]Terrestrial Ecology Unit (TEREC), Ghent University, K.L. Ledeganckstraat 35, 9000 Gent, Belgium; [4]Houston Museum of Natural Science, Department of Vertebrate Biology, Houston, TX 77019, USA

Citation: Huysentruyt, F., Callaghan, C.T., Strubbe, D., Winston, K., Adriaens, T. and Brooks, D.M. (2020) Egyptian Goose (*Alopochen aegyptiaca* Linnaeus, 1766). In: Downs, C.T. and Hart, L.A. (eds) *Invasive Birds: Global Trends and Impacts*. CAB International, Wallingford, UK, pp. 206–212.

28.1 Common Names

Egyptian Goose, Nile Goose, Phoenician Goose, African Goose, Egyptian Shelgoose (Avibase, 2018).

28.2 Distribution

The Egyptian Goose (*Alopochen aegyptiaca* Linnaeus, 1766) has a widespread natural distribution across sub-Saharan Africa and the upper Nile Valley, where population levels are estimated to be stable at around 500,000 individuals (Fig. 28.1) (Brown *et al.*, 1982; Madge and Burn, 1988; Maclean, 1997; Banks *et al.*, 2008). The species has established non-native breeding populations in Belgium, the Czech Republic, Denmark, France, Germany, Israel, Luxembourg, Mauritius, the Netherlands, Sweden, Switzerland, United Arab Emirates, the UK and the USA (Banks *et al.*, 2008; Kampe-Persson 2010; Gyimesi and Lensink, 2012; van Daele *et al.*, 2012; Ries *et al.*, 2014; Callaghan and Brooks, 2017; Jaška and Řepa, 2017; CAB International, 2018; DAISIE, 2009).

28.3 Description

The Egyptian Goose is one of two species of African shelgeese, with an overall brownish appearance, including a yellowish-brown head, neck and breast, a whitish belly and dark brown upper parts (Fig. 28.2). A pinkish bill with dark edges and long pink legs give the species, in combination with a chestnut neck ring and chocolate brown eye and breast patches, a distinctive appearance. Egyptian Geese have clear white wing coverts, forming white flank streaks when on the ground. In flight, these coverts are visible as large white patches on broad wings, highly contrasting with the surrounding chestnut tertials, metallic green and purple secondaries, and black primaries. Young birds are dull and darker overall, lacking the distinctive head, neck and breast markings. Greater coverts are grey-brown instead of clear white, and secondaries lack metallic shine. Both sexes are similar in plumage, but male birds on average are larger than females. Wing length, tarsus length and body weight on average are, respectively, 39 cm, 85 mm and 2.45 kg in males, and 37 cm, 80 mm and 1.94 kg in females. Within a pair, males are almost always larger than females (Madge and Burn 1988; Baker 1993).

28.4 Diet

The Egyptian Goose is a predominantly herbivorous species, with grass, seeds, leaves, grain, crop seedlings, aquatic rhizomes, tubers and plant detritus making up the main part of the

*Corresponding author: frank.huysentruyt@inbo.be

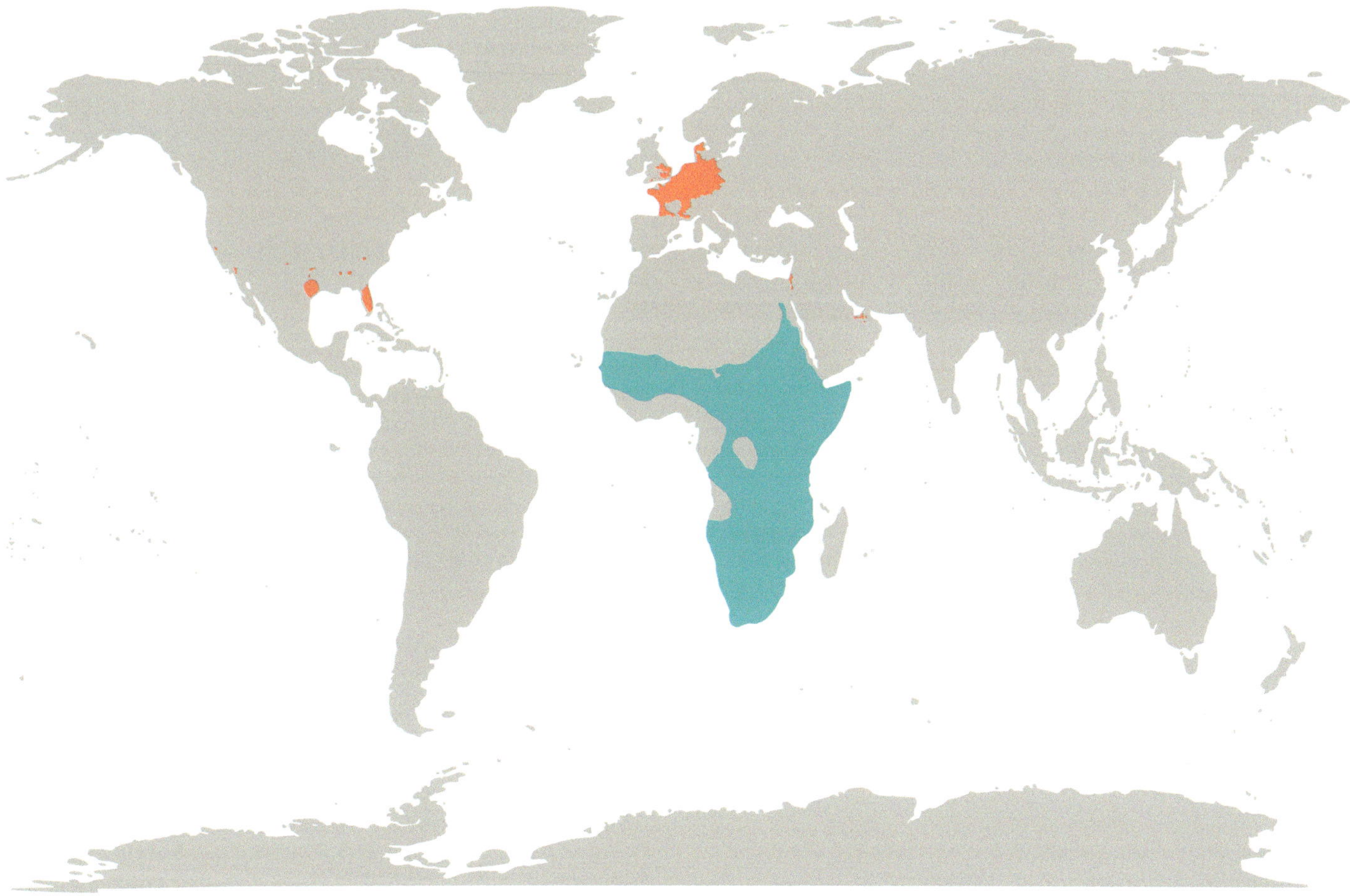

Fig. 28.1. Global distribution of the Egyptian Goose (*Alopochen aegyptiaca*) showing the native (blue) and invasive (red) ranges.

Fig. 28.2. Egyptian Goose adults. (© Photograph: Yves Adams, Vildaphoto.)

diet. Occasionally, they also feed on small insects, terrestrial worms and frogs (Halse, 1984; Maclean, 1997; Mangnall and Crowe, 2002). In Texas, 12% of reports documented some type of anthropogenic feeding, including bread, maize, bird seed and cat food (Callaghan and Brooks, 2016). The comprehensive diet of Egyptian Geese allows seasonal variation in the actual diet composition (Halse, 1984). For example, during the flightless moulting period and brood rearing, when geese seek protection on and alongside various types of water bodies, the diet largely shifts to aquatic and waterside vegetation. Egyptian Geese also rely heavily on body reserves during this period, and potentially lose up to 25% of their body mass (Halse, 1984; Gyimesi and Lensink, 2010).

28.5 Introduction and Invasion Pathways

The establishment of local populations of Egyptian Geese outside the native range generally occurs through escaped individuals from ornamental birds kept in captivity. The species' presence in Western Europe dates to the 17th century, when it was introduced into parks in the UK (Sutherland and Allport, 1991; Rehfisch *et al.*, 2010). Similar introductions in the second half of the 20th century made the species a widespread ornamental bird introduced at various locations across Europe (Banks *et al.*, 2008). Documented escape events in The Hague and Brussels led to birds breeding in the wild from 1967 in the Netherlands and from 1982 in Belgium (Segers, 1989; Vangeluwe and Roggeman, 2002; Anselin, 2004; Anselin and Vermeersch, 2005; Gyimesi and Lensink, 2012). In contrast to the UK population, populations on mainland Europe showed rapid growth immediately after first breeding in the wild (Gyimesi and Lensink, 2012). The initial spread of these populations was slow, however, leading to additional breeding populations in France and Germany from the mid-1980s and in Luxembourg, Denmark, Switzerland and the Czech Republic in the first decade of the 21st century (Kampe-Persson, 2010; Gyimesi and Lensink, 2012, Ries *et al.*, 2014; Jaška and Řepa, 2017). More recent occasional observations of individual birds, small flocks and isolated breeding events in countries such as Sweden, Poland, Austria and Spain indicate a dispersive potential for further future establishment of breeding populations (Kampe-Persson, 2010; CAB International, 2018). Thus, whereas the first Egyptian Geese populations in Europe originated from introductions for ornamental reasons, current range expansion and new populations most likely stem from natural dispersal from previously established populations.

In the USA, the earliest known records were from the late 1800s, probably from captive birds (Akhurst, 1877; Kirkwood, 1900). Egyptian Geese remained common in avicultural collections throughout the 20th century (Wilbur and Yocom, 1971). Breeding in nature was first documented in 1967 in California, and in the mid-1980s in Florida (Callaghan *et al.*, 2017). There are now currently significant populations in the states of Florida, Texas and California, among other regions. Similar to European birds, some Egyptian Geese appear to disperse either long or short distances, accounting for records in south-eastern states (Callaghan and Brooks, 2016, 2017).

Egyptian Goose dispersal in non-native ranges, once established, corroborates the knowledge on the dispersive abilities of this species in their native range. Within the native range, the species is thought to be a partial migrant, able to disperse over large distances of up to 1000 km (Toms, 2002). In Europe, the population in the UK shows little movement, and remains confined to eastern England, without reported exchanges with the continent (van Dijk and Majoor, 2011). In mainland Europe, more movement is observed, with individuals regularly dispersing more than 100 km from the ringing site, with occasional maxima of over 350 km (van Dijk and Majoor, 2011; BeBirds: http://odnature.naturalsciences.be/bebirds, accessed 1 November 2019). Higher dispersion is suggested to occur during severe winters (Gyimesi and Lensink, 2012).

28.6 Breeding Behaviour

In their native range, Egyptian Geese usually rear a single brood within a breeding season in which most eggs are laid in March and April although laying can occur throughout the year (Cramp *et al.*, 1984; Maclean, 1997; Callaghan and Brooks, 2016). In Florida, broods have been observed in every month (Pranty and Ponzo, 2014). In the European invasive range, year-round breeding events have also been observed, but the main part of the breeding season in Western Europe, comprising nest-building, egg-laying and breeding, is from March to June (Lensink, 1996, 1999; van Daele *et al.*, 2012). This seems to differ somewhat with the breeding season on the eastern front of the invasive range, where, for example, in the Czech Republic, breeding only starts in mid-April and lasts until August, which probably relates to differences in local climate (Jaška and Řepa, 2017). Nest site selection is highly variable, and nests can be located on the ground (Fig. 28.3), in holes or on cliff edges but often in trees (Cramp *et al.*, 1984; Callaghan and Brooks, 2016). Ground nests are usually confined to islands, whereas tree nests can be found in old trees with holes or epicormic shoots, such as various *Salix* spp. (Sutherland and Allport, 1991). Occasionally, nests are constructed on buildings, and various cases of Egyptian Goose nests in highly elevated nest boxes for Peregrine Falcons (*Falco peregrinus*) and Kestrels (*Falco tinnunculus*) have been reported (Beck *et al.*, 2002). Although the rearing habitat for the young consists of waterside grazing pastures, nest sites can be located at larger distances from water, as birds have been shown to walk their broods over distances of more than 1 km from a nest site to suitable rearing locations (Sutherland and Allport, 1991). Sexual maturity is reached after 1 year, at which point sustained, monogamous pair-bonds are formed, which will actively defend a relatively large territory of variable size throughout the breeding and rearing season (Cramp *et al.*, 1984; Lensink, 1999; Beck *et al.*, 2002).

The Egyptian Goose clutch size averages eight to nine eggs but can often be as high as 12 (Cramp *et al.*, 1984). Brooding is done exclusively by the female and incubation takes 28–30 days (Cramp *et al.*, 1984). After the fledging stage, which lasts 70–75 days, the young stay with their parents for several weeks to months (Cramp *et al.*, 1984; Callaghan and Brooks, 2016). In the invasive as well as parts of the native range, overall breeding success can be low (1.1 fledglings per pair in England) (Sutherland and Allport, 1991; Gyimesi and Lensink 2012). In other parts of the invasive and natural range, reproductive success can be much higher but variable, with success rates ranging from 1.5 to 5.7 fledglings per pair (Lensink, 1996, 1999; Vangeluwe and Roggeman, 2002; Jaška and Řepa, 2017). The high success in the Netherlands can probably be attributed to low predation pressure and the abundance of outstanding habitats for herbivorous waterfowl with plenty of highly nutritious

Fig. 28.3. An Egyptian Goose nest located on the ground in between brushes on an island in a small freshwater lake in Flanders, Belgium (eggs were marked within the context of a nest success study). (© Photograph: F. Huysentruyt, INBO.)

grasslands situated adjacent to fresh water (Gyimesi and Lensink, 2012).

28.7 Habitat

Egyptian Geese can be found within a wide range of habitats but all within the vicinity of various freshwater systems such as rivers, lakes, reservoirs, pools, sewage ponds and marshes (Cramp *et al.*, 1984; Carboneras, 1992; Callaghan and Brooks, 2016). The species generally avoids coastal regions and densely forested areas, but in mountainous regions within the species' native range, it can occur up to relatively high altitudes (Cramp *et al.*, 1984). The preferred habitat offers sufficient open grassland with a few trees in close proximity to open freshwater bodies (Cramp *et al.*, 1984; Carboneras, 1992; Gyimesi and Lensink, 2012). In the invasive and native range, this prime habitat is often available in urbanized areas including city parks and various green landscape structures that have water bodies such as lakes, golf courses and swimming pools (Gyimesi and Lensink, 2010; Callaghan and Brooks, 2016; C.T. Downs, unpublished data).

28.8 Impacts

The impact of Egyptian Geese on other (waterfowl) species is probably limited to interspecific aggression at certain locations and during specific periods throughout the year. Prior to and during the reproductive season, breeding Egyptian Geese are highly territorial and are spread across the landscape. During breeding, Egyptian Geese will actively and fiercely defend these territories, possibly preventing smaller native species, such as ducks and coots, from establishing territories at these locations (Anselin and Devos, 2007; Callaghan and Brooks, 2016). However, most territorial aggression is targeted towards congeners, and the impact on other native waterfowl could be low, as Egyptian Geese are often observed in association with other waterfowl species (Cramp *et al.*, 1984; Callaghan and Brooks, 2016). Occasional reports of Egyptian Geese drowning other bird species are indeed known, but generally, aspects such as nesting success of native birds is not affected by the presence of breeding Egyptian Geese (Gyimesi and Lensink, 2010, 2012; Callaghan and Brooks, 2016).

Negative effects on native birds during the breeding season are mostly limited to nest site competition with birds of prey. In South Africa, Egyptian Geese have been shown to compete with several raptor species for nest sites (McPherson *et al.*, 2016; Wreford *et al.*, 2017) and negatively influence Black Sparrowhawk (*Accipiter melanoleucus*) nest success (Curtis *et al.*, 2007). In the UK, Egyptian Geese can outcompete Ospreys (*Pandion haliaetus*) and Barn Owls (*Tyto alba*) for artificial nest boxes (Rehfisch *et al.*, 2010). As the use of nest boxes for Peregrine Falcons and Kestrels has been reported in mainland Europe (Beck *et al.*, 2002), possible competition effects can also be expected for these species but have not been reported to date.

In contrast to the breeding season, during the moulting period, when large flocks are formed, interspecific aggression may have a much higher impact on native waterfowl. Aggression towards other species may cause avoidance behaviour, limiting the availability of optimal foraging and moulting areas for these other species, and increasing physiological stress during an already vulnerable stage (Gyimesi and Lensink, 2010). In parts of the USA, these large congregations and flocks of Egyptian Geese can occur at any time of the year (Callaghan *et al.*, 2017), suggesting that these impacts can be persistent throughout the year.

Egyptian Geese are known to hybridize with native species such as Barnacle Geese (*Branta leucopsis*), Greylag Geese (*Anser anser*) and various duck species, and with other introduced anatid species such as Ruddy Shelducks (*Tadorna ferruginea*) and Canada Geese (*Branta canadensis*), but hybrids are usually infertile (Lensink, 1996; Banks *et al.*, 2008; Rehfisch *et al.*, 2010; Gyimesi and Lensink, 2010, 2012; Avibase, 2018).

Different strains of avian influenza virus have been identified in Egyptian Geese in their native range (Cumming *et al.*, 2011). The presence of this virus outside the native range is thus possible, but given the absence of actual long-distance migratory behaviour in these populations, the possibility of Egyptian Geese serving as a vector for avian influenza is expected to be minimal (Gyimesi and Lensink, 2010).

Within the native range, Egyptian Geese are recognized locally as an agricultural pest, both by direct grazing and by trampling of crops (Mangnall and Crowe, 2002; Gyimesi and Lensink, 2010; Callaghan and Brooks, 2017). As far as crops are concerned, the species is known to mainly cause damage to grasslands, cornfields and wheat but occasionally other crops such as lettuce, peas and lucerne are also targeted (Gyimesi and Lensink, 2010; BIJ12: https://monitorfaunaschade.bij12.nl/, accessed 1 November 2019). The high dependency of the species on grassland habitats and its ability to forage on other crop types also raise concern in the region in which the species is introduced (Sutherland and Allport, 1991; Beck *et al.*, 2002; Gyimesi and Lensink, 2010). High local grazing pressure on grasslands can even result in competition with livestock for food, in addition to direct damage and yield loss. However, as in all cases of goose damage on growing crops, moderate grazing pressure can exert positive effects through additional fertilization via goose droppings and plant growth stimulation as a response to short grazing periods (Kear, 1970; Mangnall and Crowe, 2002). As Egyptian Geese are mainly non-migratory or limited to short-distance movements within the invasive range, crop grazing often occurs repeatedly and at high pressure, so that negative effects often outcompete the positive influences.

The presence of large flocks of Egyptian Geese may also cause eutrophication of standing water, which can heavily impact oligotrophic water systems of high biological value (Anselin and Devos, 2007; Callaghan and Brooks, 2016, 2017). Within other areas such as parks or swimming ponds, pollution by defaecation can cause nuisance to the public and interfere with the intended use of these areas (Callaghan and Brooks, 2016, 2017). Nuisance and damage caused by Egyptian Geese on golf courses in both the native and invasive ranges are well documented (Mackay *et al.*, 2014). Finally, the presence of large flocks of Egyptian Geese around airports poses the risk of aircraft collisions (Rehfisch *et al.*, 2010).

28.9 Control

Egyptian Geese experience little impact by predators in their invasive range, although kills by various raptor species are sometimes reported. Red foxes (*Vulpes vulpes*) and brown rats (*Rattus norvegicus*) are the most common predators of nests, but generally nests are fiercely defended, and predation will mostly be limited to ill-attended or deserted nests.

In most of its invasive range, Egyptian Geese are either a huntable species or can be shot within the context of invasive species or crop damage management. However, the species is not a very popular game species and appears to be difficult to hunt due to its vigilant nature and its large territorial spread during much of the year, as well as the general urban habitats it often resides in. In addition, there typically is no requirement to report the number of birds shot, making it difficult to assess the effectiveness of shooting as a management policy. The best evidence on shooting effectiveness comes from the Netherlands (where reporting on shot birds is mandatory) (Visser *et al.*, 2015). Gyimesi and Lensink (2012) used demographic models to assess the feasibility of population management of shooting to control Egyptian Geese numbers and found that about 28% of the population needs to be culled annually to prevent the populations from growing, and credited culling by shooting as a main reason behind the (near) stabilization of the Dutch Egyptian Goose population. Visser *et al.* (2015) also report a very high shooting effectiveness for Egyptian Geese in the Zuid-Holland region, and data from the native range also shows that large numbers of Egyptian Geese can be shot in organized shoots, directly removing hundreds of birds (Mangnall and Crowe, 2002). In general, although bag data on the species are mostly limited within the invasive range and hunting pressure put on Egyptian Geese can vary locally, overall the effect is expected to be limited in both Europe and the USA.

The Egyptian Goose has also been shown to be largely unaffected by a moult trapping approach as is often used for other goose species such as the Canada Goose and Barnacle Goose. Egyptian Geese, although often present at the same locations as moulting Canada and Barnacle Geese, are more vigilant and will not easily leave the water. Additionally, in contrast to other goose species, Egyptian Geese will also dive easily, and escape being herded from the water into a land-set trap in contrast to other species.

The Egyptian Goose does not generally nest in colonies and regularly uses nesting sites in trees, making the nests less accessible for viability control through egg pricking or oiling with liquid paraffin (Baker *et al.*, 1993). Viability control through nest destruction, pricking or oiling eggs has also been shown to be ineffective at the population level for goose control (Klok *et al.*, 2010). Visser *et al.* (2015) report that in exceptional cases (i.e. when it is impossible to catch or shoot the adult birds), egg oiling can be used to manage Egyptian Geese populations. For example, around Schiphol Airport in the Netherlands, where shooting is difficult for obvious reasons, up to 19% of Egyptian Goose nests could be targeted by oiling eggs in the nest (Visser *et al.*, 2015).

From ringing efforts in their native range, it has long been known that Egyptian Geese can be trapped with the use of baited walk-in traps with live decoy birds (Siegfried, 1967). Therefore, as an alternative to shooting, which may be opposed by parts of the general public or can be unfeasible in certain sites (e.g. strongly urbanized areas, or natural reserves harbouring species vulnerable to disturbance), trapping methods have been trialled. Floating or land-based Larsen traps have been used to reduce Egyptian Geese numbers across Flanders in northern Belgium (van Daele *et al.*, 2012). At low Egyptian Geese densities, trapping can be effective, as a field trial conducted in Belgium showed that a single Larsen trap was able to remove all breeding pairs present in a site (typically only one or two pairs) within a timespan of 1–9 days. Catching success with this method is best achieved with tamer, docile decoy birds, which exhibit no stress behaviour when placed in the trap. In general, the best results are obtained using a calm, frequently calling male as the decoy bird, which seems to trigger territorial behaviour in local birds most effectively. The highest success with these trap types is reached at nesting sites during the breeding season (Adriaens and Huysentruyt, 2014). This implies that traps need to be frequently moved between breeding sites (although pairs are reported usually to be caught within hours at a specific site). Experience with field trials in Flanders have shown that this approach can be successful in trapping territorial birds, leading to approximately 100 birds/year for each trapper. While effective and promising, it remains unclear whether this approach is cost-effective for removing larger populations spread across a larger geographical extent, especially as it requires daily inspection of all cage-traps installed.

28.10 Uses

As is common in waterfowl species, Egyptian Geese are hunted or shot in derogation and the meat is consumed. The species is a popular ornamental bird, often kept in captivity, but with pinioned birds also kept in open park settings.

28.11 References

Adriaens, T. and Huysentruyt, F. (2014) RINSE Partner Report: Field trial Egyptian goose. Instituut voor Natuur – en Bosonderzoek, Brussels.

Akhurst, J. (1877) Capture of the Egyptian goose on Long Island. *Bulletin of the Nuttall Ornithological Club* 2, 52–53.

Anselin, A. (2004) Nijlgans. In: Vermeersch, G., Anselin, A., Devos, K., Herremans, M., Stevens, J., *et al.* (eds) *Atlas van de Vlaamse broedvogels 2000–2002.* Instituut voor Natuurbehoud, Brussels, pp. 130–131.

Anselin, A. and Devos, K. (2007) Review of the status of introduced non-native waterbirds and their effects in Flanders, Belgium. Instituut voor Natuur – en Bosonderzoek, Brussels. Available at: https://pureportal.inbo.be/portal/files/3655910/INBO.A.2007.115.pdf (accessed 1 November 2019).

Anselin, A. and Vermeersch, G. (2005) Numbers and distribution of breeding feral geese in Flanders. *Oriolus* 71, 111–120.

Avibase (2018) Egyptian Goose: Alopochen aegyptiaca (Linnaeus, 1766). Avaiable at: https://avibase.bsc-eoc.org/species.jsp?lang=EN&avibaseid=DA-2F24E310CF72A6 (accessed 1 November 2019).

Baker, K. (1993) *Identification Guide to European Non-Passerines: BTO Guide 24*. British Trust for Ornithology, Thetford, UK.

Baker, S., Feare, C., Wilson, C., Malam, D. and Sellars, G. (1993) Prevention of breeding of Canada geese by coating eggs with liquid paraffin. *International Journal of Pest Management* 39, 246–249.

Banks, A.N., Wright, L.J., Maclean, I.M.D., Hann, C. and Rehfisch, M.M. (2008) Review of the status of introduced non-native waterbird species in the area of the African-Eurasian Waterbird Agreement: 2007 update. BTO Research Report No. 489. British Trust for Ornithology, Thetford, UK.

Beck, O., Anselin, A. and Kuijken, E. (2002) Beheer van verwilderde watervogels in Vlaanderen. Instituut voor Natuurbehoud, Brussels.

Brown, L.H., Urban, E.K. and Newman, K. (1982) *The Birds of Africa*, Vol. 1. Academic Press, London.

CAB International (2018) *Alopochen aegyptiaca* (Egyptian goose). Available at: https://www.cabi.org/isc/datasheet/94205 (accessed 1 April 2020).

Callaghan, C.T. and Brooks, D.M. (2016) Ecology, behaviour, and reproduction of invasive Egyptian Geese (*Alopochen aegyptiaca*) in Texas. *Bulletin of the Texas Ornithological Society* 49, 37–45.

Callaghan, C.T. and Brooks, D.M. (2017) History, current distribution and status of the Egyptian Goose (*Alopochen aegyptiaca*) in the contiguous United States. *Southwestern Naturalist* 62, 296–300.

Callaghan, C.T., Brooks, D.M. and Pyle, P. (2017) Egyptian Goose (*Alopochen aegyptiaca*). In: Rodewald, P.G. (ed.) *The Birds of North America*. The Cornell Lab of Ornithology, Ithaca, New York. Available at: https://doi.org/10.2173/bna.egygoo.01 (accessed 1 November 2019).

Carboneras, C. (1992) Family Anatidae (ducks, geese and swans). In: Del Hoyo, J., Elliott, A. and Sargatal, J. (eds) *Handbook of Birds of the World, Vol. 1: Ostrich to Ducks*. Lynx Edicions, Barcelona, Spain, pp. 536–629.

Cramp, S., Simmons, K.E.L., Ferguson-Lees, I.J., Gillmor, R., Hollom, P.A.D., *et al.* (1984) *Handbook of the Birds of Europe, the Middle East, and North Africa: The Birds of the Western Palearctic, Vol. 1: Ostrich to Ducks*. Oxford University Press, Oxford.

Cumming, G.S., Caron, A., Abolnik, C., Cattoli, G., Bruinzeel, L.W., *et al.* (2011) The ecology of influenza A viruses in wild birds in southern Africa. *EcoHealth* 8, 4–13.

Curtis, O.E., Hockey, P.A.R. and Koeslag, A. (2007) Competition with Egyptian geese *Alopochen aegyptiaca* overrides environmental factors in determining productivity of black sparrowhawks *Accipiter melanoleucus*. *Ibis* 149, 502–508.

DAISIE (2009) *Handbook of Alien Species in Europe*. Invading Nature – Springer Series in Invasion Ecology, INNA, Vol. 3. Springer Science+Business Media, Dordrecht, Netherlands.

Gyimesi, A. and Lensink, R. (2010) Risk analysis of the Egyptian Goose in the Netherlands. Bureau Waardenburg BV, Culemborg, Netherlands.

Gyimesi, A. and Lensink, R. (2012) Egyptian Goose *Alopochen aegyptiaca*: an introduced species spreading in and from the Netherlands. *Wildfowl* 62, 128–145.

Halse, S.A. (1984) Diet, body condition, and gut size of Egyptian Geese. *Journal of Wildlife Management* 48, 569–573.

Jaška, P. and Řepa, P. (2017) Breeding of the Egyptian Goose (*Alopochen aegyptiaca*) in the Czech Republic in 2006–2016 and a detailed analysis of its occurrence in the Karlovy Vary and Plzeň regions. *Sylvia* 53, 21–40.

Kampe-Persson, H. (2010) Occurrence of Egyptian Goose *Alopochen aegyptiaca* in Europe. *Goose Bulletin* 10, 34–37.

Kear, J. (1970) The experimental assessment of goose damage to agricultural crops. *Biological Conservation* 2, 206–212.

Kirkwood, F.C. (1900) The occurrence of Egyptian goose in North America. *Auk* 17, 64–65.

Klok, C., van Turnhout, C., Willems, F., Voslamber, B., Ebbinge, B. and Schekkerman, H. (2010) Analysis of population development and effectiveness of management in resident greylag geese *Anser anser* in the Netherlands. *Animal Biology* 60, 373–393.

Lensink, R. (1996) De opkomst van exoten in de Nederlandse avifauna; verleden, heden en toekomst [The rise of exotic bird species in the Dutch avifauna; past, present and future]. *Limosa* 69, 103–130.

Lensink, R. (1999) Aspects of the biology of Egyptian Goose *Alopochen aegyptiacus* colonizing The Netherlands. *Bird Study* 46, 195–204.

Mackay, B., Little, R.M., Amar, A. and Hockey, P.A.R. (2014) Incorporating environmental considerations in managing Egyptian Geese on golf courses in South Africa. *Journal of Wildlife Management* 78, 671–678.

Maclean, G.L. (1997) Egyptian Goose. In: Harrison, J.A., Allan, D.G., Underhill, L.G., Herremans, M., Tree, A.J. *et al.* (eds) *The Atlas of Southern African Birds, Vol. 1: Non-passerines*. BirdLife South Africa, Johannesburg, pp. 122–123.

Madge, S. and Burn, H. (1988) *Waterfowl: An Identification Guide to the Ducks, Geese, and Swans of the World*. Houghton Mifflin, Boston, Massachusetts.

Mangnall, M. and Crowe, T. (2002) Population dynamics and the physical and financial impacts to cereal crops of the Egyptian Goose *Alopochen aegyptiacus* on the Agulhas Plain, Western Cape, South Africa. *Agriculture, Ecosystems & Environment* 90, 231–246.

McPherson, S., Brown, M. and Downs, C.T. (2016) Crowned Eagle nest sites in an urban landscape: requirements of a large eagle in the Durban Metropolitan Open Space System. *Landscape and Urban Planning* 146, 43–50.

Pranty, B. and Ponzo, V. (2014) Status and distribution of Egyptian Geese (*Alopochen aegyptiaca*) in Southeastern Florida. *Florida Field Naturalist* 42, 91–107.

Rehfisch, M.M., Allan, J.R. and Austin, G.E. (2010) The effect on the environment of Great Britain's naturalized Greater Canada *Branta canadensis* and Egyptian Geese *Alopochen aegyptiacus*. In: *British Ornithologists' Union Proceedings – The Impacts of Non-native Species*. Available at: www.bou.org.uk/bouprocnet/impacts-of-non-native-species/ (accessed 1 November 2019).

Ries, C., Pfeiffenschneider, M., Engel, E., Heidt, J.-C. and Lauff, M. (2014) Environmental impact assessment and black, watch and alert list classification after the ISEIA Protocol of vertebrates in Luxembourg. *Bulletin de la Société des Naturalistes Luxembourgeois* 115, 195–201.

Segers, M. (1989) Nijlgans. In: *Vogels in Vlaanderen: Voorkomen en Verspreiding*. Vlaamse Avifauna Commissie, Bornem, Belgium, pp. 84–85.

Siegfried, W.R. (1967) Trapping and ringing of Egyptian Geese and African Shellduck at Vogelvlei, Cape. *Ostrich* 38, 173–178.

Sutherland, W.J. and Allport, G. (1991) The distribution and ecology of naturalized Egyptian Geese *Alopochena egyptiacus* in Britain. *Bird Study* 38, 128–134.

Toms, M. (2002) Egyptian Goose. In: Wernham, C.V., Toms, M.P., Marchant, J.H., Clark, J.A., Siriwardena, G.M. and Baillie, S.R. (eds) *The Migration Atlas: Movements of the Birds of Britain and Ireland*. T. & A.D. Poyser, London, p. 688.

van Daele, P., Adriaens, T., Devisscher, S., Huysentruyt, F., Voslamber, B., *et al.* (2012) Beheer van Zomerganzen in Vlaanderen en Zeeuws-Vlaanderen. Instituut voor Natuur- en Bosonderzoek, Brussels. Available at: https://pureportal.inbo.be/portal/files/6824472/INBO.R.2012.58web.pdf (accessed 1 November 2019).

van Dijk, K. and Majoor, F. (2011) Geboortedispersie van Nijlganzen van Nederland naar Duitsland en vice versa. *Limosa* 84, 82–84.

Vangeluwe, D. and Roggeman, W. (2002) Dynamique d'expansion d'une population exotique d'ouettes d'Egypte (*Alopochen aegyptiacus*). *Bulletin van het Koninklijk Belgisch instituut voor Natuurwetenschappen* 72, 229–230.

Visser, A., Keuper, D. and Guldemond, A. (2015) Faunabeheerplan ganzen Zuid-Holland 2015–2020. Bijlage Regioplan Zuid-Holland-Noord. Publicatienummer CLM-879. CLM, Culemborg, Netherlands. Available at: https://www.clm.nl/uploads/pdf/866-CLMrapport-Faunabeheerplan_Ganzen_ZH2015-2020-web.pdf (accessed 1 November 2019).

Wilbur, S.R. and Yocom, C.F. (1971) Unusual geese in the Pacific Coast states. *Murrelet* 52, 16–19.

Wreford, E., Hart, L., Brown, M. and Downs, C.T. (2017) Black sparrowhawk (*Accipiter melanoleucus*) breeding behaviour and reproductive success in KwaZulu-Natal, South Africa. *Ostrich* 88, 287–290

29 Greylag Goose (*Anser anser* Linnaeus, 1758)

Lorinda A. Hart[1,2]* and Colleen T. Downs[1]

[1]*Centre for Excellence in Invasion Biology, School of Life Sciences, University of KwaZulu-Natal, Pietermaritzburg 3209, South Africa;* [2]*Department of Biological Sciences, University of Namibia, Windhoek 10023, Namibia*

Citation: Hart, L.A. and Downs, C.T. (2020) Greylag Goose (*Anser anser* Linnaeus, 1758). In: Downs, C.T. and Hart, L.A. (eds) *Invasive Birds: Global Trends and Impacts*. CAB International, Wallingford, UK, pp. 213–222.

29.1 Common Names

Graylag Goose, Eastern Greylag Goose, Feral Goose, Grey Goose, Grey Lag-Goose, Greylag, Grey-lag Goose, Western Graylag Goose (English), Graugans (German), Ansar Comun, Ganso común (Spanish), Oie cendrée (French), Oca selvatica (Italian), Burung Angsa Kelabu (Malay) (Lepage, 2004).

29.2 Nomenclature

The Greylag Goose (*Anser anser* Linnaeus, 1758) is a member of the family Anatidae and has two subspecies recognized. The Western Greylag Goose (*A. a. anser*; European) breeds in north-western Eurasia including Iceland and north and central Europe, and its winter range spans the area encompassing Scotland, North Africa, Turkey and Iran (Clements *et al.*, 2018; Carboneras and Kirwan, 2019). It is considered partially migratory (Blair *et al.*, 2000) as this species can further be divided into four biogeographical groups found in Iceland, Great Britain and Ireland (where they remain resident), north-west and south-west Europe, and central Europe (Madsen *et al.*, 1999; Powolny *et al.*, 2018). The Eastern Greylag Goose (*A. a. rubrirostris*; Siberian) breeds in north-eastern Eurasia including Romania, Turkey and Russia eastwards to north-eastern China and winters in Anatolia, India and northern Indochina (Clements *et al.*, 2018; Carboneras and Kirwan, 2019). It is fully migratory (Blair *et al.*, 2000). This subpecies forms two populations located at the Black Sea and Caspian Sea (Madsen *et al.*, 1999). The migratory patterns of Greylag Geese can vary and are discussed in detail by Polowny *et al.* (2018).

A. a. domesticus/*A. a. forma domestica* were bred for produce and are thus larger, heavier and flightless (Delacour, 1964). Their appearance may resemble that of wild geese, or they can have variable amounts of white plumage (Fig. 29.1), and can even be completely white. They typically have orange bills and legs, while their wild counterparts have pink legs (Delacour, 1964). Their close resemblence to their wild ancestors, despite a long period of domestication was noted by Charles Darwin (Darwin, 1890).

29.3 Distribution

The Greylag Goose has a large native range extending across the Palearctic (Fig. 29.2) (BirdLife International, 2019). Its native range includes the following countries and islands: Afghanistan, Albania, Algeria, Armenia, Austria, Azerbaijan, Bahrain, Bangladesh, Belarus, Belgium, Bhutan, Bosnia and Herzegovina, Bulgaria, China, Croatia, Cyprus, Czech Republic, Denmark, Estonia, Faroe Islands, Finland, France, Georgia, Germany, Gibraltar, Greece, Hungary, Iceland, India, Iran, Iraq, Ireland, Israel, Italy, Japan, Jordan, Kazakhstan, Republic of Korea, Kuwait, Kyrgyzstan, Laos, Latvia, Liechtenstein, Lithuania, Luxembourg, Macedonia, Malta, Moldova, Mongolia, Montenegro, Morocco, Myanmar, Nepal, the Netherlands, Norway (including Svalbard and Jan Mayen), Oman, Pakistan, Poland, Portugal, Qatar, Romania, Russian Federation, Saudi Arabia, Serbia, Slovakia, Slovenia, Spain, Sri Lanka, Sweden, Switzerland, Syria, Tajikistan, Tunisia, Turkey, Turkmenistan, Ukraine, the UK, Uzbekistan, United Arab Emirates and Vietnam.

Vagrant Greylag Goose sightings, probably of wild birds, have been reported in Egypt, Gibraltar, Greenland, Hong Kong, Libya, Lebanon, Canary Islands (Spain), Thailand and Yemen.

*Corresponding author: lorinda.hart.research@gmail.com

Fig. 29.1. Colour variation of the domestic Greylag Goose (*Anser anser*). (©Photograph: Filippo Salamone from Settimo Milanese (MI), Italia; CC BY-SA 2.0, https://creativecommons.org/licenses/by-sa/2.0.)

In terms of their invasive range, feral and established Greylag Geese are most probably A. a. domesticus and are reported in Australia, New Zealand, the Falkland Islands, South Africa, Namibia, Canada, the USA and Columbia (Voller and McNay, 2007; Salaman *et al.*, 2008; Florida Fish and Wildlife Conservation Commission, 2019; L. Hart personal communication).

29.4 Description

The Greylag Goose is a large goose (76–98 cm in length) with a thick neck (Porter and Aspinall, 2013; Carboneras and Kirwan, 2019). Male *A. a. anser* weigh 2600–4560 g and females weigh 2070–3960 g, while *A. a. rubrirostris* have narrower weight ranges that fall within these sizes (Carboneras and Kirwan, 2019). On average, a Greylag Goose weighs 3270 g, but variation is observed among seasons, ages and sexes (Matthews and Campbell, 1969). For example, geese are lightest after completing a migration and heaviest when feeding on grains and tubers during winter, after which they again lose weight on a summer grass diet (Matthews and Campbell, 1969). Non-migratory feral individuals are also heavier than migratory birds (Matthews and Campbell, 1969). Greylag Geese have pink (*A. a. rubrirostris*) or orange (*A. a. anser*) bills, pink legs and a pink-orange ring around brown eyes (Fig. 29.3) (Johnsgard, 2010). Overall, Greylag Geese have brown-grey plumage, except for the posterior abdomen, which is white (Fig. 29.3) (Johnsgard, 2010). The head and upper chest is uniform grey-brown, while the neck area has a furrowed appearance (Johnsgard, 2010). There is slight barring on the lower chest with patterning becoming more prominent towards the flanks, which are darker (Johnsgard, 2010). Mantle feathers and wing coverts have a light buff edging, with the wing darkening towards the primaries. They have a distinctive pale grey forewing in flight (Porter and Aspinall, 2013). Dark brown tail feathers have white edges and tips (Johnsgard, 2010). Juveniles generally have a more uniform appearance, lacking well-defined markings and have pale grey legs (Fig. 29.3) (Porter and Aspinall, 2013).

29.5 Diet

The Greylag Goose is predominantly vegetarian and feeds on a variety of plant materials including grass, leaves, roots, seedling

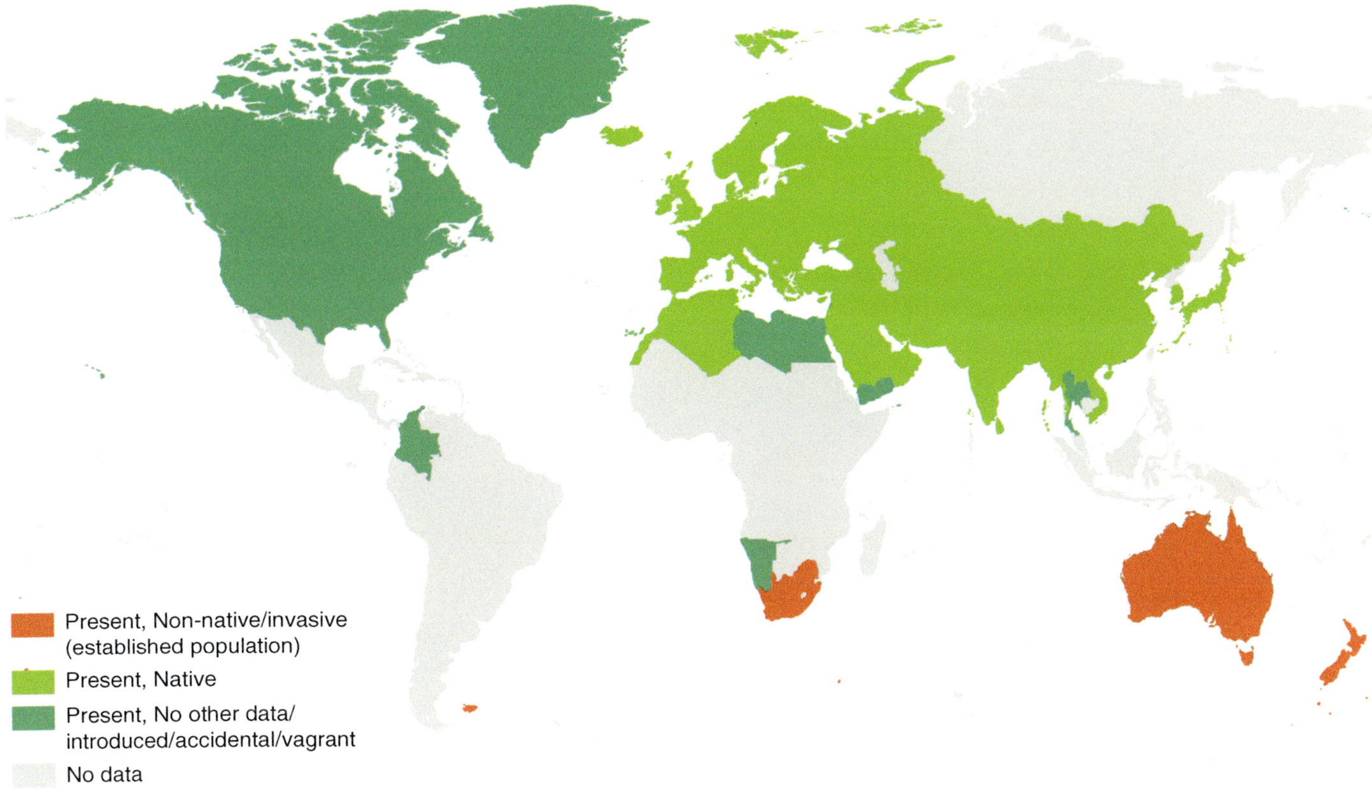

Fig. 29.2. Global distribution of the Greylag Goose showing native (green) and invasive (red) ranges.

sprouts and seeding flower heads (Carboneras and Kirwan, 2019). Their diets are sometimes supplemented by people, particularly in urban parks (Hutton, 2015). In some of their native range (British Isles), they feed almost exclusively on grass and clover during egg-laying, with ripening grain crops added to their diet during the late summer (Delacour, 1964). Russian geese incorporate aquatic plants over and above a predominantly grass diet, which also forms an important component of goslings' diets (Delacour, 1964). During migration and at wintering grounds, they feed on cereal crops, coastal grasses, tubers and bulbs (Dementiev and Gladkov, 1952; Delacour 1964; Newton and Kerbes, 1974). They also occasionally eat snails and insects (Southey, 2013). Non-migratory geese in Germany prefer grains and cereals in winter (Kleinhenz and Koenig, 2018), which in part has facilitated increasing goose populations by providing high-energy, concentrated food sources (Buij *et al.*, 2017).

29.6 Introduction and Invasion Pathways

Throughout much of its native range, Greylag Goose populations were once declining. They almost became extinct in Scotland in the 1940s due to hunting, persecution and the outbreak of World War II (Bainbridge, 2017). Similarly, Great Britain saw the geese heading to extinction during the 19th century as a result of hunting and land drainage for cultivation (Mitchell *et al.*, 2012). Other negative impacts to goose populations include egg collecting, increased predation, recreational disturbance and vegetation changes due to overgrazing (Mitchell *et al.*, 2012). Since then, favourable agriculturural practices, changes in land use and the implementation of protective policies have seen the populations increase dramatically (Bainbridge, 2017). On Orkney Island, populations have increased 60-fold in the last 24 years, with around 19% annual growth (Mitchell *et al.*, 2014; Tulloch *et al.*, 2017). In the Netherlands, populations increased by 20% annually between 1961 and 2009 (Voslamber *et al.*, 2010).

Introductions of both domestic and wild Greylag Geese have occurred in their native range, including: United Arab Emirates (1980s), Austria (1985–1995), Finland and France (1990s), Germany (1950s, possibly earlier), Great Britain (Middle Ages and 1930s–1960s), Ireland (1730s), Israel (1990s), Italy (1978–mid-1980s), Switzerland, Lithuania (1970s), the Netherlands (1972) and Ukraine (Blair *et al.*, 2000; Banks *et al.*, 2008). In France, the non-native *A. a. rubrirostris* was introduced, and in Great Britain, Scottish and possibly Icelandic birds were released (Boyd and Matthews, 1963; Owen and Salmon, 1988; Rehfisch *et al.*, 2002; Banks *et al.*, 2008; Mitchell *et al.*, 2012). Some introductions were motivated and carried out by hunting communities and wildfowl associations (Owen and Salmon, 1988). In a few instances, escaped birds and natural range expansion have also occurred (Blair *et al.*, 2000). Today, feral and captive populations occur throughout Europe (Carboneras and Kirwan, 2019).

Outside of Eurasia, acclimatization societies repeatedly introduced geese during the 18th and 19th centuries as a food

Fig. 29.3. Identity composite of Greylag Geese. The front left bird is an adult and front right is a juvenile. (From Crossley and Couzens, 2011; ©Photograph: Richard Crossley; CC BY-SA 3.0, https://creativecommons.org/licenses/by-sa/3.0.)

source and as a reminder of their homelands (Blair *et al.*, 2000). Captain Cook unsuccessfully tried to introduce five birds from South Africa to New Zealand, and later the acclimatization societies also repeatedly tried to establish them there (Southey, 2013). Today, Greylag Geese are recognized as an established species in New Zealand (Carboneras and Kirwan, 2019), although it is suspected that they originated from farm strays (Southey, 2013). Similarly, in other countries where more recent populations are arising, Greylag Geese are probably domestic escapees or releases.

There has been much debate about the inclusion of Greylag Geese on official bird lists, for example in Australia (Palliser, 2014) and the Falkland Islands (Donegan, 2007). While the latter has included them (Remsen *et al.*, 2019), Australia has not. They are also not recognized as 'feral', as Australian Greylag Geese are probably domestic or potentially hybrid individuals (i.e. not recognized as a true species) and although breeding in 'the wild', they are not necessarily free flying (Palliser, 2014; Hutton, 2015).

Introduction to the Falkland Islands started with one pair in 1935 and another documented release in 1944 (Woods, 1988). In 1955, around 50 birds were reported, and by 1983, more than 130 Greylag Geese were counted and had spread beyond the point of introduction (Woods, 1988). Their mixed plumage containing varying amounts of white supports the conclusion that these were from domestic geese (Woods, 1988). In 2007, the population was still considered small with 300–600 pairs reported (Donegan, 2007).

29.7 Breeding Behaviour

The Greylag Goose starts breeding in March and April in Europe and Iceland (Carboneras and Kirwan, 2019). Commencement of egg-laying varies; for example, in France it begins in February (Schricke, 2018, cited by Powolny *et al.*, 2018), in Europe and Iceland predominantly from late April, in the UK from late March (Carboneras and Kirwan, 2019), and in the Hebrides off the west coast of Scotland in mid-April (Newton and Kerbes, 1974). In their invasive range in New Zealand, laying occurs during August and September (Southey, 2013). The timing of breeding has been linked to breeding success (Newton and Kerbes, 1974). Nest construction is variable with geese generally using materials found within a 2 m radius of the nest site (Young, 1972). Typically, a twig base is laid to which dead leaves, moss and/or lichen is added, with a final grass and down lining (Young, 1972). Each nest is unique and can be a simple depression in vegetation, a scrape on a dry nesting site or a specific assemblage of plant species used in construction (Young, 1972; Brown *et al.*, 1987). Nests typically have a 41 cm diameter with a 6 cm deep bowl, which is added to during incubation, gaining 6–8 cm (Young, 1972). Nests are reused annually, with some nests in Scotland used for up to five breeding seasons (Delacour, 1964). Nest sites are close to water and grazing areas, usually with some wooded cover (Delacour, 1964), dense emergent vegetation (Powolny *et al.*, 2018) or long grass (Southey, 2013). Nests can be densely packed, particularly on islands, ranging from 2–12 m apart, averaging approximately 11 m (Delacour, 1964; Young, 1972). Geese pair for life, with both sexes building the nest, but only the female incubates the eggs while the male remains nearby to defend it (Delacour, 1964; Powolny *et al.*, 2018).

Greylag Geese lay an average of 5–6 eggs, which are incubated for approximately 30 days, after which there is a 50–60 day rearing period (Delacour, 1964; Southey, 2013). Eggs are cream-white and approximately 85 × 58 mm, weighing 160 g (Fig. 29.4) (Johnsgard, 2010). Young females may parasitize experienced females' nests (Kotrschal *et al.*, 2006). Young are brooded for 2 weeks (Kotrschal *et al.*, 2006) and are fully feathered by 12–16 weeks (Serjeantson, 2002). They remain near the parents until the next breeding season (Southey, 2013). Greylag Goose familes form social bonds, with both active and passive benefits, clan structures, dominance heirarchies and individual recognition (Frigerio *et al.*, 2005; Weiß *et al.*, 2010; Scheiber *et al.*, 2011).

Fig. 29.4. A Greylag Goose nest with eggs. (©Photograph: Lämpel; CC BY-SA 3.0, https://creativecommons.org/licenses/by-sa/3.0.)

29.8 Habitat

The Greylag Goose inhabits open or low-lying landscapes with water bodies that are usually bordered by vegetation (Carboneras and Kirwan, 2019). They occupy a wide range of wetlands from boreal to subarctic (Powolny *et al.*, 2018). They occur on arable fields, agricultural land, river flood plains, grasslands, islets, marshes (salt and freshwater), estuaries, reed beds, boggy thickets, and damp or flooded meadows (Delacour, 1964; Porter and Aspinall, 2013; Powolny *et al.*, 2018). Foraging geese prefer improved grass, semi-natural grass and arable fields to natural moorland habitats (Mitchell *et al.*, 2014). In New Zealand, they are generally seen in urban parks, on grazing fields and in association with ponds on farms (Southey, 2013). In Germany and the Netherlands, they are also becoming more common in urban areas (Powolny *et al.*, 2018).

Habitat use varies seasonally. Greylag Geese respond negatively to extremely cold conditions, and migrating flocks are larger during milder winters at some locations (Esselink *et al.*, 1997). During the non-breeding season, Greylag Goose reliance on water is predominantly for communal roosting after foraging afield (Powolny *et al.*, 2018). These water bodies range from freshwater to protected marine bays (Powolny *et al.*, 2018). Non-migratory individuals forage near water and maintain small home ranges when breeding, but following moult, foraging areas increase around these waterbodies, only to contract again during winter (Kleinhenz and Koenig, 2018). Some flocks may remain in urban areas during winter, taking advantage of food from people (Kleinhenz and Koenig, 2018).

29.9 Impacts

29.9.1 Negative impacts

Greylag Geese impacts, like other invasive geese species, include eutrophication of water bodies, damage to agricultural crops, animal and human health impacts, damage/fouling of buildings or recreational areas, hybridization, competition and an increased risk of bird strikes (Blair *et al.*, 2000, Dolbeer and Seubert, 2009; Maragakis, 2009; van Ham *et al.*, 2013; see also

Chapter 30, this volume). Additionally, in urban parks, aggressive behaviour and biting occurs towards people (Southey, 2013; Hutton, 2015). However, there are concerns that the evidence for some of the impacts are often anecdotal observations from relatively small areas (Strubbe *et al.*, 2011; see Chapter 30, this volume).

Migratory birds have the potential to disperse propagules over great distances. Greylag Geese passively disperse both alien and native plant seeds and bryozoans (Figuerola *et al.*, 2004; García-Álvarez *et al.*, 2015). They also carry and transmit a variety of viruses, bacteria and parasites (Buij *et al.*, 2017; Elmberg *et al.*, 2017). In 2015, more than 90% of domestic geese died in Taiwan from avian influenza (Chang *et al.*, 2016).

Throughout Europe, there has been an increase in Greylag Goose populations due to re-establishment programmes and more controlled hunting regulations (Blair *et al.*, 2000). Considering a mean growth rate of 8.9%, the north-western and south-western European populations could reach 5.1 million–8.3 million birds in 2037, from around 750,000 birds in 2012 (Powolny *et al.*, 2018). However, feral and introduced Greylag Geese can complicate accurate species counts of wild populations (Banks *et al.*, 2008). The number of non-migratory geese in north-western Europe has increased by 20% annually over the last 20 years, adding pressure to agricultural and natural systems (Kleinhenz and Koenig, 2018). Examples of agricultural damage include crop loss, pasture fouling, soil puddling and ground compaction, prompting government reimbursement schemes (van Eerden, 1990; MacMillan *et al.*, 2004; Tulloch *et al.*, 2017).

Moulting geese are rendered flightless for 28 days (Loonen *et al.*, 1991) and undergo both physical and physiological changes that are energetically demanding. They lose weight and flight muscle tone (Fox and Kahlert, 2005), and their heart rate reaches a maximum (Wascher *et al.*, 2018). Food availability is essential for this period, and birds may change moult sites, or even not moult, should food stocks be limited (Loonen *et al.*, 1991). Geese selectively feed on plants with higher protein content when moulting (Fox *et al.*, 1998), and concentrate feeding efforts near water bodies during this flightless period (Schwartz and Woog, 2012). Vigilance increases and foraging occurs predominantly at night (Kahlert *et al.*, 1996). This concentrated pressure from selectively feeding geese at moult sites has adverse long-term effects on the nutrient dynamics and growth of plants (van den Wyngaert *et al.*, 2003). It also stimulates more rapid methane emission by plants, with five times more methane being emitted compared with non-grazed vegetation (Dingemans *et al.*, 2011). Grubbing for rhizomes and tubers and intensive grazing can alter the plant community and upper soil layer, particularly where birds concentrate at wetland edge habitats (Esselink *et al.*, 1997; Mulder and Ruess, 2001). Grazing pressure by geese can result in competition with other grazing species (Loonen *et al.*, 1991; van der Wal *et al.*, 1998). As Greylag Goose numbers increase, there may be reduced tolerance of smaller waterbird species (Blair *et al.*, 2000), but data on competitive interactions are lacking (Banks *et al.*, 2008). The end product of grazing, excessive defaecation, can render fields and water bodies unfit for use by other animals (Hahn *et al.*, 2008; Southey, 2013; Hutton, 2015).

Globally, the greatest impacts of non-native Anseriformes (ducks, geese and swans) are hybridization and herbivory (Evans *et al.*, 2016; see also Chapter 30, this volume). Greylag Geese hybridize with other Anatidae (23 species in captivity), with more than 1500 instances recorded since 1970 (Blair *et al.*, 2000). In the wild, feral domestic Greylag Geese have produced approximately 18,000–20,000 hybrids in Germany and the Netherlands (Blair *et al.*, 2000). Some species that hybridize with Greylag Geese include Canada Geese (*Branta canadensis*), Barnacle Geese (*B. leucopsis*), Egyptian Geese (*Alopochen aegyptiacus*), Greater White-fronted Geese (*Anser albifrons*), Bar-headed Geese (*A. indicus*), Snow Geese (*A. caerulescens*) and Swan Geese (*A. cygnoides*) (Blair *et al.*, 2000; Ottenburghs *et al.*, 2016). The most common wild hybirds are Canada Geese × Greylag Geese (Ottenburghs *et al.*, 2016). Canada Geese are larger than Greylag Geese, and while they display greater levels of intraspecies aggression at nest sites, attacks on Greylag Geese do occur, with evidence suggesting that they can prevent their nesting (Fabricius *et al.*, 1974). This could be due to Greylag Geese performing both inter- and intraspecies nest parasitism (Beauchamp, 1998; Ottenburghs *et al.*, 2016). In an experiment, 14% of male Greylag Geese raised by Canada Geese imprinted on them for life, later choosing them as mates (Fabricius, 1991). Such behaviour therefore increases the risk of hybridization (Ottenburghs *et al.*, 2016).

29.9.2 Positive impacts

Buij *et al.* (2017) provide detailed impacts of geese. Some positive effects include: (i) their faeces can increase soluble nitrogen available to plants (nutrient cycling); (ii) other animals may eat goose droppings as a concentrated source of grass; (iii) they disperse seeds and aquatic organisms; (iv) intermediate grazing may stimulate plant growth; (v) geese are an important food source for people and wild animals; (vi) they can be used in disease detection/surveillance; and (vii) their grazing and foraging may provide a preferred habitat for some organisms (Buij *et al.*, 2017). They are also culturally recognized as part of our history, and today generate income through hunting, birdwatching and farming for various products (Serjeantson, 2002; Buij *et al.*, 2017).

29.10 Control

The increasing populations of Greylag Geese have given rise to the need for revised policies that balance economic, social and scientific opinions (Bainbridge, 2017). An action plan has recently been drawn up for Greylag Geese in Europe. Some of its objectives include: maintaining population numbers (regulating hunting levels), limiting agricultural damage, reducing health and air-strike risks, and reducing impacts on native plants and animals (Powolny *et al.*, 2018). Members of the European Union can hunt Greylag Geese, but it is protected in the Netherlands and Walloon Region of Belgium, with temporary protection enforced in Finland and northern Norway (Powolny *et al.*, 2018). However, derogations are granted where there is a risk of serious damage by geese (Powolny *et al.*, 2018). Falkland Islands legislation,

Conservation of Wildlife and Nature Ordinance (1999), allows Greylag Geese to be killed throughout the year (Woods *et al.*, 2009). Across Europe, where records are available, it is estimated that 450 000 geese have been harvested between 2012 – 2016 (Powolny et al., 2018). Modelling suggests that increasing the amount of unfavourable habitat or culling would effectively reduce goose numbers (Tulloch *et al.*, 2017).

As with Canada Geese, Greylag Geese can be actively managed through fertility reduction or through culling, which involves shooting and/or capturing flocks of geese during the moult period when they are flightless (Allan *et al.*, 1995; Smith *et al.*, 1999; Reyns *et al.*, 2018). Additionally, prevention of hatching by pricking eggs or coating them can be successful, but this depends on the effort, and model-based approaches show it must be sustained over several years (Beston *et al.*, 2016; Beaumont *et al.*, 2018; see also Chapter 30, this volume), and up to 88% of the nests must be found and treated (Klok *et al.*, 2010). Culling is also effective; for example, 60% of the national hunting bag of the Greylag Goose wintering population was culled through hunting (Aubry *et al.*, 2016).

29.11 Uses

Wild Greylag Geese are used predominantly for hunting and birdwatching, although wild geese may be captured and fattened or become integrated into captive flocks (Zeuner, 1963; Serjeantson, 2002; Buij *et al.*, 2017). Most goose products are sourced from commercially farmed domestic geese, but similar products can be harvested from hunted geese. Domestic Greylag Geese are used for meat, fat, eggs, down, controlling weeds, as pets and as guards (Southey, 2013). Today, commercial goose farming is a specialized industry in the west, while it is most prolific in China where traditional practices are mostly used (Farrell, 2004). In 2004, China was reported as the biggest producer of goose meat, with 200 million birds producing 1.93 million of the 2.10 million tonnes produced globally (Farrell, 2004). They are also the biggest supplier of goose down (Serjeantson, 2002). Geese are the fastest-growing poultry birds and their meat is the fattiest (Farrell, 2004). Their fatty livers are most valuable and are used to make pâté foie gras, a delicacy consumed predominantly in France where 800,000 geese are harvested annually (Farrell, 2004). Geese readily lay down fat deposits, and are fattened on carbohydrates, allowed limited exercise and kept warm to obtain livers of 600–1000 g (del Hoyo *et al.*, 1992; Serjeantson, 2002; Farrell, 2004). The French developed a breed that is easily force-fed, a practice recorded in Egypt 4500 years ago and also in ancient Rome, where a mixture of milk, flour and honey was used (Zeuner, 1963; Serjeantson, 2002; Farrell, 2004). This practice is now banned in several countries (Farrell, 2004).

The harvesting of down feathers is also ethically challenged by many animal rights groups. While many countries only harvest down from slaughtered geese, some countries still practice live plucking (Farrell, 2004). Goose down was most valuable during the 18th–19th centuries but, given its light weight and thermal properties, it is still sought after today (Serjeantson, 2002). Eggs are only available for a limited time of the year, with generally low laying rates, and therefore they are not a major commercial product (Serjeantson, 2002).

Historical uses of feathers (other than down products) include quill pens, fletching of arrows, feathering shuttlecocks, hollow quills used to hold gold dust, paint brushes and instrument picks, to name a few (Crawford, 1984; Serjeantson, 2002; Cocker and Mabey, 2005). Goose fat was used for lubrication and as an ointment base (Serjeantson, 2002; Cocker and Mabey, 2005). Geese were also historically used as sacrifices (Zeuner, 1963). They have also been bred for fighting in Russia, a sport that was banned in 1906 (Serjeantson, 2002).

29.12 Notes

Geese are easily domesticated due to their imprinting tendencies as goslings (Zeuner, 1963), and it is also this behaviour that made them the subject of Konrad Lorenz's behavioural research (Lorenz, 1935). The history of goose domestication is unclear, and genetic studies are being undertaken to gain a better understanding of this (Honka *et al.*, 2018), but they were kept by humans from the Neolithic period (Zeuner, 1963). There are also early records of domestic geese dating back to the 2nd millennium bc in Egypt and pre-historic Greece (Boessneck, 1988; Serjeantson, 2002). Some suggest that domestication occurred within their breeding range in south-eastern Europe around 3000 bce (Crawford, 1984), while others suggest that domestication occurred in Egypt based on iconographic depictions of geese dating from 2686 to 1991 bce (Zeuner, 1963).

They are representatives of gods and deities in Indian, Egyptian, Sumerian, Greek and Roman beliefs (Zeuner, 1963; Cocker and Mabey, 2005), commonly associated with fertility (Zeuner, 1963). For example, in ancient Rome and Greece, they were associated with Aphrodite and their fat was used as an aphrodisiac (Cocker and Mabey, 2005). As sacred animals, they were kept in the temple of Juno on Capitoline Hill in Rome, when in 390 bc they alerted guards to an attack by Gaul (Riddell, 1943; Zeuner, 1963; Cocker and Mabey, 2005). Reminders of their historical importance persist in western societies as nursery rhymes (e.g. Goosey goosey gander), fairy tales, fables (e.g. the goose and the golden egg) and superstitions (e.g. pulling of the wishbone) (Cocker and Mabey, 2005).

29.13 References

Allan, J.R., Kirby, J.S. and Feare, C.J. (1995) The biology of Canada Geese Branta canadensis in relation to the management of feral populations. *Wildlife Biology* 1, 129–143.

Aubry, P., Anstett, L., Ferrand, Y., Reitz, F., Klein, F., *et al.* (2016) Enquête nationale sur les tableaux de chasse à tir. Saison 2013–2014 – Résultats nationaux. *Faune Sauvage* 310, Suppl. Available at: http://www.oncfs.gouv.fr/IMG/file/publications/revue%20faune%20sauvage/FS_310_enquete_tableau_de_chasse.pdf (accessed 1 November 2019).

Bainbridge, I. (2017) Goose management in Scotland: an overview. *Ambio* 46, 224–230.

Banks, A.N., Wright, L.J., Maclean, I.M.D., Hann, C. and Rehfisch, M.M. (2008) Review of the status of introduced non-native waterbird species in the area of the African–Eurasian Waterbird Agreement: 2007 Update. BTO Research Report No. 489. British Trust for Ornithology, Thetford, UK. Available at: https://www.unep-aewa.org/sites/default/files/document/tc8_25_review_non_native_species_0.pdf (accessed 1 November 2019).

Beauchamp, G. (1998) The relationship between intra- and interspecific brood amalgamation in waterfowl. *Condor* 100, 153–162.

Beaumont, M., Rodrigue, J., Pilotte, C., Chalifour, É. and Giroux, J.F. (2018) Behavioral response of Canada Geese to egg-oiling and nest removal. *Journal of Wildlife Management* 82, 1359–1366.

Beston, J.A., Williams, C.K., Nichols, T.C. and Castelli, P.M. (2016) A population model for management of Atlantic flyway resident population Canada Geese. *Wildlife Society Bulletin* 40, 106–111.

BirdLife International (2019) Greylag Goose, *Anser anser*. Available at: http://datazone.birdlife.org/species/factsheet/greylag-goose-anser-anser (accessed 1 November 2019).

Blair, M.J., McKay, H., Musgrove, A.J. and Rehfisch, M.M. (2000) Review of the status of introduced non-native waterbird species in the agreement area of the African-Eurasian Waterbird Agreement. BTO Research Report No. 229. British Trust for Ornithology, Thetford, UK. Available at: http://www.unep-aewa.org/sites/default/files/document/inf2_17_non-native_0.pdf (accessed 1 November 2019).

Boessneck, J. (1988) *Die Tierwelt des Alten Ägypten*. Beck, Munich, Germany.

Boyd, H. and Matthews, J. (1963) The control of wildfowl stocks. In: Atkinson-Willes, G.L. (ed.) *Wildfowl in Great Britain*. HMSO, London, pp. 329–333.

Brown, R., Ferguson, J., Lawrence, M. and Lees, D. (1987) *Tracks and Signs of the Birds of Britain and Europe*. Christopher Helm, London.

Buij, R., Melman, T.C., Loonen, M.J. and Fox, A.D. (2017) Balancing ecosystem function, services and disservices resulting from expanding goose populations. *Ambio* 46, 301–318.

Carboneras, C. and Kirwan, G.M. (2019) Greylag Goose (*Anser anser*). In: *Handbook of the Birds of the World*. Lynx Edicions, Barcelona, Spain. Available at: www.hbw.com/species/greylag-goose-anser-anser (accessed 4 November 2019).

Chang, C.-F., King, C.-C., Wan, C.-H., Chang, Y.-C., Chan, T.-C., *et al.* (2016) Lessons from the largest epidemic of Avian Influenza Viruses in Taiwan, 2015. *Avian Diseases* 60, 156–171.

Clements, J.F., Schulenberg, T.S., Iliff, M.J., Roberson, D., Fredericks, T.A., *et al.* (2018) The Clements Checklist of Birds of the World: v2018. Available at: www.birds.cornell.edu/clementschecklist/download/ (accessed 4 November 2019).

Cocker, M. and Mabey, R. (2005) *Birds Britannica*. Chatto & Windus, London.

Crawford, R.D. (1984) Goose. In: Mason, I.L. (ed.) *Evolution of Domesticated Animals*. Longman, London, pp. 345–349.

Crossley, R. and Couzens, D. (2011) *The Crossley ID Guide: Britain and Ireland*. Princeton University Press, Princeton, New Jersey.

Darwin, C. (1890) *The Variation of Animals and Plants Under Domestication*. Cambridge University Press, London.

del Hoyo, J., Elliot, A. and Sargatal, J. (1992) *Handbook of the Birds of the World*. Lynx Edicions, Barcelona, Spain.

Delacour, J. (1964) *The Wildfowl of the World*, Vol. 4. Hamlyn Publishing Group, London.

Dementiev, G. and Gladkov, N. (1952) *Birds of the Soviet Union*, Vol. IV: *Moscow*. Israel Program for Scientific Translations, Jerusalem.

Dingemans, B.J.J., Bakker, E.S. and Bodelier, P.L.E. (2011) Aquatic herbivores facilitate the emission of methane from wetlands. *Ecology* 92, 1166–1173.

Dolbeer, R. A. and Seubert, J.L. (2009) Canada Goose populations and strikes with civil aircraft, 1990–2008: challenging trends for aviation industry. Special report, US Department of Agriculture, Wildlife Services, Airport Wildlife Hazards Program, Washington, DC.

Donegan, T. (2007) Proposal (353) to South American Classification Committee: Move *Anser anser* (Graylag Goose) to main list. Available at: www.museum.lsu.edu/~Remsen/SACCprop353.htm (accessed 4 November 2019).

Elmberg, J., Berg, C., Lerner, H., Waldenström, J. and Hessel, R. (2017) Potential disease transmission from wild geese and swans to livestock, poultry and humans: a review of the scientific literature from a One Health perspective. *Infection Ecology and Epidemiology* 7, 1300450.

Esselink, P., Helder, G.J., Aerts, B.A. and Gerdes, K. (1997) The impact of grubbing by Greylag Geese (*Anser anser*) on the vegetation dynamics of a tidal marsh. *Aquatic Botany* 55, 261–279.

Evans, T., Kumschick, S. and Blackburn, T.M. (2016) Application of the Environmental Impact Classification for Alien Taxa (EICAT) to a global assessment of alien bird impacts. *Diversity and Distributions* 22, 919–931.

Fabricius, E. (1991) Interspecific mate choice following cross-fostering in a mixed colony of Greylag Geese (*Anser anser*) and Canada Geese (*Branta canadensis*). A study on development and persistence of species preferences. *Ethology* 88, 287–296.

Fabricius, E., Bylin, A., Fernö, A. and Radesäter, T. (1974) Intra- and interspecific territorialism in mixed colonies of the Canada Goose *Branta canadensis* and the Greylag Goose *Anser anser*. *Ornis Scandinavica* 5, 25–35.

Farrell, D. (2004) Management, nutrition and products of domestic geese: a review. In: *Proceedings of the Australian Poultry Science Symposium*, Sydney, Australia, 9–11 February 2004, pp. 139–144.

Florida Fish and Wildlife Conservation Commission (2019) Waterfowl. Available at: http://myfwc.com/wildlifehabitats/nonnatives/birds/ (accessed 4 November 2019).

Figuerola, J., Green, A.J., Black, K. and Okamura, B. (2004) Influence of gut morphology on passive transport of freshwater bryozoans by waterfowl in Donana (southwestern Spain). *Canadian Journal of Zoology* 82, 835–840.

Fox, A.D. and Kahlert, J. (2005) Changes in body mass and organ size during wing moult in non-breeding Greylag Geese *Anser anser*. *Journal of Avian Biology* 36, 538–548.

Fox, A.D., Kahlert, J. and Ettrup, H. (1998) Diet and habitat use of moulting Greylag Geese *Anser anser* on the Danish island of Saltholm. *Ibis* 140, 676–683.

Frigerio, D., Weiß, B.M., Scheiber, I.B.R. and Kotrschal, K. (2005) Active and passive social support in families of Greylag Geese (*Anser anser*). *Behaviour* 142, 1535–1557.

García-Álvarez, A., van Leeuwen, C.H., Luque, C.J., Hussner, A., Vélez-Martín, A., *et al.* (2015) Internal transport of alien and native plants by geese and ducks: an experimental study. *Freshwater Biology* 60, 1316–1329.

Hahn, S., Bauer, S. and Klaassen, M. (2008) Quantification of allochthonous nutrient input into freshwater bodies by herbivorous waterbirds. *Freshwater Biology* 53, 181–193.

Honka, J., Heino, M., Kvist, L., Askeyev, I., Shaymuratova, D., *et al.* (2018) Over a thousand years of evolutionary history of domestic geese from Russian archaeological sites, analysed using ancient DNA. *Genes* 9, 367.

Hutton, K. (2015) Aggressive, noisy interlopers... domestic geese. *The Beast*, Australia. Available at: https://thebeast.com.au/other/aggressive-noisy-interlopers-domestic-geese/# (accessed 4 November 2019).

Johnsgard, P.A. (2010) Ducks, geese, and swans of the world: tribe Anserini (swans and true geese). DigitalCommons@University of Nebraska–Lincoln, Nebraska. Available at: https://digitalcommons.unl.edu/biosciducksgeeseswans/5/ (accessed 4 November 2019).

Kahlert, J., Fox, A.D. and Ettrup, H. (1996) Nocturnal feeding in moulting Greylag Geese *Anser anser* – an anti-predator response. *Ardea* 84, 15–22.

Kleinhenz, A. and Koenig, A. (2018) Home ranges and movements of resident Graylag Geese (*Anser anser*) in breeding and winter habitats in Bavaria, South Germany. *PLoS One* 13, e0202443.

Klok, C., van Turnhout, C., Willems, F., Voslamber, B., Ebbinge, B. and Schekkerman, H. (2010) Analysis of population development and effectiveness of management in resident Greylag Geese *Anser anser* in the Netherlands. *Animal Biology* 60, 373–393.

Kotrschal, K., Hemetsberger, J. and Weiß, B.M. (2006) Making the best of a bad situation: homosociality in male Greylag Geese. In: Sommer, V. and Vasey, P.L. (eds) *Homosexual Behaviour in Animals. An Evolutionary Perspective*. Cambridge University Press, New York, pp. 45–76.

Lepage, D. (2004) Avibase: The World Bird Database, Bird Studies Canada. Available at: https://avibase.bsc-eoc.org (accessed 4 November 2019).

Loonen, M., Zijlstra, M. and van Eerden, M.R. (1991) Timing of wing moult in Greylag Geese *Anser anser* in relation to the availability of their food plants. *Ardea* 79, 253–260.

Lorenz, K. (1935) Der kumpan in der umwelt des vogels. *Journal of Ornithology* 83, 289–413.

MacMillan, D., Hanley, N. and Daw, M. (2004) Costs and benefits of wild goose conservation in Scotland. *Biological Conservation* 119, 475–485.

Madsen, J., Cracknell, G. and Fox, A.D.E. (1999) *Goose Populations of the Western Palearctic: A Review of Status and Distribution*. Wetlands International, Wageningen, The Netherlands/National Environmental Research Institute, Rønde, Denmark.

Maragakis, I. (2009) Bird population trends and their impact on aviation safety 1999–2008. Executive Directorate-Safety Analysis and Research Department, European Aviation Safety Agency. Available at: https://www.skybrary.aero/index.php/Bird_Population_Trends_and_Impact_on_Aviation_Safety (accessed 4 November 2019).

Matthews, G.V.T. and Campbell, C.R.G. (1969) Weights and measurements of Greylag Geese in Scotland. *Wildfowl* 20, 86–93.

Mitchell, C., Hearn, R. and Stroud, D. (2012) The merging of populations of Greylag Geese breeding in Britain *British Birds* 105, 498–505.

Mitchell, C., Leitch, A. and Meek, E. (2014) The abundance and distribution of British Greylag Geese in Orkney, August 2014. A report by the Wildfowl & Wetlands Trust to Scottish Natural Heritage, Slimbridge, UK. Available at: https://monitoring.wwt.org.uk/wp-content/uploads/2013/06/OrkneyGJ_2016report.pdf (accessed 4 November 2019).

Mulder, C.P. and Ruess, R. (2001) Long-term effects of changes in goose grazing intensity on arrowgrass populations: a spatially explicit model. *Journal of Ecology* 89, 406–417.

Newton, I. and Kerbes, R. (1974) Breeding of Greylag Geese (*Anser anser*) on the Outer Hebrides, Scotland. *Journal of Animal Ecology* 43, 771–783.

Ottenburghs, J., van Hooft, P., van Wieren, S.E., Ydenberg, R.C. and Prins, H.H.T. (2016) Hybridization in geese: a review. *Frontiers in Zoology* 13, 20.

Owen, M. and Salmon, D. (1988) Feral Greylag Geese *Anser anser* in Britain and Ireland, 1960–86. *Bird Study* 35, 37–45.

Palliser, T. (2014) Submission No 809: Greylag Goose *Anser anser* Nofolk Island. Free ranging and breeding for at least 20 years prior to January 2014. Birdlife Australia Rarities Committee, Australia. Available at: http://birdlife.org.au/documents/barc/SUMM809.htm (accessed 4 November 2019).

Porter, R. and Aspinall, S. (2013) *Birds of the Middle East*. Christopher Helm, London.

Powolny, T., Jensen, G.H., Nagy, S., Czajkowski, A., Fox, A.D., *et al.* (2018) International single species management plan for the Greylag Goose (*Anser anser*): northwest/southwest European population. AEWA Technical Series. African-Eurasian Migratory Waterbird Agreement (AEWA), Bonn, Germany. Available at: https://www.unep-aewa.org/en/publication/international-single-species-management-plan-greylag-goose-northwestsouthwest-european (accessed 4 November 2019).

Rehfisch, M.M., Austin, G.E., Holloway, S.J., Allan, J.R. and O'Connell, M. (2002) An approach to the assessment of change in the numbers of Canada Geese *Branta canadensis* and Greylag Geese *Anser anser* in southern Britain: population change in geese was assessed using an approach that requires a relatively small sampling effort. *Bird Study* 49, 50–59.

Remsen, J.V. Jr, Areta, J.I., Cadena, C.D., Claramunt, S., Jaramillo, A., *et al.* (2019) A classification of the bird species of South America. American Ornithological Society, Chicago, Illinois. Available at: www.museum.lsu.edu/~Remsen/SACCBaseline.htm (accessed 4 November 2019).

Reyns, N., Casaer, J., de Smet, L., Devos, K., Huysentruyt, F., *et al.* (2018) Cost–benefit analysis for invasive species control: the case of greater Canada Goose *Branta canadensis* in Flanders (northern Belgium) *PeerJ* 6, e4283.

Riddell, W.H. (1943) The domestic goose. *Antiquity* 17, 148–155.

Salaman, P., Bayly, N., Burridge, R., Grantham, M., Gurney, M., *et al.* (2008) Sixteen bird species new for Colombia. *Conservación Colombiana* 5, 80–85.

Scheiber, I.B.R., Hohnstein, A., Kotrschal, K. and Weiss, B.M. (2011) Juvenile Greylag Geese (*Anser anser*) discriminate between individual siblings. *PloS One* 6, e22853.

Schricke, V. (2018) L'oie cendrée *Anser anser*. In: *La Conservation et la Gestion des Oiseaux Chassables en Europe*. Le Statut des Espèces de l'Annexe IIA de la Directive Oiseaux. Paris, France.

Schwartz, K. and Woog, F. (2012) Limits of grazing area use by feral Greylag Geese *Anser anser* during moult. *Wildfowl* 62, 117–127.

Serjeantson, D. (2002) Goose husbandry in Medieval England, and the problem of ageing goose bones. *Acta Zoologica Cracoviensia* 45, 39–54.

Smith, V.H., Tilman, G.D. and Nekola, J.C. (1999) Eutrophication: impacts of excess nutrient inputs on freshwater, marine, and terrestrial ecosystems. *Environmental Pollution* 100, 179–196.

Southey, I. (2013) Greylag Goose. In: Miskelly, C.M. (ed.) *New Zealand Birds*. Available at: www.nzbirdsonline.org.nz/ (accessed 4 November 2019).

Strubbe, D., Shwartz, A. and Chiron, F. (2011) Concerns regarding the scientific evidence informing impact risk assessment and management recommendations for invasive birds. *Biological Conservation* 144, 2112–2118.

Tulloch, A.I., Nicol, S. and Bunnefeld, N. (2017) Quantifying the expected value of uncertain management choices for over-abundant Greylag Geese. *Biological Conservation* 214, 147–155.

van den Wyngaert, I.J.J., Wienk, L., Sollie, S., Bobbink, R. and Verhoeven, J.T.A. (2003) Long-term effects of yearly grazing by moulting Greylag Geese (*Anser anser*) on reed (*Phragmites australis*) growth and nutrient dynamics. *Aquatic Botany* 75, 229–248.

van der Wal, R., Kunst, P. and Drent, R. (1998) Interactions between hare and Brent Goose in a salt marsh system: evidence for food competition? *Oecologia* 117, 227–234.

van Eerden, M.R. (1990) The solution of goose damage problems in the Netherlands, with special reference to compensation schemes. *Ibis* 132, 253–261.

van Ham, C., Genovesi, P. and Scalera, R. (2013) *Invasive Alien Species: The Urban Dimension. Case Studies on Strengthening Local Action in Europe*. IUCN, Brussels.

Voller, J. and McNay, R.S. (2007) Problem analysis: Effects of invasive species on species at risk in British Columbia. FORREX Series 20, Forest Research Extension Society, Kamloops, Canada. Available at: http://wildlifeinfometrics.com/wp-content/uploads/2018/02/Voller_and_McNay_2007_Effects_of_invasive_species_on_species_at_risk.pdf (accessed 4 November 2019).

Voslamber, B., Knecht, E. and Kleijn, D. (2010) Dutch Greylag Geese *Anser anser*: migrants or residents. *Ornis Svecica* 20, 207–214.

Wascher, C.A.F., Kotrschal, K. and Arnold, W. (2018) Free-living Greylag Geese adjust their heart rates and body core temperatures to season and reproductive context. *Scientific Reports* 8, 2142.

Weiß, B.M., Kehmeier, S. and Schloegl, C. (2010) Transitive inference in free-living Greylag Geese, *Anser anser*. *Animal Behaviour* 79, 1277–1283.

Woods, R., Ingham, R. and Brown, A. (2009) Falkland Islands (Malvinas). In: Devenish, C., Díaz Fernández, D.F., Clay, R.P., Davidson, I. and Yépez Zabala, I. (eds) *Important Bird Areas Americas: Priority Sites for Biodiversity Conservation*. BirdLife Conservation Series No. 16. BirdLife International, Quito, Ecuador, pp. 205–212.

Woods, R.W. (1988) *Guide to Birds of the Falkland Islands*. Anthony Nelson, Oswestry, UK.

Young, J.G. (1972) Breeding biology of feral Greylag Geese in south-west Scotland. *Wildfowl* 23, 83–87.

Zeuner, F.E. (1963) *A History of Domesticated Animals*. Harper & Row Publishers, New York.

30 Canada Goose (*Branta canadensis* Linnaeus, 1758)

Tim Adriaens[1]*, Andrew Close[2], Peter A. Robertson[2], Jean-François Maillard[3], Matthieu Guillemain[4], Claire A. Pernollet[5] and Frank Huysentruyt[1]

[1]*Research Institute for Nature and Forest (INBO), Havenlaan 88 Bus 73, B-1000 Brussels, Belgium;* [2]*School of Natural and Environmental Sciences, Newcastle University, Newcastle NE1 7RU, UK;* [3]*Office Français de la Biodiversité (OFB), 44300 Nantes, France;* [4]*Office Français de la Biodiversité (OFB), Unité Avifaune Migratrice, La Tour du Valat, Le Sambuc, 13200 Arles, France;* [5]*Conservatoire d'espaces naturels de Provence-Alpes-Côte d'Azur, 4 avenue Marcel Pagnol, 13100 Aix-en-Provence, France*

Citation: Adriaens, T., Close, A., Robertson, P., Maillard, J.-F., Guillemain, M., Pernollet, C.A. and Huysentruyt, F. (2020) Canada Goose (*Branta canadensis* Linnaeus, 1758). In: Downs, C.T. and Hart, L.A. (eds) *Invasive Birds: Global Trends and Impacts*. CAB International, Wallingford, UK, pp. 223–231.

30.1 Common Names

Canada Goose, Greater Canada Goose.

30.2 Nomenclature

The Canada Goose (*Branta canadensis* Linnaeus, 1758) has several subspecies including *B. c. canadensis*, *B. c. interior*, *B. c. parvipes*, *B. c. moffiti*, *B. c. maxima*, *B. c. occidentalis* and *B. c. fulva* (Avibase, 2018). The subspecies of Canada Goose present in north-west Europe, New Zealand and Australia is *B. c. canadensis*.

30.3 Distribution

The Canada Goose has a large native range, breeding across tundra in much of Canada, Alaska and the USA, and parts of the northern USA, and wintering in southern North America, including Mexico (Fig. 30.1) (BirdLife International, 2019). Natural populations are migratory, wintering in the southern USA states and along the coasts of North America, while feral populations are mostly sedentary with limited dispersal movements and exchange between subpopulations. In much of the USA, introduced populations are now resident south of the normal breeding range. This resident component of the North American population increased dramatically during the 1990s (Dolbeer *et al.*, 2014). The Canada Goose has been widely introduced as an ornamental species and for hunting purposes. In Europe, the Canada Goose was originally introduced in the UK in the 17th century (Lever, 2005), and established populations now occur across much of Europe (Shirley, 2009). The species is mostly sedentary in Europe, although in Scandinavian and Dutch populations, seasonal migratory behaviour does occur, for example during moult (Cooleman *et al.*, 2005; Tanger and Voslamber, 2011; Voslamber, 2011).

Banks *et al.* (2008) provided a detailed account of the population status and trends in Europe and reported a general increase in numbers in all countries throughout northern and western Europe where established populations exist. There are signs of the species spreading gradually into southern and eastern Europe (Banks *et al.*, 2008). The largest numbers are present in the UK, Scandinavia (especially Sweden), the Netherlands, Germany and Belgium. Smaller populations are present in Ireland, Russia, Ukraine, Denmark, Latvia, France and Switzerland. The European breeding population, including the UK, has increased significantly over the past decades and was estimated at roughly 90,000–100,000 breeding pairs (based on Banks *et al.*, 2008). In the UK, assessments of population change in the Canada Goose are undertaken annually in the form of the Breeding Bird Survey (Harris *et al.*, 2018), and the annual wintering Wetland Bird Survey (Frost *et al.*, 2018) carried out by the British Trust for Ornithology. The Breeding Bird Survey

*Corresponding author: tim.adriaens@inbo.be

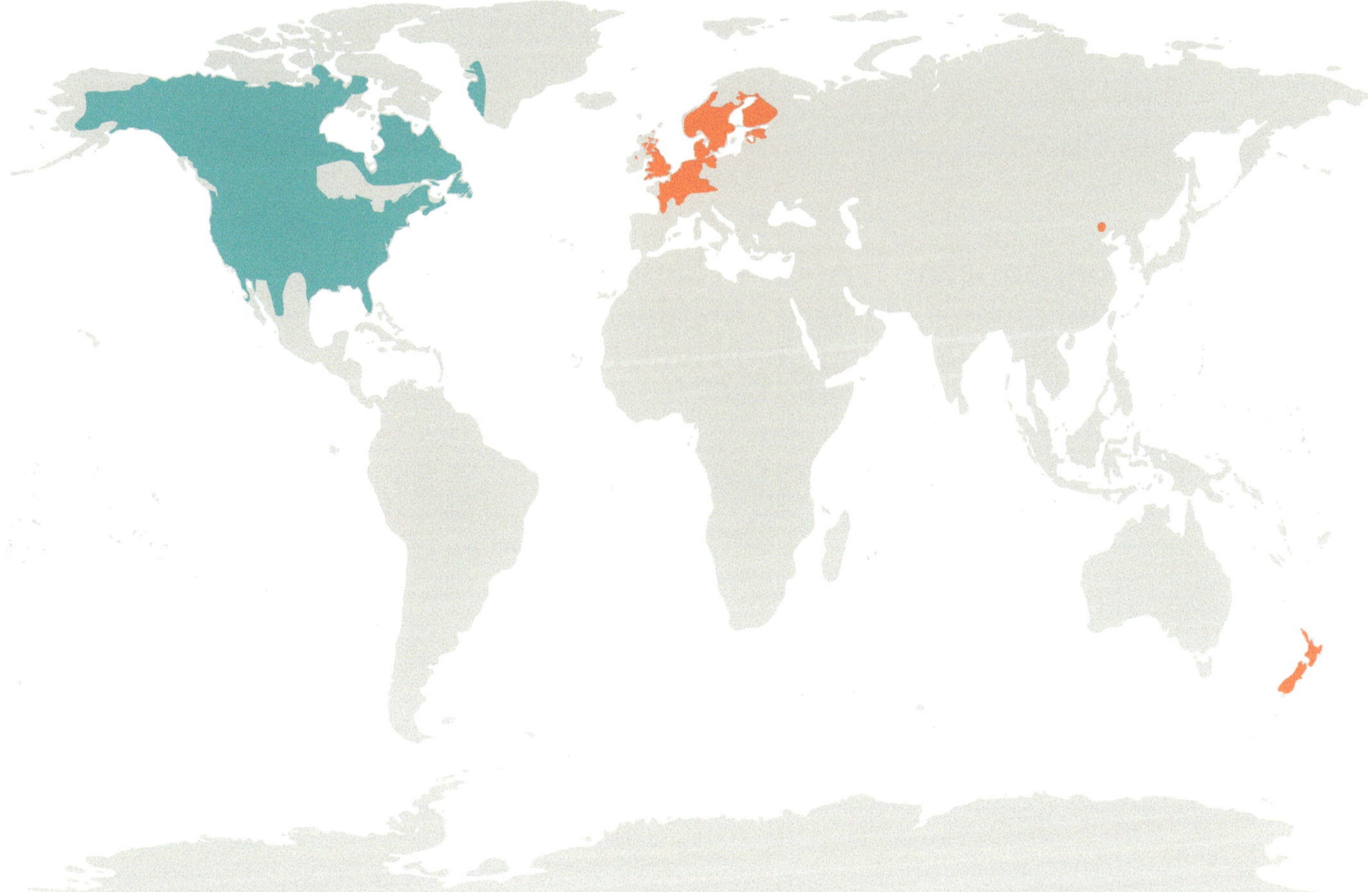

Fig. 30.1. Global distribution of the Canada Goose (*Branta canadensis canadensis*) showing its native (blue) and invasive (red) ranges.

examines trends in breeding bird numbers annually, at 10-year and 21-year temporal scales. During the 2016–2017 survey, the Canada Goose demonstrated a 34% increase in numbers. In contrast, an 8% decrease in breeding numbers was estimated during 2006–2016. However, longer-term population estimates describing the 21-year period 1995–2016 indicated an overall increase in breeding numbers of approximating 82% (Harris *et al.*, 2018). The wintering Wetland Bird Survey examines trends in bird numbers at 10-year and 25-year temporal scales (Frost *et al.*, 2018). At each scale, the numbers of wintering Canada Geese increased by 12% and 56%, respectively. Musgrove *et al.* (2013) used extant and published data sources and estimated the UK breeding and wintering populations to be 62,000 pairs and 190,000 individuals, respectively. In France, the number of wintering Canada Geese increased by 119% between 2009 and 2016, representing a 15% yearly increase, although the situation varies spatially from one department to another. Most wildlife conflicts with Canada Geese are observed in departments bordering Belgium where the increase in numbers was most considerable (Guillemain *et al.*, 2018).

Established populations also occur in Russia, Ukraine and New Zealand where the species was introduced in 1905 and is currently subject to a management plan (Spurr and Coleman, 2005). In Asia, the species is considered a vagrant (i.e. recorded sporadically, but not native to the area) to Japan and North and South Korea, and an introduced population is present in Beijing (BirdLife International, 2019).

30.4 Description

The Canada Goose is a large (55–110 cm in length, wingspan 122–183 cm) goose with an unmistakable dark grey-brown body, black head and neck, and white throat patch extending from under the chin up to behind the eye, white lower belly and tailcoverts, dark flight feathers and tail, a large black bill and black legs (Fig. 30.2) (Cramp, 1977). Juveniles are similar in appearance but have a greyer throat patch in their first year, while second-year birds are morphologically indistinguishable from adults. Its deep resonant honking call is very loud and easily recognizable. Chicks have olive-brown down above and are buffish below (del Hoyo *et al.*, 2011). A number of subspecies (*B. c. canadensis*, *B. c. interior*, *B. c. parvipes*, *B. c. moffiti*, *B. c. maxima*, *B. c. occidentalis*, *B. c. fulva*) have been described (Avibase, 2018) that differ in size, proportion, coloration and distribution area. The Canada Goose is now considered to be distinct from the Cackling Goose (*Branta hutchinsii*) and its subspecies (Sangster *et al.*, 2005), which are usually smaller and thinner and only rarely occur in the wild in Europe. The subspecies of Canada Goose present in north-west Europe is *B. c. canadensis*, a large bird that can weigh up to 5 kg. The sexes are similar, although males are generally larger and heavier (Cramp, 1977). Canada Geese are mostly gregarious, especially during moult, when large groups flock together on post-breeding moulting sites, mostly water bodies with adjacent foraging (grassland) territory and with at least some sloping shore where the chicks can get out.

30.5 Diet

The Canada Goose is essentially vegetarian, feeding on plant materials including grasses, roots, stems, rhizomes, leaves, fruits, seeds, and the green parts of aquatic plants and sedges. It also feeds on crops (grain such as wheat and maize, and legumes) and seaweed (del Hoyo *et al.*, 2011; Shirley, 2009). It feeds mostly by day by grazing on grassland and crops. On water, it feeds on rooted aquatic plants by dipping its head and upending. It will readily consume food provided by people in city parks (e.g. grain, bread).

30.6 Introduction and Invasion Pathways

The initial introductions of the Canada Goose in Europe occurred in 1665 where it formed a constituent part of the royal collection of Charles II (Lever, 2005). Birds were also introduced to New Zealand from about 1876. Although numbers in continental Europe remained low for a long time, during the 20th century there were multiple, poorly documented new releases and translocations to many parts of Europe, especially in Scandinavia, accompanied by natural range expansion as the new populations continued to grow (Banks *et al.*, 2008). Whereas early introductions were for ornamental or sentimental reasons, those during the 20th century were mainly for hunting purposes (CAB International, 2018). At a country level within Europe, pathways were a combination of importation from neighbouring countries, local unintentional escapes from zoos and collections, and natural dispersal across national borders (CAB International, 2018). Translocations within the native range in Canada and the USA have resulted in a massive expansion of urbanized, mainly resident populations, and have been largely responsible for a fourfold increase in the population in North America between 1970 and 2008 (Dolbeer and Seubert, 2009). Pooling the wild and introduced/feral individuals, the Canada Goose is now considered to be the most abundant goose species in the world (Rehfisch *et al.*, 2002). Based on ringing data, in north-west Europe, Canada Geese can undertake long-distance dispersal, such as during moult (Voslamber, 2011), or in winter from Scandinavia to Germany. Within the UK, non-breeding birds demonstrate a high degree of fidelity to their choice of moult-site (White and Combs, 2004). However, a proportion of the national population undertakes moult migration, mimicking the seasonal dispersal behaviours observed within some native North American populations (Davis *et al.*, 1985). Nevertheless, many subpopulations (e.g. in Belgium and the UK) are relatively sedentary, and evidence from North America suggests that site fidelity and the configuration of subpopulation groups are important mechanisms of social structuring among Canada Geese (Christensen *et al.*, 2004; Reed and Hughes, 2004; White and Combs, 2004). For instance, in the Netherlands, where most populations originated from escapes from waterfowl collections in parks and gardens (Lensink, 1996), birds primarily move locally for foraging, breeding and moulting, and their home ranges seldom exceed a 50 km radius (Cooleman *et al.*, 2005). Similarly, dispersal behaviour of UK birds varies both temporally and geographically, and ranges between 4 and 29 km/year (Austin *et al.*, 2004).

30.7 Breeding Behaviour

The Canada Goose breeds in parks, gardens, recreational areas, rural areas, often on islands, in single pairs or in breeding colonies (Fig. 30.3) (Cramp, 1977; Allan *et al.*, 1995; Smith *et al.*, 1999a).

Fig. 30.2. An adult Canada Goose. (©Photograph: Vildaphoto, Yves Adams.)

Fig. 30.3. A Canada Goose nest with chicks. (©Photograph: Vildaphoto, Yves Adams.)

Breeding colonies are often located on islands that are inaccessible to predators such as the red fox (*Vulpes vulpes*). The birds are territorial and defend the nesting zone against other geese, waterfowl and even people who approach too close. Nests from the previous year are often reused (Cramp, 1977; Converse and Kennelly, 1994; Smith *et al.*, 1999a) and many Canada Geese return to their natal area homelands to breed during later years (Allan *et al.*, 1995). Canada Geese lay one clutch per season and clutches usually contain four to seven eggs (range one to 12, more than eight usually by dump laying) (Cramp, 1977; Allan *et al.*, 1995; Smith *et al.*, 1999a). Breeding starts in spring (onset of breeding usually March–April) and the incubation period is 26–28 days (Bellrose, 1976; Cramp, 1977; Gosser *et al.*, 1997; Smith *et al.*, 1999a). In New Zealand, where the species breeds in loose colonies in lowland areas and is more dispersed in the high country, the clutch size is most commonly four to five (up to 10) eggs, incubation is 28 days and fledging occurs after about 80 days (Holloway *et al.*, 1997; Spurr and Coleman, 2005). Following the early loss of a nest, a replacement nest is often started in the vicinity of the former one (Cramp 1977; Christens *et al.*, 1995; Smith *et al.*, 1999a). Elsewhere, juveniles are usually ready to fly after about 6–9 weeks and stay with their parents during the first autumn and winter (Harrison, 1977; Allan *et al.*, 1995, Gosser *et al.*, 1997). About 20% of males start breeding at 2 years old, and almost all males are breeding by 3 years old. On average, the first clutch is laid by females of 3 years of age (Bellrose 1976). However, sparse data from the Netherlands suggest that Canada Geese could start breeding at a younger age (from 1 year old; Voslamber, 2011). Mortality is 45% in the first year and 30% annually thereafter (Williams, 1981): 20% from hunting and 10% from natural mortality (Heather and Robertson, 1996). Powell *et al.* (2004) calculated a survival based on ringing data for a Nebraska population of 62% in the first year and 70% for adult birds. Survival also depends on the areas of origin. In urban environments in the USA, a survival rate of 77% (starting from first moulting episode) to 90% after the first hunting season was reported (Johnson and Sibly, 1991; Smith *et al.*, 1999a), in contrast to a mean survival rate of 59% in the first year for animals from rural areas (25–84%; Samuel *et al.*, 1990; Smith *et al.*, 1999a). Balmer *et al.* (1997) estimated the annual adult and juvenile survivorship rates of UK Canada Geese to be 72.4 % and 75.4%, respectively.

30.8 Habitat

The Canada Goose is a very adaptable species, occurring in a diversity of habitats. It thrives well in (peri-)urban areas and agricultural areas that provide food and have few natural predators. It is generally bound to grasslands, pastures and arable land for foraging, and to water bodies for foraging, roosting and moulting. Canada Geese breed on the ground in all kinds of wetlands, usually less than 50 m away from water (Bellrose, 1976). The nests are well hidden and are often located on islands where there is better protection against predators (e.g. foxes, mink).

30.9 Impacts

The Canada Goose is listed among the worst invasive alien species threatening biodiversity in Europe. Impact scoring, using established impact assessment protocols, has shown Canada Geese to have the highest actual environmental impact and also the highest impact on the economy of 26 established non-native bird species in Europe (Kumschick and Nentwig, 2010). In many countries, Canada Geese rank among the top in human–wildlife conflict (e.g. Hughes *et al.*, 1999). Impacts of Canada Geese include eutrophication of water bodies, damage to agriculture, animal and human health impacts, damage (fouling) to recreational areas and an increased risk of bird strikes (Dolbeer and Seubert, 2009; Maragakis, 2009; van Ham *et al.*, 2013). Worldwide, non-native Anseriformes (ducks, geese and swans) mostly have an impact through hybridization and herbivory (Evans *et al.*, 2016), although concerns have been raised that the evidence for some impacts is often based on anecdotal observations relating to small areas (Strubbe *et al.*, 2011).

Impacts and damage by Canada Geese are multifaceted. They include crop damage, food competition with livestock, eutrophication of ponds and fens, overgrazing, fouling and trampling of vegetation such as reed beds and meadows, soil and water pollution, pathogen transmission (Converse *et al.*, 1999) and spread of wildlife pathogens (Garmyn *et al.*, 2012). Competition and hybridization with native species can also be an issue, although there is little evidence that this is affecting indigenous species (Lever, 2005). The presence of flocks of geese can also hamper costly nature restoration projects because of nutrient enrichment through their excreta/faeces, notably nitrogen and phosphorous (Smith *et al.*, 1999b; van Ham *et al.*, 2013).

In rural areas, Canada Geese show a strong preference for pasture, especially newly sown grass, and compete for it directly with livestock. They also foul pasture with their droppings, and these deter grazing by sheep and cattle. In urban areas, goose droppings foul parks, water bodies and sports fields, and may constitute a public disease risk. Their presence around airports considerably increases the risk of bird strikes. Because of their large size and flocking behaviour (Allan *et al.*, 1995) the species has become of primary concern for airport avifauna management in north-west Europe (Maragakis, 2009; Rehfisch *et al.*, 2010). For example, in the UK between 2008 and 2013, the estimated cost of Canada Goose strikes was £107,000/year (corresponding to about four strikes per year), but this is probably underestimated as it does not include costs to military aviation, and costs for bird-strike prevention such as investments to control local populations around airports or disperse birds from adjacent areas, or to pay farmers to discontinue agricultural practices that attract these birds to the airport vicinity (Dave Cowan, UK Animal and Plant Health Agency, personal communication).

Canada Geese are a host for the highly pathogenic avian influenza virus H5N1 (Pasick *et al.*, 2007) and are therefore a potential public health threat (Feare, 2010). Concerns have also been raised over the dissemination and transmission of several other viral and bacterial diseases of human or agricultural importance by Canada Geese, including Newcastle disease virus of concern to poultry, and coliform bacteria such as *Escherichia coli* (Clark, 2003). However, currently there seems to be no conclusive evidence that Canada Geese transmit these pathogens to humans, and in general, more data on excreta/faecal pathogenic loadings (Feare *et al.*, 1999) and transmission and exposure of humans and wildlife are needed to better assess the risks (Allan *et al.*, 1995).

The economic impact of Canada Geese occurs mainly through crop damage, either directly by feeding on crops, or by feeding on crops and pasture intended for other animals. In Belgium, agricultural damage is especially evident on parcels with winter wheat and on temporary or permanent grasslands (Huysentruyt *et al.*, 2010). Because of their size and tendency to aggregate, consumption of crops is often combined with trampling of vegetation and soil, which can cause soil compaction and impact on soil quality and the germination of crops through the formation of a 'hard pan' that prevents new growth (Conover, 1991). It is generally acknowledged that calculating the total cost of damage by geese is difficult, as individual farmers can be disproportionately affected due to the high degree of gregariousness of the birds and a tendency to repeatedly utilize individual fields (Kirby *et al.*, 1998). The level of damage is also dependent on the timing of grazing (Huysentruyt *et al.*, 2010). However, Williams *et al.* (2010) stated that, in the EU, Canada Geese cause yield losses of 0–56% on cereals and 0–40% on grass, and they used a weighted average of damage per goose of £14.90 as a damage cost for all geese species, which resulted in an estimated £1,324,103 of damage caused annually by Canada Geese based on a UK population of 88,866 in 2000. Simpson (1991) cited instances of yield losses in the UK on winter cereals continuously grazed by Canada Geese at 20%.

Soil and water pollution also cause management costs for maintaining areas suitable for recreation. Some people enjoy seeing geese in urban areas, while others consider them a nuisance. Large flocks leave behind large amounts of excreta/faecal material. Defecations can reduce the water quality of (swimming) ponds and can spoil lawns of (sub)urban parks, golf courses, apartment complexes, etc. In addition, geese are attracted by open expanses of grasses, such as runways of airports. Because they are large and flocking, Canada Geese represent a particular human safety hazard by increasing the possibility of goose–plane collisions, and the financial and safety risks associated with their presence near airports are high (Allan *et al.*, 1995; Allan, 2000a,b; Allan and Orosz, 2001; Rehfisch *et al.*, 2010).

In Belgium, several case studies have shown that the presence of Canada Geese can compromise the outcome of nature restoration projects (e.g. Anselin and Devos, 2007). In particular, restoration efforts of mesotrophic and oligotrophic fens, transition mires, lowland hay meadows, natural eutrophic lakes, Natura 2000 habitats in Europe, are affected. Although rarely backed by scientific data, an impact on local fauna has also been suggested through competition for food and space. More specifically, concern has been raised about the direct aggression of Canada Geese towards other breeding bird species through their strong territorial behaviour. Where high breeding concentrations occur, this could locally prevent smaller water birds from establishing territories. Circumstantial evidence also suggests that meadow birds such as Black-tailed Godwits (*Limosa limosa*), for which specific conservation schemes exist in many countries, might be affected (G. Spanoghe, personal communication).

30.10 Control

Canada Geese can be actively managed through fertility reduction or through culling, which involves shooting and/or capturing flocks of geese during the moult when they are flightless (Allan *et al.*, 1995; Smith *et al.*, 1999a; Reyns *et al.*, 2018). Although Canada Goose problems are often tackled locally, especially in urban areas, few management programmes exist on a regional or higher (national or international) spatial scale.

Prevention of hatching by pricking eggs or coating them with paraffin is practised in the UK (Baker *et al.*, 1993), as well as on the European mainland (van Daele *et al.*, 2012). Several good best practices are available to apply this method (e.g. USDA/APHIS, 2009; Titchenell and Lynch, 2010; Food and Environmental Research Agency, 2011; Maillard and Hurel, 2017). Although the method may be practical and desirable to slow down population growth locally, for highly developed populations of long-living species such as the Canada Goose, reducing the number of eggs has a small effect on the increase in numbers (Rockwell *et al.*, 1997; Schekkerman *et al.*, 2000; van der Jeugd *et al.*, 2006; Klok *et al.*, 2010). The success of this measure depends to a great extent on the effort, and model-based approaches show that it must be sustained over several years (Beston *et al.*, 2016; Beaumont *et al.*, 2018) and up to 88% of the nests must be found and treated (cf. Klok *et al.*, 2010, for the Greylag Goose, *Anser anser*). In practice, the method is difficult to apply at higher spatial scales and the actual implication can be difficult to harmonize. For pricking, the timing of egg treatment is essential as the species can produce replacement clutches, for example when eggs are destroyed or removed from the nest too early in incubation or when the entire nest is destroyed. Nest destruction can also result in females dispersing away from the nesting site (Beaumont *et al.*, 2018). The problem of replacement clutches can partly be avoided using egg oiling as a method. Mineral oil stops embryo development through asphyxia. Beaumont *et al.* (2018) showed that egg oiling at 3 ml of oil per egg or 20 ml for a mean clutch of six eggs was efficient in reducing the annual reproductive output of Canada Geese on breeding islands in Quebec, Canada. Oiling resulted in females that remained on their nests after treatment and even after the expected hatching date for an average of 18 days. Thus, whereas nest removal may promote moult migration, egg oiling does not. This can be more practical when planning a moult capture at the same location, and maintaining the birds at their nest sites may lessen the risk of nuisance problems.

Canada Geese populations can be managed in an adaptive management approach, using an integrated strategy that involves hunting, fertility reduction (egg pricking, oiling, nest removal) and moult capturing (Fig. 30.4) (French and Parkhurst, 2009). Presumably, although management methods might influence each other (e.g. by inducing migratory or sedentary behaviour), combining different methods and managing

Fig. 30.4. Moult capture of Canada Geese. (©Photograph: Jan Rodts.)

at different scales and impact levels is a more effective management strategy. Canada Geese are highly susceptible to moult captures. For example, in Flanders, in north Belgium, and Zeeland in the southern part of the Netherlands, management through moult capturing was intensified and upscaled to the level of a 9000 km^2 region, with on average 2000 geese (or approximately 50% of the summer population) caught per year in the period 2009–2012. Although other goose species (Greylag Goose, Barnacle Goose and Upland Goose (*Chloephaga picta*)) were also caught, Canada Geese represented 87% of the birds caught in such captures (van Daele *et al.*, 2012). Summer census data of the regional population have shown a significant decrease in Canada Goose numbers since 2010 as a consequence (Huysentruyt *et al.*, 2013; Adriaens *et al.*, 2014). The estimated decline in the population was on average 11%/year in the period 2010–2018 (unpublished data). As a result, the estimated number of Canada Goose at the municipal level in Flanders has dropped by 50% since 2010. Cost–benefit analysis has shown this strategy of supplemental moult captures to be cost-effective (Reyns *et al.*, 2018). In France, the Canada Goose population increased from around 150 birds at the beginning of the 1990s (Dubois, 2007) to 6506 wintering birds in 2009 (ONCFS, 2010), and consequently became a legal game species in 2012 (Guillemain *et al.*, 2018). Over four hunting seasons, the national reported hunting bag ranged between 1768 birds during the 2012–2013 season and 2884 birds in 2014–2015 (mean ±SD of 2272 ± 482 birds); thus, 20% of the national wintering population were culled through hunting (by contrast, this was approximately one-third of the national Greylag Goose hunting bag) (Aubry *et al.*, 2016). The latest field surveys suggest a total of 13,490 wintering individuals in 2016, so culling alone was insufficient to decrease population numbers. In 2017, Luxembourg, Belgium and north-east French authorities convened to draft a common strategy to better collaborate on goose management. Following the adaptive management example of Canada Geese in Flanders, a pilot project was initiated by France and Wallonia, Belgium. A first moult capture was performed in June 2018 in the border region between France and Belgium (Guillemain *et al.*, 2018).

In general, when implementing an adaptive management strategy with stakeholders and interest groups, the debate on management choices needs information regarding expected population trends and what measures would have the most impact. Population models should be informed with data on breeding success, recruitment, mortality and survival, and high-quality data on the applied management, such as the number of birds culled through shooting and moult captures. Any strategy involving the culling of birds needs to be accompanied by measures to raise awareness and increase public and stakeholder support towards controlling goose problems.

Non-lethal preventive strategies to mitigate the impact of Canada Geese locally include discouraging and redistributing geese to alternative foraging sites, scaring, chemical anti-feedants and various forms of habitat management (e.g. Conover, 1992;

Melman *et al.*, 2011, Maillard and Hurel, 2017). Although these methods can mitigate damage locally (e.g. locally moving birds can be an effective way of reducing bird-strike risk), they do not represent population management and are mostly poorly effective in reducing damage as they simply shift the goose problems to other areas (Melman *et al.*, 2011; Tombre *et al.*, 2013; Nolet *et al.*, 2016; Simonsen *et al.*, 2016).

30.11 Uses

The Canada Goose is kept in wildfowl collections for ornamental purposes. Canada Geese also generate benefits through wildlife observation and hunting. In many countries, Canada Geese are a game species. The meat is used for consumption and several recipe books exist (e.g. Invexo, 2012).

30.12 References

Adriaens, T., Huysentruyt, F., Devisscher, S., Devos, K. and Casaer, J. (2014) Integrated management of invasive Canada geese populations in an international context: a case study. In: Uludağ, A., Yazlik, A., Jabran, K., Türkseve, S. and Starfinger, U. (eds) NEOBIOTA 2014, *Proceedings of the 8th International Conference on Biological Invasions: From Understanding to Action*, 3–8 November, Antalya, Turkey. Available at: www.cabi.org/isc/FullTextPDF/2017/20173063613.pdf (accessed 5 November 2019).

Allan, J. (2000a) A protocol for bird strike risk assessment at airports. *International Bird Strike Proceedings* 25, 29–46.

Allan, J.R. (2000b) The costs of bird strikes and bird strike prevention. In: *USDA National Wildlife Research Center Symposia: Human Conflicts with Wildlife: Economic Considerations*, pp. 147–153

Allan, J.R., and Orosz, A.P. (2001) The costs of birdstrikes to commercial aviation. In: *2001 Bird Strike Committee-USA/Canada, Third Joint Annual Meeting, Calgary, AB*, p. 2.

Allan, J.R., Kirby, J.S. and Feare, C.J. (1995) The biology of Canada geese *Branta canadensis* in relation to the management of feral populations. *Wildlife Biology* 1, 129–143.

Anselin, A. and Devos, K. (2007) Review of the status of introduced non-native waterbirds and their effects in Flanders, Belgium. INBO.A.2007.115. Instituut voor Natuur- en Bosonderzoek, Brussels.

Aubry, P., Anstett, L., Ferrand, Y., Reitz, F., Klein, F., *et al.* (2016) Enquête nationale sur les tableaux de chasse à tir. Saison 2013–2014 – Résultats nationaux. *Faune Sauvage* 310, I–VIII.

Austin, G.E., Belman, P.J. and McMeeking, J. (2004) Canada Goose. In: Wernham, C., Toms, M., Marchant, J., Clark, J., Siriwardena, G. and Baillie, S. (eds) *The Migration Atlas: Movements of the birds of Britain and Ireland*. British Trust for Ornithology. T. and A.D. Poyser, London.

Avibase (2018) Canada Goose: *Branta canadensis* (Linnaeus, 1758). Available at: https://avibase.ca/B59E1863 (accessed 5 November 2019).

Baker, S., Feare, C., Wilson, C., Malam, D. and Sellars, G. (1993) Prevention of breeding of Canada geese by coating eggs with liquid paraffin. *International Journal of Pest Management* 39, 246–249.

Balmer, D.E., Peach, W.J. and Norfolk, I.P. (1997) Review of natural avian mortality rates. BTO Research Report No. 175, British Trust for Ornithology, Thetford, UK.

Banks, A.N., Wright, L.J., Maclean, I.M.D., Hann, C. and Rehfisch, M.M. (2008) Review of the status of introduced non-native waterbird species in the area of the African-Eurasian Waterbird Agreement: 2007 update. BTO Research Report No. 489, British Trust for Ornithology, Thetford, UK.

Beaumont, M., Rodrigue, J., Pilotte, C., Chalifour, É. and Giroux, J.F. (2018) Behavioral response of Canada geese to egg-oiling and nest removal. *Journal of Wildlife Management* 82, 1359–1366.

Bellrose, F.C. (1976) *Ducks, Geese, and Swans of North America*. Stackpole Books, Harrisburg, Pennsylvania.

Beston, J.A., Williams, C.K., Nichols, T.C. and Castelli, P.M. 2016. A population model for management of Atlantic flyway resident population Canada geese. *Wildlife Society Bulletin* 40, 106–111.

BirdLife International (2019) Species factsheet: *Branta canadensis*. Available at: http://www.birdlife.org (accessed 6 November 2019).

CAB International (2018) *Branta canadensis* (Canada Goose) [original text by John Marchant]. In: *Invasive Species Compendium*. CAB International, Wallingford, UK. Available at: www.cabi.org/isc/datasheet/91754 (accessed 5 November 2019).

Christens, E., Blokpoel, H., Rason, G. and Jarvie, S.W.D. (1995) Spraying white mineral oil on Canada Goose eggs to prevent hatching. *Wildlife Society Bulletin* 23, 228–230

Christensen, C.J., Combs, D.L., Sykes, R.E. and Mukherjee, K. (2004) Social biology in two flocks of Giant Canada Geese in Tennessee. In: Moser, T.J., Lien, R.D., ver Cauteren, K.C., Abraham, K.F., Andersen, D.E., *et al.* (eds) *Proceedings of the 2003 International Canada Goose Symposium*, Madison, Wisconsin.

Clark, L. (2003) A review of pathogens of agricultural and human health interest found in Canada geese. USDA National Wildlife Research Center-Staff Publications: 205. Available at: https://digitalcommons.unl.edu/icwdm_usdanwrc/205 (accessed 5 November 2019).

Conover, M.R. (1991) Herbivory by Canada geese: diet selection and effect on lawns. *Ecological Applications* 1, 231–236.

Conover, M.R. (1992) Ecological approach to managing problems caused problems caused by urban Canada geese. In: *Proceedings of the 15th Vertebrate Pest Conference 1992*. Available at: https://digitalcommons.unl.edu/vpc15/19/ (accessed 5 November 2019).

Converse, K., Wolcott, M., Douchety, D. and Cole, R. (1999) Screening for potential human pathogens in fecal material deposited by resident Canada Geese on areas of public utility. USGS National Wildlife Health Center, Madison, Wisconsin.

Converse, K.A. and Kennelly, J.J. (1994) Evaluation of Canada goose sterilization for population control. *Wildlife Society Bulletin* 22, 265–69.

Cooleman, S., Anselin, A., Beck, O., Kuijken, E. and Lens, L. (2005) Verplaatsingen en mortaliteit van Canadese Ganzen *Branta canadensis* in Vlaanderen. *Natuur Oriolus* 71, 152–160.

Cramp, S. (1977) *Handbook of the Birds of Europe, the Middle East, and North Africa: The birds of the Western Palearctic*, Vol. 1: *Ostrich to Ducks*. Oxford University Press, Oxford.

Davis, R.A., Jones, R.N., MacInnes, C.D. and Pakulak, A.J. (1985) Moult migration of large Canada Geese on the West Coast of Hudson Bay. *Wilson Bulletin* 97, 296–305.

del Hoyo, J., Elliott, A. and Sargatal, J. (2011) *Handbook of the Birds of the World*, Vol. 1: *Ostrich to Ducks*. Lynx Edicions, Barcelona, Spain.

Dolbeer, R. A. and Seubert, J.L. (2009) Canada goose populations and strikes with civil aircraft, 1990–2008: challenging trends for aviation industry. Special report, US Department of Agriculture, Wildlife Services, Airport Wildlife Hazards Program, Washington, DC.

Dolbeer, R.A., Seubert, J.L. and Begier, M.J. (2014) Population trends of resident and migratory Canada geese in relation to strikes with civil aircraft. *Human–Wildlife Interactions* 8, 88–99..

Dubois, J.P. (2007) Les oiseaux d'eau allochtones en France: statut et interactions avec les espèces indigènes. *Ornithos* 14, 329–364.

Evans, T., Kumschick, S. and Blackburn, T.M. (2016) Application of the Environmental Impact Classification for Alien Taxa (EICAT) to a global assessment of alien bird impacts. *Diversity and Distributions* 22, 919–931.

Feare, C.J. (2010) Role of wild birds in the spread of highly pathogenic avian influenza virus H5N1 and implications for global surveillance. *Avian Diseases* 54, 201–212.

Feare, C.J., Sanders, M., Blasco, R. and Bishop, J. (1999) Canada goose (*Branta canadensis*) droppings as a potential source of pathogenic bacteria. *Journal of the Royal Society for the Promotion of Health* 119, 146–155.

Food and Environmental Research Agency (2011) The management of problems caused by Canada geese: a guide to best practice. Natural England Technical Information Note TIN009. Food and Environmental Research Agency, York, UK.

French, L. and Parkhurst, J. (2009) Managing wildlife damage: Canada goose (*Branta canadensis*). Virginia State University, Petersburg, Virginia. Available at: www.pubs.ext.vt.edu/420/420-203/420-203.html (accessed 5 November 2019).

Frost, T.M., Austin, G.E., Calbrade, N.A., Mellan, H.J., Hearn, R.D., *et al.* (2018) Waterbirds in the UK 2016/17: The Wetland Bird Survey. BTO, RSPB and JNCC, in association with WWT, British Trust for Ornithology, Thetford, UK.

Garmyn, A., van Rooij, P., Pasmans, F., Hellebuyck, T., van den Broeck, W., *et al.* (2012) Waterfowl: potential environmental reservoirs of the chytrid fungus *Batrachochytrium dendrobatidis*. *PLoS One* 7, e35038.

Gosser, A.L., Conover, M.R. and Mesmer, T.A. (1997) Managing problems caused by urban Canada geese. Berryman Institute, Utah State University, Logan, Utah. Available at: www.berrymaninstitute.org/files/geese.pdf (accessed 5 November 2019).

Guillemain, M., Maillard, J.F., Pernollet, C.A. and Arnauduc, J.P. (2018) La bernache du Canada en France: des pistes pour limiter ses nuisances et ses populations. *Faune Sauvage* 321, 64–68.

Harris, S.J., Massimino, D., Gillings, S., Eaton, M.A., Noble, D.G., *et al.* (2018) The Breeding Bird Survey 2017. BTO Research Report No. 706, British Trust for Ornithology, Thetford UK.

Harrison, C. (1977) *Les Nids, les Oeufs et les Poussins d'Europe en Couleurs*. Elsevier Sequoia, Paris.

Heather, B.D. and Robertson, H.A. (1996) *The Field Guide to the Birds of New Zealand*. Penguin Books, Auckland, New Zealand.

Holloway, J.D.R., Scott, R.R. and Wilson, K.-J. (1997) Management of the Canada goose in the South Island of New Zealand. Unpublished report, Lincoln University, Lincoln, New Zealand.

Hughes, B., Kirby, J. and Rowcliffe, J. (1999) Waterbird conflicts in Britain and Ireland: ruddy ducks *Oxyura jamaicensis*, Canada geese *Branta canadensis*, and cormorants *Phalacrocorax carbo*. *Wildfowl* 50, 77–99.

Huysentruyt, F., Adriaens, T., Devisscher, S. and Casaer, J. (2013) Evaluation of a largescale management strategy for summering geese in Flanders and Zealand (Belgium/The Netherlands). In: *20th Annual Conference of the Wildlife Society*, Milwaukee, Wisconsin (poster).

Huysentruyt, F., Devos, K. and Casaer, J. (2010) Het bepalen van mogelijke herkomstgebieden bij landbouwschade door overzomerende ganzen: Een eerste aanzet voor een modelmatige benadering. Rapporten van het Instituut voor Natuur- en Bosonderzoek, 2010(9) Instituut voor Natuur- en Bosonderzoek, Brussels.

Invexo (2012) 'Gans en Frietje'. Recipe booklet produced by Inagro vzw in the framework of the Interreg IVa-project INVEXO [in Dutch].

Johnson, L.P. and Sibly, R.M. (1991) Pre-breeding behaviour affects condition, assessed by abdominal profile, and hence breeding success of Canada geese *Branta canadensis*. *Wildfowl* 42, 91–99.

Kirby, J.S., Owen, M. and Rowcliffe, J.M. (1998) *Geese and their Interactions with Agriculture and the Environment*. The Wildfowl and Wetlands Trust, UK.

Klok, C., van Turnhout, C., Willems, F., Voslamber, B., Ebbinge, B. and Schekkerman, H. (2010) Analysis of population development and effectiveness of management in resident greylag geese *Anser anser* in the Netherlands. *Animal Biology* 60, 373–393.

Kumschick, S. and Nentwig, W. (2010) Some alien birds have as severe an impact as the most effectual alien mammals in Europe. *Biological Conservation* 143, 2757–2762.

Lensink, R. (1996) De opkomst van exoten in de Nederlandse avifauna; verleden, heden en toekomst. *Limosa* 69, 103–130.

Lever, C. (2005) *Naturalised Birds of the World*. T. and A.D. Poyser, London.

Maillard, J.F. and Hurel, P. (2017) *Réduire les Nuisances Engendrées par la Bernache du Canada en Milieu Urbain*. Office National de la Chasse et de la Faune Sauvage, Paris.

Maragakis, I. (2009) Bird population trends and their impact on Aviation safety 1999–2008. Executive Directorate-Safety Analysis and Research Department, European Aviation Safety Agency, Cologne, Germany.

Melman, T.C.P., de Lange, H.J. and Clerkx, A.P.P.M. (2011) QuickScan effectiviteit van het weren en verjagen van ganzen. Wageningen Environmental Research (Alterra) Report No. 2251, Wageningen, Netherlands.

Musgrove, A.J., Aebischer, N.J., Eaton, M.A., Hearn, R.D., Newson, S.E., *et al.* (2013) Population estimates of birds in Great Britain and the United Kingdom. *British Birds* 106, 64–100.

Nolet, B.A., Kölzsch, A., Elderenbosch, M. and Noordwijk, A.J. (2016) Scaring waterfowl as a management tool: how much more do geese forage after disturbance? *Journal of Applied Ecology* 53, 1413–1421.

ONCFS (2010) Note Technique. La Bernache du Canada *Branta canadensis*. Office National de la Chasse et de la Faune Sauvage, Paris. Available at: www.oncfs.gouv.fr/IMG/Bernache.pdf (accessed 5 November 2019).

Pasick, J., Berhane, Y., Embury-Hyatt, C., Copps, J., Kehler, H., *et al.* (2007) Susceptibility of Canada geese (*Branta canadensis*) to highly pathogenic avian influenza virus (H5N1). *Emerging Infectious Diseases* 13, 1821–1827.

Powell, L.A., Vrtiska, M.P. and Lyman, N. (2004) Survival rates and recovery distributions of Canada geese banded in Nebraska. In: Moser, T.J., Lien, R.D., Vercauteren, K.C., Abraham, K.F., Andersen, D.E., *et al.* (eds). *Proceedings of the 2003 International Canada Goose Symposium*, Madison, Wisconsin.

Reed, E.T. and Hughes, R.J. (2004) Survival and breeding philopatry of Atlantic population Canada Geese. In: Moser, T.J., Lien, R.D., ver Cauteren, K.C., Abraham, K.F., Andersen, D.E., *et al.* (eds) *Proceedings of the 2003 International Canada Goose Symposium*, Madison, Wisconsin.

Rehfisch, M.M., Allan, J.R. and Austin, G.E. (2010) The effect on the environment of Great Britain's naturalized Greater Canada *Branta canadensis* and Egyptian Geese *Alopochen aegyptiacus*. In: *Proceedings of the British Ornithologists' Union 2008 – The Impacts of Non-native Species.*

Rehfisch, M.M., Austin, G.E., Holloway, S.J., Allan, J.R. and O'Connell, M. (2002) An approach to the assessment of change in the numbers of Canada Geese *Branta canadensis* and Greylag Geese *Anser anser* in southern Britain: population change in geese was assessed using an approach that requires a relatively small sampling effort. *Bird Study* 49, 50–59.

Reyns, N., Casaer, J., de Smet, L., Devos, K., Huysentruyt, F., *et al.* (2018) Cost–benefit analysis for invasive species control: the case of greater Canada goose *Branta canadensis* in Flanders (northern Belgium). *PeerJ* 6, e4283.

Rockwell, R.F., Cooch, E.G. and Brault, S. (1997) High goose populations: causes, impacts and implications. In: Batt B.D.J. (ed.) *Arctic Ecosystems in Peril.* Arctic Goose Joint Venture Special Publication, US Fish and Wildlife Service, Washington, DC, and Canadian Wildlife Service, Ottawa, Ontario, pp. 73–100.

Samuel, M.D., Rusch, D.H. and Craven, S. (1990) Influence of neck bands on recovery and survival rates of Canada geese. *Journal of Wildlife Management* 54, 45–54.

Sangster, G., Collinson, J.M., Helbig, A.J., Knox, A.G. and Parkin, D.T. (2005) Taxonomic recommendations for British birds: third report. *Ibis* 147, 821–826.

Schekkerman, H., Klok, C., Voslamber, B., van Turnhout, C., Willems, F. and Ebbinge, B.S. (2000) Overzomerende Grauwe Ganzen in het noordelijk Deltagebied. Een modelmatige benadering van de aantalsontwikkeling bij verschillende beheerscenario's. Wageningen Environmental Research (Alterra) Report No. 139/ SOVON-onderzoeksrapport 2000/06. SOVON Vogelonderzoek Nederland, Beek-Ubbergen, Netherlands.

Shirley, S. (2009) *Branta canadensis* (Linnaeus), Canada goose (Anatidae, Aves). In: *DAISIE (Delivering Alien Invasive Species Inventories for Europe). Handbook of Alien Species in Europe*, Invading Nature – Springer Series in Invasion Ecology, Vol. 1. Springer Netherlands, Dordrecht, Netherlands.

Simonsen, C.E., Madsen, J., Tombre, I.M. and Nabe-Nielsen, J. (2016) Is it worthwhile scaring geese to alleviate damage to crops? An experimental study. *Journal of Applied Ecology* 43, 916–924.

Simpson, W. (1991) Agricultural damage and its prevention. In: *Canada Geese Problems and Management Needs.* British Association for Shooting and Conservation, Rossett, UK, pp. 21–24.

Smith, A.E., Craven, S.R. and Curtis, P.D. (1999a) *Managing Canada Geese in Urban Environments.* Jack Berryman Institute and Cornell University Cooperative Extension, Ithaca, New York.

Smith, V.H., Tilman, G.D. and Nekola, J.C. (1999b) Eutrophication: impacts of excess nutrient inputs on freshwater, marine, and terrestrial ecosystems. *Environmental Pollution* 100, 179–196.

Spurr, E.B. and Coleman, J.D. (2005) Review of Canada Goose population trends, damage, and control in New Zealand. Landcare Research Science Series No. 30, Lincoln, Canterbury, New Zealand.

Strubbe, D., Shwartz, A. and Chiron, F. (2011) Concerns regarding the scientific evidence informing impact risk assessment and management recommendations for invasive birds. *Biological Conservation* 144, 2112–2118.

Tanger, D. and Voslamber, B. (2011) Ruitrek van Canadese ganzen over Nederland in relatie tot de veiligheid rond vliegvelden. *De Grauwe Gors* 39, 135–137.

Titchenell, M.A. and Lynch, W.E. Jr (2010) Coping with Canada geese: conflict management and damage prevention strategies. Ohio State University Extension, Columbus, Ohio. Available at: https://ohioline.osu.edu/factsheet/W-3 (accessed 04 February 2014).

Tombre, I.M., Eythórsson, E. and Madsen, J. (2013) Towards a solution to the goose-agriculture conflict in North Norway, 1988–2012: the interplay between policy, stakeholder influence and goose population dynamics. *PLoS One* 8, e71912.

USDA/APHIS (2009) Management of Canada goose nesting. USDA Wildlife Services, Washington, DC. Available at: www.aphis.usda.gov/wildlife_damage/downloads/canada_goose.pdf (accessed 5 November 2019).

van Daele, P., Adriaens, T., Devisscher, S., Huysentruyt, F., Voslamber, B., *et al.* (2012) Beheer van Zomerganzen in Vlaanderen en Zeeuws-Vlaanderen – Rapport opgesteld in het kader van het INVEXO INTERREG project. INBO.R.2012.58. Instituut voor Natuur- en Bosonderzoek, Brussels.

van der Jeugd, H., Voslamber, B., van Turnhout, C., Sierdsema, H., Feige, N., *et al.* (2006) Overzomerende ganzen in Nederland: grenzen aan de groei? Sovon-onderzoeksrapport 2006/02, Sovon Vogelonderzoek, Beek-Ubbergen, Netherlands.

van Ham, C., Genovesi, P. and Scalera, R. (2013) Invasive alien species: the urban dimension – case studies on strengthening local action in Europe. International Union for Conservation of Nature, Brussels.

Voslamber, B. (2011) Canadese Ganzen in Groningen: herkomst ruiende vogels. *De Grauwe Gors* 39, 128–134.

White, H.B. and Combs, D.L. (2004) Molt-site fidelity and sub-flocking in giant Canada Geese in Tennessee. In: Moser, T.J., Lien, R.D., ver Cauteren, K.C., Abraham, K.F., Andersen, D.E., *et al.* (eds) *Proceedings of the 2003 International Canada Goose Symposium*, Madison, Wisconsin.

Williams, F., Eschen, R., Harris, A., Djeddour, D., Pratt, C., *et al.* (2010) The economic cost of invasive non-native species on Great Britain. CAB International Project No. VM10066, pp. 1–99.

Williams, M. (1981) *The Duckshooter's Bag. An Understanding of New Zealand's Wetland Gamebirds.* Wetland Press, Wellington.

31 Mute Swan (*Cygnus olor* Gmelin, 1789)

Guillaume Gayet[1]*, Matthieu Guillemain[2], Eileen C. Rees[3], Kevin A. Wood[3] and Mike W. Eichholz[4]

[1]Unite Mixte de Service Patrimoine Naturel Agence Française pour la Biodiversité, Centre National de la Recherche Scientifique, Muséum national d'Histoire naturelle, 75005 Paris, France; [2]Office Français de la Biodiversité (OFB), Unité Avifaune Migratrice, La Tour du Valat, Le Sambuc, 13200 Arles, France; [3]Wildfowl and Wetlands Trust, Slimbridge Wetland Centre, Gloucestershire GL2 7BT, UK; [4]Department of Zoology, Southern Illinois University, Carbondale, IL 62901, USA

Citation: Gayet, G., Guillemain, M., Rees, E., Wood, K.A. and Eichholz, M. (2020) Mute Swan (*Cygnus olor*, Gmelin, 1789). In: Downs, C.T. and Hart, L.A. (eds) *Invasive Birds: Global Trends and Impacts*. CAB International, Wallingford, UK, pp. 232–242.

31.1 Common Names

Mute Swan, Tame Swan, Knobbelzwaan (Dutch), Cygne tuberculé (French), Hökerschwan (German).

31.2 Nomenclature

The Mute Swan (*Cygnus olor* Gmelin, 1789) belongs to the family Anatidae, subfamily Anserinae and tribe Cygnini, where swan species are placed in two genera: *Cygnus* and *Coscoroba*. *Olor* is sometimes separated as a third genus but is more usually classed as *Cygnus* with six other species and subspecies: the Whooper Swan (*C. cygnus*) and Bewick's Swan (*C. columbianus bewickii*) of Eurasia, the Whistling Swan (*C. columbianus columbianus*) and Trumpeter Swan (*C. buccinator*) of North America, the Black-necked Swan (*C. melanocoryphus*) of South America and the Black Swan (*C. atratus*) of Australia. The subspecies *C. c. columbianus* and *C. c. bewickii* are also known collectively as the Tundra Swan. The evolutionary relationship between the Coscoroba Swan (*Coscoroba coscoroba*) from South America and the *Cygnus* swans continues to be a matter of debate (Callaghan *et al.*, 2005).

31.3 Distribution

The range of Mute Swans is the largest of the *Cygnus* spp. (Fig. 31.1). It is considered native to Eurasia, but captive birds have also been bred and released in several countries in central and Western Europe from the 16th century onwards, and it is not always possible to distinguish between wild and introduced stocks (Cramp *et al.*, 1977; Delany, 2005). In the 1960s, its breeding range included only Great Britain, Ireland, the Netherlands, wide strips along the Baltic Sea and the Black Sea, and areas from the Caspian Sea to the East of China (Voous, 1960). More recently, increasing numbers and range expansion have seen the species become established across Europe, and the formerly discrete populations of the Ural and Volga deltas in Russia and Kazakhstan have joined (Snow and Perrins, 1998). It has the most southerly breeding range among all Eurasian swans (Rowell and Spray, 2004), and it is likely that the introductions have contributed to the regional expansion of the Mute Swan in Eurasia. The species is also quite mobile, however, and since the 1990s, it has increasingly been sighted as a summer visitor to the Nenetskiy region of the Russian Arctic. Pairs with up to six cygnets have also been observed in this area (on the Korovinskaya Gulf) in recent years, although the precise locations of their nests in the Pechora River delta are not yet known (A. Glotov, personal communication).

The expansion of Mute Swans' geographical range has occurred in conjunction with an increase in numbers in the different populations. For the Black Sea population (approximately 45,000 individuals) and the West and Central Asia/Caspian population (approximately 250,000 individuals), coverage is

*Corresponding author: ggayet@mnhn.fr

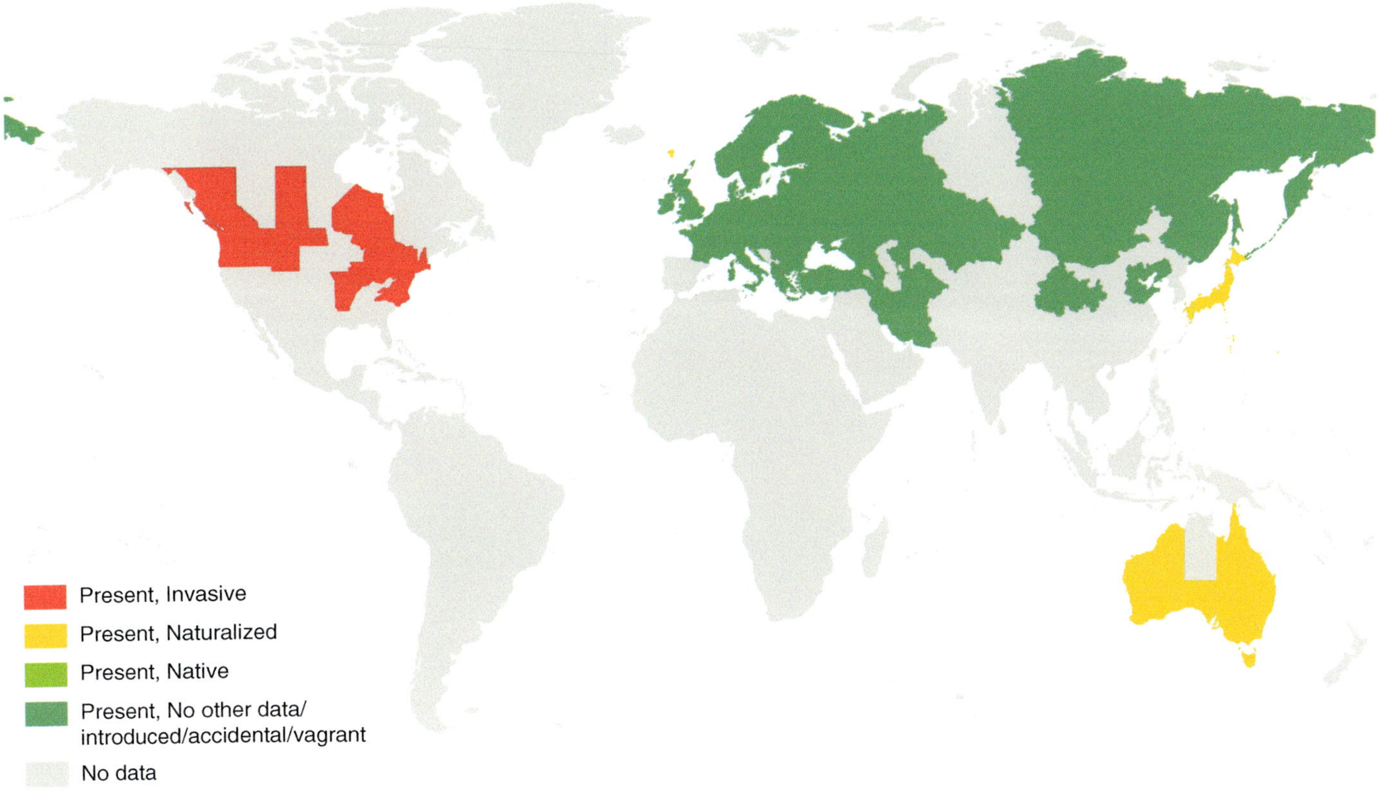

Fig. 31.1. Global distribution of the Mute Swan (*Cygnus olor*) (based on Seabrook-Davison 2013, BirdLife International 2020, Atlas of Living Australia 2019, Invasive species of Japan 2020).

patchy and some counts date back to the 1990s, so trends in numbers into the 21st century remain unclear (Wetlands International, 2017). Elsewhere, the north-west mainland and central European population has grown by 2.3% annually over the period 1972–2015, reaching around 200,000 individuals to date (Wetlands International, 2017; see also Wieloch, 1991; Kirby *et al.*, 1994; Musil and Fuchs, 1994; Fouque *et al.*, 2007). The British population has seen an even greater increase, with numbers doubling between the late 1980s and early 2000s before becoming relatively stable during 2000–2010 (Wood *et al.*, 2019). The most recent estimate was 74,000 individuals by around 2009 (Musgrove *et al.*, 2013). In Ireland, however, the population declined slightly by 2.9% between 1999 and 2009, to 9180 individuals, following major increases during the second half of the 20th century (Crowe and Holt, 2013). The increases both in numbers and distribution have been attributed to several factors, which may affect the birds to varying degrees across their range, most notably legal protection from persecution, bans on the sale or use of lead shot and lead fishing weights in some countries including the UK and Denmark (which result in lead poisoning when ingested as grit), creation of suitable habitats (e.g. gravel pits, ponds), intensification of agriculture (providing food during winter) and milder winters (Wieloch, 1991; van Eerden *et al.*, 1996; Rowell and Spray, 2004; Wood *et al.*, 2019).

Release and/or accidental escapes have created non-native Mute Swan populations outside its original range, in Mauritius (Banks *et al.*, 2008), Japan (Albertsen and Kanazawa, 2002) and South Africa, Australia, New Zealand, Iceland and the United Arab Emirates (Avibase: https://avibase.bsc-eoc.org). In North America, Mute Swans are especially present in British Columbia, California and Michigan, with small flocks scattered around the other states and provinces, totalling around 50,000–60,000 individuals (Petrie and Francis, 2003; Banks *et al.*, 2008; Baldassarre, 2014). In most such countries, Mute Swans have experienced high growth rates in suitable areas (increasing approximately 10% annually in North America when no control efforts are enacted; Petrie and Francis, 2003), while population size has remained lower where availability of breeding sites, predation or winter climate are less favourable (e.g. Japan; Albertsen and Kanazawa, 2002). Birds formerly introduced to Iceland, South Africa and the United Arab Emirates are now considered to have vanished (Banks *et al.*, 2008), although Mute Swans occasionally occur in Iceland as vagrants in Whooper Swan flocks (Nielsen, 1985; Ó. Einarsson, personal communication).

31.4 Description

The Mute Swan is one of the largest of the swan species and is one of the biggest flying birds on Earth. Its wing beats make a characteristic creaking sound in flight, audible at up to 1–2 km, caused by the movement of its flight feathers. The two sexes look generally similar (but see sexual differences below) and like several other swan species, adult birds have white plumage throughout the year. It can, however, readily be distinguished from other swans by its red-orange bill outlined in black, with a black bill knob (Fig. 31.2). Its feet and legs are usually black in adulthood.

Fig. 31.2. Adult Mute Swans with a cygnet. (©Photograph: Guillaume Gayet.)

Adult Mute Swan body mass ranges from 6.6 to 15 kg (del Hoyo *et al.*, 1992), and males are larger on average than females (Cramp *et al.*, 1977), with a larger bill knob, especially during breeding (Horrocks *et al.*, 2009). Immature swans are brown above and whitish below, lack a bill knob, and have a grey bill and grey feet. In a leucistic morph (the 'Polish' phase), which lacks melanin, immature swans have white instead of brown plumage, and retain pink or yellowish legs into adulthood (Delany, 2005). The morphological appearance of immature swans becomes very close to that of adults after their first winter and spring (Cramp *et al.*, 1977), so that precise age cannot generally be determined in fully grown individuals.

31.5 Diet

The Mute Swan is primarily an aquatic herbivore (del Hoyo *et al.*, 1992) and consumes 3–4 kg of fresh aquatic vegetation daily. It relies on submerged aquatic macrophytes when dipping or upending (up to 1 m deep; Owen and Cadbury, 1975), and feeds on emergent plants or seeds when grazing at the water edge (Cramp *et al.*, 1977). The most common plant species in the diet of the Mute Swan, both within its native range and in North America, include *Potamogeton*, *Myriophyllum*, *Chara*, *Najas*, *Elodea*, *Ceratophyllum*, *Vallisneria* and *Zizania* spp. in freshwater, as well as *Zostera* and *Ruppia* spp. and green algae in brackish or salt waters (Perry *et al.*, 2004; Rowell and Spray, 2004; Bailey *et al.*, 2008). Mute Swans preferentially consume plant stems and leaves compared with other swan species (e.g. Whooper and Tundra Swans), which exposes roots and tubers when digging into the substrate and trampling with their feet (Källander, 2005; see also Bailey *et al.* 2008 for a similar behaviour in North America). Mute Swan parents commonly pull up vegetation and bring it to the water surface for the cygnets to feed on (Cramp *et al.*, 1977). In addition to feeding in wetlands, Mute Swans may also rely on grazing in agricultural lands (e.g. oilseed rape, grasses, cereals, sometimes young vegetables) to a significant extent (e.g. Sears, 1989). Supplementary feeding by humans can contribute to its diet, especially in urban areas (Rowell and Spray, 2004; Bailey *et al.*, 2008).

31.6 Introduction and Invasion Pathways

There is a relatively long history of farming Mute Swans as a source of food. In the UK, the birds have been kept for their down, feathers and particularly meat since mediaeval times (Birkhead and Perrins, 1986; Kear, 1990), and former swan farms may have been the main reason behind the establishment of local swan populations. The Fleet in Dorset, UK, where up to 1000 individuals occur (Frost *et al.*, 2018), is adjacent to Abbotsbury Manor where swan farming was recorded as early as 1393 (Perrins and Ogilvie, 1981), and the swans continue to breed colonially to this day (Birkhead and Perrins, 1986). In the Netherlands, the breeding bird atlas similarly shows that Mute Swan concentrations still occur in areas where swan farms were once located (Banks *et al.*, 2008).

In addition to being farmed for meat, Mute Swans were frequently released at sites not only within but also outside their native range for ornamental purposes. Their large size, white plumage, elegant shape, breeding behaviour (with cygnets accompanying their parents after hatching) and impressive raised-wing displays make it easy and pleasant for the public to observe these birds in urban parks and on country estates. As the Mute Swan can thrive in a range of environments, it became established following introduction in many regions, although where the birds are thought to be non-native, the introduced populations often remained relatively limited in numbers. Only a couple of hundred birds occur in Japan, New Zealand and Australia (del Hoyo *et al.*, 1992; Rees *et al.*, 2019), and no negative impact of the swans has been reported in these particular countries. The population seems to be in decline in New Zealand because of predation and botulism outbreaks (Seabrook-Davison, 2013), and several formerly introduced populations (e.g. in Iceland, South Africa and the United Arab Emirates) have now disappeared (Banks *et al.*, 2008).

One major exception to this pattern is in North America, where Mute Swan numbers and their geographical range increased substantially in the late 20th century. Although the precise year is not known, Mute Swans were first brought to North America during the late 1800s for ornamental purposes, and were first recorded as breeding in the wild in 1910 along the lower Hudson River (Baldassarre, 2014). Thereafter, the species was imported multiple times and to multiple locations for private collections, zoos and parks, or to reduce the abundance of Canada Geese (*Branta canadensis*), which are also perceived as a pest in urban areas. All Mute Swans in North America originated from the escape or release of swans from these introductions (Ciaranca *et al.*, 1997).

The Mute Swan is now present in high densities along the North Atlantic Coast of North America, the Great Lakes region, the Chesapeake Bay area and the Pacific Coast, with small sporadic populations found in many of the states and provinces of all four flyways (Hindman and Tjaden, 2014). In the lower Great Lakes, Atlantic Coast and Pacific Coast, the rate of increase is such that, in the absence of population control, numbers are expected to double every 7–8 years (Petrie and Francis, 2003).

While the Mute Swan is migratory in some parts of its native range, it is sedentary or only a partial migrant in others (Snow and Perrins, 1998), and in these areas the rate of increase in population size is generally greater than that of its geographical range. In the UK, most ringed Mute Swans are found to breed in the general area where they were raised, and permanent movements of over 50 km are not common (Birkhead and Perrins, 1986; Rowell and Spray, 2004). In contrast, birds caught and ringed in the Russian Arctic during summer migrate more than 1000 km to more temperate wintering areas (Wildfowl & Wetlands Trust, Nenetskiy Zapovednik and Moscow Birds Ringing Centre, unpublished data). Although introduced Mute Swans generally do not disperse widely, and hence tend to build multiple sets of local populations rather than covering wide geographical areas, recent data from radio transmitter-fitted swans in Michigan, USA, indicate that some individuals, especially juveniles, move as far as 300 km from their natal wetland (R. Knapik, unpublished data).

31.7 Breeding Behaviour

The Mute Swan is gregarious during autumn and winter, although pairs breeding on high-quality habitat may remain on their territory throughout the winter (Scott, 1984). During the rest of the year, non-breeders tend to form temporary groups, while breeding pairs remain on their territories (Scott, 1984). Mute Swans often pair at the age of 2–4 years (Minton, 1968) and are monogamous; repairing ('divorce') occurred in just 3.7% of pair bonds recorded where both members of the original pair were still alive (Rees *et al.*, 1996). The nest is commonly located close to extended stands of vegetation, with easy access to and from the water (Figs 31.3 and 31.4) (Rowell and

Fig. 31.3. A typical Mute Swan nest. (©Photograph; Guillaume Gayet.)

Fig. 31.4. A Mute Swan nesting in (A) open habitat and (B) more closed habitat. (©Photographs: Maurice Benmergui (A) and Guillaume Gayet (B).)

Spray, 2004). It is often made of the local emergent vegetation, and is 1–2 m in diameter with a height of 60–80 cm above the water surface (Cramp *et al.*, 1977).

Food availability for Mute Swan cygnets does not seem to influence nesting site selection (Rowell and Spray, 2004). The first eggs of the season are laid in early March, and laying may continue into June. Eggs are laid at 48 h intervals, clutch sizes are of 5.5–7 eggs on average (depending on the study area), and the eggs are incubated for 36 days. Smaller replacement clutches may be produced in case of clutch failure, generally 2–4 weeks after the initial eggs have been lost (Cramp *et al.*, 1977). Parental duties are often split between the parents, with males defending their territory and their brood after hatching, especially against conspecifics (Lind, 1984), while females incubate the eggs and later provide most care during the cygnets' feeding activity (Wlodarczyk and Minias, 2015). Mute Swans have a reputation for being highly territorial during the breeding period, but aggressive behaviour is also observed during winter in order to ensure access to food resources and secure future breeding sites (Scott, 1984). Colonial nesting also occurs, however, where pairs defend only the nest site and breed as little as 2 m apart (Snow and Perrins, 1998). These dense breeding colonies are established at just a few sites, in Denmark, England and Estonia (Birkhead and Perrins, 1986; Bacon and Andersen-Harild, 1987; Luigujõe, 2018). Wieloch (1991) also mentions colonies in Sweden, Germany, Russia and formerly in Poland. The reason for such variation in territorial behaviour among locations

Fig. 31.5. Mute Swans on water. (©Photograph: G. Gayet.)

remains unclear. Cygnet development is relatively slow, taking 120–150 days from hatching to fledging. Vandalism during incubation and starvation of cygnets (especially during the 2 weeks after hatching) are the most common causes of breeding failure (Cramp *et al.*, 1977), and newly fledged young are susceptible to flying accidents, particularly collisions with powerlines (Perrins and Sears, 1991). Failed breeders may join moulting flocks of non-breeders in May–June (Fig. 31.5).

31.8 Habitat

The Mute Swan can use a wide variety of wetland habitats: slow-flowing rivers, fluvial canals, lakes, fishponds, coastal marine habitats, gravel pits and urban ponds, with farmland also important for grazing (Kirby *et al.*, 1994; Rees *et al.*, 1997; Petersen, 2000; Rowell and Spray, 2004; Fouque *et al.*, 2007). Although habitat use may vary markedly with the time of year and the swans' breeding status (Gayet *et al.*, 2011a, 2013; Wood *et al.*, 2013), Mute Swans demonstrate a high level of ecological plasticity, which explains their ability to colonize these diverse habitats within their native range in recent decades (Wieloch, 1991; Fouque *et al.*, 2007). Indeed, these birds have few strict requirements for establishing breeding sites (Gayet *et al.*, 2011b), and can readily rely on agricultural crops for feeding during winter in addition to their more traditional use of wetland habitats (e.g. Sears, 1989; Chisholm and Spray, 2002).

31.9 Impacts

The growth of Mute Swan populations, both in areas where the species has been introduced and within their native geographical range, has raised concerns about their interactions with other waterbird communities, the extent to which they deplete food resources such as aquatic macrophyte beds, and hence their general impacts on other wildlife and human recreational activities. The extent of the concern varies among regions, and may be influenced by whether the species is native or non-native, but only for the latter can the expansion of the Mute Swan numbers sometimes be considered a biological invasion (Gayet *et al.*, 2014).

Being such large herbivorous waterbirds, Mute Swans can potentially overgraze aquatic plants, generating conflicts with various stakeholders such as those managing habitats for other species, including hunters and fish farmers. Mute Swan feeding activity can reduce plant biomass substantially, in some cases leaving only bare sediment, and can also influence species dominance relationships within plant communities (Tatu *et al.*, 2006; O'Hare *et al.*, 2007; Sandsten and Klaassen, 2008; Gayet *et al.*, 2011c, 2012; Stafford *et al.*, 2012). For instance, aquatic plant cover was on average 79% lower, mean shoot density was 76% lower and canopy height was 40% lower in areas of the Chesapeake Bay grazed by swans (Tatu *et al.*, 2006). In mid-continental American wetlands, Mute Swan effects on above-ground plant biomass could not be demonstrated after only 2 years of swan exclusion, but they were found to reduce below-ground plant biomass significantly, probably forecasting a future effect on above-ground biomass (Stafford *et al.*, 2012).

In other areas, emergent plants could potentially be uprooted by the swans (Chaichana *et al.*, 2011). The magnitude of grazing impacts and the vegetation parameters affected by Mute Swans appear to depend on the density of birds, the time of year and the swans' migratory behaviour, and also to vary among wetland ecosystems (Wood *et al.*, 2012). As Mute Swans may remain on or be faithful to the same waterbodies over successive years (e.g. as breeding or flocking sites), it has been postulated that their effects on plants could be cumulative, although a recent study found no evidence for Mute Swan grazing impacts carrying over into the following year (Wood *et al.*, 2018). It is additionally suggested that the swans could cause chronic disturbance, potentially affecting the whole ecosystem through trophic cascades, given the critical role of plants on biogeochemical functioning and as habitat for a range of animals (Tatu *et al.*, 2006; Gayet *et al.*, 2011a), but such potential effects have been poorly studied. It should also be noted, however, that Mute Swan grazing does not invariably have negative consequences for the vegetation: some studies found no evidence for a significant impact on the plants (e.g. Conover and Kania, 1994), and Mute Swans can in fact partially promote some plant species of conservation concern through their grazing of other, dominant plant species (Gayet *et al.*, 2012).

Interspecific interactions among waterbirds including swans have been reported to range from exclusion (Conover and Kania, 1994) to commensalism (e.g. Källander, 2005), with Mute Swans affecting other waterbirds indirectly through competition for food resources (Bailey *et al.*, 2008; Gyimesi *et al.*, 2011), and also directly through territorial behaviour on breeding sites. While territorial behaviour is directed mainly towards conspecifics (reviewed by Wood *et al.*, 2017), breeding pairs have been observed behaving aggressively towards other waterbirds, sometimes killing or badly injuring the young, and anecdotal evidence suggests that removal of Mute Swans is facilitating the propagation of reintroduced native Trumpeter Swans in North America (e.g. Conover and Kania, 1994). Large moulting Mute Swan flocks can also disrupt the nesting of threatened species (e.g. Least Terns (*Sterna antillarum*) and Black Skimmers (*Rynchops niger*) in Tar Bay, Dorchester County, South Carolina; see Hindman and Tjaden, 2014). However, breeding Mute Swans were found to lack common agonistic behaviour in some areas. For instance, in French fishponds, waterbird abundance was actually higher on Mute Swan breeding lakes, resulting presumably from shared habitat preferences (Gayet *et al.*, 2011c), and waterbirds generally do not avoid Mute Swan breeding territories (Gayet *et al.*, 2016). Commensalism with foraging ducks and coots has been reported, where birds with lower diving or foraging abilities benefited from the underwater vegetation brought back to the surface of the water by feeding adult swans (Oksanen *et al.*, 1979; Källander, 2005; Phillips, 2010). In Europe, such a lack of agonistic interactions has been postulated to result from swans occupying previously vacant ecological niches while (re)colonizing new habitats (Oksanen *et al.*, 1979; Pöysä and Sorjonen, 2000). Wood *et al.* (2017) demonstrated that Mute Swans do not behave more aggressively than other waterbird species. Its displays are easy to observe where swans occur, so its territorial behaviour should simply be more noticeable, explaining an exaggerated aggressive reputation. Resource availability, and genetic or other innate factors, may explain variation in aggressive behaviour between Mute Swan populations, as well as between individual birds.

Beside potential impacts on natural aquatic plants and waterbirds in aquatic ecosystems, and the consequences of these for socio-economic activities (e.g. hunting, fish farming), Mute Swans can cause concern to farmers where they congregate on agricultural land from late winter to early spring (Rowell and Spray, 2004). Damage may result from the consumption and trampling of crops, puddling of fields and perhaps soil compaction (Rowell and Spray, 2004; Parrott and Watola, 2007a,b). There is no apparent evidence for farmland being adversely affected by swan droppings (e.g. Kear, 1963), but these may reduce the public's enjoyment of riparian areas and parklands used for leisure activities if large numbers of Mute Swans are in the vicinity. Individual birds can also be problematic in urban areas if a breeding pair threatens people and pets that approach close to the nest, in defence of eggs or cygnets, making some places difficult for the public to access during the swans' breeding season (Conover and Kania, 1994; Hindman and Tjaden, 2014). Attacks by territorial Mute Swans towards small watercraft and people have led to two reported cases of drowning in North America (Hindman and Tjaden, 2014). The risk of swans colliding with air transport, undertaken as part of broader bird strike management programmes, is controlled both through the design of the aircraft and by managing any presence of swans near airports. While collision risk is relatively low in comparison with other large avian species as the swans' flight height is generally well below that of aircraft outside the airport's perimeter, there were eight collisions between civil aircraft and Mute Swans in the USA during 1990–2012 (of which two resulted in damage to the aircraft; Dolbeer *et al.*, 2013), and three collisions have been reported at JFK International Airport in New York since 2010. The birds' size means that the consequences of a collision with the fuselage and particularly the engine can be severe.

Like other Anatidae species, Mute Swans have been suspected of being a vector of avian influenza viruses because of the relatively large numbers of dead swans found during some outbreaks (e.g. Olsen *et al.*, 2006). Virus surveillance has indeed demonstrated that Mute Swans can carry and shed the virus, but they may be infected by other migratory birds instead of or in addition to transporting the viruses themselves. The prominence of Mute Swans among the casualties of avian influenza virus outbreaks may largely be an artefact caused by the greater detectability of relatively large white swan carcasses compared with those of other species (EFSA Panel on Animal Health and Welfare, 2017).

31.10 Control

The charismatic nature of the Mute Swan is the reason why it has been introduced as an ornamental bird in so many areas. This may also be a hurdle to the control of non-native populations, which may face vigorous public opposition. Several groups have formed to defend the protection status of the Mute Swan in North America, sometimes contesting whether the

Mute Swan is historically non-native in areas where it was introduced on the continent from the 19th century onwards. Plans to reduce or eliminate Mute Swans in parts of North America have therefore resulted in strong public reactions, including through petitions (in New Jersey, New York and Toronto) and legal challenges (Craves and Susko, 2010; Hindman and Tjaden, 2014).

In Europe, the Mute Swan is listed in Appendix III of the Bern Convention, so any exploitation must maintain the populations in a favourable conservation status. It is also listed in Annex II of the European Bird Directive, which permits hunting if allowed under the national legislation of Member States and if conservation efforts are not being jeopardized, but the species is in fact protected in most European countries. Some localities, however, prohibit feeding the swans when they pose a threat to humans, fencing is used to keep the birds away from some areas, and various measures are put in place to deter the swans from using fields where crop damage may occur (Parrott and Watola, 2007b). Culling is also practised under licence, again in relation to crop damage, although the damage is not always proven; for instance, approximately 2000–3000 individuals are shot annually in the Netherlands (Esselink and Beekman, 1991; Wieloch, 1991; C. Perrins personal communication in Rowell and Spray, 2004). In France, local permits have been granted to undertake several control measures within a management programme, such as deterring swans from feeding on fields with young vegetable crops, oiling Mute Swan eggs and culling fully grown birds in the northern part of the country. This has had a significant impact on the number of Mute Swans locally (Parc Naturel Régional des Caps et Marais d'Opale, 2016). In regions where regulation is not permitted, stakeholders (either individually or as a group) sometimes attempt to regulate swan populations illegally, mostly through oiling, needling or shaking the eggs (G. Gayet, France, personal observation). Illegal culling of Mute Swans has also happened in the UK because of perceived damage to valuable salmonid fisheries (because of overgrazing of the aquatic plants that form critical foraging habitats and refugia for the salmon and trout) (Wood *et al.*, 2015).

Non-native Mute Swans are protected in several countries where the birds have been introduced, including in Canada, Japan and New Zealand. The protection level in New Zealand was reduced in 2010 (from absolute protection under the Wildlife Act 1953) so that it can now be culled or held in captivity at the discretion of the Minister of Conservation, but there have been no incidences of populations becoming invasive or destructive as in the USA, and Mute Swan numbers are declining in New Zealand through predation and botulism rather than control measures (Seabrook-Davison, 2013).

The legal status of Mute Swans in the USA has been variable and controversial. The USA Migratory Bird Treaty Act (MBTA) of 1918 ratified the 1916 treaty between the USA and Canada (later including Mexico, Japan and Russia), which protected migratory birds from exploitation. Although it was not specified in the Act, until 2001 the US Fish and Wildlife Service had interpreted the protection to extend only to migratory birds native to North America. In the legal case of Hill v. Norton, 275 F. 3d 98 (D.C. Cir. 2001), a Federal Appeals Court ruled that the MBTA did not specifically exclude exotic species; thus, Mute Swans were covered by the act and protected. Between 2001 and 2004, when the Migratory Bird Treaty Reform Act was passed, which specified that the treaty covered only species native to North America, Mute Swans were treated as native migratory birds under US Federal Law and a special depredation permit or order was required for direct control of Mute Swans that were causing damage to crops, livestock or wildlife, or when causing a health hazard or other nuisance. Mute Swans in the USA are currently unprotected by federal law and their protected status varies among individual states, from being given the same protective status as native migratory birds to being designated as a 'deleterious exotic wildlife' or 'prohibited invasive species', making it illegal to own or import Mute Swans within the state's borders. These regulations are not well followed, however, or enforced.

Mute Swans are unprotected in many areas of the USA, so legal hunting should theoretically be possible, but no USA state has so far proposed a hunting season for the species, and the very small numbers taken are misidentifications with other, hunted, swan species. Instead, the increase in numbers in some parts of North America resulted in calls for more focused population control in these areas, on the basis that social conflicts, depletion of aquatic vegetation and conflicts with other species should be alleviated. Non-lethal humane control measures are often preferred (e.g. the State of New York has been planned to extirpate Mute Swans mainly by prioritizing non-lethal management techniques: New York State Department of Environmental Conservation, 2017), so that egg addling or oiling has often been the most practised management technique (e.g. Hindman *et al.*, 2014). However, Mute Swans are long-lived species, so controlling the eggs may have limited or no impact on population growth rate, and only culling adult birds is considered an efficient technique to reduce population size in some cases (Ellis and Elphick, 2007; Baldassarre, 2014). A combination of population control actions, including public education, the culling of adult swans using shooting and live capture with euthanasia, along with egg oiling, has therefore been used to reduce the number of Mute Swans at Chesapeake Bay in Maryland from 3995 individuals in 1999 to 41 in 2014 (Hindman *et al.*, 2016, 2018), and to stabilize abundance at approximately 12,000 birds in Michigan.

31.11 Uses

Historically, Mute Swans were hunted or farmed for food, and their feathers used as quills for writing. At one time, all swans in England were in the possession of the Crown, but as time passed, rights to swans were granted to local noblemen in certain areas, and no mediaeval feast was complete without a roast swan. The earliest statement of ownership was in the year 966 when King Edgar gave the Abbots of Croyland in Lincolnshire the rights over stray swans in their area, and the Benedictine monks at Abbotsbury in Dorset farmed swans over many centuries following the establishment of the monastery during the 1040s (Birkhead and Perrins, 1986; Kear, 1990). Marks, usually made by notching the bill to create permanent scar tissue, were used to denote ownership, and by the 16th century there were about 900 different swan marks registered. Any unmarked swan remained the property of the Crown, giving the Mute Swan its status as a royal bird. The annual ceremony of swan upping, in which Mute Swans on the River Thames are caught and fitted with leg rings, is a continuation of the historic tradition where the Dyers' and the Vintners' Companies in the City of London were granted permission to own swans on the Thames, and the birds were rounded up and marked during their annual moult (Birkhead and Perrins, 1986; Kear, 1990). These days, people's main engagement with the Mute Swan is observing birds in the wild and as an ornamental bird in parks and country estates. In the USA, the perceived aggressive behaviour of Mute Swans is exploited by commercial entrepreneurs, with Mute Swans being raised and sold commercially to be placed on water bodies as a mechanism to reduce the abundance of local Canada Geese.

31.12 Notes

One main problem faced by managers willing to control non-native Mute Swan populations is the very reason why they were introduced to these regions: these large, attractive and generally tame birds are much appreciated by the public, which may be opposed to any kind of regulation of their numbers.

31.13 References

Albertsen, J.O. and Kanazawa, Y. (2002) Numbers and ecology of swans wintering in Japan. *Waterbirds* 25, 74–85.

Atlas of living australia (2020) Available at: https://www.ala.org.au/ (accessed 23 July 2020).

Bacon, P.J. and Andersen-Harild, P. (1987) Colonial breeding in Mute Swans associated with an allozyme of lactate dehydrogenase. *Biological Journal of the Linnean Society* 30, 193–220.

Bailey, M., Petrie, S.A. and Badzinski, S.S. (2008) Diet of mute swans in Lower Great Lakes Coastal Marshes. *Journal of Wildlife Management* 72, 726–732.

Baldassarre, G. (2014) *Ducks, Geese and Swans of North America. Revised and Updated Edition*, Vol. 1. John Hopkins University Press, Baltimore, Maryland.

Banks, A.N., Wright, L.J., Maclean, I.M.D., Hann, C. and Rehfisch, M.M. (2008) Review of the status of introduced non-native waterbird species in the area of the African-Eurasian Waterbird Agreement: 2007 update. BTO Research Report No. 489, British Trust for Ornithology, Thetford, UK.

BirdLife International (2020) Species factsheet: Cygnus olor. Downloaded from http://www.birdlife.org on 23/07/2020.

Birkhead, M. and Perrins, C. (1986) *The Mute Swan*. Croom Helm, London.

Callaghan, D., Rees, E. and Harshman, J. (2005) The swans: taxonomy. In: Kear, J. (ed.) *Ducks, Geese and Swans*, Vol. 1: *General Chapters and Species Accounts (*Anhima *to* Salvadorina*)*. Oxford University Press, Oxford, pp. 218–219.

Chaichana, R., Leah, R., and Moss, B. (2011) Seasonal impact of waterfowl on communities of macrophytes in a shallow lake. *Aquatic Botany* 95, 39–44.

Chisholm, H. and Spray, C. (2002) Habitat usage and field choice by Mute and Whooper Swans in the Tweed Valley, Scotland. *Waterbirds* 25, 177–182.

Ciaranca, M. A., Allin, C.C. and Jones, G.S. (1997) Mute Swan (*Cygnus olor*). In: Poole, A. (ed.) *The Birds of North America*. The Cornell Lab of Ornithology, Ithaca, New York. Available at: https://birdsna.org/Species-Account/bna/species/mutswa/introduction (accessed 5 November 2019).

Conover, M.R. and Kania, G.S. (1994) Impact of interspecific aggression and herbivory by mute swans on native waterfowl and aquatic vegetation in New England. *Auk* 11, 744–748.

Cramp, S., Simmons, K.E.L., Ferguson-Lees, I.J., Gillmor, R., Hollom, P.A.D., *et al.* (eds) (1977) *Handbook of the Birds of Europe, the Middle East and North Africa: The Birds of Western Paleartic*. Oxford University Press, Oxford.

Craves, J.A. and Susko, D.J. (2010) Mute Swans: an ecological overview with an emphasis on the lower Detroit River. Report to the Friends of the Detroit River, Taylor, Michigan.

Crowe, O. and Holt, C. (2013) Estimates of waterbird numbers wintering in Ireland, 2006/07–2010/11. *Irish Birds* 9, 545–552.

del Hoyo, J., Andrew, E. and Sargatal, J. (1992) *Handbook of the Birds of the World*, Vol. 1: *Ostrich to Ducks*. Lynx Editions, Barcelona, Spain.

Delany, S. (2005) Mute Swan *Cygnus olor*. In: Kear, J. (ed.) *Ducks, Geese and Swans, Vol. 1: General Chapters and Species Accounts (*Anhima *to* Salvadorina*)*. Oxford University Press, Oxford, pp. 231–234.

Dolbeer, R.A., Wright, S.E., Weller, J. and Begier, M.J. (2013) Wildlife strikes to civil aircraft in the United States 1990–2012. Serial Report No. 19, Federal Aviation Administration National Wildlife Strike Database, Washington, DC.

EFSA Panel on Animal Health and Welfare (2017) Scientific opinion on avian influenza. *EFSA Journal* 15, 4991.

Ellis, M.M. and Elphick, C.S. (2007) Using a stochastic model to examine the ecological, economic and ethical consequences of population control in a charismatic invasive species: Mute Swans in North America. *Journal of Applied Ecology* 44, 312–322.

Esselink, H. and Beekman, J.M. (1991) Between year variation and causes of mortality in the non-breeding population of the Mute Swan *Cygnus olor* in the Netherlands, with special reference to hunting. *Wildfowl*, Suppl. No. 1, 110–119.

Fouque, C., Guillemain, M., Benmergui, M., Delacour, G., Mondain-Monval, J.-Y. and Schricke, V. (2007) Mute swan (*Cygnus olor*) winter distribution and numerical trends over a 16-year period (1987/1988–2002/2003) in France. *Journal of Ornithology* 148, 477–487.

Frost, T.M., Austin, G.E., Calbrade, N.A., Mellan, H.J., Hearn, R.D., *et al.* (2018) *Waterbirds in the UK 2016/17: The Wetland Bird Survey*. British Trust for Ornithology, Thetford, UK.

Gayet, G., Guillemain, M., Benmergui, M., Mesleard, F., Boulinier, T., *et al.* (2011a) Effects of seasonality, isolation and patch quality for habitat selection processes by mute swans *Cygnus olor* in a fishpond landscape. *Oikos* 120, 801–812.

Gayet, G., Eraud, C., Benmergui, M., Broyer, J., Mesleard, F., *et al.* (2011b) Breeding mute swan habitat selection when accounting for detectability: a plastic behaviour consistent with rapidly expanding populations. *European Journal of Wildlife Research* 57, 1051–1056.

Gayet, G., Guillemain, M., Fritz, H., Mesleard, F., Begnis, C., *et al.* (2011c) Do mute swan (*Cygnus olor*) grazing, swan residence and fishpond nutrient availability interactively control macrophyte communities? *Aquatic Botany* 95, 110–116.

Gayet, G., Croce, N., Grillas, P., Nourry, C., Deschamps, C. and Defos du Rau, P. (2012) Expected and unexpected effects of waterbirds on Mediterranean aquatic plants. *Aquatic Botany* 103, 98–105.

Gayet, G., Guillemain, M., Mesleard, F., Fritz, H., Curtet, L. and Broyer, J. (2013) Annual use of man-made wetlands by the mute swan (*Cygnus olor*). *Journal of Environmental Management* 120, 120–126.

Gayet, G., Guillemain, M., Defos du Rau, P. and Grillas, P. (2014) Effects of mute swans on wetlands: a synthesis. *Hydrobiologia* 723, 195–204.

Gayet, G., Calenge, C., Broyer, J., Mesléard, F., Vaux, V., *et al.* (2016) Analysis of spatial point pattern shows no desertion of breeding Mute Swan areas by the other waterbirds within fishpond. *Acta Ornithologica* 51, 151–162.

Gyimesi, A., de Vries, P.P., de Boer, T. and Nolet, B.A. (2011) Reduced tuber banks of fennel pondweed due to summer grazing by waterfowl. *Aquatic Botany* 94, 24–28.

Hindman, L.J. and Tjaden, R.L. (2014) Awareness and opinions of Maryland citizens toward Chesapeake Bay Mute Swans *Cygnus olor* and management alternatives. *Wildfowl* 64, 167–185.

Hindman, L.J., Harvey, W.F. IV and Conley, L.E. (2014) Spraying corn oil on Mute Swan *Cygnus olor* eggs to prevent hatching. *Wildfowl* 64, 186–196.

Hindman, L.J., Harvey, W.F., Walbridge, H.R., Hooper, M. and Driscoll, C.P. (2016) An efficient method of capture and field euthanasia of flightless Mute Swans. In: Conner, L.M. and Smith, M.D. (eds) *Proceedings of the 16th Wildlife Damage Management Conference*, Auburn University, Auburn, Alabama, pp. 55–64.

Hindman, L.J., Harvey, W.F., Walbridge, H.R., Hooper, M. and Driscoll, C.P. (2018) The capture and field euthanasia of non-native Mute Swans at Chesapeake Bay, Maryland, USA. *Swan News* 14, 9–10.

Horrocks, N., Perrins, C. and Charmantier, A. (2009) Seasonal changes in male and female bill knob size in the mute swan *Cygnus olor*. *Journal of Avian Biology* 40, 511–519.

Invasive species of Japan (2020) Available at: https://www.nies.go.jp/biodiversity/invasive/DB/detail/20010e.html

Källander, H. (2005) Commensal association of waterfowl with feeding swans. *Waterbirds* 28, 326–330.

Kear, J. (1963) The agricultural importance of wild goose droppings. *Wildfowl* 14, 72–77.

Kear, J. (1990) *Man and Wildfowl*. T. & A.D. Poyser, London.

Kirby, J., Delany, S. and Quinn, J. (1994) Mute swans in Great Britain: a review, current status and long-term trends. *Hydrobiologia* 279/280, 467–482.

Lind, H. (1984) The rotation display of the mute swan *Cygnus olor*: synchronised neighbour responses as instrument in the territorial defence strategy. *Ornis Scandinavia* 15, 98–104.

Luigujõe, L. (2018) Kühmnokk-luik – Rmt: Linnuatlas. Eesti haudelindude levik ja arvukus. Eesti Ornitoloogiaühing, Tartu, Estonia.

Minton, C.D.T. (1968) Pairing and breeding of mute swans. *Wildfowl* 19, 41–60.

Musgrove, A., Aebischer, N., Eaton, M., Hearn, R., Newson, S., *et al.* (2013) Population estimates of birds in Great Britain and the United Kingdom. *British Birds* 106, 64–100.

Musil, P. and Fuchs, R. (1994) Changes in abundance of water bird species in southern Bohemia (Czech Republic) in the last 10 years. *Hydrobiologia* 279–280, 511–519.

New York State Department of Environmental Conservation (2017) Mute Swans in New York: a management plan to prevent population growth and minimize impacts of a non-native invasive species. Third draft. New York State Department of Environmental Conservation Report, New York.

Nielsen, Ó.K. (1985) The Mute Swan in Iceland 1958–77, its introduction and history. *Bliki* 4, 2–7 [in Icelandic].

O'Hare, M., Stillman, R.A., McDonnell, J. and Wood, L.R. (2007) Effects of mute swan grazing on a keystone macrophyte. *Freshwater Biology* 52, 2463–2475.

Oksanen, L., Fretwell, S.D. and Järvinen, O. (1979) Interspecific agression and the limiting similarity of close competitors: the problem of size gaps in some community arrays. *American Naturalist* 114, 117–129.

Olsen, B., Munster, V.J., Wallensten, A., Waldenström, J., Osterhaus, A.D.M.E. and Fouchier, R.A.M. (2006) Global patterns of influenza A virus in wild birds. *Science* 312, 384–388.

Owen, M. and Cadbury, C. (1975) The ecology and mortality of swans at the Ouse Washes, England. *Wildfowl* 25, 31–42.

Parc Naturel Régional des Caps et Marais d'Opale (2016) L'expérience de régulation du cygne dans le marais Audomarois. PNR des Caps et Marais d'Opale report, Le Wast, France.

Parrott, D. and Watola, G. (2007a) Variation in yield loss of oilseed rape due to grazing by mute swans *Cygnus olor*. *Crop Protection* 26, 1794–1800.

Parrott, D. and Watola, G. (2007b) Deterring mute swans from fields of oilseed rape using suspensed high visibility tape. *Crop Protection* 27, 632–637.

Perrins, C.M. and Ogilvie, M.A. (1981) A study of the Abbotsbury Mute swans. *Wildfowl* 32, 35–47.

Perrins, C.M. and Sears, J. (1991) Collisions with overhead wires as a cause of mortality in Mute Swans *Cygnus olor*. *Wildfowl* 42, 5–11.

Perry, M.C., Osenton, P.C. and Lohnes, E.J.R. (2004) Food habits of Mute Swans in the Chesapeake Bay. In: Perry, M.C. (ed.) *Mute Swans and their Chesapeake Bay Habitats: Proceedings of a Symposium*. US Geological Survey, Biological Resources Discipline Information and Technology Report USGS/BRD/ITR, pp. 31–36.

Petersen, B.M. (2000) A quantitative model of the distribution of mute swans wintering in Danish waters. *Wildfowl* 51, 17–32.

Petrie, S.A. and Francis, C.M. (2003) Rapid increase in the lower Great Lakes population of feral mute swans: a review and a recommendation. *Wildlife Society Bulletin* 31, 407–416.

Phillips, A. (2010) Mute swan impacts on native waterbirds and submersed aquatic vegetation in Illinois. MS thesis, Southern Illinois University Carbondale, Carbondale, Illinois.

Pöysä, H. and Sorjonen, J. (2000) Recolonization of breeding waterfowl communities by the whooper swan: vacant niches available. *Ecography* 23, 342–348.

Rees, E.C., Kirby, J.S. and Gilburn, A. (1997) Site selection by swans wintering in Britain and Ireland; the importance of habitat and geographic location. *Ibis* 139, 337–352.

Rees, E.C., Lievesley, P., Pettifor, R.A. and Perrins, C. (1996) Mate fidelity in swans: an interspecific comparison. In: Black, J.M. (ed.) *Partnerships in Birds; the Study of Monogamy*. Oxford University Press, Oxford, pp. 118–136.

Rees, E.C., Cao, L., Clausen, P., Coleman, J., Cornely, J., et al. (2019) Conservation status of the world's swan populations, Cygnus sp. and Coscoroba sp.: a review of current trends and gaps in knowledge. Wildfowl, Special Issue No. 5, 35–72.

Rowell, H. and Spray, C. (2004) Mute swan *Cygnus olor* (Britain and Ireland population) in Britain and Northern Ireland 1960/61 – 2000/01. Waterbird Review Series, The Wildfowl and Wetlands Trust/Joint Nature Conservation Committee, Slimbridge, UK.

Sandsten, H. and Klaassen, M. (2008) Swan foraging shapes spatial distribution of two submerged plants, favouring the preferred prey species. *Oecologia* 156, 569–576.

Scott, D.K. (1984) Winter territoriality of Mute Swans *Cygnus olor*. *Ibis* 126, 168–176.

Seabrook-Davison, M. (2013) Mute swan. In: Miskelly, C.M. (ed.) *New Zealand Birds Online*. Available at: http://nzbirdsonline.org.nz/species/mute-swan (accessed 6 November 2019).

Sears, J. (1989) Feeding activity and body condition of Mute Swans *Cygnus olor* in rural and urban areas of a lowland river system. *Wildfowl* 40, 88–98.

Snow, D.W. and Perrins, C.M. (1998) *The Birds of Western Paleartic: Concise Edition*, Vol. 1: *Non-Passerines*. Oxford University Press, Oxford.

Stafford, J.D., Eichholz, M.W. and Phillips, A.C. (2012) Impacts of Mute Swans (*Cygnus olor*) on submerged aquatic vegetation in Illinois River Valley Backwaters. *Wetlands* 32, 851–857.

Tatu, K.S., Anderson, J.T., Hindman, L.J. and Seidel, G. (2006) Mute swan's impact on submerged aquatic vegetation in Chesapeake Bay. *Journal of Wildlife Management* 71, 1431–1439.

van Eerden, M.R., Zijlstra, M., van Roomen, M. and Timmerman, A. (1996) The response of Anatidae to changes in agricultural practice: long-term shifts in the carrying capacity of wintering waterfowl. *Gibier Faune Sauvage* 13, 681–707.

Voous K.H. (1960) *Atlas of European Birds*. Elsevier, Amsterdam.

Wetlands International (2017) Flyway trend analyses based on data from the African–Eurasian Waterbird Census from the period of 1967–2015. Wetlands International, Ede, Netherlands. Available at: http://iwc.wetlands.org/static/files/0-IWC-trend-analysis-report-2017-final.pdf (accessed 6 November 2019).

Wieloch, M. (1991) Population trends of the mute swan *Cygnus olor* in the Palearctic. *Wildfowl*, Special Issue No. 1, 22–32.

Wlodarczyk, R. and Minias, P. (2015) Division of parental duties confirms a need for bi-parental care in a precocial bird, the mute swan *Cygnus olor*. *Animal Biology* 65, 163–176.

Wood, K.A., Stillman, R.A., Clarke, R.T., Daunt, F. and O'Hare, M.T. (2012) Understanding plant community responses to combinations of biotic and abiotic factors in different phases of the plant growth cycle. *PLoS One* 7, e49824.

Wood, K.A., Stillman, R.A., Coombs, T., McDonald, C., Daunt, F. and O'Hare, M.T. (2013) The role of season and social grouping on habitat use by Mute Swans (*Cygnus olor*) in a lowland river catchment. *Bird Study* 60, 229–237.

Wood, K.A., Stillman, R.A., Daunt, F. and O'Hare, M.T. (2015) The swan grazing conflict in chalk rivers. In: Redpath, S.M., Gutierrez, R.J., Wood, K.A. and Young, J.C. (eds) *Conflicts in Conservation: Navigating Towards Solutions*. Cambridge University Press, Cambridge, pp. 134–136.

Wood, K.A., Ponting, J., d'Costa, N., Newth, J.L., Rose, P.E., *et al.* (2017) Understanding intrinsic and extrinsic drivers of aggressive behaviour in waterbird assemblages: a meta-analysis. *Animal Behaviour* 126, 209–216.

Wood, K. A., Stillman, R. A., Clarke, R. T., Daunt, F. and O'Hare, M. T. (2018) Water velocity limits the temporal extent of herbivore effects on aquatic plants in a lowland river. *Hydrobiologia* 812, 45–55.

Wood, K.A., Brown, M.J., Cromie, R.L., Hilton, G.M., Mackenzie, C., *et al.* (2019) Regulation of lead fishing weights results in mute swan population recovery. *Biological Conservation* 230, 67–74.

32 Grey-headed Swamphen (*Porphyrio poliocephalus* Latham, 1801)

Bill Pranty[1]* and Corey T. Callaghan[2]
[1]*Bayonet Point, Florida, USA;* [2]*Centre for Ecosystem Science, School of Biological, Earth and Environmental Sciences, UNSW Sydney, Sydney, NSW 2052, Australia*

Citation: Pranty, B. and Callaghan, C.T. (2020) Grey-headed Swamphen (*Porphyrio poliocephalus* Latham, 1801). In: Downs, C.T. and Hart, L.A. (eds) *Invasive Birds: Global Trends and Impacts*. CAB International, Wallingford, UK, pp. 243–247.

32.1 Common Names

Grey-headed Swamphen, Purple Swamphen, Pukeko (New Zealand).

32.2 Nomenclature

The taxonomic status of *Porphyrio* swamphens is under refinement. Although taxonomic uncertainties exist in the eastern portion of the swamphen's range (A.W. Kratter, personal communication), most authorities now consider the Purple Swamphen complex to consist of six species: the Western Swamphen (*P. porphyrio*) of south-western Europe and north-western Africa; the African Swamphen (*P. madagascariensis*) of Egypt, sub-Saharan Africa and Madagascar; the Grey-headed Swamphen (*P. poliocephalus*) from Turkey to Thailand; the Black-backed Swamphen (*P. indicus*) from Burma to Borneo; the Philippine Swamphen (*P. pulverulentus*) of the Philippines; and the Australasian Swamphen (*P. melanotus*) from Sulawesi through Australia, New Guinea and New Zealand to American Samoa (Sangster, 1998; Garcia-R and Trewick, 2015). For this chapter, we supplement information on the non-native Grey-headed Swamphen population found in Florida with information on Purple Swamphens throughout their range, even if not necessarily specific to Grey-headed Swamphens.

32.3 Distribution

Purple Swamphens are widespread in the Old World, occurring from Western Europe and Africa east through southern Asia, Australia, New Zealand and islands in the Pacific Ocean to American Samoa. Swamphens are non-migratory throughout their range, but are known to disperse long distances (over 1000 km; e.g. Grussu, 1999; Sánchez-Lafuente *et al.*, 2001) in response to local wetland conditions. Grey-headed Swamphens are native from Turkey and Saudi Arabia east through India and southern Asia to northern Thailand (Sangster, 1998; Fig. 32.1). The first record of a Grey-headed Swamphen in North America was an adult photographed at Wilmington, Delaware, in December 1990; this individual was considered to be ship-assisted, perhaps from an American naval vessel returning from the Persian Gulf (Boyle *et al.*, 1991; Mumford, 2013; Pranty, 2013a, b). Around 1996, a population of Grey-headed Swamphens was discovered in south-eastern Florida, 1580 km farther south; this population began from an unintended release and is now well established (Pranty and Schnitzius, 1998; Pranty *et al.*, 2000; Pranty, 2001, 2012, 2013a, b; Hardin *et al.*, 2011; Callaghan *et al.*, 2017). The Florida population of Grey-headed Swamphens is believed to be the only non-native population of the Purple Swamphen complex found anywhere in the world.

32.4 Description

Purple Swamphens are huge rails with a stocky body, a large bill and prominent frontal shield, and long legs and feet (Fig. 32.2). Swamphen measurements differ by populations; adult measurements include body lengths of 38–50 cm and wingspans of 90–100 cm. Adult body mass ranges are 480–737 g in African Swamphens, 679–1310 g in Australasian Swamphens, and 505–850 g in Grey-headed Swamphens (Taylor, 1996; Callaghan and Gawlik, 2016). Adult plumage varies but is characterized by bluish or purplish bodies, often with greenish feathering on the back and/or upperwings, reddish bills, frontal shields and legs (the latter often with darker joints), reddish or orangish irides and white undertail coverts. The sexes may be indistinguishable

*Corresponding author: billpranty@hotmail.com

Fig. 32.1. Global distribution of the Grey-headed Swamphen (*Porphyrio poliocephalus*) showing its native (green) and non-native (red) ranges. The population of Grey-headed Swamphens in Florida, USA, is believed to represent the only non-native population of the Purple Swamphen complex found anywhere in the world. (Map created by Corey T. Callaghan.)

Fig. 32.2. Adult Grey-headed Swamphen feeding on its preferred food item in Florida, Gulf coast spikerush *Eleocharis cellulosa*. Swamphens feed by grasping tubers in the toes of one foot while balancing on the other foot. (Location: Wakodahatchee Wetlands, Delray Beach, Palm Beach County, Florida; 15 November 2007.) (©Photograph: Bill Pranty.)

by plumage, but males are larger and heavier than females. The wings are rather broad and rounded, and the tail is short, as is typical of rails. Juveniles have duller plumage and dusky bare parts. The chicks are downy black with a whitish bill and frontal shield and pink legs and feet. True to their name, adult Grey-headed Swamphens are distinguished from other swamphens by their pale heads and upper necks, along with a generally duller purple coloration in their plumage. Vocalizations include a variety of croaks and moans typical of rail species, including a 'creek' and a 'bark' (Pranty 2013a; Callaghan *et al.*, 2017), frequently calling in flight. Although strong fliers, swamphens often run away to escape danger, and they rarely swim. Callaghan *et al.* (2017) has provided detailed plumage descriptions, including of juveniles, as well as detailed vocalization descriptions of Purple Swamphens.

32.5 Diet

Purple Swamphens have a varied diet but feed primarily on the soft shoots of reeds (Poaceae) and rushes (Juncaceae) (Ripley, 1977). Some individuals also feed opportunistically on insects, arthropods, molluscs, fish, amphibians, reptiles, birds and rodents (Taylor, 1996). Tubers are plucked off with the bill and held with the toes of one foot, while the swamphen balances on the other foot. Sometimes the tubers are eaten whole, while at other times the stem is opened to reveal the soft inner core, and the hard outer shell is discarded (Pranty, 2013a). Swamphens also feed by climbing tall (over 1 m) reeds and eating from the top (Hamling, 1949). A diet study at three sites in southern Florida (Lake Okeechobee in Glades County, Water Conservation Area 2B in Broward County, and Stormwater Treatment Area 1W in Palm Beach) County found that Grey-headed Swamphens feed primarily on the stems of Gulf coast spikerush (*Eleocharis cellulosa*), but the study also found grit, insects, Lepidoptera, and molluscs in swamphen stomachs (Callaghan and Gawlik, 2016). Additionally, swamphens in Florida have been observed feeding on rice (*Oryza sativa*) planted in conservation areas and the Everglades Agricultural Area, and on earthworms (Oligochaeta), bird seed, grass shoots and human foods such as peas, melon rinds and cooked pasta in suburban areas (Pranty *et al.*, 2000; Pranty, 2013a).

32.6 Introduction and Invasion Pathways

The population of Grey-headed Swamphens in Florida, USA, was founded inadvertently by one or two aviculturists at Pembroke Pines, Broward County, who allowed their captive swamphens to roam freely. Evidently, some of these swamphens found suitable wetlands nearby and remained to breed. The first swamphen was photographed in May 1998, a period thought to be around 18 months after the first sighting; thus, many references list the date of discovery as 1996 or December 1996 (e.g. Pranty *et al.*, 2000; Pranty, 2001, 2012; Callaghan *et al.*, 2017). The founding population must have been small, as neither aviculturist noted any reduction in their captive flocks, but additional immigrants may have supplemented the population over several years (Pranty, 2013a). The non-captive swamphen population at Pembroke Pines numbered 84 individuals in October 1998, 135 individuals in February 1999 and smaller numbers up to February 2003 when the surveys ended (Pranty and Schnitzius, 1998; Pranty *et al.*, 2000; Pranty, 2001, 2012, B. Pranty, personal observation).

By 2000, Grey-headed Swamphens in Florida had begun to disperse, with movements noted as far as 83 km from Pembroke Pines (Pranty, 2001). Based on the swamphen's increasing numbers and expanding range, staff of two state agencies undertook an eradication programme. Begun in 2006, the programme was deemed a failure and was discontinued in 2008 (see section 32.10). By 2011, the Grey-headed Swamphen population in Florida was considered established: swamphens had been breeding for at least 15 years; they occurred at 30 discrete sites; four dispersals between 250 and 350 km from Pembroke Pines had been documented; their core area in Broward, Hendry and Palm Beach counties encompassed 2840 km^2; and a polygon drawn around all swamphen locations in the state exceeded 35,000 km^2 (Pranty, 2012). Two Grey-headed Swamphens photographed at Gainesville, Alachua County, in 2015 and 2016 currently represent the northern-most record in Florida (Callaghan *et al.*, 2017; B. Pranty, personal observation); this site is less than 112 km from the Florida/Georgia border. Colonization by swamphens of additional wetlands in Florida, and perhaps beyond, is expected if not already under way. An adult Grey-headed Swamphen photographed at Bull Island, Cape Romain National Wildlife Refuge in Charleston County, South Carolina, in October 2018 (Wagner-McLean, 2018), probably dispersed from the Florida population. If so, it would represent a dispersal of more than 775 km.

32.7 Breeding Behaviour

The breeding ecology of Grey-headed Swamphens in Florida is poorly known and is based exclusively on opportunistic observations, so most information provided here refers to studies of Purple Swamphens elsewhere in their range. Swamphens have variable mating systems, ranging from monogamous to promiscuous to communal. The mating system of Grey-headed Swamphens in Florida is unknown, but their propensity for occurring in groups suggests that they may breed communally (Callaghan *et al.*, 2017). Some Purple Swamphen populations build several nests and then select one of these for egg-laying. Swamphen nests are built close to, above or on top of shallow water and are composed of the leaves and stems of various aquatic plants (Fig. 32.3). Both sexes bring material to the nest, but it is the female that largely builds the nest. Nest construction continues through incubation (Callaghan *et al.*, 2017). Western Swamphen nests in Italy contain ramps, and other swamphen nests have had stems bent over the top to form a canopy (Callaghan *et al.*, 2017); these behaviours have not been noted in Florida.

Purple Swamphen clutch size varies by population; communal clutches are larger than single clutches. Grey-headed Swamphen clutch size in India was 4.5 eggs (range three to seven, $n = 25$; Doss *et al.*, 2009). Incubation begins after the

Fig. 32.3. An incubated Grey-headed Swamphen clutch in Florida. Information on the breeding biology of swamphens in Florida is poorly known and based entirely on opportunistic observations; this may be the only active swamphen nest found (Location: Pembroke Pines, Broward County, Florida, 25 July 1999). (©Photograph: Bill Pranty.)

penultimate egg is laid. The incubation period varies among populations but is 19–22 days for Grey-headed Swamphens in India (Doss *et al.*, 2009). From about day 2, chicks can feed themselves but are reliant on adults for most of their food until they reach about 2 months of age, after which the juveniles feed themselves exclusively (Craig, 1980). Because of the subtropical climate in Florida, Grey-headed Swamphens appear to breed year-round; downy chicks have been observed in all months except December (Pranty, 2012; Callaghan *et al.*, 2017; eBird: www.ebird.org, accessed 7 November 2019). Year-round breeding was also reported for Western Swamphens in Italy (Grussu, 1999). In contrast, egg-laying in Western Swamphens was restricted to March and April in Algeria (Mouslim *et al.*, 2014) and in Grey-headed Swamphens to January–April in southern India (Doss *et al.*, 2009).

32.8 Habitat

Throughout their range, Purple Swamphens are habitat generalists, found in and along the margins of slow-flowing or stagnant fresh- or brackish-water marshes, lakes, ponds, rivers and floodplains, as well as artificial habitats such as sewage-treatment facilities, agricultural fields, parks, golf courses and other grasslands (Taylor, 1996, 1998). Swamphens use their long legs to wade in shallow water and their long toes to walk across floating vegetation. In Florida, Grey-headed Swamphens inhabit shallowly flooded fresh-water wetlands with open or semi-open, emergent vegetation such as Gulf coast spikerush, arrowhead (*Sagittaria* spp.), pickerelweed (*Pontederia cordata*), water lily (*Nymphaea* spp.) and sawgrass (*Cladium jamaicense*) interspersed with patches of cattail (*Typha domingensis* and *T. latifolia*) and willow (*Salix caroliniana*) (Pranty *et al.*, 2000; Hardin *et al.*, 2011; Pranty, 2012). Swamphens in Florida are commonly observed in agricultural areas growing rice or sugarcane, and they often forage on residential lawns, dikes and other grassy or weedy areas adjacent to marshes (Hardin *et al.*, 2011; Callaghan *et al.*, 2017; B. Pranty and C.T. Callaghan, personal observation). In contrast, swamphens in Florida appear to avoid brackish and saltwater habitats, such as some habitats in the Everglades National Park.

32.9 Impacts

The attempted Grey-headed Swamphen eradication programme (see section 32.10) was initiated based on the concerns that swamphens in Florida would negatively impact native species such as Purple Gallinules (*Porphyrio martinica*), Common Gallinules (*Gallinula galeata*) and American Coots (*Fulica americana*), and/or substantially damage vegetation in native and human-modified wetlands (Hardin *et al.*, 2011). The swamphen population was also thought to be small and limited in range, and thus susceptible to a targeted eradication effort (Hardin *et al.*, 2011). However, the impacts of swamphen competition on native bird species have yet to be documented, and negative impacts of swamphens foraging on wetland vegetation presently appear to be localized and minor (e.g. photographs in Pranty, 2013a), but can sometimes encompass areas of approximately 500 m^2. None the less, additional studies on the potential impacts of swamphens on native flora and fauna are needed (Callaghan *et al.*, 2017).

32.10 Control

From October 2006 to December 2008, 3187 Grey-headed Swamphens were culled, mostly by shot-gunning from airboats, at seven sites in the Lake Okeechobee to Everglades region in south-eastern Florida (Hardin *et al.*, 2011). The eradication programme was discontinued because it was not reducing the size of the swamphen population, which proved to be much larger than had been anticipated, and because some swamphens had moved into wetlands where control efforts were not taking place (Hardin *et al.*, 2011). After the attempted eradication programme ended, state authorities discussed making the swamphen a game bird in Florida (United Press International, 2010), but, to date, no hunting season has been enacted. In addition to humans, predators of Grey-headed Swamphens and their eggs in Florida undoubtedly include reptiles such as snakes and American Alligators (*Alligator mississippiensis*), birds, and native and domestic mammals, but few depredations have been observed (Pranty, 2013a). One swamphen chick at Pembroke Pines was taken by a Great Blue Heron (*Ardea herodias*), and other swamphens have been traffic casualties (Pranty *et al.*, 2000; Pranty, 2013a).

32.11 Uses

The Grey-headed Swamphen in Florida has proven to be a popular species for study, from the perspectives of research, management and observation (e.g. Pranty *et al.*, 2000; Hardin *et al.*, 2011; Pranty 2012, 2013a; Callaghan and Gawlik, 2016). Through August 2018, birders in Florida had submitted 11,891 checklists to eBird (www.ebird.org) that contain Grey-headed Swamphen observations. These data originated from the following counties (with the number of reports from each in parentheses): Alachua (141), Brevard (217), Broward (1231), Collier (36), Glades (133), Hendry (1512), Indian River (42), Lee (753), Martin (33), Miami-Dade (652), Okeechobee (34), Orange (178), Osceola (8) and Palm Beach (6921). Another potential use for swamphens in Florida, as noted above, is as a game bird.

32.12 Acknowledgements

Andrew W. Kratter provided information on the swamphen's range. Valeri Ponzo improved drafts of the manuscript.

32.13 References

Boyle, W.J. Jr, Paxton, R.O. and Cutler, D.A. (1991) Hudson–Delaware region [winter 1990–1991 regional bird report]. *American Birds* 45, 253–257.

Callaghan, C.T. and Gawlik, D.E. (2016) Diet and selectivity in the *Porphyrio porphyrio* (Purple Swamphen) in Florida. *Southeastern Naturalist* 15, 1–14.

Callaghan, C.T., Pranty, B., Pyle, P. and Patten, M.A. (2017) Gray-headed Swamphen (*Porphyrio poliocephalus*), version 1.0. In: Rodewald, G. (ed.) *The Birds of North America*. The Cornell Lab of Ornithology, Ithaca, New York. Available at: https://doi.org/10.2173/bna.purswa3.01 (accessed 6 November 2019).

Craig, J.L. (1980) Pair and group breeding behavior of a communal gallinule, the Pukeko, *Porphyrio porphyrio*. *Animal Behaviour* 28, 593–603.

Doss, D.P.S., Gopukumar, N. and Sripathi, K. (2009) Breeding biology of the Purple Swamphen (*Porphyrio porphyrio*) at Tirunelveli, South India. *Wilson Journal of Ornithology* 121, 796–800.

Garcia-R, J.C. and Trewick, S.A. (2015) Dispersal and speciation in Purple Swamphens (Rallidae: Porphyrio). *Auk* 132, 140–155.

Grussu, M. (1999) Status and breeding ecology of the Purple Swamphen in Italy. *British Birds* 92, 183–192.

Hamling, H.H. (1949) Ling Reed-Hen or Purple Gallinule. *Ostrich* 20, 91–94.

Hardin, S., Donlan, E., Ward, M. and Eggeman, D. (2011) Attempted eradication of *Porphyrio porphyrio* Linnaeus in the Florida Everglades. *Management of Biological Invasions* 2011, 47–55.

Mouslim, B., Eddine, M.S., Rassim, K., Zihad, B. and Moussa, H. (2014) Aspects of the breeding ecology of the Purple Swamphen (*Porphyrio porphyrio*) in the wetland complex of Guerbes-Sanhadja, northeast Algeria. *Ostrich* 85, 185–191.

Mumford, B. (2013) The Delaware swamphen. *Birding* 45(5), 13.

Pranty, B. (2001) Purple Swamphens on the move. *Winging It* 13, 1–3.

Pranty, B. (2012) Population growth, spread, and persistence of Purple Swamphens (*Porphyrio porphyrio*) in Florida. *Florida Field Naturalist* 40, 1–12.

Pranty, B. (2013a) Introducing the Purple Swamphen: Management, taxonomy, and natural history. *Birding* 45(3), 38–45.

Pranty, B. (2013b) The Delaware swamphen: response. *Birding* 45(5), 13.

Pranty, B. and Schnitzius, K. (1998) Purple Swamphens found in Florida. *Winging It* 10(11), 7.

Pranty, B., Schnitzius, K., Schnitzius, K. and Lovell, H.W. (2000) Discovery, origin, and current distribution of the Purple Swamphen (*Porphyrio porphyrio*) in Florida. *Florida Field Naturalist* 28, 1–11.

Ripley, S.D. (1977) *Rails of the World: A Monograph*. David R. Godine, Boston, Massachusetts.

Sánchez-Lafuente, A.M., Valera, F., Godino, A. and Muela, F. (2001) Natural and human-mediated factors in the recovery and subsequent expansion of the Purple Swamphen *Porphyrio porphyrio* L. (Rallidae) in the Iberian Peninsula. *Biodiversity and Conservation* 10, 851–867.

Sangster, G. (1998) Purple Swamphen is a complex of species. *Dutch Birding* 20, 13–22.

Taylor, B. (1998) *A Guide to the Rails, Crakes, Gallinules, and Coots of the World*. Yale University Press, New Haven, Connecticut.

Taylor, P.B. (1996) Family Rallidae (Rails, Gallinules, and Coots). In: del Hoyo, J., Elliott, A. and Sargatal, J. (eds) *Handbook of Birds of the World*, Vol. 3: *Hoatzins to Auks*. Lynx Edicions, Barcelona, Spain.

United Press International (2010) Invasive swamphen thrives in Florida. United Press International, Science News. Available at: www.upi.com/Invasive-swamphen-thrives-in-Florida/17111274109827/ (accessed 16 October 2018).

Wagner-McLean, N. (2018) Post in Facebook group: ABA Rare Bird Alert. www.facebook.com/groups/ABArare/permalink/2091122050948427 (accessed 20 November 2018).

33 African Sacred Ibis (*Threskiornis aethiopicus* Latham, 1790)

Jean-François Maillard[1]*, Carlos Gutiérrez-Expósito[2] and Pierre Yésou[3]
[1]*Office Français de la Biodiversité, 8 Boulevard Albert Einstein, 44300 Nantes, France;* [2]*Estación Biológica de Doñana (CSIC), Americo Vespucio 26, Isla de la Cartuja, Sevilla 41032, Spain;* [3]*3 rue de Châteaulin, 44000 Nantes, France*

Citation: Maillard, J.-F., Gutiérrez-Expósito, C. and Yésou, P. (2020) African Sacred Ibis (*Threskiornis aethiopicus* Latham, 1790). In: Downs, C.T. and Hart, L.A. (eds) *Invasive Birds: Global Trends and Impacts*. CAB International, Wallingford, UK, pp. 248–251.

33.1 Common Names

African Sacred Ibis, Sacred Ibis (English), Ibis sacré (French), Ibis Sagrado (Spanish).

33.2 Distribution

The African Sacred Ibis (*Threskiornis aethiopicus* Latham, 1790) is a common and widespread bird throughout large parts of sub-Saharan Africa (Fig. 33.1) (Barnes, 2005). A small isolated population persists in southern Iraq (Fazaa *et al.*, 2017). The species still bred in the Nile Valley, Egypt, in the early 19th century but disappeared within a few decades (del Hoyo *et al.*, 1992). All African Sacred Ibises found in the wild outside the African continent have a captive origin, having been deliberately or accidentally released. Breeding outside the native range has occurred in the USA, the Netherlands and Spain, and currently there are still wild populations breeding in France, Italy, Taiwan and the United Arab Emirates. Currently, free-flying African Sacred Ibises are known in Belgium, Germany, the UK, the Netherlands, Portugal and Spain without evident breeding (Yésou *et al.*, 2017).

33.3 Description

The nominate form of the African Sacred Ibis is a large, mainly white bird with a long decurved dark bill, a wholly bare dark head and neck, and a black trailing edge to the wings (Figs 33.2 and 33.3) (Barnes, 2005). Its length is about 65–75 cm, its wingspan 112–124 cm and its weight about 1500 g (del Hoyo *et al.*, 1992). The iris is dark brown and the legs are black. Breeding adults show a 'dirty' yellow-toned plumage, with areas of bare red skin on the underwing and prominent ornamental black feathers on the rear back. Juveniles show a feathered neck and head; their tertials and the tip to their remiges are brownish-black. First-winter birds are intermediate, with the head and neck still well feathered and some blackish ornamental feathers. They also show a variable amount of black along the centre of the tertials and greater coverts and, in a few birds, on the median coverts (Barnes, 2005).

33.4 Diet

In their native and introduced ranges, African Sacred Ibises feed by day, mainly in flocks and often in mixed-species flocks (Barnes, 2005; Yésou *et al.*, 2017). They feed on locusts, grasshoppers, aquatic beetles, crustaceans, molluscs, and a variety of other aquatic and terrestrial invertebrates (Barnes, 2005). In addition, they can feed on fish, amphibians, lizards, birds and small mammals. Breeding birds sometimes raid other species' nests for eggs and chicks (Barnes, 2005; Yésou *et al.*, 2017). They sometimes scavenge at refuse sites, abattoirs or feedlots (Barnes, 2005; Yésou *et al.*, 2017). In urban areas, their diets sometimes include vegetable matter and anthropogenic foods (Barnes, 2005; Yésou *et al.*, 2017).

33.5 Introduction and Invasion Pathways

Escapee African Sacred Ibis remained a rare sight until it became more common for zoological collections to house free-flying

*Corresponding author: jean-francois.maillard@ofb.gouv.fr

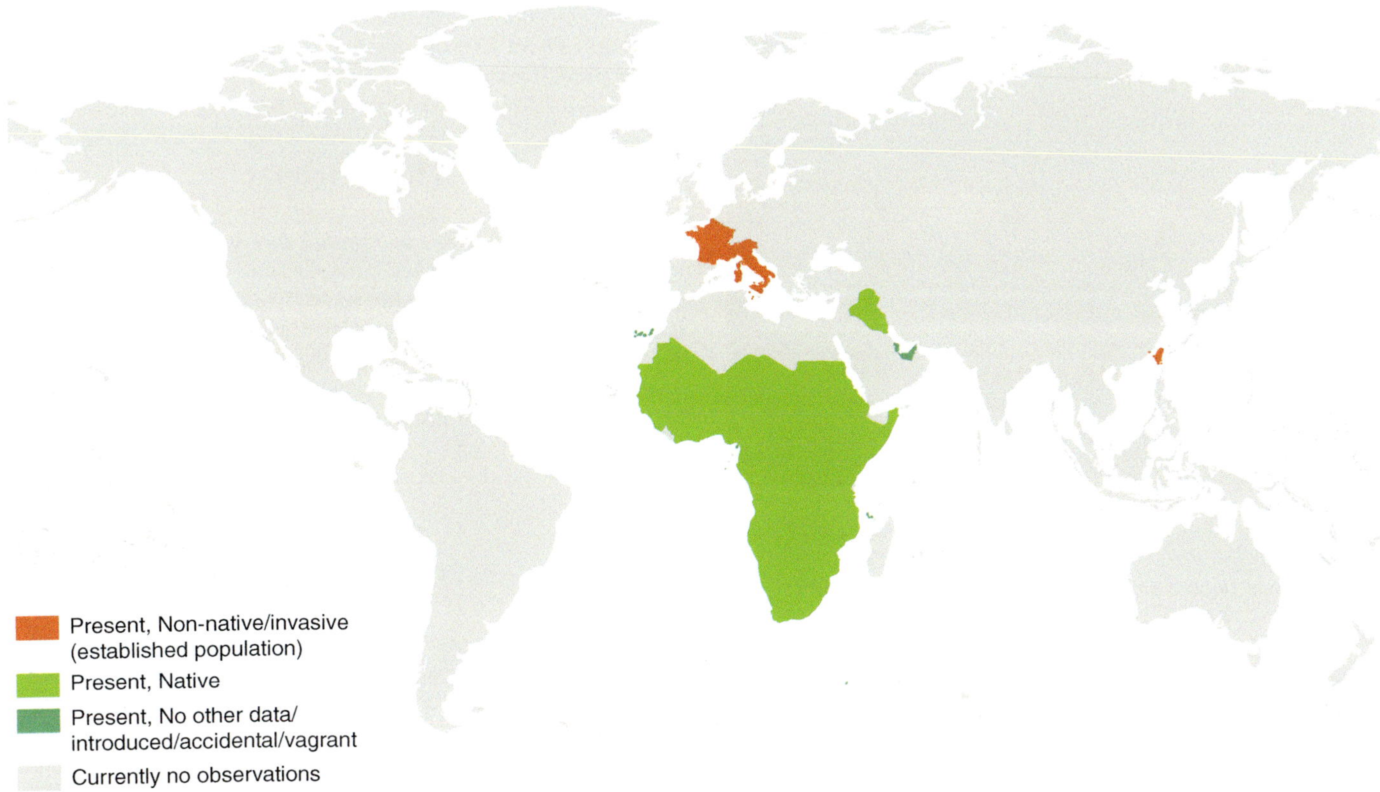

Fig. 33.1. Global distribution of the African Sacred Ibis (*Threskiornis aethiopicus*) showing its native (green) and non-native (red) ranges.

Fig. 33.2. Adult African Sacred Ibises flying in Doñana National Park, Spain (spring 2008). These birds were extirpated immediately after they started showing breeding behaviour at a mixed heronry. (©Photograph: José Antonio Sencianes.)

groups of ibises; the increasing number of escapees then led to the establishment of significant non-native populations in some areas (Fàbregas *et al.*, 2010; Yésou *et al.*, 2017).

In France, 20 African Sacred Ibises were introduced in the 1970s to a private park in southern Brittany. One decade later, the young, left free to fly, moved beyond the park, wandering hundreds of kilometres away along the Atlantic Coast. There were about 1400–1800 pairs in the wild in 2007, totalling more than 5000 individuals before control measures were employed, resulting in about 250 birds remaining in January 2020 (OFB - unpublished data).

In southern France, a similar case occurred leading to about 300 individuals and around 100 pairs in the wild in 2007; this population has since been fully extirpated. In Italy, breeding in the upper Po Valley, Piedmont, began in 1989, and a census coordinated in late 2016 by the Piemontese Ornithological Group found 4068 individuals in 13 roosts in north-west Italy. In Spain, free-ranging ibises have nested in the Canary Islands (Fuerteventura and Lanzarote) since at least 1997, with a few pairs nesting sporadically to date (Lorenzo and González, 2015). A colony formerly established in a park in Barcelona was controlled in 2001 (Clavell, 2003). In spring 2008, a group of 11 birds of wild French origin (including one bird colour-ringed as a pullus at Lac Grand Lieu, near Nantes, in 2006) was settled at Doñana National Park in a mixed heronry, and nest building began: culling action was immediately carried out by National Park staff before laying could take place (Gutiérrez-Expósito, 2008).

Since 1989, a small introduced population persists on Sir Bani Yas island in the United Arab Emirates (Yésou and Clergeau, 2005) and probably disperses to Bahrain and Qatar. A dozen birds that escaped from a zoological garden in Taiwan in the early 1990s rapidly established a thriving population, with hundreds of birds living in the wild by 2002 (Ding *et al.*, 2017).

In the Netherlands, free-flying African Sacred Ibises that escaped from a zoological park attempted to breed in 2001, and successful breeding occurred in 2002, but eradication measures

Fig. 33.3. An adult African Sacred Ibis feeding with two Cattle Egrets (*Bubulcus ibis*) at El Rocío, Doñana National Park, Spain. This bird was tagged as a chick in the colony at Lac Grand Lieu, France, and had dispersed up to 1200 km; it was one of the members of the colonizing group found in the area in spring 2008 that were later extirpated. (©Photograph: Carlos Gutiérrez-Expósito.)

reduced the number to a few birds with possibly new escapees or birds arriving from Germany (Smits, 2010; Robert *et al.*, 2013).

In the USA, five African Sacred Ibises escaped from Miami Zoo, Florida, in 1992 when Hurricane Andrew destroyed their enclosure. Local government and state and federal administrations joined forces to eradicate the species, and now all African Sacred Ibises living in the wild in southern Florida, approximately 75 birds, have been removed (Herring and Gawlik, 2008).

33.6 Breeding Behaviour

The African Sacred Ibis is a colonial breeding species. Nests comprise a platform, usually built on islands or near the water on a variety of supports, mostly trees and bushes but sometimes on the ground or among rocks, using sticks, reeds, twigs or driftwood collected close to the site (Hancock *et al.*, 1992). A fidelity to nesting sites is observed year after year, and pairs lay quickly after arrival at the colony. In the native range, breeding usually starts during or just after the rain season. In the Loire estuary, western France, the first clutches are laid in late March or early April, and laying peaks in the second half of April and occurs up to the last week of July; whether late nests are related to late breeders or to replacement clutches is not known. Females lay two to four eggs, and both parents alternately incubate the eggs and share the feeding duties once the eggs have hatched. The survival rate for African Sacred Ibis chicks is low, and it is rare that more than one chick leaves the nest alive in its native range. Parents continue to feed the young until they are fully fledged. In France, the average (± SD) clutch size (2.41 ± 0.68 eggs per nest, $n = 58$) and the productivity (1.46 young fledged per pair, $n = 486$) were higher than most values from the African natural range of the species (Yésou *et al.*, 2006). Yésou *et al.* (2006) suggested that both relatively low predation rates and rich food resources, including rubbish dumps, helped higher breeding success.

33.7 Habitat

The African Sacred Ibis is adapted to a wide range of mainly wet habitats including inland marshes and freshwater wetlands, mud flats, meadows and cultivated fields. The species also inhabits coastal lagoons, intertidal areas, offshore islands (del Hoyo *et al.*, 1992) and mangroves. It may also visit rubbish dump or dung heaps. It feeds by wading in very shallow wetlands or slowly stomping in wet pastures with soft soil. In France, it also forages in waste-water plants, in ploughed fields or in open-air poultry farms.

33.8 Impacts

In its native range, the African Sacred Ibis can raid nests for eggs and/or chicks of a variety of bird species including pelicans, herons, spoonbills and cormorants, gulls and terns on offshore islands, and two endangered species, the African Penguin (*Spheniscus demersus*) and the Cape Cormorant (*Phalacrocorax capensis*). They can even take eggs of the Nile Crocodile (*Crocodylus niloticus*; Williams and Ward, 2006). In introduction areas, such predation behaviour has been observed on a variety of bird species, including the Mallard (*Anas platyrhynchos*), Garganey (*Anas querquedula*), European Shag (*Phalacrocorax aristotelis*), Cattle Egret (*Bubulcus ibis*), Little Egret (*Egretta garzetta*), Night Heron (*Nycticorax nycticorax*), Squacco Heron (*Ardeola ralloides*), Black-winged Stilt (*Himantopus himantopus*), Northern Lapwing (*Vanellus vanellus*), Common Redshank (*Tringa tetanus*), Black Tern (*Chlidonias niger*), Sandwich Tern (*Sterna sandvicensis*), Common Tern (*Sterna hirundo*), Black-headed Gull (*Chroicocephalus ridibundus*) and European Herring Gull (*Larus argentatus*). Predation is sometimes opportunistic, but there are documented instances of African Sacred Ibises deliberately and systematically searching for eggs or chicks (Yésou *et al.*, 2017).

In France, farmers, particularly those rearing cattle and ducks, expressed concern that the growing numbers of African Sacred Ibises might be significant carriers of disease. African Sacred Ibises regularly feed at rubbish dumps and can also commute to farmland, which may exacerbate the problem. Studies have shown that the risk of the African Sacred Ibis to farm animals is only moderate, except in cases where specific outbreaks of an avian disease are causing concern (Bastian *et al.*, 2010), although Vorimore *et al.* (2013) described a pathogen new to science from non-native African Sacred Ibises.

33.9 Control

Apart from wild boars (*Sus scrofa*) destroying clutches in reedbeds, the African Sacred Ibis has no obvious natural predators in Europe. In France, the national wildlife service (Office National de la Chasse et de la Faune Sauvage), renamed from 2020 Office Français de la Biodiversité has been carrying out an eradication programme of the African Sacred Ibis since 2007, culling adults and subadults with firearms. Killing adults is the best way to significantly reduce a bird population with a long lifespan. Culling gives better results during the rearing period of chicks, as adults use specific paths from the nest to the foraging sites and back. The use of flocks of artificial African Sacred Ibis decoys in the early morning, imitating birds foraging on meadows, improves the shooting efficiency. In winter, the birds do not have the same feeding pattern and are more dispersed, reducing the efficiency of shooting. Shooting on breeding sites is not recommended when African Sacred Ibises nest in mixed colonies with other species. In nature reserves where shooting is prohibited, a programme of egg sterilization ensures that most eggs do not hatch. In western France, about of 9150 African Sacred Ibises were culled between 2007 and 2019, and 3500 clutches were sterilized at Lac de Grand Lieu nature reserve colonies in 2009–2019. The population declined from 5000 birds in 2006 to around around 250 birds in January 2020. However, the population still disperse over much of the area occupied before the eradication programme began, now in much smaller groups at a much lower density, which makes completion of the programme difficult (personal observation). In recent years on the Mediterranean Coast, 485 ibises were removed from the wild when coming back to a zoological park to feed; 395 of them were killed and 90 were captured alive and placed in an aviary.

33.10 Uses

No specific use has been reported apart from being used as an ornamental species in zoological gardens.

33.11 References

Barnes, K.N. (2005) African Sacred Ibis. In: Hockey, P.A.R., Dean, W.R.J. and Ryan, P.G. (eds) *Roberts Birds of Southern Africa*, 7th edn. Trustees of the John Voelcker Bird Book Fund, Cape Town, pp. 611–612.

Bastian, S., Passet, A., Lagrange, P., Laroucau, K., Pellerin, J.L., *et al.* (2010) Contact of livestock with an invasive species, the African Sacred Ibis (*Threskiornis aethiopicus*) – a multidisciplinary approach for risk assessment. In: *9th EWDA Conference*, 13–16 September 2010, Vlieland, Netherlands.

Clavell, J. (2003) Especies introducidas, no establecidas. In: Martí, R. and del Moral, J.C. (eds) *Atlas de las Aves Reproductoras de España*. Dirección General de Conservación de la Naturaleza-Sociedad Española de Ornitología, Madrid, pp. 39–47.

del Hoyo, J., Elliott, A. and Sargatal, J. (eds) (1992) *Handbook of the Birds of the World*, Vol. 1. Lynx Edicions, Barcelona, Spain.

Ding, T.-S., Juan, C.-S., Lin, R.-S., Pan, C.-Y., Tsai, Y.-J., *et al.* (2017) *The 2017 CWBF Checklist of the Birds of Taiwan*. Chinese Wild Bird Federation, Taipei.

Fàbregas, M.C., Guillén-Salazar, F. and Garcés-Narro, C. (2010) The risk of zoological parks as potential pathways for the introduction of non-indigenous species. *Biological Invasions* 12, 3627–3636.

Fazaa, N.A., Dunn, J.C. and Whittingham, M.J. (2017) Distributions and community composition of birds in Iraq's Central Marsh. *International Journal of Biodiversity* 2017, 1–28.

Gutiérrez-Expósito, C. (2008) Propuesta de seguimiento y control de las poblaciones de Ibis sagrado (*Threskiornis aethiopicus*) en Doñana. Espacio Natural Doñana, Spain.

Hancock, J.A., Kushlan, J.A. and Kiahl, M.P. (1992) Sacred Ibis. In: *Storks, Ibises and Spoonbills of the World*. Academic Press, London. pp. 211–215.

Herring, G. and Gawlik, D.E. (2008) Potential for successful population establishment of the nonindigenous sacred ibis in the Florida Everglades. *Biological Invasions* 10, 969–976.

Lorenzo, J.A. and González, C. (2015) Qué sabemos de las aves exóticas en Canarias? Memoria técnica. SEO/BirdLife y Dirección General de Protección de la Naturaleza del Gobierno de Canarias, Santa Cruz, Tenerife, Spain

Robert, H., Lafontaine, R.-M., Delsinne, T. and Beudels-Jamar, R.C. (2013) Risk analysis of the Sacred Ibis *Threskiornis aethiopicus* (Latham 1790). Risk analysis report of non-native organisms in Belgium from the Royal Belgian Institute of Natural Sciences for the Federal Public Service Health, Food Chain Safety and Environment. Royal Belgian Institute of Natural Sciences, Brussels.

Smits, I.R. (2010) A risk analysis of the sacred ibis in The Netherlands. Bureau Waardenburg bv, Culemborg, Netherlands. Available at: www.buwa.nl/en/sacred-ibis-in-the-netherlands.html (accessed 6 November 2019).

Vorimore, F., Hsia, R.C., Huot-Creasy, H., Bastian, S., Deruyter, L., *et al.* (2013) Isolation of a new *Chlamydia* species from the feral Sacred Ibis (*Threskiornis aethiopicus*): *Chlamydia ibidis*. *PLoS One* 8, e74823.

Williams, A.J. and Ward, V.L. (2006) Sacred Ibis and Gray Heron predation of Cape Cormorant eggs and chicks; and a review of ciconiiform birds as seabird predators. *Waterbirds* 29, 321–327.

Yésou, P. and Clergeau, P. (2005) Sacred Ibis: a new invasive species in Europe. *Birding World* 18, 517–526.

Yésou, P., Cabelguen, J. and Potiron, J.-L. (2006) Quelques aspects de la reproduction de l'Ibis sacré *Threskiornis aethiopicus* dans l'estuaire de la Loire. *Alauda* 74, 421-427.

Yésou, P., Clergeau, P., Bastian, S., Reeber, S. and Maillard, J-F. (2017) The Sacred Ibis in Europe: ecology and management. *British Birds* 110, 197–212.

34 Great Horned Owl (*Bubo virginianus* Gmelin, 1788)

Alice Cibois[1]*, Jean-Claude Thibault[2] and Jean-Yves Meyer[3]

[1]*Natural History Museum of Geneva, CP 6434, CH 1211 Geneva 6, Switzerland;* [2]*Muséum National d'histoire Naturelle, F-75005 Paris, France;* [3]*Délégation à la Recherche, Gouvernement de la Polynésie Française, BP 20981, 98713 Papeete, Tahiti, French Polynesia*

Citation: Cibois, A., Thibault, J.-C. and Meyer, J.-Y. (2020) Great Horned Owl (*Bubo virginianus* Gmelin, 1788). In: Downs, C.T. and Hart, L.A. (eds) *Invasive Birds: Global Trends and Impacts*. CAB International, Wallingford, UK, pp. 252–254.

34.1 Common Names

Great Horned Owl (English), Grand Duc de Virginie, Grand Duc d'Amérique (French), Búho Americano (Spanish).

34.2 Nomenclature

The Great Horned Owl (*Bubo virginianus* Gmelin, 1788) has 15 subspecies currently considered (Holt *et al.*, 2018). It belongs to the order Strigiformes and family Strigidae.

34.3 Distribution

The Great Horned Owl is native to North America (from Alaska and the Northwest Territories to Mexico) and South America (Andes and lowlands from Colombia to Peru, Brazil and Argentina) (Fig. 34.1). It has been introduced to the island of Hiva Oa, Marquesas (French Polynesia, South Pacific).

34.4 Description

The Great Horned Owl is a large and bulky owl, 45–60 cm tall, with a wingspan of 91–152 cm and body mass between 0.9 and 2.5 kg with stout ear-tufts (Fig. 34.2). Its plumage is greyish-brown, mottled and vermiculated above and barred below, with a white throat. Clinal variation in plumage exists, with the eastern populations (nominate subspecies) rufous below with a tawny-orange face, western populations darker and northern subspecies greyer. Females are on average larger and more heavily marked than males (Holt *et al.*, 2018).

34.5 Diet

In its native range, small mammals represent the bulk of the Great Horned Owl's diet (around 90%) and include lagomorphs, rodents, voles and ground squirrels. They also feed on birds, amphibians, reptiles, fish, insects, other invertebrates and sometimes carrion (Holt *et al.*, 2018).

On Hiva Oa, the diet of the Great Horned Owl has not been studied but probably includes seabirds (nocturnal petrels), native and introduced land birds such as Fruit Doves (*Ptilinopus* spp.), Kingfishers (*Todiramphus* spp.) and Chickens (*Gallus* spp.), as well as rats (*Rattus* spp.) (Thibault and Cibois, 2017).

34.6 Introduction and Invasion Pathways

The Great Horned Owl was probably introduced only once, with the release of eight birds bought in San Francisco, USA, in December 1927 by the Catholic Mission (Bishop David Lecadre), with the objective of eliminating rats from Hiva Oa (Holyoak and Thibault, 1984). Surveys in the 2000s and 2010s suggested that the species has not yet reached the nearby island of Tahuata (only 4 km from Hiva Oa) (Thibault and Cibois, 2017). The numbers of Great Horned Owls on Hiva Oa are currently not known, but they are regularly recorded by local people, although rarely seen or photographed by ornithologists.

*Corresponding author: alice.cibois@ville-ge.ch

Fig. 34.1. Map of the native (green) and non-native (red dot) distribution ranges of the Great Horned Owl. (Modified from *Handbook of the Birds of the World Alive*, www.hbw.com/.)

Fig. 34.2. An adult Great Horned Owl on Hiva Oa, French Polynesia. (©Photograph: Thomas Ghestemme, SOP Manu.)

34.7 Breeding Behaviour

In its native range, the Great Horned Owl builds its nest in a variety of locations. Examples include in old raptor or corvid nest, in a hollow in a snag, on a human-made platform, in a depression on the ground or on a cliff, and on a cave entrance. Their clutch size varies from one to three eggs. Generally, incubation lasts approximately 28–30 days. The fledglings leave the nest at 6–7 weeks and are fed by the parents for up to 5 months (Holt *et al.*, 2018). There is no breeding behaviour recorded in its introduced range.

34.8 Habitat

The Great Horned Owl, in its native range, is found in a wide array of habitat types with open woodlands and groves (Sibley, 2000; Holt *et al.*, 2018). It occurs locally in desert or rocky areas with woodlands. It generally avoids dense rainforests and cloud forests. In its introduced range, it probably hunts and occupies all habitats from sea level up to 1000 m. It is also found in gardens and plantations near villages, and in dense forests at low and mid-elevations (Anon., 1994; Thibault and Cibois, 2017).

34.9 Impacts

Predation by the Great Horned Owl is possibly a factor of the decline of two endemic landbirds, the Marquesan Kingfisher (*Todiramphus godeffroyi*) and the White-capped Fruit-dove (*Ptilinopus dupetithouarsii*) (Thibault and Cibois, 2017). It probably hastened the extinction of the Red-moustached Fruit-dove (*Ptilinopus mercierii*). As with the Swamp Harrier (*Circus approximans*) in the Society Islands, the introduction of this raptor has had no assessable impact on the control of rats.

34.10 Control

The Great Horned Owl has no natural predators. Control methods have not yet been tested.

34.11 Uses

Injured Great Horned Owls are sometimes kept as pets in aviaries (personal observation).

34.12 Notes

Introduction of the Great Horned Owl to an avifauna devoid of native raptors, like that of the Swamp Harrier in the Society Islands, has had major consequences on the native birds.

34.13 References

Anon. (1994) Rapaces introduits: éperviers-hiboux. Te Manu No. 9, December 1994. Bulletin de la Société d'Ornithologie de Polynésie, French Polynesia.

Holt, D.W., Berkley, R., Deppe, C., Enríquez Rocha, P., Petersen, J.L., *et al.* (2018) Great Horned Owl (*Bubo virginianus*). In: del Hoyo, J., Elliott, A., Sargatal, J., Christie, D.A. and de Juana, E. (eds) *Handbook of the Birds of the World Alive*. Lynx Edicions, Barcelona, Spain. Available at: www.hbw.com/species/great-horned-owl-bubo-virginianus (accessed 6 November 2019).

Holyoak, D.T. and Thibault, J.-C. (1984) *Contribution à l'Etude des Oiseaux de Polynésie Orientale*. Mémoires du Muséum National d'Histoire Naturelle, Éditions du Muséum, Paris.

Sibley, D.A. (2000) *The Sibley Guide to Birds*. Alfred A. Knopf, New York.

Thibault, J.C. and Cibois, A. (2017) *Birds of Eastern Polynesia. A Biogeographic Atlas*. Lynx Edicions, Barcelona, Spain.

35 Swamp Harrier (*Circus approximans* Peale, 1848)

Alice Cibois[1]*, Jean-Claude Thibault[2] and Jean-Yves Meyer[3]

[1]*Natural History Museum of Geneva, CP 6434, CH 1211 Geneva 6, Switzerland;* [2]*Muséum National d'Histoire Naturelle, 55 rue Buffon, F-75005 Paris, France;* [3]*Délégation à la Recherche, Gouvernement de la Polynésie Française, BP 20981, 98713 Papeete, Tahiti, French Polynesia*

Citation: Cibois, A., Thibault, J.-C. and Meyer, J.-Y. (2020). Swamp Harrier (*Circus approximans* Peale, 1848). In: Downs, C.T. and Hart, L.A. (eds) *Invasive Birds: Global Trends and Impacts*. CAB International, Wallingford, UK, pp. 255–257.

35.1 Common Names

Swamp Harrier, Australasian Harrier, Pacific Harrier (English), Busard de Gould (French), Manu'amu moa (Tahitian).

35.2 Nomenclature

The Swamp Harrier (*Circus approximans* Peale, 1848) was previously considered a subspecies of the Western Marsh Harrier (*Circus aeruginosus* Linnaeus, 1758). Two subspecies are currently recognized: *C. a. approximans* Peale, 1848 and *C. approximans gouldi* Bonaparte, 1850 (Debus and Kirwan, 2018). The Swamp Harrier belongs to the family Accipitridae and is one of a few raptors considered invasive.

35.3 Distribution

The Swamp Harrier is native to Australia (including Tasmania), New Zealand (also resident in the Chatham Islands), Fiji, New Caledonia, Tonga and Vanuatu (Fig. 35.1). It is a regular visitor to islands at the margin of its range including the Kermadec Islands, Norfolk and Lord Howe Islands, and Wallis and Futuna Islands. It is considered a vagrant in the Subantarctic Islands and isolated islands in western Polynesia (Debus and Kirwan, 2018).

The Swamp Harrier has been introduced and is considered an invasive on the Society Islands (French Polynesia, South Pacific), including the high volcanic islands of Tahiti, Moorea, Raiatea, Tahaa, Bora Bora, Huahine and Maupiti, and the atolls of Tetiaroa and Tupai (it is not recorded on Maio and the small uninhabited island of Mehetia). It is a vagrant on the Cook Islands and the Austral Islands (French Polynesia) (Thibault and Cibois, 2017).

35.4 Description

The Swamp Harrier is a brown harrier, 48–61 cm long, with a wingspan of 118–145 cm and a body mass of between 0.4 and 10 kg) (Fig. 35.2). It has round-tipped and well-fingered wings, and a long narrow tail. The females are darker and more rufous below, compared with males (Debus and Kirwan, 2018).

35.5 Diet

In its native range, the Swamp Harrier feeds mostly on small mammals and carrion. It also feeds on birds and eggs, reptiles, amphibians, fish and large insects (Debus and Kirwan, 2018). In its introduced range, it feeds mainly on native birds (seabirds and land birds), including endemic species (Leopold, 1965; Anon., 1994; Wilcox and Spotswood, 2011; Faulquier, 2015).

35.6 Introduction and Invasion Pathways

The Swamp Harrier was first introduced to Tahiti *c.*1885 by the German Consulate to control rats (Holyoak and Thibault, 1984). It rapidly colonized the other islands of the Society Archipelago during the 20th century (found in Bora Bora in 1922) and was considered 'noxious' (as it eats chickens, thus its Tahitian name) and was hunted between the 1900s and 1930s. Vagrant birds are not uncommon, and its spread to the archipelagos close to the

*Corresponding author: alice.cibois@ville-ge.ch

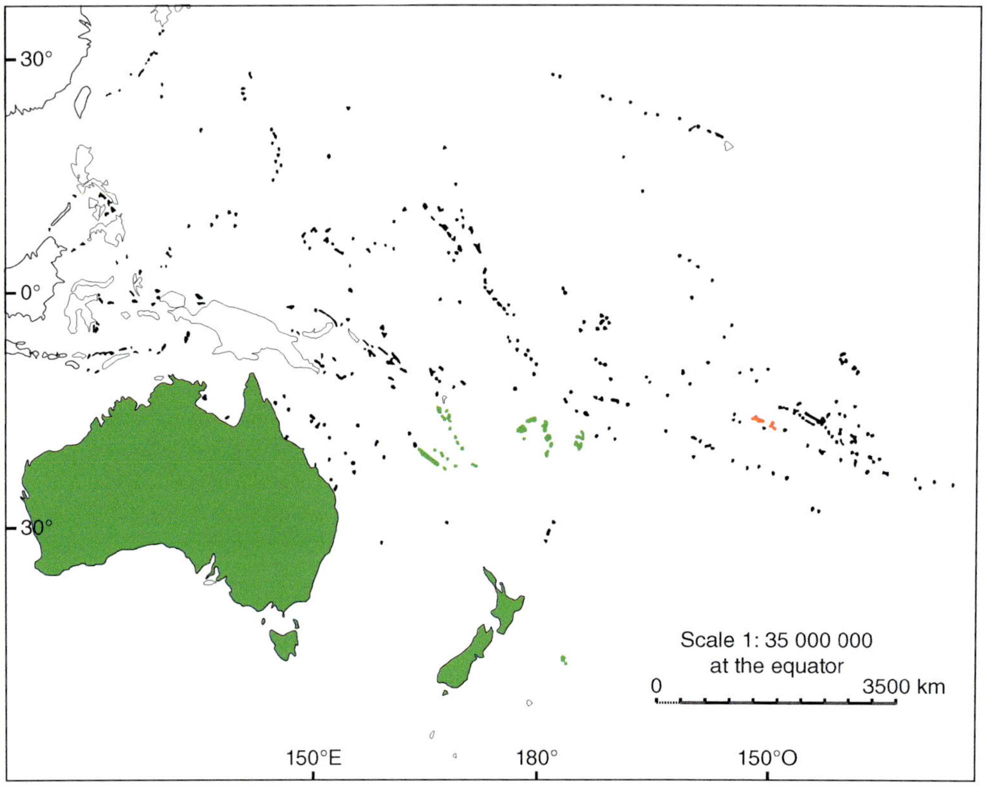

Fig. 35.1. Distribution of the Swamp Harrier (*Circus approximans*) showing its native (green) and non-native (red) ranges.

Fig. 35.2. An adult Swamp Harrier. (©Photograph: Frédéric Jacq.)

Society Islands is possible in the near future (Thibault and Cibois, 2017).

35.7 Breeding Behaviour

In its native range, Swamp Harrier breeding takes place between September and February. Their nests are built on the ground or in low trees. Their clutch size is three to four eggs and their incubation is between 31 and 34 days. Their fledglings leave the nest at 6–7 weeks and are fed by the parents for up to 6 weeks (Debus and Kirwan, 2018). No breeding behaviour has been recorded in its introduced range.

35.8 Habitat

The native habitat of the Swamp Harrier includes forests, shrublands, wetlands, marshes, grasslands, croplands and pastures, from sea level up to 1700 m. In its introduced range on Tahiti, Moorea and Raiatea, it forages more often over open ridges (up to 1900 m on Tahiti) than in valleys. On the lower islands, it is also seen foraging on the shore and frequently near villages. It also occasionally forages in the understorey of large-canopy trees, and has been observed in urban areas of Tahiti (personal observation).

35.9 Impacts

The Swamp Harrier has had strong negative impacts on seabird colonies (including petrels, e.g. *Pseudobulweria rostrata*) and on native land birds. It is probably one of the drivers leading to the extinction of the Polynesian Imperial-pigeon (*Ducula aurorae*) in the Society Islands. The low density of endemic fruit doves (*Ptilinopus* spp.) could also be due to Swamp Harrier predation (Thibault and Cibois, 2017). As with the Great Horned Owl (*Bubo virginianus*) in the Marquesas Islands, the introduction of this raptor has had little apparent impact on the control of rats.

35.10 Control

The Swamp Harrier has no natural predators in its introduced range except for humans. Few control methods have been tested, except using guns to shoot them (personal observation).

35.11 Notes

The introduction of the Swamp Harrier to areas with avifauna devoid of raptors, like that of the Great Horned Owl to the Marquesas Islands, has had major consequences on the native birds. Additionally, the Swamp Harrier's dispersal capacity represents a high risk for the avifauna of nearby archipelagos.

35.12 References

Anon. (1994) Rapaces introduits: éperviers-hiboux. Te Manu No. 9, December 1994. Bulletin de la Société d'Ornithologie de Polynésie, French Polynesia.

Debus, S. and Kirwan, G.M. (2018) Swamp Harrier (*Circus approximans*). In: del Hoyo, J., Elliott, A., Sargatal, J., Christie, D.A. and de Juana, E. (eds) *Handbook of the Birds of the World Alive*. Lynx Edicions, Barcelona, Spain. Available at: www.hbw.com/species/swamp-harrier-circus-approximans (accessed 6 November 2019).

Faulquier, L. (2015) Etude du Pétrel de Tahiti sur les plateaux Te Mehani (Raiatea). *Te Manu* 86, 4–8.

Holyoak, D.T. and Thibault, J.-C. (1984) Contribution à l'étude des oiseaux de Polynésie orientale. *Bulletin du Muséum National d'Histoire Naturelle, Paris* 127, 1–209.

Leopold, A.S. (1965) Harrier observed catching a fairy tern in Tahiti. *Condor* 67, 91.

Thibault, J.C. and Cibois, A. (2017) *Birds of Eastern Polynesia. A Biogeographic Atlas*. Lynx Edicions, Barcelona, Spain.

Wilcox, R.C. and Spotswood, E. (2011) Introduced predators and seabird predation on Mo'orea, French Polynesia. *Notornis* 58, 39–42.

36 Continental Analysis of Invasive Birds: Australia and New Zealand

David A. Ehlers Smith*, Yvette C. Ehlers Smith and Colleen T. Downs

Centre for Excellence in Invasion Biology, School of Life Sciences, University of KwaZulu-Natal, Pietermaritzburg 3209, South Africa

Citation: Ehlers Smith, D.A., Ehlers Smith, Y.C. and Downs, C.T. (2020) Continental analysis of invasive birds: Australia and New Zealand. In: Downs, C.T. and Hart, L.A. (eds) *Invasive Birds: Global Trends and Impacts*. CAB International, Wallingford, UK, pp. 258–264.

36.1 Introduction

The term 'Australasia' has several classifications, depending on the basis for interpretation: geopolitical, physiographical or ecological. For our purpose of exploring avian invasions, we focus on Australia and New Zealand, arguably the most significant land masses of the Australasian region.

36.1.1 Australia

Australia is considered the 'island continent' because of its area – some 7692 million km^2, including 34,218 km of coastline – and relative isolation. Owing to this, the climate is heavily dictated by oceanic currents and varied topography. As the sixth-largest country by total area, Australia spans climatic zones from the tropics at its northern extremity (9°S) to the temperate at the southern extremity of Tasmania (44°S). Its expansive longitudinal spread (112–154°E) similarly facilitates a varied climate, from tropical rainforests in the north-east, sprawling mountain ranges with alpine heaths in the south-east, xeric shrub- and woodland on the periphery of arid desert in the interior, broad plateaus to the west and spectacular sandy coastlines for much of the country. The desert landscape, popularly known as the Outback, represents the greatest land-cover classification, and the mean annual rainfall for Australia is less than 500 mm; as such, it is the driest inhabited continent on Earth. At 2.8 individuals/km^2, the human population density is also relatively very low; however, most inhabitants occupy the more mesic coastline regions, particularly the temperate south-east.

The history of human occupation of Australia is a rich and fascinating – if somewhat turbulent – account. Australian Aboriginal culture is one of the Earth's most ancient: recent studies estimate that the earliest occupation of the Australian continent occurred 65,000–70,000 years ago (Clarkson *et al.*, 2017) via land bridges and short sea voyages from South-east Asia. From that time, Aboriginal culture organized into complex hunter–gatherer societies, and population estimates at the time of first contact with European explorers is predicted to be 750,000–1,000,000, with over 250 distinct languages spoken (Australian Bureau of Statistics, 2012; www.abs.gov.au/, accessed 7 November 2019). Dutch navigators in 1606 were the first to sight, make landfall and meet with Aboriginal peoples at the Cape York Peninsular in the north-east. After that time, Spanish, Dutch and British navigators increasingly explored and mapped the eastern coastline, culminating in the claiming of the east coast for Britain by James Cook in 1770, under the name New South Wales. After Britain lost the American colonies in the early 1770s, colonization of the Australian continent began in earnest. Over the next century, colonial expansion and transportation of convicts from Britain saw the European population of Australia and Tasmania swell, while the indigenous Aboriginal population declined because of infectious European diseases and frontier conflicts.

36.1.2 New Zealand

The sovereignty of New Zealand comprises three main islands (Te Ika-a-Māuithe or the North Island, Te Waipounamu or the South Island, and the much smaller Rakiura or Stewart Island) and approximately 600 smaller islands; here, we focus mainly on the three main islands. Like Australia, New Zealand is similarly isolated in the south-western Pacific Ocean, lying approximately 2000 km to the south-east of Australia and approximately 1000 km south of other Pacific Island nations such as Fiji. The North and South Islands are separated by the Cook Strait, 22 km

*Corresponding author: ehlerssmithd@ukzn.ac.za

at its narrowest point, and the South and Stewart Islands are separated by 30 km of the Foveaux Strait. The North Island occupies some 113,729 km^2 and its latitude spans 34–41°S. The South Island is 150,437 km^2 in area and spans 40–46°S, while Stewart Island is 1746 km^2, spanning 46–47°S. Both main islands are long and narrow, with collectively around 15,000 km of coastline. The South Island is mountainous, bisected by the Southern Alps containing 18 peaks over 3000 m above sea level. Vast fiords and glaciers cover the south-western portion of the South Island. Both islands host many lakes, but the North Island is less mountainous, with more volcanic activity, including the Lake Taupo plateau, which contains the nation's largest lake. The islands straddle the Pacific and Indo-Australasian tectonic plates, responsible for New Zealand's volcanic activity and mountainous topography. The islands are climatically varied, ranging from subtropical in the north, high rainfall in the central and western regions, semi-arid in the central South Island to cold, snowy winters at the southern extreme and at altitude. Prior to human colonization, it is predicted that around 80% of New Zealand was forested; volcanic and alpine regions were the only landscapes inhospitable to tree cover (Forestry Department, 1997). Eighty million years of isolation resulted in a high level of biological endemism, including 82% of all New Zealand's vascular plants (de Lange *et al.*, 2006).

New Zealand was one of the last major land masses to be colonized by humans, owing to this isolation. It is estimated that Eastern Polynesian peoples concluded the series of voyages and migrations that populated Polynesia by arriving in New Zealand between 1250 and 1300 CE. A distinct Polynesian culture, Māori, evolved and developed into *iwi* (tribes or nations) and *hapū* (subtribes or political units) until the Māori population was estimated at 100,000–120,000 by the time of first contact with European explorers some 350–400 years later (Poole and Kukutai, 2011). Like Australia, the following relationship between the Māori and European colonizers was violent and uneasy. After a hostile initial contact by Dutch explorers in 1642, Europeans would not revisit New Zealand until James Cook mapped the majority of its coastline in 1769. A 40-year 'Musket War' ensued, and immigration of European colonizers and Christian missionaries increased, until Māori population numbers had fallen to 40% of their pre-contact levels (Brailsford, 1972; Lange, 1999). Britain continued to send immigrants and declared a separate Colony of New Zealand in 1841; New Zealand remains a constitutional monarchy with the British Crown representing its Head of State today.

36.1.3 Avian invasion pathways and invasion vulnerability

Australia and New Zealand's relative isolation leave them at a distinct vulnerability to invasion by exotic species (Denslow, 2003; Allen *et al.*, 2006). New Zealand is further disadvantaged because of its smaller land area, a key component of invasion vulnerability (Denslow, 2003). The socio-political history of these colonized countries exacerbated the susceptibility to invasive infestations: the European colonizers of Australia and New Zealand were of the broad opinion that the native flora and fauna of these colonies were inferior to those of their points of origin (McLintock, 1966). Additionally, the familiarity and nostalgia for European species helped the newly arrived settlers to feel more 'at home' in their adopted countries. As such, so-called acclimatization societies became popular and highly active all over Australia and New Zealand, organizing mass importations by ship of European flora and fauna during the 19th and early 20th centuries (Wilson, 2004). Individual accounts of species and their introduction history are detailed in section 36.2.2.

The anthropogenic history of New Zealand in particular represents a further confounding factor in its invasion vulnerability: since the arrival of humans to the islands around 700 years ago, the native avifauna has undergone mass extinction events, resulting in 131 species, or 31% of the total predicted avian assemblage, having become extinct on the North and South Islands alone (Allen *et al.*, 2006). Pre-European extinctions were of larger-bodied bird species such as the six Moa species, and Haast's Eagle (*Harpagornis moorei*), while post-European extinctions were of smaller-bodied birds lacking anti-predator responses to introduced carnivores (Duncan and Blackburn, 2004). Thus, the availability of unoccupied ecological niches allowed invasive avian species to thrive (Boast *et al.*, 2018).

36.2 Species of the Global Invasive Species Database

Of the 31 avian species listed on the International Union for Conservation of Nature (IUCN) Invasive Species Specialist Group's (ISSG) Global Invasive Species Database (GISD), 15 occur in the region of New Zealand and Australia (Table 36.1); 13 occur in Australia and ten in New Zealand. Only the Chukar Partridge (*Alectoris chukar*) occurs in New Zealand but not Australia; the House Crow (*Corvus splendens*), the Mute Swan (*Cygnus olor*), the Red Junglefowl (*Gallus gallus*), the Monk Parakeet (*Myiopsitta monachus*) and the Red-whiskered Bulbul (*Pycnonotus jocosus*) occur in Australia but not New Zealand. The Australian Magpie (*Gymnorhina tibicen*) is an Australian species invasive in New Zealand. Three species are listed in the ISSG's 100 worst invasive species on Earth: the Common/Indian Myna (*Acridotheres tristis*), the Common/European Starling (*Sturnus vulgaris*) and the Red-vented Bulbul (*Pycnonotus cafer*). Each occurs in both Australia and New Zealand.

36.2.1 ISSG GISD species native and introduced within the region

Australian Magpies were introduced into New Zealand in the 1860s and 1870s from Victoria and Tasmania into Canterbury, Otago (South Island), Auckland (Kawau Island), Hawke's Bay and Wellington (North Island) (Troup 2008). They have expanded their range and are now found in most regions of the country, except for the Fiordland and open plains of Otago (Troup, 2008). Australian Magpies were introduced into New

Table 36.1. The 15 species of the International Union for Conservation of Nature's Global Invasive Species Database that occur in New Zealand and Australia. Thirteen occur in Australia (AUS) and ten in New Zealand (NZ).

Species	Order	Family	System	Location
Acridotheres tristis	Passeriformes	Sturnidae	Terrestrial	AUS, NZ
Alectoris chukar	Galliformes	Phasianidae	Terrestrial	NZ
Anas platyrhynchos	Anseriformes	Anatidae	Freshwater terrestrial	AUS, NZ
Branta canadensis	Anseriformes	Anatidae	Freshwater terrestrial	AUS, NZ
Bubulcus ibis	Ciconiiformes	Ardeidae	Terrestrial	AUS, NZ
Columba livia	Columbiformes	Columbidae	Terrestrial	AUS, NZ
Corvus splendens	Passeriformes	Corvidae	Terrestrial	AUS
Cygnus olor	Anseriformes	Anatidae	Freshwater terrestrial	AUS
Gallus gallus	Galliformes	Phasianidae	Terrestrial	AUS
Gymnorhina tibicen	Passeriformes	Cracticidae	Terrestrial	NZ
Myiopsitta monachus	Psittaciformes	Psittacidae	Terrestrial	AUS
Passer domesticus	Passeriformes	Passeridae	Terrestrial	AUS, NZ
Pycnonotus cafer	Passeriformes	Pycnonotidae	Terrestrial	AUS, NZ
Pycnonotus jocosus	Passeriformes	Pycnonotidae	Terrestrial	AUS
Sturnus vulgaris	Passeriformes	Sturnidae	Terrestrial	AUS, NZ

Zealand to control crop pest species, affording them protection until 1951; however, they are reputed to prey on the eggs and nestlings of native avifauna (Troup, 2008). The Laughing Kookaburra (*Dacelo novaeguineae*) was introduced into New Zealand from south-east Australia in the late 19th century on to Kawau Island, where a small population (fewer than 500 individuals) has since spread to the Whāngārei and Waitākere Ranges (Troup, 2008). Sulfur-crested Cockatoos (*Cacatua galerita*) were also introduced to the same region during the same time from northern Australia, and the wild population is estimated to be stable at approximately 1000 individuals (Troup, 2008). Furthermore, two other Psittacidae from Australia have been introduced into New Zealand – the Galah (*Eolophus roseicapillus*) and the Eastern Rosella (*Platycercus eximius*), but population numbers remain very small (Troup, 2008).

Conversely, the avifauna of New Zealand has had little impact on Australia; however, the Kea (*Nestor notabilis*), the Antipodes Green Parakeet (*Cyanoramphus unicolor*), the Yellow-fronted Parakeet (*Cyanoramphus auriceps*), the Red-fronted Parakeet (*Cyanoramphus novaezelandiae*), the New Zealand Pigeon (*Hemiphaga novaeseelandiae*), the New Zealand Scaup (*Aythya novaeseelandiae*) and the Paradise Shelduck (*Tadorna variegata*) are all species of bird from New Zealand that are kept in captivity in Australia. Given the established proficiency of the Psittacidae, Columbidae and Anatidae for successful invasion (Brook, 2004), these species pose a potential future invasion risk in the case of escape or release.

36.2.2 ISSG GISD species introduced to the region

Of the avian species that have invaded Australia, the House Crow (also known as the Indian or Ceylon Crow) was introduced by ships and has successfully established in human settlements of Western Australia, particularly because of the abundance of sea ports. There are also smaller populations in the coastal, south-eastern state of Victoria (Western Australian Agriculture Authority, 2015). The Mute Swan was introduced in the 19th Century and has established a wild colony on the Avon River of Western Australia. Smaller populations have also been sustained in eastern Tasmania, eastern South Australia and Victoria, New South Wales and coastal southern Queensland (Atlas of Living Australia: www.ala.org.au, accessed 6 November 2019). First domesticated 5000 years ago in Asia and the ancestor of the domestic chicken, the Red Junglefowl is widely distributed across Australia and is present in each state. It is particularly successful along the eastern coastlines and southern Western Australia (Western Australian Agriculture Authority, 2015).

As with most invasive birds in Australia, the Red-whiskered Bulbul has established in the temperate and densely populated eastern states of Queensland and New South Wales, and the south-eastern state of Victoria (Mo, 2015). It is an attractive bird that was popular in the caged-bird trade and was briefly introduced into South Australia but did not establish and is no longer present (Mo, 2015).

The Monk Parakeet, or Quaker Parrot, has established in the state of Queensland and is predicted to be one of the worst potential invaders in Australia, because of the favourable climate and thriving pet trade present in the eastern states (Csurhes, 2016). As such, the Western Australia government has banned the ownership of Monk Parakeets in captivity (Western Australian Agriculture Authority, 2015).

The Chukar Partridge was first introduced into New Zealand from India in 1926, with a later introduction from Iran into the Marlborough region. The two populations apparently hybridized and are now confined to the South Island's highlands region. However, the population is reported to be declining, and is a popular hunting species (Wilson, 2004).

One of the most successful invasive bird species globally, the Common or Indian Myna, has successfully invaded both Australia and New Zealand. Common Mynas were introduced to Melbourne, Australia, from South-east Asia in the 1860s (Hone, 1978). Later introductions to control grasshoppers and cane beetles resulted in the species now dominating urban areas in the east of the country, Tasmania and urban centres such as Perth on the west coast (Martin, 1996). Research on its ecological impact is limited to a region part of its invasive range; it has had varying levels of deleterious impact in Australia, depending on the habitat in which it has established (see Chapter 42, this volume). Although often cited as outcompeting native avifauna for food resources, little documented evidence exists to support this (Haythorpe *et al.*, 2014; Sol *et al.*, 2012). Mynas may be more aggressive than native species in the case of occupying the scarce resource of nest cavities; however, ecological partitioning has been recorded whereby mynas prefer cavities in modified habitats, while native bird species tend toward cavities in natural settings (Grarock *et al.*, 2013). In New Zealand, introductions to both the North and South Islands in the 1870s to control insect pests saw populations proliferate in the subtropical northern North Island, but further south the cooler climate prevented the Common Myna from successfully establishing colonies (Troup, 2008).

The Mallard Duck (*Anas platyrhynchos*) was one of the species chosen for introduction into the entirety of New Zealand by various acclimatization societies over a period of almost 100 years, ending as late as the 1970s with as many as 30,000 individuals released (Dyer and Williams, 2010). Similarly, Mallards dominate Tasmania, east and south-east Australia, the northern reaches of the Northern Territories and south-western Western Australia (Guay and Tracy, 2009). They pose significant threats to native waterfowl via hybridization in both New Zealand and Australia (Guay and Tracy, 2009; Dyer and Williams, 2010).

Canada Geese (*Branta canadensis*) were introduced into the South Island of New Zealand in 1905 as a game bird, and were protected under legislation until 2011, when it was recognized that the large population sizes – some 40,000 by 1996 – posed a threat to crops and livestock via food competition and infections through faecal deposits (Spurr *et al.*, 2005; New Zealand Government, 2011). In Australia, Canada Geese are considered more of a potential emerging threat, after several unsuccessful introduction attempts into Western Australia and Victoria in the 1920s and 1930s (Western Australian Agriculture Authority, 2015). More recently, several incursions of geese thought to be from New Zealand into New South Wales have alerted wildlife authorities of the potential for migration and establishment of populations in Australia (Dawes, 2008).

Rock Doves or 'feral pigeons' (*Columba livia*) were brought to New Zealand and Australia by the earliest European settlers for food, messenger vectors or simply as pets, whereupon they quickly established populations across the New Zealand islands (Troup, 2008) and the Australian landscape, particularly in urban and coastal areas.

The House Sparrow (*Passer domesticus*) was introduced to Melbourne, Australia, in 1863 and quickly dominated the anthropogenic landscape of eastern and southern Australia, the Northern Territory and Tasmania (Summers-Smith, 1990). There are isolated populations in urban centres of southern Western Australia; however, eradication efforts have proven successful in stopping the spread of the species in Western Australia (Western Australian Agriculture Authority, 2015). The invasion process was similar after introduction to New Zealand in 1859, and this population is considered to have spread to other islands of Oceania, including Hawaii (Anderson, 2006).

The Red-vented Bulbul has had limited impact in New Zealand, and the small number of sightings in Auckland are considered not to be of a self-sustaining population (Miskelly, 2015); however, it is one of the most successful invaders in tropical regions of Oceana (Thibault *et al.*, 2018) and its invasion risk is not to be underestimated. Similarly, in Australia, the species has had limited impact after introduction in Melbourne, where numbers remain small (Atlas of Living Australia: www.ala.org.au).

Arguably one of the most successful avian invaders of the region is the Common or European Starling. Approximately 1000 individuals were originally introduced into New Zealand in the 1860s to control insect pests for farming purposes; the species now covers the landscape in all but the most densely forested and mountainous regions (Troup, 2008). In Australia, the motivation for introduction was the same as in New Zealand, and the first individuals were introduced to Melbourne in 1857 and to Sydney two decades later. However, by 1895, Western Australia had banned the introduction and led a successful eradication programme (Woolnough *et al.*, 2005). The remaining states of Australia are heavily infested with Common Starlings, despite considering the species a pest by the 1920s (Higgins *et al.*, 2006).

Finally, the Eastern Cattle Egret (*Bubulcus ibis*) is of particular interest as a more recent self-colonizer of Australia in 1948 and New Zealand in 1963 (Arent, 1988). It is a non-breeding migrant and occupies both the North and South Islands of New Zealand, Tasmania and the majority of Australia with the exception of the arid interior. As a self-colonizer, it is accepted as a native bird and is fully protected under New Zealand law.

36.3 Other Invasive Species and Potential Emerging Species of Concern

Owing to the activities of acclimatization societies, as well as accidental transportations, escapes and release of captive birds, since European colonization of New Zealand and Australia it is estimated that some 242 species of bird have been introduced to the region; of these, approximately 32% have established viable populations (Table 36.2) (Brook, 2004). Indeed, 32% of New Zealand's terrestrial bird species are alien (Clout, 1999). Furthermore, there are 225 exotic bird species known to be kept in captivity in Australia alone, posing a risk for potential invasions through escape or release of individuals (Olsen *et al.*, 2006). Of concern is the high numbers of species from the six most invasive bird families, particularly Psittacidae (106 species), Fringillidae (49 species), Phasianidae (22 species), Columbidae (11 species), Anatidae (eight species) and Passeridae (two species).

36.4 Synthesis and Conclusions

The impact likelihood is highly dependent on the functional traits of the individual species (Shirley and Kark, 2009; Kumschick and Nentwig, 2010). Competition is likely to be greatest between native and invasive avifauna in modified habitats, in which native birds often are at a disadvantage (see Chapter 42, this volume). Anatidae and Psittacidae have the most negative impact on native biodiversity (Kumschick and Nentwig, 2010), of which several species have successfully invaded New Zealand and Australia. Numerous studies have also concluded that competition with

Table 36.2. Invasive species with established populations in Australia (AUS) and New Zealand (NZ).

Scientific name	Common name	Native origin	Invaded location
Aix sponsa	Wood Duck	North America	NZ
Alauda arvensis	Eurasian Skylark	Eurasia	AUS, NZ
Alectoris rufa	Red-legged Partridge	South-western Europe	AUS, NZ
Anas acuta	Northern Pintail	North America, Eurasia	NZ
Anas penelope	Eurasian Wigeon	Eurasia, Africa	NZ
Anser anser	Greylag Goose	Eurasia	AUS, NZ
Anser caerulescens	Snow Goose	North America	NZ
Athene noctua	Little Owl	Eurasia, North Africa	NZ
Aythya fuligula	Tufted Duck	Eurasia	NZ
Callipepla californica	California Quail	North and Central America	AUS, NZ
Carduelis cannabina	Common Linnet	Eurasia, North Africa	AUS, NZ
Carduelis carduelis	European Goldfinch	Eurasia, North Africa	AUS, NZ
Carduelis chloris	European Greenfinch	Eurasia, North Africa	AUS, NZ
Carduelis flammea	Common Redpoll	North America, Eurasia	NZ
Carduelis spinus	Eurasian Siskin	Eurasia	AUS, NZ
Colinus virginianus	Northern Bobwhite	North and Central America, Caribbean	NZ
Corvus frugilegus	Rook	Eurasia	NZ
Corvus monedula	Western Jackdaw	Eurasia	NZ
Emberiza citrinella	Yellowhammer	Eurasia	AUS, NZ
Emberiza cirlus	Cirl Bunting	Eurasia	NZ
Emberiza schoeniclus	Common Reed Bunting	Eurasia	NZ
Erithacus rubecula	European Robin	Europe, North Africa	AUS, NZ
Fringilla coelebs	Common Chaffinch	Eurasia, North Africa	NZ
Fringilla montifringilla	Brambling	Eurasia	AU, NZ
Lagopus lagopus	Willow Ptarmigan	North America, Eurasia	NZ
Lonchura punctulata	Scaly-breasted Munia	South-east Asia	AUS
Luscinia megarhynchos	Common Nightingale	Eurasia, North Africa	AUS, NZ
Passer montanus	Eurasian Tree Sparrow	Eurasia	AUS
Perdix perdix	Grey Partridge	Eurasia	AUS, NZ
Phasianus colchicus	Common Pheasant	Asia	AUS, NZ
Prunella modularis	Dunnock	Eurasia	NZ
Pyrrhula pyrrhula	Eurasian Bullfinch	Eurasia	AUS, NZ
Spilopelia chinensis	Spotted Dove	Asia	AUS, NZ
Spilopelia senegalensis	Laughing Dove	Africa	AUS
Streptopelia risoria	Barbary Dove	Africa	NZ
Streptopelia turtur	European Turtle-Dove	Eurasia, Africa	AUS
Struthio camelus	Common Ostrich	Africa	AUS
Sylvia atricapilla	Eurasian Blackcap	Eurasia, Africa	NZ
Sylvia communis	Common Whitethroat	Eurasia, Africa	NZ
Tetrao tetrix	Black Grouse	Eurasia	NZ
Turdus merula	Common Blackbird	Eurasia, N Africa	AUS, NZ
Turdus philomelos	Song Thrush	Eurasia	AUS, NZ
Tympanuchus phasianellus	Sharp-tailed Grouse	North America	NZ
Tyto alba	Barn Owl	Eurasia, Africa, North and South America, Australia	NZ
Vanellus vanellus	Northern Lapwing	Eurasia, North Africa	NZ

native bird species for food sources and nesting sites is a primary negative effect of invasive bird species, as is hybridization, herbivory, predation and transmission of disease (Martin-Albarracin *et al.*, 2015). Fewer studies have focused on how the functional traits of invasive species may contribute to ecosystem functioning in natural habitats, and whether the negative impacts of invasive species consistently outweigh the positive impacts, if any exist (Kelly *et al.*, 2015). For example, the concept that unoccupied niches may be filled by introduced species has long been suggested (e.g. Herbold and Moyle, 1986), and certain studies have indicated that Common or European Blackbirds (*Turdus merula*) and European Song Thrushes (*Turdus philomelos*) may disperse the seeds of New Zealand's native plants in the absence of medium-bodied frugivores (Williams, 2006; Burns, 2012). However, further studies indicate in this case that such introduced frugivores are more adept at spreading the seeds of invasive plants (MacFarlane *et al.*, 2015), hinting at an invasive seed-dispersal and frugivore feedback loop.

Invasive species are prominent in the consciousness of the governments, scientists and citizens of New Zealand and Australia. New Zealand has recently become globally prominent in invasive species control through ambitious initiatives such as the Department of Conservation's 'Predator Free 2050' and 'Battle for our Birds' (New Zealand Department of Conservation, 2019). These drives seek to eradicate all invasive predators by 2050, such as rats, mustelids and possums. A range of methods, including trapping, ground-laid toxins and the application of Aerial 1080 (an aerially applied pesticide) have already seen positive results, with estimates of native bird populations doubling in the last 20 years in some regions (O'Donnell and Monks, 2012).

Similarly, Australia has an Invasive Species Council, which aims to detect, prevent and eradicate invasive species from establishing or spreading into uninvaded regions of the country (Invasive Species Council, 2016). Western Australia, in particular, has had success in controlling the spread of and eradicating House Sparrows and Common Mynas from the state, but at a significant economic cost to the Department of Agriculture and Food (Western Australian Agriculture Authority, 2015). The Invasive Species Council has a 7-year *Strategic Plan 2016–2022* based on 'Eradication, Containment and Control', through which it plans to enhance biosecurity and lobby for government-led containment strategies for emerging threats (Invasive Species Council, 2016). The chance for successful control of invasive species and the realization of the ambitious goals established by various councils and governmental departments in New Zealand and Australia are stronger with a combination of scientific study and consensus, governmental support and economic funding, and consistent education drives to keep the threat from invasive species in the media.

36.5 References

Allen, R.B., Duncan, R.P. and Lee, W.G. (2006) Updated perspective on biological invasions in New Zealand. In: Allen, R.B. and Lee, W.G. (eds) *Biological Invasions in New Zealand*. Springer, Heidelberg, pp. 435–448.

Anderson, T.R. (2006) *Biology of the Ubiquitous House Sparrow: from Genes to Populations*. Oxford University Press, Oxford.

Arent, W.J. (1988) Range expansion of the Cattle Egret (*Bubulcus ibis*) in the Greater Caribbean Basin. *Colonial Waterbirds* 11, 252–262.

Boast, A.P., Weyrich, L.S., Wood, J.R., Metcalf, J.L., Knight, R. and Cooper, A. (2018) Coprolites reveal ecological interactions lost with the extinction of New Zealand birds. *Proceedings of the National Academy of Sciences USA* 115, 1546–1551.

Brailsford, B. (1972) *Arrows of Plague*. Hick Smith & Sons, Wellington.

Brook, B.W. (2004) Australasian bird invasions: accidents of history? *Ornithological Science* 3, 33–42.

Burns, K.C. (2012) Are introduced birds unimportant mutualists? A case study of frugivory in European blackbirds (*Turdus merula*). *New Zealand Journal of Ecology* 36, 171–176.

Clarkson, C., Jacobs, Z., Marwick, B. and Fullagar, R. (2017) Human occupation of northern Australia by 65,000 years ago. *Nature* 547, 306–310.

Clout, M.N. (1999) Biodiversity conservation and the management of invasive animals in New Zealand. In: Sandlund, O.T., Schei, P.J. and Viken, Å. (eds) *Invasive Species and Biodiversity management*. Kluwer Academic Publishers, Dordrecht, pp. 349–361.

Csurhes, S. (2016) Invasive animal risk assessment: Monk/Quaker Parakeet (*Myiopsitta monachus*). Department of Agriculture and Fisheries. Biosecurity Queensland, Queensland Government, Australia. Available at: www.daf.qld.gov.au/__data/assets/pdf_file/0005/65597/IPA-Monk-Parakeet-Risk-Assessment.pdf (accessed 6 November 2019).

Dawes, J. (2008) *Canada Geese – A New Incursion into Australia*. Pestat Pty, Bruce, Australia. Available at: www.pestsmart.org.au/wp-content/uploads/2010/03/CanadaGeese0803.pdf (accessed 6 November 2019).

de Lange, P.J., Sawyer, J.W.D. and Rolfe, J. (2006) New Zealand indigenous vascular plant checklist. New Zealand Plant Conservation Network, Wellington. Available at: https://www.nzpcn.org.nz/publications/documents/new-zealand-indigenous-vascular plant-checklist-2010/ (accessed 10 April 2020).

Denslow, J.S. (2003) Weeds in paradise: thoughts on the invasibility of tropical islands. *Annals of the Missouri Botanical Gardens* 90, 119–127.

Duncan, R.P. and Blackburn, T.M. (2004) Extinction and endemism in the New Zealand avifauna. *Global Ecology and Biogeography* 13, 509–517.

Dyer, J. and Williams, M. (2010) An introduction most determined: Mallard (*Anas platyrhynchos*) to New Zealand. *Notornis* 57, 178–195.

Grarock, K., Lindenmayer, D.B., Wood, J.T. and Tidemann, C.R. (2013) Does human-induced habitat modification influence the impact of introduced species? A case study on cavity-nesting by the introduced common myna (*Acridotheres tristis*) and two Australian native parrots. *Environmental Management* 52, 958–970.

Guay, P.-J. and Tracy, J.P. (2009) Feral Mallards: a risk for hybridisation with wild Pacific Black Ducks in Australia? *Victorian Naturalist* 126, 87–91.

Haythorpe, K.M., Burke, D. and Sulikowski, D. (2014) The native versus alien dichotomy: relative impact of native noisy miners and introduced common mynas. *Biological Invasions* 16, 1659–1674.

Herbold, B. and Moyle, P.B. (1986) Introduced species and vacant niches. *American Naturalist* 128, 751–760.

Higgins, P.J., Peter, J.M. and Cowling, S.J. (eds) (2006) *Handbook of Australian, New Zealand, and Antarctic Birds*, Vol. 7: *Boatbill to Starlings*. Oxford University Press, Melbourne, Australia.

Hone, J. (1978) Introduction and spread of the Common Myna in New South Wales. *Emu* 78, 227–230.

Invasive Species Council (2016) *Strategic Plan 2016–2022*. Invasive Species Council, Victoria, Australia.

Kelly, D., MacFarlane, A.F. and Briskie, J.V. (2015) European Blackbirds (*Turdus merula*) in New Zealand: useful substitutes for lost mid-sized native frugivores, or weed vectors? In: *6th International Symposium of the Frugivores and Seed Dispersal*, Drakensberg, South Africa, 21–26 June.

Kumschick, S. and Nentwig, W. (2010) Some alien birds have as severe an impact as the most effectual alien mammals in Europe. *Biological Conservation* 143, 2757–2762.

Lange, R. (1999) *May the People Live: A History of Māori Health Development 1900–1920*. Auckland University Press, Auckland.

MacFarlane, A.F., Kelly, D. and Briskie, J.V. (2015) Introduced blackbirds and song thrushes: useful substitutes for lost mid-sized native frugivores, or weed vectors? *New Zealand Journal of Ecology* 40, 80–87.

Martin, W.K. (1996) The current and potential distribution of the Common Myna *Acridotheres tristis* in Australia. *Emu* 96, 169–170.

Martin-Albarracin, V.L., Amico, G.C., Simberloff, D. and Nuñez, M.A. (2015) Impact of non-native birds on native ecosystems: a global analysis. *PLoS One* 10, e0143070.

McLintock, A.H. (1966) *Acclimatisation Societies and their Activities*. In: McLintock, A.H. (ed.) *An Encyclopedia of New Zealand*. New Zealand Ministry for Culture and Heritage, Wellington. Available at: https://teara.govt.nz/en/1966/acclimatisation-of-animals/page-7 (accessed 6 November 2019).

Miskelly, C.M. (2015) Red-vented bulbul. In: Miskelly, C.M. (ed.) *New Zealand Birds Online*. Available at: www.nzbirdsonline.org.nz/species/red-vented-bulbul (accessed 19 March 2019).

Mo, M. (2015) The Red-whiskered Bulbul (*Pycnonotus jocosus*) in Australia – a global perspective, history of introduction, current status and potential impacts. *Australian Zoologist* 37, 461–471.

New Zealand Department of Conservation (2019) Predator Free 2050 programme. Predator Free NZ. Available at: http://predatorfreenz.org (accessed 9 March 2019).

New Zealand Government (2011) Canada geese protection status changed. Beehive.govt.nz. Available at: www.beehive.govt.nz/release/canada-geese-protection-status-changed (accessed 19 March 2019).

O'Donnell, C.F.J. and Monks, A. (2012) Quantifying the benefits of long-term integrated pest control for forest bird populations in a New Zealand temperate rainforest. *New Zealand Journal of Ecology* 36, 131–140.

Olsen, P., Silcocks, A. and Weston, N. (2006) The state of Australia's birds 2006: invasive species. *Wingspan* 16, 1–17.

Poole, R. and Kukutai, T. (2011) Building ethnic boundaries in New Zealand: representations of Maori identity in the census. In: Axelsson, P. and Sköld, P. (eds) *Indigenous Peoples and Demography. The Complex Relation Between Identity and Statistics*. Berghahn Books, New York, pp. 33–54.

Shirley, S.M. and Kark, S. (2009) The role of species traits and taxonomic patterns in alien bird impacts. *Global Ecology and Biogeography* 18, 450–459.

Sol, D., Bartomeus, I. and Griffin, A.S. (2012) The paradox of invasion in birds: competitive superiority or ecological opportunism? *Oecologia* 169, 553–564.

Spurr, E.B, Coleman, J.D. and Whenua, M. (2005) *Review of Canada Goose Population Trends, Damage, and Control in New Zealand*. Manaaki Whenua Press, Landcare Research, Lincoln, New Zealand.

Summers-Smith, D. (1990) Changes in the distribution and habitat utilisation by members of Passer. In: Pinowski, J. and Summers-Smith, J.D. (eds) *Proceedings of the General Meetings of the Working Group on Granivorous Birds*. PWN, Warszawa, pp 110–120.

Thibault, M., Vidal, E., Potter, M.A., Sanchez, T. and Brescia, F. (2018) The invasive Red-vented bulbul (*Pycnonotus cafer*) outcompetes native birds in a tropical biodiversity hotspot. *PLoS One* 13, e0192249.

Troup, C. (2008) Introduced land birds. In: McLintock, A.H. (ed.) *An Encyclopedia of New Zealand*. New Zealand Ministry for Culture and Heritage, Wellington. Available at: www.TeAra.govt.nz/en/introduced-land-birds/page-12 (accessed 20 March 2019).

Western Australian Agriculture Authority (2015) *Invasive Species Plan for Western Australia 2015–2019*. Department of Agriculture and Food. Government of Western Australia, Australia.

Williams, P.A. (2006) The role of blackbirds (*Turdus merula*) in weed invasion in New Zealand. *New Zealand Journal of Ecology* 30, 285–291.

Wilson, K. (2004) *Flight of the Huia: Ecology and Conservation of New Zealand's Frogs, Reptiles, Birds and Mammals*. Canterbury University Press, Christchurch, New Zealand.

Woolnough, A.P., Massam, M.C., Payne, R.L. and Pickles, G.S. (2005) Out on the border: keeping starlings out of Western Australia. In: *Proceedings of the 13th Australasian Vertebrate Pest Conference*, Te Papa, Wellington, New Zealand, 2–6 May., pp. 183–189. Available at: www.pestsmart.org.au/wp-content/uploads/2011/06/VPC2005.pdf (accessed 6 November 2019).

37 Continental Analysis of Invasive Birds: Africa

Lorinda A. Hart[1,2]* and Colleen T. Downs[1]

[1]Centre for Excellence in Invasion Biology, School of Life Sciences, University of KwaZulu-Natal, Pietermaritzburg 3209, South Africa; [2]Department of Biological Sciences, University of Namibia, Windhoek 10023, Namibia

Citation: Hart, L.A. and Downs, C.T. (2020) Continental analysis of invasive birds: Africa. In: Downs, C.T. and Hart, L.A. (eds) *Invasive Birds: Global Trends and Impacts*. CAB International, Wallingford, UK, pp. 265–277.

37.1 Introduction

Most of Africa is characterized by hot and humid conditions, due to its inter-tropical location. The equator is almost centrally located on the continent, with the Tropic of Cancer and Capricorn intercepting the north and south, respectively. Africa is broadly divided into six climate zones. The equatorial region is predominantly characterized by rainforest in western and central Africa. Towards the Indian Ocean, the eastern equatorial area consists of tropical and then desert conditions. The equatorial forest zone is bordered by humid tropical conditions. This in turn is surrounded by a narrow tropical band to the north and extends as the dominant climate zone to the south, consisting mostly of savannah. North of this tropical zone, a narrow semi-desert zone transitions into the Sahara, the largest hot desert in the world which characterizes most of North Africa. A smaller desert zone is in south-west Africa, of which Namibia makes up the largest portion. This desert is also bordered by a semi-desert zone on its eastern boundary, while the Atlantic Ocean borders the west. The final and smallest zone is the Mediterranean, which is located on the northern- and southern-most fringes of Africa. Unlike northern Africa, southern Africa is more diverse in terms of climate zones, transitioning from desert in the west to more forested areas in the east, and has a Mediterranean zone in the south. South Africa has one of the richest biodiversies in the world, consisting of nine biomes, three of which are global biodiversity hotpots (Driver *et al.*, 2005).

Humans originated from Africa, and the first foreign species introductions were probably from livestock trade and movements with humans on the continent (Deacon and Deacon, 1999; Richardson *et al.*, 2003). The number of species introductions to the continent took flight in the 17th century with colonization by Europeans (Richardson *et al.*, 2003). The 'Scramble for Africa' was driven by what has become known as the three Cs of colonization: Christianity, civilization and commercialism. From 1870 to 1914, Africa shifted from 10% to 90% European occupation, and by the early 20th century, only Ethiopia and Liberia remained independent (New World Encyclopedia, 2015). Early settler colonies included Algeria, Tunisia, South Africa, Namibia, Angola, Mozambique and central African areas such as Zimbabwe and Zambia. The scrambling powers included Britain, France, Germany, Belgium, Italy, Portugal and Spain. Britain had the greatest influence, with colonies spanning from Cairo, Egypt, in the north to Cape Town, South Africa, in the south, with a few scattered colonies in western Africa (New World Encyclopedia, 2015). The early colonizers' intentions to 'improve' African nations were coupled with a sense of nostalgia, and as part of this, several avian species were deliberately introduced. Most bird introductions coincide with this colonization period in the late 19th century and are of British origin (Blackburn and Duncan, 2001; Blackburn *et al.*, 2009). More recently, social transformation and the ending of wars and civil unrest have further opened intra- and inter-African trade routes (and therefore potential invasion pathways), driven by the need for economic growth (Richardson *et al.*, 2003). In 2016, 21 million people entered South Africa alone (van Wilgen and Wilson, 2018).

Within Africa, there is some disparity in the levels of urbanization, with southern and northern Africa being most urbanized, while East Africa is least urbanized (Chen *et al.*, 1998). However, urbanization rates are estimated to be the fastest in Africa, particularly in the east (Chen *et al.*, 1998). Africa is also the only major region that continues to observe moderate population growth within the rural sector (Chen *et al.*, 1998). These increasing populations will no doubt lead to continued land transformation on the continent. In South Africa, the expanding cities of Pretoria and Johannesburg (the second biggest mega-urban expanse in sub-Saharan Africa)

*Corresponding author: lorinda.hart.research@gmail.com

have led to an increase in the number of indigenous and exotic bird species in the area (Symes *et al.*, 2017). Although more diverse, there have been losses of specialist grassland species (Symes *et al.*, 2017). This altered bird community is a result of the greening of suburbs, increased tree cover and water sources, and a variety of habitats ranging from highly urbanized city centres to more natural environments on the outskirts (Symes *et al.*, 2017). In Africa, introduced birds stem predominantly from accidental escapees and deliberate releases, with a few arriving as ship stowaways or through natural expansion (Richardson *et al.*, 2003; Picker and Griffiths, 2013; Dyer *et al.*, 2017). Today, the caged-bird trade is the primary source of bird introductions, with 37% of extant bird species being traded globally (Butchart, 2008), and is spurred on by globalized transport systems (Hulme, 2009). Of course, not all species become invasive. Most never form breeding populations and a few remain localized with relatively stable populations.

Fourteen of the 31 avian species listed on the Invasive Species Specialist Group's (ISSG) Global Invasive Species Database (GISD) of the International Union for Conservation of Nature (IUCN) are not present in Africa or its islands (note that Réunion is not included in this review). These include the Canada Goose (*Branta canadensis*), Swamp Harrier (*Circus approximans*), Great Horned Owl (*Bubo virginianus*), House Finch (*Carpodacus mexicanus*), Australian Magpie (*Gymnorhina tibicen*), Pekin Robin (*Leiothrix lutea*), Brown-headed Cowbird (*Molothrus ater*), Shiny Cowbird (*Molothrus bonariensis*), Green Junglefowl (*Gallus varius*), Red Junglefowl (*Gallus gallus*), Derby Flycatcher (*Pitangus sulphuratus*), Red-vented Bulbul (*Pycnonotus cafer*), Jungle Myna (*Acridotheres fuscus*) and Warbling White-eye (*Zosterops japonicus*). While some of these species are imported by African countries, they have not been consistently observed outside of captivity and are therefore not considered to be introduced species in Africa. A summary of the 17 species that do occur in Africa is provided in Table 37.1. Ten of these are native to Africa, predominantly to the north. Northern, western and central Africa have comparatively fewer introduced ISSG GISD listed species than eastern and southern Africa and its islands (Table 37.1).

Although the Green and Red Junglefowl are not listed as invasive in Africa, domestic chickens (*Gallus gallus domesticus*), descendants of wild Red Junglefowl, do occur throughout the continent and its islands. The introduction of chickens to Africa is still under debate; however, multiple introduction pathways and events have been identified (Mwacharo *et al.*, 2013). In the Seychelles, domestic chickens are known to breed in the wild on the uninhabited Marianne Island (Nature Seychelles, 2018). Currently, the genetic integrity of the existing population of Red Junglefowl is questioned, due to introgression and hybridization with domestic and feral *G. g. domesticus* (Peterson and Brisbin, 1998).

37.2 ISSG GISD Species Native and Introduced Within Africa

In Africa, Rock Doves (*Columba livia*) are native to the north and central regions (BirdLife International, 2018). Given their use as carrier pigeons, food, and for sport and breeding by pigeon fanciers, they have been widely traded and dispersed across the world and are one of the most globally invasive bird species today (Dyer *et al.*, 2017). In countries where this species occurs naturally, captive birds have been introduced or accidentally escaped, and it is difficult to discern between native and foreign birds in these areas. They consume 2–3 t of grain daily in Cape Town alone and feed on crop shoots (Picker and Griffiths, 2013). They also cause fouling of buildings and statues with faeces and nest sites, and cause diseases. Numbers of known diseases in pigeons have risen from 50 (Weber, 1979; Long, 1981) to 60 identified pathogens, although only five pathogens routinely infect humans (Haag-Wackernagel and Moch, 2004). This figure is likely to be even higher as the study by Haag-Wackernagel and Moch (2004) was not a global review and Newcastle disease virus, for example, which is known to be carried by Rock Doves, was absent from their list (Meulemans *et al.*, 2002). People suffering from immunodeficiency (such as those with HIV/AIDS) are at a 1000-fold greater risk of infection from avian-borne diseases (Glaser *et al.*, 1994), which is particularly concerning for the African population.

North Africa makes up part of the native range of Eurasian Collared-doves (*Streptopelia decaocto*), which are not only expanding their invasive range, but in the last century have extended their native range extensively as well (Rocha-Camarero and de Trucios, 2002; Bled *et al.*, 2011). North Africa also hosts non-breeding European Greylag Goose (*Anser anser*) migrants, which are otherwise absent from the wild in Africa. Feral Greylag Geese reported in South Africa and Namibia have not established any notable populations (Banks *et al.*, 2008; L. Hart personal observation).

The Purple Swamphen (*Porphyrio porphyrio*) nominate subspecies *P. p. porphyrio* is indigenous to the northern tip of Africa. From Egypt and south of the Sahara, the subspecies *P. p. madagascariensis* is recognized, and is present in most African countries including Madagascar, except for forested areas and arid locations in north-east and south-west Africa (BirdLife International, 2018). For brevity, we have only listed *P. porphyrio* in Table 37.1. The contemporary distribution of *P. porphyrio* is questioned given its historical trading and captive breeding, particularly in the Mediterranean, and its ability to hybridize with native populations (Mlíkovský, 2011; Lopes *et al.*, 2016).

Mallard Ducks (*Anas platyrhynchos*) are native to the far north and north-eastern parts of Africa, where they occur as breeding and non-breeding populations, respectively. They have been introduced around the world, primarily for hunting, but are also kept as attractive domestic ducks (Reaser *et al.*, 2007). Although Mallard Ducks are considered vagrant when they are observed in the Seychelles (Sinclair, 2013), occasional breeding is reported (Banks *et al.*, 2008). In Lesotho, South Africa and Mauritius populations are breeding, and probably increasing (Banks *et al.*, 2008). They are potential carriers of diseases such as avian influenza virus and West Nile virus (Lipkind *et al.*, 1979; Rappole *et al.*, 2000) and hybridize with native species throughout their global invasive range (Reaser *et al.*, 2007). Hybridization compromises the genetic integrity of a species and can ultimately cause its extinction (Thymer and Simberloff, 1996; Mank *et al.*, 2004). At least 18 introduced waterbird species have been observed to hybridize, but cases are generally rare and can result in infertile young (Banks *et al.*, 2008). In South Africa, Mallard Ducks hybridize with Yellow-billed Ducks (*Anas undulata*), Cape Teals (*A. capensis*), Egyptian Geese (*A. aegyptiaca*) and Cape Shovelers (*A. smithii*) (Stafford, 2010; Gaertner *et al.*, 2016). Hybridization has also been observed with the globally endangered Meller's Duck (*A. melleri*) in its introduced range in Mauritius (Rhymer, 2006). This raises concerns for this species in its endemic

Table 37.1. Invasive avian species listed on the Invasive Species Specialist Group's Global Invasive Species Database that are present in Africa and its islands[a].

	Species	*Anas platyrhynchos*	*Anser anser*	*Cygnus olor*	*Oxyura jamaicensis*	*Porphyrio porphyrio*	*Bubulcus ibis*	*Columba livia*	*Streptopelia decaocto*	*Alectoris chukar*	*Acridotheres tristis*	*Sturnus vulgaris*	*Corvus splendens*	*Estrilda astrild*	*Passer domesticus*	*Pycnonotus jocosus*	*Myiopsitta monachus*	*Psittacula krameri*
	Common Name	Mallard Duck	Greylag Goose	Mute swan	Ruddy Duck	Purple Swamphen	Cattle Egret	Pigeon	Eurasian Collared-dove	Chukar Partridge	Common Myna	European Starling	House Crow	Common Waxbill	House sparrows	Red-whiskered Bulbul	Monk Parakeet	Rose-ringed Parakeet
Region	**Country**																	
Northern Africa	Algeria	N	N		V	N	N	N				N			N			P
	Egypt	N	V				N	N	N		P	N	P		N			P
	Libya	N	V					N				N			N			
	Morocco	N	N		P	N	N	N	N			N	V		N		V	V
	South Sudan						N	P						N	N			N
	Sudan	N					N	N					P	V	N			N
	Tunisia	N	N		P	N	N	N	N			N			N			V
	Western Sahara	N					N	N	N						N			
Western Africa	Benin						N	P					V					N
	Burkina Faso						N	N						N				N
	Cote d`Ivoire						N	LN						N	P			N
	Gambia						N	N							P			N
	Ghana						N	N			V			N	P			N
	Guinea						N	N						N				N
	Guinea-Bissau						N	LN							P			N
	Mali	P					N	N						V				N
	Mauritania						N	N	N						P			N
	Senegal						N	N							P			N
	Sierra Leone						N	P						N				V
	Togo						N	N										N
Central Africa	Angola						N	P						N	P			
	Cameroon						N	P						N	V			N
	Central African Republic						N	P						N	V			N
	Chad						N	N							P			N
	Republic of the Congo						N	P						N	P			
	Democratic Republic of the Congo						N	P						N	P			
	Equatorial Guinea						N	P						N				
	Gabon						N	P						N				
	Liberia						N	P						N	P			N
	Niger						N	N							V			N
	Nigeria						N	P						N	V			N
Eastern Africa	Burundi						N	P						N	P			
	Djibouti						N	p					P		P			N
	Eritrea	N					N	N					P		N			N
	Ethiopia	N					N	P				V	V	N	V			N
	Kenya	P					N	P					P	N	P		?	P
	Malawi						N	P						N	P			
	Mozambique	R					N	P			P		P	N	P			V
	Rwanda						N	P						N	P			
	Somalia						N	P					P	N	P			V
	Tanzania						N	P					P	N	P			
	Uganda						N	P						N	N			N
	Zambia						N	P			P			N	P			
	Zimbabwe	P					N	P			P			N	P			

Table 37.1. Continued.

		Species	*Anas platyrhynchos*	*Anser anser*	*Cygnus olor*	*Oxyura jamaicensis*	*Porphyrio porphyrio*	*Bubulcus ibis*	*Columba livia*	*Streptopelia decaocto*	*Alectoris chukar*	*Acridotheres tristis*	*Sturnus vulgaris*	*Corvus splendens*	*Estrilda astrild*	*Passer domesticus*	*Pycnonotus jocosus*	*Myiopsitta monachus*	Psittacula krameri
		Common Name	Mallard Duck	Greylag Goose	Mute swan	Ruddy Duck	Purple Swamphen	Cattle Egret	Pigeon	Eurasian Collared-dove	Chukar Partridge	Common Myna	European Starling	House Crow	Common Waxbill	House sparrows	Red-whiskered Bulbul	Monk Parakeet	Rose-ringed Parakeet
Southern Africa		Botswana						N	P			P	V		N	P			
		Lesotho	P					N	P			P	P		N	P			
		Namibia	V	V				N	P			P	P	V	N	P			
		South Africa	P	V	P			N	P			P	P	P	N	P	V		P
		Eswatini						N	P			P			N	P			
African Islands	**Oceanic Islands**	Cape Verde						N	P	N					P	P			V
		Comoros						N	P			P				P			
		Equatorial Guinea						N	P						N				
		Madagascar	P					N	P			P		V	P	P			
		Mauritius	P		P			V	P			P		P	P	P	P		P
		São Tomé and Príncipe						N	P						N				
		Seychelles	V					N/V	P			P		P	P	P	X		X
	Continental Islands	Mozambique: Inhaca Island										P		P	N	P			V
		South Africa: Robben Island						N	P		P		V		V	P			
		Tanzania: Zanzibar & Pemba Island						N	P					P	N	P			P

Madagascan range where domestic Mallard Ducks are kept (Banks *et al.*, 2008).

Common Starlings (*Sturnus vulgaris*) are non-breeding North African natives as well. In 1889, 18 birds were released in Cape Town by Cecil John Rhodes, a British businessman and former Prime Minister of South Africa (Craig, 2005; Picker and Griffiths, 2013). After an initial lag period, the species rapidly spread and currently occupies most of the country, with continued trends of expansion (Ivanova and Symes, 2018). It was also during this time that House Sparrows (*Passer domesticus*) were brought with Indian labourers, probably as pets, to the sugarcane farms in KwaZulu-Natal, South Africa (Dean, 2005a; Picker and Griffiths, 2013). House Sparrows are also breeding and non-breeding natives to northern and north-eastern Africa, respectively, but are invasive throughout most of the remaining African continent and its islands, except for equatorial forest and northern desert areas (BirdLife International, 2018). Following 50 years of relatively slow dispersal, this expansion rate increased to more than 80 km/year (Msimanga and Slotow, 2000). These sparrows spread northwards to Malawi, Zambia, the Democratic Republic of Congo and Angola (Dean, 2005a). In 1955, House Sparrows were introduced to Maputo, Mozambique (Pinto, 1959), with additional individuals colonizing from the South African population (Dyer *et al.*, 2017). In 1957 they were introduced to Harare, Zimbabwe (Harwin and Irwin, 1966). House Sparrows in Kenya, Somalia, Senegal and Guinea-Bissau are thought to originate from ship stowaways (Dyer *et al.*, 2017). The Tanzanian population is attributed to birds from Zanzibar and colonizers from Zambia (Dyer *et al.*, 2017). Invasion pathways are diverse for this species, even within a single country.

Ring-necked Parakeets (*Psittacula krameri*) are native to the subtropical belt spanning sub-Saharan Africa and southern Asia, with two subspecies occurring on each continent (Juniper and Parr, 1998; Pithon and Dytham, 2001). They are globally recognized as the most widespread invasive parrot species (Butler, 2003). Previously absent from Australia, today several feral flocks occur there (Anon., 2018), making them present on six continents. In Africa, invasive populations occur in the north and south of the continent and on some of its islands (Table 37.1). The Ring-necked Parakeet is the only recognized established invasive parakeet in Africa and is expanding its range in South Africa (Hart and Downs, 2014; Symes, 2014).

The Cattle Egret (*Bubulcus ibis*) is native to Africa occurring in all continental countries. Although reported as absent in Libya (BirdLife International, 2018), it has been recorded to breed there (Herring and Fuchs, 2010). Cattle Egrets were apparently released in Mauritius but did not establish successfully (Watson, 1963; Lever, 2005). In the Seychelles, a separate, endemic subspecies is recognized, *B. i. seychellarum* (Sinclair, 2013), although some argue that data supporting this taxonomic split are lacking (Ahmed, 2011).

Another 'African-born invasive export' listed by the ISSG GISD is the Common Waxbill (*Estrilda astrild*). It is an afro-tropical native that has successfully invaded four of Africa's islands (Table 37.1), as well as countries in Europe, South and North America, and several international islands (BirdLife International, 2018). In the Seychelles, they were considered agricultural pests feeding on rice in the 19th century, but today as rice is no longer grown there, their populations have declined and they pose no major threats (Nature Seychelles, 2018).

37.3 ISSG GISD Species Introduced to Africa

A population of Chukar Partridges (*Alectoris chukar*) is only present on Robben Island off the coast of South Africa, where six confiscated imported birds were introduced in 1964 (Ratcliffe, 2005; Picker and Griffiths, 2013). Over the next 10 years, the population rose to approximately 500 but dropped to around 300 in the following decade (Picker and Griffiths, 2013). Despite several attempts to introduce them to mainland South Africa, they did not establish (Ratcliffe, 2005).

The Mute Swan (*Cygnus olor*) was introduced to Mauritius during the last 30–40 years at a bird park and at hotels as tourist attractions and are now observed breeding in the surrounding areas (Banks *et al.*, 2008). Mute Swans were first introduced to South Africa 100 years ago in 1918 (Dean, 2000), and while populations initially increased, most have reportedly become extinct or declined (Banks *et al.*, 2008; Shaw and Waller, 2017). Mute Swans are still reported in South Africa, although it is believed that these are probably new introductions (Banks *et al.*, 2008).

The Ruddy Duck (*Oxyura jamaicensis*) is only reported in three North African countries (Algeria, Tunisia and Morocco) where it hybridizes with the endangered White-headed Duck (*O. leucocephala*) (Hughes *et al.*, 2006). Morocco has implemented some control of this species, but extensive control in Spain was linked to a population reduction in Morocco, suggesting that Moroccan birds are migrants from Spain (Hughes *et al.*, 2006; Banks *et al.*, 2008). Caution should be taken to prevent future introductions and spread of this species as it has been observed to hybridize with other African species, such as the South African Shelduck (*Tadorna cana*) and Common Shelduck (*T. tadorna*) where they are introduced (Blair *et al.*, 2000; Banks *et al.*, 2008).

House Crow (*Corvus splendens*) populations are present in several African countries but are absent from western and central Africa (Table 37.1) (Dean, 2005b). However, a single bird was observed in Benin in 2010 (Demey, 2010) and in Namibia in 2011 (Ryall, 2016). Additionally, a small group of 15 birds was discovered in 2014 in Madagascar (Linders and Langrand, 2014). This highlights the continuous process of new introductions and the need for monitoring and controlling potentially harmful founder populations with immediate effect. In Egypt, founder birds were escapees from a zoo (Lever, 2005), and in Zanzibar, birds were deliberately released to assist with refuse and crop pests in the late 1800s (Ryall, 2016). In many cases they were ship stowaways (Ryall, 2016). They have the potential to form dense flocks and breed near one another. In Durban, South Africa, 52–62 breeding pairs of House Crows were observed within a 1 km^2 area (Allan and Davies, 2005). House Crows are a pest to both natural ecosystems and humans (see Chapter 24, this volume). They feed on many vertebrate species, mob people and animals, damage subsistence gardens and transmit disease (Picker and Griffiths, 2013).

The invasive stronghold of the Common Myna (*Acridotheres tristis*) in Africa is predominantly in the south of the continent and its islands (Table 37.1) (Craig, 2005; SABAP2, 2018; see also Chapter 3, this volume). A member of the starling family (Sturnidae), it is an aggressive and highly adaptable species. Unlike many invasive birds, which are predominantly urban dwellers, Common Mynas occur in urban, suburban and semi-natural areas, although they are more common in urban zones (van Rensburg *et al.*, 2009). It is estimated that escaped birds established the initial population in Durban, South Africa, in 1902, with a second potential introduction occurring during the 1930s in Johannesburg (Peacock *et al.*, 2007). After an initial lag period, their rate of spread has increased, and they have become established throughout most of southern Africa (Peacock *et al.*, 2007).

The Red-whiskered Bulbul (*Pycnonotus jocosus*) is currently only present in a one-quarter degree grid in South Africa (van Wilgen and Wilson, 2018) and is not currently an invader of continental Africa. Due to its global invasiveness, it is a regulated species in South Africa and requires a permit for trade, outside of which it should be controlled (van Wilgen and Wilson, 2018). It is invasive in Mauritius (Sinclair, 2013) and has been successfully eradicated from the Seychelles (Uranie, 2015).

Although the ISSG (and subsequent sources that cite the ISSG) lists the Monk Parakeet (*Myiopsitta monachus*) as invasive in Kenya, bird checklists (Lepage, 2004; IGoTerra, 2018), record apps (Sullivan *et al.*, 2009), atlases (Baker and Baker, 2018; Kenya Bird Map, 2018; SABAP2, 2018), and field guides (Zimmerman *et al.*, 1996; Hockey *et al.*, 2005; Picker and Griffiths, 2013; Sinclair, 2013) for the country and subregion have no records for this species. As Ring-necked Parakeets do occur there, it is possible that misidentification has occurred between these two green parakeets, or that feral birds were observed and have since died out. Of concern is that in September 2018, three sightings of Monk Parakeets were reported in Morocco (Sullivan *et al.*, 2009). These could be from Spanish birds expanding their range southwards (or escapees), and potentially mark the beginning of the species' expansion into Africa. Given the global invasive trend for this species, it is certainly one species that Africa should aim to eradicate before founder populations become too large and widespread to control.

37.4 Invasive Species Not on IUCN List and Potential Emerging Species of Concern

Compiling lists of introduced species for African countries is often challenging, due to the lack of accurate record keeping and historical unrest and wars in many countries. Additionally, some databases include indigenous species introductions and reintroductions to native areas. One example is the Seychelles Magpie-robin (*Copsychus sechellarum*) where populations have successfully been re-established on islands where they had previously become extinct (Watson *et al.*, 1992). In some instances, birds may be native to one area yet be introduced elsewhere in the same country. Ostriches (*Struthio camelus*) are native to the north-western portion of South Africa but are introduced over much of their current range, spanning the country, including Robben Island and various game reserves (Brook and Prins, 1986; Macdonald, 1988). Such species are typically listed as native to the country. Additionally, some databases list species that are recognized as global invaders simply as present in a country, without identifying native areas, creating further confusion. Thus, species reports should be carefully considered and cross-checked using reputable resources.

Faulkner *et al.* (2017) reported 87 bird species introduced to Africa, while Dean (2000) reported 48 for South Africa alone. More recently, Dyer *et al.* (2017) reported 79 species for South Africa. Other African mainland countries are dwarfed in comparison, with Botswana reporting the second highest number of introductions with 12 species (Dyer *et al.*, 2017). Diverse species introductions have also been documented on African islands. For example, Mauritius reports 67 species, with the remaining islands ranging from 21 to two species (Dyer *et al.*, 2017). An amalgamation of resources (including, but not limited to, Zimmerman *et al.*, 1996; Lepage, 2004; Hockey *et al.*, 2005; Sullivan *et al.*, 2009; Picker and Griffiths, 2013; Sinclair, 2013; Dyer *et al.*, 2017; Baker and Baker, 2018; BirdLife International, 2018; IGoTerra, 2018; IUCN, 2019; Kenya Bird Map, 2018; Lack, 2018; SABAP2, 2018) produced a list of 123 species for mainland Africa and 94 species for its islands, with 34 species introduced to both islands and the mainland (Table 37.2). Combined, mainland Africa and its islands have had 192 avian species introduced. This list excludes native reintroductions, hybrids, vagrant records (where possible to identify the case as a natural disperser) and any species native to any part of a country or a disputed population origin for a given country. This value is probably still an underestimate, as several additional species are known imports for the pet trade, and almost certainly escape at some stage, but are subsequently undetected in the wild. This has been predicted for traded parrot species in South Africa (Symes, 2014). According to the Global Avian Invasions Atlas (GAVIA) developed by Dyer *et al.* (2017), there have been 218 introduced bird species in sub-Saharan Africa and its islands. However, this database includes some records excluded here. The high diversity of introduced species in the south of Africa could reflect its historical and current importance as a trade post, or could be an artefact of more accurate record keeping and reporting rates by a very active birding community. The establishment success of more invasive species at least in part reflects the diverse ecological zones present there.

In South Africa, approximately 30 parrot species have been observed out of their natural range, probably due to intentional and unintentional releases (Symes, 2014). Black-cheeked Lovebirds (*Agapornis nigrigenis*) had previously been reported as escapees in the Gauteng province, but were not recognized as an established species (Symes, 2014). Today, they are regularly seen there, with a small breeding population also reported in the Eastern Cape province (SABAP2, 2018, C.T. Downs, personal communication). This species is listed as vulnerable, due to its isolated and declining population based predominantly in south-western Zambia (BirdLife International, 2018). Some may suggest that this warrants the preservation of these charismatic, exotic birds outside their native range.

The Egyptian Goose (*Alopochen aegyptiaca*) is native to sub-Saharan Africa but has established invasive populations in Israel, United Arab Emirates, the UK and throughout Europe, where, in the Netherlands, it is one of the fastest-expanding invasive birds (Banks *et al.*, 2008; Gyimesi and Lensink, 2012; see also Chapter 28, this volume). More recently, North America is reporting established and increasing populations (Callaghan and Brooks, 2017; see also Chapter 28, this volume). In the mid-1950s, pinioned adults were released in a private park in Mauritius; however, subsequent breeding has led to a steadily growing population (Banks *et al.*, 2008). Populations of Egyptian Geese have also increased in their native ranges and

Table 37.2. List of species introduced to African islands, the African continent or both[a]. Indigenous species introductions to a country have been excluded.

Species	Common name	Region introduced
Acridotheres albocinctus	Collared Myna	Both
Acridotheres grandis	White-vented Myna	Islands
Acridotheres tristis	Common Myna	Islands
Agapornis canus	Grey-headed Lovebird	Both
Agapornis fischeri	Fischer's Lovebird	Mainland
Agapornis lilianae	Lilian's Lovebird	Mainland
Agapornis nigrigenis	Black-cheeked Lovebird	Mainland
Agapornis personatus	Yellow-collared Lovebird	Mainland
Agapornis pullarius	Red-headed Lovebird	Both
Agapornis roseicollis	Rosy-faced Lovebird	Both
Aix galericulata	Mandarin Duck	Mainland
Aix sponsa	Wood Duck	Mainland
Alauda arvensis	Eurasian Skylark	Islands
Alectoris chukar	Chukar	Both
Alectoris graeca	Rock Partridge	Both
Alectoris melanocephala	Arabian Partridge	Mainland
Alectoris rufa	Red-legged Partridge	Both
Alopochen aegyptiaca	Egyptian Goose	Islands
Amandava amandava	Red Avadavat	Both
Amazona aestiva	Blue-fronted Parrot	Mainland
Anas acuta	Northern Pintail	Mainland
Anas discors	Blue-winged teal	Mainland
Anas erythrorhyncha	Red-billed Duck	Mainland
Anas melleri	Meller's Duck	Islands
Anas platyrhynchos	Mallard Duck	Both
Anas querquedula	Garganey	Mainland
Anas rubripes	American Black Duck	Mainland
Anas undulata	Yellow-billed Duck	Mainland
Anser anser	Greylag Goose	Mainland
Ara ambiguus	Great Green Macaw	Mainland
Ara ararauna	Blue-and-yellow Macaw	Islands
Aratinga jandaya	Jandaya Parakeet	Mainland
Aratinga pertinax	Brown-throated Parakeet	Mainland
Aratinga weddellii	Dusky-headed Parakeet	Mainland
Aythya ferina	Common Pochard	Mainland
Aythya fuligula	Tufted Duck	Mainland
Aythya nyroca	Ferruginous Duck	Mainland
Branta canadensis	Canada Goose	Mainland
Bubulcus ibis	Cattle Egret	Islands
Cacatua goffiniana	Tanimbar Cockatoo	Islands
Cacatua sulphurea	Yellow-crested Cockatoo	Both
Cairina moschata	Muscovy Duck	Both
Callipepla californica	California Quail	Mainland
Callonetta leucophrys	Ringed Teal	Mainland
Carduelis carduelis	European Goldfinch	Both
Carduelis chloris	European Greenfinch	Islands
Chrysolophus pictus	Golden Pheasant	Both
Coccycolius iris	Emerald Starling	Mainland
Colinus virginianus	Northern Bobwhite	Mainland
Columba livia	Rock Pigeon	Both
Columbina inca	Inca Dove	Mainland
Coracias cyanogaster	Blue-bellied Roller	Mainland
Coracopsis nigra	Lesser Vasa Parrot	Islands
Coracopsis vasa	Vasa Parrot	Islands

Continued

Table 37.2. Continued.

Species	Common name	Region introduced
Corvus albus	Pied Crow	Islands
Corvus frugilegus	Rook	Mainland
Corvus monedula	Eurasian Jackdaw	Mainland
Corvus splendens	House Crow	Both
Coturnix chinensis	Blue Quail	Both
Coturnix coturnix	Common Quail	Islands
Coturnix japonica	Japanese Quail	Islands
Crinifer piscator	Western Grey Plaintain-eater	Mainland
Criniferoides leucogaster	White-bellied Go-away-bird	Mainland
Cyanocitta cristata	Blue Jay	Mainland
Cyanoliseus patagonus	Burrowing Parrot	Mainland
Cygnus atratus	Black Swan	Both
Cygnus olor	Mute Swan	Both
Dendrocitta vagabunda	Rufous Treepie	Mainland
Dendrocygna autumnalis	Black-bellied Whistling-duck	Mainland
Dendrocygna bicolor	Fulvous Whistling-duck	Mainland
Dendrocygna viduata	White-faced Whistling-duck	Islands
Eos bornea	Red Lory	Islands
Estrilda astrild	Common Waxbill	Islands
Estrilda melpoda	Orange-cheeked Waxbill	Mainland
Eudocimus ruber	Scarlet Ibis	Mainland
Euodice malabarica	Indian Silverbill	Mainland
Euplectes aureus	Golden-backed Bishop	Islands
Euplectes capensis	Yellow Bishop	Islands
Falco columbarius	Merlin	Mainland
Falco tinnunculus	Common Kestrel	Islands
Forpus passerinus	Green-rumped Parrotlet	Mainland
Foudia madagascariensis	Madagascar Red Fody	Islands
Francolinus afer	Red-necked Spurfowl	Islands
Francolinus capensis	Cape Francolin	Islands
Francolinus hildebrandti	Hildebrandt's Francolin	Islands
Francolinus pintadeanus	Chinese Francolin	Islands
Francolinus pondicerianus	Grey Francolin	Islands
Fregilupus varius	Réunion Starling	Islands
Fringilla coelebs	Eurasian Chaffinch	Mainland
Fulica americana	American Coot	Mainland
Gallinula comeri	Gough Moorhen	Mainland
Gallinula nesiotis	Tristan Moorhen	Mainland
Gallus gallus	Red Junglefowl	Both
Geopelia cuneata	Diamond Dove	Both
Geopelia striata	Barred Ground Dove	Islands
Glareola ocularis	Madagascar Pratincole	Mainland
Guaruba guarouba	Golden Conure	Mainland
Halcyon leucocephala	Grey-headed Kingfisher	Islands
Hypsipetes borbonicus	Olivaceous Bulbul	Islands
Lagonosticta senegala	Red-billed Firefinch	Mainland
Lamprotornis iris	Emerald Starling	Mainland
Lamprotornis purpuroptera	Rüppell's Starling	Mainland
Lamprotornis superbus	Superb Starling	Mainland
Leiothrix argentauris	Silver-eared Mesia	Mainland
Lonchura cantans	African Silverbill	Mainland
Spermestes cucullata	Bronze Munia	Islands
Lonchura malabarica	White-throated Munia	Mainland
Lonchura punctulata	Scaly-breasted Munia	Islands
Lophura nycthemera	Silver Pheasant	Mainland

Continued

Table 37.2. Continued.

Species	Common name	Region introduced
Luscinia megarhynchos	Common Nightingale	Mainland
Margaroperdix madagascariensis	Madagascar Partridge	Islands
Melanocorypha bimaculata	Bimaculated Lark	Mainland
Melopsittacus undulatus	Budgerigar	Both
Merops malimbicus	Rosy Bee-eater	Mainland
Musophaga violacea	Violet Turaco	Mainland
Myiopsitta monachus	Monk Parakeet	Mainland
Nandayus nenday	Nanday Parakeet	Mainland
Nesoenas picturata	Malagasy Turtle-dove	Islands
Netta peposaca	Rosy-billed Pochard	Mainland
Netta rufina	Red-crested Pochard	Mainland
Numida meleagris	Helmeted Guineafowl	Both
Nymphicus hollandicus	Cockatiel	Mainland
Oxyura jamaicensis	Ruddy Duck	Mainland
Oxyura maccoa	Maccoa Duck	Mainland
Padda oryzivora	Java Sparrow	Both
Paroaria dominicana	Red-cowled Cardinal	Both
Passer domesticus	House Sparrow	Both
Passer euchlorus	Arabian Golden Sparrow	Mainland
Passer hispaniolensis	Spanish Sparrow	Islands
Pavo cristatus	Indian Peafowl	Both
Pelecanus crispus	Dalmatian Pelican	Islands
Perdicula argoondah	Rock Bush-quail	Islands
Perdicula asiatica	Jungle Bush-quail	Islands
Phasianus colchicus	Common Pheasant	Both
Platycercus elegans	Crimson Rosella	Islands
Platycercus eximius	Eastern Rosella	Islands
Ploceus capensis	Cape Weaver	Islands
Ploceus cucullatus	Village Weaver	Islands
Ploceus manyar	Streaked Weaver	Mainland
Ploceus melanocephalus	Black-headed Weaver	Islands
Ploceus nigerrimus	Vieillot's Black Weaver	Mainland
Ploceus velatus	Southern Masked Weaver	Islands
Poicephalus cryptoxanthus	Brown-headed Parrot	Mainland
Poicephalus meyeri	Meyer's Parrot	Mainland
Poicephalus rueppellii	Rüppell's Parrot	Mainland
Poicephalus rufiventris	African Orange-bellied Parrot	Mainland
Poicephalus senegalus	Senegal Parrot	Mainland
Poicephalus suahelicus	Grey-headed Parrot	Mainland
Porphyrio porphyrio	Purple Swamphen	Islands
Pseudeos fuscata	Dusky Lory	Islands
Psittacula cyanocephala	Plum-headed Parakeet	Mainland
Psittacula krameri	Ring-necked Parakeet	Both
Psittacus erithacus	Grey Parrot	Both
Psittrichas fulgidus	Vulturine Parrot	Mainland
Pycnonotus jocosus	Red-whiskered Bulbul	Both
Pyrrhula pyrrhula	Eurasian Bullfinch	Islands
Pyrrhura rupicola	Black-capped Conure	Mainland
Serinus alario	Black-headed Canary	Mainland
Serinus canaria	Island Canary	Islands
Serinus canicollis	Cape Canary	Islands
Serinus mozambicus	Yellow-fronted Canary	Islands
Spilopelia chinensis	Eastern Spotted Dove	Islands
Stigmatopelia senegalensis	Laughing Dove	Islands

Continued

Table 37.2. Continued.

Species	Common name	Region introduced
Streptopelia capicola	Ring-necked Dove	Islands
Streptopelia decaocto	Eurasian Collared-dove	Both
Streptopelia risoria	Barbary Dove	Islands
Streptopelia roseogrisea	African Collared-dove	Islands
Streptopelia semitorquata	Red-eyed Dove	Mainland
Streptopelia turtur	European Turtle-dove	Mainland
Struthio camelus	Ostrich	Both
Sturnus roseus	Rosy Starling	Islands
Sturnus vulgaris	Common Starling	Both
Tadorna cana	South African Shelduck	Mainland
Tadorna tadorna	Common Shelduck	Mainland
Taeniopygia guttata	Zebra Finch	Mainland
Turdus merula	Eurasian Blackbird	Mainland
Turdus philomelos	Song Thrush	Mainland
Turnix nigricollis	Madagascar Buttonquail	Islands
Turtur tympanistria	Tambourine Dove	Islands
Tyto alba	Barn Owl	Islands
Uraeginthus angolensis	Blue Waxbill	Islands
Uraeginthus bengalus	Red-cheeked Cordon-bleu	Both
Zenaida macroura	Mourning Dove	Mainland

[a]See Table 37.1 footnote for key references.

Fig. 37.1. Invasive Zebra Doves (*Geopelia striata*), Madagascar Fodys (*Foudia madagascariensis*) and Turtle Doves (probably hybrids of *Streptopelia picturata rostrata* × *S. p. picturata*) (left) and a Common Myna (*Acridotheres tristis*) (right) feeding on restaurant scraps on Praslin Island, Seychelles. (©Photographs: L. Hart.)

have thrived with urbanization (McKenzie, 2011). In South Africa, cereal crop farmers suffer just over 60% annual mean yield losses due to Egyptian Geese, which feed on sown seeds, newly emerging shoots and drying harvested crops (Mangnall and Crowe, 2002). They are also considered a nuisance species on golf courses where they frequently graze (Little and Sutton, 2013; Mackay *et al.*, 2014). Additionally, they are known nest disturbers and usurpers for many bird species (Tate and Amar, 2012; McPherson 2015; Thompson *et al.* 2017; Wreford *et al.*, 2017) and have been associated with disease transfer to domestic fowl (Thompson *et al.*, 2008). Given the gregarious and competitive nature of this species, any introductions or expansions should be of concern.

Common/Indian Peafowl (*Pavo cristatus*) have established feral populations in South Africa and on Robben Island (Picker and Griffiths, 2013). Populations are still relatively small and their potential impact as an invasive species remains to be determined in South Africa (Shaw and Waller, 2017). However, in Australia, they are known to eat seabird eggs and chicks (Picker and Griffiths, 2013). A self-sustaining and contained population is also reported for Dar es Salaam, Tanzania (N. Baker, personal communication).

Not all introduced and established species pose a threat to native species. For example, in South Africa, the Common Chaffinch (*Fringilla coelebs*) is an established alien species, introduced in the 1890s by Cecil Rhodes, but has remained within a small area in Cape Town (Picker and Griffiths, 2013). It is estimated that the population is at most 2000 birds (Picker and Griffiths, 2013). The removal of alien conifer vegetation seems to negatively impact this species and, given its confined population, there is no need to actively control this species (Picker and Griffiths, 2013). Similarly, in the Seychelles, seed-eating species such as the

Barred Ground Dove (*Geopelia striata*) and Madagascar Fody (*Foudia madagascariensis*) prefer open grassy areas associated with humans, and competition with native species is therefore not of concern (Fig. 37.1) (Nature Seychelles, 2018).

37.5 Conclusions

There is a paucity of literature from central, western and northern Africa (see Bendjoudi *et al.*, 2013). In 2003, the *Invasive Alien Species in Southern Africa: National Reports and Directory of Resources* was published by the Global Invasive Species Programme in Cape Town. In these reports, only Mauritius, Mozambique, Tanzania and South Africa reported invasive bird species, with most countries strongly focused on invasive plants and insect pests (Macdonald *et al.*, 2003). Indeed, Smit and Steenkamp (2003) highlighted the lack of knowledge regarding the status of invasive species in Namibia. Control programmes were largely ad hoc and isolated, with a strong need to develop and implement legislation and education programmes. Generally, countries require import permits for bird species; unfortunately, in some cases the only restrictions are for species listed on CITES (Convention on International Trade in Endangered Species) Appendices I and II (Mauremootoo *et al.*, 2003).

South Africa potentially has one of the best regulatory mandates in Africa in terms of managing invasive and alien species, namely the National Environmental Management: Biodiversity Act (NEMBA) (Department of Environmental Affairs, 2014). This places the onus on both the government and private parties to manage invasive species (Gaertner *et al.*, 2016). Indeed, many species have been proactively identified, some of which are listed by ISSG GISD but are not yet present in the wild in South Africa. However, 'regulated' does not equate to 'restricted'. For example, duck and falcon species with the potential to hybridize with native species are still imported, as are species with known invasive tendencies (Richardson *et al.*, 2003). In 2016, 81 permits were granted, and none were refused for the importation of Ring-necked Parakeets (van Wilgen and Wilson, 2018). While species restrictions and owner responsibilities for importing and owning or trading certain species are in place, enforcing laws once species are imported remains challenging. Furthermore, provinces differ in permitting requirements and stringency and thus national norms and standards are recommended for permitting (Richardson *et al.*, 2003). Informal trade between aviculturalists compounds the invasive risks of imported species, as birds are moved into new microclimates where they could pose greater invasive threats (Richardson *et al.*, 2003).

Given the restricted range, stable populations, and non-destructive presence of species such as the Chukar Partridge on Robben Island and Common Chaffinch in Cape Town, no control measures are warranted (Shaw and Waller, 2017). For some species, the possibility of eradication is not plausible. However, although no formal control programmes are in place for highly invasive species such as Common Starlings, House Sparrows and Rock Doves, localized control measures are taken when these species become a nuisance (Shaw and Waller, 2017). Reducing flock sizes will reduce their damage and disease risk, as well as their impact on native species (Picker and Griffiths, 2013).

One species that is routinely targeted in many African countries is the House Crow (Dean, 2005b). Unfortunately, Mozambique has been unsuccessful in eradicating House Crows, brought to Inhaca Island via ship in the 1970s, using poisons and mechanical control methods (Jone, 2003). In South Africa, control efforts have been very successful in reducing numbers (Ryall, 2016; Shaw and Waller, 2017) from 10,000 birds in 2009 to fewer than 400 birds in 2016 in Cape Town and only five birds reported for Durban in 2016 (van Wilgen and Wilson, 2018). These efforts must be maintained, especially as sightings of House Crows at new coastal locations arise (Demey, 2010). Crows are also recognized for their intelligence, and staff implementing control measures have had to change uniforms and vehicles to target remaining birds that associate staff with poisoned baits (L. Stafford, personal communication). House Crows were successfully eradicated from the Seychelles in the 1980s before populations were too large (Nature Seychelles, 2018), and it is promising that when a population of House Crows was established in Madagascar, similar eradication efforts were planned (Ryall, 2016). This is especially important given the island's high level of endemism and the devastating effect that crows would have on subsistence farmers in impoverished communities (Ryall, 2016). House Crow control was also planned for Mauritius using poisoned baits at nest sites (Mauremootoo *et al.*, 2003). Unfortunately, House Crow programmes have been suspended in Kenya and Tanzania due to unavailability of the preferred poison and funding (Ryall, 2016). Tanzania's programme saw the destruction of 1.2 million birds, which have since bounced back (Ryall, 2016). With the collapse of control efforts, the inland eastward spread of crows to densely populated Uganda and Rwanda seems inevitable (Ryall, 2016).

Barn Owls (*Tyto alba*) have a natural global distribution but were deliberately introduced to the Seychelles in the 1950s to control rats (Nature Seychelles, 2018). However, they have broad diets and they soon spread to smaller islands where rats were absent and instead fed primarily on birds, some of which are endangered (Nature Seychelles, 2018). As such, the government offers rewards for any owls caught and killed, and today they have successfully been removed from some islands (Nature Seychelles, 2018). The Seychelles has also been successful in effectively eradicating Ring-necked Parakeets, with a total of 548 birds killed primarily by shooting (Karapetyan, 2017). However, new birds may arise, and a bounty is offered for any parakeet reports, as they pose a threat to the endemic Black Parrot (*Coracopsis barklyi*) (Karapetyan, 2017). The Seychelles have also successfully eradicated the Red-whiskered Bulbul from Assumption island using mist-netting and shooting, with a total of 5279 bulbuls killed over 3 years (Uranie, 2015). The birds originated from Mauritius, with six individuals released in 1997 (Uranie, 2015).

The Common Myna on mainland Africa is not formally controlled. However, on some Seychelles Islands eradication has been successful. The motivation for their removal, particularly from important bird-supporting islands, is their detrimental effect on endangered and vulnerable species such as Seychelles Magpie-robins and Seychelles White-eyes (*Zosterops modestus*) (Payet, 2007; Canning, 2011). They have also been observed to predate seabird eggs on Bird Island (Feare *et al.*, 2015). Following a combination of methods including various trapping techniques, nest disturbance and shooting, a total of 745 birds and 42 eggs were removed, ridding Fregate Island of this invader (Canning, 2011).

The control of Ruddy Ducks and their hybrids in their global invasive ranges has led to declines in their populations (Banks *et al.*, 2008), and similar trends can be expected for Mallard Ducks with co-ordinated and consistent control efforts. However, unlike House Crows, which are more easily vilified given their aggressive nature and direct negative impact, particularly towards people, lethal Mallard Duck removal has been met with resistance in South Africa. Control methods have been uncoordinated, inconsistent, and ultimately unsuccessful and a waste of resources (Stafford, 2010). The detrimental effect of hybridization is more challenging to motivate for the removal of a species, particularly to the public and policy makers, and thus Mallard Ducks are currently being tolerated (Gaertner *et al.*, 2016). However, a strategy for their control has been formulated (Stafford, 2010), and continued education and discussion with stake holders is proposed as a part thereof.

The success of an eradication programme has many considerations, but most importantly it should be undertaken as an open-ended task that requires the implementation of several methods and dedicated staff (Feare, 2010). On islands, particularly isolated ones, total eradication is possible and should be the end goal of any invasive control programme, especially given their high levels of endemism (Feare 2010). Due to the commensal nature of most invasive bird species, implementing control and eradication programmes is more challenging as there are often conflicting stakeholder views (Gaertner *et al.* 2016; see also Chapter 43, this volume). Strategies must therefore be adjusted to account for these challenges, particularly when faced in an urban environment. Invasive species can be categorized by ranking their socio-economic benefits and negative environmental impacts to gauge the potential for controversial species control (Zengeya *et al.* 2017). Additionally, restriction policies have in the past been amended to accommodate conflicts that arose. For example, Rock Doves were initially proposed to be banned from the trade list for South Africa given their highly invasive nature and health risk to humans (Zengeya *et al.*, 2017). However, racing pigeon and pigeon fancier groups opposed this, and the policies were amended accordingly (Zengeya *et al.*, 2017). Similarly, caged-bird species regulations have also been amended to accommodate trade and pet ownership. Unfortunately, this means that the risk of introduction is ever present.

Moving forward, control efforts and education need to be co-ordinated and consistent. There also needs to be an involvement of citizen scientists (see Chapter 44, this volume), which will improve public support. These efforts should not be constrained by country borders, as consistent immigration and emigration will render control efforts futile. Efforts should also be concentrated on islands. These often have high species endemicity and, given their relative size and isolation, generally have more successful outcomes (Feare, 2010). On the mainland, perhaps managing population levels rather than total eradication for prolific species should be considered to reduce their impact and slow their spread. Ultimately, none of these efforts will succeed without government and public support and collaboration.

37.6 References

Ahmed, R. (2011) Subspecific identification and status of Cattle Egret. *Dutch Birding* 33, 294–304.

Allan, D.G. and Davies, G.B. (2005) Breeding biology of House Crows (*Corvus splendens*) in Durban, South Africa. *Ostrich* 76, 21–31.

Anon. (2018) Indian Ringneck Parakeet: Animal pest alert. Department of Primary Industries and Regional Development, Australia. Available at: www.agric.wa.gov.au/birds/indian-ringneck-parakeet-animal-pest-alert (accessed December 2018).

Baker, L. and Baker, N. (2018) *Tanzania Bird Atlas*. Iringa, Tanzania. Available at: http://tanzaniabirdatlas.net (accessed 8 November 2019).

Banks, A.N., Wright, L., Maclean, I.M., Hann, C. and Rehfisch, M. (2008) Review of the status of introduced non-native waterbird species in the area of the African-Eurasian Waterbird Agreement: 2007 update. British Trust for Ornithology, Thetford, UK.

Bendjoudi, D., Chenchouni, H., Doumandji, S. and Voisin, J.-F. (2013) Bird species diversity of the Mitidja Plain (northern Algeria) with emphasis on the dynamics of invasive and expanding species. *Acrocephalus* 34, 13–26.

BirdLife International (2018) IUCN Red List for birds. Available at: www.birdlife.org/news/tag/iucn-red-list (accessed 8 November 2019).

Blackburn, T.M. and Duncan, R.P. (2001) Establishment patterns of exotic birds are constrained by non-random patterns in introduction. *Journal of Biogeography* 28, 927–939.

Blackburn, T.M., Lockwood, J.L. and Cassey., P. (2009) *Avian Invasions: The Ecology and Evolution of Exotic Birds*. Oxford University Press, Oxford.

Blair, M.J., McKay, H., Musgrove, A.J. and Rehfisch, M.M. (2000) Review of the status of introduced non-native waterbird species in the agreement area of the African-Eurasian Waterbird Agreement. British Trust for Ornithology, Thetford, UK.

Bled, F., Royle, J.A. and Cam, E. (2011) Hierarchical modeling of an invasive spread: the Eurasian Collared-Dove *Streptopelia decaocto* in the United States. *Ecological Applications* 21, 290–302.

Brook, R.K. and Prins, A.J. (1986) Review of alien species on South African offshore islands. *South African Journal of Antarctic Research* 16, 102–109.

Butchart, S.H. (2008) Red List Indices to measure the sustainability of species use and impacts of invasive alien species. *Bird Conservation International* 18, S245–S262.

Butler, C.J. (2003) Population biology of the introduced Rose-ringed Parakeet *Psittacula krameri* in the UK. PhD thesis, University of Oxford, Oxford.

Callaghan, C.T. and Brooks, D.M. (2017) History, current distribution, and status of the Egyptian Goose (*Alopochen aegyptiaca*) in the contiguous United States. *Southwestern Naturalist* 62, 296–300.

Canning, G.J.P. (2011) Eradication of the invasive Common Myna, *Acridotheres tristis*, from Fregate Island, Seychelles. *Phelsuma* 19, 43–53.

Chen, N., Valente, P. and Zlotnik, H. (1998) What do we know about recent trends in urbanization? In: Bilsborrow, R. (ed.) *Migration, Urbanization, and Development: New Directions and Issues*. United Nations Populatuion Fund and Kluwer Academic Publishers, Norwell, Massachusetts, pp. 59–88.

Craig, A.J.F.K. (2005) Common Myna. In: Hockey, P.A.R., Dean, W.R.J. and Ryan, P.G. (eds) *Roberts Birds of Southern Africa*, 7th edn. Trustees of the John Voelker Bird Book Fund, Cape Town, pp. 972.

Deacon, H.J. and Deacon, J. (1999) *Human Beginnings in South Africa. Uncovering the Secrets of the Stone Age*. David Philip, Cape Town.

Dean, W.R.J. (2000) Alien birds in southern Africa: what factors determine success? *South African Journal of Science* 96, 9–14.

Dean, W.R.J. (2005a) House Sparrow. In: Hockey, P.A.R., Dean, W.R.J. and Ryan, P.G. (eds) *Roberts Birds of Southern Africa*, 7th edn. Trustees of the John Voelker Bird Book Fund, Cape Town, pp. 1082–1083.

Dean, W.R.J. (2005b) House Crow. In: Hockey, P.A.R., Dean, W.R.J. and Ryan, P. (eds) *Roberts Birds of Southern Africa*, 7th edn. Trustees of the John Voelker Bird Book Fund, Cape Town, pp. 721–722.

Demey, R. (2010) Recent reports. *Bulletins of the African Bird Club* 17, 239–252.

Department of Environmental Affairs (2014) Government Gazette No. 37885. National Environmental Management: Biodiversity Act, 2004 (Act No. 10 of 2004) Alien and Invasive Species Regulations, Vol. 590. Department of Environmental Affairs, Pretoria, South Africa.

Driver, A., Maze, K., Rouget, M., Lombard, A.T., Nel, J., *et al.* (2005) National Spatial Biodiversity Assessment 2004: priorities for biodiversity conservation in South Africa. South African National Biodiversity Institute (SANBI), Pretoria.

Dyer, E.E., Redding, D.W. and Blackburn, T.M. (2017) The global avian invasions atlas, a database of alien bird distributions worldwide. *Scientific Data* 4, 170041.

Faulkner, K.T., Hurley, B.P., Robertson, M.P., Rouget, M. and Wilson, J.R. (2017) The balance of trade in alien species between South Africa and the rest of Africa. *Bothalia* 47, 1–16.

Feare, C.J. (2010) Invasive bird eradication from tropical oceanic islands. *Aliens: The Invasive Species Bulletin* 30, 12–19.

Feare, C.J., Lebarbenchon, C., Dietrich, M. and Larose, C.S. (2015) Predation of seabird eggs by Common Mynas *Acridotheres tristis* on Bird Island, Seychelles, and its broader implications. *Bulletin of the African Bird Club* 22, 162–170.

Gaertner, M., Larson, B.M., Irlich, U.M., Holmes, P.M., Stafford, L., *et al.* (2016) Managing invasive species in cities: a framework from Cape Town, South Africa. *Landscape and Urban Planning* 151, 1–9.

Glaser, C.A., Angulo, F.J. and Rooney, J.A. (1994) Animal-associated opportunistic infections among persons infected with the human immunodeficiency virus. *Clinical Infectious Diseases* 18, 14–24.

Gyimesi, A. and Lensink, R. (2012) Egyptian Goose *Alopochen aegyptiaca*: an introduced species spreading in and from the Netherlands. *Wildfowl* 62, 128–145.

Haag-Wackernagel, D. and Moch, H. (2004) Health hazards posed by feral pigeons. *Journal of Infection* 48, 307–313.

Hart, L.A. and Downs, C.T. (2014) Public surveys of Rose-ringed Parakeets, *Psittacula krameri*, in the Durban Metropolitan area, South Africa. *African Zoology* 49, 283–289.

Harwin, R.M. and Irwin, M.P.S. (1966) The spread of the House Sparrow, *Passer domesticus*, in south-central Africa. *Arnoldia* 2, 1–17.

Herring, J. and Fuchs, E. (2010) Mixed breeding colony of Little Egret *Egretta garzetta* and Cattle Egret *Bubulcus ibis* in Benghazi, Libya. *Alauda* 78, 149–152.

Hockey, P.A.R., Dean, W.R.J. and Ryan, P.G. (2005) *Roberts Birds of Southern Africa*, 7th edn. Trustees of the John Voelker Bird Book Fund, Cape Town.

Hughes, B., Robinson, J., Green, A.J., Li, D. and Mundkur, T. (2006) International single species action plan for the White-headed Duck *Oxyura leucocephala*. CMS Technical Series No. 13 and AEWA Technical Series No.8. Bonn, Germany.

Hulme, P.E. (2009) Trade, transport and trouble: managing invasive species pathways in an era of globalization. *Journal of Applied Ecology* 46, 10–18.

IGoTerra (2018) IGoTerra species checklists. Available at: www.igoterra.com (accessed 8 November 2019).

IUCN (2019) The IUCN Red List of Threatened Species, version 2019-2. Available at: www.iucnredlist.org (accessed 8 November 2019).

Ivanova, I.M. and Symes, C.T. (2018) Common Starling *Sturnus vulgaris* expansion in South Africa. *Biodiversity Observations* 9, 1–6.

Jone, A. (2003) Mozambique. In: Macdonald, I.A.W., Reaser, J.K., Bright, C., Neville, L.E., Howard, G.W., *et al.* (eds) *Invasive Alien Species in Southern Africa: National Reports and Directory of Resources*. Global Invasive Species Programme, Cape Town, pp. 38–39.

Juniper, T. and Parr, M. (1998) *Parrots: A Guide to the Parrots of the World*. Pica Press, Sussex.

Karapetyan, S. (2017) Death of last known Ring-necked Parakeet a big win in Seychelles' invasive species fight. Seychelles News Agency. Available at: www.seychellesnewsagency.com/articles/7878/Death+of+last+known+ringnecked+parakeet+a+big+win+in+Seychelles%27+invasive+species+fight (accessed 8 November 2019).

Kenya Bird Map (2018) Kenya Bird Map Project. Animal Demography Unit, University of Cape Town, Cape Town. Available at: http://kenyamap.adu.org.za (accessed 8 November 2019).

Lack, P. (2018) African Bird Club: African Checklist. African Bird Club. Available at: www.africanbirdclub.org/resources/checklist/intro (accessed 8 November 2019).

Lepage, D. (2004) Avibase: The World Bird Database. Bird Studies Canada. Available at: https://avibase.bsc-eoc.org (accessed 8 November 2019).

Lever, C. (2005) *Naturalised Birds of the World*. T. & A.D. Poyser, London.

Linders, T.E.W. and Langrand, O. (2014) First record of House Crow *Corvus splendens* for Madagascar – potential impacts and suggested management of an invasive bird species. *Bulletins of the African Bird Club* 21, 216–222.

Lipkind, M., Weisman, Y., Shihmanter, E. and Shoham, D. (1979) Identification of avian influenza viruses isolated from wild Mallard Ducks in Israel. *Veterinary Record* 105, 558–558.

Little, R.M. and Sutton, J.L. (2013) Perceptions towards Egyptian Geese at the Steenberg Golf Estate, Cape Town, South Africa. *Ostrich* 84, 85–87.

Long, J.L. (1981) *Introduced Birds of the World: The Worldwide History, Distribution and Influence of Birds Introduced to New Environments*. Universe Books, New York.

Lopes, R.J., Gomez, J.A., Andreotti, A. and Andreoni, M. (2016) Purple Swamphen or Gallinule (*Porphyrio porphyrio*) and humans. *Society & Animals* 24, 574–595.

Macdonald, I.A.W. (1988) The invasion of introduced species into nature reserves in tropical savannas and dry woodlands. *Biological Conservation* 44, 67–93.

Macdonald, I.A.W., Reaser, J.K., Bright, C., Neville, L.E., Howard, G.W., *et al.* (2003) *Invasive Alien Species in Southern Africa: National Reports and Directory of Resources*. Global Invasive Species Programme, Cape Town, South Africa.

Mackay, B., Little, R.M., Amar, A. and Hockey, P.A.R. (2014) Incorporating environmental considerations in managing Egyptian Geese on golf courses in South Africa. *Journal of Wildlife Management* 78, 671–678.

Mangnall, M.J. and Crowe, T.M. (2002) Population dynamics and the physical and financial impacts to cereal crops of the Egyptian Goose *Alopochen aegyptiacus* on the Agulhas Plain, Western Cape, South Africa. *Agriculture, Ecosystems & Environment* 90, 231–246.

Mank, J.E., Carlson, J.E. and Brittingham, M.C. (2004) A century of hybridization: decreasing genetic distance between American Black Ducks and Mallards. *Conservation Genetics* 5, 395–403.

Mauremootoo, J.R., Leckraz, N.R., Puttoo, M., Bellouard, E., Ganeshan, S. and Beni Madhu, S.P. (2003) Mauritius. In: Macdonald, I.A.W., Reaser, J.K., Bright, C., Neville, L.E., Howard, G.W., *et al.* (eds) *Invasive Alien Species in Southern Africa: National Reports and Directory of Resources*. Global Invasive Species Programme, Cape Town, pp. 12–37.

McKenzie, D. (2011) Reporting rate comparisons for birds in the Nelspruit area – SABAP1 vs SABAP2. *Biodiversity Observations* 2, 22–31.

McPherson, S.C. (2015) Urban ecology of the Crowned Eagle *Stephanoaetus coronatus* in KwaZulu-Natal, South Africa. PhD thesis, University of KwaZulu-Natal, Pietermaritzburg, South Africa.

Meulemans, G., van den Berg, T.P., Decaesstecker, M. and Boschmans, M. (2002) Evolution of pigeon Newcastle disease virus strains. *Avian Pathology* 31, 515–519.

Mlíkovský, J. (2011) Records of the Purple Swamphen (*Porphyrio porphyrio*) from the Czech Republic: a review. *Sylvia* 47, 123–129.

Msimanga, A. and Slotow, R. (2000) The House Sparrow *Passer domesticus* (Passeriformes: Ploceidae) invasion of southern Africa: dispersal rates and the Allee effect. *Durban Museum Novitates* 25, 5–11.

Mwacharo, J., Bjørnstad, G., Han, J. and Hanotte, O. (2013) The history of African village chickens: an archaeological and molecular perspective. *African Archaeological Review* 30, 97–114.

Nature Seychelles (2018) Introduced land birds. Available at: www.natureseychelles.org/component/content/article?id=171:introduced-land-birds (accessed 8 November 2019).

New World Encyclopedia Contributors (2015) Scramble for Africa. New World Encyclopedia. Available at: www.newworldencyclopedia.org/p/index.php?title=Scramble_for_Africa& oldid=988092 (accessed 8 November 2019].

Payet, H.E. (2007) The introduced Seychelles White-eye (*Zosterops modestus*) population of Fregate Island, Seychelles: territories and aspects of population dynamics, biology and ecology. MSc thesis, University of Reunion, Saint Denis, France.

Peacock, D.S., van Rensburg, B.J. and Robertson, M.P. (2007) The distribution and spread of the invasive alien Common Myna, *Acridotheres tristis* L. (Aves: Sturnidae), in southern Africa. *South African Journal of Science* 103, 465–474.

Peterson, A.T. and Brisbin, I.L. (1998) Genetic endangerment of wild Red Junglefowl *Gallus gallus*? *Bird Conservation International* 8, 387–394.

Picker, M. and Griffiths, C. (2013) *Alien and Invasive Animals: A South African Perspective*. Penguin Random House, Milnerton, South Africa.

Pinto, A.D.R. (1959) Alguns novos 'Records' de aves para o Sul do Save e Moçambique, incluindo o de um género novo para a sub-regiáo da África do Sul, com a descrição de novas sub-espécies. *Boletim da Sociedade de Estudos da Provincia de Mocambique* 118, 15–28.

Pithon, J.A. and Dytham, C. (2001) Determination of the origin of British feral Rose-ringed Parakeets. *British Birds* 94, 74–79.

Rappole, J.H., Derrickson, S.R. and Hubálek, Z. (2000) Migratory birds and spread of West Nile virus in the Western Hemisphere. *Emerging Infectious Diseases* 6, 319–328.

Ratcliffe, C.S. (2005) Chukar Partridge. In: Hockey, P.A.R., Dean, W.R.J. and Ryan, P.G. (eds) *Roberts Birds of Southern Africa*, 7th edn. Trustees of the John Voelker Bird Book Fund, Cape Town, pp. 61–62.

Reaser, J.K., Meyerson, L.A., Cronk, Q., de Poorter, M., Eldrege, L., *et al.* (2007) Ecological and socioeconomic impacts of invasive alien species in island ecosystems. *Environmental Conservation* 34, 98–111.

Rhymer, J.M. (2006) S33-4 Extinction by hybridization and introgression in anatine ducks. *Acta Zoologica Sinica* 52, 583–585.

Richardson, D.M., Cambray, J.A., Chapman, R.A., Dean, W.R.J., Griffiths, C.L., *et al.* (2003) Vectors and pathways of biological invasions in South Africa: Past, present and future. In: Ruiz, G.M. and Carlton, J.T. (eds) *Invasive Species: Vectors and Management Strategies*. Island Press, Washington, DC, pp. 292–349.

Rocha-Camarero, G. and de Trucios, S.J.H. (2002) The spread of the Collared Dove *Streptopelia decaocto* in Europe: colonization patterns in the west of the Iberian Peninsula. *Bird Study* 49, 11–16.

Ryall, C. (2016) Further records and updates of range expansion in House Crow *Corvus splendens*. *Bulletin of the British Ornithologists' Club* 136, 39–45.

SABP2 (2018) The Southern African Bird Atlas Project. Animal Demography Unit, University of Cape Town, Cape Town. Available at: http://sabap2.adu.org.za (accessed 8 November 2019).

Shaw, K.A. and Waller, L.J. (2017) Avifauna. In: Turner, A.A. (ed) *Western Cape Province State of Biodiversity*. Cape Nature Scientific Services, Stellenbosch, South Africa, pp. 153–190.

Sinclair, I. (2013) *Birds of the Indian Ocean Islands*. Penguin Random House, Milnerton, South Africa.

Smit, P. and Steenkamp, C. (2003) Namibia. In: Macdonald, I.A.W., Reaser, J.K., Bright, C., Neville, L.E., Howard, G.W., *et al.* (eds) *Invasive Alien Species in Southern Africa: National Reports and Directory of Resources*. Global Invasive Species Programme, Cape Town, pp. 40–44.

Stafford, L. (2010) Mallard strategy for South Africa. L. Stafford in collaboration with The National Problem Bird Forum and The CAPE Invasive Animal Working Group, Cape Town. Available at: www.wingshooters.co.za/pdf/NationalMallard%20Strategy-May10.pdf (accesed 8 November 2019).

Sullivan, B.L., Wood, C.L., Iliff, M.J., Bonney, R.E., Fink, D. and Kelling, S. (2009) eBird: a citizen-based bird observation network in the biological sciences. *Biological Conservation* 142, 2282–2292.

Symes, C.T. (2014) Founder populations and the current status of exotic parrots in South Africa. *Ostrich* 85, 235–244.

Symes, C.T., Roller, K., Howes, C., Lockwood, G. and van Rensburg, B.J. (2017) Grassland to urban forest in 150 years: avifaunal response in an African metropolis. In: Murgui, E. and Hedblom, M. (eds) *Ecology and Conservation of Birds in Urban Environments*. Springer, Cham, Switzerland, pp. 309–341.

Tate, G. and Amar, A. (2012) Possessed: Black Sparrowhawks and Egyptian Geese vie for nest sites. *Africa Birds and Birding* 17, 12–13.

Thompson, L.J., Davies, J.P., Gudehus, M., Botha, A.J., Bildstein, K.L., *et al.* (2017) Visitors to nests of Hooded Vultures *Necrosyrtes monachus* in northeastern South Africa. *Ostrich* 88, 155–162.

Thompson, P.N., Sinclair, M. and Ganzevoort, B. (2008) Risk factors for seropositivity to H5 avian influenza virus in Ostrich farms in the Western Cape Province, South Africa. *Preventive Veterinary Medicine* 86, 139–152.

Thymer, J. and Simberloff, D. (1996) Extinction by hybridisation. *Annual Review of Ecology, Evolution, and Systematics* 27, 83–109.

Uranie, S. (2015) Eradication success – Seychelles wins war against invasive Red-whiskered Bulbul. Seychelles News Agency. Available at: http://www.seychellesnewsagency.com/articles/2332/Eradication+success++Seychelles+wins+war+against+invasive+red-whiskered+bulbul (accessed 8 November 2019).

van Rensburg, B.J., Peacock, D.S. and Robertson, M.P. (2009) Biotic homogenization and alien bird species along an urban gradient in South Africa. *Landscape and Urban Planning* 92, 233–241.

van Wilgen, B.W. and Wilson, J.R. (eds) (2018) *The Status of Biological Invasions and their Management in South Africa in 2017*. South African National Biodiversity Institute, Kirstenbosch, and DST-NRF Centre of Excellence for Invasion Biology, Stellenbosch, South Africa.

Watson, G.E. (1963) *Preliminary Field Guide to the Birds of the Indian Ocean: For Use During the International Indian Ocean Expedition*. Smithsonian Institution, Washington, DC.

Watson, J., Warman, C., Todd, D. and Laboudallon, V. (1992) The Seychelles Magpie Robin *Copsychus sechellarum*: ecology and conservation of an endangered species. *Biological Conservation* 61, 93–106.

Weber, W.J. (1979) *Health Hazards from Pigeons, Starlings and English Sparrows: Diseases and Parasites Associated with Pigeons, Starlings, and English Sparrows which Affect Domestic Animals*. Thomson Publications, Fresno, California.

Wreford, E., Hart, L., Brown, M. and Downs, C.T. (2017) Black Sparrowhawk (*Accipiter melanoleucus*) breeding behaviour and reproductive success in KwaZulu-Natal, South Africa. *Ostrich* 88, 287–290.

Zengeya, T., Ivey, P., Woodford, D.J., Weyl, O., Novoa, A., *et al.* (2017) Managing conflict-generating invasive species in South Africa: challenges and trade-offs. *Bothalia* 47, 1–11.

Zimmerman, D.A., Turner, D.A., Pearson, D.J., Willis, I. and Pratt, H.D. (1996) *Birds of Kenya and Northern Tanzania*. Princeton University Press, Princeton, New Jersey.

Continental Analysis of Invasive Birds: North America

Jessica L. Burnett[1] and Craig R. Allen
Nebraska Cooperative Fish and Wildlife Research Unit, School of Natural Resources, University of Nebraska–Lincoln, Lincoln, NE 68583, USA

Citation: Burnett, J.L. and Allen, C.R. (2020) Continental analysis of invasive birds: North America. In: Downs, C.T. and Hart, L.A. (eds) *Invasive Birds: Global Trends and Impacts*. CAB International, Wallingford, UK, pp. 278–294.

38.1 Introduction

The USA began to monitor imports of non-native wildlife through ports beginning in the early 1900s; however, what happened to the birds afterwards was usually not monitored (Phillips, 1928). Caged-bird releases, intentional release for harvest and biocontrol are among the most common reasons for bird introductions to North America. Despite the estimated hundreds of species documented as introduced to North America, relatively few have successfully acclimatized/established and persisted.

Only a small percentage of the approximately 1000 bird species introduced globally (Dyer *et al.*, 2017) are currently established in North America (see Appendix 38.1). There are documented introductions of birds as early as the 1800s in North America, at a time when acclimatization societies were active, but despite changing pathways, the rate of introductions has continued to increase (Dyer *et al.*, 2017). The lack of knowledge of the outcomes of bird introductions to North America, particularly with respect to unsuccessful introductions, makes analysis of less recent introductions difficult (Moulton *et al.*, 2010).

Introductions of successfully established birds in North America are largely attributed to European settlement and nostalgia, multiple introduction events and climate-matching introductions. Additional pathways for introductions include biocontrol and the pet trade. One of the earliest successful introductions to North America was that of the House Sparrow (*Passer domesticus*), with introductions documented as early as 1851, when 16 birds were introduced to Brooklyn, New York (Barrows, 1889). Facilitated by railway cars, a high density of horses (which provided undigested grain in droppings), waste grains and additional introduction events led to the House Sparrow spreading across the contiguous USA in only a few decades (Barrows, 1889).

38.2 Biogeography and Climate

The geographical breadth of mainland North America comprises a contiguous land mass encompassing numerous climate zones, ranging from tropical and wet to arctic tundra and taiga plains. The North American Bird Conservation Initiative (NABCI) delineates 66 bird conservation regions, or regions with distinct bird communities, coinciding with biotic and abiotic landscape features (Fig. 38.1). Despite the variation in habitat availability and climate across North America, non-native species richness is highest in California, Florida (bird conservation region 32) and Hawaii (Dawson *et al.*, 2017). Due largely to the subtropical climate and human population size, Florida has a high non-native bird species richness, conservatively estimated at 15 species (Table 38.1).

38.3 Introduction Hotspots

Introduction hotspots of North America have occurred largely in the Hawaiian Archipelago and Florida, and to a lesser extent in southern California and New York City. Florida and Hawaii have more established non-native species relative to other regions of North America, and the number of species introduced to each of these regions is approximately equal to the number introduced to mainland North America (not including Florida) (Fig. 38.2).

38.3.1 Florida

Florida is a peninsular region of the USA and is primarily in a subtropical climate zone. A steady climate without winter and

[1]Corresponding author: jessicaleighburnett@gmail.com

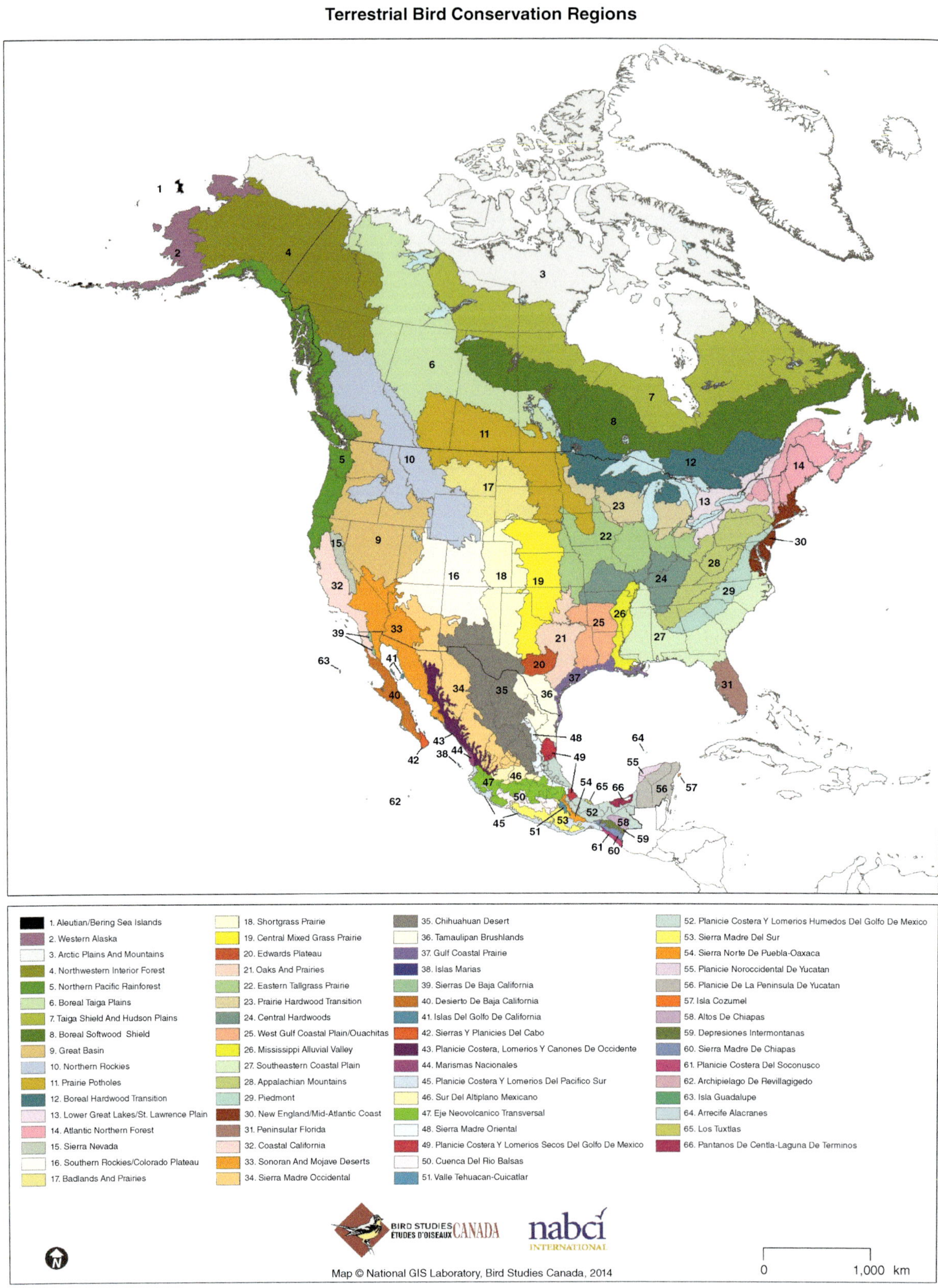

Fig. 38.1. Terrestrial regions from the North American Bird Conservation Initiative (NABCI) bird conservation regions in North America.

Table 38.1. Non-native species in Florida, USA. (Data from the Florida Ornithological Society Checklist: www.fosbirds.org/florida-bird-list.html, accessed 11 November 2019.)

Common name	Species	Established exotic	Disestablished exotic
Egyptian Goose	*Alopochen aegyptiacus*	✓	
Muscovy Duck	*Cairina moschata*	✓	
Rock Pigeon	*Columba livia*	✓	
Eurasian Collared-dove	*Streptopelia decaocto*	✓	
Purple Swamphen	*Porphyrio porphyrio*	✓	
Monk Parakeet	*Myiopsitta monachus*	✓	
Nanday Parakeet	*Aratinga nenday*	✓	
White-winged Parakeet	*Brotogeris versicolurus*	✓	
Budgerigar	*Melopsittacus undulatus*		✓
Red-whiskered Bulbul	*Pycnonotus jocosus*	✓	
Common/ European Starling	*Sturnus vulgaris*	✓	
Common Myna	*Acridotheres tristis*	✓	
House Sparrow	*Passer domesticus*	✓	
House Finch	*Haemorhous mexicanus*	✓	
Spot-breasted Oriole	*Icterus pectoralis*	✓	

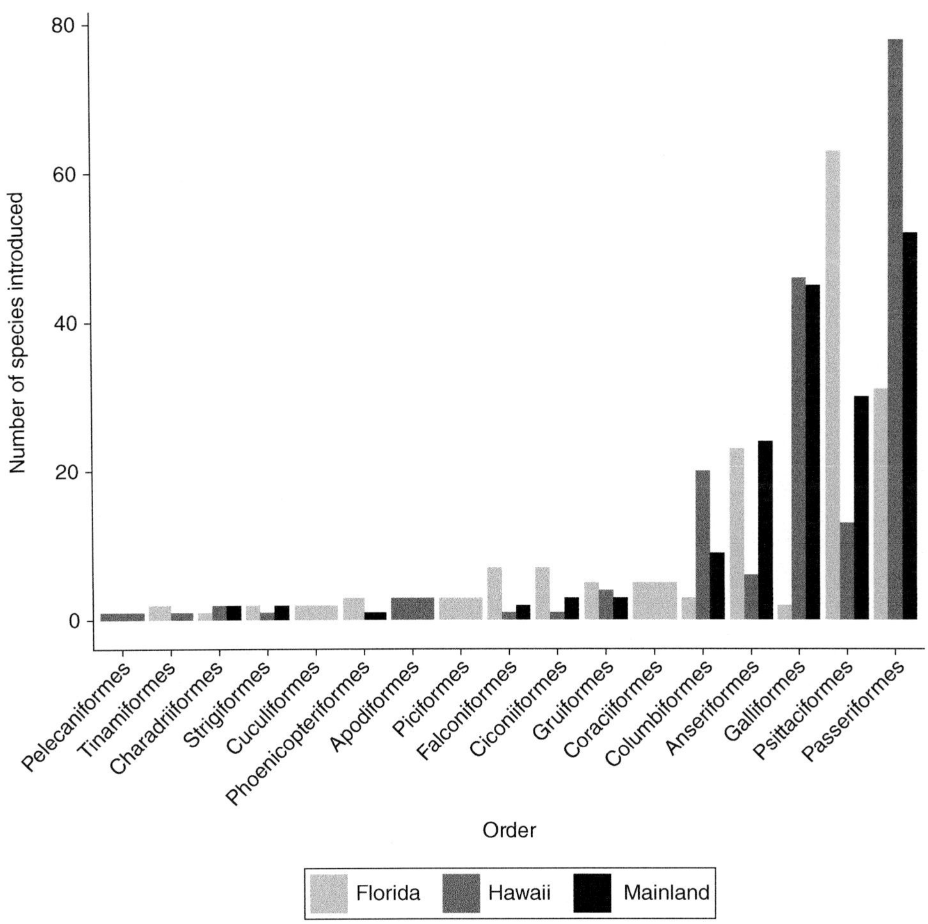

Fig. 38.2. Number of species introduced (successfully and unsuccessfully) to Florida (n = 159), Hawaii (n = 177), and mainland North America (n = 173; does not include Florida) by taxonomic order. Estimates were obtained from the Global Avian Invasions Atlas (GAVIA) database.

busy ports (especially southern Florida) has made this region a hotspot for introductions of most taxa (especially birds, reptiles and plants). Of the 31 species listed as non-native and established in the Global Invasive Species Database (GISD), 15 species are currently considered 'exotic and established' in Florida (Table 38.1). It is worth noting that although the Florida Ornithological Society's current list may be conservative (e.g. see species list for Florida in Allen, 2006), it contains three species that are not currently listed in the GISD: the Nanday Parakeet (*Aratinga nenday*), the Egyptian Goose (*Alopochen aegyptiacus*) and the House Finch (*Haemorhous mexicanus*). Several species common to these lists are currently exhibiting spread, notably the Monk Parakeet (*Myiopsitta monachus*) and Eurasian Collared-dove (*Streptopelia decaocto*). Others, such as the House Sparrow and Common or European Starling (*Sturnus vulgaris*) have been so long established that they are considered a naturalized and permanent part of the fauna.

38.3.2 Hawaiian islands

The study of non-native bird impacts on native bird communities is more pronounced in island ecosystems relative to studies of mainland communities (Simberloff, 1995), probably fuelled by the high rate of endemism and the proportion of threatened or endangered species relative to mainland communities. The native avifauna of the Hawaiian archipelago have been subjected to numerous stressors, which have greatly influenced native community composition, endemic species populations and functional diversity (Boyer and Jetz, 2014). Although the primary stressors on native avifauna are the introduction of non-native predators (e.g. snakes, mongeese) and habitat loss, non-native birds are considered to influence native bird community dynamics (Moulton and Pimm, 1987; Boyer and Jetz, 2014).

The Warbling or Japanese White-eye (*Zosterops japonicus*) is considered invasive in Hawaii, as it is one of the most commonly occurring non-native birds of the Hawaiian forests, yet the direction of its impacts on native wildlife are variable (van Riper, 2000). For example, this species may serve as a buffer for disease transmission of avian malaria, as it is currently immune to existing strains of this disease (Samuel *et al.*, 2011), yet it is also suspected of influencing native bird morphologies by exploitative competition for food (Freed and Cann, 2009). The impacts of the Common Myna (*Acridotheres tristis tristis*), a species introduced to Hawaii to control a pest worm, are also contestable (Funasaki *et al.*, 1988; Grarock *et al.*, 2012). It is clear, however, that the Common Myna is a known disperser of an invasive weed, *Lantana camara* (Pimentel *et al.*, 2000).

38.4 Spread and Distribution Patterns of Successful Introductions

The success of acclimatization efforts for introduced birds is influenced by multiple factors, predominantly local abiotic conditions, propagule pressure and species traits. Establishment of birds varies by hotspot. North American hotspots for introduced birds are all subtropical (Florida, Hawaii and southern California), and most of the species established are likewise from subtropical conditions, which allows species with a low tolerance for cold to thrive (e.g. Monk Parakeet; see Chapters 10 and 43, this volume).

Many avian species that have successfully established but whose rate of spread is limited, such as the Grey Partridge (*Perdix perdix*) and Monk Parakeet, are often more *K*-selected, while species capable of reproducing and surviving in more harsh and unstable environments are often generalists, and are arguably 'more' *r*-selected, such as the House Sparrow and Common Starling. The most successful avian invaders of mainland USA are species that are neophilic, or are bolder in new environments, and these species are often at high density in urban environments, relative to their native counterparts. In Florida, established non-native birds were generalists with large native ranges (Allen *et al.*, 2013), and predictors of establishment and spread differed, and predictors of both varied with taxa. However, Florida may be an outlier due to the high number of caged-bird releases with low propagule pressure. Predictors of spread in Florida include a positive association with reproductive output and propagule pressure, and negatively associated with a species' body mass.

Introductions to mainland North America have occurred primarily because of the pet trade and hunting industries, with introductions from the pet trade occurring primarily in Florida, and game bird introductions spanning much of the eastern and central USA. An estimated 46 species of the orders Galliformes (game birds), 94 Passeriformes and 80 Psittaciformes have been introduced to North America to date (see Appendix 38.1), but only 31 are listed in the GISD (Figs 38.2 and 38.3).

38.4.1 Upland game birds

Game birds require special attention, because as a group they have been subject to regular, frequent and intensive efforts to establish species outside their native range. Regular (i.e. semi-annually or annually) and documented introductions of non-native game birds to North America date back to at least the early 1900s with the goal of sustaining or creating local populations for harvest. Introductions for some game birds led to local establishment across much of central North America. For example, by approximately the 1980s, populations of Grey Partridge thrived as far west as Utah, and as far north-east as the Atlantic Provinces of Canada (Dumke *et al.*, 1980; Potts and Aebischer, 1995). Like many facultative and obligate grassland game bird species, however, Grey Partridge populations are negatively impacted by large-scale agricultural monocultures, which have led to a loss of vegetative cover and poor diet for some. Future range size, population persistence and local population size of other non-native game birds may be predicted by agricultural land use and intensity. Species regularly introduced across much of North America to create hunting opportunities include the Ring-necked Pheasant (*Phasianus colchicus*), the Grey Partridge and the Chukar (*Alectoris chukar*).

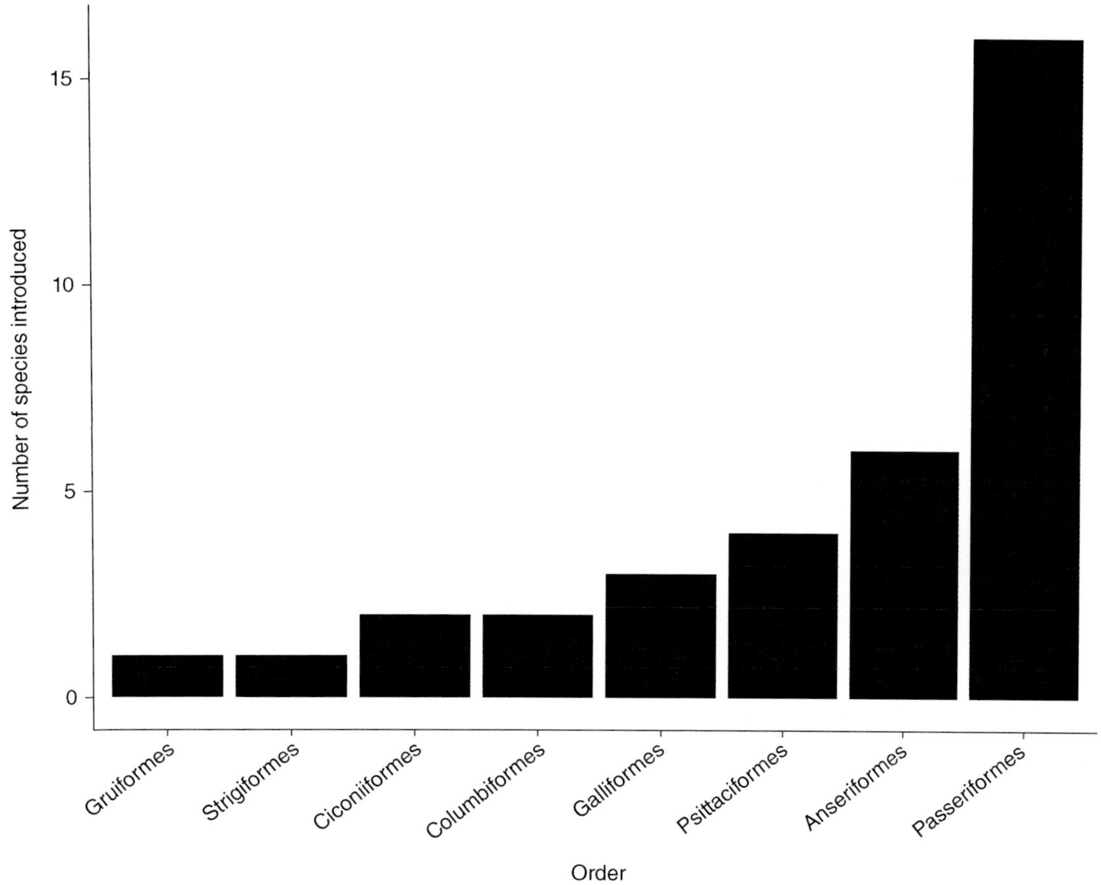

Fig. 38.3. Number of species currently established and breeding in Florida. (Data from the Florida Ornithological Society official species list at www.fosbirds.org/florida-bird-list.html and the Global Invasive Species Database at www.iucngisd.org/gisd/, accessed 12 November 2019.)

These regular introduction events, coupled with a decline in hunting interest, may reduce the modern spread of some game bird species relative to the recent past.

38.4.2 Pet trade

The pet trade threatens the ecological integrity of both the systems from which the species was collected and the habitat to which it is introduced (Carrete and Tella, 2008). Despite the potential ecological consequences of selling and exchange of non-native birds, the pet trade industry provides modern sources of non-native parrots and passerine individuals. Although few species that are introduced through the pet trade by way of escapes or intentional releases establish successfully, those that have have caused major economic (e.g. Monk Parakeet; see Chapters 10 and 43, this volume) and ecological (e.g. Common Myna) harm. Introductions because of the pet trade are largely restricted to Florida, with species such as the Monk Parakeet proving an economic burden in some communities (see Chapter 10, this volume). The unique introduced species in North America are predominantly songbirds (Passierformes) and parrots (Psittaciformes) (Fig. 38.4).

38.5 Data Sources for Avian Introductions, Invasions and Population Trends

There are five major data sources for avian introductions, invasions and population trends in North America.

38.5.1 Global Register of Introduced and Invasive Species

The Global Register of Introduced and Invasive Species (GRIIS; Pagad *et al.*, 2018) comprises country-specific and edited databases. It does not currently contain datasets for the USA, but does contain records for Canada and Mexico. It is a simple species checklist denoting whether the source contains verified records and whether impacts are known. However, it does not appear to be comprehensive for birds.

38.5.2 IUCN Global Invasive Species Database

The International Union for Conservation of Nature (IUCN) Global Invasive Species Database (GISD) is a register of invasive

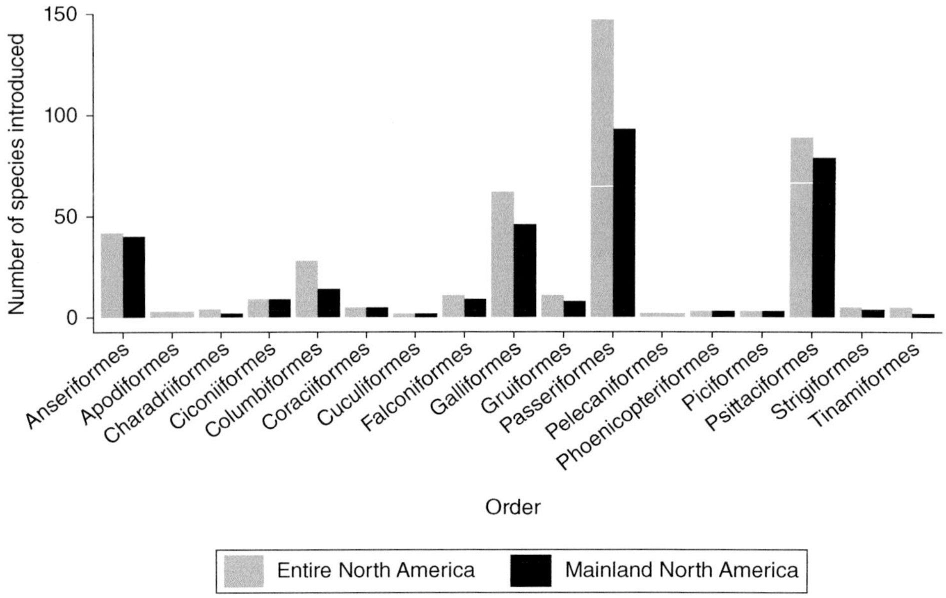

Fig. 38.4. Number of species introduced to North America by taxonomic order using the Global Avian Invasions Atlas (GAVIA) and Global Invasive Species Database (GISD). Mainland North America comprises Belize, Canada, Costa Rica, El Salvador, Guatemala, Honduras, Mexico, Nicaragua, Panama and the USA.

species with documented or predicted threat to native biodiversity. It is managed by the Invasive Species Specialist Group (ISSG) of the IUCN Species Survival Commission (ISSG Invasive Species Specialist Group, 2015).

38.5.3 Global Avian Invasions Atlas

The Global Avian Invasions Atlas (GAVIA; Dyer *et al.*, 2017) is the most comprehensive, publicly-available meta-database for non-native bird introductions. Contains known historic and modern introduction events with associated references. Records are never removed; therefore, it is useful for historic analysis.

38.5.4 North American Breeding Bird Survey

The North American Breeding Bird Survey (NABBS; Pardieck *et al.*, 2017) is an annual roadside survey of more than 4100 locations across Canada and continental USA. Roadside surveys consist of 5 min point counts by trained citizen scientists, conducted at 50 points along an approximately 39 km route. The scheme is designed to gain inference on long-term population trends. Data collection began in 1966. Routes are designed to avoid urban areas, so are useful mostly for non-native species occurring in rural areas.

38.5.5 Christmas Bird Count

The Christmas Bird Count (CBC; National Audubon Society, 2010) is an annual winter count conducted by citizen scientists within 2 weeks of 25 December. Data collection began in the early 1900s. Survey areas comprise a circle with a diameter of 24 km and are conducted in both terrestrial and aquatic habitats. These data may be more useful than NABBS when estimating trends of non-native, urban-dwelling birds, as the circles typically include an urban centre.

38.6 Population Trend Analyses

We modelled species population trends of 24 of the 31 GISD species using the NABBS data (Pardieck *et al.*, 2017) using generalized additive models (GAMs; Hastie *et al.*, 2015). GAMs are a flexible implementation of generalized linear models in cases where species populations exhibit non-linear trends. GAMs optimize the predictability of the relationship between the response and predictor variable(s) while accounting for the noise associated with year-to-year fluctuations in species counts. Should the species trend be linear, the GAM will not force a non-linear fit.

We fit GAMs to estimate population trends of individual species using data from the Breeding Bird Survey using the equation:

$$g(E(Y)) = \beta_0 f(\text{year})$$

Species counts (Y) on the log-scale were modelled as a function of time, f(year) (i.e. a smooth function is produced for each species over time). We assumed the response variable (Y) distribution as quasi-Poisson and therefore we modelled the expected value of species counts $E(Y)$ using a log-link function, $g(E(Y))$, where β_0 is the intercept.

The rate of change (derivative) of the population trend (with 95% confidence intervals) estimates obtained from the GAMs are described in Fig. 38.5 for the House Sparrow. A trend (solid line) and 95% confidence interval (shaded area) below zero (red dashed line) indicates a declining trend, and above zero an increasing trend for the House Sparrow in North America (Fig. 38.5).

38.7 Knowledge Gaps

Three major gaps exist in our understanding of non-native birds in North America: (i) modern range shifts; (ii) drivers of population declines; and (iii) impacts of non-native species on native wildlife.

38.7.1 Range shifts

Globalization and climate change are impacting local and regional populations and species distributions worldwide, and accounts of these phenomena are increasing (Veech *et al.*, 2011). The rate at which non-native species ranges are shifting or non-native populations are increasing or declining because of widespread changes in biotic and abiotic conditions is relatively understudied. Most successful non-native species in North America have dense populations in urban areas; however, whether the rate at which shift, expansion or contraction of non-native species ranges is accelerated by urbanization remains a question (MacGregor-Fors and Schondube, 2011). Two non-native passerines occurring primarily in developed areas in North America, the Great Tit (*Parus major*) and the Eurasian Tree Sparrow (*Passer montanus*), have exhibited range shifts in the last 10 or 20 years (Burnett *et al.*, 2017). However, studies of these species are quite limited, such that the potential mechanisms driving these shifts have not yet been identified.

38.7.2 Population declines of non-native GISD species

Although the decline of a non-native species is not necessarily of concern to management entities, the ecological significance of some species declines may be an indication of general system health and resilience. The House Sparrow has exhibited declines across much of its non-native and native range worldwide, so much so that this species is now difficult to find in many regions. Coinciding with changes in agricultural practices in North America, the House Sparrow first exhibited widespread declines in North America in the 1980s in rural and agricultural areas, although the drivers of modern declines in urbanized areas worldwide are largely undetermined (Fig. 38.5) (Summers-Smith, 2003; Burnett and Moulton, 2015).

The Common Starling was originally introduced to North America in the late 1800s (Linz *et al.*, 2007) and then quickly spread and is thought to have caused significant economic harm in the USA (Pimentel *et al.*, 2000). Our analysis suggests that the starling is steadily declining in much of North America (Fig. 38.6), which agrees with the IUCN population trend declining estimate (BirdLife International, 2018).

The House Finch is native to western North America but was introduced to the north-eastern USA in the mid-1900s and has since established across much of continental USA, Mexico and some parts of southern Canada. Despite the rapid rate of spread of House Finches towards the central USA, its populations appear to be steadily declining (Fig. 38.7). Declines in this population are strongly linked to epizootic conjunctivitis (*Mycoplasma gallisepticum*; Dhondt *et al.*, 1998; Altizer *et al.*, 2004) and less so to competition with House Sparrows (Cooper

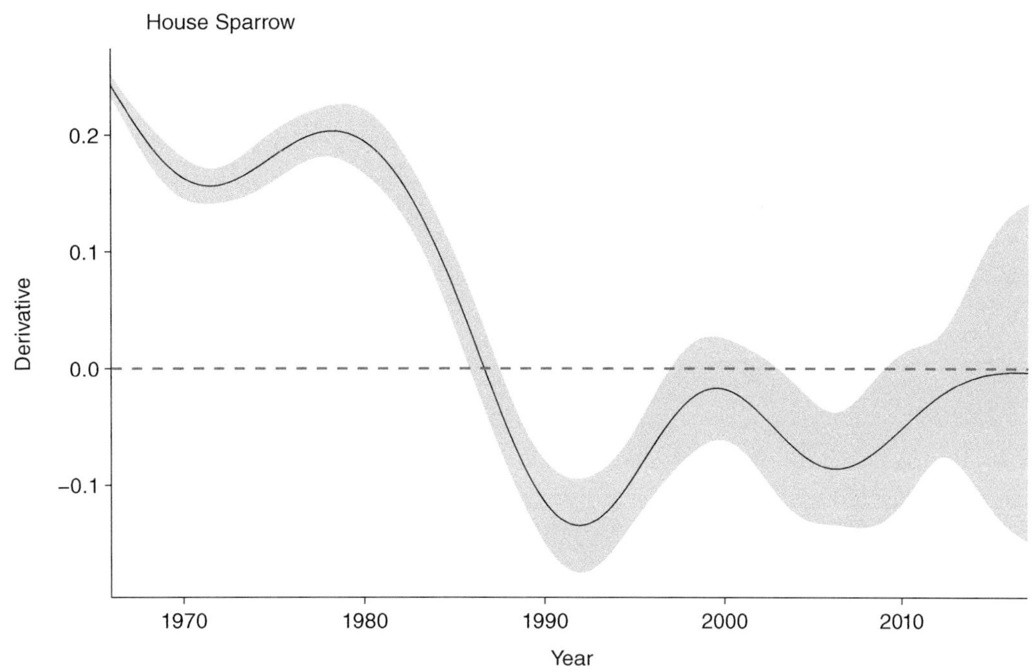

Fig. 38.5. A generalized linear model of the North American Breeding Bird Survey (NABBS) data suggesting that House Sparrows (*Passer domesticus*) are generally declining across North America.

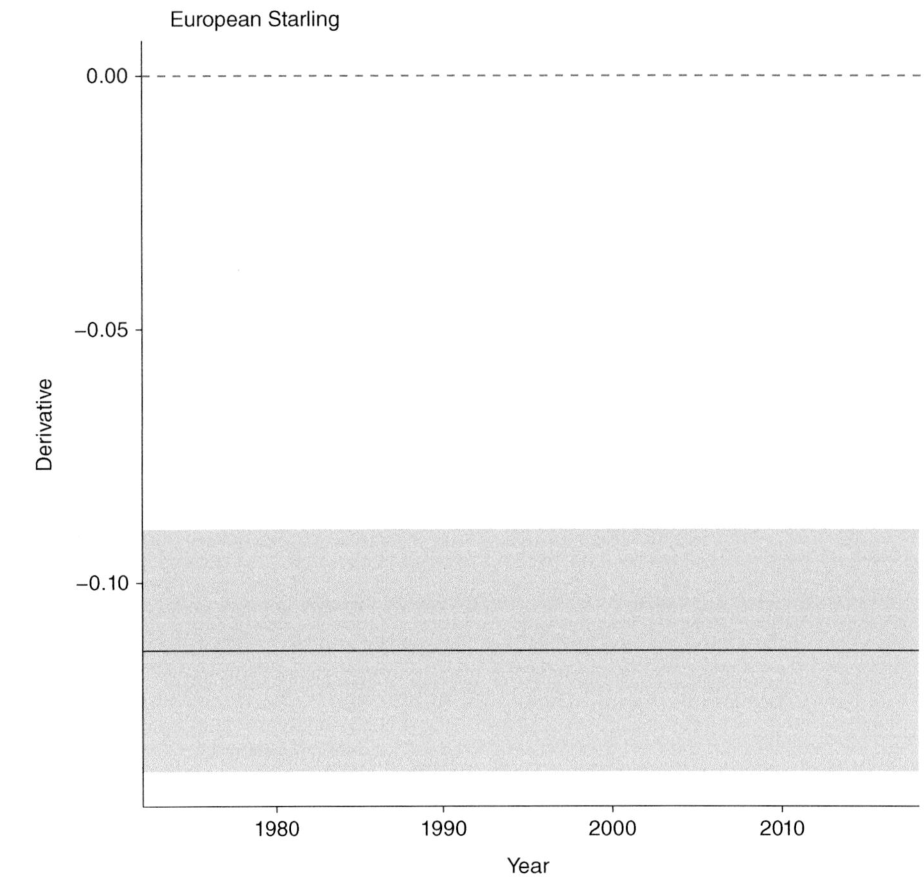

Fig. 38.6. North American Breeding Bird Survey (NABBS) data suggest that the Common Starling (*Sturnus vulgaris*) population has declined in recent years in North America. Note the constant negative predicted values of the derivative, indicating constant decline over time.

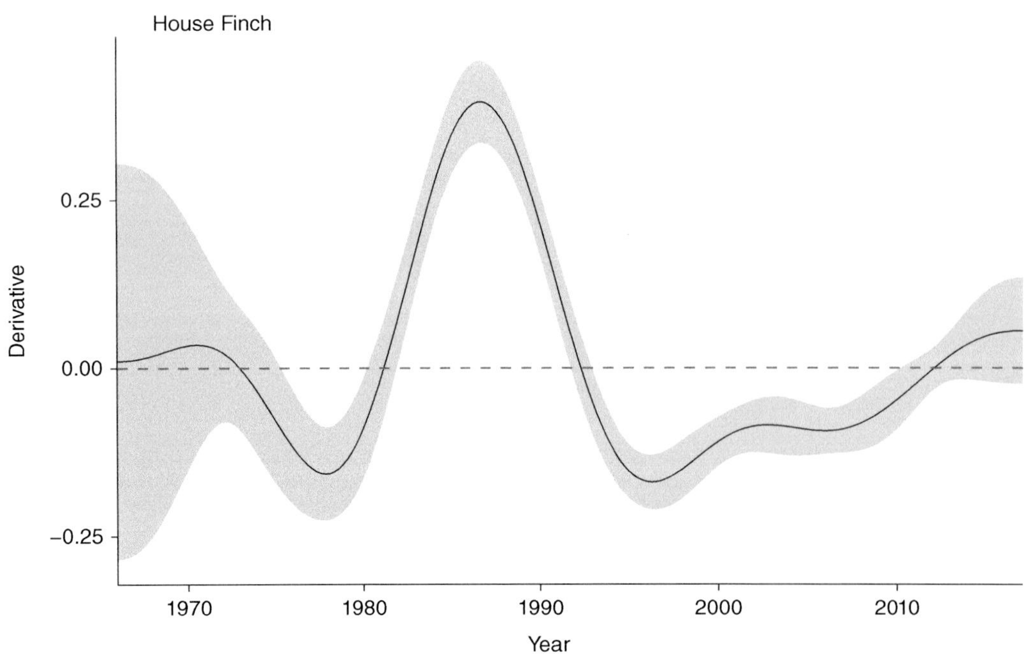

Fig. 38.7. North American Breeding Bird Survey (NABBS) data trends for the House Finch (*Haemorhous mexicanus*) in North America.

et al., 2007). Interestingly, and unlike the sedentary western (i.e. California) population of House Finches, which exhibit no migratory behaviour, the eastern population is seasonally migratory (Able and Belthoff, 1998).

The Cattle Egret (*Bubulcus ibis*), native to Africa, spread from its original introduction sites of South America, first breeding in North America in 1953 (Crosby, 1972). Much like the Common Starling, House Finch and House Sparrow, the Cattle Egret spread rapidly across much of Mexico, central and southern USA, and most islands of the Caribbean Sea in a matter of decades. Since its spread, however, it has exhibited increasing rates of declines since the mid-1980s (Fig. 38.8).

38.7.3 Non-native birds in the urban environment

Numerous studies suggest that the suburban habitat may be a point of colonization for non-native, urban-philic birds, and a point of local extirpation for native, urban-avoiding or urban-sensitive birds (Marzluff, 2001; Blair and Johnson, 2008; Lepczyk *et al.*, 2008). These patterns are quantified using simple composite community metrics (e.g. species richness, evenness); however, the role of non-native species in functional homogenization remains unknown (Devictor *et al.*, 2007). Furthermore, the study of biotic homogenization in the urban ecosystem, especially as it relates to the influence of non-native birds, '...raises more questions than [it] answers...' (Marzluff, 2016).

Non-native birds are generally denser in highly urbanized areas in mainland North America, comprising in many cases a large proportion of the avifaunal biomass (Faeth *et al.*, 2011). Although decline and eradication of non-native species are typically considered positively by ecological managers, it is not clear whether the niches occupied by non-native species would be filled by native species upon eradication. This is primarily due to the relatively high behavioural flexibility and stress tolerance levels required of successful urban exploiters and urban adapters. Urban biodiversity has generally been found to have a positive influence on the well-being (mental health) of its human occupants (Dallimer *et al.*, 2012), and the non-native species influencing urban biodiversity provide important ecosystem services, such as pollination (e.g. Warbling or Japanese White-eye; Cox, 1983) and seed dispersal (Foster and Robinson, 2007). Successful eradication of non-native birds from cities may lead to undesirable consequences including an increase in pest insects, a decline in native plant biodiversity, and a decline in human mental health and well-being.

38.7.4 Impacts of non-native birds on native wildlife

Although the population trends and distributions of many non-native birds listed in the GISD are captured in the large-scale monitoring efforts of either the NABBS or the CBC, the impacts and trends of many of the non-native species are not well understood. Most 'impact' studies either focus on the biology of individual species, or make population and community-level inference based on limited empirical evidence. For example, most studies of non-native birds in the urban setting are limited to correlative studies of population-level (e.g. trends) and community-level metrics, including species richness, alpha diversity, and community composition (MacGregor-Fors *et al.*, 2010), while studies in more natural habitats often include

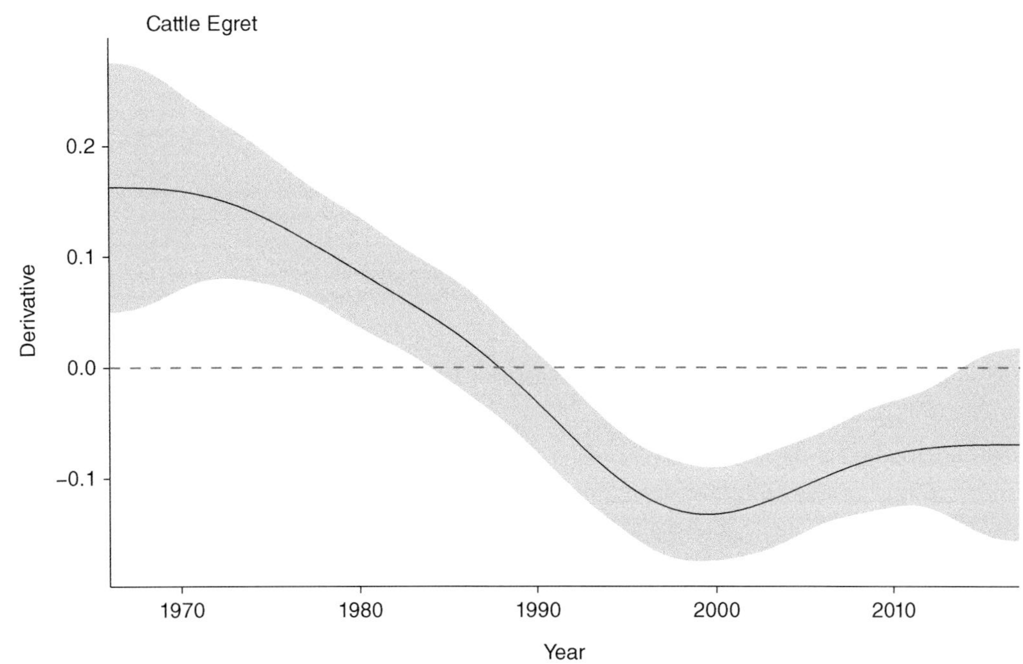

Fig. 38.8. A generalized linear model of North American Breeding Bird Survey (NABBS) data suggesting that the Cattle Egret (*Bubulcus ibis*) has been declining in recent years in North America.

crude pre- and post-invasion studies of community composition and population trends.

The impacts of non-native birds on native wildlife is contestable in the case of many species (e.g. Common Myna; Grarock *et al.*, 2012), but there is a clear lack of understanding regarding the impacts of many of the 31 GISD species on native wildlife. In fact, a recent review and global analysis of studies of non-native bird impacts on native birds yielded only 23 scientific articles containing empirical evidence for impacts of any naturalized bird in North America (Martin-Albarracin *et al.*, 2015). Given the paucity of studies of non-native species in North America (approximately 10% of the 80 or so naturalized species considered have been studied; Martin-Albarracin *et al.*, 2015), it is not surprising that our understanding of the effects of non-native species on native wildlife and ecosystems is limited.

38.8 Conclusions

Florida will remain the hotspot for introduced avian species in North America for the foreseeable future. The large geographical extent of North America is occupied by few introduced species, and these tend to be global species associated with anthropogenic habitats, or game species introduced into agricultural and grassland landscapes. Global change, including climate change, will continue to shift the mosaic of available habitat leading to additional spread, and the rate of introductions globally has been increasing. Therefore, it is likely that the species richness of introduced species will continue to climb. Some introduced species are viewed positively (e.g. pheasant), and most have negligible or local impacts. Few species are subject to intensive management, and these tend to be species with negative economic impacts (e.g. Monk Parakeet) or with perceived impacts on native fauna (e.g. those species that compete with native species for nest cavities. International trade is currently a driving force for North American introductions. Even species with known deleterious impacts such as the Monk Parakeet, and that are subject to expensive control methods, are still available in the pet trade; conflicting policies make successful control unlikely. However, as a particular taxon, avian introductions have been arguably less harmful than introductions of vertebrates such as fish and mammals, and much less harmful than introduced plants and insects. Vigilance in biosecurity and the development of stronger policies preventing imports of potential invaders are warranted.

38.9 References

Able, K.P. and Belthoff, J.R. (1998) Rapid 'evolution' of migratory behaviour in the introduced house finch of eastern North America. *Proceedings of the Royal Society of London B: Biological Sciences* 265, 2063–2071.

Allen, C.R. (2006) Predictors of introduction success in the south Florida avifauna. *Biological Invasions* 8, 491–500.

Allen, C.R., Nemec, K.T., Wardwell, D.A., Hoffman, J.D., Brust, M., *et al.* (2013) Cross-taxa invasions. *Global Ecology and Biogeography* 22, 889–899.

Altizer, S., Hochachka, W.M., and Dhondt, A.A. (2004) Predictors of regional establishment success and spread of introduced non-indigenous vertebrates. *Journal of Animal Ecology* 73, 309–322.

Barrows, W.B. (1889) The English sparrow (*Passer domesticus*) in North America: especially in its relations to agriculture. Division of Economic Ornithology; Mammalogy, US Department of Agriculture, Washington, DC.

BirdLife International (2018) The IUCN Red List of Threatened Species. Available at: www.iucnredlist.org/ (accessed 11 November 2019).

Blair R.B. and Johnson, E.M. (2008) Suburban habitats and their role for birds in the urban–rural habitat network: points of local invasion and extinction? *Landscape Ecology* 23, 1157–1169.

Boyer, A.G. and Jetz, W. (2014) Extinctions and the loss of ecological function in island bird communities. *Global Ecology and Biogeography* 23, 679–688.

Burnett, J.L. and Moulton, M.P. (2015) Recent trends in house sparrow (*Passer domesticus*) distribution and abundance in Gainesville, Alachua County, Florida. *Florida Field Naturalist* 43, 167–172.

Burnett, J.L., Roberts, C.P., Allen, C.R., Brown, M.B. and Moulton, M.P. (2017) Range expansion by *Passer montanus* in North America. *Biological Invasions* 19, 5–9.

Carrete, M. and Tella, J. (2008) Wild-bird trade and exotic invasions: a new link of conservation concern? *Frontiers in Ecology and the Environment* 6, 207–211.

Cooper, C.B., Hochachka, W.M. and Dhondt, A.A. (2007) Contrasting natural experiments confirm competition between house finches and house sparrows. *Ecology* 88, 864–870.

Cox, P.A. (1983) Extinction of the Hawaiian avifauna resulted in a change of pollinators for the ieie, *Freycinetia arborea*. *Oikos* 41, 195–199.

Crosby, G.T. (1972) Spread of the cattle egret in the Western Hemisphere. *Bird Banding* 43, 205–212.

Dallimer, M., Irvine, K.N., Skinner, A.M., Davies, Z.G., Rouquette, J.R., *et al.* (2012) Biodiversity and the feel-good factor: understanding associations between self-reported human well-being and species richness. *BioScience* 62, 47–55.

Dawson, W., Moser, D., van Kleunen, M., Kreft, H., Pergl, J., *et al.* (2017) Global hotspots and correlates of alien species richness across taxonomic groups. *Nature Ecology and Evolution* 1, 0186.

Devictor, V., Julliard, R., Couvet, D., Lee, A. and Jiguet, F. (2007) Functional homogenization effect of urbanization on bird communities. *Conservation Biology* 21, 741–751.

Dhondt, A.A., Tessaglia, D.L. and Slothower, R.L. (1998) Epidemic mycoplasmal conjunctivitis in house finches from eastern North America. *Journal of Wildlife Diseases* 34, 265–280.

Dumke, R., Dahlgren, R., Peterson, S., Schulz, J., Wiegand, J. and Wishart, W. (1980) A gray partridge management/research plan for North America. In: *Proceedings of PERDIX II: Gray Partridge Workshop*. Forest, Wildlife and Range Experiment Station, University of Idaho, Moscow, Idaho, pp. 165–197.

Dyer, E.E., Redding, D.W. and Blackburn, T.M. (2017) The global avian invasions atlas, a database of alien bird distributions worldwide. *Scientific Data* 4, 170041.

Faeth, S.H., Bang, C. and Saari, S. (2011) Urban biodiversity: patterns and mechanisms. *Annals of the New York Academy of Sciences* 1223, 69–81.

Foster, J.T. and Robinson, S.K. (2007) Introduced birds and the fate of Hawaiian rainforests. *Conservation Biology* 21, 1248–1257.

Freed, L.A. and Cann, R.L. (2009) Negative effects of an introduced bird species on growth and survival in a native bird community. *Current Biology* 19, 1736–1740.

Funasaki, G.Y., Lai, P.-Y., Nakahara, L.M., Beardsley, J.W. and Ota, A.K. (1988) A review of biological control introductions in Hawaii: 1890 to 1985. *Proceedings of the Hawaiian Entomological Society* 28, 105–160.

Grarock, K., Tidemann, C.R., Wood, J. and Lindenmayer, D.B. (2012) Is it benign or is it a pariah? Empirical evidence for the impact of the common myna (*Acridotheres tristis*) on Australian birds. *PLoS One* 7, e40622.

Hastie, T., Tibshirani, R. and Wainwright, M. (2015) *Statistical Learning with Sparsity: The Lasso and Generalizations.* Chapman & Hall/CRC Press, New York.

ISSG Invasive Species Specialist Group (2015) The Global Invasive Species Database. Available at: www.iucngisd.org/gisd/ (accessed 12 November 2019).

Lepczyk, C.A., Flather, C.H., Radeloff, V.C., Pidgeon, A.M., Hammer, R.B. and Liu, J. (2008) Human impacts on regional avian diversity and abundance. *Conservation Biology* 22, 405–416.

Linz, G.M., Homan, H.J., Gaulker, S.M., Penry, L.B. and Bleier, W.J. (2007) European starlings: a review of an invasive species with far-reaching impacts. In: Witmer, G.W., Pitt, W.C. and Fagerstone, K.A. (eds) *Managing Vertebrate Invasive Species: Proceedings of an International symposium.* USDA/APHIS Wildlife Services, National Wildlife Research Center, Fort Collins, Colorado, pp. 378–386.

MacGregor-Fors, I. and Schondube, J.E. (2011) Gray vs. green urbanization: relative importance of urban features for urban bird communities. *Basic and Applied Ecology* 12, 372–381.

MacGregor-Fors, I., Morales-Pérez, L., Quesada, J. and Schondube, J.E. (2010) Relationship between the presence of house sparrows (*Passer domesticus*) and neotropical bird community structure and diversity. *Biological Invasions* 12, 87.

Martin-Albarracin, V.L., Amico, G.C., Simberloff, D. and Nuñez, M.A. (2015) Impact of non-native birds on native ecosystems: a global analysis. *PLoS One* 10, e0143070.

Marzluff, J.M. (2001) Worldwide urbanization and its effects on birds. In: Marzluff, J., Bowman, R. and Donnelly, R. (eds) *Avian Ecology and Conservation in an Urbanizing World.* Springer Science+Business Media, New York, pp. 19–47.

Marzluff, J.M. (2016) A decadal review of urban ornithology and a prospectus for the future. *Ibis* 159, 1–13.

Moulton, M.P. and Pimm, S. (1987) Morphological assortment in introduced Hawaiian passerines. *Evolutionary Ecology* 1, 113–124.

Moulton, M.P., Cropper, W.P., Avery, M.L. and Moulton, L.E. (2010) The earliest house sparrow introductions to North America. *Biological Invasions* 12, 2955–2958.

National Audubon Society (2010) The Christmas bird count historical results. Available at: http://netapp.audubon.org/cbcobservation/historical/resultsby-count.aspx (accessed 11 November 2019).

Pagad, S., Genovesi, P., Carnevali, L., Schigel, D. and McGeoch, M.A. (2018) Introducing the global register of introduced and invasive species. *Scientific Data* 5, 170202.

Pardieck, K.L., Ziolkowski, D.J., Lutmerding, M. and Hudson, M.A.R. (2017) North American breeding bird survey dataset 1966–2017, version 2017.0. US Geological Survey, Patuxent Wildlife Research Center. Available at: https://doi.org/10.5066/F76972V8 (accessed 11 November 2019).

Phillips, J.C. (1928) Wild birds introduced or transplanted in North America. *United States Department of Agriculture Technical Bulletin* 61, 1–66.

Pimentel, D., Lach, L., Zuniga, R. and Morrison, D. (2000) Environmental and economic costs of nonindigenous species in the United States. *BioScience* 50, 53–66.

Potts, G. and Aebischer, N.J. (1995) Population dynamics of the grey partridge *Perdix perdix* 1793–1993: monitoring, modelling and management. *Ibis* 137, S29–S37.

Samuel, M.D., Hobbelen, P.H., DeCastro, F., Ahumada, J.A., LaPointe, D.A., *et al.* (2011) The dynamics, transmission, and population impacts of avian malaria in native Hawaiian birds: a modeling approach. *Ecological Applications* 21, 2960–2973.

Simberloff, D. (1995) Why do introduced species appear to devastate islands more than mainland areas? *Pacific Science* 49, 1, 87–97

Summers-Smith, J.D. (2003) The decline of the house sparrow: a review. *British Birds* 96, 439–446

van Riper, S.G. (2000) Japanese white-eye (*Zosterops japonicus*). In: Poole, A.F. and Gill, F.B. (eds) *The Birds of North America.* The Cornell Lab of Ornithology, Ithaca, New York. Available at: https://doi.org/10.2173/bna.487 (accessed 11 November 2019).

Veech, J.A., Small, M.F. and Baccus, J.T. (2011) The effect of habitat on the range expansion of a native and an introduced bird species. *Journal of Biogeography* 38, 69–77.

Appendix 38.1. List of non-native species introduced to or invasive in mainland North America[a] and whether they are present in Global Invasive Species Database (GISD), Global Avian Invasions Atlas (GAVIA) or Florida Ornithological Society (FOS) databases.

Order	Species	FOS	GISD	GAVIA
Anseriformes	*Aix galericulata*			✓
	Aix sponsa			✓
	Alopochen aegyptiaca			✓
	Alopochen aegyptiacus	✓		
	Anas bahamensis			✓
	Anas crecca			✓
	Anas formosa			✓
	Anas fulvigula			✓
	Anas hottentota			✓
	Anas luzonica			✓
	Anas platyrhynchos		✓	✓
	Anas poecilorhyncha			✓
	Anas rhynchotis			✓
	Anas rubripes			✓
	Anser anser		✓	✓
	Anser cygnoides			✓
	Anser fabalis			✓
	Anser indicus			✓
	Aythya americana			✓
	Branta canadensis		✓	✓
	Branta leucopsis			✓
	Branta ruficollis			✓
	Cairina moschata	✓		✓
	Callonetta leucophrys			✓
	Coscoroba coscoroba			✓
	Cygnus atratus			✓
	Cygnus buccinator			✓
	Cygnus cygnus			✓
	Cygnus melancoryphus			✓
	Cygnus olor		✓	✓
	Dendrocygna arborea			✓
	Dendrocygna autumnalis			✓
	Dendrocygna bicolor			✓
	Dendrocygna viduata			✓
	Neochen jubata			✓
	Netta peposaca			✓
	Netta rufina			✓
	Nettapus coromandelianus			✓
	Oxyura jamaicensis		✓	
	Tadorna ferruginea			✓
	Tadorna tadorna			✓
Charadriiformes	*Scolopax minor*			✓
	Vanellus chilensis			✓
Ciconiiformes	*Bubulcus ibis*		✓	✓
	Ciconia abdimii			✓
	Ciconia ciconia			✓
	Ciconia episcopus			✓
	Circus approximans		✓	
	Ephippiorhynchus asiaticus			✓
	Eudocimus ruber			✓
	Platalea leucorodia			✓
	Threskiornis aethiopicus			✓

Continued

Appendix 38.1. Continued.

Order	Species	FOS	GISD	GAVIA
Columbiformes	*Columba livia*	✓	✓	✓
	Columba palumbus			✓
	Columbina inca			✓
	Gallicolumba luzonica			✓
	Geopelia cuneata			✓
	Ocyphaps lophotes			✓
	Pterocles exustus			✓
	Pterocles orientalis			✓
	Stigmatopelia chinensis			✓
	Streptopelia decaocto	✓	✓	✓
	Streptopelia risoria			✓
	Streptopelia roseogrisea			✓
	Syrrhaptes paradoxus			✓
	Zenaida asiatica			✓
Coraciiformes	*Aceros undulatus*			✓
	Bucorvus abyssinicus			✓
	Bycanistes brevis			✓
	Megaceryle torquata			✓
	Tockus nasutus			✓
Cuculiformes	*Musophaga violacea*			✓
	Tauraco schalowi			✓
Falconiformes	*Buteogallus anthracinus*			✓
	Buteogallus urubitinga			✓
	Chondrohierax uncinatus			✓
	Falco peregrinus			✓
	Falco tinnunculus			✓
	Geranospiza caerulescens			✓
	Gymnogyps californianus			✓
	Parabuteo unicinctus			✓
	Sarcoramphus papa			✓
Galliformes	*Alectoris barbara*			✓
	Alectoris chukar		✓	✓
	Alectoris graeca			✓
	Alectoris rufa			✓
	Ammoperdix griseogularis			✓
	Bambusicola thoracicus			✓
	Bonasa bonasia			✓
	Bonasa umbellus			✓
	Callipepla californica			✓
	Callipepla douglasii			✓
	Callipepla gambelii			✓
	Callipepla squamata			✓
	Catreus wallichi			✓
	Chrysolophus amherstiae			✓
	Chrysolophus pictus			✓
	Colinus virginianus			✓
	Coturnix chinensis			✓
	Coturnix coturnix			✓
	Coturnix japonica			✓
	Dendragapus canadensis			✓
	Dendragapus obscurus			✓
	Francolinus erckelii			✓
	Francolinus francolinus			✓
	Francolinus pondicerianus			✓

Continued

Appendix 38.1. Continued.

Order	Species	FOS	GISD	GAVIA
	Gallus gallus		✓	✓
	Gallus varius		✓	
	Lagopus lagopus			✓
	Lagopus leucura			✓
	Lophura leucomelanos			✓
	Lophura nycthemera			✓
	Meleagris gallopavo			✓
	Meleagris ocellata			✓
	Numida meleagris			✓
	Oreortyx pictus			✓
	Ortalis vetula			✓
	Pavo cristatus			✓
	Perdix perdix			✓
	Phasianus colchicus			✓
	Phasianus versicolor			✓
	Syrmaticus reevesii			✓
	Syrmaticus soemmerringii			✓
	Tetrao tetrix			✓
	Tetrao urogallus			✓
	Tetraogallus himalayensis			✓
	Tympanuchus cupido			✓
	Tympanuchus phasianellus			✓
Gruiformes	*Aramides cajanea*			✓
	Balearica pavonina			✓
	Balearica regulorum			✓
	Crex crex			✓
	Grus americana			✓
	Grus antigone			✓
	Grus virgo			✓
	Porphyrio porphyrio	✓	✓	✓
Passeriformes	*Acridotheres cristatellus*			✓
	Acridotheres fuscus		✓	✓
	Acridotheres tristis	✓	✓	✓
	Alauda arvensis			✓
	Amandava amandava			✓
	Amandava subflava			✓
	Cardinalis cardinalis			✓
	Carduelis cannabina			✓
	Carduelis carduelis			✓
	Carduelis chloris			✓
	Carduelis psaltria			✓
	Carduelis spinus			✓
	Carpodacus mexicanus		✓	✓
	Cinclus cinclus			✓
	Corvus corax			✓
	Corvus splendens		✓	✓
	Cyanocorax caeruleus			✓
	Cyanocorax dickeyi			✓
	Cyanocorax yncas			✓
	Emberiza bruniceps			✓
	Emberiza citrinella			✓
	Erithacus rubecula			✓
	Estrilda astrild		✓	✓
	Estrilda melpoda			✓

Continued

Appendix 38.1. Continued.

Order	Species	FOS	GISD	GAVIA
	Euplectes afer			✓
	Euplectes ardens			✓
	Euplectes aureus			✓
	Euplectes franciscanus			✓
	Euplectes orix			✓
	Fringilla coelebs			✓
	Garrulax caerulatus			✓
	Garrulax canorus			✓
	Garrulax pectoralis			✓
	Gracula religiosa			✓
	Gymnorhina tibicen		✓	✍
	Haemorhous mexicanus	✓		✍
	Hypsipetes madagascariensis			✓
	Icterus icterus			✓
	Icterus pectoralis	✓		✓
	Irena puella			✓
	Lagonosticta rubricata			✓
	Leiothrix lutea		✓	✓
	Lonchura atricapilla			✓
	Lonchura cantans			✓
	Lonchura maja			✓
	Lonchura malabarica			✓
	Lonchura malacca			✓
	Lonchura nana			✓
	Lonchura punctulata			✓
	Loxia pytyopsittacus			✓
	Loxigilla violacea			✓
	Lullula arborea			✓
	Luscinia megarhynchos			✓
	Melopyrrha nigra			✓
	Mimus gilvus			✓
	Mimus polyglottos			✓
	Molothrus ater		✓	✓
	Molothrus bonariensis		✓	✓
	Padda oryzivora			✓
	Paroaria coronata			✓
	Paroaria gularis			✓
	Parus caeruleus			✓
	Parus major			✓
	Passer domesticus	✓	✓	✓
	Passer luteus			✓
	Passer montanus			✓
	Pica pica			✓
	Pitangus sulphuratus		✓	✓
	Pitta guajana			✓
	Ploceus cucullatus			✓
	Ploceus velatus			✓
	Prunella modularis			✓
	Psarocolius montezuma			✓
	Pycnonotus cafer		✓	✓
	Pycnonotus jocosus	✓	✓	✓
	Pyrrhula pyrrhula			✓
	Serinus canaria			✓
	Serinus mozambicus			✓

Continued

Appendix 38.1. Continued.

Order	Species	FOS	GISD	GAVIA
	Sicalis flaveola			✓
	Sporophila torqueola			✓
	Sturnus vulgaris	✓	✓	✓
	Sylvia atricapilla			✓
	Thraupis episcopus			✓
	Tiaris canorus			✓
	Turdus grayi			✓
	Turdus iliacus			✓
	Turdus merula			✓
	Turdus philomelos			✓
	Turdus plumbeus			✓
	Turdus rufopalliatus			✓
	Uraeginthus bengalus			✓
	Vidua macroura			✓
	Zosterops japonicus		✓	✓
	Zosterops palpebrosus			✓
Phoenicopteriformes	*Phoenicopterus chilensis*			✓
	Phoenicopterus roseus			✓
	Phoenicopterus ruber			✓
Piciformes	*Ramphastos sulfuratus*			✓
	Ramphastos toco			✓
	Ramphastos vitellinus			✓
Psittaciformes	*Agapornis fischeri*			✓
	Agapornis personatus			✓
	Agapornis roseicollis			✓
	Amazona aestiva			✓
	Amazona albifrons			✓
	Amazona amazonica			✓
	Amazona auropalliata			✓
	Amazona autumnalis			✓
	Amazona barbadensis			✓
	Amazona farinosa			✓
	Amazona festiva			✓
	Amazona finschi			✓
	Amazona ochrocephala			✓
	Amazona oratrix			✓
	Amazona pretrei			✓
	Amazona ventralis			✓
	Amazona viridigenalis			✓
	Amazona xantholora			✓
	Anodorhynchus hyacinthinus			✓
	Ara ararauna			✓
	Ara chloropterus			✓
	Ara macao			✓
	Ara militaris			✓
	Ara severus			✓
	Aratinga acuticaudata			✓
	Aratinga aurea			✓
	Aratinga canicularis			✓
	Aratinga chloroptera			✓
	Aratinga erythrogenys			✓
	Aratinga finschi			✓
	Aratinga holochlora			✓
	Aratinga leucophthalma			✓

Continued

Appendix 38.1. Continued.

Order	Species	FOS	GISD	GAVIA
	Aratinga mitrata			✓
	Aratinga nenday	✓		✍
	Aratinga pertinax			✓
	Aratinga wagleri			✓
	Aratinga weddellii			✓
	Brotogeris chiriri			✓
	Brotogeris jugularis			✓
	Brotogeris pyrrhoptera			✓
	Brotogeris sanctithomae			✓
	Brotogeris versicolurus	✓		✓
	Cacatua alba			✓
	Cacatua galerita			✓
	Cacatua goffiniana			✓
	Cacatua moluccensis			✓
	Cacatua roseicapilla			✓
	Cyanoliseus patagonus			✓
	Diopsittaca nobilis			✓
	Eclectus roratus			✓
	Eos bornea			✓
	Lorius garrulus			✓
	Melopsittacus undulatus	✓		✓
	Myiopsitta monachus	✓	✓	✓
	Nandayus nenday			✓
	Nymphicus hollandicus			✓
	Pionites melanocephalus			✓
	Pionus senilis			✓
	Platycercus elegans			✓
	Platycercus eximius			✓
	Poicephalus crassus			✓
	Poicephalus rueppellii			✓
	Poicephalus senegalus			✓
	Primolius auricollis			✓
	Psephotus haematonotus			✓
	Pseudeos fuscata			✓
	Psittacula alexandri			✓
	Psittacula columboides			✓
	Psittacula cyanocephala			✓
	Psittacula eupatria			✓
	Psittacula krameri		✓	✓
	Psittacula roseata			✓
	Psittacus erithacus			✓
	Pyrrhura frontalis			✓
	Pyrrhura molinae			✓
	Rhynchopsitta pachyrhyncha			✓
	Rhynchopsitta terrisi			✓
	Trichoglossus chlorolepidotus			✓
	Trichoglossus haematodus			✓
	Trichoglossus ornatus			✓
Strigiformes	*Bubo scandiaca*			✓
	Bubo virginianus		✓	
	Pulsatrix perspicillata			✓
	Strix varia			✓
Tinamiformes	*Nothura maculosa*			✓
	Rhynchotus rufescens			✓

[a]Mainland North America comprises Belize, Canada, Costa Rica, El Salvador, Guatemala, Honduras, Mexico, Nicaragua, Panama and the USA.

39 Continental Analysis of Invasive Birds: South America

Karina L. Speziale*

Grupo de Investigaciones en Biología de la Conservación, Laboratorio Ecotono, INIBOMA (CONICET-UNCOMA), Bariloche, Rio Negro 8400, Argentina

Citation: Speziale, K.L. (2020) Continental analysis of invasive birds: South America. In: Downs, C.T. and Hart, L.A. (eds) *Invasive Birds: Global Trends and Impacts*. CAB International, Wallingford, UK, pp. 295–302.

39.1 Introduction

As one of the most biodiverse realms, South America is considered a conservation priority to protect this invaluable biodiversity hotspot (Myers *et al.*, 2000). Its wide latitudinal gradient creates a high diversity of biomes, some unique to this subcontinent, including several areas containing the highest species diversity in the world (Myers *et al.*, 2000; Kier *et al.*, 2005). South America has 3540 bird species, with Brazil and Peru hosting 232 and 107 endemic bird species, respectively (BirdLife International, 2004). Despite this importance, biodiversity is threatened by global change drivers. Among these invasive alien species are considered an important driver of threats to bird species in South America (Díaz *et al.* 2019). Furthermore, some native bird species are being introduced to new areas within the continent and often become invasive themselves, even impacting their congeners (BirdLife International, 2018).

The negative impacts of invasive species in South America outnumber the positive ones (Jaksic, 1998; Jaksic *et al.*, 2002; Novillo and Ojeda, 2008; Lambertucci *et al.*, 2009). These impacts include the emergence of infectious diseases, extinction of endemic species and large-scale habitat changes (Jaksic, 1998; Rodríguez, 2001; Novillo and Ojeda, 2008). Despite the number of introduced bird species recorded for South America is relatively low, both the number and the impacts of these introduced species are probably under-recorded, given the paucity of research effort in biological invasions in South American countries (Quiroz *et al.*, 2009; Speziale *et al.*, 2012).

39.2 History of the Bird Invasion Process

The threat that South America is facing by invasive species has only recently been considered a problem, at least for biodiversity (Jaksic, 1998; Rodríguez, 2001; Pauchard *et al.*, 2004). Research interest regarding biological invasions has also been delayed compared with first-world countries and is less intense (Nuñez and Pauchard, 2010; Speziale *et al.*, 2012). Information on invasive species in South America varies greatly among countries but is mainly important in just three of them, Argentina, Brazil and Chile, probably as a correlation with higher levels scientific research in all areas (Pauchard *et al.*, 2011; Speziale *et al.*, 2012). This relatively low level of interest in the study of introduced species may be based on the culture and idiosyncrasies of South American people and colonizers (Speziale *et al.*, 2012). Species introductions were conducted by acclimatization societies in colonies such as New Zealand, Australia and the USA, the same societies that changed their roles about species introductions, some of them in less than 20 years, to prevent new ones after realizing their negative impacts (Dunlap, 1997; Ritvo, 2012). In contrast, South America did not experience an early shift in its people's attitudes towards introduced species (Speziale *et al.*, 2012). The lack of acclimatization societies in South America also influenced the lack of systematic registers on the introduction history of the various invasive species, or their fate.

Despite this general lack of historical introduction records for most of the introduced bird species in South America, a few can be traced back. The first avian introductions probably occurred with Polynesian colonizers before the Europeans', arrival between AD 1321 and 1407, such as the domesticated descendant of the Red Jungle Fowl (*Gallus gallus*; Storey *et al.*, 2007). The Rock Pigeon (*Columba livia*) was then registered as an introduced species, arriving in the 16th century with the Spanish conquerors. No other records are found till the end of the 19th century and the beginning of the 20th. The Shiny

*Corresponding author: karina.speziale@crub.uncoma.edu.ar

Cowbird (*Molothrus bonariensis*) and California Quail (*Callipepla californica*) were recorded in Chile before 1870 (Marin, 2000), and the Common Waxbill (*Estrilda astrild*) in Brazil in 1870 (Sick, 1965). The House or English Sparrow (*Passer domesticus*) was first introduced in Argentina in 1872, in Brazil in 1903 and in Chile in 1904 (Sick, 1959; Summers-Smith, 1963, 1988). The Cattle Egret (*Bubulcus ibis*) was first recorded in Suriname and Guyana in 1877 as a natural arrival from Africa (Crosby, 1972). In addition, at the beginning of the 20th century, the European Goldfinch (*Carduelis carduelis*) was found in Argentina and Uruguay (1913 for the latter) and the European Greenfinch (*Carduelis chloris*) was first recorded in Uruguay in 1908. In the same period, the Silver Pheasant (*Lophura nycthemera*) was first recorded in Argentina (1902–1911).

By the end of the 20th century, two more introduced avian species were recorded in Argentina; the Crested Myna (*Acridotheres cristatellus*) and the Common or European Starling (*Sturnus vulgaris*) in October 1988. Most species introductions, however, were not recorded, particularly when they were unintentional escapees, intentional release of captive individuals or changes in the distribution of native species. Yet introductions did not stop, and even in 2015 new species introductions were still being registered (Arbeláez-Cortés *et al.*, 2016). This was the case for many pet species that were released and were first recorded during the 21st century (e.g. Ibañez *et al.*, 2014).

39.3 Current Knowledge on Introduced Bird Species

39.3.1 Reasons for introductions and vectors

The first avian introductions in South America from the end of the 19th to the beginning of the 20th century were aimed mostly at introducing species useful for hunting, or as food resources. This was typical during the human colonization and immigration era when migrants wanted to reproduce their missed homeland and to feed on food items they knew how to hunt and cook (Dunlap, 1997; Simberloff and Rejmánek, 2010; Ritvo, 2012). Currently, introductions occur mainly because of escapees or intentional release of pets. This is the case for most native species introductions (Fig. 39.1), in particular parrots. Furthermore, environmental organizations for the welfare of animals often promote the release of captive individuals. This is the primary route of introductions in Brazil where species recorded in new areas beyond their native range correspond to escaped or released individuals from captivity (Fontoura *et al.*, 2013). These two clearly distinct periods that led to the introduction of different types of birds in South America have also been observed in other parts of the world (Cassey *et al.*, 2015).

Once introduced, bird species that spread from the original introduction points were generally aided by human disturbance of the environment that promoted conditions that benefitted them. Of the species introduced in Brazil that currently have stable populations, two, the European Goldfinch and European Greenfinch, were introduced to the neighbouring countries of Argentina and Uruguay, and subsequently expanded their distributions into Brazil, where the European Goldfinch was first recorded in 1994 (Fontoura *et al.*, 2013).

Two contrasting examples are the Pied Crow (*Corvus albus*) and the Cattle Egret. The Pied Crow was probably a ship-assisted vagrant that reached Brazil (Fontoura *et al.*, 2013). The Cattle Egret in contrast, is a regular migrator from Africa (Crosby, 1972). Since its first register in 1877 it was not recorded again until 1965 in Brazil and 1969 in Argentina (Sick, 1965; Olrog, 1972). It may have spread to Brazil because of the anthropogenic modifications of the environment. In the case of Argentina, given its discovery in the south of the country, far away from the arrival point first it is considered a new arrival from Africa (Olrog, 1972). Native introduced

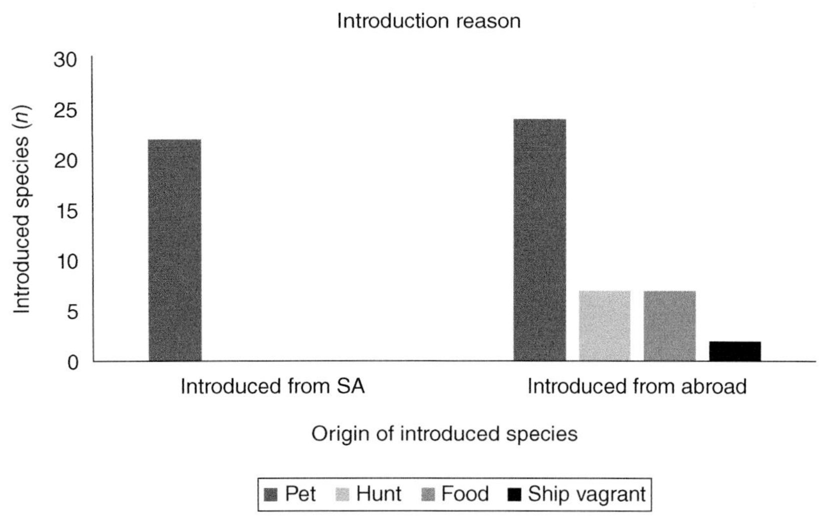

Fig. 39.1. Number of bird introduced species in South America by reason of introduction and species origin. 'Introduced from SA' refers to native species from South America that were introduced to new areas in the continent by human means. 'Introduced from abroad' refers to species that arrived from other continents.

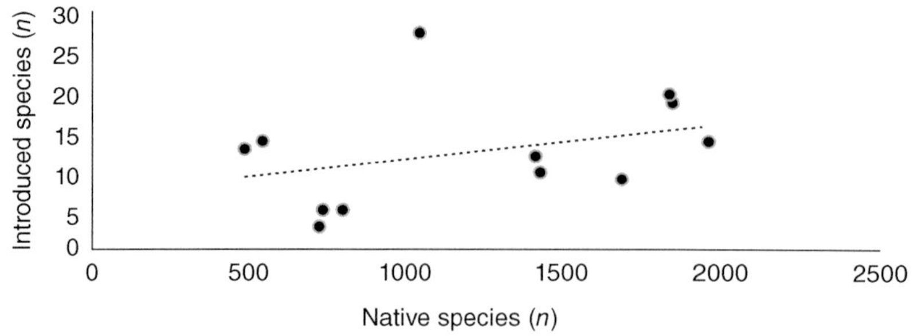

Fig. 39.2. Correlation between the total number of native bird species per country in South America and the number of bird introduced species. Correlation = 0.38 (t = 1.28, d.f. = 10, p = 0.23).

species can also spread aided by human habitat disturbance (see section 39.3.4).

39.3.2 Species numbers

Of the 213 introduced bird species known worldwide, 16% (34 species) were present in South America by 2015 (Martin-Albarracin *et al.*, 2015). By 2018, that number had increased by 210% if native species that were introduced outside their natural ranges intentionally or unintentionally by humans were considered. A total of 71 species belonging to 22 families from nine orders are registered as introduced by human means and naturalized in South America according to Avibase, the Invasive Species Specialist Group (ISSG) and BirdLife International (BirdLife International, 2004) and published papers searched for in Scopus and Scholar Google. Passeriformes, Galliformes and Psittaciformes are the most represented orders with 11, three and two families each. The rest of the orders are represented by just one family. Regarding families, Psittacidae, Phasianidae, Columbidae and Anatidae are the most speciose ones with 23, eight, six and six species, respectively. These families are among those that have also had the highest impact on biodiversity, both in Europe (Kumschick and Nentwig, 2010) and at a global scale (Martin-Albarracin *et al.*, 2015).

Several hypotheses have been proposed to understand biological invasions. Among these, 'the rich get richer' one proposes that environments with high resource availability will receive and sustain a higher number of introduced species (Stohlgren *et al.*, 2003). In contrast, the biotic resistance hypothesis posits that the higher the number of native species, the fewer introduced species will succeed in establishing, given that they will encounter competitors, predators, parasites and pathogens (Simberloff, 2013). However, in South America, there is little correlation between the total number of native bird species and the number of introduced species (Fig. 39.2), neither positively supporting 'the rich get richer' hypothesis nor negatively supporting the biotic resistance one. This may reflect research effort where countries that are not the most diverse ones have had greater scientific research intensity in terms of biological invasions, such as Argentina and Chile (Pauchard *et al.*, 2011; Speziale *et al.*, 2012). However, it may also indicate that introduced species have not yet occupied all the possible niches and that the invasion process may continue with available resources to sustain more introduced species (Fig. 39.3).

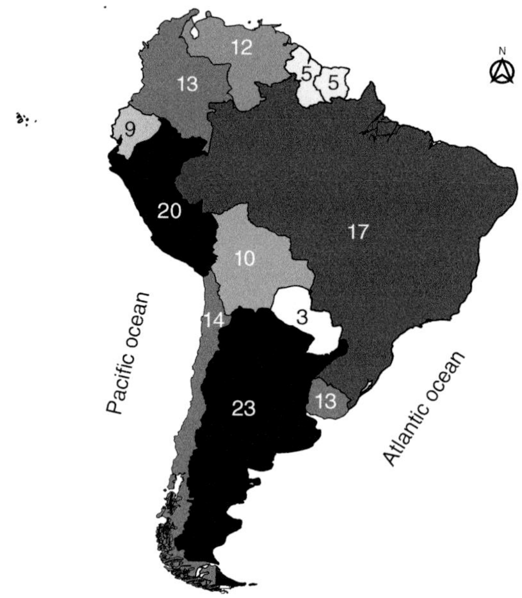

Fig. 39.3. Number of introduced avian species per country in South America. Increasing colour intensity reflects an increased number of introduced species.

Other hypotheses have tried to find the perfect invader by identifying particular species traits that favour invasion (Catford *et al.*, 2009; Evans *et al.*, 2014). Among these characteristics, bird species with relatively short generation lengths (Simberloff, 2013) and species that have a wide native geographical range (e.g. a broad niche and adaptability capacity or plasticity) (Pyšek *et al.*, 2009; Evans *et al.*, 2014) are generally prone to succeed. Other variables not related to species traits such as residence time (Pyšek and Jarošík, 2005) could also play a role in the invasion process and are mostly related to the end of the lag time (Simberloff, 2013; Lockwood *et al.*, 2016).

According to available data, the number of South American countries occupied by each introduced bird species (a surrogate of invasion potential) is better explained positively by residence

Table 39.1. General linear model (Poisson distribution) analyses of the effect of time since a bird species was first introduced to South America (years since primary introduction), and generation length, altitudinal native range, native range and body mass of each introduced bird species on the number of invaded South American countries. The analysis was conducted in R (R Core Team, 2018).

	Estimate	SE	*z* value	Pr(>|z|)
Intercept	1.53	1.29	4.38	<0.001***
Years since primary introduction	3.78	0.15	4.45	<0.001***
Generation length (years)	−1.51	0.66	−3.06	0.002**
Altitudinal range (m)	−7.89	0.02	−0.68	0.50
Native range (km²)	5.93	1.57	4.65	<0.001***
Body mass (g)	−1.400	0.03	−0.90	0.37

**$p < 0.05$; ***$p < 0.01$.

time and distribution area (in km^2) in the native range, while it is negatively related to generation length (Table 39.1). Over time, species may have had the chance to undergo mutational changes that may have facilitated population explosions, experienced abiotic or biotic conditions favouring them, or merely had enough time to increase their population numbers (Simberloff, 2013). In contrast, wider native distributions may represent a higher tolerance of a wide range of climates acquired in the region of origin that could favour their adaptation to the new environment (Pyšek *et al.*, 2009). Among potential causal drivers of introduced bird impacts in Europe and Australia habitat breadth and body mass affect introduced bird species impacts (Kumschick *et al.*, 2013; Evans *et al.*, 2014). Currently, in South America, available data do not allow such an analysis of the variables that determine the impacts of introduced bird species. However, data suggest that species traits including widely distributed species and short generation length, but also residence time, are drivers of invasion. This agrees in part with analyses for Europe and Australia, showing that invasive species with wider geographical native ranges produce higher impacts (Kumschick *et al.*, 2013; Evans *et al.*, 2014). However, the parameters are low, so other variables could also be playing a role (Table 39.1).

39.3.3 Origin and current distritubtion of introduced bird species

Argentina, Brazil, Peru, Colombia, Chile and Uruguay are the countries that host the most known bird species introductions in South America (Table 39.2). Most species that were introduced from outside South America are of Palearctic origin, followed by species from the Afrotropical and Nearctic realms (Fig. 39.4). The pattern of higher numbers of Palearctic species has also been shown for Australia, probably reflecting Australia and South America's former status as European colonies (Cassey *et al.*, 2015; Speziale *et al.*, 2012). However, more than half of the introduced species are native to South America, particularly the recently introduced ones. Of these, six have arrived or have been introduced from nearby countries, whereas 29 are registered as introduced in non-native areas within the same country where they are natives. The 37 new introduced species since 2015 included only eight arriving from outside South America, from Afrotropical (three species), Palearctic (two species), Australasia (two species) and Nearctic (one species) realms. This may also reflect the change in the monopoly of commerce and interchange of goods among colonizers and colonized countries towards an increased globalization economical exchange.

The Rock Pigeon, House Sparrow and Cattle Egret are already present in all South American countries, whereas the Shiny Cowbird has not yet been recorded in Paraguay, although this may simply reflect the relatively low research effort on biological invasions in Paraguay (Speziale *et al.*, 2012). These four widely spread species are those that were introduced or naturally arrived earlier, which could support residence time as an important factor of the invasion process (Pyšek and Jarošík, 2005; Simberloff, 2013). However, the case of the California Quail differs. The California Quail was introduced as early as these four species but has spread only to neighbouring countries, indicating that, together with the time since first introduction and the impact of introduced species, species traits or local conditions may also be important in affecting distribution (Catford *et al.*, 2009; Pyšek *et al.*, 2009). The rest of the introduced species are present in a maximum of six South American countries (e.g. Common Starling and Monk Parakeet (*Myiopsitta monachus*), with most of them present in only one (n = 35) or a few countries. Given the lack of information for most species, no clear conclusions can be drawn.

39.3.4 Native species introduced beyond their natural ranges

Among South American introduced native species, most records relate to native species that were kept as pets and then escaped or were intentionally released, or to native species that increased their distribution because of anthropogenic habitat alteration. In the case of former pets, most species are parrots, which generally have an adaptation capacity that facilitates their colonization of urban and rural areas outside their native ranges once they are released (Guerrero-Peláez, 2016). The Psittacidae family represents one of the more successful bird groups as non-native species (Lockwood, 1999). However, their impact when introduced in new areas within the same native country remains unstudied. In South America, the number of native parrot species that are present beyond their native ranges is 13, and it is the family most represented in this category. All are recorded as having been either intentionally released or having escaped confinement, and are already naturalized in

Table 39.2. Number of introduced avian species by country in South America.

Country	Number of species
Argentina	36
Bolivia	10
Brazil	20
Chile	13
Colombia	14
Ecuador	9
Guyana	5
Paraguay	3
Peru	20
Suriname	5
Uruguay	13
Venezuela	12

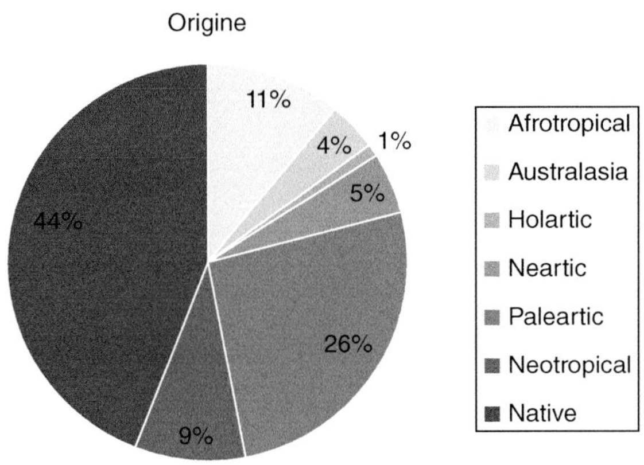

Fig. 39.4. Percentages of introduced avian species in South America according to biogeographical origin.

urban areas. However, it is not known whether they are confined to urban places or whether they are spreading in more natural environments. In Colombia, 73% of captive native parrots are released outside their native distribution areas (Restrepo-Rodas and Pulgarín-Restrepo, 2017). Moreover, native species can be introduced and increase their populations outside their native ranges while being threatened in their original ranges. The threats parrots face include overexploitation and illegal traffic for pet commerce (Restrepo-Rodas and Pulgarín-Restrepo, 2017) and destruction of their natural environment.

For example, the Turquoise-fronted Amazon (*Amazona aestiva*), a parrot native to Bolivia, Brazil, Paraguay and north of Argentina, has been found in some South American cities outside its native range. In Argentina, introduced populations have been recorded in Buenos Aires since 1990 and in Córdoba Province since 2002 (Mori *et al.*, 2017). In Brazil, individuals have been observed in São Paulo and Porto Alegre (Fontoura *et al.*, 2013; Mori *et al.*, 2017). Turquoise-fronted Amazon flocks are generally composed of a small number of individuals, although they have been established for a relatively long period (Mori *et al.*, 2017). It is the psittacine species most recorded as introduced outside the native range (Mori *et al.*, 2017). However, this species, and other native parrot species escaping captivity, are currently recorded only in urban or human-modified areas, and no record or study has yet shown impacts in natural environments.

Another mechanism by which native species are aided by humans to reach new distribution areas is through human land-use change (Medan *et al.*, 2011). In Argentina's Pampas, native avifauna changed with the history of agricultural expansion. Despite reported extinctions, the pampean avifauna richness has increased since the onset of agriculture and now includes 20 more species (Medan *et al.*, 2011). This process was favoured by the planting of trees as windbreaks and shade refuge for cattle or houses, the extension of wire fence networks and increased food availability as a result of afforestation and the production of crops (Narosky and Giacomo, 1993; Ghersa and León, 2001). Species such as the Rufous Hornero (*Furnarius rufus*), Rufous-bellied Thrush (*Turdus rufiventris*) and Monk Parakeet, whose original habitat was restricted to the xeric woodlands, have expanded to the whole region due to of the change in habitat structure caused by anthropogenic land-use changes (Ghersa and León, 2001). Other species including the Picazuro Pigeon (*Columba picazuro*) and White-throated hummingbird (*Leucochloris albicollis*) were positively affected because of increased food availability (Montaldo, 1984; Narosky and Giacomo, 1993). However, the impacts of these native species have not been fully studied beyond their impacts on crops.

39.3.5 Impacts

By 2012, 41 species out of the 100 most invasive species of the world were already established in South America (IUCN-ISSG, 2000; Speziale *et al.*, 2012) but only one species belonged to the class Aves. Given that South America is among the regions where the impacts of non-native species are less studied (Martin-Albarracin *et al.*, 2015), these numbers are probably underestimated. Only ten bird species (14%) out of those registered as naturalized till 2018 are recorded as invasive, and three have been declared pests (Bruggers *et al.*, 1998), although these categorizations are not supported by research, nor do they indicate which invasive species definition has been used. The impacts of only nine species (11%) have been studied in South America: Cattle Egret, Rock Pigeon, Shiny Cowbird, Monk Parakeet, English Sparrow, Great Kiskadee (*Pitangus sulphuratus*), Common Starling, Picazuro Pigeon and Saffron Finch (*Sicalis flaveola*). Among these, most species were studied because of their economic impacts on crops (89%, all but the Saffron Finch). Five species (55%) produce ecological impacts in terms of competition, parasitism or habitat alteration, whereas two species (22%), the Rock Pigeon and the Common Starling, are known to also produce health problems for humans. This contrasts with data at a global scale, where only 17% of naturalized bird species produce ecological impacts (Martin-Albarracin *et al.*, 2015). However, these results are based on only the few species that were studied on the continent. Despite so, most of the introduced bird species in South America are widespread distributed and their impacts are known at least for Europe and Australia (Martin Albarracin *et al.*, 2015).

In South America, the recorded impacts of the Cattle Egret are contrasting. In Peru (Capurro, 2017) and Brazil, it poses a threat to aviation in Fernando de Noronha Island, but information on ecological impacts remains anecdotal (Nunes *et al.*, 2010). In Argentina and Colombia, it is considered a biological control important for agriculture given that they prey on Orthoptera, Lepidoptera (Torres and Gutiérrez, 1999) and slugs (Constantino, 2013). However, in Colombia it has been also recorded as forming large colonies that can destroy trees and displace native species.

The Common Starling is one of the 100 most invasive species worldwide. In Argentina, beyond the place of their first observation, in Buenos Aires they are found in two more provinces to the north (Jensen, 2008). It has extended its distribution at a rate of 7.5 km/year since the first records in 1987 (Peris *et al.*, 2005). This rate is much lower than in the USA where it was introduced as early as 1890 and was found to be spreading at a rate of 91.6 km/year 100 years after its introduction (van den Bosch *et al.*, 1992). Thus, the spreading rate found for Argentina just 18 years after the first detection can be expected to increase with time.

The Rock Pigeon is perhaps the most studied introduced bird species in South America. Most research has been conducted in relation to their parasites and potential zoonosis by contact with humans. In Chile, pigeons were found to suffer from salmonellosis, aspergillosis, staphylococcosis and clamidiosis (González-Acuña *et al.*, 2007). In Argentina, pigeons aid the mobility of the vector of Chagas disease, *Trypanosoma cruzi* (Carrizo Páez *et al.*, 2008). In Colombia, pigeons are considered a pest owing to the transmission of zoonotic diseases, food contamination with excrement and the noise level produced (de la Ossa *et al.*, 2017). In urban areas of Envigado, Colombia, they have high percentages of oocysts compatible with *Eimeria* spp. and enterobacteria such as *Escherichia coli* (Pérez-García *et al.*, 2015). Beyond health issues, rock pigeons also affect infrastructure, and despite people generally do not perceive them as problematic, they agree with control measures (Bernal *et al.*, 2012).

39.4 Introduction Trends

At a global scale, the number of introduced species recorded shows that the introduction rate is increasing worldwide, and new invasions may emerge (Essl *et al.*, 2011; Lockwood *et al.*, 2016), although the purpose of introduction may have changed (Cassey *et al.*, 2015). These trends are also seen in South America given the high percentage of change seen since 2015 (see section 39.3.2) and the release of native pets. Despite current regulations that limit the importation of non-native species, and given that most species exhibit detectable lag times (Aagaard and Lockwood 2014), it is possible that non-native bird species already present in South America will be detected or declared invasive in coming years. Moreover, non-native species populations are not being monitored, so population numbers and trends are not known for most species and thus the invasion processes cannot be traced or prevented. Observations show that House Sparrows, Rock Pigeons and Common Starlings are spreading from urban areas to agroecosystems in South America (Peris *et al.*, 2005; Garaffa *et al.*, 2009). The case of the Common Starling is particularly important in terms of monitoring, given that it is expected to increase its impacts to of other ecosystems and countries as a result of its invasion capability. The negative impacts recorded in other introduced areas include competition with native cavity-nesting birds and damage to agriculture estimated at US$800 million annually (Pimentel *et al.*, 2001).

39.5 Knowledge Gaps

Despite the increasing scientific research conducted in South America, relatively little research has been carried out on introduced species (Pauchard *et al.*, 2011; Speziale *et al.*, 2012) especially regarding bird species. Until 2015, only three naturalized bird species had been studied in South America (Martin-Albarracin *et al.*, 2015). Additionally, global databases are not updated, and most information on naturalized birds for South America is found in local journals or is reported in Spanish. This language bias is positive, as the information is available for local managers. However, given that most articles just report new records for particular places, most often associated with research centres or universities (e.g. Guerrero-Peláez, 2016), little information is produced to aid management. Particularly important is the study and monitoring of those introduced bird species that have already shown negative impacts in other countries.

The fact that few studies were carried out on introduced bird species in South America provides an advantage. There is still time to design standardized methodologies to produce comparable results across the continent. This is particularly important given the patterns of spread of introduced species from nearby countries and that of native species naturalization outside their natural ranges, both aided by the anthropogenic disturbance of native environments. Collaboration among scientists from each South American country combined with standard methodologies is a priority to deal with introduced species on this continent (Pauchard *et al.*, 2011).

39.6 Conclusions

With the increase in studies focusing on introduced species, new introductions and their impacts are being recorded for South America (Pauchard *et al.*, 2011). However, this information is relatively sparce, at least for introduced bird species and their actual impacts across South America or at local scale. To fill this gap is a need for increased funds that can be devoted to the generation of strategies to prevent and deal with the impacts of invasive bird species. However, most South American countries' budget is relatively little for scientific research and conservation activites. Given the importance of birds in ecosystem functioning as seed dispersers, regulators of other animal populations and indicators of human impacts on the environment (Wheland *et al.*, 2015), it is essential to fully understand the impacts of introduced bird species in South America to protect this invaluable biodiversity realm.

39.7 References

Aagaard, K. and Lockwood, J. (2014) Exotic birds show lags in population growth. *Diversity and Distributions* 20, 547–554.

Aikio, S., Duncan, R.P. and Hulme, P.E. (2010) Lag-phases in alien plant invasions: separating the facts from the artefacts. *Oikos* 119, 370–378.

Arbeláez-Cortés, E., Marín-Gómez, O.H., Duque-Montoya, D., Cardona-Camacho, P.J., Renjifo, L.M. and Gómez, H.F. (2016) Birds, Quindío Department, Central Andes of Colombia. *Check List* 7, 227–247.

Bernal, L., Rivas, M., Rodríguez, C., Vásquez, C. and Vélez, M.P. (2012) Nivel de impacto de la sobrepoblación de palomas (*Columba livia domestica*) en los habitantes del perímetro del parque Principal del Municipio de Envigado en el año 2011. Available at: http://marthanellymesag.weebly.com/uploads/6/5/6/5/6565796/palomas.pdf (accessed 11 November 2019).

BirdLife International (2004) Most threatened birds are deteriorating in status. Available at: http://datazone.birdlife.org/sowb/casestudy/most-threatened-birds-are-in-decline (accessed 11 November 2019).

BirdLife International (2018) *State of the World's Birds: Taking the Pulse of the Planet*. BirdLife International, Cambridge.

Bruggers, R.L., Rodriguez, E. and Zaccagnini, M.E. (1998) Planning for bird pest problem resolution: a case study. *International Biodeterioration and Biodegradation*, 42, 173–184.

Capurro, V.P. (2017) Peligro aviar en aeropuertos del Perú. Estrategias Cognitivas En Adolescentes Limeños: Factores De Riesgo Clínico Y Social. Revista Exégesis, Universidad Inca Garcilazo de la Vega, Lima, pp. 25–33.

Carrizo Páez, R., Pickenhayn, J. and Carrizo Páez, M. (2008) Chagas urbano en San Juan: diagnóstico, revisión y propuesta para un sistema integrado de ataque. *Revista Argentina de Cardiología* 76, 480–487.

Cassey, P., Vall-Llosera, M., Dyer, E. and Blackburn, T.M. (2015) The biogeography of avian invasions: history, accident and market trade. In: Canning-Clode, J. (ed.) *Biological Invasions in Changing Ecosystems*. de Gruyter Open, Warsaw, pp. 37–50.

Catford, J.A., Jansson, R. and Nilsson, C. (2009) Reducing redundancy in invasion ecology by integrating hypotheses into a single theoretical framework. *Diversity and Distributions* 15, 22–40.

Constantino, L.M. (2013) Descripción y daños causados por las babosas Colosius pulcher y Sarasinula plebeya en el cultivo del café en Colombia. Cenicafé, Avances técnicos 392. Available at: www.cenicafe.org/es/publications/avt0392.pdf (accessed 11 November 2019).

Crosby, G.T. (1972) Spread of the Cattle Egret in the Western Hemisphere. *Bird-Banding* 43, 205.

de la Ossa, V., la Ossa-Lacayo, D. and Monroy-Pineda, M. (2017) Abundance of domestic dove (*Columbia livia domestica* Gmelin, 1789) in Santiago de Tolú, Sucre, Colombia. *Revista MVZ Córdoba* 22, 5718–5727.

Diamond, A.W. and Filion, F.L. (1987) *The Value of Birds: Based on the Proceedings of a Symposium and Workshop Held at the XIX World Conference of the International Council for Bird Preservation*, June 1986, Queens University, Kingston, Ontario, Canada.

Díaz *et al.* (2019) pervasive human-driven decline of life on Earth points to the need for transformative change. *Science* 366, 1327.

Dunlap, T.R. (1997) Remaking the land: the acclimatization movement and anglo ideas of nature. *Journal of World History* 8, 303–319.

Essl, F., Dullinger, S., Rabitsch, W., Hulme, P.E., Hülber, K., *et al.* (2011) Socioeconomic legacy yields an invasion debt. *Proceedings of the National Academy of Sciences USA* 108, 203–207.

Evans, T., Kumschick, S., Dyer, E. and Blackburn, T. (2014) Comparing determinants of alien bird impacts across two continents: implications for risk assessment and management. *Ecology and Evolution* 4, 2957–2967.

Fontoura, P.M., Dyer, E., Blackburn, T.M. and Orsi, M.L. (2013) Non-native bird species in Brazil. *Neotropical Biology and Conservation* 8, 165–175.

Garaffa, P.I., Filloy, J. and Bellocq, M.I. (2009) Bird community responses along urban–rural gradients: does the size of the urbanized area matter? *Landscape and Urban Planning* 90, 33–41.

Ghersa, C.M. and León, R.J.C. (2001) Ecología del paisaje pampeano: consideraciones para su manejo y conservación. Ecología de Paisajes, Teorías y Aplicaciones. Facultad de Agronomía de la Universidad de Buenos Aires, Buenos Aires, pp. 471–512.

González-Acuña, D., Silva, F., Moreno, L., Cerda, F., Donoso, S., *et al.* (2007) Detección de algunos agentes zoonóticos en la paloma doméstica (*Columba livia*) en la ciudad de Chillán, Chile. *Revista Chilena de Infectología* 24, 199–203.

Guerrero-Peláez, S. (2016) Registro de dos especies de loros (Psittacidae) e híbrido en zonas rurales y urbanas de Armenia, Quindío, Colombia. *Acta Biológica Colombiana* 21, 653–656.

Ibañez, L.M., Girini, J.M., Palacio, F.X. and Montalti, D. (2014) Nidificación y alimentación de psittácidos introducidos en el noreste de la provincia de Buenos Aires, Argentina. *El Hornero* 29, 13–22.

IUCN-ISSG (2000) *100 of the World's Worst Invasive Alien Species: A Selection from the Global Invasive Species Database*. IUCN (International Union for Conservation of Nature)/Invasive Species Specialist Group (ISSG), Auckland, New Zealand. Available at: http://www.issg.org/pdf/publications/worst_100/english_100_worst.pdf (accessed 13 November 2019).

Jaksic, F.M. (1998) Vertebrate invaders and their ecological impacts in Chile. *Biodiversity and Conservation* 7, 1427–1445.

Jaksic, F.M., Iriarte, J.A., Jiménez, J.E. and Martínez, D.R. (2002) Invaders without frontiers: cross-border invasions of exotic mammals. *Biological Invasions* 4, 157–173.

Jensen, R.F. (2008) Nuevos registros de estornino pinto (*Sturnus vulgaris*) para el sureste de la provincia de Entre Ríos, Argentina. *Nuestras Aves* 53, 22.

Kier, G., Mutke, J., Dinerstein, E., Ricketts, T.H., Küper, W., *et al.* (2005) Global patterns of plant diversity and floristic knowledge. *Journal of Biogeography* 32, 1107–1116.

Kumschick, S. and Nentwig, W. (2010) Some alien birds have as severe an impact as the most effectual alien mammals in Europe. *Biological Conservation* 143, 2757–2762.

Kumschick, S., Bacher, S. and Blackburn, T.M. (2013) What determines the impact of alien birds and mammals in Europe? *Biological Invasions* 15, 785–797.

Lambertucci, S.A., Trejo, A., Di Martino, S., Sánchez-Zapata, J.A., Donázar, J.A. and Hiraldo, F. (2009) Spatial and temporal patterns in the diet of the Andean condor: ecological replacement of native fauna by exotic species. *Animal Conservation* 12, 338–345.

Lockwood, J.L. (1999) Using taxonomy to predict success among introduced avifauna: relative importance of transport and establishment. *Conservation Biology* 13, 560–567.

Lockwood, J.L., Hoopes, M.F., and Marchetti, M.P. (2016) *Invasion Ecology*. Wiley-Blackwell, London.

Marin, M. (2000) The Shiny Cowbird (*Molothrus bonariensis*) in Chile: introduction or dispersion – its hosts and parasitic trends. *Ornitologia Neotropical* 11, 285–296.

Martin-Albarracin, V.L., Amico, G.C., Simberloff, D. and Nuñez, M.A. (2015) Impact of non-native birds on native ecosystems: a global analysis. *PLoS One* 10, e0143070.

Medan, D., Torretta, J.P., Hodara, K., Elba, B. and Montaldo, N.H. (2011) Effects of agriculture expansion and intensification on the vertebrate and invertebrate diversity in the Pampas of Argentina. *Biodiversity and Conservation* 20, 3077–3100.

Montaldo, N.H. (1984) Asociación de dos especies de picaflores con árboles del género *Eucalyptus* (Myrtaceae) en la provincia de Buenos Aires. *El Hornero* 12, 159–162.

Mori, E., Grandi, G., Menchetti, M., Tella, J.L., Jackson, H.A., *et al.* (2017) Worldwide distribution of non-native Amazon parrots and temporal trends of their global trade. *Animal Biodiversity and Conservation* 40, 49–62.

Myers, N., Mittermeier, R.A., Mittermeier, C.G., da Fonseca, G.A. and Kent, J. (2000) Biodiversity hotspots for conservation priorities. *Nature* 403, 853.

Narosky, T. and Giacomo, A.D. (1993) *Las Aves de la Provincia de Buenos Aires: Distribución y Estatus.* AOP – Vázquez Mazzini, Buenos Aires.

Novillo, A. and Ojeda, R.A. (2008) The exotic mammals of Argentina. *Biological Invasions* 10, 1333–1344.

Nunes, M., Langeloh Roos, A., Barbosa Filho, R. and Mestre, L. (2010) The Cattle Egret (*Bubulcus ibis*) on Fernando de Noronha Archipelago: history and population trends. *Revista Brasileira de Ornitologia* 18, 315–327.

Nuñez, M.A. and Pauchard, A. (2010) Biological invasions in developing and developed countries: does one model fit all? *Biological Invasions* 12, 707–714.

Olrog, C.C. (1972) Adiciones a la avifauna Argentina. *Acta Zoológica Lilloana* 26, 255–266.

Pauchard, A., Cavieres, L., Bustamante, R., Becerra, P. and Rapoport, E. (2004) Increasing the understanding of plant invasions in southern South America: first symposium on alien plant invasions in Chile. *Biological Invasions* 6, 255–257.

Pauchard, A., Quiroz, C.L., García, R., Anderson, C.B. and Kalin, M.T. (2011) Invasiones biológicas en América Latina y El Caribe: tendencias en investigación para la conservación. In: *Conservación Biológica: Perspectivas desde América Latina.* Editorial Universitaria, Universidad de Chile, Santiago de Chile, pp. 79–94.

Pérez-García, J., Monsalve-Arcila, D. and Márquez-Villegas, C. (2015) Presence of parasites and enterobacterial infection in feral pigeons (*Columba livia*) in urban areas in Envigado, Colombia. *Revista Facultad Nacional de Salud Pública* 33, 370–376.

Peris, S., Soave, G.E., Camperi, A., Darrieu, C. and Aramburu, R. (2005) Range expansion of the European Starling *Sturnus vulgaris* in Argentina. *Ardeola* 52, 359–364.

Pimentel, D., McNair, S., Janecka, J., Wightman, J., Simmonds, C., *et al.* (2001) Economic and environmental threats of alien plant, animal, and microbe invasions. *Agriculture, Ecosystems and Environment* 84, 1–20.

Pyšek, P. and Jarošík, V. (2005) Residence time determines the distribution of alien plants. In: Inderjit, S. (ed.) *Invasive Plants: Ecological and Agricultural Aspects.* Birkhäuser Basel, Basel, Switzerland, pp. 77–96.

Pyšek, P., Jarošík, V., Pergl, J., Randall, R., Chytrý, M., *et al.* (2009) The global invasion success of Central European plants is related to distribution characteristics in their native range and species traits. *Diversity and Distributions* 15, 891–903.

Quiroz, C., Pauchard, A., Lohegren, A.C. and Anderson, C.B. (2009) Análisis cuantitativo de la investigación en invasiones biológicas en Chile: tendencias y desafíos. *Revista Chilena de Historia Natural* 82, 497–505.

R Core Team (2018) R: a language and environment for statistical computing. Available at: www.R-project.org/ (accessed 11 November 2019).

Restrepo-Rodas, D.C. and Pulgarín-Restrepo, P.C. (2017) Dinámicas de los loros en cautiverio en Colombia: tráfico, mortalidad y liberación. *Ornitología Colombiana* 16, eA06.

Ritvo, H. (2012) President's lecture: going forth and multiplying: animal acclimatization and invasion. *Environmental History* 17, 404–414.

Rodríguez, J.P. (2001) Exotic species introductions into South America: an underestimated threat? *Biodiversity and Conservation* 10, 1983–1996.

Sick, H. (1959) A invasão de América Latina pelo pardal (*Passer domesticus*). *Boll Museo Nacional Brazil* 207, 1–31.

Sick, H. (1965) *Bubulcus ibis* (L.) na Ilha de Marajo, Para: Garca ainda nao registrada no Brasil. *Anais da Academia Brasileira de Ciências* 37, 567–570.

Simberloff, D. (2013) *Invasive Species: What Everyone Needs to Know.* Oxford University Press, New York.

Simberloff, D. and Rejmánek, M. (2010) *Encyclopedia of Biological Invasions.* University of California Press, California.

Speziale, K.L., Lambertucci, S.A., Carrete, M. and Tella, J.L. (2012) Dealing with non-native species: what makes the difference in South America? *Biological Invasions* 14, 1609–1621.

Stohlgren, T.J., Barnett, D.T. and Kartesz, J.T. (2003) The rich get richer: patterns of plant invasions in the United States. *Frontiers in Ecology and the Environment* 1, 11–14.

Storey, A.A., Ramírez, J.M., Quiroz, D., Burley, D.V., Addison, D.J., *et al.* (2007) Radiocarbon and DNA evidence for a pre-Columbian introduction of Polynesian chickens to Chile. *Proceedings of the National Academy of Sciences USA* 104, 10335–10339.

Summers-Smith, J.D. (1963) *The House Sparrow.* Collins, London.

Summers-Smith, J.D. (1988) *The Sparrows: A Study of the Genus* Passer. T. & A.D. Poyser, London.

Torres, R. and Gutiérrez, M. (1999) Dieta de los pichones de la Garcita Bueyera (*Bubulcus ibis*) en la Laguna Mar Chiquita (Córdoba, Argentina). *Natura Neotropicalis* 1, 35–41.

van den Bosch, F., Hengeveld, R. and Metz, J.A.J. (1992) Analysing the velocity of animal range expansion. *Journal of Biogeography* 19, 135–150.

Whelan, C.J., Şekercioğlu, Ç.H. and Wenny, D.G. (2015) Why birds matter: from economic ornithology to ecosystem services. *Journal of Ornithology* 156, 227–238. https://doi.org/10.1007/s10336-015-1229.

40 Continental Analysis of Invasive Birds: Europe and the Middle East

Yvette C. Ehlers Smith* and David A. Ehlers Smith

Centre for Excellence in Invasion Biology, School of Life Sciences, University of KwaZulu-Natal, Pietermaritzburg 3209, South Africa

Citation: Ehlers Smith, Y.C. and Ehlers Smith D.A. (2020) Continental analysis of invasive birds: Europe and the Middle East. In: Downs, C.T. and Hart, L.A. (eds) *Invasive Birds: Global Trends and Impacts*. CAB International, Wallingford, UK, pp. 303–313.

40.1 Introduction

The Aichi Biodiversity Target 9 states, 'By 2020, invasive alien species and pathways are identified and prioritized, priority species are controlled or eradicated, and measures are in place to manage pathways to prevent their introduction and establishment.' Cardador *et al.* (2019) showed the effectiveness of the European wild bird trade restrictions in preventing avian invasions by limiting introduction of invaders at early stages of the invasion process. However, avian invasions still remain within Europe and the Middle East, and some species have been highlighted as priority species to enable targeted programmes to manage and mitigate further spread. Here, we provide a continental assessment of Europe and the Middle East to establish the history of avian invasion on the continent, explore the invasion pathways and species of concern, and highlight research gaps and challenges that will aid in achieving this target by 2020.

40.2 Invasion Pathways

Hulme *et al.* (2008) described the arrival of an alien species into a new region through six different pathways: release, escape, contaminant, as a stowaway, through a corridor and unaided. Introduced species that are incapable of self-maintaining sustainable populations, but on occasion exist beyond 'aided propagation' are considered casual species. If an introduced species is capable of reproducing in its novel environment/region unaided, then it is considered an established species. When an established species naturally extends its distribution and causes damage or becomes a pest species, then it is referred to as an invasive species (Keller *et al.*, 2011). However, only a subset of those introduced species become established, as they must overcome ecological, climatic and stochastic influences that pose barriers to their success (Zenni and Nuñez, 2013; Lockwood, 2017). Subsequently, only certain species spread and become invasive (Blackburn *et al.*, 2011; Lockwood, 2017). This subset, or filtering effect (Lockwood, 2017), was described by Williamson (1996) as the 'tens rule', which estimates that about 10% of all introduced species survive the four filtration stages: (i) transportation of live specimens (passage); (ii) an effective introduction process (release); (iii) formation of a non-native breeding population (establishment); and (iv) the natural range expansion (spread) of an invasive species (Williamson, 1996; Lockwood *et al.*, 2013). Abellán *et al.* (2016) suggested that the degree to which the colonized region resembles a species' native range is a critical factor regulating the spread of an invasive species. Similarly, the availability of a niche that the invasive species is capable of filling in its new non-native range is also a limitation (Strubbe and Matthysen, 2014).

40.3 An Accident of Geography?

Europe lies entirely within the northern hemisphere between the Tropic of Cancer and the Arctic Circle. The continent is partially surrounded by the Atlantic and Arctic Oceans and separated from the African continent by the Mediterranean and Red Seas. The divide between the Eurasian continents is largely arbitrarily through geographical features referred to as the Caucasian barrier comprising the Caucasus Mountains, the Ural River, and the Caspian and Black Seas, where Europe comprises the western-most part of Eurasia (Cunliffe, 2015).

*Corresponding author: EhlersSmithY@ukzn.ac.za

The climate of Europe is temperate overall but can be categorized into six broad classifications: boreal, Mediterranean, polar, temperate continental, temperate transitional and temperate oceanic. Geography, ocean currents and wind patterns keep Europe warmer than other landmasses at similar latitudes. The polar zone, characterized by subzero winter and cold summer months, falls within Northern Europe spanning the Arctic Circle. The boreal climate has long cold winters and short cool wetter summers. The temperate zone is subdivided into oceanic, transitional and continental climates. The continental zone is characterized by hot dry summers and very cold wet winters. The temperate oceanic climate is buffered by oceans with a narrow annual temperature range consistent with warm summers and mild winters, with rainfall all year round.

The transition zone between the continent and oceanic climates is also characterized by warm summers, cold winters and year-round precipitation. The Mediterranean climatic zone is buffered by the sea and is subsequently characterized by hot and dry summers and mild wet winters. The climate, geological features and orientation of the Eurasian continent along the east–west axis was pivotal in the domestication of plants and animals within the Fertile Crescent, which was also the driving force behind the establishment of trade routes and movement of people and commodities along the latitudes (Diamond, 1999). As Diamond (1999) stated, much of the status quo of modern society stemmed from accidents of geography and biogeography, where Europe became a centre for international trade for many centuries and consequently led to the spread of exotic or alien species of peoples and wildlife (Keller *et al.*, 2011).

40.4 The East–West Roads to Invasion

The German geographer Ferdinand von Richthofen described the 'silk roads', the network of trade routes between China and the West from the 2nd century BC to the 13th century AD (Cunliffe, 2015). However, some 300 years prior to the opening of the Silk Road, the transportation of goods and services was already established along the Royal Road. The Royal Road spanned approximately 1600 miles from east to west, connecting Susa (modern-day Iran) and Sardis, near the Mediterranean Sea (present-day Turkey). This trade road was established by the Persians, under the rule of Darius I. The Royal Road also incorporated smaller routes that connected Mesopotamia to the Indian subcontinent and Egypt (North Africa). Alexander the Great (Alexander III, died *c.*323 BC), ruler of the Ancient Greek Kingdom of Macedonia, used the Royal Road to expand his rule into Persia (Asia Minor, Syria and Egypt), incorporating the Silk Road. His empire stretched from Greece in the West, India in the East and the Danube River to the north. He transported birds from India to numerous European regions.

Dyer *et al.* (2017) found two distinct peaks in the rate at which birds have been introduced globally, one occurring at the end of the 19th century and another occurring in the past two decades. The first peak came from the actions of acclimatization societies and other groups that purposefully transported and intentionally introduced birds as commodities to supplement local hunting opportunities (game birds) or to improve the natural aesthetics of their local avifauna, so called 'fauna improvement' (Blackburn *et al.*, 2009; Brochier *et al.*, 2010; Cassey *et al.*, 2015; Dyer *et al.*, 2017). Early explorers and settlers were responsible for many of the exotic bird introductions within the 18th and 19th centuries globally (Crosby, 1986; McDowall, 1994).

The second peak came as a result of the aviculture market where birds are sold as captive companion animals, for their singing ability or as ornamental birds for displays in zoos or bird parks, which then enter introduction pathways either as intentional release or accidental escapees (Hulme *et al.*, 2008; Kark *et al.*, 2009; Brochier *et al.*, 2010; Souviron-Priego *et al.*, 2018). The demand for exotic species and the number of subsequent introductions depends on political and economic factors, cultural practices and human population size (Mulliken *et al.*, 1996; Kark *et al.*, 2009; Chiron *et al.*, 2010). Trade deals between America and Western Europe meant an influx of North American birds during the Cold War. Similarly, immigration from Africa was responsible for a peak in the number of African birds imported into Western Europe during this period (Mulliken *et al.*, 1996). The Iron Curtain proved to be ecologically beneficial to the Eastern Bloc countries in that the international trade embargoes in place helped to limit the import and subsequent population establishment of exotic species (Chiron *et al.*, 2010). Overall, an increase in the number of deliberate and accidental releases of exotic/non-native birds in the wild in Europe during the 20th century was linked to increased human immigration rate (Jeschke and Strayer, 2005) and to both the legal and illicit international trade (Jenkins and Jenkins, 1996; Kark *et al.*, 2009).

40.5 Exotics Introduced into the Region

Kark *et al.* (2009) estimated that 193 bird species have historically been introduced into the European and Middle East region, and out of these 140 remained by 2007. However, only 77 species (40%) could be considered established (Kark *et al.*, 2009). Chiron *et al.* (2009), using the Delivering Alien Invasive Species Inventories for Europe (DAISIE) research consortium database considered 175 as being introduced as exotic bird species into Europe. They estimated that 75 species (43%) had already established populations by the year 2000. However, Abellán *et al.* (2016) distinguished 377 exotic species in Spain and Portugal alone, whereas the DAISIE database only recorded 73 species within Spain and Portugal. The Abellán *et al.* (2016) dataset thus highlights the shortcomings of large-scale databases and notes that inventories must continuously be updated at a national level.

Sixteen of the 31 Global Invasive Species Database (GISD: www.iucngisd.org/gisd/, accessed 13 November 2019) species are present in Europe and the Middle East: the Common or Indian Myna (*Acridotheres tristis*), Chukar Partridge (*Alectoris chukar*), Mallard Duck (*Anas platyrhynchos*), Greylag Goose (*Anser anser*), Canada Goose (*Branta canadensis*), Cattle Egret (*Bubulcus ibis*), Rock Dove (*Columba livia*), House Crow (*Corvus splendens*), Mute Swan (*Cygnus olor*), Common Waxbill (*Estrilda *astrild*), Quaker or Monk Parakeet (*Myiopsitta monachus*), Ruddy Duck (*Oxyura jamaicensis*), House Sparrow (*Passer domesticus*), Western Swamphen (*Porphyrio porphyrio*), Ring-necked or Rose-ringed Parakeet (*Psittacula krameri*) and Eurasian Collared-Dove (*Streptopelia decaocto*). The database is managed by the Invasive Species Specialist Group (ISSG) of the International Union for Conservation of Nature (IUCN) Species Survival Commission.

40.6 Invasive Species Native to the Region and Natural Colonizations

The legacies of colonizations by European settlers have left their mark on the natural histories of the continents that were conquered. Many of the successful avian invaders are of European descent and were either intentionally (through acclimatization societies) or accidentally introduced into non-native regions as escaped pets (Crosby, 1986; McDowall, 1994; Blackburn *et al.*, 2009; Cassey *et al.*, 2015).

The Common, or European Starling (*Sturnus vulgaris*) is native to the Eurasian continent and has a broad east–west natural distribution, but has been successfully introduced into numerous countries globally, where it has gained invasive status (Craig and Feare, 1998). Throughout parts of its native range, it is considered an agricultural pest species, and measures to manage the population are being taken (Feare *et al.*, 1992). The House Sparrow and Mute Swan, native to the Eurasian continent, are some of the invasive species with the largest non-native range globally. The House Sparrow was deliberately introduced into Africa, Australia and North and South America as a pest control or as part of faunal 'improvement' programmes (McDowall, 1994; Cassey *et al.*, 2015). The Western Swamphen is native to Europe and the Middle East but has become invasive within the USA through accidental introduction of free-roaming birds. These birds may exert negative impacts on native Gallinule species through competition, and cause damage to native vegetation. The distribution of Mute Swans is partly a result of introductions and reintroductions, but some natural range extensions also occurred once populations became established. Their historic (pre-1960s) breeding range was considered to be only within Great Britain, Ireland and the Netherlands, along with patchy distributions along the Baltic and the Black Seas, and from the Caspian Sea to eastern China (Voous and Thomson, 1960). The Mute Swan's success was attributed to legal protection status in parts of their range, increasing artificial water bodies (e.g. ponds and lakes) and an increase in year-round food availability through agricultural intensification (Wieloch, 1991). Tomialojc (2017) showed that Mute Swans are highly adaptable and are associated with habitats within urban land, with breeding populations established within cities.

The Rock Dove is native to Eurasia and is considered the ancestral stock of homing and messenger pigeons (*Columba livia domestica*), renowned for their ability to find their way home. The species has been studied since the turn of the century and their methods for navigation have been hypothesized for decades (Schiffner *et al.*, 2018). The species were used to convey messages from as early as approximately 700 BC (for Ancient Olympic announcements) and they were even donned with medals of honour for wartime services (Blechman, 2007). However, feral populations have now established globally; they are considered 'flying rats' and vermin (Jerolmack, 2008). Al-Barwari and Saeed (2012) in their examination of the parasitic communities of the rock pigeons within Iraq recommended the eradication of pigeons from poultry-breeding facilities and food and water handling facilities to counteract disease dissemination. This species breeds all year round and forms large flocks, which have a great presence in public open spaces in cities worldwide. For many, the sheer numbers of individuals are a major influencer of public perception as a pest species, and make their number difficult to control.

The Mallard Duck, a species synonymous with duck-feeding toddlers worldwide (i.e. supplementary feeding), is native to the northern hemisphere and is present across Europe. This species, however, is one of the most widely distributed ducks globally, with invasive populations present on all continents. The species was originally introduced into its non-native range as an ornamental species or as hunting stock but spread locally.

Some colonization events exist where birds have spread to other regions through natural range extensions. European Collared-doves colonized anthropogenic habitats in a westerly direction across Europe from their native range in North Africa and Asia. The spread of grain crops throughout Europe facilitated their colonization efforts (Robertson, 1990). At the end of the 19th century the expanded species distribution was limited to Turkey and the Balkans (Fisher, 1953; Hudson, 1965), but by 1955, the species had already established a breeding population in Britain (Richardson *et al.*, 1957). Colonization of new areas is ongoing. In recent years, the species colonized the region of Extremadura, situated in the centre-west of the Iberian Peninsula, where they first established populations in urban areas and then slowly expanded towards rural farmland. The species is highly commensal with humans, who provide it with advantages against certain predators (Rocha-Camarero and de Trucios, 2002). They are considered agricultural pests throughout their newly established range, as they eat stored grain crops (Robertson, 1990). They were originally considered capable of hybridization with other doves in the genus *Streptopelia* and subfamilies Zenaidinae and Columbinae (Fisher, 1953), which may have conservation implications for similar species. As the species is highly adaptable and their range is continuously expanding, it could pose a threat to other resident columbiform species (Rocha-Camarero and Hidalgo de Trucios, 2002). The species is, however, controlled in areas where they cause financial losses (Robertson, 1990).

The Cattle Egret, native to Africa and more recently parts of southern Europe (mainly France and the Iberian Peninsula), has spread naturally throughout the Eurasian continent (Ahmed, 2011). Parts of Spain and Portugal remained uncolonized until the 1960s, and in parts of France they did not establish breeding populations until the 1980s (Voisin, 1991). The first breeding record within the Netherlands was in 1998 (Messemaker and Veldkamp, 1999) and in Britain in 2008 (Hill, 2008; Hiley *et al.*, 2013). The GISD database notes that Cattle Egrets, which are communal nesters and roosters (Voisin, 1991), are potentially able to outcompete native birds for nesting areas. Where they nest or roost in urban areas, they could be considered a nuisance due to noise and fouling, but there are currently no studies looking at their impacts specifically in Europe and the Middle East. Other than as a potential nuisance, there are no publications on the invasive impacts of Cattle Egret in Europe.

40.7 ISSG GISD Species Introduced to the Region: Current Distribution and Origin of Introduced Bird Species

40.7.1 Indian or Common Myna

The Common Myna hails from tropical central and southern Asia (Craig and Feare, 1998). The species is classified as an extinct alien within France and Great Britain. They are considered established aliens in the Canary Islands, Russia and Israel and as 'non-established aliens' within Italy, Spain, the Balearic Islands and Turkey (DAISIE, 2009), although according to Saavedra (2010), Common Mynas have been eradicated from the Islands of Tenerife, Gran Canaria and Mallorca and are currently being controlled on Fuerteventura, and St Helena and Ascension Islands. Unregulated sales of Common Mynas remain common practice in Portuguese and Dutch pet shops and are also available online. They are imported into Spain without legislative control (Saavedra *et al.*, 2015).

The IUCN/SSC ISSG (ISSG, 2011) noted them as present within Israel, Kuwait, Qatar, Iraq, Oman, Saudi Arabia and the United Arab Emirates. Mynas were first noted within the Gulf States of the Middle East from the 1970s onwards. The species was introduced as a biological pest control (Porter *et al.*, 1996), although it is unclear specifically which pests were targeted. The Common Myna is a popular caged bird in Israel and other Middle Eastern countries, and some of the earlier observations of the birds were probably from captive populations. Within Israel, the species utilizes buildings and man-made structures for nesting and roosting. They forage within parks and irrigated grass lawns (Holzapfel *et al.*, 2006). Although evidence for interference competition with native species exists beyond Europe (see Baker *et al.*, 2014), the actual invasive impact of Common Mynas within Europe and the Middle East remains uncertain (Shirley and Kark, 2006). However, Cohen *et al.* (2019), using species distribution models, showed the potential extensive expansion in the Common Myna distribution, putting greater areas at risk of Common Myna invasion. They also highlighted the species' broad environmental tolerance and human commensalism within urbanized areas.

40.7.2 Chukar Partridge

The Chukar Partridge is only partially native to Europe, with populations existing from the Aegean Sea through to Central and East Asia (Barbanera *et al.*, 2009). The species is classed as an established alien within Austria, France, Greece and Italy, and as an extinct alien in Germany, Great Britain and Sicily. Within Croatia, the Czech Republic and Spain, they are classified as non-established aliens. Their current status in Norway is unknown (DAISIE, 2009). The GISD database, however, only considers them as invasive within Italy, Spain and the UK. The Chukar Partridge was introduced into Spain, France and Italy with the aim of establishing a hunting population to supplement the declining native Red-legged Partridge (*Alectoris rufa*; Barbanera *et al.*, 2005; Blanco-Aguiar *et al.*, 2008) and the Rock Partridge (*Alectoris graeca*; Barilani *et al.*, 2007a).

40.7.3 Greylag Goose

Although the Greylag Goose is native to certain parts of Northern and Eastern Europe and the Middle East, it is listed as an alien within Austria, Belgium, Germany and Switzerland (DAISIE, 2009). The main threat that Greylag Geese pose is the potential hybridization with domestic and other Anatidae species, which has been recorded to cause feral hybrid populations in the Netherlands and Germany (Blair *et al.*, 2000). Bakker *et al.* (2018) found that herbivory by the Greylag Goose negatively influences the persistence and expansion of riparian vegetation, particularly *Phragmites australis* reed beds, habitats that are in decline throughout Europe. The genetic vulnerability of wild native species and the eutrophication of waterbodies may become a problem (Blair *et al.* 2000), although the evidence is currently lacking for Europe and the Middle East.

40.7.4 Canada Goose

The Canada Goose is a native of North America, but established alien populations are present within Belgium, Denmark, Finland, France, Germany, Great Britain, Ireland, Lithuania, the Netherlands, Norway, Poland and Sweden. They are classed as non-established aliens in Austria, Czech Republic, Italy, Latvia, Sardinia, Switzerland and Poland (DAISIE, 2009; Brochier *et al.*, 2010). The species was originally introduced into Europe for hunting purposes and then spread naturally to neighbouring countries (Josefsson and Andersson, 2001; Brochier *et al.*, 2010; Vall-Llosera, 2015; Carpio *et al.*, 2017).

40.7.5 House Crow

The native range of the House Crow spans a wide variety of climatic conditions including the wet (monsoon), semi-arid and arid zones of the tropical and subtropical regions and temperate alpine regions throughout the Indian subcontinent (Krzemińska *et al.*, 2018). The House Crow is considered native to southern Asia and parts of East and South-east Asia, including Afghanistan, Bangladesh, Bhutan, China, India, Iran, Nepal, Pakistan, Sri Lanka and Thailand (Meininger *et al.*, 1980; Ali, 2002; Nyári *et al.*, 2006; Ryall, 2010). However, phylogenetic analyses have shown that northern South Asia, specifically Nepal, Bangladesh and possibly North India, is the centre of origin for the species (Krzemińska *et al.*, 2018).

In certain places, the House Crow was deliberately introduced to deal with refuse and crop pests (Ryall, 2016). The

GISD database notes that the species is present within Bahrain, Denmark, France, Ireland, Israel, Jordan, the Netherlands, Qatar, Saudi Arabia, Spain (and Gibraltar), United Arab Emirates, the UK and Yemen (although the DAISIE database only recognizes them as established aliens within Israel). Ryall (2016) noted the first individuals present on the island of Cyprus in 2011 and in Turkey in 2015. The ecological niche model of Krzemińska *et al.* (2018) predicted the suitability of all known geographical areas where they have successfully colonized (Nyári *et al.*, 2006; Ryall, 2016). They also showed that House Crow distribution is influenced by rainfall and precipitation. The Arabian Peninsula was not predicted as suitable by their ecological niche model, and they suggested that House Crows might have evolved adaptations to much drier environments (Krzemińska *et al.*, 2018). However, an earlier study by Nyári *et al.* (2006) found that anthropogenic associations (human habitation, augmented food and water availability) have enabled them to spread into otherwise intolerable areas, particularly areas such as the southern rim of the Arabian Peninsula.

40.7.6 Common Waxbill

Common Waxbills are attractive small granivores native to sub-Saharan Africa and imported into Europe as caged birds in the 1960s (Reino, 2005; Cardoso and Reino, 2018). The species has adapted well to the Iberian Peninsula's Mediterranean climes. Weather conditions seem to be a limiting factor for their current non-native distribution. The species is sensitive to cooler and wetter temperatures below 15°C (Steinbacher and Wolters, 1965; Nicolai and Steinbacher, 2007). In the Iberian Peninsula, Common Waxbills occupy a marginal human-modified niche, which native species have not yet filled. Interspecific competition between Common Waxbills and their native counterparts and harmful impacts on agriculture are yet to be demonstrated (Reino, 2005; Cardoso and Reino, 2018). However, Batalha *et al.* (2013) suggested that studying interactions between Common Waxbills and the south-western subspecies of endangered Reed Buntings (*Emberiza schoeniclus*) is warranted, as their ecological nearest neighbour.

40.7.7 Monk or Quaker Parakeet

The Monk Parakeet is a popular caged bird native to South America, with patchy distributions within open savannas, scrub forests and palm groves in Argentina, Bolivia, Brazil, Paraguay and Uruguay (Souviron-Priego *et al.*, 2018). It is a highly adaptable species that has subsequently established and spread as an alien invasive species throughout Europe. According to the DAISIE (2009), their non-native range includes Austria (although considered extinct), Belgium (established), Czech Republic (established but not breeding; Hudec, 2015), Denmark (single population; Fox *et al.*, 2015), France (cities of Toulon and Marseilles; Dubois and Cugnasse, 2015), Germany (non-breeding), Greece (Athens; Kalodimos, 2013), Italy (country wide; Mori *et al.*, 2013), Israel (Tel-Aviv; Postigo *et al.*, 2017), the Netherlands, Portugal (Lisbon and Porto), Spain (largest number of CITES imports; Mori *et al.*, 2013) and the UK (non-established alien). Monk Parakeets can exist in colder climes in Europe in comparison with their native range, but this results from their introduction history and not because of range expansion towards colder areas (Strubbe and Matthysen, 2014). Strubbe and Matthysen (2014) suggested that this cold 'adaptation' is facilitated by their association with human-dominated landscapes and the availability of anthropogenic food sources.

40.7.8 Ruddy Duck

The Ruddy Duck is native to North America but was imported into wildfowl collections in the UK in the 1940s. The species was first observed in Belgium in 1979, assumed to be a natural expansion of the population in the UK. In 2013, the established breeding population was estimated at around 15 individuals (Lafontaine *et al.*, 2013).

They are classed as established alien species in France, Great Britain, Ireland, Norway and Sweden. They are present, but not established, in Austria, Belgium, Denmark, Finland, Germany, Hungary, Iceland, Israel, Italy, the Netherlands, Portugal, Sardinia, Slovenia, Spain and Turkey (Muñoz-Fuentes *et al.*, 2006), with breeding populations in the UK, Iceland, Ireland, France, Belgium, the Netherlands, Germany and Morocco (Muñoz-Fuentes *et al.*, 2006). Research by Muñoz-Fuentes *et al.* (2006) suggested that the population in Europe has undergone a genetic bottleneck, associated with the 'founder effect', probably associated with the small number of birds brought into the UK. The species is being controlled within the UK, Belgium, the Netherlands, France and Spain, but self-sustaining breeding populations remain in France and the Netherlands (Lafontaine *et al.*, 2013; Robertson *et al.*, 2015).

40.7.9 Ring-necked or Rose-ringed Parakeet

Although native to Africa and Asia (Mauritania, Guinea, Senegal, Uganda, South Sudan, Afghanistan, West Pakistan, the Indian subcontinent and Myanmar; Forshaw, 1978), the Ring-necked Parakeet has established populations throughout Europe (Strubbe and Matthysen, 2009). Pârâu *et al.* (2016) estimated, as of 2015, a European invasion of at least 90 breeding populations, consisting of at least 85,000 individuals across ten countries. The species was traded as caged birds, particularly during the late 1960s and 1970s, and non-native populations established from accidental and deliberate escapees. They have subsequently spread naturally into human-dominated landscapes (Brochier *et al.*, 2010). Like the Monk Parakeet, their spread has been facilitated by anthropogenic associations within their native range (Strubbe and Matthysen, 2007, 2009, 2014).

40.8 Emerging Species of Concern

The DAISIE project highlighted the Canada Goose, Ruddy Duck and Sacred Ibis (*Threskiornis aethiopicus*) as among the 100 worst invasive species in Europe (DAISIE, 2009; Brochier *et al.*, 2010). However, the Sacred Ibis is not noted on the GISD species list for Europe. Lowe *et al.* (2000) listed the Common Myna, Red-vented Bulbul (*Pycnonotus cafer*) and Common Starling among the '100 of the World's Worst Invasive Alien Species', a subset of GISD. The Red-vented Bulbul is currently not classed as invasive within the region. Roy *et al.* (2015) compiled a list totalling 95 species across different taxonomic groups that represents high- or very high-risk species with the potential of arriving, establishing, spreading and posing a threat to biodiversity and ecosystem functioning within the European Union by 2025. Six bird species are featured within this list, all of which are already present within Europe (Table 40.1) (Roy *et al.*, 2015). The Common Myna is already highlighted as a highly problematic invasive species within Europe. DAISIE (2009) also states that the Egyptian goose (*Alopochen aegyptiaca*) has established non-native breeding populations within several European countries as well as in the United Arab Emirates, and their competitive effects have been shown for certain birds of prey species on the continent (Rehfisch *et al.*, 2010).

Perhaps one of the most famous species brought into Europe by Alexander the Great was the Alexandrine Parakeet (*Psittacula eupatria*), which was native to the Asian continent and naturally present between Afghanistan and Vietnam, through India and Indochina, and north to Nepal and Bhutan (Juniper and Parr, 1998). Several feral populations can be found across Europe and the Middle East, including Spain, England, Belgium, the Netherlands, Germany, Turkey, Saudi Arabia, Bahrain, Qatar, United Arab Emirates and Iran. It is classed as an established alien within Belgium, Germany, Italy, Turkey, Great Britain, Israel and the Netherlands. As of 2015, sightings of this parakeet had increased and populations were rapidly growing in Belgium, the Netherlands and Germany, with an estimate of 100–300 birds present in each county (Ancillotto *et al.*, 2016). Many of these countries have climates much colder than habitats within its native range. Ancillotto *et al.* (2016) found interspecific facilitation with established populations of Ring-necked Parakeets, which may contribute to niche expansion and invasion success of the congeneric Alexandrine Parakeets. They are also known to hybridize with Ring-necked Parakeets (Postigo, 2016), although the exact implications of this are not yet known.

The species' distribution models also predict a global high invasion risk that indicates the potential for further invasion and range expansion, particularly if the species is tolerant to colder weather conditions beyond its native range. Ironically, the species is classed as 'near threatened' in its native range due to habitat loss, and the capture of live birds for trade is likely to negatively influence its persistence in the future (BirdLife International, 2017).

40.9 Impacts of Introductions

40.9.1 Economic impacts

Kumschick and Nentwig (2010) highlighted the Canada Goose, Sacred Ibis, Ruddy Duck, Monk Parakeet and Ring-necked Parakeet as the avian alien species with the greatest economic impact. Kumschick *et al.* (2013) found that the Canada Goose was the only bird species that has a combined environmental and economic impact score greater than the average of invasive mammals tested. However, Strubbe *et al.* (2011) highlighted the shortcomings of Kumschick and Nentwig's (2010) study. Strubbe *et al.* (2011) showed that Kumschick and Nentwig assumed that the abundance and impact of Canada Geese would be comparative and relevant throughout the entire invaded range. Population booms within their native range (Wichita, Kansas) are thought to be linked to increased hunting restrictions, the increased availability of artificial water sources and supplementary feeding (Maccarone and Cope, 2004); none the less, these factors will be context dependent and area specific. Some of the impacts such as aircraft collisions in areas where high volumes of air traffic and high-density Canada Goose populations exist can be extrapolated, but some impacts can be ecologically context specific (Baxter and Robinson, 2007). Alderisio (1999) found that the excreta/faecal coliform concentrations excreted by the Canada Geese could have a negative effect on the microbiological quality of water in Westchester County, New York, while Keller *et al.* (2011) stated that their droppings can cause human health hazards and algal blooms; however, they did not reference specific studies. Within its native range, the species is considered a crop pest (e.g. flowers, grains, fruits and maize; Brochier *et al.*, 2010).

40.9.2 Disease vectors

The companion animal trade is indirectly a pathway for the introduction of infectious diseases (Daut *et al.*, 2016). Fifty-three species of alien bird have been identified as linked with human health impacts (Schindler *et al.*, 2015). West Nile virus is transmitted via mosquitoes, but birds serve as amplifying hosts. West Nile virus arrived in Mesopotamia around 300 BC, killing indigenous birds and people (Marr and Calisher, 2003). The virus has been isolated from some species of actively migrating birds in Western Europe (Daut *et al.*, 2016), but could the trade of exotic birds from Mesopotamia and Babylon (present-day Baghdad) to the West be considered an additional pathway for the spread of West Nile virus? Given that Alexander the Great was renowned for procuring exotic birds, ironically it is now thought that he died of a West Nile virus infection (Marr and Calisher, 2003).

In a study conducted in the USA, two isolates of the H5 2.3.4.4 clade of avian influenza virus were tested, namely GYRF/14 H5N8, a Eurasian virus, and NOPI/14 H5N2, in Ruddy Ducks (Spackman *et al.*, 2017). The H5N8 strain has been detected in wild birds in Europe, and this isolate could be maintained by wild bird populations, of which migrating wildfowl could potentially be spreading the virus. In 2017, this virus infected and killed large numbers of wildfowl, including Tufted Ducks (*Aythya fuligula*). Spackman *et al.* (2017) showed that Ruddy Ducks were susceptible to infection with the H5 2.3.4.4 clade. However, a lack of clinical symptoms of the disease suggested that they could act as reservoirs of the virus.

Table 40.1. Six bird species extracted from the dataset composed by Roy *et al*. (2015), including the threat that each species poses to native biodiversity.

Rank/95	Species	Common name	Pathway	Origin	Risk category	Competition	Predation	Hybridization	Disease	Parasitism
11	*Pycnonotus cafer*	Red-vented Bulbul	Pet, escapee	Asia	Very high	✓	✓	✓	✓	
12	*Acridotheres tristis*	Common Myna	Unaided through other pathways, Pet, zoo	Asia	Very high	✓	✓	✓	✓	✓
44	*Acridotheres cristatellus*	Crested Myna	Pet, escapee	Asia	High	✓				
49	*Pycnonotus jocosus*	Red-whiskered Bulbul	Pet, escapee	Asia	High	✓	✓	✓	✓	✓
53	*Rhea americana*	Greater Rhea	Zoo	South America	High	✓	✓	✓		
54	*Psittacula eupatria*	Alexandrine Parakeet	Zoo, unaided through other pathways	Asia	High	✓	✓	✓	✓	✓

They also found that the Ruddy Duck would not be the most efficient species for maintaining and disseminating the virus. The potential for this invasive species as a vector for avian influenza virus within Europe therefore needs to be explored further.

40.9.3 Genetic pollution

Barbanera *et al.* (2005) highlighted that within the UK, France and Italy, hybridization has occurred naturally between the Chukar Partridge and Red-legged Partridge in the wild. Captive-bred hybrids between these two species produce more offspring compared with the wild native counterparts, making them more profitable to be released for hunting within Spain (Blanco-Aguiar *et al.*, 2008). The Red-legged Partridge and the Rock Partridge are also known to hybridize in the wild, where their distributions overlap in the French Alps (Barilani *et al.*, 2007b). Barilani *et al.* (2007a) similarly found introgressive hybridization between Chukar and Rock Partridges, suggesting that released captive-bred partridges have reproduced and hybridized causing genetic pollution of wild Rock Partridge populations in Greece. There is thus potential for interbreeding between the hybrids of Chukar × Red-legged Partridge descent with those of Chukar × Rock Partridge descent, but the evidence for this is yet to be obtained. In Spain, the Ruddy Duck is known to hybridize with the native White-headed Duck (*Oxyura leucocephala*), a species that is classed as endangered globally (BirdLife International, 2017). The mallard is known to hybridize with the Pacific Black Duck (*Anas wyvilliana*) and the Hawaiian Duck (*Anas superciliosa*) within their native range (Baker *et al.*, 2014), but relatively little evidence of hybridization with native species within Europe has been collected.

40.9.4 Competition and other impacts on native bird species

Much of the research on competitive exclusion of many of the invasive species within Europe stems from studies outside Europe. However, Strubbe *et al.* (2010) found that Nuthatches (*Sitta europaea*) compete with Ring-necked Parakeets for nesting cavities, with a moderate competition strength. They found no compelling evidence indicating that parakeets pose a threat large enough to justify eradication campaigns. Others have found that this species may reduce the resources available for the House Sparrow, Stock Dove (*Columba oenas*) and Common Starlings, which are considered pests or invasive throughout much of their distribution (Brochier *et al.*, 2010).

40.10 Working Towards the Aichi Biodiversity Ttargets

The European Union Regulation 1143/2014 on Invasive Alien Species is the legal instrument that has been implemented to meet Target 9 of the Aichi Biodiversity Targets. This regulation highlights prevention, early detection and management as crucial measures to control or eradicate priority alien invasive species. Some research is now focusing on those species that disappeared at each stage progressing to invasive status to be able to better understand which factors influence successful invasion to prevent future invasions (Zenni and Nuñez, 2013; Abellán *et al.*, 2016; Lockwood, 2017). Species distribution models are being utilized to predict range expansions or areas vulnerable to invasions as early detections measures, but these methods need to be explored for a greater range of species encompassing broader regions. Relatively few invasive bird species are targets for eradication campaigns in Europe (Kumschick and Nentwig, 2010), partly because of lack of feasibility studies and impact assessments that encompass all aspects of biological invasion (Strubbe *et al.*, 2011), including ecological impacts, public perception (Vane and Runhaar, 2016) and cost-effectiveness (Reyns *et al.*, 2018). Vanderhoeven *et al.* (2017) also highlighted the need for transparency and repeatability of assessments.

40.10.1 What is good for the goose is not necessarily good for the gander

Despite heavy focus on the impacts of invasive species on the continent, much of the research relating to invasive birds is being generated outside Europe, although the outcomes are being applied within the region. Ten years on from the study of Kark *et al.* (2009), there is relatively little empirical evidence of impacts on biodiversity, human health and competition with native species for various GISD-listed species within Europe. Strubbe *et al.* (2011) critically assessed the impact risk assessment and management recommendations for invasive birds and highlighted that assessments are often largely based on anecdotal observations or tenuous links to research elsewhere that are likely to mislead management recommendations. More focus needs to be applied to the implications of invasions, particularly where management campaigns are potentially ineffective, time consuming and expensive. For example, there are still some countries within Europe and the Middle East that are not party to the Nagoya Protocol. The European Commission has highlighted the need for a harmonized response to tackling alien invasive species as the measures taken by one member state will be ineffective if neighbouring countries do not act or respond in a coordinated manner (European Commission, 2008).

40.11 References

Abellán, P., Carrete, M., Anadón, J.D., Cardador, L. and Tella, J.L. (2016) Non-random patterns and temporal trends (1912–2012) in the transport, introduction and establishment of exotic birds in Spain and Portugal. *Diversity and Distributions* 22, 263–273.

Ahmed, R. (2011) Subspecific identification and status of Cattle Egret. *Dutch Birding* 33, 294–304.

Al-Barwari, S. and Saeed, I. (2012) The parasitic communities of the rock pigeon *Columba livia* from Iraq: component and importance. *Türkiye Parazitolojii Dergisi* 36, 232–239.

Alderisio, K.A. (1999) Seasonal enumeration of fecal coliform bacteria from the feces of Ring-Billed Gulls (*Larus delawarensis*) and Canada Geese (*Branta canadensis*). *Applied and Environmental Microbiology* 65, 5628–5630.

Ali, S. (2002) *The Book of Indian Birds*, 13th edn. Oxford University Press, New Dehli.

Ancillotto, L., Strubbe, D., Menchetti, M. and Mori, E. (2016) An overlooked invader? Ecological niche, invasion success and range dynamics of the Alexandrine parakeet in the invaded range. *Biological Invasions* 18, 583–595.

Baker, J., Harvey, K.J. and French, K. (2014) Threats from introduced birds to native birds. *Emu* 114, 1–12.

Bakker, E.S., Veen, C.G.F., ter Heerdt, G.J.N., Huig, N. and Sarneel, J.M. (2018) High grazing pressure of geese threatens conservation and restoration of reed belts. *Frontiers in Plant Science* 9, 1–12.

Barbanera, F., Negro, J.J., Di Giuseppe, G., Bertoncini, F., Cappelli, F. and Dini, F. (2005) Analysis of the genetic structure of red-legged partridge (*Alectoris rufa*, Galliformes) populations by means of mitochondrial DNA and RAPD markers: a study from central Italy. *Biological Conservation* 122, 275–287.

Barbanera, F., Guerrini, M., Khan, A.A., Panayides, P., Hadjigerou, P., *et al.* (2009) Human-mediated introgression of exotic chukar (*Alectoris chukar*, Galliformes) genes from East Asia into native Mediterranean partridges. *Biological Invasions* 11, 333–348.

Barilani, M., Bernard-Laurent, A., Mucci, N., Tabarroni, C., Kark, S., *et al.* (2007a) Hybridisation with introduced chukars (*Alectoris chukar*) threatens the gene pool integrity of native rock (*A. graeca*) and red-legged (*A. rufa*) partridge populations. *Biological Conservation* 137, 57–69.

Barilani, M., Sfougaris, A., Giannakopoulos, A., Mucci, N., Tabarroni, C. and Randi, E. (2007b) Detecting introgressive hybridisation in rock partridge populations (*Alectoris graeca*) in Greece through Bayesian admixture analyses of multilocus genotypes. *Conservation Genetics* 8, 343–354.

Batalha, H.R., Ramos, J.A. and Cardoso, G.C. (2013) A successful avian invasion occupies a marginal ecological niche. *Acta Oecologica* 49, 92–98.

Baxter, A.T. and Robinson, A.P. (2007) Monitoring and influencing feral Canada goose (*Branta canadensis*) behaviour to reduce birdstrike risks to aircraft. *International Journal of Pest Management* 54, 341–346.

BirdLife International (2017) The IUCN Red List of Threatened Species. Available at: www.iucnredlist.org/ (accessed 11 November 2019).

Blackburn, T., Lockwood, J. and Cassey, P. (2009) *Avian Invasions: The Ecology and Evolution of Exotic Birds*. Oxford University Press, Oxford.

Blackburn, T.M., Pyšek, P., Bacher, S., Carlton, J.T., Duncan, R.P., *et al.* (2011) A proposed unified framework for biological invasions. *Trends in Ecology and Evolution* 26, 333–339.

Blair, M.J., Mckay, H., Musgrove, A.J. and Rehfisch, M.M. (2000) Review of the status of introduced non-native waterbird species in agreement area of the African-Eurasian Waterbird Agreement research contract CR0219. *British Trust for Ornithology Research Report* 229, 1–129.

Blanco-Aguiar, J.A., González-Jara, P., Ferrero, M.E., Sánchez-Barbudo, I., Virgós, E., *et al.* (2008) Assessment of game restocking contributions to anthropogenic hybridization: the case of the Iberian red-legged partridge. *Animal Conservation* 11, 535–545.

Blechman, A.D. (2007) *Pigeons: The Fascinating Saga of the World's Most Revered and Reviled Bird*. Grove Press, New York.

Brochier, B., Vangeluwe, D. and van den Berg, T.P. (2010) Alien invasive birds. *Scientific and Technical Review of the Office International des Epizooties* 29, 217–226.

Cardador, L., Tella, J.L., Anadón, J.D., Abellán, P. and Carrete, M. (2019) The European trade ban on wild birds reduced invasion risks. *Conservation Letters* 2019, e12631.

Cardoso, G.C. and Reino, L. (2018) Ecologically benign invasions: the invasion and adaptation of Common Waxbills (*Estrilda astrild*) in Iberia. In: Quiroz, A. and Pooley, S. (eds) *Histories of Bioinvasions in the Mediterranean*. Environmental History, Vol. 8. Springer, Cham, Switzerland, pp. 149–169.

Carpio, A.J., Guerrero-Casado, J., Barasona, J.A., Tortosa, F.S., Vicente, J., *et al.* (2017) Hunting as a source of alien species: a European review. *Biological Invasions* 19, 1197–1211.

Cassey, P., Vall-Llosera, M., Dyer, E. and Blackburn, T.M. (2015) The biogeography of avian invasions: history, accident and market trade. In: Canning-Clode, J. (ed.) *Biological Invasions in Changing Ecosystems*. De Gruyter Open, Warsaw. pp. 37–54.

Chiron, F., Shirley, S. and Kark, S. (2009) Human-related processes drive the richness of exotic birds in Europe. *Proceeding of the Royal Society B: Biological Sciences* 276, 47–53.

Chiron, F., Shirley, S.M. and Kark, S. (2010) Behind the Iron Curtain: Socio-economic and political factors shaped exotic bird introductions into Europe. *Biological Conservation* 143, 351–356.

Cohen, T.M., McKinney, M., Kark, S. and Dor, R. (2019) Global invasion in progress: modeling the past, current and potential global distribution of the common myna. *Biological Invasions* 21, 11–15.

Craig, A. and Feare, C. (1998) *Starlings and Mynas*. A&C Black Publishers, London.

Crosby, A.W. (1986) *Ecological Imperialism: The Biological Expansion of Europe, 900–1900*. Cambridge University Press, Cambridge.

Cunliffe, B. (2015) *By Steppe, Desert and Ocean: The Birth of Eurasia*. Oxford University Press, Oxford.

DAISIE (2009) *Handbook of Alien Species in Europe*. Springer, Berlin.

Daut, E.F., Lahodny, G., Peterson, M.J. and Ivanek, R. (2016) Interacting effects of Newcastle disease transmission and illegal trade on a wild population of White-Winged Parakeets in Peru: a modeling approach. *PLoS One* 11, e0147517.

Diamond, J. (1999) *Guns, Germs, and Steel: The Fates of Human Societies*. W.W. Norton and Co., New York.

Dubois, P.J. and Cugnasse, J.-M. (2015) Les populations doiseaux allochtones en France en 2014 (3e enquête nationale) [Third invasive bird species survey in France in 2014]. *Ornithos* 22, 72–91.

Dyer, E.E., Redding, D.W. and Blackburn, T.M. (2017) The global avian invasions atlas, a database of alien bird distributions worldwide. *Scientific Data* 4, 170041.

European Commission (2008) Commission presents policy options for EU strategy on invasive species. Press release database IP/08/1890. Available at: http://europa.eu/rapid/press-release_IP-08-1890_en.htm (accessed 11 November 2019).

Feare, C.J., de Franssu, P.D. and Peris, S.J. (1992) The starling in Europe: multiple approaches to a problem species. In: Borrecco, J. and Marsh, R. (eds) *Proceedings of the Fifteenth Vertebrate Pest Conference*. University of California, California, pp 28. Available at: https://digitalcommons.unl.edu/vpc15/28/ (accessed 11 November 2019).

Fisher, J. (1953) The collared turtle dove in Europe. *British Birds* 46, 41–62.

Forshaw, J.M. (1978) *Parrots of the World*, 2nd edn. David & Charles, London.

Fox, A., Heldbjerg, H. and Nyegaard, T. (2015) Invasive alien birds in Denmark. *Dansk Ornitologisk Forenings Tidsskrift* 109, 193–205.

Hiley, J.R., Bradbury, R.B., Holling, M. and Thomas, C.D. (2013) Protected areas act as establishment centres for species colonizing the UK. *Proceedings of the Royal Society B: Biological Sciences* 280, 20122310.

Hill, B. (2008) Cattle Egrets breed in Somerset. Bird Guides. Available at: www.birdguides.com/news/cattle-egrets-breed-in-somerset/ (accessed 11 November 2019).

Holzapfel, C., Levin, N., Hatzofe, O. and Kark, S. (2006) Colonisation of the Middle East by the invasive Common Myna *Acridotheres tristis* L., with special reference to Israel. *Sandgrouse* 28, 44–51.

Hudec, K. (2015) Non-native species of birds in the Czech Republic. *Zoo Report*, June. Available at: www.zoobrno.cz/img/UK profi cerven 2015 K03.pdf (accessed 11 November 2019).

Hudson, R. (1965) The spread of the Collared Dove in Britain and Ireland. *British Birds* 58, 139–155.

Hulme, P.E., Bacher, S., Kenis, M., Klotz, S., Kühn, I., *et al.* (2008) Grasping at the routes of biological invasions: a framework for integrating pathways into policy. *Journal of Applied Ecology* 45, 403–414.

ISSG (2011) Global Invasive Species Database (GISD). Invasive Species Specialist Group of the IUCN Species Survival Commission. Available at: www.issg.org/database (accessed 13 November 2019).

Jenkins, P.T. (1996) Free trade and exotic species introductions. *Conservation Biology* 10, 300–302.

Jerolmack, C. (2008) How pigeons became rats: the cultural-spatial logic of problem animals. *Social problems* 55, 72–94.

Jeschke, J.M. and Strayer, D.L. (2005) From the cover: invasion success of vertebrates in Europe and North America. *Proceedings of the National Academy of Sciences USA* 102, 7198–7202.

Josefsson, M. and Andersson, B. (2001) The Environmental Consequences of Alien Species in the Swedish Lakes Mälaren, Hjälmaren, Vänern and Vättern. *AMBIO* 30, 514–521.

Juniper, T. and Parr, M. (1998) *Parrots: A Guide to Parrots of the World*, 1st edn. Yale University Press, New Haven, Connecticut.

Kalodimos, N. (2013) First account of a nesting population of monk parakeets, *Myiopsitta monachus* with nodule-shaped bill lesions in Katehaki, Athens, Greece. *Bird Populations* 12, 1–6.

Kark, S., Solarz, W., Chiron, F., Clergeau, P. and Shirley, S. (2009) Alien birds, amphibians and reptiles of Europe. In: Drake, J.A. (ed.) *Handbook of Alien Species in Europe*. Springer, Heidelberg, Germany, pp. 105–118.

Keller, R.P., Geist, J., Jeschke, J.M. and Kühn, I. (2011) Invasive species in Europe: ecology, status, and policy. *Environmental Sciences Europe* 23, 23.

Krzemińska, U., Morales, H.E., Greening, C., Nyári, Á.S., Wilson, R., *et al.* (2018) Population mitogenomics provides insights into evolutionary history, source of invasions and diversifying selection in the House Crow (*Corvus splendens*). *Heredity* 120, 296–309.

Kumschick, S. and Nentwig, W. (2010) Some alien birds have as severe an impact as the most effectual alien mammals in Europe. *Biological Conservation* 143, 2757–2762.

Kumschick, S., Bacher, S. and Blackburn, T.M. (2013) What determines the impact of alien birds and mammals in Europe? *Biological Invasions* 15, 785–797.

Lafontaine, R.-M., Robert, H., Delsinne, T., Adriaens, T., Devos, K. and Beudels-Jamar, R.C. (2013) Risk analysis of the Ruddy Duck *Oxyura jamaicensis* (Gmelin, 1789), Risk analysis report of non-native organisms in Belgium. Royal Belgian Institute of Natural Sciences for the Federal Public Service Health, Food Chain Safety and Environment. Available at: http://biodiversite.wallonie.be/servlet/Repository/?ID=33503 (accessed 11 November 2019).

Lockwood, J., Hoopes, M., and Marchetti, M. (2013) *Invasion Ecology*, 2nd edn. Wiley-Blackwell, London.

Lockwood, J.L. (2017) Exotic birds provide unique insight into species invasions. *Proceedings of the National Academy of Sciences USA*, 114, 9237–9239.

Lowe, S., Browne, M., Boudjelas, S. and De Poorter, M. (2000) *100 of the World's Worst Invasive Alien Species: A Selection from the Global Invasive Species Database*. The Invasive Species Specialist Group (ISSG), Auckland, New Zealand. Available at: www.issg.org/pdf/publications/worst_100/english_100_worst.pdf (accessed 11 November 2019).

Maccarone, A.D. and Cope, C.H. (2004) Recent trends in the winter population of Canada Geese (*Branta canadensis*) in Wichita, Kansas: 1998–2003. *Transactions of the Kansas Academy of Science* 107, 77–82.

Marr, J.S. and Calisher, C.H. (2003) Alexander the Great and West Nile Virus Encephalitis. *Emerging Infectious Diseases* 9, 1599–1603.

McDowall, R. (1994) *Gamekeepers for the Nation: The Story of New Zealand's Acclimatisation Societies, 1861–1990*. Canterbury University Press, Christchurch, New Zealand.

Meininger, P.L., Mullie, W.C. and Bruun, B. (1980) The spread of the House Crow, *Corvus splendens*, with special reference to the occurrence in Egypt. *Le Gerfaut* 70, 245–250.

Messemaker, R. and Veldkamp, R. (1999) Eerste Nederlandse broedgeval van Koereiger in De Wieden in 1998. *Dutch Birding* 21, 206–263.

Mori, E., Di Febbraro, M., Foresta, M., Melis, P., Romanazzi, E., *et al.* (2013) Assessment of the current distribution of free-living parrots and parakeets (Aves: Psittaciformes) in Italy: a synthesis of published data and new records. *Italian Journal of Zoology* 80, 158–167.

Mulliken, T.A, Broad, S.R. and Thomsen, J.B. (1996) A global review of worldwide bird trade. I: Leader-Williams, N. and Tibanyenda, R.K. (eds) *The Live Bird Trade in Tanzania*. IUCN, Gland, Switzerland, and Cambridge, pp. 5–26.

Muñoz-Fuentes, V., Green, A.J., Sorenson, M.D., Negro, J.J. and Vilà, C. (2006) The ruddy duck *Oxyura jamaicensis* in Europe: natural colonization or human introduction? *Molecular Ecology* 15, 1441–1453.

Nicolai, J. and Steinbacher, J. (2007) *Prachtfinken Afrika (Handbuch der Vogelpflege)*, 3rd edn. Eugen Ulmer, Stuttgart, Germany.

Nyári, A., Ryall, C. and Townsend Peterson, A. (2006) Global invasive potential of the house crow *Corvus splendens* based on ecological niche modelling. *Journal of Avian Biology* 37, 306–311.

Pârâu, L.G., Strubbe, D., Mori, E., Menchetti, M., Ancillotto, L., *et al.* (2016) Rose-ringed Parakeet populations and numbers in Europe: a complete overview. *Open Ornithology Journal* 9, 1–13.

Porter, R.F., Christensen, S. and Schiermacker-Hansen, P. (1996) *Field Guide to the Birds of the Middle East*. Helm Field Guides, London.

Postigo, J.-L. (2016) New records of invasive Parakeet hybrids in Spain. A great opportunity to apply the rapid response mechanism. *European Journal of Ecology* 2, 19–22.

Postigo, J., Shwartz, A., Strubbe, D. and Muñoz, A.-R. (2017) Unrelenting spread of the alien monk parakeet *Myiopsitta monachus* in Israel. Is it time to sound the alarm? *Pest Management Science* 73, 349–353.

Rehfisch, M.M., Allan, J.R. and Austin, G.E. (2010) The effect on the environment of Great Britain's naturalized Greater Canada *Branta canadensis* and Egyptian Geese *Alopochen aegyptiacus*. In: *British Ornithologists' Union Proceedings – The Impacts of Non-native Species*. BOU 2008 Autumn Scientific Meeting. Available at: www.bou.org.uk/bouprocnet/impacts-of-non-native-species/ (accessed 11 November 2019).

Reino, L. (2005) Variation partitioning for range expansion of an introduced species: the common waxbill *Estrilda astrild* in Portugal. *Journal of Ornithology* 146, 377–382.

Reyns, N., Casaer, J., de Smet, L., Devos, K., Huysentruyt, F., *et al.* (2018) Cost-benefit analysis for invasive species control: the case of greater Canada goose *Branta canadensis* in Flanders (northern Belgium). *PeerJ* 6, e4283.

Richardson, R.A., Seago, M.J. and Church, A. (1957) Collared Doves in Norfolk: a bird new to the British list. *British Birds* 50, 239–246.

Robertson, H.A. (1990) Breeding of Collared Doves *Streptopelia decaocto* in rural Oxfordshire, England. *Bird Study* 37, 73–83.

Robertson, P.A., Adriaens, T., Caizergues, A., Cranswick, P.A., Devos, K., *et al.* (2015) Towards the European eradication of the North American ruddy duck. *Biological Invasions* 17, 9–12.

Rocha-Camarero, G. and Hidalgo de Trucios, S.J. (2002) The spread of the Collared Dove *Streptopelia decaocto* in Europe: colonization patterns in the west of the Iberian Peninsula. *Bird Study* 49, 11–16.

Roy, H.E., Adriaens, T., Aldridge, D.C., Bacher, S., Bishop, J.D.D., *et al.* (2015) Invasive alien species – prioritising prevention efforts through horizon scanning. Technical Report No. ENV.B.2/ETU/2014/0016. European Commission.

Ryall, C. (2010) Further records and updates of range extension in House Crow *Corvus splendens*. *Bulletin of British Ornithologists' Club* 130, 246–254.

Ryall, C. (2016) Further records and updates of range expansion in House Crow *Corvus splendens*. *Bulletin of the British Ornithologists' Club* 136, 39–45.

Saavedra, S. (2010) Eradication of invasive Mynas from islands. Is it possible? *Aliens: The Invasive Species Bulletin* 29, 40–47.

Saavedra, S., Maraver, A., Anadón, J.D. and Tella, J.L. (2015) A survey of recent introduction events, spread and mitigation efforts of mynas (*Acridotheres* sp.) in Spain and Portugal. *Animal Biodiversity and Conservation* 38, 121–128.

Schiffner, I., Fuhrmann, P., Reimann, J. and Wiltschko, R. (2018) Behavioural traits of individual homing pigeons, *Columba livia* f. *domestica*, in their homing flights. *PLoS One* 13, e0201291.

Schindler, S., Staska, B., Adam, M., Rabitsch, W. and Essl, F. (2015) Alien species and public health impacts in Europe: a literature review. *NeoBiota* 27, 1–23.

Shirley, S.M. and Kark, S. (2006) Amassing efforts against alien invasive species in Europe. *PLoS Biology* 4, e279.

Souviron-Priego, L., Muñoz, A.R., Olivero, J., Vargas, J.M. and Fa, J.E. (2018) The legal international wildlife trade favours invasive species establishment: the Monk and Ring-Necked Parakeets in Spain. *Ardeola* 65, 233–246.

Spackman, E., Prosser, D.J., Pantin-Jackwood, M.J., Berlin, A.M. and Stephens, C.B. (2017) The pathogenesis of clade 2.3.4.4 h5 highly pathogenic avian influenza viruses in ruddy duck (*Oxyura jamaicensis*) and lesser scaup (*Aythya affinis*). *Journal of Wildlife Diseases* 53, 832–842.

Steinbacher, J. and Wolters, H.E. (1965) *Vogel in Kafig und Voliere. Ein Handbuch fu r Vogelliebhaber. Prachtfinken*, Vol. 1: *Astrilde*, 2nd edn. Hans Limberg, Aachen, Germany.

Strubbe, D. and Matthysen, E. (2007) Invasive ring-necked parakeets *Psittacula krameri* in Belgium: habitat selection and impact on native birds. *Ecography* 30, 578–588.

Strubbe, D. and Matthysen, E. (2009) Establishment success of invasive ring-necked and monk parakeets in Europe. *Journal of Biogeography* 36, 2264–2278.

Strubbe, D. and Matthysen, E. (2014) Patterns of niche conservatism among non-native birds in Europe are dependent on introduction history and selection of variables. *Biological Invasions* 16, 759–764.

Strubbe, D., Matthysen, E. and Graham, C.H. (2010) Assessing the potential impact of invasive ring-necked parakeets *Psittacula krameri* on native nuthatches *Sitta europeae* in Belgium. *Journal of Applied Ecology* 47, 549–557.

Strubbe, D., Shwartz, A. and Chiron, F. (2011) Concerns regarding the scientific evidence informing impact risk assessment and management recommendations for invasive birds. *Biological Conservation* 144, 2112–2118.

Tomialojc, L. (2017) Human initiation of synurbic populations of waterfowl, raptors, pigeons and cage birds. In: Murgui, E. and Hedblom, M. (eds) *Ecology and Conservation of Birds in Urban Environments*. Springer, New York, pp. 1–525.

Vall-Llosera, M. (2015) Avian invasions: from basic to applied research. *Revista Catalana d'Ornitologia* 31, 28–37.

Vanderhoeven, S., Branquart, E., Casaer, J., Dhondt, B., Hulme, P.E., *et al.* (2017) Beyond protocols: improving the reliability of expert-based risk analysis underpinning invasive species policies. *Biological Invasions* 19, 2507–2517.

Vane, M. and Runhaar, H.A.C.C. (2016) Public support for invasive alien species eradication programs: insights from the Netherlands. *Restoration Ecology* 24, 743–748.

Voisin, C. (1991) *The Herons of Europe*. A&C Black, London.

Voous, K.H. and Thomson, S.A.L. (1960) *Atlas of European Birds*. Elsevier, Amsterdam.

Wieloch, M. (1991) Population trends of the mute swan *Cygnus olor* in the Palearctic. *Wildfowl* 1, 22–32.

Williamson, M. (1996) *Biological Invasions*, 1st edn. Springer, Amsterdam.

Zenni, R.D. and Nuñez, M.A. (2013) The elephant in the room: the role of failed invasions in understanding invasion biology. *Oikos* 122, 801–815.

41 Continental Analysis of Invasive Birds: Asia

Richard T. Corlett[1]*, Michael R. Leven[2], Ding Li Yong[3], James A. Eaton[4] and Philip D. Round[5]

[1]Xishuangbanna Tropical Botanical Garden, Chinese Academy of Sciences, Menglun, Mengla, Yunnan 666303, China; [2]AEC Ltd, Palm Springs, Yuen Long, Hong Kong, China; [3]BirdLife International (Asia), Tanglin International Centre, Singapore 247672; [4]Birdtour Asia Ltd, Littleover, Derby DE23 4EE, UK; [5]Department of Biology, Mahidol University, Bangkok, Thailand

Citation: Corlett, R.T., Leven, M.R., Yong, D.L., Eaton, J.A. and Round, P.D. (2020) Continental analysis of invasive birds: Asia. In: Downs, C.T. and Hart, L.A. (eds) *Invasive Birds: Global Trends and Impacts*. CAB International, Wallingford, UK, pp. 314–339.

41.1 Introduction

There is no universally accepted geographical definition of Asia, with disagreements about the western, northern and eastern boundaries. For example, in the definition used by the Intergovernmental Science-Policy Platform on Biodiversity and Ecosystem Services (IPBES), 37 countries are included, but Central Asia, Asian Russia and Papua New Guinea are excluded, while the Intergovernmental Panel on Climate Change (IPCC) has 51 countries in Asia, including all those omitted by IPBES. In the broadest definition, Asia includes one-third of the world's total land area and nearly two-thirds of the world's human population, while narrower definitions reduce the area and, to a lesser extent, the population. In this chapter, we will focus on tropical and subtropical Asia – the Oriental region – because it forms a distinct zoogeographical unit recognized by all vertebrate biologists, while temperate northern Asia is part of the Palearctic region, western Asia is Palearctic and/or Afrotropical, and New Guinea is Australian. However, as some Asian countries – particularly China and Indonesia – span two zoogeographical regions, with consequences for the bird trade and thus for invasive birds, we have included non-Oriental Asia where relevant. Accordingly, in the north-east we have included Palearctic China, Russia, Korea and Japan, while in the south-east we have included all of Wallacea east to Lydekker's Line but excluding New Guinea.

The Oriental region includes all of Sri Lanka, Bangladesh, Myanmar, Thailand, Laos, Vietnam, Cambodia, Malaysia, Singapore, Brunei and the Philippines, plus Pakistan, India, Nepal and Bhutan below the Himalayan treeline, China south of approximately 35°N, the Ryukyu Islands of Japan and western Indonesia. It has a total area of around 10 million km^2 (approximately 7% of the Earth's total land surface) and a human population of around 2.8 billion (approximately 37% of the total), depending on the precise boundaries chosen. Natural ecosystems range from aseasonal lowland rainforests to hot deserts. More than half of the region has been transformed by agriculture, and most natural ecosystems that have not been completely transformed have been more or less strongly impacted by hunting, logging, grazing or other forms of exploitation. Only a small proportion of the region still supports an intact native avifauna.

Despite this diversity in natural ecosystems and the recent anthropogenic changes, the Oriental land-bird avifauna is dominated by forest-dependent or -associated bird families and species, reflecting the primeval post-glacial landscape, which was largely forest, especially in the east, where forest once covered most of the land from the Siberian taiga to the tropical rainforests of the Thai–Malay Peninsula, and then east through the Indonesian Archipelago to New Guinea. If the taxonomic treatment of Gill and Donsker (2019) is followed, as we have in this chapter, the most species-rich Oriental land-bird families are almost all composed of forest-associated passerines, including: (i) the endemic tit-babblers and scimitar-babblers (Timaliidae); (ii) the fulvettas and ground-babblers (Pellorneidae), shared with the Afrotropical region but mostly Oriental; (iii) the near-endemic laughingthrushes and allies (Leiothrichidae), with just a few species in the southern Palearctic and Afrotropical regions; (iv) the bulbuls (Pycnonotidae), shared with the Afrotropical region; (v) the widespread Old World

*Corresponding author: corlett@xtbg.org.cn

chats and flycatchers (Muscicapidae); and (vi) the sunbirds (Nectariniidae), in which most of the other species are Afrotropical. Among the non-passerines, the terrestrial forest pheasants and allies (Phasianidae) reach their greatest diversity in the region, woodpeckers (Picidae) are highly diverse, and the large frugivore niche is dominated by the endemic Asian Barbets (Megalaimidae), hornbills (Bucerotidae) and, in the east of the region, pigeons (Columbidae).

In addition to the diverse and endemic Timaliidae and Megalaimidae, smaller endemic families include the ioras (Aegithinidae), the cupwings (Pnoepygidae), the leafbirds (Chloropseidae), two species of fairy-bluebirds (Irenidae) and four single-species families – the Bornean Bristlehead (*Pityriasis gymnocephala*; Pityriasidae), Rail-babbler (*Eupetes macrocercus*; Eupetidae), Hylocitrea (*Hylocitrea bonensis*; Hylocitreidae) and Spotted Elachura (*Elachura formosa*; Elachuridae) – with a fifth single-species family, the Crested Jay (*Platylophus galericulatus*) also recognized by some authorities (e.g. Winkler *et al.*, 2015). All except some ioras are forest dependent. Especially in the north of the region, land-bird species diversity increases markedly in the boreal winter with an influx of species breeding in the eastern Palearctic.

41.2 Invasion Pathways

Globally, the international bird trade is considered to be the main source of bird invasions (Cassey *et al.*, 2004; Souviron-Priego *et al.*, 2018). In most of the developing world, however, the domestic bird trade is much larger, in terms of numbers of individuals and species, than the international one. This can also lead to bird invasions where species are moved outside their native ranges, and this is particularly likely in large countries that span two or more zoogeographical regions, such as China, or that consist of an island archipelago, such as the Philippines, or both, as in Indonesia, which includes both Wallace's and Lydekker's zoogeographical lines.

The Asian bird trade is huge and diverse, involving hundreds of species and millions of individuals every year (BirdLife International, 2018). It is also old: 900 years ago, the Chinese Emperor Huizong painted an Ornate Lorikeet (*Trichoglossus ornatus*) from Sulawesi in his Palearctic capital Kaifeng (Lindholm, 1995). In the 1860s, ornithologist Robert Swinhoe recorded the Black-throated Laughingthrush (*Pterorhinus chinensis*), Red-billed Blue Magpie (*Urocissa erythrorhyncha*) and Red-billed Leiothrix (*Leiothrix lutea*) in the bird shops in Hong Kong and commented, 'I could send a fine collection of so-called Chinese birds, if I only had an agent at Hong Kong or Canton to watch the bird shops' (Swinhoe, 1862).

Excluding domesticated poultry, live birds are now sold in this region mainly as pet caged birds but also for singing competitions (Chng and Eaton, 2016a,b; Shepherd *et al.*, 2016; Eaton *et al.*, 2017a,b), for food and medicines (Webster, 1975; Chan, 2004; Gilbert *et al.*, 2012), for bird-fighting contests (Dai and Zhang, 2017) and for merit release, mostly by Buddhists (Severinghaus and Chi, 1999; Chan, 2006; Gilbert *et al.*, 2012). Japan has a tradition of falconry supplied by imported birds (Vall-Llosera and Su, 2019), and other species have been imported for ornamental collections and hunting.

Bird markets in Japan (Vall-Llosera and Su, 2019), Hong Kong (Chan 2006) and Singapore (Eaton *et al.*, 2017a) are dominated by imported birds, reflecting a combination of effective enforcement of laws against domestic bird trapping and high average incomes, which make imported birds affordable. In contrast, the birds sold in markets in Vietnam (Edmunds *et al.*, 2011; Eaton *et al.*, 2017b), Cambodia (Gilbert *et al.*, 2012) and Indonesia (Chng *et al.*, 2015; Chng and Eaton, 2016b; Chng *et al.*, 2018; Rentschlar *et al.*, 2018) are largely native to the country, while those sold in China (Dai and Zhang, 2017), Taiwan (Su *et al.*, 2014) and Thailand (Chng and Eaton, 2016a) are a mixture of native and non-native species. Javan markets sell birds from across the Indonesian archipelago (Chng *et al.*, 2015; Chng and Eaton, 2016b; Shepherd *et al.*, 2016) and markets in China sell birds from across the latitudinal range of the country (Chan, 2006; Dai and Zhang, 2017).

The most widely sold wild-caught birds are common, open-country or forest-edge 'species of least concern' in the International Union for Conservation of Nature (IUCN) Red List (IUCN, 2019), including munias, bulbuls, white-eyes and doves. However, most markets in the region also include a proportion of forest specialists and of 'threatened' species. Parrots dominate the non-native bird trade into and within the region. A relatively few widely available species are consistently captive bred, including budgerigars, cockatiels, canaries and lovebirds, but some other claims for captive breeding are dubious.

Captive birds can only form invasive populations if they escape or are released in sufficient numbers. Accidental escapes are presumably as frequent in Asia as elsewhere, and this is likely also to be true for the deliberate release of unwanted birds by traders and pet owners. In contrast, merit release – the freeing of captive animals for religious reasons – is an Asian phenomenon, found outside Asia only among emigrant communities. It is typically carried out by Buddhists and Taoists, and ranges from individual acts to organized mass releases of thousands of birds and other animals by religious organizations (Chan, 2006; Severinghaus and Chi, 1999; Gilbert *et al.*, 2012; Su *et al.*, 2016). Most individual birds released are small passerines sold specifically for this purpose, including munias (*Lonchura* spp.), white-eyes (*Zosterops* spp.), bulbuls (Pycnonotidae), swallows (Hirundinidae), sparrows (*Passer* spp.), weavers (*Ploceus* spp.) and, at least historically, the now critically endangered Yellow-breasted Bunting (*Emberiza aureola*), but any captive bird may be released, and some practitioners favour birds sold as pets to avoid the obvious paradox of birds being captured specifically for release. The numbers released are huge – an estimated 500,000–1,000,000/year in Hong Kong alone 15 years ago (Chan, 2006).

Where the released birds are native species from surrounding areas, as they are in Cambodia (Gilbert *et al.*, 2012), then the risks of them becoming invasive are minimized, but all birds released in Hong Kong, Taiwan and Singapore are imported, and many of these, such as the popular Mongolian Lark (*Melanocorypha mongolica*) in Hong Kong, are released outside their native ranges (Chan, 2006). Moreover, these birds are typically released in groups of multiple individuals, favouring establishment of wild populations if suitable habitats are available. In Hong Kong, of 19 bird species that were not present in the 19th century and that appear to have established wild populations as a result of human agency, populations of all but the House Crow of the 16 passerine species were considered by Leven and Corlett (2004) to most probably have originated from deliberate releases. Strikingly, most of these are babblers (Timaliidae and Leiothrichidae) that have become established in shrublands and/or secondary forests rather than the anthropogenic open habitats that are normally occupied by alien introductions. A similar pattern is apparent in Japan and Taiwan where at least six forest and shrubland species, including four babblers, have become established in recent years (Eguchi and Amano, 2004; Brazil, 2009).

The practice of merit release appears to have begun as a Taoist ritual in China at least as early as the 3rd century AD and was adopted by Chinese Buddhists soon after Buddhism's arrival in China, becoming popular during the Tang Dynasty (618–907 AD), and subsequently spreading, as a Buddhist practice, to Japan, Korea, Sri Lanka, Thailand and, more recently, Taiwan, where it is now also practised by followers of Christianity and other religions (Shiu and Stokes, 2008). The antiquity of release practices, especially in China, begs the question as to how much the supposed natural ranges of some species, especially those that are largely confined to anthropogenic habitats, have been extended through this agency before the beginning of the historical record which is patchy, at best, before the mid-19th century. The earliest documented deliberate introduction in the region appears to be that of the Oriental Magpie (*Pica serica*) to Kyushu, Japan, in the 16th century (Eguchi and Amano, 2004); in view of the magpie's importance in Buddhist culture, this might also have been an early instance of merit release. Leven and Corlett (2004) speculated that the wide distributions of the (frequently released) Red-whiskered Bulbuls (*Pycnonotus jocosus*) in south-east China and White-rumped Munias (*Lonchura striata*) in Hong Kong were likely to have resulted from released birds. Similarly, while the initial spread of the Eurasian Tree Sparrow (*Passer montanus*) from its presumed origin in the Huang He (Yellow River) valley through eastern China may have followed the spread of rice and millet cultivation (Summers-Smith, 1988), its later spread into South-east Asia seems likely to have been at least assisted by merit releases. Other frequent merit release species, such as Oriental Magpie-Robins (*Copsychus saularis*) and several munias, which are largely confined to anthropogenic habitats in much of the region, may also have had their distributions extended by merit releases, but this may be impossible to prove.

The choice of species involved in merit releases usually appears to be driven largely by their availability and cost, and some species, such as Azure-winged Magpies (*Cyanopica cyanus*) in China, are chosen because they are associated with good luck. As a consequence, other than captive-bred Budgerigars (*Melopsittacus undulatus*), which are often included in merit releases in Hong Kong where they are inexpensive (M.R. Leven, personal observation), parrots are not usually purchased for merit release, as they are too expensive. It seems likely, therefore, that the non-native parrot populations in the region, which are largely confined to urban centres (Gibson and Yong, 2017), are derived from similar sources to those in other parts of the world – accidental escapes and owners tiring of the responsibility (and perhaps noise) of looking after their pets and thus releasing them, although the prevailing culture of merit release may encourage owners to let them go rather than returning them to shops or animal welfare organizations.

Non-native bird species may also be introduced as free-living individuals transported on ships. Birds can easily fly back to shore, so only a few species are carried this way, but two apparently ship-assisted species are important in the region. The House Crow (*Corvus splendens*), the population of which in Hong Kong is centred on a relatively busy container port, almost certainly reached here by ship (Leven and Corlett, 2004). The spread of Tree Sparrows through the Philippines and the Indonesian Archipelago is also likely to have been facilitated by individuals travelling on ships, although there is little firm evidence.

House Crows were deliberately introduced to Malaysia to control caterpillars, and the introduction of Pied Mynas (*Gracupica contra*) in rubber plantations in Peninsular Malaysia seems to have had a similar objective (Yap and Sodhi, 2004). In contrast to Europe and North America, few species of waterfowl and pheasants have been deliberately introduced for hunting or ornamental purposes, except in Japan.

41.3 Invasive Birds

41.3.1 Species on the Global Invasive Species Database occurring in Asia

Out of the 31 species listed in the Invasive Species Specialist Group (ISSG) Global Invasive Species Database (GISD), seven have not been recorded in the wild in Asia. These are the Ruddy Duck (*Oxyura jamaicensis*), Great Horned Owl (*Bubo virginianus*), Great Kiskadee (*Pitangus sulphuratus*), Australian Magpie (*Gymnorhina tibicen*), House Finch (*Carpodacus mexicanus*), and Shiny Cowbird (*Molothrus bonariensis*) and Brown-headed Cowbird (*Molothrus ater*). In addition, the 'Canada Goose' (*Branta canadensis/hutchinsii*) occurs only as a natural vagrant and deliberate releases in northern Asia (Japan and Beijing), and the Swamp Harrier (*Circus approximans*) only as a natural vagrant. All but two of the remaining species on the database – the Monk Parakeet (*Myiopsitta monachus*) and Common Waxbill (*Estrilda astrild*) – are native somewhere in Asia and also in the Oriental region, although the Common Starling (*Sturnus vulgaris*) is only present as a winter visitor in the north of the Oriental region, and the status of the possibly native Rock Dove (*Columba livia*) populations in the Indian subcontinent and Myanmar is unclear.

Eleven of the species that are native in Asia also have invasive populations in Asia outside their natural ranges, although distinguishing native and introduced ranges is not always possible in long-established species. These invasions are localized in six species: the Mute Swan (*Cygnus olor*; Japan), Eurasian Collared Dove (*Streptopelia decaocto*; Hong Kong, Japan), Ring-necked Parakeet (*Psittacula krameri*; Thailand, Hong Kong, Japan, Taiwan, Singapore), Red-whiskered Bulbul (Japan, Taiwan, Singapore, southern Peninsular Malaysia), Red-billed Leiothrix (Japan, Hong Kong) and Warbling White-eye (*Zosterops japonicus*; Singapore).

The Red Junglefowl (*Gallus gallus*) was a prehistoric introduction to the Indonesian archipelago east of Bali and perhaps the Philippines. The Rock Dove occurs in most substantial urban areas in the region, although most individuals seen in urban areas are free-flying domestic birds kept by hobbyists, and truly wild populations are relatively small compared with, for example, those in Western Europe. The House Crow is established in Hong Kong, Thailand, Singapore, Peninsular Malaysia and Kota Kinabalu, the Andaman Islands, and Batam Island and Medan in Indonesia. The Common Myna (*Acridotheres tristis*) occurs outside its (uncertain) native range in southern China, Sumatra, Borneo and probably elsewhere. The status of several House Sparrow (*Passer domesticus*) populations outside their core range is unclear: most probably reflect natural spread, but they were deliberately introduced in the Andaman Islands.

41.3.2 Species not on the GISD

The great majority of invasive bird species and individuals in Asia and the Oriental region are not currently on the GISD (Table 41.1). Most of these species are native somewhere in Asia, and the few established species from Australia, Africa and the Neotropics are all localized. Indeed, invasive populations of most species, Asian or not, are currently localized, although some of these are locally abundant and many are increasing. Most of the more successful invasive Asian bird species are doves, bulbuls, 'babblers' (Timaliidae and Leiothrichidae) or mynas. The babblers are almost all weak flyers, and most are forest dependent, so their spread across even fairly minor barriers probably depends on human assistance. Most of the doves, bulbuls and mynas, in contrast, are relatively strong flyers. Outside these major groups, the Tree Sparrow is the most widespread and abundant invasive bird species in the region, and is still extending its range, but its precise area of origin and the history of its spread are unclear.

41.3.3 Impacts

The serious impact of captures for the caged-bird trade on wild-bird populations in the region has been well documented (Harris *et al.*, 2017; BirdLife International, 2018), but the impact of invasive bird populations on wild species, ecosystems and people has received much less attention. Most evidence of impacts in Asia is anecdotal and/or very local (Table 41.1). On the basis of studies outside Asia, potential negative impacts include: predation, competition for nest sites (Charter *et al.*, 2016) and food (Le Louarn *et al.*, 2016), spread of invasive plant species (Martin-Albarracin *et al.*, 2018; Thibault *et al.*, 2018), acting as reservoirs of diseases of birds and people (Gutiérrez *et al.*, 2011; Fogell *et al.*, 2018), hybridization with native relatives (Robertson *et al.*, 2015), damage to crops (Brochier *et al.*, 2010), and noise and droppings, particularly from colonial roosting species.

In the Oriental region, predation by the invasive Indian Peafowl (*Pavo cristatus*) may have contributed to the decline of an endemic lizard, *Plestiodon kishinouyei*, in the southern Ryukyu Islands, and competition for nest sites with native species has been suspected for invasive barbets, parrots, mynas and other hole-nesting species. Competition for food, nesting space and other resources between other invasives and related native species is widely suspected, but with only anecdotal support. The potential for transmission of avian influenza virus by captive and released birds has been highlighted in multiple studies (Chan, 2004; Brooks-Moizer *et al.*, 2009, Ellis *et al.*, 2009; Edmunds *et al.*, 2011; Gutiérrez *et al.*, 2011; Gilbert *et al.*, 2012), but there is no evidence that this has actually occurred. Health surveys of captive birds reveal a diversity of other diseases, some of which may be acquired in captivity and transmitted to wild birds after escape or release (Gilbert *et al.*, 2012; Sa-ardta *et al.*, 2019), although again there is no evidence that this has happened. Hybridization with closely related native species or races has been documented in Japan in geese, pheasants and stilts (*Himantopus* spp.), and in Taiwan in hwameis (*Garrulax canorus* and *G. taewanus*), bulbuls, the Chestnut Munia (*Lonchura atricapilla* ssp.), and locally elsewhere. In Singapore, the local population of the endangered Milky Stork (*Mycteria cinerea*), which originated from escaped birds, has suffered several generations of admixture with Painted Storks (*Mycteria leucocephala*) and hybrids that have escaped from captive collections (Baveja *et al.*, 2019), while in Thailand, hybridization is known to have taken place between the native near-threatened Black-headed Ibis (*Threskiornis melanocephalus*) and the increasing population of escaped African Sacred Ibises (*Threskiornis aethiopicus*) (P.D. Round, personal observation). Urban nuisance species include Rock Doves, House Crows and several species of mynas.

Invasive species may also have positive impacts where they substitute for extinct native species as pollinators (Aslan *et al.*, 2014) or seed-dispersal agents. In Singapore, for example, where several native barbet species are now extirpated, the introduced Lineated Barbet (*Psilopogon lineatus*) is now the dominant barbet species in shrubland. Similarly, in Hong Kong, most large-gaped native frugivorous bird species have been extirpated, but the invasive Greater Necklaced Laughingthrush (*Pterorhinus pectoralis*) has colonized most areas of forest and shrubland since the 1970s and may provide a replacement dispersal agent for large-fruited (less than 2 cm diameter) plant species (Corlett, 2011). As with several other species that have become established in Hong Kong since the 19th century, the Greater Necklaced Laughingthrush was probably present there before deforestation, and could thus be viewed as an informal reintroduction, but relying on the vagaries of the bird trade and Buddhist release practices to rebuild depleted native avifaunas is not a sensible conservation strategy.

41.4 Discussion

The invasive avifauna of Asia – and of the Oriental region, which is the focus of this chapter – is characterized by its diversity (over 100 species), the dominance of species native to the region and the significant representation of forest-dependent species. These characteristics, in turn, reflect the influence of the massive regional bird trade – mostly within individual

Table 41.1. Summary of invasive birds established in Asia. Taxomomic sequence and scientific and English names follow Gill and Donsker (2019).

Species name	Common name	Invaded region	Origin (region)	Date established	Source/ reason	Population size	Current trend	Habitats	Interactions and effects	Control	Sources
Casuarius casuarius	Southern Cassowary	Seram, Indonesia	New Guinea, Northern Australia	Prehistory	Food	?	?	Forest	Cassowaries are important dispersers of seeds	Hunted for food	Eaton *et al.* (2016)
Branta canadensis[a]	Canada Goose	Honshu, Japan; Beijing, China	North America, Western Europe	20th century	Deliberate release for ornamental purposes?	?	Increasing and spreading in Japan; status in Beijing uncertain	Wetlands including city parks	Hybridization with native populations of Cackling Goose (*Branta hutchinsii*)	None?	Brazil (2009)
Cygnus olor[a]	Mute Swan	Hokkaido and Honshu, Japan	Europe, Middle East, Central Asia, east to eastern China	20th century	Deliberate release for ornamental purposes?	?	Increasing and slow spread?	Wetlands	?	None?	Brazil (2009)
Alopochen aegyptiaca[a]	Egyptian Goose	Honshu, Japan	Sub-Saharan Africa	?	Deliberate release for ornamental purposes?	?	?	Wetlands	?	None	Kawakami and Kanouchi (2012)
Aix sponsa[a]	American Wood Duck	Honshu, Japan	Eastern North America	1900	Deliberate release for ornamental purposes?	?	?	Wetlands	?	None	Kawakami and Kanouchi (2012)
Colinus virginianus[a]	Northern Bobwhite	Shikoku and Honshu, Japan	Eastern North America, Central America	20th century	Sport (hunting)	?	?	Forest, shrubland, riparian areas	Competition with native pheasants	None	Kawakami and Kanouchi (2012)
Bambusicola sonorivox[a]	Taiwan Bamboo Partridge	Honshu, Japan	Taiwan	1931; 20th century	Sport (hunting)	Localized	?	Forest edge, shrubland, farmland	?	None	Kawakami and Kanouchi (2012)
Bambusicola thoracicus[a]	Chinese Bamboo Partridge	Kyushu, Shikoku and Honshu, Japan	Southern China, mainland South-east Asia	1915; early 20th century	Sport (hunting)	?	?	Forest, shrubland	Competition with native birds (linked to the decline of Izu Thrush (*Turdus celaenops*) on Mikurajima)	None (hunted for food)	Eguchi and Amano (2004)

Gallus gallus	Red Junglefowl	Prehistoric introduction of birds resembling ancestral wild form to Indonesian Archipelago east of Bali and perhaps Philippines; domestic chickens in eastern India, Myanmar, Thailand, Malaysia, Indonesia and Philippines	Indian subcontinent, South-east Asia	Prehistory for wild population; ongoing for domestic chickens	Food, free-ranging domestic and feral chickens	Unknown	Decreasing	Forest, shrubland	Hybridization with domestic chickens is considered to be a threat to the genetic integrity of wild birds throughout the range of the species	None	Dickinson *et al.* (1991); Eaton *et al.* (2016)
Phasianus colchicus[a]	Common Pheasant	Japan	Widely across Eurasia, east to China	*c.*1920	Sport (hunting)	?	?	Forest, farmland	Hybridization with native endemic Green Pheasant (*P. versicolor*)	Releases ended when hybridization issue identified, except in Hokkaido where Green Pheasants are absent	Eguchi and Amano (2004); Brazil (2009)
Syrmaticus soemmeringii[a]	Copper Pheasant	Hokkaido, Japan	Honshu, Japan	20th century	Sport (hunting)	?	?	Forest, farmland	?	?	Kawakami and Kanouchi (2012)
Pavo cristatus[b]	Indian Peafowl	Ryukyu Islands and Honshu, Japan; Singapore	Indian subcontinent	20th century	Deliberate release for ornamental purposes?	?	?	?	Modification of plant community; predation of vulnerable endemic lizard (*Plestiodon kishinouyei*) in the southern Ryukyu Islands	None	Eguchi and Amano (2004); Brazil (2009); Lim (2009); Kawakami and Kanouchi (2012)

Continued

Table 41.1. Continued.

Species name	Common name	Invaded region	Origin (region)	Date established	Source / reason	Population size	Current trend	Habitats	Interactions and effects	Control	Sources
Threskiornis aethiopicus	Sacred Ibis	Taiwan; Thailand	Sub-Saharan Africa, Middle East	20th century	Accidental release from collection; free-flying birds from collections in Thailand	300–350 pairs in Taiwan (2009)	Increasing and spreading in Taiwan; increasing in Thailand	Wetland	Hybridization with native Black-headed Ibis (*T. melanocephalus*) confirmed in Thailand where listed as invasive; predation of nests of waterbirds a problem in Europe	None?	Brazil (2009); Bird Conservation Society of Thailand (2018); P.D. Round (pers. obs.)
Mycteria leucocephala[b]	Painted Stork	Singapore; Peninsular Malaysia; Japan	Indian subcontinent, South-east Asia	20th century	Accidental release from collection	?	Localized, but increasing (breeding in Singapore)	Wetland	Hybridization with Milky Stork (see below)	None, but recommended	Lim (2009); Baveja *et al.* (2019)
Mycteria cinerea	Milky Stork	Singapore; Peninsula Malaysia	South-east Asia	20th century	Accidental release from collection	?	Localized, but increasing in Singapore, Johor Bahru, Klang Valley area (also breeding in Singapore)	Wetland	Hybridization with introduced Painted Stork in Singapore and Malay Peninsula	None, but recommended	Lim (2009); Baveja *et al.* (2019)
Vanellus miles	Masked Lapwing	Singapore	New Guinea, Australia	1994	Accidental release from collection	?	Localized but increasing (with breeding observed)	Parkland and open grass (golf courses)	Potentially competing with Red-wattled Lapwing (*Vanellus indicus*)	None	Lim (2009)
Himantopus mexicanus[a]	Black-necked Stilt	Honshu, Japan	Eastern North America, south to South America	2001	Deliberate release from collection	?	?	Wetlands	Hybridization with Black-winged Stilt (*Himantopus himantopus*)	None	Kawakami and Kanouchi (2012)

Columba livia[b]	Rock Dove	Urban areas throughout region	Widely across Eurasia, east to China	13th century in Japan	Caged birds, food?	Common but localized	Stable	Urban	Urban nuisance; often considered probable disease vector for (e.g.) avian influenza but no firm evidence?	Passive deterrence in some urban centres	Carey *et al.* (2001); Gibbs *et al.* (2001)
Ducula bicolor	Pied Imperial Pigeon	Singapore	South-east Asia, east to New Guinea	20th century	Accidental release from collection	Localized	Increasing; spreading	Parkland, shrubland	?	None	Yong *et al.* (2017)
Streptopelia decaocto[b]	Eurasian Collared Dove	Honshu, Japan; Hong Kong	Widely across Eurasia, east to China	Late 20th century	Caged birds; escaped pets	Localized in Japan; ~1000 in Hong Kong	Localized population in Hong Kong slowly spreading/ increasing	Suburbs, villages, fish ponds	Decrease in Spotted Dove (*Spilopelia chinensis*) observed in Hong Kong where *Streptopelia decaocto* abundant	None	Brazil (2009); M.R. Leven (pers. obs.)
Streptopelia tranquebarica	Red Turtle Dove	Sulawesi, Indonesia; Peninsular Malaysia; Singapore	Indian subcontinent, South-east Asia, eastern China	Late 20th century?	Caged birds	Common in Sulawesi	Increasing; spreading	Farmland?	?	None	Robson (2002); Eaton *et al.* (2016)
Spilopelia chinensis	Spotted Dove	Philippines; Sulawesi and Moluccas, Indonesia	Indian subcontinent, South-east Asia, eastern China	20th century	Caged birds	?	Probably native in Palawan, but introduced on other Philippine islands, where increasing and spreading	?	?	None	Dickinson *et al.* (1991); Eaton *et al.* (2016)

Continued

Table 41.1. Continued.

Species name	Common name	Invaded region	Origin (region)	Date established	Source / reason	Population size	Current trend	Habitats	Interactions and effects	Control	Sources
Geopelia striata	Zebra Dove	Thailand; Laos; Philippines; Borneo; Sulawesi and Moluccas, Indonesia	South-east Asia	20th century	Caged birds	Common in some areas	Increasing, spreading	Urban areas, farmland	?	None	Dickinson *et al.* (1991); Robson (2002); Eaton *et al.* (2016)
Geopelia maugeus	Barred Dove	Tukangbesi, Indonesia	South-east Asia; Lesser Sundas	20th century?	Caged birds	?	?	?	?	None	Eaton *et al.* (2016)
Aerodramus fuciphagus	Edible-nest Swiftlet	Thailand; Malaysia; South-west Sulawesi and Manado, Indonesia	South-east Asia	20th/21st century	Food (nests)	Populations within natural range of species but source populations for 'swift house' artificial colonies often not local	Increasing	Urban areas, farmland	Loss of genetic integrity of wild populations due to translocations	None	Eaton *et al.* (2016)
Psilopogon lineatus	Lineated Barbet	Singapore	South-east Asia, to northern Malay Peninsula	21st century	Caged birds	Widespread	Increasing	Forest, shrubland	Presumed to compete with native *Psilopogon* spp., especially for nest holes	None	Yong *et al.* (2017)
Psilopogon pyrolophus	Fire-tufted Barbet	West Java, Indonesia	Sumatra, Malay Peninsula	Late 20th century	Caged birds	Small, localized	Increasing		Presumed nest site and other competition, especially with native *Psilopogon* spp.	None	Eaton *et al.* (2016); R. Hutchinson (pers. com.)
Probosciger aterrimus	Palm Cockatoo	Kai Islands, Indonesia	New Guinea, northern Australia	?	Caged birds	?	?	?	?	None	Eaton *et al.* (2016)

Cacatua goffiniana	Tanimbar Corella	Taiwan; Singapore; Kai Islands, Indonesia	Wallacea: Tanimbar Islands	20th century in Taiwan and Singapore, unknown for Kai Islands	Caged birds	Localized in Taiwan and Singapore	?	?	?	None	Brazil (2009); Eaton *et al.* (2016)
Cacatua galerita	Sulphur-crested Cockatoo	Taiwan; Kai Islands, Indonesia; Singapore	New Guinea, Australia	20th century in Taiwan, unknown in Kai Islands?	Caged birds	Localized in Taiwan and Singapore	?	?	Presumed nest site competition	None	Brazil (2009); Lim (2009); Eaton *et al.* (2016)
Cacatua sulphurea	Yellow-crested Cockatoo	Hong Kong; Singapore	Across Wallacea, especially Lesser Sundas	1961	Caged birds	~100–200 in Hong Kong; <30 in Singapore	Stable or slow decrease in Hong Kong; rapid decrease in Singapore	Urban	Tree damage in urban parks; presumed nest site competition	Nest hole blocking, no feeding in parks	Carey *et al.* (2001); Leven and Corlett (2004); Yong *et al.* (2017); Allcock *et al.* (2018); M.R. Leven (pers. obs.)
Cacatua alba	White Cockatoo	Taiwan, Obi and Morotai, Indonesia	Wallacea: Halmahera	20th century	Caged birds	Localized	?	?	?	None	Brazil (2009); C. Bocos (pers. com. to J.A. Eaton from pers. obs.)
Psittacula alexandri	Red-breasted Parakeet	Tokyo, Japan; Mumbai, India; Singapore	Himalayas, South-east Asia	20th century	Caged birds	Localized; widespread across Singapore	Increasing in Singapore?	Urban	Nest site competition with native parrots	None?	Brazil (2009); Lim (2009); Yong *et al.* (2017)
Psittacula eupatria[b]	Alexandrine Parakeet	Honshu, Japan; Hong Kong; Mumbai, India	Indian subcontinent, South-east Asia	20th century	Caged birds	Localized in Japan; localized and very small population in Hong Kong (~20)	Slow increase in Hong Kong but not definitely self-sustaining	Urban, farmland	Presumed nest site competition	None in Hong Kong; unknown in Japan	Brazil (2009); M.R. Leven (pers. obs.)

Continued

Table 41.1. Continued.

Species name	Common name	Invaded region	Origin (region)	Date established	Source/ reason	Population size	Current trend	Habitats	Interactions and effects	Control	Sources
Psittacula krameri[b]	Ring-necked Parakeet	Japan; Hong Kong; Taiwan; Thailand; Singapore	Sub-Saharan Africa, east to Indian subcontinent	20th century	Caged birds	Populations small and localized?	Hong Kong population much declined from peak of >200 to ~20 birds; increasing in Thailand and Singapore	Urban	Presumed nest site competition; listed as invasive in Thailand	Passive (no feeding in parks) in Hong Kong	Carey *et al.* (2001); Robson (2002); Brazil (2009); Lim (2009); P.D. Round (pers. obs.)
Tanygnathus lucionensis	Blue-naped Parrot	Kota Kinabalu, Malaysia	Philippines	20th century	Caged birds	Populations small and localized	Stable?	Shrubland, parkland	?	None	Sheldon *et al.* (2001)
Melopsittacus undulatus[b]	Budgerigar	Honshu, Shikoku, and Kyushu, Japan; Taiwan	Australia	20th century	Caged birds	Frequent observa-tions but uncertain if any estab-lished popula-tions	?	?	?	None?	Eguchi and Amano (2004); Brazil (2009)
Trichoglossus haematodus	Coconut Lorikeet	Singapore	Wallacea, east to New Guinea and western Pacific Islands	21st century	Caged birds	Populations localized, but wide-spread	Stable	Urban and woodland	Presumed nest site competition	None	Lim (2009)
Eos bornea	Red Lory	Taiwan; Singapore	Wallacea: Moluccas	20th century	Caged birds	Localized	?	Urban	Presumed nest site competition	None	Eaton *et al.* (2016); D.L. Yong (pers. obs.)
Myiopsitta monachus[b]	Monk Parakeet	Singapore; Honshu, Japan; South Korea	Southern and central South America	20th century	Caged birds	Localized	?	Urban and woodland		None	Kawakami and Kanouchi (2012)
Pachycephala nudigula	Bare-throated Whistler	Lombok, Indonesia	?	20th/21st century	Caged birds	?	?	Forest	?	None	J.A. Eaton (pers. obs.)

Cyanopica cyanus	Azure-winged Magpie	Hong Kong	Eastern China, Mongolia, south-eastern Russia	2003	Caged birds; merit release	Localized (500–1000)	Increase and slow spread	Suburbs, Village, fish ponds	Possible parasitism by Asian Koel (*Eudynamys scolopaceus*)	None	Allcock *et al.* (2018); M.R. Leven (pers. obs.)
Urocissa caerulea[a]	Taiwan Blue Magpie	Honshu, Japan	Taiwan	Before 1976	Caged birds	Localized	?	Woodland, shrubland	?	None	Kawakami and Kanouchi (2012)
Urocissa erythrorhyncha	Red-billed Blue Magpie	Taiwan; Singapore	Himalayas, South-east Asia, southern China	20th/21st century	Caged birds, accidental release from collection	Localized (breeding in Singapore)	?	Forest edge, shrubland, farmland	Potential competition and even hybridization with endemic native Taiwan Blue Magpie	None?	Lim (2009)
Cissa hypoleuca	Indochinese Green Magpie	South-east Asia	Hong Kong	21st century	Caged birds (merit release)	Localized (<100)	Recently established; increasing	Forest	?	None	Allcock *et al.* (2018)
Pica sericea[a]	Oriental Magpie	Kyushu, Japan	China, Korea, south to northern South-east Asia	16th century	Deliberately introduced, religious reasons?	Common but has not spread to other islands?	?		?	None	Eguchi and Amano (2004); Brazil (2009)
Corvus splendens	House Crow	Hong Kong; Phuket and Chonburi, Thailand; Peninsular and Kota Kinabalu, Malaysia; Singapore; Batam Island and Sumatra (Medan), Indonesia; Andamans	Indian subcontinent, east to Myanmar	Introduced *c.*1903 to Peninsular Malaysia; Singapore since *c.*1948; Hong Kong since 1990s; late 20th century in Indonesia	Deliberately introduced Malaysia to control caterpillars; presumed that this population spread to Singapore; presumed ship-assisted to Hong Kong and Indonesia, perhaps Thailand	Abundant but localized in Peninsular Malaysia; small in Singapore; <100 in Hong Kong	In Hong Kong: sharp increase to ~250 in early 21st century; decrease with control to <100	Urban	Urban 'nuisance'. Dead wild individual found in Hong Kong proven to be avian influenza (H5N1) positive. Presumed predation on native species. Frequent host species for Asian Koel (*Eudynamys scolopaceus*) in Malaysia	Poisoning of adults, and nest destruction in Hong Kong; shooting, and trapping in Singapore and Malaysia	Agriculture, Fisheries and Conservation Department (2007); Wells (2007); Eaton *et al.* (2016); P.D. Round (pers. obs.)

Continued

Table 41.1. Continued.

Species name	Common name	Invaded region	Origin (region)	Date established	Source / reason	Population size	Current trend	Habitats	Interactions and effects	Control	Sources
Corvus levaillantii	Eastern Jungle Crow	Andamans	South-east Asia	?	?	?	?	?	?	?	J.A. Eaton (pers. obs.)
Rhipidura dryas	Arafura Fantail	Lombok, Indonesia	Eastern Lesser Sundas to Australia	20th/21st century	Caged birds; escaped pets	?	?	?	?	?	J.A. Eaton (pers. obs.)
Machlolophus spilonotus	Yellow-cheeked Tit	Hong Kong	Himalayas, South-east Asia, to southern and central China	1988 (but probably present before deforest-ation)	Caged birds; escaped pets	<100; localized	Stable	Natural forest, plant-ations	None known; possible competition with other small cavity nesters	None	Leven and Corlett (2004)
Pycnonotus jocosus[b]	Red-whiskered Bulbul	Japan; Taiwan; distribution in China and South-east Asia may be partly a result of 19th century introductions	Indian subcontinent, South-east Asia, to eastern China	20th century in Japan and Taiwan	Caged birds; escaped pets	Localized	?		Important seed disperser	None	Leven and Corlett (2004); Brazil (2009)
Pycnonotus sinensis	Light-vented Bulbul	Eastern Taiwan; Okinawa, Japan	Northern South-east Asia, eastern and southern China	20th/21st century century	Caged birds; escaped pets	Localized	Increased incidence of mass releases	Forest, shrubland, farmland, urban areas	Mass releases in south-east Taiwan within range of endemic native Taiwan Bulbul (*P. taivanus*), threaten the latter species through hybridization	Passive: education effort to reduce caged-bird release in range of *P. taiva-nus*?	Brazil (2009)
Pycnonotus aurigaster	Sooty-headed Bulbul	Sumatra, Borneo, Sulawesi, Lesser Sundas, and Moluccas, Indonesia; Singapore	Southern China, South-east Asia	20th century	Caged birds; escaped pets	Generally common in suitable habitats	Continuing increase and spread	Farmland and other anthropo-genic habitats	?	None	Eaton *et al.* (2016)
Pycnonotus bimaculatus	Orange-spotted Bulbul	Lombok, Indonesia	Sumatra, Java	201th/21st century	Caged birds; escaped pets	?	?	Forest	?	None	J.A. Eaton (pers. obs.)

Pycnonotus goiavier	Yellow-vented Bulbul	Sulawesi	South-east Asia, east to Philippines	20th century	Caged birds; escaped pets	Generally common in suitable habitats	Continuing increase and spread	Farmland and other anthropogenic habitats	?	None	Eaton *et al.* (2016)
Pycnonotus flaviventris	Black-crested Bulbul	Singapore	Himalayas, South-east Asia	20th century	Caged birds; escaped pets	Small and localized	Decreasing	Forest		None	Yong *et al.* (2017)
Prinia familiaris	Bar-winged Prinia	Eastern Kalimantan, Indonesia	Sumatra, Java	20th/21st century	Caged birds; escaped pets	?	?	?	?	?	J.A. Eaton (pers. obs.)
Prinia inornata	Plain Prinia	Bali, Indonesia	South and East Asia	20th/21st century	Caged birds; escaped pets	?	?	?	?	?	J.A. Eaton (pers. obs.)
Orthotomus sepium	Olive-backed Tailorbird	Sumba, Indonesia	Java, Bali	20th/21st century	Caged birds	Small, recently discovered localized population considered to be probably introduced but study required	Not known	Shrubland, forest edge	?	None	Eaton *et al.* (2016)
Pomatorhinus montanus	Chestnut-backed Scimitar Babbler	Lombok, Indonesia	Sumatra, Java	20th/21st century	Caged birds; escaped pets	?	?	?	?	?	J.A. Eaton (pers. obs.)
Pomatorhinus ruficollis	Streak-breasted Scimitar Babbler	Hong Kong	Himalayas, northern South-east Asia, central and southern China	1986 (but probably present before deforestation)	Caged birds (merit release)	Widespread, 500–1000	Stable or slow increase and spread	Natural forest, plantation, shrubland	Seed disperser	None	Leven and Corlett (2004)
Stachyridopsis ruficeps	Rufous-capped Babbler	Hong Kong	Himalayas, northern South-east Asia, central and southern China	1985 (but probably present before deforestation)	Caged birds (merit release)	Widespread, 1000–5000	Moderate increase and spread	Natural forest, plantation, shrubland	Frequent flock member	None	Leven and Corlett (2004)
Alcippe hueti	Huet's Fulvetta	Hong Kong	Southern south-east China	1992 (but probably present before deforestation)	Caged birds (merit release)	Localized, ~100–200	Slow increase and spread	Natural forest, plantations	Frequent flock member	None	Allcock *et al.* (2018)

Continued

Table 41.1. Continued.

Species name	Common name	Invaded region	Origin (region)	Date established	Source / reason	Population size	Current trend	Habitats	Interactions and effects	Control	Sources
Minla cyanouroptera	Blue-winged Minla	Hong Kong	Himalayas, South-east Asia, central and southern China	1992	Caged birds (merit release)	Localized, 500–1000	Marked increase in 1990s, now stable?	Forest, shrubland	Frequent flock member	None	Leven and Corlett (2004)
Leiothrix lutea[b]	Red-billed Leiothrix	Honshu, Kyushu, and Shikoku, Japan; Hong Kong	Himalayas, South-east Asia, central and southern China	20th century (but probably present before deforestation)	Caged birds; escaped pets	Widespread and common in Japan; localized in Hong Kong, 100–500	Increase and spread in Japan; decline since 1990s in Hong Kong, now stable but doubtfully self-sustaining	Forest, shrubland, urban parks	Competes with other native birds in Japan; see *L. argentauris*	Import, transport and keeping are prohibited by the Invasive Alien Species Act in Japan	Leven and Corlett (2004)
Leiothrix argentauris	Silver-eared Mesia	Hong Kong	Himalayas, South-east Asia, southern and south-east China	1987	Caged birds; escaped pets	Localized, 500–1000	Marked increase in 1990s, now stable?	Forest, shrubland	Possibly dominates *L. lutea* but only circumstantial evidence	None	Leven and Corlett (2004)
Garrulax canorus[b]	Chinese Hwamei	Japan; Taiwan; Singapore	Southern and central China, to northern South-east Asia	20th century; 1980s in Japan	Caged birds; escaped pets	Common in parts of Japanese range; localized in Singapore	Increasing; no trend information from Singapore	Forest, shrubland	Hybridization with native endemic Taiwan Hwamei (*G. taewanus*) in Taiwan, placing this species at risk; in Japan, competes with native species	Taiwan: education effort to reduce caged-bird release in range of *G. taewanus*; Japan: import, transport and keeping are prohibited by the Invasive Alien Species Act	Eguchi and Amano (2004); Brazil (2009); Lim (2009)

Garrulax leucolophus	White-crested Laughingthrush	Singapore; Peninsular Malaysia	Himalayas, South-east Asia, southern and south-east China	20th century	Caged birds; escaped pets	Widespread and common in Singapore; localized in Penang, Malaysia	Increasing	Forest, shrubland	Competition with native babbler species	None	Lim (2009); Wong (2014)
Garrulax perspicillatus[a]	Masked Laughingthrush	Honshu, Japan	Northern South-east Asia, southern and eastern, China	?	Caged birds; escaped pets	?	?	Forest, shrubland	Competition with native bird species	Import, transport and keeping are prohibited by the Invasive Alien Species Act	Kawakami and Kanouchi (2012)
Ianthocincla cineracea[a]	Moustached Laughingthrush	Shikoku, Japan	Himalayas, southern, south-east and central China	1998	Caged birds	Localized	?	Forest, shrubland		None?	Brazil (2009)
Pterorhinus chinensis	Black-throated Laughingthrush	Hong Kong; Taiwan; Singapore	South-east Asia, southern and south-east China	Early 20th century in Hong Kong (but probably present before deforestation); late 20th century in Taiwan; 20th century in Singapore	Caged birds; escaped pets	Common and widespread in Hong Kong; localized in Singapore	Stable or slow increase and spread in Hong Kong; increasing in Taiwan; unknown in Singapore	Forest, urban parks	Seed disperser in Hong Kong; possible interactions with other forest and shrubland laughingthrushes	None	Leven and Corlett (2004); Brazil (2009); Lim (2009)
Pterorhinus pectoralis	Greater Necklaced Laughingthrush	Hong Kong; Singapore	Himalayas, South-east Asia, southern and south-east China	1969 (but probably present before deforestation)	Caged birds; escaped pets	1000–5000 in Hong Kong; numbers unknown in Singapore	Steady increase and spread; localized in Singapore	Forest, shrubland	Seed disperser	None	Leven and Corlett (2004); Lim (2009)

Continued

Table 41.1. Continued.

Species name	Common name	Invaded region	Origin (region)	Date established	Source / reason	Population size	Current trend	Habitats	Interactions and effects	Control	Sources
Pterorhinus sannio[b]	White-browed Laughingthrush	Hong Kong; Honshu, Japan	South-east Asia, southern, central and south-east China	1976 (but probably present before deforestation); 1994 in Japan	Caged birds; escaped pets	<100 in Hong Kong; unknown in Japan, but localized	Slow decrease	Shrubland, farmland, village, formerly urban	Seed disperser; possible interactions with other farmland and shrubland *Garrulax* spp. (*G. canorus* and *G. perspicillatus*)	Import, transport and keeping are prohibited in Japan by the Invasive Alien Species Act	Leven and Corlett (2004)
Sinosuthora webbiana	Vinous-throated Parrotbill	Hong Kong	Northern South-east Asia, widely across China, to Korea and south-eastern Russia	1980	Caged birds (merit release)	<100	Stable or slow decrease	Montane shrubland and dwarf bamboo	None known	None	Leven and Corlett (2004)
Heleia crassirostris	Thick-billed Heleia	Lombok, Indonesia	Eastern Lesser Sundas	20th/21st century	Caged birds; escaped pets	?	?	?	?	None?	J.A. Eaton (pers. obs.)
Zosterops japonicus	Warbling (Japanese) White-eye	Singapore	Widely across East Asia, and northern South-east Asia	?	Caged birds; escaped pets	?	?	?	?	None?	
Sitta frontalis	Velvet-fronted Nuthatch	Hong Kong	Indian subcontinent, South-east Asia, to southern and south-eastern China	1989	Caged birds; escaped pets	Localized, 100–500	Stable or slow increase and spread	Forest, plantations, parkland	None known; possible competition with other small cavity nesters; cavity enlargement may benefit Hainan Blue Flycatcher (*Cyornis hainanus*)?	None	Leven and Corlett (2004)

Aplonis panayensis	Asian Glossy Starling	Taiwan	South-east Asia	20th/21st century	Caged birds; escaped pets	Locally numerous	?	Urban areas	Potential effects probably similar to other *Acridotheres* spp.	?	Brazil (2009)
Scissirostrum dubium	Grosbeak Starling	Java	Sulawesi	20th/21st century	Caged birds; escaped pets	Several estab-lished popula-tions	Increasing?	Forest	?	None	Eaton *et al.* (2016)
Acridotheres grandis	Great Myna	Thailand, Taiwan; Kuala Lumpur, Malaysia	South-east Asia, to northern Malay Peninsula	20th/21st century	Natural spread in anthropo-genic habitats; caged birds; escaped pets	Widespread in southern Thailand, localized elsewhere	Much increased in Thailand in the past 30 years	Urban areas	Potential effects probably similar to other *Acridotheres* spp.	?	Brazil (2009); Jeyara-jasingam (2012); P.D. Round (pers. obs.)
Acridotheres cristatellus[b]	Crested Myna	Beijing, China; Japan; Luzon and Negros, Philippines; Penang, Sarawak, and Sabah, Malaysia	South-east Asia, to southern, central and south-eastern China	20th century in China and Japan, *c.*1850 in Philippines	Caged birds; escaped pets	Native in south China, but intro-duced in Beijing; Tokyo, Japan	?	Urban, farmland	Potential effects probably similar to other *Acridotheres* spp.	?	Dickinson *et al.* (1991); Brazil (2009)
Acridotheres javanicus[b]	Javan Myna	Southern Thailand; Japan; Taiwan; Sabah, Malaysia; Singapore; Sumatra, Kalimantan, Timor, Sumba, Sulawesi, Flores, and Simeulue, Indonesia; Thailand	Java, Bali	20th century	Caged birds; escaped pets	Common in some areas, notably Taiwan and Singapore	Likely increase and spread in most areas; stable in Singapore with control measures	Urban, farmland	More success-ful in urban areas than *A. tristis* where both species are present; urban nuisance; presumed to predate nests of native species but no firm evidence in the region	Ongoing control in Singa-pore, largely by trapping; unknown else-where?	Brazi (2009); Eaton *et al.* (2016)
Acridotheres cinereus	Pale-bellied Myna	Sabah, Malaysia	South Sulawesi	20th/2st century	Caged birds; escaped pets	Localized	?	Farmland?	?	None?	Eaton *et al.* (2016)

Continued

Table 41.1. Continued.

Species name	Common name	Invaded region	Origin (region)	Date established	Source / reason	Population size	Current trend	Habitats	Interactions and effects	Control	Sources
Acridotheres fuscus	Jungle Myna	Thailand, Taiwan	Indian subcontinent, South-east Asia, southern China	20th/21st century	Natural spread in anthropogenic habitats; Caged birds; escaped pets	Localized	Marked recent decline in Thailand	Farmland	Declining in Thailand due to exclusion by other invasive *Acridotheres* spp.	?	Brazil (2009), P.D. Round (pers. obs.)
Acridotheres tristis[b]	Common Myna	Thailand; Peninsular Malaysia; Singapore; widespread records in Greater Sundas but status unclear; Andaman and Nicobar Islands; Hong Kong; Taiwan; perhaps elsewhere in eastern China; central Japan	Central Asia, across Indian subcontinent to mainland South-east Asia	Early 20th century in Southeast Asia; in Hong Kong since 1952; 20th century in north and east Asia	Probably natural spread aided by releases in South-east Asia, deliberate introduction to Hong Kong, Andaman and Nicobar Islands; presumed merit release in North and East Asia	Common to abundant in South-east Asia, Andaman and Nicobar Islands; locally common in Japan, Hong Kong, Taiwan and elsewhere in China	Stable?	Urban and suburban areas; farmland	Urban nuisance; excluding (also non-native) *A. fuscus* in Thailand; presumed to predate nests of native species but no firm evidence in the region	Ongoing control in Singapore, largely by trapping; no control measures in Hong Kong or elsewhere	Carey *et al.* (2001); Wells (2007); Eaton *et al.* (2016)
Acridotheres melanopterus	Black-winged Starling	Singapore	Java, Bali	Mid-20th century	Caged birds; escaped pets	Very localized, now probably extinct	Probably extinct	Shrubland	?	None	Lim (2009)
Acridotheres burmannicus	Vinous-breasted Starling	Southern Thailand	South-western China, Northern Thailand, Indochina	20th/21st century	Caged birds; escaped pets	Locally common	Increasing	Farmland	Potential effects probably similar to other *Acridotheres* spp.	None	P.D. Round (pers. obs.)

Gracupica contra	Pied Myna	Mumbai, India. Malay Peninsula	South and East Asia; Java (perhaps extinct)	20th century	Malay Peninsula in rubber plantation (presumably to combat pests); also caged birds, escaped pets	Common in Mumbai	?	Urban areas, farmland	?	None?	del Hoyo *et al*. (2009); Eaton *et al*. (2016); J.A. Eaton (pers. com.)
Passer domesticus[b]	House Sparrow	Eastern Russia (*P. d. domesticus*); Andaman Islands; Thailand; Cambodia; Laos; Vietnam (*P. d. indicus*); perhaps western China (*P. d. bactrianus*); Singapore	Widely across Eurasia	19th/ 20th/ 21st century	Northern population (*P. d. domesticus*) natural spread from western Russia; southern population (*P. d. indicus*) natural spread from Myanmar; deliberately introduced to Andaman Islands; perhaps some natural spread of migratory *P. d. bactrianus* in western China	Locally common	Increase and continuing spread; localized in Singapore	Urban and suburban areas; farmland	Displacing native Plain-backed Sparrow (*P. flaveolus*) in both villages and farmland	?	Summers-Smith (1988); Lim (2009); J.A. Eaton (pers. obs.); M.R. Leven (pers. obs.); P.D. Round (pers. obs.)
Passer montanus	Eurasian Tree Sparrow	Malaysia; Singapore; Indonesia; Philippines	Widely across Eurasia	16th–20th century	Largely natural spread, but perhaps also ship-assisted and merit releases	Common to abundant	Probably still spreading with expansion of agriculture	Urban and suburban areas; farmland	None known	?	Summers-Smith (1988); Dickinson *et al*. (1991)

Continued

Table 41.1. Continued.

Species name	Common name	Invaded region	Origin (region)	Date established	Source / reason	Population size	Current trend	Habitats	Interactions and effects	Control	Sources
Ploceus intermedius[b]	Lesser Masked Weaver	Honshu, Japan; Taiwan	East, Central and South Africa	20th/21st century	Caged birds (merit release)	?	?	Farmland, open woodland, parks	?	None?	Brazil (2009)
Ploceus jacksoni	Golden-backed Weaver	Singapore	East Africa	21st century	Caged birds (merit release)	Localized	?	Open woodland, grassland	Possibly competition with native weavers and munias	None	D.L. Yong (pers. obs.)
Ploceus hypoxanthus	Asian Golden Weaver	Thailand; Singapore, Bali	Mainland South-east Asia, Java	21st century	Caged birds (merit release), perhaps natural spread due to farming changes in Thailand	Localized	Increasing in Thailand and Bali, decreasing in Singa-pore	Open woodland, grassland	Possibly competition with native weavers and munias	None	J.A. Eaton (pers. obs.); P.D. Round (pers. obs.)
Ploceus manyar[b]	Streaked Weaver	Japan; Singa-pore; Kaliman-tan, Lombok, Sumbawa, and south-west Sulawesi, Indonesia	Indian subcon-tinent, mainland South-east Asia, Java	20th/21st century	Caged birds (merit release)	?	Decreasing in Singapore	Farmland, wetlands	?	None?	Eaton *et al.* (2016)
Ploceus phillippinus	Baya Weaver	Sabah, Malaysia; southwest Sulawesi, Lombok, and Sumbawa, Indonesia	Indian subcon-tinent, South-east Asia	20th/21st century	Caged birds (merit release)	?	?	Farmland, wetlands	?	None?	Eaton *et al.* (2016); Myers (2016); J.A. Eaton (pers. obs.)
Euplectes franciscanus[a]	Northern Red Bishop	Japan	Sub-Saharan Africa	20th/21st century	Caged birds (merit release)	Perhaps estab-lished?	?	Farmland, wetland?	?	None?	Brazil (2009)
Euplectes orix	Southern Red Bishop	Singapore	East and South Africa	21st century	Caged birds (merit release)	Localized	?	Grassland, shrubland	?	None	Yong *et al.* (2017)
Estrilda melpoda[b]	Orange-cheeked Waxbill	Honshu, Japan; Singapore	West and Central Africa	20th/21st century; recent decade in Singapore	Caged birds (merit release)	Localized?	?	Grassland, wetlands	?	None?	Brazil (2009)
Estrilda astrild	Common Waxbill	Taiwan; Singapore	Sub-Saharan Africa	20th/21st century; recent decade in Singapore	Caged birds (merit release)	Localized	Increasing in Singapore	Woodland edge, grassland	?	None?	Brazil (2009)

Amandava amandava[b]	Red Avadavat	Honshu, Kyushu, and Shikoku, Japan; Taiwan; Singapore; Sabah, Malaysia; Kalimantan and Sumatra, Indonesia	Indian subcontinent, South-east Asia	20th/21st century	Caged birds (merit release); escaped pets	Localized	Decreasing in Singapore	Grassland, farmland, wetland	?	None?	Brazil (2009); Myers (2016)
Euodice malabarica	Indian Silverbill	Taiwan	Indian subcontinent	20th/21st century	Caged birds (merit release)	?	?	Grassland, farmland	?	None?	Brazil (2009)
Lonchura striata	White-rumped Munia	Ryukyu Islands	Indian subcontinent, South-east Asia, southern, central and south-eastern China	20th/21st century	Uncertain if natural colonists and/or released birds	?	?	Shrubland, grassland, farmland	?	None?	Brazil (2009)
Lonchura leucogastroides	Javan Munia	Singapore; Johor, Malaysia; Sumatra and East Kalimantan, Indonesia	Java, Bali	1920s	Caged birds (merit release)	Localized	Decreasing	Shrubland, grassland, farmland	?	None	Yong *et al.* (2017); J.A. Eaton (pers. obs.)
Lonchura punctulata[a]	Scaly-breasted Munia	South Japan	Indian subcontinent, across South-east Asia east to Australia	20th/21st century (1990s)	Uncertain if natural colonists and/or released birds	?	?	Shrubland, grassland, farmland	?	None?	Brazil (2009)
Lonchura atricapilla[b]	Chestnut Munia	Japan; Taiwan; Ambon, Indonesia	Indian subcontinent, South-east Asia, southern China	20th/21st century	Caged birds (merit release)	?	?	Grassland, wetlands, farmland	Non-native race(s) introduced to Taiwan threaten genetic integrity of endemic native *L.a. formosana*	None?	Brazil (2009)
Lonchura malacca[a]	Tricoloured Munia	Honshu and Kyushu, Japan	Southern and south-western India	1968	Cagebird (merit release)/ escaped pets	?	?	Grassland, wetlands, farmland		None?	Kawakami and Kanouchi (2012)
Lonchura maja[a]	White-headed Munia	Honshu, Japan	South-east Asia	20th/21st century	Caged birds (merit release)	Localized	?	Wetlands, farmland	?	None?	Brazil (2009)

Continued

Table 41.1. Continued.

Species name	Common name	Invaded region	Origin (region)	Date established	Source / reason	Population size	Current trend	Habitats	Interactions and effects	Control	Sources
Lonchura oryzivora[b]	Java Sparrow	Japan; Shanghai; Taiwan; Philippines; Borneo; Sulawesi, Indonesia; Ipoh, Sabah, Malaysia	Java, Bali	20th/21st century	Caged birds (merit release)	?	Uncertain if some populations self-sustaining	Urban areas, farmland	?	None?	Kennedy *et al.* (2000); Brazil (2009); Eaton *et al.* (2016); Myers (2016)
Vidua paradisaea[a]	Eastern Paradise Whydah	Honshu, Japan	East and South Africa	20th/21st century	Caged birds (merit release)	Localized	?	Farmland, wetlands	?	None?	Brazil (2009)
Vidua macroura[b]	Pin-tailed Whydah	Singapore; Honshu, Japan	Sub-Saharan Africa	2oth/21st century	Caged birds (merit release)	Localized	Increasing	Shrubland	?	None?	Lim (2009)
Paroaria coronata[a]	Red-crested Cardinal	Honshu, Japan	Central and southern South America	20th century	Escaped pets	Localized		Shrubland, parks	?	None?	Kawakami and Kanouchi (2012)
Crithagra mozambica	Yellow-fronted Canary	Taiwan; Hong Kong	Sub-Saharan Africa	20th/21st century	Caged birds (merit release)	Localized	Uncertain if Hong Kong population self-sustaining	Farmland	?	None?	Brazil (2009); M.R. Leven (pers. obs.)

[a]All introduced populations are in Palearctic East Asia outside the Oriental region.
[b]Some introduced populations are in Palearctic East Asia outside the Oriental region.

countries – and the widespread occurrence of merit releases by Buddhists and others. As a consequence, the GISD is not a useful guide to invasive bird problems in the region.

The invasive birds highlighted in this chapter are, in some ways, just the visible tip of the iceberg. Bird lists and guide books for human-dominated parts of this region include as native many species that could not have survived in the primeval habitats that existed there before people. This 'shifted baseline' reflects the long-term, pervasive influence of anthropogenic pressures in Asia dating back millennia – in particular, the impacts of habitat destruction, agriculture and urbanization. Whether species that have expanded their ranges following deforestation without direct human assistance should be considered as invasive is a moot point, but it is often difficult or impossible to determine how much assistance (from the bird trade, shipping or other means) a particular species has received. In any case, without this ability of human-tolerant bird species to expand their ranges, most anthropogenic habitats in the region would have fewer birds and the most extreme (e.g. urban centres) might have none. Few people would welcome this.

Invasive birds in anthropogenic habitats can be pest or nuisance species and may also pose a potential threat to human health, but they are rarely a conservation issue. Threats to native species and ecosystems are most likely to arise where the ranges of invasive birds overlap with native species that are confined to natural and semi-natural habitats. Asia has few habitats that can be considered pristine, but the fragmented and more or less degraded semi-natural areas that remain are home to most of the region's native biodiversity. Most of these areas do not (yet) contain invasive bird species, but the exceptions – most strikingly in Hong Kong (Leven and Corlett, 2004; Kwok, 2007) and, to a lesser extent, Singapore and Taipei – suggest a potential for future problems. Invasive birds are not yet a significant conservation issue in the region, but they could become so in the future.

These currently localized threats could become globalized through the international bird trade, as has already happened with the Asian species on the GISD. We hope that this chapter will serve as a warning that Asia is not just a source of invasive open-country mynas and bulbuls, but also of attractive, forest-adapted species that present a potentially much greater conservation threat. Moreover, the release practices that can introduce expensive caged birds to suitable natural habitats are also found among the Asian diaspora in cities worldwide.

In Asia, the potential conservation threat from the release of birds pales in comparison with the immediate, well-documented threat from their capture in the wild (Eaton *et al.*, 2015; Harris *et al.*, 2017; BirdLife International, 2018). Controlling this would also reduce the threat from invasives by reducing the diversity and abundance of forest-adapted species in the regional trade. This needs to be the first priority. Moreover, most countries in the region have clearer, more enforceable laws on wild captures and the associated trade than they do on releases. Controlling merit releases is clearly a close second in terms of priority actions. Regional governments focus on economic development, and most have been reluctant to address issues associated with livelihoods and cultural practices that are seen as traditional. However, there are multiple reasons for stopping merit releases, including animal welfare and public health, as well as conservation concerns. Public education, as carried out currently in Hong Kong, Taiwan and Singapore, can be effective locally, but clear laws, consistently enforced, could go a long way in reinforcing this.

Control of established invasive bird species is a practical option for very few species and then only if started while the population is still small and localized (Table 41.1). Control measures for the House Crow in Hong Kong have prevented spread beyond a small part of the urban area, and control has also been carried out in Singapore and Malaysia. Common and Javan Mynas (*Acridotheres javanicus*) are the subject of ongoing control measures in Singapore.

41.5 References

Agriculture, Fisheries and Conservation Department (2007) Two birds tested positive for H5N1 virus. Government of the Hong Kong Special Administrative Region. Available at: www.afcd.gov.hk/english/publications/publications_press/pr676.html (accessed 19 March 2019).

Allcock, J.A., Diskin, D., Holmes, J., Lee, L. and Lewthwaite, R. (2018) Systematic List 2016. *Hong Kong Bird Report* 2016, 20–2017.

Aslan, A., Hart, P., Wu, J. and Aslan, C.E. (2014) Evaluating the qualitative effectiveness of a novel pollinator: a case study of two endemic Hawaiian plants. *Biotropica* 46, 732–739.

Baveja, P., Tang, Q., Lee, J.G.H. and Rheindt, F.E. (2019) Impact of genomic leakage on the conservation of the endangered Milky Stork. *Biological Conservation* 229, 59–66.

Bird Conservation Society of Thailand (2018) Revised checklist of Thai birds. BCST Bird Data and Reports. Available at: www.bcst.or.th/report-archives/ (accessed 19 March 2019).

BirdLife International (2018) *State of the World's Birds: Taking the Pulse of the Planet.* BirdLife International, Cambridge.

Brazil, M. (2009) *Birds of East Asia: China, Taiwan, Korea, Japan, and Russia.* Princeton University Press, Princeton, New Jersey.

Brochier, B., Vangeluwe, D. and van den Berg, T. (2010) Alien invasive birds. *Revue Scientifique et Technique-office International Des Epizooties* 29, 217–225.

Brooks-Moizer, F., Roberton, S.I., Edmunds, K. and Bell, D. (2009) Avian influenza H5N1 and the wild bird trade in Hanoi, Vietnam. *Ecology and Society* 14, 28.

Carey, G.J., Chalmers, M.L., Diskin, D.A., Kennerley, P.R., Leader, P.J., *et al.* (2001) *The Avifauna of Hong Kong.* Hong Kong Bird Watching Society, Hong Kong.

Cassey, P., Blackburn, T.M., Sol, D., Duncan, R.P. and Lockwood, J.L. (2004) Global patterns of introduction effort and establishment success in birds. *Proceedings of the Royal Society of London B: Biological Sciences* 271, S405–S408.

Chan, S. (2004) Yellow-breasted Bunting *Emberiza aureola*. *Birding ASIA* 1, 16–17.

Chan, S.W. (2006) Religious release of birds in Hong Kong. MPhil thesis, University of Hong Kong, Hong Kong.

Charter, M., Izhaki, I., Ben Mocha, Y. and Kark, S. (2016) Nest-site competition between invasive and native cavity nesting birds and its implication for conservation. *Journal of Environmental Management* 181, 129–134.

Chng, S., Eaton, J. and Shepherd, C.R. (2018) In the market for extinction: birds for sale at selected outlets in Sumatra. *TRAFFIC Bulletin* 30, 15–22.

Chng, S.C.L. and Eaton, J.A. (2016a) Snapshot of an on-going trade: an inventory of birds for sale in Chatuchak weekend market, Bangkok, Thailand. *Birding Asia* 25, 24–29.

Chng, S.C.L. and Eaton, J.A. (2016b) *In the Market for Extinction: Eastern and Central Java*. TRAFFIC, Petaling Jaya, Malaysia.

Chng, S.C.L., Eaton, J.A., Krishnasamy, K., Shepherd, C.R. and Nijman, V. (2015) *In the Market for Extinction: An Inventory of Jakarta's Bird Markets*. TRAFFIC, Petaling Jaya, Malaysia.

Corlett, R.T. (2011) Seed dispersal in Hong Kong, China: past, present and possible futures. *Integrative Zoology* 6, 97–109.

Dai, C.Y. and Zhang, C.L. (2017) The local bird trade and its conservation impacts in the city of Guiyang, Southwest China. *Regional Environmental Change* 17, 1763–1773.

del Hoyo, J., Elliott, A. and Christie, D.A. (eds) (2009) *Handbook of the Birds of the World*, Vol. 14: *Bush-shrikes to Old World Sparrows*. Lynx Edicions, Barcelona, Spain.

Dickinson, E.C., Kennedy, R.S. and Parkes, K.C. (1991) *Birds of the Philippines: An Annotated Checklist*. British Ornithologists' Union, Peterborough, UK.

Eaton, J.A., Leupen, B.T.C. and Krishnasamy, K. (2017a) *Songsters of Singapore: An Overview of the Bird Species in Singapore Pet Shops*. TRAFFIC, Petaling Jaya, Malaysia.

Eaton, J.A., Nguyen, M.D.T., Willemsen, M., Lee, J. and Chng, S.C.L. (2017b) *Caged in the City: An Inventory of Birds for Sale in Ha Noi and Ho Chi Minh City, Viet Nam*. TRAFFIC, Petaling Jaya, Malaysia.

Eaton, J.A., Shepherd, C.R., Rheindt, F.E., Harris, J.B.C., van Balen, S.B., *et al.* (2015) Trade-driven extinctions and near-extinctions of avian taxa in Sundaic Indonesia. *Forktail* 31, 1–12.

Eaton, J.A., van Balen, B., Brickle, N.W. and Rheindt, F.E. (2016) *Birds of the Indonesian Archipelago*. Greater Sundas and Wallacea. Lynx Edicions, Barcelona, Spain.

Edmunds, K., Roberton, S.I., Few, R., Mahood, S., Bui, P.L., *et al.* (2011) Investigating Vietnam's ornamental bird trade: implications for transmission of zoonoses. *Ecohealth* 8, 63–75.

Eguchi, K. and Amano, H.E. (2004) Spread of exotic birds in Japan. *Ornithological Science* 3, 3–11.

Ellis, T.M., Dyrting, K.C., Wong, C.W., Chadwick, B., Chan, C., *et al.* (2009) Analysis of H5N1 avian influenza infections from wild bird surveillance in Hong Kong from January 2006 to October 2007. *Avian Pathology* 38, 107–119.

Fogell, D.J., Martin, R.O., Bunbury, N., Lawson, B., Sells, J., *et al.* (2018) Trade and conservation implications of new beak and feather disease virus detection in native and introduced parrots. *Conservation Biology* 32, 1325–1335.

Gibbs, D., Barnes, E. and Cox, J. (2001) *Pigeons and Doves: A Guide to the Pigeons and Doves of the World*. Pica Press, Crowborough, UK.

Gibson, L. and Yong, D.L. (2017) Saving two birds with one stone: solving the quandary of introduced threatened species. *Frontiers in Ecology and the Environment* 15, 35–41.

Gilbert, M., Sokha, C., Joyner, P.H., Thomson, R.L. and Poole, C. (2012) Characterizing the trade of wild birds for merit release in Phnom Penh, Cambodia and associated risks to health and ecology. *Biological Conservation* 153, 10–16.

Gill, F. and Donsker, D. (eds) (2019) IOC World Bird List, version 9.2. Available at: www.worldbirdnames.org/ (accessed 11 November 2019).

Gutiérrez, R.A., Sorn, S., Nicholls, J.M. and Buchy, P. (2011) Eurasian Tree Sparrows, risk for H5N1 virus spread and human contamination through Buddhist ritual: an experimental approach. *PLoS One* 6, e28609.

Harris, J.B.C., Tingley, M.W., Hua, F.Y., Yong, D.L., Adeney, J.M., *et al.* (2017) Measuring the impact of the pet trade on Indonesian birds. *Conservation Biology* 31, 394–405.

IUCN (2019) The IUCN Red List of threatened species. Available at: www.iucnredlist.org (accessed 14 November 2019).

Jeyarajasingam, A. (2012) *A Field Guide to the Birds of Peninsular Malaysia and Singapore*. Oxford University Press, Oxford.

Kawakami, K. and Kanouchi, T. (2012) *The Handbook of Introduced Birds in Japan*. Bun-ichi Sogo Shuppan, Tokyo.

Kennedy, R.S., Gonzales, P.C., Dickinson, E.E., Miranda, H.C. and Fisher, T.H. (2000) *A Guide to the Birds of the Philippines*. Oxford University Press, Oxford.

Kwok, H.-K. (2007) Changes of a forest bird community in Hong Kong of China in 10 years. *Acta Ecologica Sinica* 27, 3993–4001.

Le Louarn, M., Couillens, B., Deschamps-Cottin, M. and Clergeau, P. (2016) Interference competition between an invasive parakeet and native bird species at feeding sites. *Journal of Ethology* 34, 291–298.

Leven, M.R. and Corlett, R.T. (2004) Invasive birds in Hong Kong, China. *Ornithological Science* 3, 43–55.

Lim, K.S. (2009) *The Avifauna of Singapore*. Nature Society (Singapore), Singapore.

Lindholm, J.H. (1995) Lories may be hazardous... (a cautionary tale). *AFA Watchbird* September/October, 22–27.

Martin-Albarracin, V.L., Nunez, M.A. and Amico, G.C. (2018) Non-redundancy in seed dispersal and germination by native and introduced frugivorous birds: implications of invasive bird impact on native plant communities. *Biodiversity and Conservation* 27, 3793–3806.

Myers, S. (2016) *Birds of Borneo: Sabah, Sarawak, Brunei and Kalimantan*. Christopher Helm, London.

Rentschlar, K.A., Miller, A.E., Lauck, K.S., Rodiansyah, M., Bobby, *et al.* (2018) A silent morning: the songbird trade in Kalimantan, Indonesia. *Tropical Conservation Science* 11, 1940082917753909.

Robertson, P.A., Adriaens, T., Caizergues, A., Cranswick, P.A., Devos, K., *et al.* (2015) Towards the European eradication of the North American ruddy duck. *Biological Invasions* 17, 9–12.

Robson, C. (2002) *A Field Guide to the Birds of South-east Asia*. New Holland, London.

Sa-ardta, P., Rinder, M., Sanyathitiseree, P., Weerakhun, S., Lertwatcharasarakul, P., *et al.* (2019) First detection and characterization of *Psittaciform* bornaviruses in naturally infected and diseased birds in Thailand. *Veterinary Microbiology* 230, 62–71.

Severinghaus, L.L. and Chi, L. (1999) Prayer animal release in Taiwan. *Biological Conservation* 89, 301–304.

Sheldon, F.H., Kennard, J. and Moyle, R. (2001) Ornithology of Sabah: History, Gazetteer, Annotated Checklist, and Bibliography. *Ornithological Monographs* 52, 1–285.

Shepherd, C.R., Eaton, J.A. and Chng, S.C.L. (2016) Nothing to laugh about – the ongoing illegal trade in laughingthrushes (*Garrulax* species) in the bird markets of Java, Indonesia. *Bird Conservation International* 26, 524–530.

Shiu, H. and Stokes, L. (2008) Buddhist animal release practices: historic, environmental, public health and economic concerns. *Contemporary Buddhism* 9, 181–196.

Souviron-Priego, L., Munoz, A.R., Olivero, J., Vargas, J.M. and Fa, J.E. (2018) The legal international wildlife trade favours invasive species establishment: the monk and ring-necked parakeets in Spain. *Ardeola* 65, 233–246.

Su, S., Cassey, P. and Blackburn, T.M. (2014) Patterns of non-randomness in the composition and characteristics of the Taiwanese bird trade. *Biological Invasions* 16, 2563–2575.

Su, S., Cassey, P. and Blackburn, T.M. (2016) The wildlife pet trade as a driver of introduction and establishment in alien birds in Taiwan. *Biological Invasions* 18, 215–229.

Summers-Smith, J.D. (1988) *The Sparrows: A Study of the Genus* Passer. T. and A.D. Poyser, Calton, UK.

Swinhoe, R. (1862) Letter to the editor. *Ibis* 4, 304–307.

Thibault, M., Masse, F., Pujapujane, A., Lannuzel, G., Bordez, L., *et al.* (2018) "Liaisons dangereuses": the invasive red-vented bulbul (*Pycnonotus cafer*), a disperser of exotic plant species in New Caledonia. *Ecology and Evolution* 8, 9259–9269.

Vall-Llosera, M. and Su, S. (2019) Trends and characteristics of imports of live CITES-listed bird species into Japan. *Ibis* 161, 12653.

Webster, M.A. (1975) *An Annotated Checklist of the Birds of Hong Kong*. Hong Kong Bird Watching Society, Hong Kong.

Wells, D.R. (2007) *The Birds of the Thai-Malay Peninsula*, Vol. 2: *Passerines*. Christopher Helm, London.

Winkler, D.W., Billerman, S.W. and Lovette, I.J. (2015) *Bird Families of the World: A Guide to the Spectacular Diversity of Birds*. Lynx Edicions, Barcelona, Spain.

Wong, S.H.F. (2014) The spread and relative abundance of the non-native White-crested Laughingthrush *Garrulax leucolophus* and Lineated Barbet *Megalaima lineata* in Singapore. *Forktail* 30, 90–95.

Yap, C.A. and Sodhi, N.S. (2004) Southeast Asian invasive birds: ecology, impact and management. *Ornithological Science* 3, 57–67.

Yong, D.L., Lim, K.C. and Lee, T.K. (2017) *A Naturalist's Guide to the Birds of Singapore*. John Beaufoy Publishing, Oxford.

42 Competition and Invasive Species Impacts on Native Communities

Andrew M. Rogers* and Salit Kark

The Biodiversity Research Group, The School of Biological Sciences, Centre for Biodiversity and Conservation Science, The University of Queensland, Brisbane, QLD 4072 Australia

Citation: Rogers, A.M. and Kark, S. (2020) Competition and invasive species impact on native communities. In: Downs, C.T. and Hart, L.A. (eds) *Invasive Birds: Global Trends and Impacts*. CAB International, Wallingford, UK, pp. 340–348.

42.1 Introduction

Competition between birds can influence species' access to resources, altering species persistence within an environment and ultimately community structure. However, predicting where and when invasive species significantly compete with native species can be challenging due to a lack of information on interspecies interactions around resources in different environments. The extent to which critical resources, such as habitat space, territories, nesting sites and food, vary across different environments changes levels of competition and makes the importance of invasive–native species competition context specific. In order to better understand how competition between invasive and native species impacts invaded communities, it is important to identify critical resources, the species interacting over the resource and the functional traits that influence interaction frequency and outcome. Such an approach will allow a more mechanistic understanding of competition at the community level and facilitate better predictions of invasive bird impacts.

42.2 Competition as a Key Factor

Competitive interactions within bird communities can influence species access to resources and ultimately structure community composition at landscape scales (Minot and Perrins, 1986; Alatalo and Moreno, 1987; Montague-Drake *et al.*, 2011; Farwell and Marzluff, 2013; Peck *et al.*, 2014). In global assessments of invasive bird impacts, competition between invasive and native birds is generally considered to be of little ecological importance relative to other impacts such as predation and disease transmission (Blackburn *et al.*, 2009). However, increasingly, studies have shown that invasive species can significantly impact native species through competitive interactions, especially when species are competing over resources critical to breeding such as nest sites and breeding territories (Ingold, 1998; Ghilain and Belisle, 2008; Brazill-Boast *et al.*, 2010; Yosef *et al.*, 2016). While invasive–native competition and aggression are often invoked as the mechanism driving patterns of reduced species richness or changes in breeding dynamics (e.g. nest site choice and breeding success), the mechanism of that competition is rarely quantified. Furthermore, much more work needs to be done on how competition is likely to vary in different landscapes or across different communities, as competitive interactions will often occur alongside other environmental changes, changing the relative importance of competitive impacts (Grarock *et al.* 2012).

Some of the best examples of significant competition in birds occur where species require the same critical resources (e.g. foraging space, breeding sites or territories) (Rusterholz, 1981; Remm *et al.*, 2008; Menchetti and Mori, 2014; Sanz-Aguilar *et al.*, 2015). However, identifying the form of the competitive impacts (e.g. direct aggression or more efficient resource exploitation; Griffin *et al.*, 2012) for each member of an invaded community requires data on interspecies interactions among all species that require that resource. These interactions are often hard to observe, as they may occur infrequently or be habitat specific (e.g. for cavity-nesting species, interactions often occur high in tree canopies around tree hollows) (Davis, 2003; Kéfi *et al.*, 2015). Novel use of technology and large spatial databases on species occurrence provide opportunities to explore competition and the patterns it creates in community structure. For example, Davis *et al.* (2013) used remote motion-activated camera traps to monitor birds and other species visiting nesting sites in tree hollows, demonstrating high levels of aggressive interactions around urban tree hollows and revealing the extent to which aggression is a driver of competition for a limited resource in woodland fragments in Sydney, Australia. At broader spatial scales, Cooper *et al.* (2007) found significant evidence of competition between introduced House Finches (*Haemorhous mexicanus*) and native House

*Corresponding author: a.munro.rogers@gmail.com

Sparrows (*Passer domesticus*) in the North American Christmas bird counts. As more landscape-scale databases on bird presence continue to grow with the aid of citizen science, more opportunities to explore individual species trends and test the importance of competition and its drivers at the species and community levels will be possible (Joyce *et al.*, 2018).

Recent work has shown the importance of quantifying invasive species impacts on whole communities (Hui *et al.*, 2016; Hui and Richardson, 2019). Where invasive–native interactions have been mapped at the community level, complex interaction graphical representations or webs have revealed both direct and indirect competitive interactions, with important implications for invasive species management (Orchan *et al.*, 2013; Goldshtein *et al.*, 2018). Describing the community-wide interaction network (where species are represented as nodes and connected by pairwise interactions or links between nodes; Fath *et al.*, 2007) is critical for managing invaded communities to avoid perverse outcomes and optimally achieve efforts to mitigate invasive species impacts, especially in communities with multiple invasive species (Bode *et al.*, 2015). For example, Orchan *et al.* (2013) found in the cavity breeding bird community (invaded by multiple invasive species) in a large urban park in Israel, that managing the invasive Common Myna (*Acridotheres tristis*) alone could lead to the competitive release of a second invasive species, the Venus-breasted Starling (*Acridotheres burmannicus*), which was more dominant over the native Syrian Woodpecker (*Dendrocopos syriacus*). In a study of interspecies fighting around bird feeders across North America, Miller *et al.* (2017) described the dominance hierarchy over food resources. This hierarchy revealed that, while the competitive dominance of species in general was predicted by larger body mass, some species were notable outliers that were more aggressively dominant than would be predicted by functional traits alone. In these studies, species interactions and functional traits were quantified at the community level and revealed more complex relationships between species than would have been found in studies of single species or species pairs.

In global assessments of invasive–native bird impacts, there is relatively little information on most invasive bird interactions with local species, and very few examples (relative to the number of introductions) of direct impacts of a species on native populations (Blackburn *et al.*, 2009; Baker *et al.*, 2014; Evans *et al.*, 2016). Where impacts do occur, they are often the result of competition. Impacts from predation are considered more significant but are less common (Blackburn *et al.*, 2009; Batalha *et al.*, 2013). The lack of data on interspecies interactions is a limiting factor in assessing the global importance of invasive–native competition and hampers predictions of where significant impacts are likely to occur. Invasive species do not always significantly compete with native species where resources are not limited, or where there are differences in habitat preference (e.g. native species avoiding heavily modified environments that are often favoured by introduced species), or when a species can shift resources use (even slight differences in food preferences or foraging strategy, allowing species to reduce levels of competition) (Griffin *et al.*, 2012; Batalha *et al.*, 2013). Even where competitive impacts have been identified for an invasive species in one part of its invasive range, it is often hard to generalize these impacts across their global invasive range. For example, the Common or European Starling (*Sturnus vulgaris*) is invasive and widespread in both North America and Australia (among other regions), yet a lack of research focus on the drivers of competition in North America and a lack of interaction data in Australia make generalizations about its impacts globally uncertain. The impacts of this species across North America (mostly competition for nesting sites with other cavity-nesting species) varies significantly based on the local community and the habitat type invaded (Koenig, 2003; Linz *et al.*, 2007). Similar variation is likely within and between continents, and more information on competitive interactions between starlings and native Australian bird species (as well as other cavity-dependent wildlife) from across its Australian range would provide a better context for understanding the potential impact across its global distribution.

Other factors that are likely to change interspecies interactions and therefore competitive impacts include invasive species abundance, environmental variation in resources, and the fluctuation in resource abundance or importance over time (Parker *et al.*, 1999; Fogarty *et al.*, 2011; Grarock *et al.*, 2013; Simberloff *et al.*, 2013). These factors are likely to be especially important for understanding when invasive impacts are likely, in turn allowing conservation managers to work in a more informed and targeted way rather than relying on the precautionary principle to justify management actions, such as invasive species control. Accounting for competitive interactions and the factors that influence them is especially important for the management of communities with multiple invasive species and where the impacts of management are uncertain (Bode *et al.*, 2015; Baker *et al.*, 2018). Therefore, in order to generate a more mechanistic understanding (incorporation of physiological ecology and ecomorphology into analysis of community ecology) (Schoener, 1986) of invasive bird competitive impacts, data must be collected on resource use, interacting species, the strength of those interactions, the functional traits related to resource use and the context in which the impacts occur. Such an approach will improve assessments of how invasive species change community-wide interactions and allow predictions of where and when invasive species are likely to significantly compete with native species (Dick *et al.*, 2014).

42.3 A Framework for Assessing Species Interactions

To help include interactions into invasion studies and management for birds, we present a simple framework that outlines the steps required to map interactions and their drivers (Fig. 42.1). Such an approach will improve information on invasive species impacts on individual species as well as on communities (Rusterholz, 1981; Romanuk *et al.*, 2009; Green *et al.*, 2018), and will allow improved predictions of invasive species impacts beyond the immediate study area usually covered by invasive species studies. In particular, it is important to: (i) identify resources; (ii) identify the actors interacting around resources and the functional traits of those actors; (iii) identify all the relationships between interacting species; (iv) map the direction, strength and

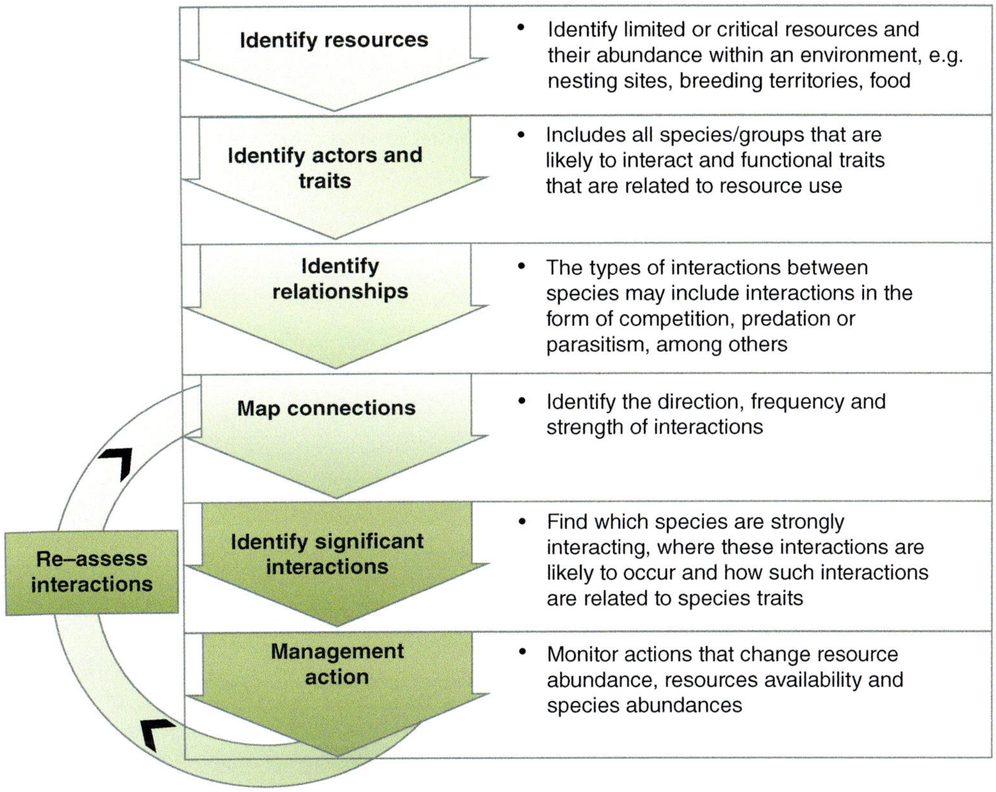

Fig. 42.1. By taking species interactions into account when evaluating the importance of competition, impacts on specific species can more easily be identified and targeted. This approach can reveal complex relationships at the community level that inform community dynamics and how species management should be undertaken.

frequency of the interactions, in multiple environmental contexts if possible; (v) identify the significant interactions within a given community; (vi) apply management actions; and finally (vii) reassess interactions between species to test how changes driven by management actions have altered species competitive relationships.

42.3.1 Identifying resources

Identifying critical resources such as nesting sites, breeding territories and foraging areas, around which species are likely to interact is one way to define a potentially interacting community (Dhondt, 2012). The habitat and environmental conditions in which the resources occur will provide important context, as variation in resource abundance and quality will change their importance for species (Le Roux *et al.*, 2016a). For example, in a study of nest box visitation by birds, Le Roux *et al.* (2016a) found that nest boxes on small trees were used at significantly lower rates compared with large trees, such that simply adding nest boxes to habitats is unlikely to increase nesting attempts by native birds. In similar studies for other nest box-using species, both the traits of the box and the surrounding habitats influenced whether the boxes were used by native species (Mänd *et al.*, 2005; Lindenmayer *et al.*, 2009; Le Roux *et al.*, 2016a,b).

42.3.2 Identifying the actors and traits

One of the primary challenges in assessing the importance of interactions in an invaded system is adequately surveying the species and their abundances within a community. Interactions should be more common between species that are more abundant, who co-occur, and who occupy more similar ecological niches. A species ecological niche includes its resource requirements and use, and can be influenced by the functional traits of the species. For example, the body size of cavity-nesting species reflects what size of nesting cavity they require, with larger species needing larger cavities in general. High overlap in the occupied niche space between species is often associated with high levels of competition (Rusterholz, 1981; Aderhold *et al.*, 2012; Green *et al.*, 2018; Reif *et al.*, 2018). Additionally, interacting species are likely to be a subset of the local species pool, creating challenges for identifying the limits of an interacting community, so efforts to quantify how completely a community has been sampled are important (Jordano, 2016).

42.3.3 Identify relationships

Species interacting around a resource will rarely, if ever, occur in isolation from the rest of the environment in which they occur. The presence of predators, parasites and other actors needs to be considered, as these can influence the risk associated with competing for a resource. Importantly, a single species may have multiple relationships with another one. For example, in cavity-nesting communities in Australia, cavity-using mammals both compete with birds for tree hollows and act as nest predators (Gibbons and Lindenmayer, 2002). Similarly, invasive birds such as the Common Myna have been known to destroy nests in some parts of its global range (Charter *et al.*, 2016), so understanding the conditions in which such nest predation occurs would be important when assessing the importance of competition for nest sites in tree hollows.

42.3.4 Map interactions between species

Mapping the interactions between species requires quantifying the interaction direction, frequency and variation in different environments. The direction of an interaction refers to the competitive impact of one species on another. For example, when fighting over food resources around bird feeders, direct aggressive interactions are observed between species pairs. For each interaction, the direction should be noted, where direction refers to the species that initiates or receives the interaction, or, if resource control is important, then the direction can reflect the species that most frequently wins the fight (i.e. the more dominant species).

42.3.5 Identify significant interactions

At this stage, it is important to test how species abundance, co-occurrence and interaction frequency within a specific context affect competition. The strength (frequency and intensity) of the interactions between two species can then be assessed relative to all the interactions among community members. Different interaction intensities can be quantified by the types of interactions; in the case of birds, physical fights could be considered a more intense interaction than alarm calls or threat displays. Interaction strength could also be quantified as how frequently one species is dominant over another, for example how many times a species disrupts the nesting attempts of another species (Edworthy, 2016) or how frequently a species excludes another species from a feeding resource (Peck *et al.*, 2014; Miller *et al.*, 2017). Both direct and indirect effects can be assessed at this point and used to inform management actions, such as which species or resource to prioritize or whether multi-species management will better mitigate invasive species impacts (Moon *et al.*, 2010; Bode *et al.*, 2015).

42.3.6 Manage the species

Activities that change resource abundance or availability, or alter species abundances, are likely to change competition over resources (Brazill-Boast *et al.*, 2013). Understanding the environmental or temporal contexts in which competitive impacts are the greatest between invasive and native species offers a more specific context in which to implement management actions. Furthermore, by incorporating the ecological niche of each species, slight differences in niche space can be exploited to reduce competition by excluding invasive species from accessing resources (Charter *et al.*, 2016; Goldshtein *et al.*, 2018).

42.3.7 Reassess interactions following management and changes in the system

When resources or species abundances have changed, interactions need to be reassessed. The strength of direct and indirect interactions will also change, with implications for follow-up management actions. After management, predictions on how interactions are likely to change (based on niche overlap, species abundance and resource use) with management can be tested by comparing levels of competition after management with pre-management conditions.

42.4 A Case Study for an Australian Cavity-dependent Species

Australia is home to some of the same invasive species that have been introduced in many other countries such as House Sparrows, Common Starlings and Common Mynas (see Chapter 36, this volume). Of these species, most research has focused on the Common Myna. The Common Myna has been shown to have significant impacts on native communities in other countries in which it is invasive, and the evidence of significant impacts on native Australian species varies depending on the habitat in which the myna occurs (although studies have only come from a small part of the invasive range in Australia). Competitive impacts of the Common Myna are especially interesting to explore in the Australian context because it is home to many cavity-nesting species, and cavities are limited resources in the modified habitats preferred by the Common Myna. Additionally, competition and aggression have a more significant role in structuring community composition in Australia compared with most places in the world.

Across the global invasive range of the Common Myna, impacts on native species include competition for habitat space, competition for nesting sites, and nest destruction and predation (see Chapter 3, this volume). Common Myna competition for habitat space has been invoked to explain declines in some bird species following the increases in myna abundance (Grarock *et al.* 2012) and increases in native bird abundance following myna eradication (Tindall *et al.*, 2007). In these studies, competition is often invoked to explain these patterns, but direct, aggressive interactions between mynas and native species are rarely quantified. When examining myna–native species competition for food in urban areas in Australia, Sol *et al.* (2012) and Haythorpe *et al.* (2014) found that Common Mynas were not more aggressive than native species around food resources, highlighting that the mechanism by which mynas impact other species use of foraging or habitat space is not well understood. Evidence of significant impacts on native species breeding is more substantial, such as aggressive interactions around nest sites and destruction of nests observed in Israel (Orchan *et al.*, 2013; Charter *et al.*, 2016) and from Pacific islands where the myna predates the nests of shorebirds (Byrd *et al.*, 1983). Despite significant impacts observed overseas, the evidence for significant impacts in Australia is mixed.

Like elsewhere, the most direct competitive impacts of the Common Myna in Australia come from studies of competition for nest sites, specifically when Common Mynas compete for nesting sites in less-disturbed habitats. In a study of nest site locations, Grarock *et al.* (2013) found that Common Mynas selected nest sites (in nest boxes) in more modified habitats, while native species selected nest sites in less modified habitats (areas with higher tree density), and concluded that the spatial segregation in nesting sites was a result of both habitat preferences and competition with Common Mynas. In a study of breeding success in nest boxes and tree hollows in open woodland, Pell and Tidemann (1997a) found that the Common Myna was the most aggressive cavity-nesting species compared to the invasive Common Starling and two native species. They concluded that this aggression could cause reduced breeding opportunities for the native species. While Common Mynas rarely nest deep into intact forests with high tree density (Pell and Tidemann, 1997b), impacts on nesting from competition are likely to be greatest where preferences for nest sites overlap. These areas include edge habitats (where urbanized and undisturbed environments meet), agricultural landscapes (particularly areas with a few large scattered trees) and native woodland with low tree density. Competition may become less important as levels of urbanization increase in areas characterized by suburban housing developments (where mynas can nest in buildings, under roofs and in rain gutters), as the effects of habitat change exclude native species more than competitive interactions.

Attempts to quantify the relative importance of additional competition from invasive species must also account for the high levels of aggression and competition among Australian native species (Sol *et al.*, 2012; Haythorpe *et al.*, 2014). Australian birds are known to be among some of the most aggressive birds globally (Low, 2014), with high levels of competition observed between species competing over foraging areas, breeding territories and nesting sites (Mac Nally and Timewell, 2005; Howes and Maron, 2009; Mac Nally *et al.*, 2012; Maron *et al.*, 2013). In particular, the native Noisy Miner (*Manorina melanocephala*) is a colony-nesting member of the honeyeater family, and so aggressively excludes other birds from its territories (in modified environments) that it is considered to impact bird communities at landscape scales (Montague-Drake *et al.*, 2011; Maron *et al.*, 2013). However, direct interactions responsible for these patterns are rarely quantified at the community level in Australia. The best examples of the importance of direct competitive interactions between species come from studies of cavity-nesting birds (Davis *et al.*, 2013), and include threatened native species such as the Forty-spotted Pardalote (*Pardalotus quadragintus*; Edworthy, 2016), Gouldian Finch (*Erythrura gouldiae*; Brazill-Boast *et al.*, 2010, 2013) and Palm Cockatoo (*Probosciger aterrimus*; Garnett *et al.*, 1999; Murphy *et al.*, 2003). Despite the importance of competition demonstrated for these species, and bird communities more generally, direct competitive interactions are not well understood for the cavity-nesting community at large, which includes 114 species of birds, and at least seven established invasive cavity-nesting species including the Common Myna.

In Australia, competition between Common Mynas and native bird species is likely to be greatest in moderately transformed environments; however, a lack of data on the relationship between functional traits (i.e. body size) and interspecific dominance prevents predictions of where the Common Myna will have an impact across its range. This, in turn, limits where managing the species would be most effective. While there are some efforts by local conservation groups to control the Common Myna through trapping, this approach is unlikely to reduce the population or reduce the impact of the Common Myna on native birds (see Chapter 3, this volume). Additionally, reducing Common Myna populations in urban areas may not improve nesting opportunities for native birds due to competition with other native urban-adapted species. In a study of urban cavity-nesting species in subtropical south-east Queensland, Rogers (2019) found a diverse community of cavity-dependent species using urban nest boxes, including the Common Myna (Fig. 42.2). While Common Mynas were found to be the most aggressive species around tree hollows, species recorded around nest boxes in the same region included several cavity-nesting birds, cavity-dependent mammals and other predators (Fig. 42.2). This work highlights that, even within the same region, interactions around different types of resources (tree hollows versus nest boxes) may be different. Additionally, the interactions between Common Mynas and native birds are occurring within a larger complex interaction web that includes predation and competition. In the case of common brushtail possums (*Trichosurus vulpecula*), this species is both a competitor for nest boxes and a nest predator of birds (Garnett *et al.*, 1999). While the interaction web described in Fig. 42.2 centres around nest boxes, similar processes are likely to be occurring around natural tree hollows but are poorly quantified. Nevertheless, like previous work, it appears that competitive interactions between Common Mynas and native species are context dependent. Competitive impacts are particularly high around natural tree hollows (Pell and Tidemann, 1997a), but native bird avoidance of nest boxes (perhaps due to interactions with native predators and native mammals) reduces the importance of competition around nest boxes.

Conservation efforts that aim to improve nesting opportunities for native cavity-nesting birds in Australia need to take into account all interactions among members of a community in order to identify which interactions may be limiting breeding opportunities (Heinsohn *et al.*, 2003; Murphy *et al.*, 2003; Brazill-Boast *et al.*, 2010; Stojanovic *et al.*, 2014). In the case of the Common Myna, competitive interactions will have a negative impact on native species in areas where Common Myna decide to nest in natural tree hollows or edge habitats. However, efforts to improve nesting opportunities for native species in modified environments in south-east Queensland through control of Common Myna populations alone are unlikely to generally succeed due to the high abundance of native competitors and predators. While nest boxes remain a popular conservation tool in Australia to increase the supply of available cavities, such projects are often of little use to species of conservation concern (Lindenmayer *et al.*, 2009, 2017). Incorporating a better

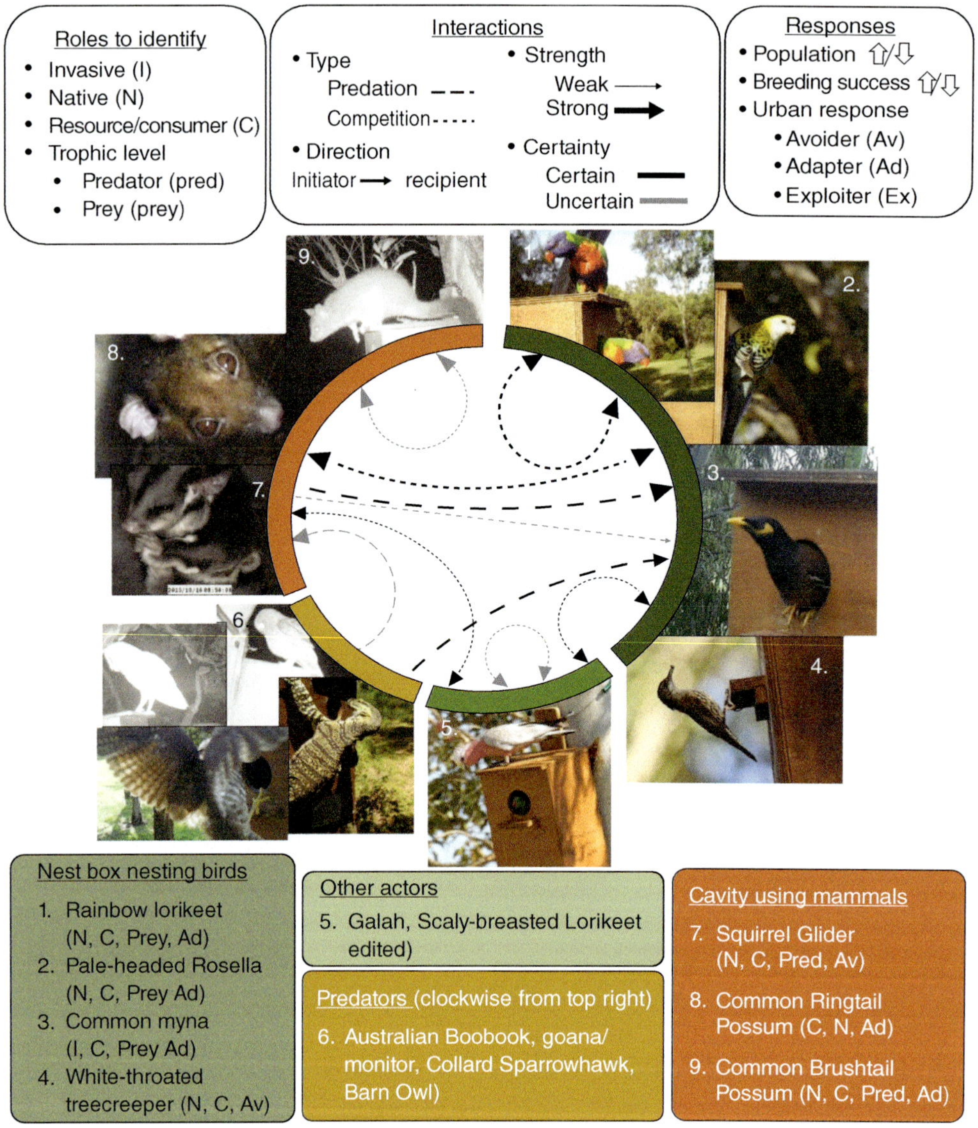

Fig. 42.2. The interaction web around nest boxes in south-east Queensland. Nest boxes were monitored with remote camera traps from October 2015 to March 2016, the peak breeding months in this region of Australia. The importance of invasive–native impacts needs to be assessed in the context of the community-wide interactions, including predation and competition among native species. The direction and strength of the interactions (arrows) between groups are essential for evaluating the importance of interactions such as competition. The response of an individual species to the total of the interactions within the network will influence its use and persistence within a habitat. In urban areas, these response groups include species that avoid urban areas (avoiders), species that adapt to use parts of urban environments (adapters) and species that can exploit urban environments (exploiters). (Photos: 1-4 & 6-9 A. Rogers; 5 Steve Gray).

understanding of interactions into such projects has been shown to increase their benefit to target species (Brazill-Boast *et al.*, 2013), and mapping all competitive and other interactions around different resources will improve management by identifying where and when invasive species interactions are most likely to impact breeding dynamics or use of habitat space.

The relative importance of invasive–native competition in the tangled interaction web around cavities in the urban environments of Australia is just starting to be quantified. The combined effects of habitat change, invasive species competition and changes in native species abundances none the less create challenges for conserving biodiversity in human-dominated environments (Kark *et al.*, 2007; Bellocq *et al.*, 2017). Species that are experts in exploiting urban areas are called 'urban exploiters', species that make some use of urban areas are called 'urban adapters' and species that avoid urban habitats are called 'urban avoiders' (Kark *et al.*, 2007; Shwartz *et al.*, 2008). Conservation efforts targeting urban adapters and avoiders within or at the edge of urban habitat fragments are likely to provide the greatest conservation benefit by increasing the resource availability in otherwise marginal habitats. In Fig. 42.2., the White-throated Treecreeper (*Cormobates leucophaea*) is an urban avoider that will use the same or similar nesting cavities as medium-sized parrots and the Common Myna in the same environment. Currently, relatively little is known about the exact nesting preferences of this treecreeper, which limits the ability of managers to exploit differences in nesting requirements and provide nest boxes that would exclude larger native or invasive birds. Nest boxes that are accessible to a wide variety of species are unlikely to be of much conservation benefit as they are frequently occupied by common species not in need of additional nesting opportunities (Lindenmayer *et al.*, 2017). For the White-throated Treecreeper, and many other less abundant cavity-nesting birds in Australia, additional work is needed on specific nest box design and reducing competition in habitats dominated by urban exploiters.

While the Common Myna is perhaps the most studied invasive cavity-nesting species in Australia, seven other invasive cavity-nesting species are already established (Gibbons and Lindenmayer, 2002), and Australia is at high risk of additional non-native species establishing invasive populations (Vall-Ilosera and Cassey, 2017). Additionally, there are 21 native Australian species that have been moved or established outside their historic ranges (Gibbons and Lindenmayer, 2002). Like the patterns for invasive species globally, the interactions between non-native and native species are poorly studied and consequently little is known about how these introductions have altered local communities. Describing the competitive interaction networks for these communities has the potential to sustainably improve native species conservation by identifying management action that can reduce the competitive impact of invasive species in targeted ways (Orchan *et al.*, 2013). Significant conservation opportunities exist in Australian cities (Garden *et al.*, 2006; Ives *et al.*, 2016), but understanding how species persist and coexist in urban areas will require more data on community-level interactions, especially as communities dynamically change in terms of ongoing environmental disturbance and novel species introductions (Mokross *et al.*, 2014; Hui *et al.*, 2016).

42.5 References

Aderhold, A., Husmeier, D., Lennon, J.J., Beale, C.M. and Smith, V.A. (2012) Hierarchical Bayesian models in ecology: Reconstructing species interaction networks from non-homogeneous species abundance data. *Ecological Informatics* 11, 55–64.

Alatalo, R.V. and Moreno, J. (1987) Body size, interspecific interactions and use of feeding sites in tits. *Ecology* 68, 1773–1777.

Baker, C.M., Holden, M.H., Plein, M., McCarthy, M.A. and Possingham, H.P. (2018) Informing network management using fuzzy cognitive maps. *Biological Conservation* 224, 122–128.

Baker, J., Harvey, K.J. and French, K. (2014) Threats from introduced birds to native birds. *Emu* 114, 1–12.

Batalha, H.R., Ramos, J.A. and Cardoso, G.C. (2013) A successful avian invasion occupies a marginal ecological niche. *Acta Oecologica* 49, 92–98.

Bellocq, M.I., Leveau, L.M. and Filloy, J. (2017) Urbanization and bird communities: spatial and temporal patterns. In: Murgui, E. and Hedblom, M. (eds) *Emerging from Southern South America. Ecology and Conservation of Birds in Urban Environments.* Springer International Publishing, Cham, Switzerland, pp. 35–54.

Blackburn, T.M., Lockwood, J.L. and Cassey, P. (2009) *Avian Invasions : The Ecology and Evolution of Exotic Birds.* Oxford University Press, Oxford.

Bode, M., Baker, C.M. and Plein, M. (2015) Eradicating down the food chain: optimal multispecies eradication schedules for a commonly encountered invaded island ecosystem. *Journal of Applied Ecology* 52, 571–579.

Brazill-Boast, J., Pryke, S.R. and Griffith, S.C. (2010) Nest-site utilisation and niche overlap in two sympatric, cavity-nesting finches. *Emu* 110, 170–177.

Brazill-Boast, J., Pryke, S.R. and Griffith, S.C. (2013) Provisioning habitat with custom-designed nest-boxes increases reproductive success in an endangered finch. *Austral Ecology* 38, 405–412.

Byrd, G.V., Moriarty, D.I. and Brady, B.G. (1983) Breeding biology of wedge-tailed shearwaters at Kilauea Point, Hawaii. *Condor* 85, 292.

Charter, M., Izhaki, I., Ben Mocha, Y. and Kark, S. (2016) Nest-site competition between invasive and native cavity nesting birds and its implication for conservation. *Journal of Environmental Management* 181, 129–134.

Cooper, C.B., Hochachka, W.M. and Dhondt, A.A. (2007) Contrasting natural experiments confirm competition between House Finches and House Sparrows. *Ecology* 88, 864–870.

Davis, A., Major, R.E. and Taylor, C.E. (2013) Housing shortages in urban regions: aggressive interactions at tree hollows in forest remnants. *PLoS One* 8, e59332.

Davis, M.A. (2003) Biotic globalization: does competition from introduced species threaten biodiversity? *Bioscience* 53, 481–489.

Dhondt, A.A. (2012) *Interspecific Competition in Birds.* Oxford University Press, Oxford.

Dick, J.T.A., Alexander, M.E., Jeschke, J.M., Ricciardi, A., MacIsaac, H.J., *et al.* (2014) Advancing impact prediction and hypothesis testing in invasion ecology using a comparative functional response approach. *Biological Invasions* 16, 735–753.

Edworthy, A.B. (2016) Competition and aggression for nest cavities between Striated Pardalotes and endangered Forty-spotted Pardalotes. *Condor* 118, 1–11.

Evans, T., Kumschick, S. and Blackburn, T.M. (2016) Application of the Environmental Impact Classification for Alien Taxa (EICAT) to a global assessment of alien bird impacts. *Diversity and Distributions* 22, 919–931.

Farwell, L.S. and Marzluff, J.M. (2013) A new bully on the block: does urbanization promote Bewick's wren (*Thryomanes bewickii*) aggressive exclusion of Pacific wrens (*Troglodytes pacificus*)? *Biological Conservation* 161, 128–141.

Fath, B.D., Scharler, U.M., Ulanowicz, R.E. and Hannon, B. (2007) Ecological network analysis: network construction. *Ecological Modelling* 208, 49–55.

Fogarty, S., Cote, J. and Sih, A. (2011) Social personality polymorphism and the spread of invasive species: a model. *American Naturalist* 177, 273–287.

Garden, J., McAlpine, C., Peterson, A., Jones, D. and Possingham, H. (2006) Review of the ecology of Australian urban fauna: a focus on spatially explicit processes. *Austral Ecology* 31, 126–148.

Garnett, S.T., Pedler, L.P. and Crowley, G.M. (1999) The breeding biology of the Glossy Black-Cockatoo *Calyptorhynchus lathami* on Kangaroo island, South Australia. *Emu* 99, 262–279.

Ghilain, A. and Belisle, M. (2008) Breeding success of tree swallows along a gradient of agricultural intensification. *Ecological Applications* 18, 1140–1154.

Gibbons, P. and Lindenmayer, D.B. (2002) *Tree Hollows and Wildlife Conservation in Australia.* CSIRO Publishing, Canberra.

Goldshtein, A., Markman, S., Leshem, Y., Puchinsky, M. and Charter, M. (2018) Nest-site interference competition with House Sparrows affects breeding success and parental care in Great Tits. *Journal of Ornithology* 159, 667–673.

Grarock, K., Tidemann, C.R., Wood, J. and Lindenmayer, D.B. (2012) Is it benign or is it a pariah? Empirical evidence for the impact of the common myna (*Acridotheres tristis*) on Australian birds. *PLoS One* 7, e40622.

Grarock, K., Lindenmayer, D.B., Wood, J.T. and Tidemann, C.R. (2013) Does human-induced habitat modification influence the impact of introduced species? A case study on cavity-nesting by the introduced common myna (*Acridotheres tristis*) and two Australian native parrots. *Environmental Management* 52, 958–970.

Green, D.S., Matthews, S.M., Swiers, R.C., Callas, R.L., Yaeger, J.S., *et al.* (2018) Dynamic occupancy modelling reveals a hierarchy of competition among fishers, grey foxes and ringtails. *Journal of Animal Ecology* 87, 813–824.

Griffin, A.S., Sol, D. and Bartomeus, I. (2012) The paradox of invasion in birds: competitive superiority or ecological opportunism? *Oecologia* 169, 553–564.

Haythorpe, K.M., Burke, D. and Sulikowski, D. (2014) The native versus alien dichotomy: relative impact of native noisy miners and introduced common mynas. *Biological Invasions* 16, 1659–1674.

Heinsohn, R., Murphy, S. and Legge, S. (2003) Overlap and competition for nest holes among eclectus parrots, palm cockatoos and sulphur-crested cockatoos. *Australian Journal of Zoology* 51, 81–94.

Howes, A.L. and Maron, M. (2009) Interspecific competition and conservation management of continuous subtropical woodlands. *Wildlife Research* 36, 617–626.

Hui, C. and Richardson, D.M. (2019) How to invade an ecological network. *Trends in Ecology and Evolution* 34, 121–131.

Hui, C., Richardson, D.M., Landi, P., Minoarivelo, H.O., Garnas, J. and Roy, H.E. (2016) Defining invasiveness and invasibility in ecological networks. *Biological Invasions* 18, 971–983.

Ingold, D.J. (1998) The influence of starlings on flicker reproduction when both naturally excavated cavities and artificial nest boxes are available. *Wilson Bulletin* 110, 218–225.

Ives, C.D. Lentini, P.E., Threlfall, C.G., Ikin, K., Shanahan, D.F., *et al.* (2016) Cities are hotspots for threatened species. *Global Ecology and Biogeography* 25, 117–126.

Jordano, P. (2016) Sampling networks of ecological interactions. *Functional Ecology* 30, 1883–1893.

Joyce, M., Barnes, M.D., Possingham, H.P. and van Rensburg, B.J. (2018) Understanding of avian assemblage change within anthropogenic environments using citizen science data. *Landscape and Urban Planning* 179, 81–89.

Kark, S., Iwaniuk, A., Schalimtzek, A. and Banker, E. (2007) Living in the city: can anyone become an 'urban exploiter'? *Journal of Biogeography* 34, 638–651.

Kéfi, S., Berlow, E.L., Wieters, E.A., Joppa, L.N., Wood, S.A., *et al.* (2015) Network structure beyond food webs: mapping non-trophic and trophic interactions on Chilean rocky shores. *Ecology* 96, 291–303.

Koenig, W.D. (2003) European starlings and their effect on native cavity-nesting birds. *Conservation Biology* 17, 1134–1140.

Le Roux, D.S., Ikin, K., Lindenmayer, D.B., Bistricer, G., Manning, A.D. and Gibbons, P. (2016a) Enriching small trees with artificial nest boxes cannot mimic the value of large trees for hollow-nesting birds. *Restoration Ecology* 24, 252–258.

Le Roux, D.S., Ikin, K., Lindenmayer, D.B., Bistricer, G., Manning, A.D. and Gibbons, P. (2016b) Effects of entrance size, tree size and landscape context on nest box occupancy: considerations for management and biodiversity offsets. *Forest Ecology and Management* 366, 135–142.

Lindenmayer, D.B., Crane, M., Evans, M.C, Maron, M., Gibbons, P., *et al.* (2017) The anatomy of a failed offset. *Biological Conservation* 210, 286–292.

Lindenmayer, D.B., Welsh, A., Donnelly, C., Crane, M., Michael, D., *et al.* (2009) Are nest boxes a viable alternative source of cavities for hollow-dependent animals? Long-term monitoring of nest box occupancy, pest use and attrition. *Biological Conservation* 142, 33–42.

Linz, G.M., Homan, H.J., Gaulker, S.M., Penry, L.B. and Bleier, W.J. (2007) European starlings: a review of an invasive species with far-reaching impacts. In: *Managing Vertebrate Invasive Species. USDA National Wildlife Research Center Symposium*, University of Nebraska – Lincoln, pp. 378–386. Available at: https://digitalcommons.unl.edu/nwrcinvasive/24 (accessed 12 November 2019).

Low, T. (2014) *Where Song Began: Australia's Birds and How They Changed the World*. Viking, Sydney, Australia.

Mac Nally, R. and Timewell, C.A.R. (2005) Resource availability controls bird-assemblage composition through interspecific aggression. *Auk* 122, 1097–1111.

Mac Nally, R., Bowen, M., Howes, A., McAlpine, C.A. and Maron, M. (2012) Despotic, high-impact species and the subcontinental scale control of avian assemblage structure. *Ecology* 93, 668–678.

Mänd, R., Tilgar, V., Lõhmus, A. and Leivits, A. (2005) Providing nest boxes for hole-nesting birds – does habitat matter? *Biodiversity and Conservation* 14, 1823–1840.

Maron, M., Grey, M.J., Catterall, C.P., Major, R.E., Oliver, D.L., *et al.* (2013) Avifaunal disarray due to a single despotic species. *Diversity and Distributions* 19, 1468–1479.

Menchetti, M. and Mori, E. (2014) Worldwide impact of alien parrots (Aves Psittaciformes) on native biodiversity and environment: a review. *Ethology, Ecology and Evolution* 26, 172–194.

Miller, E.T., Bonter, D.N., Eldermire, C., Freeman, B.G., Greig, E.I., *et al.* (2017) Fighting over food unites the birds of North America in a continental dominance hierarchy. *Behavioral Ecology* 28, 1454–1463.

Minot, E.O. and Perrins, C.M. (1986) Interspecific interference competition – nest sites for blue and great tits. *Journal of Animal Ecology* 33, 1–4.

Mokross, K., Ryder, T.B., Côrtes, M.C., Wolfe, J.D. and Stouffer, P.C. (2014) Decay of interspecific avian flock networks along a disturbance gradient in Amazonia. *Proceedings of the Royal Society of London B: Biological Sciences* 281, 20132599.

Montague-Drake, R.M., Lindenmayer, D.B., Cunningham, R.B. and Stein, J.A.A. (2011) Reverse keystone species affects the landscape distribution of woodland avifauna: a case study using the Noisy Miner (*Manorina melanocephala*) and other Australian birds. *Landscape Ecology* 26, 1383–1394.

Moon, D.C., Moon, J. and Keagy, A. (2010) Direct and indirect interactions. *Nature Education Knowledge* 3, 50.

Murphy, S., Legge, S. and Heinsohn, R. (2003) The breeding biology of palm cockatoos (*Probosciger aterrimus*): a case of a slow life history. *Journal of Zoology* 261, 327–339.

Orchan, Y., Chiron, F., Schwartz, A. and Kark, S. (2013) The complex interaction network among multiple invasive bird species in a cavity-nesting community. *Biological Invasions* 15, 429–445.

Parker, I.M., Simberloff, D., Lonsdale, W.M., Goodell, K., Wonham, M., *et al.* (1999) Impact: toward a framework for understanding the ecological effects of invaders. *Biological Invasions* 1, 3–19.

Peck, H.L., Pringle, H.E., Marshall, H.H., Owens, I.P.F. and Lord, A.M. (2014) Experimental evidence of impacts of an invasive parakeet on foraging behavior of native birds. *Behavioral Ecology* 25, 582–590.

Pell, A.S. and Tidemann, C.R. (1997a) The impact of two exotic hollow-nesting birds on two native parrots in savannah and woodland in eastern Australia. *Biological Conservation* 79, 145–153.

Pell, A.S. and Tidemann, C.R. (1997b) The ecology of the Common Myna in urban nature reserves in the Australian captial territory. *Emu* 97, 141–149.

Reif, J., Reifová, R., Skoracka, A. and Kuczyński, L. (2018) Competition-driven niche segregation on a landscape scale: Evidence for escaping from syntopy toward allotopy in two coexisting sibling passerine species. *Journal of Animal Ecology* 87, 774–789.

Remm, J., Lõhmus, A. and Rosenvald, R. (2008) Density and diversity of hole-nesting passerines: dependence on the characteristics of cavities. *Acta Ornithologica* 43, 83–91.

Rogers, A.M. (2019) The role of habitat variability and interactions around nesting cavities in shaping urban bird communities. PhD thesis, University of Queensland, Brisbane, Australia.

Romanuk, T.N., Zhou, Y., Brose, U., Berlow, E.L., Williams, R.J. and Martinez Neo, D. (2009) Predicting invasion success in complex ecological networks. *Philosophical Transactions of the Royal Society B: Biological Sciences* 364, 1743–1754.

Rusterholz, K.A. (1981) Competition and the structure of an avian foraging guild. *American Naturalist* 118, 173–190.

Sanz-Aguilar, A., Carrete, M., Edelaar, P., Potti, J. and Tella, J.L. (2015) The empty temporal niche: breeding phenology differs between coexisting native and invasive birds. *Biological Invasions* 17, 3275–3288.

Schoener, T.W. (1986) Mechanistic approaches to community ecology: a new reductionism? *American Zoologist* 26, 81–106.

Shwartz, A., Shirley, S. and Kark, S. (2008) How do habitat variability and management regime shape the spatial heterogeneity of birds within a large Mediterranean urban park ? *Landscape and Urban Planning* 84, 219–229.

Simberloff, D., Martin, J.-L., Genovesi, P., Maris, V., Wardle, D.A., *et al.* (2013) Impacts of biological invasions: what's what and the way forward. *Trends in Ecology and Evolution* 28, 58–66.

Sol, D., Bartomeus, I. and Griffin, A.S. (2012) The paradox of invasion in birds: competitive superiority or ecological opportunism? *Oecologia* 169, 553–564.

Stojanovic, D., Koch, A.J., Webb, M., Cunningham, R., Roshier, D. and Heinsohn, R. (2014) Validation of a landscape-scale planning tool for cavity-dependent wildlife. *Austral Ecology* 39, 579–586.

Tindall, D.S., Ralph, J.C. and Clout, M.N. (2007) Changes in bird abundance following Common Myna control on a New Zealand island. *Pacific Conservation Biology* 13, 202–212.

Vall-llosera, M. and Cassey, P. (2017) Leaky doors: private captivity as a prominent source of bird introductions in Australia. *PLoS One* 12, e0172851.

Yosef, R., Zduniak, P. and Żmihorski, M. (2016) Invasive Ring-Necked Parakeet negatively affects indigenous Eurasian Hoopoe. *Annales Zoologici Fennici* 53, 281–287.

43 Control or eradication: problems in the management of invasive birds

Michael L. Avery[1]* and Chris J. Feare[2]

[1]*USDA National Wildlife Research Center; Gainesville, FL 32641, USA;*
[2]*WildWings Bird Management, Haslemere, Surrey, UK*

Citation: Avery, M.L. and Feare, C.J. (2020) Control or eradication: problems in the management of invasive birds. In: Downs, C.T. and Hart, L.A. (eds) *Invasive Birds: Global Trends and Impacts*. CAB International, Wallingford, UK, pp. 349–361.

43.1 Introduction

Humans have captured, transported and intentionally released wild birds for centuries (Blackburn *et al.*, 2009). Motivations for such purposeful introductions include food (West and Zhou, 2007), religion (Agoramoorthy and Hsu, 2007), sport (McDowall, 1994), biocontrol (Bennett and Hughes, 1959; Kurdila, 1995) and aesthetics (Ryan, 1906; Thomson, 1922). Many purposeful bird introductions were the work of acclimatization societies, particularly in North America, New Zealand and Australia. These societies were formed in the 19th century by European settlers to transport bird species from their homelands in efforts to establish them in the newly settled regions (Thomson, 1922; Dunlap, 1997). As a result of these efforts, the Common or European Starling (*Sturnus vulgaris*), the House Sparrow (*Passer domesticus*) and many other species are now permanently established far beyond their native ranges.

Commercial trade in captive birds is also an important introduction pathway. Non-native species are introduced through unintentional releases of cage birds and inadvertent escapes from research facilities, zoos and private collections. The international bird trade has declined gradually following adoption in the USA of the 1992 Wild Bird Conservation Act and similar European regulations restricting trade in wild birds following the westward spread from China of the highly pathogenic H5N1 avian influenza virus in the early 2000s (Cooney and Jepson, 2006). The pattern of trade in wild birds has also changed. Mexico and Asia have replaced the USA and the European Union as the principal importers in the global cage-bird market (Cardador *et al.*, 2017; Hobson *et al.*, 2017). Nevertheless, large-scale traffic in wild and captive-bred birds continues. During the 3-year period 2000–2002, global exports of live birds totalled 3,640,135 compared with 807,476 during 2015–2017 according to the Convention on International Trade in Endangered Species (CITES, 2018).

Invasive birds have major impacts throughout the world, regardless of the invasion pathway. Pimentel *et al.* (2001) examined the published data available on invasive species in the USA, the UK, Australia, South Africa, India and Brazil. They concluded that introduced birds were responsible for US$2.4 billion in damage to agriculture, human health and natural resources among these six countries.

We focus in this chapter on a subset of these impacts, namely the threats that invasive birds pose to native species and the efforts that have been made to reduce or eradicate such impacts. Specifically, we review management options and control strategies, explore what has and has not been effective, and discuss case histories of success and failure.

43.2 Management Options

Invasive species management can be viewed as a process occurring along a time continuum on which management costs and difficulty increase with time. Following a scheme developed and articulated for Australian agricultural resources (Department of Primary Industries, 2010), Harvey and Mazzotti (2014) defined four stages within the invasive species management process: prevention, eradication, containment and long-term management.

43.2.1 Prevention

By far, prevention yields the greatest management benefit per unit cost. Prevention obviates the need for subsequent

*Corresponding author (retired): gibs8387@gmail.com

management because the invasive species does not become established in the first place. Prevention procedures must be applied throughout the importation/transportation process to minimize inadvertent releases en route or at the destination port of entry. Unfortunately, prevention strategies can fail either because they are poorly conceived or because they are not implemented properly. There are now more than 200 bird species occupying areas beyond their native ranges, and at least 36 of these have some level of ecological impact, including competition, predation, hybridization and disease transmission with native taxa (Lever, 2005; Martin-Albarracin *et al.*, 2015).

43.2.2 Eradication

When a species evades the prevention phase and invades non-native space, eradication should become the primary goal. Unfortunately, eradication is not always realistic. There is, in most cases, a relatively short window, while the invasive population is small and localized, when eradication is a realistic management objective. However, if such opportunities are not seized, the prospects for eradication diminish rapidly. Eradication goes together with a policy of early detection and rapid response (EDRR). The sooner a management response can be mounted and implemented, the greater the chance of stopping the invasion before it gets out of hand. Feasibility and cost of eradication vary with the size of the population and the size of the geographical area occupied by that population. Both of these parameters can be related to time since introduction, with founder populations predicted to be smaller and more localized than established populations.

EDRR is costlier than prevention, but, in most cases, it represents the last chance for eradication. The effectiveness of an EDRR effort can be enhanced by ongoing monitoring, research into species' life histories and impacts, and development of detection and removal tools. An EDRR strategy offers opportunities for public participation in learning how to identify, detect and report invaders.

Management based on EDRR is especially warranted when there is high probability that the invasive species will cause serious impacts and, if unchecked, is likely to reproduce and disperse, thus compounding the difficulties for successful control (Simberloff, 2003). Several conditions favour a successful EDRR strategy:

1. *Public support.* Natural resource managers ideally will have foreseen the benefit of a public well-informed about invasive species, and they will have established the need for an EDRR programme supported by the public. Education and outreach are important components of such an approach and are invaluable in implementing an EDRR programme to address invasive species concerns (Temple, 1992). In some cases, active participation by the public has been essential to the eradication success (e.g. Suleiman and Taleb, 2010).
2. *Resources.* Early detection of an incipient invasion is of little value unless properly trained personnel with adequate logistical support are available to respond promptly. Public support of an EDRR programme for invasive species management implies that personnel, equipment and funds for conducting the necessary field activities have been allocated. An effective rapid response will prevent the invasive front from advancing and dispersing, and will increase likelihood of success. The rapid response will also lessen the time that the presence of the invasive species can be publicized and attract avid birders interested in adding an unusual sighting to their records. Unless resources are already allocated and available, timely responses to reports of invasives cannot be guaranteed.
3. *Regulations.* An efficient, rapid response to an invasive bird species will be greatly facilitated if regulatory procedures (e.g. permits, access to properties) are well understood and anticipated. Management actions will proceed more smoothly if the EDRR responders maintain regular contact with the relevant oversight agencies or departments so that action plans and response strategies can be discussed and agreed prior to any urgent need. Potential permitting issues related to the use of lethal control (firearms, toxicants) or animal welfare concerns should be resolved in up-front discussions with appropriate authorities.

43.2.3 Containment

If immediate action is not taken, or if the initial response is inadequate, invasive species will reproduce and spread through suitable habitats. Eradication becomes increasingly unlikely as the species increases in number and invades new areas. When populations become established, management efforts shift from eradication to controlling the spread of a species. Intense efforts are necessary to contain the core population of a species and extirpate it from new areas.

43.2.4 Resource protection and long-term management

When an invader is too widespread and abundant to control everywhere it occurs, eradication becomes impossible. Long-term management aims to reduce populations to the lowest feasible levels and to protect specific highly valued resources. Community support can be critical to ensure the success of long-term management programmes because such efforts could require sustained funding and staffing across many years.

43.3 Eradication or Control?

'Eradication is an intense, time-limited process offering perpetual freedom from the pest, its effects, and control costs. By contrast, continuing control is a recurrent activity with continuing damage and control costs' (Bomford and O'Brien, 1995). When the existence of an endangered species or other highly valued natural resource is threatened by an invasive species, managers will probably opt for eradication of the invasive population (Feare, 2010a). Eradication offers maximum, long-term protection for native fauna and flora, but practicality, cost and public acceptability affect the likelihood that eradication can be achieved. The feasibility and cost of eradication vary with the size of the population and the size of the geographical area occupied by that population. The techniques available for eradication can be limited by potential side-effects on native fauna and flora, and the effectiveness of techniques can vary among species of invasive birds, and even among different populations of the same species.

The biota on many islands is threatened or endangered, and there have been numerous eradication efforts to rid islands of invasive wildlife (e.g. rodents, Howald *et al.*, 2007; feral goats, Campbell and Donlan, 2005; feral cats and rabbits, Robinson and Copson, 2014). Non-native bird introductions are greatly biased towards islands (Blackburn *et al.*, 2009), but compared with mammals, eradication efforts seldom target invasive birds. Glen *et al.* (2013) reviewed 1068 vertebrate eradications on 749 islands worldwide. Of the successful vertebrate eradications, 1043 (98%) were of mammals; only 24 eradications targeted invasive birds.

Regardless of taxon, eradication of an invasive continental population is rare. Fleming *et al.* (2017) noted that no established invasive vertebrate or plant has been eradicated from a continent. Bomford and O'Brien (1995) proposed the following criteria for successful eradication of an invasive wildlife population:

- The rate of removal must be greater than reproductive rate at all population densities.
- Immigration must be zero.
- All reproductive animals must be at risk of control tools and strategies.
- The target species are able to be monitored at low densities.
- The discounted cost–benefit analysis favours eradication over ongoing suppression.
- The socio-political environment is suitable.

The first three are 'essential' criteria without which eradication cannot succeed. The remaining three are 'desirable' and must be met for eradication to be the preferred management option. These six criteria are rarely ever met in mainland invasive populations.

We suggest that there have been eradications of invasive bird populations from mainland areas, such as the Sacred Ibis (*Threskiornis aethiopicus*) in Barcelona (Clergeau and Yésou, 2006) and the American Black Duck (*Anas rupripes*) in British Columbia (Fenneman, 2011), but they do not occur often and always involve small, localized populations. Thus, for most mainland invasive bird populations, particularly those that are decades old, eradication could be unrealistic. If such conditions pertain, then maintaining the population at a given level through long-term management might be an acceptable option. Long-term management is not necessarily an admission of defeat but rather should be viewed as one alternative in addressing a difficult problem (Simberloff, 2009).

43.4 Control Methods

Some wild bird populations negatively impact human health and safety through agricultural losses, zoonotic disease transmission, property damage and other factors. In response, numerous commercial devices and chemicals have been developed in attempts to alleviate these impacts. Many of the commonly used bird damage control methods are non-lethal and probably have little utility in strategies to reduce or eliminate populations of invasive birds.

When eradication is the management objective, three basic types of lethal control measures are employed: (i) live trapping, followed by relocation or humane euthanasia; (ii) selective shooting; and (iii) applying toxic bait. Each of these approaches requires training and experience to implement safely and effectively.

43.4.1 Trapping

Numerous types of live traps are available to wildlife managers. The target species will determine the design and deployment of the live traps to address a given situation. Large drop nets can be used effectively to capture groups of ground-foraging birds such as Wild Turkeys (*Meleagris gallopavo*; Morrison *et al.* 2016). Corvids and other highly social species are susceptible to capture in large, baited, drop-in pen traps (e.g. Tsachalidis *et al.*, 2006). Trap success can usually be improved by maintaining two or three birds of the target species within the trap to attract conspecifics. A dripping water source inside the pen will also attract outside birds, as well as provide those already caught with drinking and bathing water. Various designs of smaller live traps with decoy birds have been employed successfully in campaigns to control Common Mynas (*Acridotheres tristis*; Tidemann, 2005; Saavedra, 2010; Canning, 2011).

Removing nestlings and eggs from nests of invasive species can augment the lethal control measures targeting the adult birds. Eradication of the House Crow (*Corvus splendens*) from Socotra, Yemen, was attributed in part to the successful efforts of school children locating nests, removing the contents and bringing the young crows to authorities to be euthanized (Suleiman and Taleb, 2010). Destruction of nests, eggs and nestlings has also contributed, albeit in a minor way, to eradication of an island myna population (Canning, 2011).

A period of pre-baiting is recommended for most trapping programmes to overcome neophobia and to create a seemingly benign, or even attractive, environment for the target animals. Pre-baiting entails setting up the trap, providing bait and then allowing target animals to come and go freely. The trap is set or activated when observations confirm reliable visitation by the target species. Trapping proceeds until the capture rate diminishes to an unacceptable level. A different site might then be established, or a new method employed.

In most cases, trapped birds will be humanely dispatched following applicable guidelines as designated by the appropriate local authorities (e.g. American Veterinary Medical Association, 2001). Euthanized birds can be sources of otherwise unobtainable information on the target species, and the specimens should be used to the greatest extent possible to increase understanding of the invasive species (Feare, 2010a).

43.4.2 Shooting

A carefully planned and expertly executed shooting campaign can prove invaluable in an eradication programme (e.g.

Suleiman and Taleb, 2010; Morrison *et al.*, 2016). Daily feeding sites, water holes, roosts and other places where target birds reliably congregate are often ideal shooting locations, but regular shooting at such sites can instil location avoidance. For eradication of invasive species, use of expert marksmen with understanding of the behaviour of the target species is encouraged. Shooting is frequently essential for removing the final few trap- or bait-shy individuals.

Statutes and regulations governing possession and use of firearms vary geographically, so it is vital that the appropriate authorities be consulted and the necessary permits obtained in advance.

43.4.3 Toxicants

Starlicide (3-chloro-*p*-toluidine hydrochloride, also known as DRC-1339) was developed over 50 years ago and has been used ever since in management of pest populations of European starlings and other problem bird species in the USA (DeCino *et al.*, 1966; Eisemann *et al.*, 2003). Common Mynas are very sensitive to this toxicant (Avery and Eisemann, 2015). It has been used in myna control programmes on several islands with varying degrees of success (Millett *et al.*, 2004; Feare, 2010b; Parkes, 2012). Deployment of toxic bait might have most utility to reduce a large population of invasive birds quickly so that trapping and shooting can then be efficiently employed (Millett *et al.*, 2004; Parkes, 2012). Population reduction with toxic bait is best viewed as a first step and should be followed promptly by determined application using other techniques to achieve the goal of eradication. This approach was used successfully in the Cook Islands (see section 43.5.2) and might be applicable in other invasive bird management situations.

Advance planning is crucial for a successful toxic baiting programme. Potential bait sites, types of bait, schedule, amount of bait deployed and possible non-target species are among the factors to be considered. Use of Starlicide, or any toxicant, in the vicinity of endangered endemics (all taxa, not just birds) will be problematic because potential impacts on other organisms such as insects, crustaceans, amphibians, reptiles and fish are poorly studied, so extreme caution is needed.

Most birds ingesting Starlicide bait do not die at the bait site, and the numbers of birds killed will be difficult to determine. Furthermore, a day-to-day lessening in visitation rates to bait sites could reflect feeding site aversion, not simply mortality in the target population.

Candidate bait sites should be provisioned with untreated baits and monitored for several days to document use by the target species and any non-target animals. This pre-baiting period should establish a feeding pattern among the local target birds. The observations will enable personnel to determine the best time of day and duration for baiting, and the optimal quantity of treated bait to deploy at the site to maximize exposure to the target species and minimize the potential risk to non-targets. Starlicide bait should be formulated to deliver a lethal dose to the target species in a single bait (Glahn and Avery, 2001). In field applications, the treated baits are mixed with untreated baits (e.g. 20:1 or 30:1, untreated:treated). This ensures that not every bird ingests lethal bait on the first day. Thus, a portion of the feeding flock will return to the bait site on subsequent days, recruiting new birds each time until maximum effect is achieved. In an ideal baiting programme, the amount of bait offered will be depleted by the target birds during their initial daily morning feeding bout, leaving nothing at the bait site for non-targets to ingest. If there is any doubt about treated bait remaining after a presentation, it should be offered in such a way that any remaining bait can be collected and destroyed.

43.4.4 Integrated management

An eradication strategy that succeeds in one situation will not necessarily be as effective under different conditions, even if the same target species is involved. Three myna eradications in the Seychelles revealed unexpected differences in the responses of Common Mynas to trap types. On Frégate Island, Canning (2011) caught 97% of the 745 mynas killed in commercially made Mini Myna traps (funnel traps that permit multiple birds to be caught: http://mynamagnet.com.au/, accessed 15 November 2019). On Denis Island, these traps were inefficient, accounting for only 0.9% of 1090 mynas captured, whereas decoy traps (cage traps that contain a live myna decoy in a central compartment surrounded by four single-catch compartments) were much more effective (Feare *et al.*, 2016). On North Island, 226 out of 1538 mynas (14.7%) were caught in Mini Myna traps, with most of the remainder being caught in decoy traps. The efficiency of Mini Myna traps on North Island appeared to be related to habitat and the associated density of feeding birds. On managed grassland, which represented the most widespread habitat used by mynas on the island, mynas were readily caught in decoy traps but not in Mini Myna traps, whereas at a site where the dumping of organic waste each morning attracted large numbers, Mini Myna traps were much more successful. A similar association between habitat type and relative trap success was found during a feasibility study of controlling Common Mynas on St Helena Island (Feare and Saavedra, 2009).

Each management situation has its own set of peculiarities that will influence the optimal course of action. Each lethal technique has advantages and disadvantages, and none is a magic bullet. It would be very unusual for an eradication programme to succeed relying on a single method. More likely, a successful programme will employ several methods, lethal and non-lethal, in an integrated effort to address a given invasive species problem (Feare, 2010a). Sometimes, a trial-and-error approach will be needed; in other cases, eradication efforts can be guided by the successes of others under similar conditions. Parkes (2012) reviewed several control programmes targeting the Common Myna and concluded that the most effective approach was sequential application of poisoning, trapping and shooting.

The recommendation of Parkes (2012) for the Common Myna might be an appropriate starting point for devising an eradication programme, but there is no prescription that will fit all situations. Numerous factors will impinge on the feasibility of carrying out a successful eradication programme, including environmental (e.g. weather, terrain), ecological (e.g. non-targets, natural history of target species), economic (e.g. equipment, personnel), governmental (e.g. permits, access to property) and human (e.g. public support, availability of volunteers). In populated areas, the human dimension is possibly the most challenging set of factors (e.g. Phillips *et al.*, 2012). Glen *et al.* (2013) recommended that in developing an invasive species eradication plan on an inhabited island, 'The local community must be engaged, involved in the planning process, and given a degree of ownership of the project'.

We feel that adequate tools and techniques for eradication or long-term management already exist. Improvements will undoubtedly be made to improve efficacy of the present methods. But trapping, baiting and shooting, in their various incarnations and combinations, are sufficient to effectively address invasive bird issues. Management of invasive species often represents a political or economic problem, not a scientific one. 'Invasive vertebrate species control or eradication has usually proved possible. Scientists have developed the methods; what is needed is the political will to use them' (Usher, 1989).

43.5 Case Histories

We have assembled information on global invasive bird species eradication attempts, separating those of founder populations from those well-established in their new environments in Tables 43.1 and 43.2. We have not included feasibility studies (e.g. control of Common Mynas on St Helena; Feare and Saavedra, 2009), ongoing control of invasive species limited to parts of their geographical range (e.g. Common Mynas in parts of Australia; Grarock *et al.*, 2014) or field tests of potential control techniques (e.g. trials of Starlicide with free-living Common Mynas; Anon., 2009; Feare, 2010b). We define 'founder' populations as those of recent origin and which occupy a localized area within a larger area of potentially available habitat. 'Established' populations are those that have spread from the area of introduction to occupy a large area of available habitat, and which were introduced decades or centuries ago.

The range of species for which eradications have been undertaken is small (Tables 43.1 and 43.2), and many attempts at eradication have failed, especially in already established populations, but these failures have proved valuable in helping to identify techniques that are most promising for different species.

Table 43.1. Eradications of founder populations of invasive birds.

Common name	Location	Methods	Status	Duration	No. killed	Reference
Sacred Ibis	Brittany, France	Shoot, sterilize eggs	Ongoing	2007–2013	6626 + 2720 eggs	Yésou *et al.* (2015)
	Doñana, Spain	Shoot	Complete	2011	16	Junta de Andalucía (2011)
	Barcelona, Spain	No details given	Complete	2001	No details given	Clergeau and Yésou (2006)
House Crow	Socotra Island, Yemen	Remove nestlings, trap, shoot	Complete	1999–2009	<300	Suleiman and Taleb (2010)
	Seychelles	Shoot, poison	Complete	1977–1994 et seq.	~40	Rocamora and Henriette (2015)
Red-whiskered Bulbul	Dzaoudzi, Comores	Trap, shoot	Complete	Not recorded	'Small numbers'	M. Louette (unpublished data)
	Aldabra, Seychelles	Shoot	Complete	2012	1	Bunbury *et al.* (2015)
Common Myna	Tarawa island, Kiribati	Shoot	Complete	2015	3, plus 1 Jungle Myna	SPREP (2016))
	Gran Canaria, Canary Islands, Spain	Trap	Complete	2006	3	Saavedra (2010)
	Tenerife, Canary Islands, Spain	Trap, shoot	Complete	1999–2000	10	Saavedra (2010)
	Mallorca, Balearic Islands, Spain	Trap	Complete	2006	13	Saavedra (2010)
Common Starling	Western Australia	Trap, shoot, net	Ongoing	1971–present	Low thousands	Campbell *et al.* (2015)
Madagascar Fody	Aldabra, Seychelles	Mist net, shoot	Complete	2012–2014	>250	Bunbury *et al.* (2015)
House Sparrow	Mahé, Seychelles	Trap, shoot, mist net, glue	Complete	2002–2003	<20	Rocamora and Henriette (2015)

Table 43.2. Eradications of established populations of invasive birds.

Species	Location	Methods	Status	Duration	No. killed	Reference(s)
Wild Turkey	Santa Cruz Island, USA	Trap, shoot, 'Judas birds'	Complete	2006–2012	>310	Morrison *et al.* (2016)
Ruddy Duck	UK	Shoot	Ongoing	1999–present	>6800	Henderson (2010)
Rock Pigeon	Santa Cruz, Galápagos Islands	α-Chloralose, shoot, captives seized by authorities	Complete	2001–2007	256	Phillips *et al.* (2012)
	San Cristóbal, Galápagos Islands	Shoot, captives seized	Complete	2001–2007	803	Phillips *et al.* (2012)
	Isabela, Galápagos Islands	Shoot, captives seized	Complete	2001–2007	418	Phillips *et al.* (2012)
Ring-necked Parakeet	Mahé, Seychelles	Shoot	Complete	2013–2017[a]	548	Seychelles Island Foundation (2018)
Red-whiskered Bulbul	Assumption Island, Seychelles	Mist net, shoot	Complete	2011–2014	>5000	Bunbury *et al.* (2015)
Common Myna	Frégate Island, Seychelles	Trap, shoot, nest trap	Complete	1993–2002, 2010–2011	>1460	Millett *et al.* (2004); Canning (2011)
	Denis Island, Seychelles	Starlicide, trap, shoot	Complete	2010–2011, 2014–2015	1112	Millett *et al.* (2004); Feare *et al.* (2016)
	North I, Seychelles	Rodenticide bycatch, Starlicide, trap, shoot	Nearly complete	2005–2009, 2012–present	>2630 (minimum 3 remain)	Rocamora and Henriette (2015); C.J. Feare (unpublished data)
	Atiu I, Cook Islands	Poison, trap, shoot	Nearly complete	2009–2018[a]	>26,000	G. McCormack (unpublished data)
Madagascar Fody	Assumption Island, Seychelles	Mist net, shoot	Complete	2011–2015	>3000	Bunbury *et al.* (2015)
House Sparrow	Round Island, Mauritius	Trap, glue, shoot, poison	Failed	2008–2009	>300	Bednarczuk *et al.* (2010)

[a]Post-eradication monitoring continuing.

43.5.1 Founder populations

Where prevention of incursions of invasive species has failed, the eradication of recently arrived individuals to prevent establishment in a new area must be a priority, as this opportunity represents the simplest and cheapest management option. This has been achieved in some instances where founder populations have been small or geographically localized (Table 43.1). Even here, however, eradication has not always been straightforward.

Sacred Ibis

In France, Sacred Ibis, imported to a zoo in Brittany, established a breeding colony from which the young could fly free (see Chapter 33, this volume). Some dispersed along the western seaboard of France, and breeding in the wild was first recorded in 1993. Subsequently, several breeding colonies have become established, totalling more than 1100 pairs by 2005 (Yésou and Clergeau, 2005), and sightings of Sacred Ibis more widely in Western Europe probably stem from these colonies. Concern over negative impacts on native wildlife prompted an eradication programme. During 2007–2013, 6626 birds were shot, and 2720 nests with eggs destroyed, leading to a population decline to 280–300 breeding pairs by 2013 (Yésou *et al.*, 2015). In southern France, 395 adult Sacred Ibis and 90 chicks were removed from the natural environment during 2007–2013. By 2013, only three Sacred Ibis remained in Camargue (Fernandez, 2015).

In the USA, Sacred Ibis escaped from private collections and zoos following the devastating effects of Hurricane Andrew throughout South Florida in 1992 (see Chapter 33, this volume). Subsequently, numerous sightings of Sacred Ibis were reported in the region (Herring and Gawlik, 2008). The Everglades Cooperative Invasive Species Management Area, a partnership of several land management agencies, initiated the Sacred Ibis Project in 2007, and developed an EDRR plan to remove free-flying ibis from known locations and to respond efficiently to any new reports. Staff at the Zoo Miami live-trapped ibis by exploiting the birds' habituation to people and open-exhibit feeding practices. Zoo personnel captured ibis by baiting existing covered holding pens, erecting enclosures around feeding stations and orally administering sedation drugs. Each ibis trapped by Zoo Miami was surgically pinioned, measured, sexed and held for placement with other accredited facilities with a signed agreement acknowledging the invasive potential of the species and guaranteeing containment. Concurrently, the US Department of Agriculture Wildlife Services began killing birds by shooting on the zoo grounds and responded to reported sightings in nearby areas. Overall, 75 Sacred Ibis were removed from the wild (45 shot, 30 trapped) by the Wildlife Services and Zoo Miami staff during 2008–2011 (South Florida Ecosystem Task Force, 2015). The project prevented Sacred Ibis range expansions and successfully controlled the population while it remained localized.

House Crow

House Crow dispersal has been aided by ship-assisted travel along trading routes, leading to their establishment at ports around the Indian Ocean and on some of its islands (see Chapter 24, this volume). There have also been deliberate introductions in Africa, notably Zanzibar where the House Crow was released in the 1890s, hoping that it would help clean up the town 'owing to its fondness for feeding on offal and refuse' (Vaughan, 1930).

In Yemen, a control programme on Socotra began in 1999 with unsuccessful attempts to live trap and to shoot crows. Schoolchildren were then encouraged to find crow nests and were paid for bringing in nests, young crows and adult birds to be humanely dispatched. During 2002–2007, 242 crows were removed. Finally, in 2008, expert marksmen were employed to shoot the remaining few adults (Suleiman and Taleb, 2010).

In the Seychelles, eradication of repeated small incursions of House Crows, some of which succeeded in breeding, was accomplished by the government's Environment Department, mainly by shooting (Rocamora and Henriette, 2015).

In Mauritius, Feare and Mungroo (1990) found that House Crows rapidly developed an aversion to bait treated with α-chloralose following the removal of a small number of narcotized birds. Narcotization causes birds to behave abnormally, as might other toxicants (Feare, 2010b). In contrast, Puttoo and Archer (2004) reported that Starlicide applied to meat baits reduced their study population by almost 80% in a 12-week control programme 'and could be safely used to control these birds in the future.' Previously, trapping had been ineffective, and shooting was deemed uneconomical and unsafe. We are not aware of any follow-up control efforts.

House Sparrow

In the Seychelles, a breeding population of ten to 20 House Sparrows was observed in the Port Victoria area on Mahé in 2002. Initial captures were made at feeding sites using mist nets, glue boards and feeding traps. Nest sites were identified at a power station, and a special trap was fitted over the nest-hole entrance, resulting in the capture of eight adults and two juveniles. After 11 months, just two sparrows remained, one of which was subsequently shot. The lone surviving male eventually died (Fanchette, 2003, as reported by Beaver and Mougal, 2009). Reinvasion via ship traffic is a continuing threat, so ongoing vigilance and reporting by port workers and the public is essential.

Common Starling

In Western Australia, trapping and shooting have predominated in the efforts to eradicate repeated incursions of Common Starlings (Woolnough *et al.*, 2005; Campbell *et al.*, 2015). Southern parts of the state have been periodically infiltrated by flocks of starlings, mainly from well-established populations in south-eastern states, from which Western Australia is separated by the arid and treeless Nullabor Plain. The Department of Agriculture of Western Australia has maintained a team of pest management practitioners and supported research in order to protect Western Australia's agriculture and environment from these incursions of potential founder populations, so far with success but at high cost (Woolnough *et al.*, 2005; Rollins *et al.*, 2009, 2011; Campbell *et al.*, 2015).

Common Myna

In Spain, Saavedra (2010) demonstrated the efficacy of traps containing a live decoy in catching free-living birds, resulting in the eradication of three founder populations (Table 43.1).

In the port village of Betio, Kiribati, residents observed three Common Mynas and one Jungle Myna (*Acridotheres fuscus*) consistently in the vicinity. The presence of people and dogs in the area meant that toxic bait and live trapping would be unsuitable control methods. Shooting was deemed the most appropriate means of eradicating this small invasive population. Authorities brought in an experienced hunter from New Zealand who took care of the job in a couple of days. A number of factors contributed to the success of the project: (i) scouting of the area before the shooter arrived to learn the birds' activity patterns and to identify possible shooting locations; (ii) the assistance of local police to escort the shooter, manage site access and ensure public safety; (iii) an environment in which the mynas were accustomed to close encounters with people; and (iv) an experienced, committed hunter (SPREP, 2016).

43.5.2 Established populations

Many current invasive bird problems date back to the 19th century, when attempts were made to introduce birds, especially of European origin, to parts of the world that European people were colonizing (Low, 2001). Most introductions were for aesthetic reasons by acclimatization societies, but species such as the Common Myna were introduced because of anticipated pest-control benefits (Feare and Craig, 1998). Zoos and the pet trade contributed further to the international movement of animals, which resulted in the widespread establishment of non-native species in new environments. Some of these established populations have been targeted for eradication, with varying results. Eradications have concentrated on smaller populations, especially on small islands where the benefits of eradicating invasives can be particularly valuable for endangered wildlife.

Common Myna

By far the largest Common Myna eradication to date has been that on Atiu (2900 ha) in the Cook Islands, where over 26,000 mynas are estimated to have been killed (G. McCormack, 2018, unpublished results). This complex project began as a control attempt, aimed at reducing the myna population initially by poisoning (Starlicide) and community trapping, using traditional chicken traps and supported by a bounty scheme. After 2 years, when the benefits of the reduced myna numbers became apparent to the island community in terms of reduced fruit damage and harassment of native birds, they decided to aim for total eradication and began shooting to augment the other control methods. Poisoning, estimated to have accounted for the deaths of approximately 11,500 birds, was stopped after 3.5 years when the reduced number of mynas no longer formed large feeding flocks that could be targeted, and shooting and trapping, now using decoy traps, became the techniques to finalize the eradication, accounting for 10,497 and 4768 (combined traditional and decoy trapping) birds, respectively. Since late 2015, post-eradication vigilance has detected a further six mynas, of which four have been shot. These are believed to have been birds missed earlier, rather than new immigrants.

Attempts to achieve a rapid knock-down of numbers of Common Mynas, using the toxicant Starlicide, on Denis (143 ha) and North (201 ha) islands in the Seychelles, did not achieve the anticipated levels of kill that might have hastened the eradications, possibly due to the development of aversion to the treated bait and bait locations (Feare, 2010b, but see Avery and Eisemann, 2015, for discussion). However, these attempted knockdowns were not followed by determined use of additional measures to achieve eradication (Millett *et al.*, 2004). On Frégate (219 ha) and Denis Islands, eradication was eventually achieved largely through trapping (Canning, 2011; Feare *et al.*, 2016), as well as on North Island, with what is believed the last myna shot in February 2019 (Green Islands Foundation, Seychelles, 2019, personal communication).

Towards the end of the Common Myna trapping programmes in Seychelles, some individuals appeared to be 'trap shy'. It is unknown whether these individuals were wary of traps throughout the programmes or whether they learned to avoid them during the programme. On Frégate Island, Canning (2011) overcame this by resorting to different trap types at the end of the eradication, whereas on Denis and North Islands, marksmen with suitable firearms were employed to dispatch the final birds.

Red-whiskered Bulbul and Madagascar Fody

Following the successes of trapping Common Mynas in the Seychelles, trapping was considered as a useful technique for eradicating two other passerines that posed a threat to indigenous avifauna in these islands. In 1976–1977, Red-whiskered Bulbuls (*Pycnonotus jocosus*) and Madagascar Fodies (*Foudia madagascariensis*) were introduced to Assumption Island, Seychelles, which lies only 27 km from Aldabra Atoll, now a World Heritage Site managed by the Seychelles Island Foundation. Aldabra is relatively undisturbed by humans and supports a wide diversity of endemic and indigenous fauna and flora, including an endemic species of fody, *Foudia aldabrana*, and an endemic subspecies of bulbul, *Hypsipetes madagascariensis rostratus*, both of which were potentially threatened by the nearby presence of the introduced relatives. To avert the threat, eradication of these species was commenced in 2011. In contrast to their commensal behaviour elsewhere (Safford and Hawkins, 2013), early trials revealed that neither species on Assumption Island was attracted to artificially offered foods (e.g. rice, bread, fruits, dried mealworms, meat/fish) or to water in small ponds created for the birds. Nor were they attracted to broadcasts of conspecific calls or to decoy birds within cages. These attributes precluded trapping and were subsequently found to apply also to these invasive species on Aldabra Atoll. When roosting communally at night, however, it proved possible to catch commuting groups of birds in mist nets set across flight lines and this method became the main tool of the eradication programmes until low bird densities led to very low capture rates. At this stage, shooting became the final eradication technique for widely dispersed survivors, and eradication of both species was achieved (Bunbury *et al.*, 2015).

Ruddy Duck

For some invasive bird eradications, shooting by competent marksmen has been the main tool. In the UK, introduced Ruddy Ducks (*Oxyura jamaicensis*) are being targeted in a government-led eradication (see Chapter 27, this volume). This is aimed at protecting the closely related endangered White-headed Duck (*Oxyura leucocephala*) population of south-western Europe from genetic introgression (Hughes *et al.*, 1999), as Ruddy Ducks have spread from the population originating in the UK through France and Spain, leading to limited hybridization with White-headed Ducks.

Ruddy Ducks are entirely aquatic, spending most of their time on the surface of freshwater bodies. Preliminary feasibility studies (Henderson, 2006) identified shooting, using shotguns and rifles, from boats to target flocks in winter and shooting from the bank at major breeding sites, targeting especially females to depress productivity, as the most promising eradication tactics (Henderson, 2009). Between 2005 and 2009, 95% of the UK Ruddy Duck population was shot (Henderson, 2009, 2010). The estimated UK population at the end of the winter of 2017–2018 was approximately 23 birds, including at least five adult females. The birds were widely scattered across the UK and mixed-sex groups occurred in Northern Ireland, West and North London, and Greater Manchester and Cheshire. The last of these is of the most immediate concern, as north-west England is the only region where breeding seems to occur annually and was the only region where the birds bred in 2017. There was no evidence of breeding anywhere in the UK during September 2018 (I. Henderson, personal communication).

This achievement was facilitated by the birds' concentration on a limited number of preferred water bodies for winter flocking and their continued use of these waters during and after shooting. Thus, a particular aspect of their behaviour rendered them vulnerable to the chosen eradication tactics.

Wild Turkey

From an initial introduction of seven birds in 1975, the Wild Turkey (*Meleagris gallopavo*) population on Santa Cruz Island, California, USA, grew to an estimated 310 in 2006. Fearing continued growth of the turkey population with consequent elevated threats to native biota, managers initiated a removal programme. Investigators exploited the tendency of turkeys to flock in the winter and devised an integrated management approach using baited drop nets, precision shooting and monitoring of surgically sterilized, radio-telemetered 'Judas turkeys'. By October 2007, the only turkeys remaining were several of the 'Judas turkeys', which were monitored until the last one died in 2012 (Morrison *et al.*, 2016).

Ring-necked Parakeet

Ring-necked Parakeets (*Psittacula krameri*) are the most widely kept and traded parrot species, and have escaped from captivity, or been released, in many countries (see Chapter 9, this volume). They are widely seen as competitors for nest sites with native hole-nesting species (Strubbe and Matthysen, 2009), including the congeneric Echo Parakeet (*Psittacula eques*) of Mauritius (Mauritian Wildlife Foundation, 2018) and the Seychelles Black Parrot (*Coracopsis barklyi*), restricted to Praslin Island (Reuleaux *et al.*, 2013) and now recognized as a species endemic to the Seychelles (Jackson *et al.*, 2015). The latter was considered to be vulnerable to nest site competition from Ring-necked Parakeets that had established a large feral population on Mahé, and of which one individual had appeared on Praslin and Silhouette Islands (Bunbury *et al.*, 2015). To mitigate this risk to the Black Parrot, the Seychelles Island Foundation initiated an eradication programme for the Ring-necked Parakeet in 2013. Most of the parakeets on Mahé roosted in clumps of bamboo in the south-east of the island, but control there was deemed unwise as it risked breaking up the roost into satellite roosts all over the island, some of which might be inaccessible. Various control tactics were explored (e.g. high-level mist netting), but shooting birds at feeding sites by marksmen proved to be the most successful eradication tool. Potential shooting sites were identified by the Seychelles Island Foundation staff and through media appeals to the public, supported by a bounty payment for notifications that led to the shooting of a parakeet. What is believed to have been the last bird was shot in August 2017. Monitoring of reports of further birds, still supported by the bounty scheme, is continuing to verify the eradication. Up to September 2018, no further parakeets have been discovered (Seychelles Island Foundation, 2018).

Monk Parakeet

In the USA, free-flying Monk Parakeets (*Myiopsitta monachus*) were first reported from the New York/New Jersey metropolitan area in 1967, and nest construction was observed in 1970. These earliest populations in the New York/New Jersey area probably originated from escaped cage birds. The Monk Parakeet was first recorded as breeding in Florida in 1969 (Owre, 1973) and has been resident there ever since. In Connecticut, Monk Parakeets were first recorded in 1971 (Neidermyer and Hickey, 1977). Import records reveal that during 1968–1972, more than 63,000 Monk Parakeets were imported into the USA from South America, mostly from Paraguay (CITES, 2018).

Concerns over possible agricultural damage, transmission of psittacosis and interspecific competition with native wildlife precipitated a nationwide Monk Parakeet retrieval programme coordinated by the US Fish and Wildlife Service (Neidermyer and Hickey, 1977). During 1970–1975, there were 367 confirmed sightings of Monk Parakeets in 30 states, and 163 birds were removed from 16 states, mostly from New York (88) and California (35). The 163 birds removed represented 44% of the estimated population at that time, and authorities considered the programme a success in reducing the growth and spread of Monk Parakeet populations (Neidermyer and Hickey, 1977). Since 1975, there has been no coordinated, large-scale Monk Parakeet control effort, and the species is currently firmly established in several states of the USA.

Several lessons can be taken from this experience:

1. Biosecurity was lax, and thousands of birds were imported before official permitting was established. Even after a Federal permitting process was in place, thousands more were imported until the practice was prohibited in the early 1990s.
2. EDRR was non-existent. Through the commercial cage-bird trade, Monk Parakeets were imported and spread around the country for years with no management response. The single management response that was organized (Neidermyer and Hickey, 1977) was insufficient, and there was no follow-up.
3. There was little strong public opposition to the one nationwide eradication effort during 1970–1975. Currently, Monk Parakeets are fiercely defended whenever and wherever any management actions are proposed. A prime opportunity for eradication was clearly missed.

Rock Pigeon

Rock Pigeons (*Columba livia*) were first recorded in the Galápagos Islands in the early 1970s, and by 2000, the total population had grown to over 600 birds on three islands: Santa Cruz, San Cristóbal and Isabela (Phillips *et al.*, 2012). To eliminate the potential of pigeons transmitting diseases to humans and native wildlife, authorities opted to implement a pre-emptive eradication programme. The integrated management approach included toxic baiting with α-chloralose (only on Santa Cruz), shooting and confiscation of captive pigeons. The programme ended in 2007 after 1477 Rock Pigeons had been removed, and the species was formally declared eradicated from the Galápagos Islands. Phillips *et al.* (2012) noted several factors that contributed to the success of the programme: (i) the pigeon populations were dependent on humans and occurred in accessible, confined areas; (ii) the pigeons' flocking and feeding behaviour combined with lack of wariness made them very vulnerable to shooting; (iii) a diverse set of methods was not needed because the shooting programme was implemented safely and efficiently, and it proved to be highly effective; and (iv) much of the field work was conducted by reliable local residents who were integrated into and trusted by the communities.

House Sparrow

On Mauritius, a variety of techniques were used in an attempt to eradicate House Sparrows from Round Island to pave the way for introducing some of Mauritius's endangered endemic birds (Bednarczuk *et al.*, 2010). During the attempt, however, it became apparent that Round Island's House Sparrow population was not closed and that immigration from other nearby islands occurred, causing the eradication attempt to be curtailed.

43.6 Duration and Cost

The samples on which to base the estimated duration and costs of eradications are small, and few data are available on the latter. Furthermore, post-eradication vigilance, involving ongoing costs, is essential to confirm that eradication has been completed and to detect any new arrivals, especially important on archipelagos where other populations of the invasive species exist.

In general, Tables 43.1 and 43.2 show that eradication of founder populations takes considerably less time than eradication of established populations. The apparent exception is the eradication of founder populations of Common Starlings in Western Australia, which requires ongoing vigilance over a large area and the repeated eradication of incursions as they are discovered.

The successful eradications of established populations of invasive birds have all been multi-year projects, involving a number of staff. On Denis and North Islands, in the Seychelles, costs have been saved by using volunteers to undertake most of the trapping, but unexpected resignations or illness of volunteers, and difficulties of recruitment of new staff at the end of short-term (often 6-month) contracts, prolonged the eradications (Feare *et al.*, 2016; C.J. Feare *et al.*, unpublished data). Canning (2011), a permanent member of staff on Frégate Island, on the other hand, took only 8 months to eradicate Common Mynas. The eradications of Red-whiskered Bulbuls and Madagascar Fodies on Assumption Island, and that of Ring-necked Parakeets on Mahé, Seychelles, were completed in less than 5 years using dedicated staff. This suggests that the Denis and North Islands myna eradications could have been achieved over shorter time scales. Whether the eradication of Common Mynas on Atiu, in the Cook Islands, could have been completed in less than the 9 years taken so far is doubtful, given the large sizes of the island, much of it forested, and of the myna population. This also applies to the Ruddy Duck eradication in UK, where, despite the dedicated team of practitioners, the birds' wide geographical distribution, open-water habitats and difficulties of access to some waters have all led to a prolonged eradication process.

Campbell *et al.* (2015) estimated that the annual expenditure on vigilance and control of Common Starling founder populations entering the southern regions of Western Australia, of approximately AUS$1.2 million (approximately US$ 864,000) up to 2009 (dropping to AUS$600,000 thereafter following budget cuts) was economically justified in view of the potential annual economic damage inflicted on Western Australian agriculture of AUS$42.8 million (US$30.8 million) by the estimated carrying capacity of starlings in Western Australia. This damage estimate referred only to agricultural produce and did not include ecosystem or human social costs. Campbell *et al.* (2015) anticipated that expenditure on technological developments in the detection and management of starlings would provide further economic benefits.

For established populations, Feare *et al.* (2016) estimated that the overall cost of the eradication of Common Mynas on Denis Island, Seychelles, was approximately US$156,950. The cost of the much larger myna eradication on Atiu, Cook Islands, has been estimated at around NZ$270,000 (US$178,200) (G. McCormack, unpublished data).

Cost estimates of further eradications of invasive birds, currently in progress or nearing completion, are needed to assist in the planning of future eradication attempts, and the lessons learned from all eradication attempts will hopefully increase operational efficiency and reduce costs in the future.

43.7 Conclusions

Successful eradications of invasive bird populations, although much fewer than for invasive mammals, have taken place mostly on islands. Demonstration of successful eradications could increase the call for more such operations. Avian eradications were carried out using various combinations of shooting, toxic baiting and trapping. In almost every case, the justification for eradication was protection of native biota. Careful, thorough, up-front planning was essential for every effective eradication programme. Coordination with and integration of local populace and authorities was vital to the planning and implementation of eradication efforts. It is doubtful that eradication of invasive bird populations on a continent-wide basis is feasible. Long-term management of established mainland populations might be a more realistic option, given the constraints of cost and public opinion.

In two of the case histories we described, authors invoked the Precautionary Principle as justification for proceeding with eradication programmes. As stated by Rogers *et al.* (1997) 'the Principle requires action to prevent serious and irreversible damage even before harm can be scientifically demonstrated or economically assessed'. This principle was applied to eradication of Rock Pigeons in the Galápagos out of concern for disease and health risks (Phillips *et al.* 2012), and also in eradication of Wild Turkeys from Santa Cruz Island, USA, to eliminate a prey base for a top-level predator (Morrison *et al.*, 2016). This principle was also applied to House Sparrow invasions in the Lesser Antilles (Clergeau *et al.* 2004). Simberloff (2003) expressed a similar point of view, arguing that an immediate management response to invasive species is more effective than spending the time and resources to study the problem while the invader proliferates, spreads and becomes increasingly more difficult and expensive to eradicate. Managers might be uncomfortable acting without complete knowledge of the impacts and costs of an invasive species, but broader application of the Precautionary Principle is worth consideration.

43.8 Acknowledgements

We are grateful to Gerald McCormack for providing us with the latest information on the eradication of Common Mynas on Atiu, Cook Islands. Michael Moulton shared unique insights into invasive bird biology and facilitated access to many original information sources.

43.9 References

Agoramoorthy, G. and Hsu, M.J. (2007) Ritual releasing of wild animals threatens island ecology. *Human Ecology* 35, 251–254.

Anon. (2009) Myna eradication programme. DRC1339 trial report. Terrestrial Resources Conservation Section, Division of Environment and Conservation, Samoa.

Avery, M.L. and Eisemann, J.D. (2015) Invasive myna control in American Samoa. *Vertebrate Pest Conference* 26, 140–144.

American Veterinary Medical Association (2001) 2000 report of the AVMA panel on euthanasia. *Journal of the American Veterinary Medical Association* 218, 669–696.

Beaver, K. and Mougal, J. (eds) (2009) Review of IAS control and eradication programmes in Seychelles. Plant Conservation Action Group. Available at: www.pcaseychelles.org/uploads/1/2/3/6/12369400/seychelles_ias_control-erad_review_pca_report_only.pdf (accessed 12 November 2019).

Bednarczuk, E., Feare, C.J., Lovibond, S., Tatayah, V. and Jones, C.G. (2010) Attempted eradication of house sparrows *Passer domesticus* from Round Island (Mauritius), Indian Ocean. *Conservation Evidence* 7, 75–86.

Bennett, F.D. and Hughes, I.W. (1959) Biological control of insect pests in Bermuda. *Bulletin of Entomological Research* 50, 423–436.

Blackburn, T.M., Lockwood, J.L. and Cassey, P. (2009) *Avian Invasions.* Oxford University Press, New York.

Bomford, M. and O'Brien, P. (1995) Eradication or control for vertebrate pests? *Wildlife Society Bulletin* 23, 249–255.

Bunbury, N., Moumou, J., Page, N., Feare, C., Crommenacker, J., *et al.* (2015) The eradication of introduced red-whiskered bulbuls and Madagascar fodies from Assumption and Aldabra. In: Rocamora, G. and Henriette, E. (eds) *Invasive Alien Species in Seychelles.* Biotope, Mèze, France, and Muséum National d'Histoire Naturelle, Paris.

Campbell, S., Roberts, E.J., Craemer, R., Pacioni, C., Rollins, L. and Woolnough, A.P. (2015) Assessing the economic benefits of starling detection and control to Western Australia. *Australasian Journal of Environmental Management* 23, 81–99.

Campbell, K. and Donlan, C.J. (2005) Feral goat eradications on islands. *Conservation Biology* 19, 1362–1374.

Canning, G. (2011) Eradication of the invasive common myna, *Acridotheres tristis*, from Frégate Island, Seychelles. *Phelsuma* 19, 43–53.

Cardador, L., Lattuada, M., Strubbe, D., Tella, J.L., Reino, L., *et al.* (2017) Regional bans on wild-bird trade modify invasion risks at a global scale. *Conservation Letters* 10, 717–725.

CITES (2018) CITES trade database. Available at: https://trade.cites.org/ (accessed 25 September 2018).

Clergeau, P. and Yésou, P. (2006) Behavioural flexibility and numerous potential sources of introduction for the sacred ibis: causes of concern in western Europe? *Biological Invasions* 8, 1381–1388.

Clergeau, P., Levesque, A. and Lorvelec, O. (2004) The precautionary principle and biological invasion: the case of the House Sparrow on the Lesser Antilles. *International Journal of Pest Management* 50, 83–89.

Cooney, R. and Jepson, P. (2006) The international wild bird trade: what's wrong with blanket bans? *Oryx* 40, 18–23.

DeCino, T.J., Cunningham, D.J. and Schafer, E.W. (1966) Toxicity of DRC-1339 to starlings. *Journal of Wildlife Management* 30, 249–253.

Department of Primary Industries (2010) Invasive plants and animals policy framework. State of Victoria, Australia. Available at: http://agriculture.vic.gov.au/agriculture/pests-diseases-and-weeds/protecting-victoria-from-pest-animals-and-weeds/invasive-plants-and-animals/invasive-plants-and-animals-policy-framework (accessed 2 October 2018).

Dunlap, T.R. (1997) Remaking the land: the acclimatization movement and Anglo ideas of Nature. *Journal of World History* 8, 303–319.

Eisemann, J.D., Pipas, P.A. and Cummings, J.L. (2003) Acute and chronic toxicity of compound DRC-1339 (3-chloro-4- methylaniline hydrochloride) to birds. In: Linz, G.M. (ed.) *Management of North American Blackbirds.* USDA APHIS National Wildlife Research Center, Fort Collins, Colorado, pp. 49–63.

Fanchette, R. (2003) Successful eradication of an IAS from Mahe island, Seychelles (House Sparrow *Passer domesticus*) 2002–2003. Unpublished report by the Government of Seychelles.

Feare, C. and Craig, A. (1998) *Starlings and Mynas.* Christopher Helm, London.

Feare, C.J. (2010a) Invasive bird eradication from tropical oceanic islands. *Aliens* 30, 12–19.

Feare, C.J. (2010b) The use of Starlicide in preliminary trials to control invasive common myna *Acridotheres tristis* populations on St Helena and Ascension islands, Atlantic Ocean. *Conservation Evidence* 7, 52–61.

Feare, C.J. and Mungroo, Y. (1990) The status and management of the house crow *Corvus splendens* in Mauritius. *Biological Conservation* 51, 63–70.

Feare, C.J. and Saavedra, S. (2009) Development of a strategy for control of common mynas (*Acridotheres tristis*) on St Helena. Unpublished report to the Royal Society for the Protection of Birds and the South Atlantic Invasive Species Project.

Feare, C.J., van der Woude, J., Greenwell, P., Edwards, H.A., Taylor, J.A., *et al.* (2016) Eradication of common mynas *Acridotheres tristis* from Denis Island, Seychelles. *Pest Management Science* 73, 295–304.

Fenneman, J. (2011) Checklist of the Birds of British Columbia. LGL Limited Environmental Research Associates, Sidney, British Columbia, Canada. Available at: http://ibis.geog.ubc.ca/biodiversity/efauna/documents/BCBirdChecklistJan2011.pdf (accessed 17 September 2018).

Fernandez, S. (2015) Eradicating the sacred ibis around the Mediterranean basin. In: Sarat, E., Mazaubert, E., Dutartre, A., Poulet, N. and Soubeyran, Y. (eds) *Invasive alien species in aquatic environments. Practical information and management insights.* Vol. 2: *Management Insights.* Knowledge for Action Series, ONEMA. Vincennes, France, pp. 189–195. Available at: www.gt-ibma.eu/wp-content/uploads/2017/08/eee-vol-2-complet-12-10-16-1.pdf (accessed 2 October 2018).

Fleming, P.J.S., Ballard, G., Reid, N.C.H. and Tracey, J.P. (2017) Invasive species and their impacts on agri-ecosystems: issues and solutions for restoring ecosystem processes. *Rangeland Journal* 39, 523–535.

Glahn, J.F. and Avery, M.L. (2001) Estimation of red-winged blackbird mortality from toxic bait application. In: Johnston, J.J. (ed.) *Pesticides and Wildlife.* American Chemical Society Symposium Series 771. American Chemical Society, Washington, DC, pp. 109–118.

Glen, A.S., Atkinson, R., Campbell, K.J., Hagen, E., Holmes, N.D., *et al.* (2013) Eradicating multiple invasive species on inhabited islands: the next big step in island restoration? *Biological Invasions* 15, 2589–2603.

Grarock, K., Tidemann, C.R., Wood, J.T. and Lindenmayer, D.B. (2014) Understanding basic species population dynamics for effective control: a case study on community-led culling of the common myna (*Acridotheres tristis*). *Biological Invasions* 16, 1427–1440.

Harvey, R.G. and Mazzotti, F.L. (2014) The invasion curve: a tool for understanding invasive species management in south Florida. IFAS Extension Report WEC347, Wildlife Ecology and Conservation Department, University of Florida, Florida. Available at: https://edis.ifas.ufl.edu/uw392 (accessed 12 November 2018).

Henderson, I. (2010) The eradication of ruddy ducks in the United Kingdom. *Aliens Newsletter* 29, 17–24.

Henderson, I.S. (2006) Recent measures to control ruddy ducks *Oxyura jamaicensis* in the United Kingdom. In: Boere, G.C., Galbraith, C.A. and Stroud, D.A. (eds) *Waterbirds around the World*. The Stationery Office, Edinburgh, UK, pp. 822–825.

Henderson, I.S. (2009) Progress of the UK ruddy duck eradication programme. *British Birds* 102, 680–690.

Herring, G. and Gawlik, D.E. (2008) Potential for successful population establishment of the non-indigenous sacred ibis in the Florida Everglades. *Biological Invasions* 10, 969–976.

Hobson, E.A., Smith-Vidaurre, G. and Salinas-Melgoza, A. (2017) History of nonnative monk parakeets in Mexico. *PLoS One* 12, e0184771.

Howald, G., Donlan, C., Galván, J.P., Russell, J.C., Parkes, J., *et al.* (2007) Invasive rodent eradication on islands. *Conservation Biology* 21, 1258–1268.

Hughes, B., Criado, J., Delany, S., Gallo-Orsi, U., Green, A.J., *et al.* (1999) The status of the North American ruddy duck *Oxyura jamaicensis*. In: *The Western Palearctic: Towards an Action Plan for Eradication, 1999–2002*. Council of Europe Publication T-PVS/Birds (99) 9. Council of Europe Publishing, Strasbourg, France.

Jackson, H.A., Bunbury, N., Przelomska, N. and Groombridge, J.J. (2015) Evolutionary distinctiveness and historical decline in genetic diversity in the Seychelles black parrot *Coracopsis nigra barklyi*. *Ibis* 158, 380–394.

Junta de Andalucía (2011) Control del ibis sagrado africano en Andalucía. *Boletin Informative Sobre Geodiversidad de Andalucía*, 18, 1–2. Available at: www.junta-deandalucia.es/medioambiente/portal_web/web/servicios/centro_de_documentacion_y_biblioteca/fondo_editorial_digital/revistas_boletines/geobio/numero_18/ibis_sagrado_ampliada.pdf (accessed 12 November 2019).

Kurdila, J. (1995) The introduction of exotic species into the United States: there goes the neighborhood. *Environmental Affairs* 16, 95–118.

Lever, C. (2005) *Naturalized Birds of the World*. T. & A.D. Poyser, London.

Low, T. (2001) *Feral Future*. Penguin Books, Ringwood, Victoria, Australia.

Martin-Albarracin, V.L., Amico, G.C., Simberloff, D. and Nuñez, M.A. (2015) Impact of non-native birds on native ecosystems: a global analysis. *PLoS One* 10, e0143070.

Mauritian Wildlife Foundation (2018) Echo parakeet *Psittacula eques*. Available at: www.mauritian-wildlife.org/projects (accessed 12 November 2018).

McDowall, R.M. (1994) *Gamekeepers for the Nation: The Story of New Zealand's Acclimatisation Societies 1861–1990*. Canterbury University Press, Christchurch, New Zealand.

Millett, J., Climo, G. and Jivan Shah, N. (2004) Eradication of the common myna *Acridotheres tristis* populations in the granitic Seychelles: successes, failures and lessons learned. *Advances in Vertebrate Pest Management* 3, 169–183.

Morrison, S.A., DeNicola, A.J., Walker, K., Dewey, D., Laughrin, L., *et al.* (2016) An irruption interrupted: eradication of wild turkeys *Meleagris gallopavo* from Santa Cruz Island, California. *Oryx* 50, 121–127.

Neidermyer, W.J. and Hickey, J.J. (1977) The monk parakeet in the United States, 1970–1975. *American Birds* 31, 273–278.

Owre, O.T. (1973) A consideration of the exotic avifauna of southeastern Florida. *Wilson Bulletin* 85, 491–500.

Parkes, J. (2012) Review of best practice management of common mynas (*Acridotheres tristis*) with case studies of previous attempts at eradication and control: a working document. Contract Report LC986, prepared for Durrell Wildlife Conservation Trust, Landcare Research, Lincoln, New Zealand.

Phillips, R.B., Cooke, B.D., Carrión, V. and Snell, H.L. (2012) Eradication of rock pigeons, *Columba livia*, from the Galápagos Islands. *Biological Conservation* 147, 264–269.

Pimentel, D., McNair, S., Janecka, J., Wightman, J., Simmonds, C., *et al.* (2001) Economic and environmental threats of alien plant, animal, and microbe invasions. *Agriculture, Ecosystems and Environment* 84, 1–20.

Puttoo, M. and Archer, T. (2004) Control and/or eradication of Indian crows (*Corvus splendens*) in Mauritius. In: Lalouette, J.A. and Bachraz, D.Y. (eds) *Proceedings of the Sixth Meeting of Agricultural Scientists*. Food and Agricultural Research Council, Réduit, Mauritius, pp. 299–309.

Reuleaux, A., Bunbury, N., Villard, P. and Waltert, M. (2013) Status, distribution and recommendations for monitoring of the Seychelles black parrot *Coracopsis* (*nigra*) *barklyi*. *Oryx* 47, 561–568.

Robinson, S.A. and Copson, G.R. (2014) Eradication of cats (*Felis catus*) from subantarctic Macquarie Island. *Ecological Management and Restoration* 15, 34–40.

Rocamora, G. and Henriette, E. (2015) *Invasive Alien Species in Seychelles*. Biotope, Mèze, France, and Muséum National d'Histoire Naturelle, Paris.

Rogers, M.F., Sinden, J.A. and De Lacy, T. (1997) The precautionary principle for environmental management: a defensive-expenditure application. *Journal of Environmental Management* 51, 343–360.

Rollins, L.A., Woolnough, A.P., Sinclair, R., Mooney, N.J. and Sherwin, W.B. (2011) Mitochondrial DNA offers unique insights into invasion history of the common starling. *Molecular Ecology* 20, 2307–2317.

Rollins, L.A., Woolnough, A.P., Wilton, A.N., Sinclair, R. and Sherwin, W.B. (2009) Invasive species can't cover their tracks: using microsatellites to assist management of starling (*Sturnus vulgaris*) populations in Western Australia. *Molecular Ecology* 18, 1560–1573.

Ryan, C.S. (1906) President's address: on European and other birds liberated in Victoria. *Emu* 5, 110–119.

Saavedra, S. (2010) Eradication of invasive mynas from islands. Is it possible? *Aliens* 29, 40–47.

Safford, R. and Hawkins, F. (2013) *The Birds of Africa*, Vol. 8: *The Malagasy Region*. Christopher Helm, London.

Seychelles Island Foundation (2018) Seychelles Island Foundation Newsletter, Issue 68, August 2018. Available at: www.sif.sc/sites/default/files/downloads/SIF%20Newsletter%20AUGUST%202018_0.pdf (accessed 13 September 2018).

Simberloff, D. (2003) How much information on population biology is needed to manage introduced species? *Conservation Biology* 17, 83–92.

Simberloff, D. (2009) We can eliminate invasions or live with them; successful management projects. *Biological Invasions* 11, 149–157.

South Florida Ecosystem Task Force (2015) EDRR case study: the sacred ibis project. Available at: https://evergladesrestoration.gov/content/ies/docs/Goal2_Case_Study_Sacred_Ibis.pdf (accessed 29 September 2018).

SPREP (2016) Manage myna birds in the Pacific. Secretariat of the Pacific Regional Environment Programme (SPREP), Apia, Samoa.

Strubbe, D. and Matthysen, E. (2009) Experimental evidence for nest-site competition between invasive ring-necked parakeets (*Psittacula krameri*) and native nuthatches (*Sitta europaea*). *Biological Conservation* 142, 1588–1594.

Suleiman, A.S. and Taleb, N. (2010) Eradication of the house crow *Corvus splendens* on Socotra, Yemen. *Sandgrouse* 32, 136–140.

Temple, S.A. (1992) Exotic birds: a growing problem with no easy solution. *Auk* 109, 395–397.

Thomson, G.M. (1922) *The Naturalisation of Animals and Plants in New Zealand*. Cambridge University Press, Cambridge.

Tidemann, C.R. (2005) Indian mynas – can the problems be controlled? *National Urban Animal Management Conference* 15, 55–57.

Tsachalidis, E., Sokos, C., Birtsas, P. and Patsikas, N. (2006) The Australian crow trap and the Larsen trap: their capture success in Greece. In: Manolas, E.I. (ed.) *Proceedings of the 2006 Naxos International Conference on Sustainable Management and Development of Mountainous and Island Areas*, Vol. II. Department of Forestry and Management of the Environment and Natural Resources, Democritus University of Thrace, Komotini, Greece, pp. 325–329.

Usher, M.B. (1989) Ecological effects of controlling invasive terrestrial vertebrates. In: Drake, J.A., Mooney, J.A., di Castri, H.A., Groves, F., Kruger, R.H., *et al.* (eds) *Biological Invasions: A Global Perspective*. Wiley, Chichester, UK, pp. 463–489.

Vaughan, J.H. (1930) The birds of Zanzibar and Pemba, part II. *Ibis* 72, 1–48.

West, B. and Zhou, B.-X. (2007) Did chickens go north? New evidence for domestication. *World's Poultry Science Journal* 45, 205–218.

Woolnough, A.P., Massam, M.C., Payne, R.L. and Pickles, G.S. (2005) Out on the border: keeping starlings out of Western Australia. *Australasian Vertebrate Pest Conference* 13, 183–189.

Yésou, P. and Clergeau, P. (2005) Sacred ibis: a new invasive species in Europe. *Birding World* 18, 517–526.

Yésou, P., Maillard, J.-F. and Simon, L. (2015) Managing the sacred ibis in western France. In: Sarat, E., Mazaubert, E., Dutartre, A., Poulet, N. and Soubeyran, Y. (eds) *Invasive Alien Species in Aquatic Environments. Practical Information and Management Insights*. Vol. 2: *Management Insights*. Knowledge for Action Series, ONEMA, Vincennes, France, pp. 196–199. Available at: www.gt-ibma.eu/wp-content/uploads/2017/08/eee-vol-2-complet-12-10-16-1.pdf (accessed 2 October 2018).

44 Using citizen science to study exotic and invasive birds

Corey T. Callaghan[1]* and Daniel M. Brooks[2]

[1]*Centre for Ecosystem Science, School of Biological, Earth and Environmental Sciences, UNSW Sydney, Sydney, NSW 2052, Australia;* [2]*Houston Museum of Natural Science, Department of Vertebrate Zoology, Houston, TX 77019, USA*

Citation: Callaghan, C.T. and Brooks, D.M. (2020) Using citizen science to study exotic and invasive birds. In: Downs, C.T. and Hart, L.A. (eds) *Invasive Birds: Global Trends and Impacts*. CAB International, Wallingford, UK, pp. 362–367.

44.1 Introduction

Citizen science, broadly defined as cooperation between a range of experts and amateurs, involving some sort of public engagement, education and data collection (Jordan *et al.*, 2015), is rapidly expanding our knowledge of biodiversity across multiple taxa (Silvertown, 2009; Dickinson *et al.*, 2012), and is useful for planning and implementing conservation strategies (Danielsen *et al.*, 2014). Thanks to citizen science data, the temporal and spatial scales of ecological questions being addressed are becoming refined. However, there is frequently a gap between the scale of data collection and the potential for conservation-oriented policies (Pocock *et al.*, 2018). The study and management of exotic and invasive birds is generally one of small scale, and targeted responses are necessary. Citizen scientists are readily being used as sentinels in a variety of projects aimed at studying invasive species (Theobald *et al.*, 2015). Table 44.1 provides some selected examples of such projects, including cane toads *Rhinella marina*, Mediterranean geckos *Hemidactylus turcicus* and invasive plants.

Ornithological research has long-relied on citizen scientists to enhance our knowledge of the ecology and life history of our native avifauna. Some of the longest-running citizen science projects are, indeed, bird focused, such as the Christmas Bird Count or Breeding Bird Surveys in Britain (Risely *et al.*, 2010) and North America (Sauer *et al.*, 2014), or bird atlassing (e.g. southern Africa; SABAP2, 2019). These projects have added to our general understanding of the distribution and population trends of many bird species, including exotic birds. Many opportunistic observations made by non-professional ornithologists are frequently used in the study of exotic birds. For example, Thibault *et al.* (2018) relied in part on opportunistic observations of the Red-vented Bulbul (*Pycnonotus cafer*) in a recent review of the potential impacts of this species. While this type of informal use of citizen scientists has traditionally been utilized in the study of exotic birds, citizen science is now 'mainstream' (Theobald *et al.*, 2015). With a few exceptions (e.g. Brooks, 2013; Conn *et al.*, 2017), the formal application of citizen science to exotic and invasive bird management has been relatively unexplored and is currently being underutilized.

In this chapter, we highlight the current usage of citizen science in furthering our knowledge of exotic and invasive bird populations. To do this, we split citizen science projects aimed at studying and tracking exotic birds into two categories: (i) those that are part of a broad-scale biodiversity collection scheme; and (ii) those that are species' focused. We summarize some of the current literature that integrates citizen science data and the effects, management and implications of introduced birds. We rely on select examples to highlight the potential of citizen science aimed at gathering various types of information about exotic birds and conclude by discussing a vision for the future of citizen science aimed at studying exotic birds.

44.2 Citizen Science and Invasive Birds

The potential impacts of non-native birds (Temple, 1992; Baker *et al.*, 2014) and the rise of 'invasive species denialism' (Russell and Blackburn, 2017) highlight the critical necessity of better understanding the role of introduced birds and their impacts. The impacts of invasive birds, in particular, are frequently debated, but indeed contribute to global biotic homogenization (McKinney and Lockwood, 1999) and are potentially a major threat on islands (Sax and Gaines, 2008; for further reviews, see the relevant chapters in this volume). In a

*Corresponding author: c.callaghan@unsw.edu.au

Table 44.1. Select examples of citizen science projects aimed at tracking invasive species populations.

Selected example	Description	Reference
GeckoWatch	A project aiming to map the fine-scale distribution of Mediterranean House Gecko in the USA	https://nhm.org/community-science-nhm/geckowatch
iMapInvasives	A broad project aiming to share information about location, search efforts and treatment outcomes of invasives. Largely focused on plants	www.imapinvasives.org/
TexasInvasives.org	A broad portal aimed at distributing information on invasive species across taxa and enlisting citizen scientists' help	https://texasinvasives.org/
FeralScan	An Australian citizen science project aimed at collecting information on Australia's pest species, including fox, cat, pig, cane toad, and Common Myna and European Starling	www.feralscan.org.au/
Cat Tracker Australia	A two-part citizen science project which first had participants fill out an online survey about their pet cats, and then had volunteers allow their cats to be GPS tracked, gaining knowledge about cat behaviour in the urban environment	https://biocollect.ala.org.au/acsa/project/index/ffa9440e-b2a0-4b89-9545-929503c750dc
Texas Invasive Bird Project	Evaluates status of six species of invasive or introduced birds in the state of Texas	www.hmns.org/invasivebirds

world of climate change, sea-level rise and a myriad of other ecological threats, monitoring of invasive bird species generally receives little attention. Of the top 100 of the world's worst alien species, only three are birds: the Common Myna (*Acridotheres tristis*), Red-vented Bulbul and Common Starling (*Sturnus vulgaris*) (Lowe *et al.*, 2000). This is potentially indicative not only of the minimal impacts introduced birds can have on their ecosystem compared with other taxa but also perhaps of the lack of knowledge on the impacts that introduced birds have on their novel environment. Furthermore, these effects may be rather local in spatial scale, and potentially undetected in broad-scale studies. Frequently, even the most basic information (e.g. life history, breeding biology, phenology, and positive or negative interactions with local avifauna) are lacking for introduced species in their introduced range.

44.2.1 Current usage of citizen science in invasive bird research

Citizen science projects vary in their design, objectives and participation. Citizen science projects can be structured or unstructured, with a continuum between these two delineations (Welvaert and Caley, 2016). Similarly, citizen science projects that have furthered the knowledge of introduced birds can be delineated into two types: (i) the use of broad-scale citizen science data, generally from an already-existing, broader-scoped project; and (ii) the use of targeted citizen science projects, focused on a specific taxon or geographical extent. We describe both below.

Use of broad-scale citizen science data to study exotic birds

Ornithology, perhaps more than any other taxon-driven science, has greatly benefited from citizen science. Indeed, one of the longest-running citizen science projects, the Christmas Bird Count (National Audubon Society, 2012), centres around counting birds annually. These types of data have been used to help understand the effects of Common Starlings on cavity-nesting species (Koenig, 2003) and to help elucidate trends in exotic species, such as the recent range expansion by the Eurasian Tree Sparrow (*Passer montanus*) in North America (Burnett *et al.*, 2017) and the Eurasian Collared-dove (*Streptopelia decaocto*) in Florida (Romagosa and Labisky, 2000).

More recently, broad-scale citizen science data are revolutionizing ecology, providing spatial and temporal scales unimaginable mere decades ago (Pocock *et al.*, 2018). These data refer to spatiotemporal coordinates of observations of a given species. Examples of broad-scale citizen science projects include iNaturalist (an app used to collect data on all taxa; iNaturalist.org, 2018), GBIF (an aggregator providing data from a variety of sources; www.gbif.org/) and eBird (a bird-specific, semi-structured project relying on birdwatchers submitting their observations; Sullivan *et al.*, 2014). Data collected from these databases can be used to track the status of specific species in a simple fashion, monitoring trends over time (e.g. Callaghan and Brooks, 2017), or provide opportunistic observations of where a species occurs (e.g. Thibault *et al.*, 2018). Given a reasonable temporal scale, these data can provide detailed information on the spread of an invasive species, habitat associations and effects on other species (Bonter *et al.*, 2010). Bonter *et al.* (2010) relied on citizen science data to investigate the colonization of the Eurasian Collared-dove throughout the USA, finding that this species was more likely to occur in landscapes that had been highly modified by human activity, and that it apparently had minimal impacts on site-level abundance of other dove species. However, these are broad macroecological patterns, potentially not revealing local-level impacts of this species.

Studies relying on broad-scale citizen science data need not be species specific. For instance, one can confirm the patterns of exotic species abundance, diversity and richness in relation to their environment (e.g. Blair, 1996; Brooks and

Page, 2012), contrasting this with patterns of native abundance, diversity and richness (C.T. Callaghan *et al.*, unpublished data). These types of studies can help by demonstrating exotic diversity 'hotspots' for birds, potentially helpful in highlighting areas where further, more detailed research should be carried out.

Another example of the benefits of broad-scale citizen science projects is relying on citizen scientists to monitor invasive species, alerting government officials on the first presence of a species, whether it be a new location or not (BirdLife International, 2015). This has helped to eradicate the Ruddy Duck (*Oxyura jamaicensis*) in the UK and has helped collect data on exotic geese species to help inform eradication and management efforts (see Chapters 27 and 28, this volume). The use of citizen scientists to detect, monitor and report invasive species has been successful with other taxa, such as plants (Gallo and Waitt, 2011; Mannino and Balistreri, 2018).

Despite the potential for these data to help inform exotic/invasive bird species management, there are several concerns that researchers should be aware of. First, there are several spatial and temporal biases associated with broad-scale biodiversity data that researchers need to account for in their study design if intending to use these data (Boakes *et al.*, 2010). Second, there are biases associated with the submission of exotic species reports, in particular, to some citizen science projects. In a survey of 804 respondents of the birding community (Callaghan, 2017), 68% of the respondents submitted their sightings to eBird. Of these, only 53% reported all exotic bird species, 36% reported only those exotic species that are 'countable' on their life lists (defined by the American Birding Association) and 11% did not report any exotic birds to eBird. These results suggest that people intending to use these datasets to gather information need to assume that they may be incomplete records when compared with native species. Furthermore, these data are rather limiting, generally restricted to biodiversity occurrence. The values of these data are predominantly focused on where species occur in their environments, and the range in which they occur, neglecting potential behavioural, ecological and life history components.

There are important considerations necessary before investigating the effects of non-native bird species, based on data derived from broad-scale citizen science projects:

1. A researcher should understand the limitations of using these types of data, generally restricted to presence-only data, but can demonstrate correlative associations with habitat and interspecific interactions.
2. A researcher should also be aware of the spatial (more sightings near human populations and from Anglophone countries) and temporal (more sightings in recent years and potential intra-annual variability) biases. At the least, analyses should account for these different levels of effort, both spatially and temporally.
3. A researcher should focus on making the results published from any data relying on citizen scientists available to the same individuals who made it possible.

Targeted citizen science projects

Contrasting with the use of broad-scale citizen science projects are targeted citizen science projects focused on a small set of taxa, or a specific taxon. Given that exotic birds generally make up a small proportion of the local avifauna, these are, by definition, targeted. They are aided by a social media push to inform participants of the goals of the project and can take a variety of forms.

For example, the Houston Museum of Natural Science has been working on a variety of local exotic species, relying on the public's input and participation (Texas Invasive Bird Project, 2018). For instance, we have used a variety of means to distribute forms to birdwatchers and interested members of the public to better understand the occurrence and ecology of introduced birds. From this, we now know specific aspects of ecology, behaviour and reproductive biology for species such as the Red-vented bulbul (Brooks, 2013; see Chapter 5, this volume), Egyptian Goose (*Alopochen aegyptiaca*; Callaghan and Brooks, 2016, 2017), Scaly-breasted Munia (*Lonchura punctulata*) and other small granivores (Brooks and Page, 2012; Conn *et al.*, 2017; see Chapters 21 and 22, this volume). These projects extend the broad-scale biodiversity data (see above) and rely on detailed questionnaires to assess various unknown aspects of exotic bird biology. For example, questions asked include a description of the habitat where the bird was seen, whether breeding was documented, whether any positive or negative interspecific interactions were observed and whether supplementary feeding was observed. Not all questions are necessarily applicable for all species. Respondents frequently send in photos, which aid in interpreting the results.

The following steps should be considered before implementing a targeted citizen science project aimed at exotic birds. Much work has been aimed at providing guidelines for the implementation of citizen science projects (e.g. https://ecsa.citizen-science.net/blog/collection-citizen-science-guidelines-and-publications, accessed 15 November 2019), but the following steps are detailed from our own experience and are aimed at exotic bird research in particular:

1. A researcher should first investigate whether a detailed citizen science research project is already under way on a potential species of interest, helping to reduce the number of citizen science projects made available to potential volunteers (Bonney *et al.*, 2014).
2. The species in question should be relatively abundant, where you would expect a reasonable number of responses. For instance, if an exotic species is only present in small numbers, then the likelihood of participants encountering it would be low, diminishing the chances of data being collected.
3. However, the species should be not so far established (e.g. House Sparrow, Common Starling) that there is: (i) already well-known data on these instances; and (ii) that management/control of the population is not viable.
4. We recommend ground-truthing 10–15% of the observations. This involves visiting the location that respondents are reporting and: (i) checking for the presence of the exotic species; (ii) checking that the habitat descriptions match; and (iii) generally making sure that the reports make sense.

5. Be sure to publish results in due course and share these results with the broader citizen science network. It is important to identify the milestones that have been completed to the participants.

44.2.2 The potential of citizen science in invasive bird research

So far, we have highlighted some previous research that has relied on citizen science to investigate the ecology and diversity of exotic birds and delineated the overlap between citizen science and exotic bird research into two categories. But how much potential is there to rely on citizen scientists in enhancing our knowledge of exotic birds in the future?

Amateur birders have incredibly acute knowledge of local natural history that is often overlooked by 'professional' ornithologists (Callaghan *et al.*, 2018). Regarding exotic birds, many birders do enjoy watching and observing them (86%; Callaghan, 2017). As many as 57% of respondents to a survey about exotic birds said that they had travelled specifically to a location to see an exotic bird. As exotic bird populations continue to increase, so does the popularity in birding globally, and most birders are armed with cameras and recording equipment. This suggests that there is potential for birders to substantially contribute to the future study of exotic/invasive species. Indeed, 'unnatural history' (Callaghan *et al.*, 2018) could play a crucial role in future monitoring of potential negative impacts of introduced birds on native ecosystem functions. For instance, interactions between introduced and native species can and should be documented and published in scientific literature – and this published literature, even if only published in local or regional journals, should be made available to the greater ornithological community. Too frequently, potential impacts of exotic birds are labelled anecdotal and speculative but have been observed (Mo, 2015). These can be captured in short research notes with relative ease. This is most likely to occur through increased connection between amateur and professional ornithologists.

Given that one of the largest concerns with invasive birds is that of economic concerns, we are surprised that we could not find any targeted citizen science projects aimed at documenting economic effects of invasive birds, such as projects that enlist volunteers to monitor crop damage by exotic species or nesting species on telephone poles. Indeed, we envision this as a particularly important area to be explored. With some foresight, researchers could design studies that aim to monitor different economic aspects of exotic birds.

Another major threat of introduced birds is genetic swamping of native species (Huxel, 1999). For instance, the native American Black Duck (*Anas rubripes*) has become genetically more similar to the Mallard (*Anas platyrhynchos*) because of hybridization with introduced domestic types (Mank *et al.*, 2004), and a similar scenario has occurred in the White-headed Duck (*Oxyura leucocephala*) because of the introduced Ruddy Duck in Europe (Muñoz-Fuentes *et al.*, 2007). Relatively little attention has been paid to the potential for citizen science to track hybridization of introduced species with native species, but we hypothesize this is a potential avenue of future citizen science projects, especially with a revolution in genetics (Allendorf *et al.*, 2010) and a willingness of the public to collect feathers for science (e.g. Brandis, 2016).

44.3 The Future of Citizen Science and Invasive Bird Research

The public, apart from birders, are often unaware of what birds are native and non-native. But birds are appreciated by the non-scientific public (Cocker *et al.*, 2013), and invasive birds are the least supported for eradication and control programmes of all taxa (Bremmer and Park, 2007). We hypothesize that an added benefit of introducing targeted citizen science programmes aimed at better understanding non-native birds could be increased understanding on the potential impacts of exotic birds, if any. Indeed, people with prior knowledge of control and eradication programmes are more likely to support control programmes (Bremmer and Park, 2007), suggesting that increased knowledge could benefit management of exotic and invasive bird populations. With other taxa, such as plants (Jordan *et al.*, 2011), citizen science programmes have been shown to translate into knowledge gain and behavioural change by the participants (Jordan *et al.*, 2011).

Citizen science projects are continuing to increase in their prevalence but are currently being underutilized in invasive bird research. Many populations of exotic species are relatively 'new'; as you will see in the preceding book chapters in this volume, documented negative interactions between exotic and native species are generally lacking. We believe this gap can potentially be filled by citizen scientists. Ultimately, professional ornithologists need to better communicate with amateur ornithologists, recognizing their knowledge of local natural history, including the role that exotic birds play in their introduced ecosystem. Both broad-scale citizen science data and targeted citizen science projects can play an important role in furthering our understanding of the role of exotic birds in the environment.

44.4 References

Allendorf, F.W., Hohenlohe, P.A. and Luikart, G. (2010) Genomics and the future of conservation genetics. *Nature Reviews Genetics* 11, 697–709.

Baker, J., Harvey, K.J. and French, K. (2014) Threats from introduced birds to native birds. *Emu* 114, 1–12.

BirdLife International (2015) Citizen scientists help track invasive species in Belgium. Available at: http://datazone.birdlife.org/sowb/casestudy/citizen-scientists-help-track-invasive-species-in-belgium (accessed 12 November 2019).

Blair, R.B. (1996) Land use and avian species diversity along an urban gradient. *Ecological Applications* 6, 506–519.

Boakes, E H., McGowan, P.J.K., Fuller, R.A., Chang-qing, D., Clark, N.E., *et al.* (2010) Distorted views of biodiversity: spatial and temporal bias in species occurrence data. *PLoS Biology* 8, e1000385.

Bonney, R., Shirk, J.L., Phillips, T.B., Wiggins, A., Ballard, H.L., *et al.* (2014) Next steps for citizen science. *Science* 343, 1436–1437.

Bonter, D.N., Zuckerberg, B. and Dickinson, J.L.(2010) Invasive birds in a novel landscape: habitat associations and effects on established species. *Ecography* 33, 494–502.

Brandis, K. (2016) Australia's waterbirds are disappearing – but nuclear physics can help save them. *The Conversation*, UK. Available at: https://theconversation.com/australias-waterbirds-are-disappearing-but-nuclear-physics-can-help-save-them-57044 (accessed 12 November 2019).

Bremmer, A. and Park, K. (2007) Public attitudes to the management of invasive non-native species in Scotland. *Biological Conservation* 139, 306–314.

Brooks, D.M. (2013) Ecology, behavior and reproduction of an introduced population of Red-vented Bulbuls (*Pycnonotus cafer*) in Houston, Texas. *Wilson Journal of Ornithology* 125, 800–808.

Brooks, D.M. and Page, G. (2012) High diversity of invasive Passerids at a park in southeast Texas. *Bulletin of the Texas Ornithological Society* 45, 23–29.

Burnett, J.L., Roberts, C.P., Allen, C.R., Brown, M.B. and Moulton, M.P. (2017) Range expansion by *Passer montanus* in North America. *Biological Invasions* 19, 5–9.

Callaghan, C.T. (2017) Birders and Exotics – aesthetics, ecology, fair play: a survey of the birding community's strongly held views about exotic bird species. *Birding* 49, 28–36.

Callaghan, C.T. and Brooks, D.M. (2016) Ecology, behavior, and reproduction of invasive Egyptian Geese (*Alopochen aegyptiaca*) in Texas. *Bulletin of the Texas Ornithological Society* 49, 37–45.

Callaghan, C.T. and Brooks, D.M. (2017) History, current distribution, and status of the Egyptian Goose (*Alopochen aegyptiaca*) in the contiguous United States. *Southwestern Naturalist* 62, 296–300.

Callaghan, C.T., Martin, J.M., Kingsford, R.T. and Brooks, D.M. (2018) Unnatural history: is a paradigm shift of natural history in 21st century ornithology needed? *Ibis* 160, 475–480.

Cocker, M., Tipling, D., Elphick, J. and Fanshawe, J. (2013) *Birds and People*. Jonathan Cape, London.

Conn, A., Lazenby, L., and Brooks, D.M. (2017) Ecology, behavior and reproduction of an invasive population of Scaly-breasted Munias (*Lonchura punctulata*) in Houston, Texas. In: Brooks, D.M. (ed.) *Half a Century of Ornithology in Texas: The Legacy of Dr. Keith Arnold*. Miscellaneous Publications of the Houston Museum of Natural Science No. 7, Houston, Texas, pp. 40–53.

Danielsen, F., Pirhofer-Walzl, K., Adrian, T.P., Kapijimpanga, D.R., Burgess, N.D., *et al.* (2014) Linking public participation in scientific research to the indicators and needs of international environmental agreements. *Conservation Letters* 7, 12–24.

Dickinson, J.L., Shirk, J., Bonter, D., Bonney, R., Crain, R.L., *et al.* (2012) The current state of citizen science as a tool for ecological research and public engagement. *Frontiers in Ecology and the Environment* 10, 291–297.

Gallo, T. and Waitt, D. (2011) Creating a successful citizen science model to detect and report invasive species. *BioScience* 61, 459–465.

Huxel, G.R. (1999) Rapid displacement of native species by invasive species: effects of hybridization. *Biological Conservation* 89, 143–152.

iNaturalist.org. (2018) iNaturalist Research-grade Observations. Occurrence Dataset. Available at: http://doi.org/10.15468/ab3s5x (accessed 12 November 2019).

Jordan, R., Crall, A., Gray, S., Phillips, T. and Mellor, D. (2015) Citizen science as a distinct field of inquiry. *BioScience* 65, 208–211.

Jordan, R.C., Gray, S.A., Howe, D.V., Brooks, W.R. and Ehrenfeld, J.G. (2011) Knowledge gain and behavioral change in citizen-science programs. *Conservation Biology* 25, 1148–1154.

Koenig, W.D. (2003) European Starlings and their effect on native cavity-nesting birds. *Conservation Biology* 17, 1134–1140.

Lowe, S., Browne, M., Boudjelas, S. and de Poorter, M. (2000) *100 of the World's Worst Invasive Alien Species: A Selection from the Global Invasive Species Database*. The Invasive Species Specialist Group (ISSG), Auckland, New Zealand. Available at: https://portals.iucn.org/library/sites/library/files/documents/2000-126.pdf (accessed 12 November 2019).

Mank, J.E., Carlson, J.E. and Brittingham, M.C. (2004) A century of hybridization: decreasing genetic distance between American Black Ducks and Mallards. *Conservation Genetics* 5, 395–403.

Mannino, A.M. and Balistreri, P. (2018) Citizen science: a successful tool for monitoring invasive alien species (IAS) in Marine Protected Areas. The case study of the Egadi Islands MPA (Tyrrhenian Sea, Italy). *Biodiversity* 19, 42–48.

McKinney, M.L. and Lockwood, J.L. (1999) Biotic homogenization: a few winners replacing many losers in the next mass extinction. *Trends in Ecology and Evolution* 14, 450–453.

Mo, M. (2015) The Red-whiskered Bulbul *Pyconotus jocosus* in Australia – a global perspective, history of introduction, current status and potential impacts. *Australian Zoologist* 37, 461–471.

Muñoz-Fuentes, V., Vilà, C., Green, A.J., Negro, J.J. and Sorenson, M.D. (2007) Hybridization between white-headed ducks and introduced ruddy ducks in Spain. *Molecular Ecology* 16, 629–638.

National Audubon Society (2012) The Christmas Bird Count historical results. National Audubon Society, New York. Available at: www.audubon.org/conservation/science/christmas-bird-count (accessed 12 November 2019).

Pocock, M.J.O., Chandler, M., Bonney, R., Thornhill, I., Albin, A., *et al.* (2018) A vision for global biodiversity monitoring with citizen science. *Advances in Ecological Research* 59, 169–223.

Risely, K., Baillie, S.R., Eaton, M.A., Joys, A.C., Musgrove, A.J., *et al.* (2010) The Breeding Bird Survey 2009. BTO Research Report 559. British Trust for Ornithology, Thetford, UK. Available at: www.bto.org/sites/default/files/u16/downloads/reports/bbsreport09.pdf (accessed 15 November 2019).

Romagosa, C.M. and Labisky, R.F. (2000) Establishment and dispersal of the Eurasian Collared-Dove in Florida. *Journal of Field Ornithology* 71, 159–166.

Russell, J.C. and Blackburn, T.M. (2017) The rise of invasive species denialism. *Trends in Ecology and Evolution* 32, 3–6.

SABAP2 (2019) Southern African Bird Atlas Project 2. Available at: http://sabap2.adu.org.za/ (accessed 12 November 2019).

Sauer, J.R., Hines, J.E., Fallon, J.E., Pardieck, K.L., Ziolkowski, D.J. Jr. and Link, W.A. (2014) The North American Breeding Bird Survey, Results and Analysis 1966–2013. Version 01.30.2015. USGS Patuxent Wildlife Research Center, Laurel, Maryland.

Sax, D.F. and Gaines, S.D. (2008) Species invasions and extinction: the future of native biodiversity on islands. *Proceedings of the National Academy of Sciences USA* 105, 11490–11497.

Silvertown, J. (2009) A new dawn for citizen science. *Trends in Ecology and Evolution* 24, 467–471.

Sullivan, B.L., Aycrigg, J.L., Barry, J.H., Bonney, R.E., Bruns, N., *et al.* (2014) The eBird enterprise: an integrated approach to development and application of citizen science. *Biological Conservation* 169, 31–40.

Temple, S.A. (1992) Exotic birds: a growing problem with no easy solution. *Auk* 109, 395–397.

Texas Invasive Bird Project (2018). Texas Invasive Bird Project. The Houston Museum of Natural Science, Houston, Texas. Available at: www.hmns.org/invasive-birds (accessed 15 November 2019).

Theobald, E.J., Ettinger, A.K., Burgess, H.K., DeBey, L.B., Schmidt, N.R., *et al.* (2015) Global change and local solutions: tapping the unrealized potential of citizen science for biodiversity research. *Biological Conservation* 181, 236–244.

Thibault, M., Vidal, E., Potter, M.A., Dyer, E. and Brescia, F. (2018) The red-vented bulbul (*Pycnonotus cafer*): serious pest or understudied invader? *Biological Invasions* 20, 121–136.

Welvaert, M. and Caley, P. (2016) Citizen surveillance for environmental monitoring: combining the effects of citizen science and crowdsourcing in a quantitative data framework. *SpringerPlus* 5, 1890.

45 Conclusions

Lorinda A. Hart[1,2]* and Colleen T. Downs[1]
[1]*Centre for Excellence in Invasion Biology, School of Life Sciences, University of KwaZulu-Natal, Pietermaritzburg 3209, South Africa;* [2]*Department of Biological Sciences, University of Namibia, Windhoek 10023, Namibia*

Citation: Hart, L.A. and Downs, C.T. (2020) Conclusion. In: Downs, C.T. and Hart, L.A. (eds) *Invasive Birds: Global Trends and Impacts*. CAB International, Wallingford, UK, pp. 368–370.

Invasive species are defined by the International Union for Conservation of Nature (IUCN) as 'Animals, plants or other organisms introduced by man into places out of their natural range of distribution, where they become established and disperse, generating a negative impact on the local ecosystem and species' (IUCN Invasive Species Specialist Group, 2015). However, Richardson *et al.* (2000) highlighted that labelling a taxon as 'naturalized', 'non-invasive' or 'invasive' can be problematic. They further suggested that 'the term 'invasive' should be used with reference to the "biogeographic/demographic" status of a species without any connotation of impact' as the impacts are not always negative (Richardson *et al.*, 2000). Criteria for the inclusion of species in the Global Invasive Species Database (GISD) focus on invasive alien species that threaten native biodiversity and natural areas, and are based on consultation and assessment by experts using available data (Pagad *et al.*, 2015; IUCN Invasive Species Specialist Group, 2015). They further describe that these species have serious impacts on biological diversity and/or human activities (Pagad *et al.*, 2015; IUCN Invasive Species Specialist Group, 2015). In Section 1 of this volume, use was made of the bird species listed in the GISD with some additional bird species included. However, as highlighted, particularly in Section 2 of this book, some of the GISD information is inadequate or not 'global', and several additional bird species are of concern while others currently listed are perhaps of less concern. As highlighted by many authors, this emphasizes that avian invasive species and invasion processes are dynamic and require ongoing monitoring and appraisal.

Continued globalization and the establishment of new trade routes mean that introductions of potentially invasive species will continue to increase (Early *et al.*, 2016). Nearly 17% of the world's land mass is at great risk of invasion (Early *et al.*, 2016). To accurately predict, prevent and manage invasive species, accurate lists must be drawn up for a region. However, such lists are often erroneous, primarily due to a lack of knowledge (McGeoch *et al.*, 2012; Evans *et al.*, 2016). Despite resources such as publicly available online invasive species forums (e.g. DAISIE (Delivering Alien Invasive Species Inventories for Europe) and GISD), information regarding these species, particularly their impacts, are often data deficient or anecdotal (McGeoch *et al.*, 2012; see various chapters in Sections 1 and 2, this volume). Additionally, their distributions and population statuses should be regularly reviewed and updated, as species expand their distributional ranges and new introductions occur (as do eradications and extirpations). Often, such observations are also reported in 'grey' literature, particularly by citizen scientists (e.g. Fig. 45.1). It is therefore predicted that the role of bird atlassing apps and similar public platforms will play an increasingly important role in monitoring and reporting invasive and emerging species. Additionally, citizen scientists can be enlisted to record not only species presence but also their impacts, such as competitive interactions, economic damage and hybridization (see Chapter 44, this volume). These impacts are essential in refining and accurately calculating the probability of a species becoming problematic and the magnitude of its effect on the environment (Blackburn *et al.*, 2014; Hawkins *et al.*, 2015; Evans *et al.*, 2016). Their role in facilitating the development of rapid pre-border risk assessments (Keller and Kumschick, 2017) further supports impact identification as a priority in invasive research.

Hui and Richardson (2017) highlighted that biological invasions are not simple actions of invaders and reactions of invaded ecosystems, but are co-evolving complex adaptive systems with emergent features of network complexity and invasibility. These require analysing their respective spread processes and/or impact dynamics, which ultimately affect management decisions (Hui and Richardson, 2017). Preventing introductions of exotic avian species is the first line of defence.

*Corresponding author: lorinda.hart.research@gmail.com

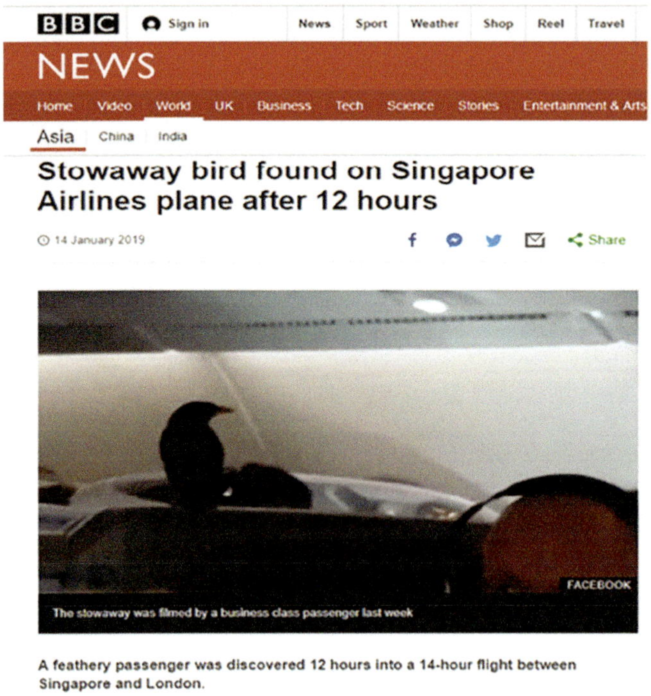

Fig. 45.1. Making the news headlines, a Common Myna (*Acridotheres tristis*) is discovered 12 h into an international flight, which departed from Singapore's Changi Airport in January 2019. (From: www.bbc.com/news/world-asia-46861136, accessed 12 November 2019.)

Should this fail, a precautionary approach should be considered, acting before populations become too large and widespread and before they demonstrate any impacts on the environment (Rogers *et al.*, 1997; see Chapter 43, this volume). Islands have been particularly successful in persevering and eradicating invasive species (Feare, 2010; Canning, 2011). Although total eradication is often the desired end goal, this is not always feasible, particularly on continents with extensive and well-established populations. In such cases, invasive species can either be contained or managed through co-ordinated actions, which can yield some success (see Chapter 43, this volume).

As has been highlighted throughout this book and in the growing body of literature on invasive species, there is a need for refined, up-to-date species data – particularly regarding their impacts. One of the main impacts of concern is the competitive effects of invasive species on native species. The competitive effects of an invading species can ultimately alter a community's structure, as native species' access to resources are limited, compromising their persistence (see Chapter 42, this volume). However, identifying these impacts remains challenging as these interactions are context specific and vary based on species, environments and the critical resources in question, as well as the number of interactions and the outcomes thereof (see Chapter 42, this volume).

These impacts should be considered from both an ecological and environmental perspective, and where possible from an economic stand point as well. The latter is more likely to motivate management decisions. More accurate resources could also ease the disagreement within the scientific community, which has been quite divided on how invasive species should be treated. Terms such as 'eco-xenophobic' or 'preservationist' are being used to describe those who favour a native species state, void of invasive and exotic species (Rotherham, 2010; Bhagwat, 2018). Those on the other side of the fence have been termed 'denialists' based on their 'rejection of scientific evidence' in favour of invasive species tolerance (Russell and Blackburn, 2017a). Pro-arguments for the presence of invasive species include: (i) that they are a product of the Anthropocene, an unstoppable consequence of human expansion; (ii) the cost of invasive control is high and rather than funding a 'losing battle', money should be put to 'better use'; and (iii) that we should learn to live with them as they are here to stay, and we should embrace this 'novel nature', which may even benefit humans (Bhagwat, 2018). Invasive birds can fulfil various beneficial ecosystem roles such as seed dispersal, pollination and pest control. While there are invasive species that pose no apparent risk to native species and provide beneficial ecosystems services and goods, one cannot ignore those that have devastating consequences. Invasive species modify the evolutionary pathway of native species genetically (by hybridization and introgression) and through niche displacement, predation and competitive exclusion (Mooney and Cleland, 2001). Wilcove *et al.*'s (1998) statement that, 'second to habitat loss, alien species pose the biggest risk to biodiversity' has been cited by many scientists (at least 3580 times according to Google Scholar, February 2020) and has since seemingly become the cornerstone in a battle of strong opinions. There is no doubt that some invasive species dramatically alter the environments they invade, but some have argued that direct evidence of total species extinction remains scant (Gurevitch and Padil, 2004). However, analysis of the IUCN Red List data, for which 170 animal species have noted causes of extinction, implicates invasive species in 91% of cases, with 20% of extinctions attributed exclusively to invasive species (Clavero and García-Berthou, 2005). However, Russell and Blackburn (2017b) are a reminder that one must not narrowly define invasive impacts as extinction but rather recognize this as the end point.

The binary standpoint of 'for' versus 'against' should perhaps be reconsidered, as nature does not always respond in such a way (Davis and Chew, 2017). While there is evidence both for and against the effects of invasive alien species, it is important that we do not forget that diverse biological systems will respond in various ways and that these responses will not necessarily be consistent (Bellard *et al.*, 2016). Hawaii, for example, has more invasive land bird species (58 species) than native species, and the implications of removing these must be carefully considered, as some fulfil important ecological niches of extinct native species (Kaplan, 2007). However, invasive birds disperse both native and invasive plants in Hawaii and do not fulfil the complete ecological role of larger native frugivores (which are becoming increasingly rare) due to their smaller gape sizes (Foster and Robinson, 2007; Pejchar, 2015; Kaushik *et al.*, 2018).

In Section 1 of this book, the authors reviewed and provided up-to-date information on invasive avian species considered to be globally problematic, with most of these listed in the GISD (IUCN Invasive Species Specialist Group, 2015). In particular, the respective bird species' diet, habitat use, breeding biology, impacts and current distributions were highlighted. It is hoped that these will provide up-to-date information that can be used by managers and to model future trends more accurately. Control strategies that have been used (successfully and unsuccessfully) are also presented, as these can reduce costs and groundwork for new eradication programmes (see Chapter 43, this volume, and other chapters in Sections 1 and 2). Currently available tools were used including bird species reporting apps, scientific publications, grey literature and personal communication with a variety of citizen scientists and experts to achieve this. The status and predicted trends of avian invaders are also presented geographically in Section 2. Introduction pathways, movement and spread differ among continents and will thus affect their prevention, management and control strategies. Indeed, impacts by invasive species can vary according to taxa, location and time (Bellard *et al.*, 2016).

This book forms part of the foundation on which it is hoped many further studies and management protocols can build, and provides examples of the data available and that which is still missing for a growing body of globally invasive bird species.

45.1 References

Bellard, C., Genovesi, P. and Jeschke, J. (2016) Global patterns in threats to vertebrates by biological invasions. *Proceedings of the Royal Society B: Biological Sciences* 283, 20152454.

Bhagwat, S. (2018) Management of nonnative invasive species in the Anthropocene. In: Dellasala, D.A. and Goldstein, M.I. (eds) *Encyclopedia of the Anthropocene*. Elsevier, Waltham, Massachusetts, pp. 409–417.

Blackburn, T.M., Essl, F., Evans, T., Hulme, P.E., Jeschke, J.M. *et al.* (2014) A unified classification of alien species based on the magnitude of their environmental impacts. *PLoS Biology* 12, e1001850.

Canning, G.J.P. (2011) Eradication of the invasive Common Myna, *Acridotheres tristis*, from Fregate Island, Seychelles. *Phelsuma* 19, 43–53.

Clavero, M. and García-Berthou, E. (2005) Invasive species are a leading cause of animal extinctions. *Trends in Ecology and Evolution* 20, 110.

Davis, M.A. and Chew, M.K. (2017) 'The denialists are coming!' Well, not exactly: a response to Russell and Blackburn. *Trends in Ecology and Evolution* 32, 229–230.

Early, R., Bradley, B.A., Dukes, J.S., Lawler, J.J., Olden, J.D. *et al.* (2016) Global threats from invasive alien species in the twenty-first century and national response capacities. *Nature Communications* 7, 12485.

Evans, T., Kumschick, S. and Blackburn, T.M. (2016) Application of the Environmental Impact Classification for Alien Taxa (EICAT) to a global assessment of alien bird impacts. *Diversity and Distributions* 22, 919–931.

Feare, C.J. (2010) Invasive bird eradication from tropical oceanic islands. *Aliens* 30, 12–19.

Foster, J.T. and Robinson, S.K. (2007) Introduced birds and the fate of Hawaiian rainforests. *Conservation Biology* 21, 1248–1257.

Gurevitch, J. and Padil, D.K. (2004) Are invasive species a major cause of extinctions? *Trends in Ecology and Evolution* 19, 470–474.

Hawkins, C.L., Bacher, S., Essl, F., Hulme, P.E., Jeschke, J.M. *et al.* (2015) Framework and guidelines for implementing the proposed IUCN Environmental Impact Classification for Alien Taxa (EICAT). *Diversity and Distributions* 21, 1360–1363.

Hui, C. and Richardson, D.M. (2017) *Invasion Dynamics*. Oxford University Press, Oxford.

IUCN Invasive Species Specialist Group (2015) Global Invasive Species Database. Available at: www.iucngisd.org/gisd/search.php. (accessed 12 November 2019).

Kaplan, M. (2007) Alien birds may be last hope for Hawaiian plants. *Nature News*, 28 September. Available at: https://www.nature.com/news/2007/070928/full/news070924-12.html (accessed 12 November 2019).

Kaushik, M., Pejchar, L. and Crampton, L.H. (2018) Potential disruption of seed dispersal in the absence of a native Kauai Thrush. *PLoS One* 13, e0191992.

Keller, R.P. and Kumschick, S. (2017) Promise and challenges of risk assessment as an approach for preventing the arrival of harmful alien species. *Bothalia* 47, 1–8.

McGeoch, M.A., Spear, D., Kleynhans, E.J. and Marais, E. (2012) Uncertainty in invasive alien species listing. *Ecological Applications* 22, 959–971.

Mooney, H.A. and Cleland, E.E. (2001) The evolutionary impact of invasive species. *Proceedings of the National Academy of Sciences USA* 98, 5446–5451.

Pagad, S., Genovesi, P., Carnevali, L., Scalera, R. and Clout, M. (2015) IUCN SSC Invasive Species Specialist Group: invasive alien species information management supporting practitioners, policy makers and decision takers. *Management of Biological Invasions* 6, 127–135.

Pejchar, L. (2015) Introduced birds incompletely replace seed dispersal by a native frugivore. *AoB Plants* 7, plv072.

Richardson, D.M., Pyšek, P., Rejmánek, M., Barbour, M., Panetta, F.D. and West, C.J. (2000) Naturalization and invasion of alien plants: concepts and definitions. *Diversity and Distributions* 6, 93–107.

Rogers, M., Sinden, J. and de Lacy, T. (1997) The precautionary principle for environmental management: a defensive-expenditure application. *Journal of Environmental Management* 51, 343–360.

Rotherham, I. (2010) Eco-xenophobia – responding to our natural aliens. *Ecos* 31, 2–10.

Russell, J.C. and Blackburn, T.M. (2017a) Invasive alien species: denialism, disagreement, definitions, and dialogue. *Trends in Ecology and Evolution* 32, 312–314.

Russell, J.C. and Blackburn, T.M. (2017b) The rise of invasive species denialism. *Trends in Ecology and Evolution* 32, 3–6.

Wilcove, D.S., Rothstein, D., Bubow, J., Phillips, A. and Losos, E. (1998) Quantifying threats to imperiled species in the United States. *Bioscience* 48, 607–615.

Index

Page numbers in **bold** type refer to figures and tables.